Lecture Notes in Electrical Engineering

Volume 444

"Lecture Notes in Electrical Engineering (LNEE)" is a book series which reports the latest research and developments in Electrical Engineering, namely:

- Communication, Networks, and Information Theory
- Computer Engineering
- Signal, Image, Speech and Information Processing
- Circuits and Systems
- Bioengineering

LNEE publishes authored monographs and contributed volumes which present cutting edge research information as well as new perspectives on classical fields, while maintaining Springer's high standards of academic excellence. Also considered for publication are lecture materials, proceedings, and other related materials of exceptionally high quality and interest. The subject matter should be original and timely, reporting the latest research and developments in all areas of electrical engineering.

The audience for the books in LNEE consists of advanced level students, researchers, and industry professionals working at the forefront of their fields. Much like Springer's other Lecture Notes series, LNEE will be distributed through Springer's print and electronic publishing channels.

More information about this series at http://www.springer.com/series/7818

Hao Liu · Cunjiang Song
Arthur Ram
Editors

Advances in Applied Biotechnology

Proceedings of the 3rd International Conference on Applied Biotechnology (ICAB2016), November 25–27, 2016, Tianjin, China

Springer

Editors
Hao Liu
College of Biotechnology
Tianjin University of Science
and Technology
Tianjin
China

Arthur Ram
Institute of Biology Leiden
Leiden University
Leiden
The Netherlands

Cunjiang Song
College of Life Sciences
Nankai University
Tianjin
China

ISSN 1876-1100 ISSN 1876-1119 (electronic)
Lecture Notes in Electrical Engineering
ISBN 978-981-10-4800-5 ISBN 978-981-10-4801-2 (eBook)
https://doi.org/10.1007/978-981-10-4801-2

Library of Congress Control Number: 2017940228

Printed on acid-free paper

This Springer imprint is published by Springer Nature
The registered company is Springer Nature Singapore Pte Ltd.
The registered company address is: 152 Beach Road, #21-01/04 Gateway East, Singapore 189721, Singapore

Contents

Part I
Microbial Genetics and Breeding

Isolation and Characterization of a Virulent Phage H6 Infecting *Lactobacillus brevis* from the Fermented Chinese Cabbage

Kechong Huang, Tian Zhang and Hongjiang Yang

1 Introduction

Nowadays, traditional vegetable fermentation products contain fermented cucumber, kimchi, sauerkraut and fermented Chinese cabbage, and etc. They are prepared by a variety of lactic acid bacteria (LAB) naturally present on surfaces of vegetables. The metabolic activities of LAB determine the quality and safety of the final fermentation products [1]. Recent studies have analyzed the diversity and dynamics of the microflora in several fermented vegetables, such as kimchi [2], sauerkraut [3] and fermented cucumber [4, 5]. Phage is the most abundant biological entity living on the earth widely distributed in nearly all natural niches. Many phage species have been discovered in the fermented vegetable products. The presence of diverse phages against LAB strains in vegetable fermentation processes can potentially cause significant mortality to LAB, causing loss or damage to the productions. Now, the phage ecology of several vegetable fermentations has been reported, including sauerkraut fermentations [6, 7], and cucumber fermentations [8]. In addition, the novel LAB phages against *Weissella cibaria* have been isolated from the commercial kimchi fermentations [9].

Fermented Chinese cabbage is one of the most popular traditional fermented Chinese foods, and it is a mainly LAB fermented vegetable product. In the cabbage fermentation, *Lactobacillus plantarum*, *Lactobacillus brevis* and *Lactobacillus pentosus* are detected as the dominant species, and others including *Lactobacillus acidophilus*, *Lactobacillus fermentum* and *Leuconostoc mesenteroides* [10].

In this study, lactic acid bacteria and the corresponding phages were isolated from the fermented Chinese cabbage samples collected in a village of Ji County,

K. Huang · T. Zhang · H. Yang (✉)
Key Laboratory of Industrial Fermentation Microbiology, Ministry of Education Tianjin, Key Laboratory of Industrial Microbiology College of Biotechnology, Tianjin University of Science and Technology, Tianjin 300457, China
e-mail: hongjiangyang@tust.edu.cn

H. Liu et al. (eds.), *Advances in Applied Biotechnology*, Lecture Notes in Electrical Engineering 444, https://doi.org/10.1007/978-981-10-4801-2_1

Tianjin. The isolated phage was further characterized, including morphology, burst size, latent time, genome size and host range. The data of this study may provide valuable information for the prevention of phage infection in the production of the fermented Chinese cabbage.

2 Materials and Methods

2.1 *Growth Conditions*

MRS medium was composed of 20 g/L glucose, 10 g/L peptone, 8 g/L beef extract, 4 g/L yeast extract, 5 g/L sodium acetate, 2 g/L ammonium citrate, 0.2 g/L $MgSO_4$, 0.05 g/L $MnSO_4$, 2 g/L KH_2PO_4, and 0.1% Tween-80 [11]. MRS (pH = 5.0) and MRS (pH = 6.4) were used for lactic acid bacteria (LAB) isolation and cultivation, respectively. Broth was used for liquid cultures, 1.5% solid agar medium was used for bacteria plating, and 0.5% semi-solid agar medium was used for phage plaque-forming assays. Both plates and liquid cultivations were statically incubated at 30 °C.

2.2 *Isolation and Identification of Lactic Acid Bacteria*

The fermented Chinese cabbage samples were collected in Ji County, Tianjin, China on March 16th, 2016. To isolate lactic acid bacteria, serial dilutions of nine fresh fermented Chinese cabbage samples were spread on MRS agar. The identity of these strains were determined by analyzing the 16S rDNA gene sequence [12].

2.3 *Isolation and Propagation of Phage*

The isolated LAB strains were used as indicator for isolating phage from the fermented Chinese cabbage samples. In brief, added 1 mL samples to 100 mL MRS medium, and incubated at 30 °C for 48 h to enrich phages. The phages were isolated by the double-layer plating technique. The propagation of phage was done as previously described in order to obtain high titer lysates [13]. Phage lysate was filtered through a 0.2 μm-pore size sterile filter and stored at 4 °C for future use.

2.4 Transmission Electron Microscopy

The phage preparations was further treated with DNase I (1 μg/mL) and RNase A (1 μg/mL) at 37 °C for 30 min. The treated lysate was washed ten times with 0.1 mol/L ammonium acetate solution (pH = 7.0) using the 100 kD Amicon filters. The retained phage solution was used directly for negative staining as described previously [14]. Photographs were taken with a JROL1011 transmission electron microscope operating at 100 kV.

2.5 Host Range Determination

As described previously [15], spotting assay was conducted to determine the sensitivity of the isolated phages to the host strains. The double-layer method was used to further confirm the above results.

2.6 Restriction Analysis of the Phage Genomic DNA

The 100 mL phage lysate was used for phage DNA extraction using the phenol-chloroform method described previously [16]. Purified phage genomic DNA was subjected to digestions with several restriction endonucleases, including *Eco*RI, *Eco*RV, *Xba*I, *Hind*III, *Sma*I, *Stu*I and *Bam*HI, respectively. The genome size was estimated by compilation of DNA fragment sizes resulting from the seven restriction enzymes digestion profiles.

2.7 SDS-PAGE Analysis of the Phage Structural Proteins

Purified phage particles were further filtered through the Amicon-100 filter, and washed three times with 0.1 mol/L ammonium acetate solution (pH = 7.0). Treated phage particles were subjected to SDS-PAGE directly, and the gel stained with Coomassie Blue R-250.

2.8 One Step Growth Curve

One-step growth experiment was carried out according to the previous descriptions [17]. In brief, 50 mL bacterial cells were incubated to mid-exponential-phase (OD_{600} = 0.5–0.6) and harvested by centrifugation at 6000 rpm for 10 min.

The pellet was resuspended in 0.5 mL fresh MRS medium and mixed with 0.5 mL phage solution (1×10^6 pfu/mL). Phage was allowed to adsorb for 1 min and the mixture was subjected to centrifuge immediately at 13,000 rpm for 30 s to remove free phage particles. The treated pellet was resuspended in 100 mL fresh MRS medium and the culture was continuously incubated at 30 °C. Samples were taken at 15 min intervals and phage titre was determined by the double-layer plate method. The latent period was deduced from the triphasic curve. The burst size of phage was calculated by dividing the phage titers at the plateau phase by the infective centers number at the latent phase.

2.9 The Lysis Curve of Host Strain by Phage

The logarithmic cultures of the strain J68 were mixed with the phage lysate at different MOIs (MOI = 100, 10, 1, 0.1, 0.01, 0). The 96-well plate was filled with the 300 μL mixture per well, and the value of OD_{600} was measured at 2 h intervals by ELISA reader (GS40A24).

3 Results

3.1 Identification of Lactic Acid Bacteria

Eight isolated strains from the fermented Chinese cabbage samples were confirmed the identity by analyzing their 16S rDNA gene sequence. The resulted sequences were deposited to GenBank and aligned to search for the most similar sequences. In final, six collected strains were validated to be *Lactobacillus brevis*, other two strains belonged to *Lactobacillus plantarum*.

3.2 Isolation and Titer of Phage

L. brevis J68 was used as indicator strain for virulent bacteriophages screening from the fermented Chinese cabbage samples. A phage was eventually isolated and named H6. Its plaques were circular, clear and transparent with smooth edge, showing 1–2 mm in diameter. At the MOI = 0.01, titer of phage H6 reached 1×10^9 pfu/mL in MRS and 5×10^9 pfu/mL in MRS with $CaCl_2$ (2.375 g/L).

3.3 Host Range Analysis

The susceptibility to phage H6 was investigated with isolated strains from fermented Chinese cabbage samples and other strains, they included six *L. brevis*, six *L. plantarum*, two *Lactobacillus vaginalis*, nine *Lactobacillus reuteri*, two *Weissella cibaria*, one *Lactobacillus curvatus* and one *Lactobacillus johnsonii*. We found that all six *L. brevis* strains were sensitive to phage H6, and other strains all were resistant to phage H6. The sequences of 16S rDNA showed that there were differences among six *L. brevis* strains. The result indicated that phage H6 might have a broad host range and was capable of infecting multiple isolates of *Lactobacillus brevis*, however, phage H6 didn't infect lactic acid bacteria from other genera.

3.4 Morphology Study by Transmission Electron Microscopy

The treated phage solution was used directly for negative staining. Images of phage H6 were developed using transmission electron microscope (Fig. 1). The obtained image showed that phage H6 had an icosahedra head of 93.3 nm in diameter and a long contractile tail about 166.6 nm in length, and it was classified as a lytic phage of *Myoviridae* in *Caudovirales*.

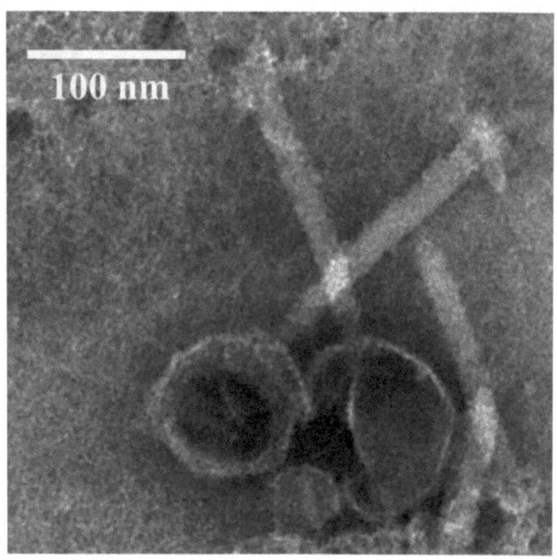

Fig. 1 Morphology of phage H6 with the transmission electron micrograph method. The *scale bar* represented 100 nm

3.5 Restriction Fragment Analysis of Genomic DNA

Phage H6 was amplified and its genomic DNA was extracted. Purified genomic DNA was digested with several restriction endonucleases, including *Eco*RI, *Eco*RV, *Xba*I, *Hind*III, *Sma*I, *Stu*I and *Bam*HI, as subsequently subjected to electrophoretic analyses (Fig. 2). Based on the digestion profiles of *Eco*RI, *Eco*RV, *Xba*I, and *Hind*III, the genome size was determined to be approximately at the range of 59.6–61.2 kb. The restriction enzymes analysis also indicated that phage H6 was a dsDNA virus.

3.6 Proteomic Analysis of Phage Structural Proteins

Purified phage particles were subjected to SDS-PAGE and proteomic patterns were obtained after Coomassie Blue R-250 staining (Fig. 3). Totally, six protein bands were displayed on the gel with the molecular weights ranging approximately from 30 to 60 kD.

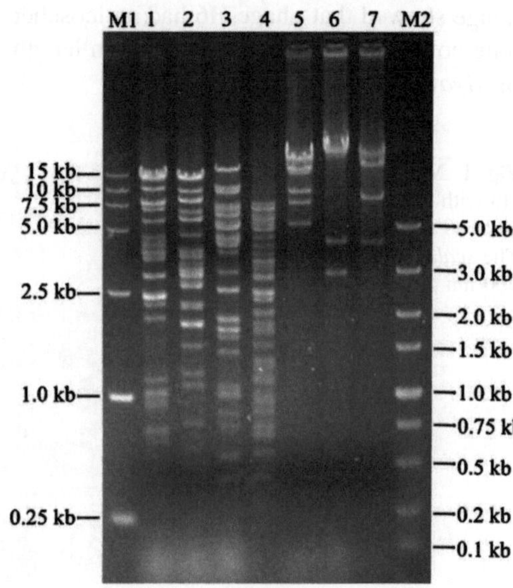

Fig. 2 Restriction fragments analysis of phage genomic DNA. Phage genomic DNA was digested with *Eco*RI (*lane 1*), *Eco*RV (*lane 2*), *Xba*I (*lane 3*), *Hind*III (*lane 4*), *Sma*I (*lane 5*), *Stu*I (*lane 6*), and *Bam*HI (*lane 7*), respectively. *M1:* 15 kb DNA marker 1; *M2:* 5 kb DNA marker 2

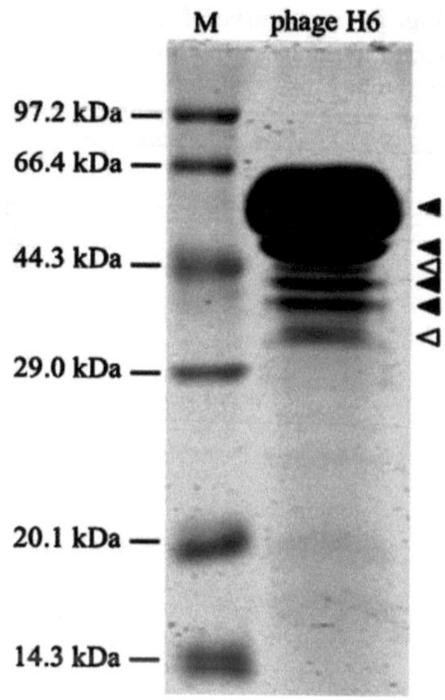

Fig. 3 SDS-PAGE of the viral structural proteins. *Solid arrows* indicated the major proteins bands; *open arrows* showed the minor proteins bands

3.7 Latent Time and Burst Size of Phage H6

One-step growth experiment was performed to determine the latent time and burst size of phage H6. As inferred from the triphasic curve (Fig. 4), the latent period was about 90 min and the burst size was about 40.4 pfu/infection center.

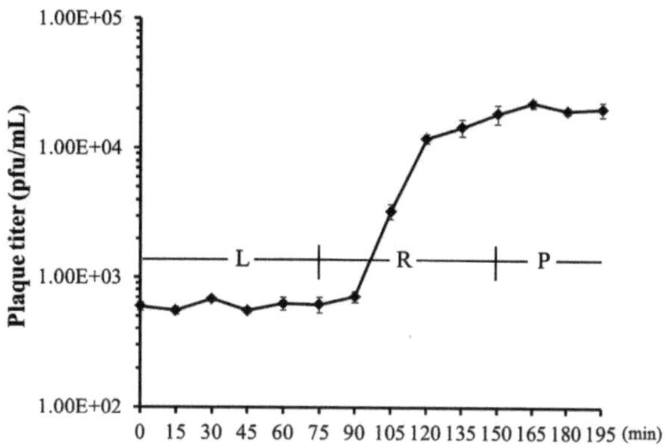

Fig. 4 One-step growth curve. Latent time and burst size of phage H6 were inferred from the triphasic curve. *L:* latent phase; *R:* rise phase; *P:* plateau phase

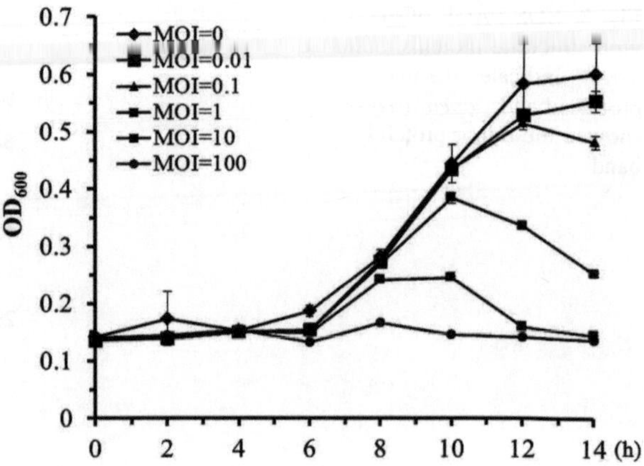

Fig. 5 Lysis curve of host strain by phage H6 with different MOIs

3.8 Lysis Curve of Host Strain by Phage H6

The lysis ability of phage H6 to the *Lactobacillus brevis* J68 was measured at different MOIs. As shown in Fig. 5, with the increase of phage titeres, the bacteriostatic ability of phage H6 gradually strengthened. At the MOI = 1, the growth of strain J68 appeared to decline in the 10th hour; the strain J68 hardly grew at the MOI = 10 or 100.

4 Discussion

To achieve high quality of fermented products, knowledge about the diversity, abundance, and property of phages in vegetable fermentation products are essential for developing phage control strategies. Lu et al. [6] investigated the ecology of phages infecting lactic acid bacteria in commercial sauerkraut fermentations, the result indicated that the phages against *Leuconostoc*, *Weissella* and *Lactobacillus* were widely existed in sauerkraut fermentations. In one recent study, a high abundance of phage DNA was found in kimchi fermentation [18]. And several LAB phages against *L. plantarum* and *Pediococcus* sp. have been isolated from commercial cucumber fermentations [7, 19]. Bacteriophage infection was a well-recognized problem in industrial food fermentations and a variety of countermeasures were employed to control the problem [20]. Therefore, the gain of more information about bacteriophages in vegetable fermentation products was essential to the phage-control strategies.

In this study, a virulent phage infecting *L. brevis* was isolated from fermented Chinese cabbage samples, and its biological characteristics were determined. The restriction analysis indicated that phage H6 was a dsDNA virus with an approximate genome size of 59.6–61.2 kb, but, it was uncertain that phage H6 genome was linear or circular. The analysis of phage H6 structural proteins showed that there was a huge protein band at 60 kD, we speculated that it contains several protein bands and this assumption can be verified by mass spectrometry. The data from this study can provide more information about *L. brevis* bacteriophages. The objectives of this study were to provide more formulation in order to prevent infection of bacteriophages in fermented Chinese cabbage.

Acknowledgements This work was partly supported by The National Natural Science Foundation of China (Grants 31370205 and 30970114).

References

1. Pederson CS, Albury MN (1969) The sauerkraut fermentation. New York State Agricultural Experiment Station Technical Bulletin 824. Geneva, New York
2. Lee JS, Heo GY, Lee JW, Oh YJ, Park JA, Park YH, Pyun YR, Ahn JS (2005) Analysis of kimchi microflora using denaturing gradient gel electrophoresis. Int J Food Microbiol 102(2):143–150
3. Plengvidhya V, Breidt F Jr, Lu Z, Fleming HP (2007) DNA fingerprinting of lactic acid bacteria in sauerkraut fermentations. Appl Environ Microbiol 73(23):7697–7702
4. Medina E, Pérez-Díaz IM, Breidt F, Hayes J, Franco W, Butz N, Azcarate-Peril MA (2016) Bacterial ecology of fermented cucumber rising pH spoilage as determined by nonculture-based methods. J Food Sci 81(1):M121–M129
5. Chen YS, Wu HC, Lo HY, Lin WC, Hsu WH, Lin CW, Lin PY, Yanagida F (2012) Isolation and characterisation of lactic acid bacteria from jiang-gua (fermented cucumbers), a traditional fermented food in Taiwan. J Sci Food Agric 92(10):2069–2075
6. Lu Z, Breidt F, Plengvidhya V, Fleming HP (2003) Bacteriophage ecology in commercial sauerkraut fermentations. Appl Environ Microbiol 69:3192–3202
7. Yoon S-S, Barrangou-Poueys R, Breidt F, Klaenhammer TR, Fleming HP (2002) Isolation and characterization of bacteriophages from fermenting sauerkraut. Appl Environ Microbiol 68(2):973–976
8. Lu Z, Perez-Diaz IM, Hayes JS, Breidt F (2012) Bacteriophage ecology in a commercial cucumber fermentation. Appl Environ Microbiol 78(24):8571–8578
9. Kleooen HP, Holo H, Jeon SR, Nes IF, Yoon SS (2012) Novel Podoviridae family bacteriophage infecting *Weissella cibaria* isolated from Kimchi. Appl Environ Microbiol 78 (20):7299–7308
10. Yan P, Chai Z, Xue W, Chang X, Kong D, Zhang H (2009) Lactic acid bacteria diversity in fermented cabbage estimated by culture-dependent and -independent. Wei Sheng Wu Xue Bao 49(3):383–388
11. De Man JC, Rogosa M, Sharpe ME (1960) A medium for the cultivation of lactobacilli. J Appl Bacteriol 23(1):130–135
12. Gürtler V, Stanisich VA (1996) New approaches to typing and identification of bacteria using the 16S–23S rDNA spacer region. Microbiology 142(Pt 1):3–16
13. Jaomanjaka F, Ballestra P, Dols-lafargue M, Le Marrec C (2013) Expanding the diversity of oenococcal bacteriophages: insights into a novel group based on the integrase sequence. Int J Food 166(2):331–340
14. Nugent KM, Cole RM (1977) Characterization of group H streptococcal temperate bacteriophage phi 227. J Virol 21(3):1061–1073

15. Yang H, Liang L, Lin S, Jia S (2010) Isolation and characterization of a virulent bacteriophage of *Acinetobacter baumannii*. BMC Microbiol 10:131, PMID: 20426877; 1471-2180-10-131
16. Sambrook J, Russell DW (2001) Molecular cloning. CSHL Press, New York
17. Chow JJ, Batt CA, Sinskey AJ (1988) Characterization of *Lactobacillus bulgaricus* bacteriophage ch2. Appl Environ Microbiol 54(5):1138–1142
18. Jung JY, Lee SH, Kim JM, Park MS, Bae JW, Hahn Y, Madsen EL, Jeon CO (2011) Metagenomic analysis of kimchi, a traditional Korean fermented food. Appl Environ Microbiol 77(7):2264–2274
19. Yoon S-S, Barrangou-Poueys R, Breidt F, Fleming HP (2007) Detection and characterization of a lytic Pediococcus bacteriophage from the fermenting cucumber brine. J Microbiol Biotechnol 17:262–270
20. Moineau S, Levesque C (2005) Control of bacteriophages in industrial fermentation. In: Kutte E, Sulakvelidze A (eds) Bacteriophage: biology and applications. CRC Press, Boca Raton, pp 286–296

Molecular Cloning and Biochemical Characterization of Oligo-1,6-Glucosidases from *Bacillus subtilis* and *Bacillus licheniformis*

Xiaoming Hao, Ke Cai and Zixing Dong

1 Introduction

Oligo-1,6-glucosidase (O-1,6-G, EC 3.2.1.10), belonging to the subfamily 31 of the glycoside hydrolase family 13 (GH13_31) [1] catalyzes the exo hydrolysis of α-1,6-glucoside bonds from the non-reducing ends of α-limit dextrin, isomaltose and other isomaltooligosaccharides (IMOs), but has no activity towards α-1,4-glucoside bonds of malto-oligosaccharides [2]. Acting together with maltase, oligo-1,6-glucosidase can completely hydrolyze α-amylase dextrins, allowing the complete digestion of starch in the gastrointestinal tract of mammals [3, 4]. Moreover, since novel oligosaccharides are finding increasing applications in biotechnological and chemical industries, the debranching enzymes containing oligo-1,6-glucosidases are also valuable [5].

In our study, two genes encoding oligo-1,6-glucosidases from *Bacillus subtilis* and *B. licheniformis* were cloned and overexpressed in *Pichia pastoris*, and their enzymatic properties were comprehensively investigated.

X. Hao · Z. Dong (✉)
Key Laboratory of Industrial Fermentation Microbiology, Ministry of Education,
School of Biotechnology, Tianjin University of Science and Technology,
Tianjin 300457, China
e-mail: dzx@tust.edu.cn

X. Hao · K. Cai · Z. Dong
School of Chemical Engineering and Materials Science, Tianjin University
of Science and Technology, Tianjin 300457, China

X. Hao · Z. Dong
Tianjin Economic-Technological Development Area,
No. 29, the 13th Street, Tianjin, China

H. Liu et al. (eds.), *Advances in Applied Biotechnology*, Lecture Notes
in Electrical Engineering 444, https://doi.org/10.1007/978-981-10-4801-2_2

2 Materials and Methods

2.1 Bacterial Strains, Plasmids

Escherichia coli JM109, *P. pastoris* strain GS115 and vector pPIC9K were obtained from Invitrogen (Carlasbad, CA).

2.2 Gene Clone

Gene encoding oligo-1,6-glucosidase from *B. subtilis* (*bsog*) was amplified by PCR using the primer sets Bs-1 5′-GTAAGTGAATGGTGGAAAGAAGCTGTC-3′ and Bs-2 5′-TGCTCTAGATCATATACTAATGCCCATCACTGCTT-3′. And primer sets Bl-1 5′-GTAAGCCAATGGTGGAAAGAGGC-3′ and Bl-2 5′-TGCTCTAGA TCATGATGTGTAATCCTTTGCCC-3′ were used for *blog*.

2.3 Enzyme Assays

The described method [6] was used to measure the activities of oligo-1,6-glucosidases from *B. subtilis* and *B. licheniformis*.

2.4 pH and Temperature Dependence of Activity

pH optima of BsOG and BlOG were analyzed by incubating them for 15 min at 37 °C as described above.

The optimal reaction temperatures of BsOG and BlOG were determined at temperatures ranging from 15 to 70 °C at pH 6.8.

2.5 Determination of Kinetic Parameters

For determination of enzyme kinetics, various amounts of *p*NPG were used as substrates, and the activities of BsOG and BlOG were measured. Isomaltose and isomaltotriose which were dissolved in 0.2 M phosphate buffer (pH 7.0). An Agilent 1200 HPLC system (Waldbronn, Germany) equipped with an evaporative light-scattering detector was used for quantification and identification of various oligosaccharides in the samples. Separation of oligosaccharides was carried out using a Prevail™ Carbohydrate ES 5μ column (GRACE, 4.6 mm × 250 mm i.d.; particle size, 5 μm).

2.6 *Specificity Towards Natural Substrates*

Isomaltose, isomaltotriose, isomaltulose, panose, maltotriose, maltose, sucrose, amylose, amylopectin and maltodextrin used as substrates to study the specificity towards natural substrates, were dissolved in 0.2 M phosphate buffer (pH 7.0).

2.7 *Hydrolysis of Isomaltotriose and IMOs by OGs*

Hydrolysis of isomaltotriose and IMOs by BsOG and BlOG were performed and analyzed as described above except that isomaltotriose were dissolved in 0.2 M phosphate buffer (pH 7.0) to a final concentration of 4 mM and that samples were withdrawn at different times for HPLC analysis.

3 Results

3.1 *Gene Cloning and Expression of Oligo-1,6-Glucosidases*

With genomic DNA as templates, respective genes encoding oligo-1,6-glucosidases from *B. subtilis* and *B. licheniformis* were amplified by PCR, and the fragments obtained were 1700 bp, consistent with their theoretical sizes (*bsog*, 1686 bp; *blog*, 1707 bp). After five days of shake flask fermentation, the activities of oligo-1,6-glucosidases of these GS115-*bsog* and GS115-*blog* were 1085 and 1037 U/mL.

3.2 *Effects of Temperature on the Activity and Stability of Recombinant Oligo-1,6-Glucosidases*

Effects of temperature on the activity and stability of recombinant oligo-l,6-glucosidases were examined. The temperature optima of BsOG and BlOG were 40 and 45 °C, respectively. BsOG retained 80% of its maximum activity at temperatures ranging from 35 to 50 °C, while BlOG had 80% of its maximum activity at temperatures between 40 and 50 °C. For determination of the thermostability,

BsOG and BlOG were pre-incubated in 0.2 M phosphate buffer (pH 7.0) at 50 °C, and the residual activities were measured at the indicated times. Incubation at 50 °C for 20 min, BsOG and BlOG activity were not detected.

3.3 *Effects of pH on the Activity and Stability of Recombinant Oligo-1,6-Glucosidases*

The relative activities of BsOG and BlOG at various pHs were measured with two different buffer systems at 37 °C. The effects of pH over a range of 4.0–10.0 on the activities of BsOG and BlOG. The optimal pH of BsOG and BlOG were 7.0 and 6.5. BsOG had a relatively broad pH optimum ranging from 6.0 to 9.5. The optimum pH range of BlOG was 5.5–7.5.

3.4 *Kinetic Parameters of Recombinant Oligo-1,6-Glucosidases*

Michaelis-Menten and Lineweaver-Burk plots were used to calculate the kinetic parameters. Although both enzymes hydrolyzed *p*NPG, isaomaltose and isomaltotriose, their kinetic parameters were different. BsOG catalyzed *p*NPG, isomaltose and isomaltotriose with K_{ms} of 0.1, 0.59 and 8.60 mM. BlOG showed K_m values of 0.27, 0.86 and 10.8 mM for *p*NPG, isomaltose and isomaltotriose.

3.5 *Substrate Specificity of Recombinant Oligo-1,6-Glucosidases*

The ability of BsOG and BlOG to hydrolyze various di- and maltooligosaccharides, as well as α-glucan polymers, such as amylose and amylopectin, was determined. As shown in Table 1, both BsOG and BlOG hydrolyzed isomaltose, isomaltotriose, isomaltulose, panose, sucrose, amylopectin and maltodextrin, and exhibited weak activity against amylose. However, no activity was observed toward maltose and maltotriose. BsOG and BlOG also exhibited α-1,2-glucosidase activity on sucrose, in accord with the substrate specificity of isomaltase from *S. cerevisiae* [2]. This restricted substrate specificity indicated that these two enzymes were oligo-1,6-glucosidases [7]. Oligo-1,6-glucosidase prefers isomaltotriose, and hyrolyzes IMOs and dextran [7]. On the other hand, *S. cerevisiae* isomaltase preferentially cleaves

Table 1 Substrate specificity of recombinant enzymes BsOG and BlOG

Substrate	BsOG	BlOG
Isomaltose	+	+
Isomaltotriose	+	+
Isomaltulose	+	+
Panose	+	+
Maltose	−	−
Maltotriose	+	+
Sucrose	+	+
Amylose	−/+	−/+
Amylopectin	+	+
Maltodextrin	+	+

Note Plus sign (+) indicates exhibiting activity toward the corresponding substrate; minus/plus sign (−/+) represents have weak activity; minus sign (−) means no activity is detected

isomaltose and methyl α-D-glucopyroside, but does not act on isomaltotriose and isomaltotetraose [8]. These differences in the specificities for substrate chain-length may be partly accounted for by the differences in the shapes of the active sites.

3.6 *Hydrolysis of Isomaltotiose by Recombinant Oligo-1,6-Glucosidases*

As shown in Fig. 1, both BsOG and BlOG hydrolyzed isomaltotriose, their hydrolysis rates were similar. Glucose increased throughout the course of the reactions, whereas isomaltotriose decreased continuously. When isomaltotriose was hydrolyzed by BsOG, isomaltose rapidly accumulated during the first hour and the relative content of isomaltotriose decreased to 6.4%. At the end of this period, the highest content of isomaltose (36%) was achieved, preceding a quick fall in this sugar level. The relative content of panose resulting from the isomerization of isomaltotriose was maintained between 6.5 and 8.0%. In contrast, isomaltotriose hydrolyzed by BlOG was the same as BsOG. The amount of isomaltose rose gradually and reached a maximum (38%) at 1 h of reaction, and slowly decreased since then. The content of panose was between 4 and 5.5% in the process. All these findings indicated that isomaltotriose can be hydrolyzed by BsOG and BlOG to glucose by cleaving single glucosyl groups from the non-reducing end. After hydrolysis by BsOG and BlOG for 18 h, isomaltotriose was completely converted into glucose, and there were no detectable panose and isomaltose (data not shown).

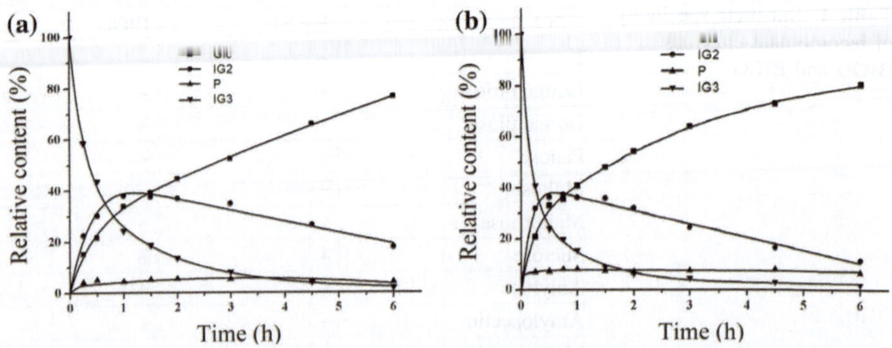

Fig. 1 The hydrolysis process of isomaltotriose by BsOG and BlOG. **a** BsOG; **b** BlOG; *down triangles* isomaltotriose; *up triangles* panose; *closed squares* glucose; *circles* isomaltose. The initial concentration of isomaltotriose was set as 100%. The *error bars* indicate standard deviations

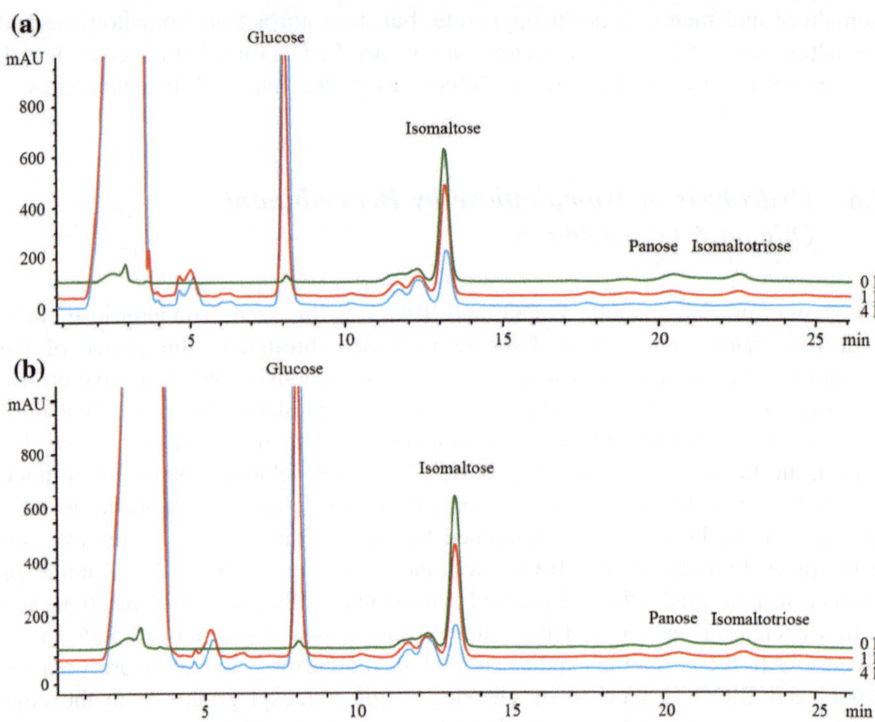

Fig. 2 HPLC profiles of IMOs hydrolyzed by BsOG and BlOG for different times. **a** BsOG; **b** BlOG; *red* and *light blue lines* represent the samples that were hydrolyzed for 1 and 4 h, respectively. *Green line* represents IMOs

3.7 Hydrolysis of IMOs by Recombinant Oligo-1,6-Glucosidases

HPLC analysis showed only three products in each digest of IMOs (Fig. 2). One was shown to be glucose, which was the sole product from isomaltose. Another was identified as maltose from panose. The third product was isomaltose from isomaltotriose. The hydrolysis modes of panose and isomaltotriose suggested that the linkage to be split in each isomaltosaccharide was at the non-reducing terminal [7]. We found that isomaltotriose in IMOs was nearly completely converted into isomaltose and glucose after hydrolyzed by BsOG for 4 h, only a small amount of panose left. When IMOs was hydrolyzed by BlOG, isomaltose and isomaltotriose were rapidly converted into glucose, while panose was hydrolyzed into maltose and glucose. At the end of the reaction, isomaltotriose was almost completely hydrolyzed and only a small quantity of panose and isomaltose left.

4 Discussion and Conclusion

In our study, the oligo-1,6-glucosidases of *B. subtilis* and *B. licheniformis* were cloned and extracellularly overexpressed with the signal peptide (α-factor) of pPIC9K in *P. pastoris* GS115. Previously, oligo-1,6-glucosidases from *B. thermoglucosidasius* and *B. cereus* ATCC7064 were expressed in *E. coli*, only 1.7 and 22.8% of recombinant enzymes were secreted into the medium [9]. The high activities of two oligo-1,6-glucosidases show their industrialization prospects.

Some characteristics of the recombinant enzymes BsOG and BlOG were then investigated. Despite BsOG and BlOG, oligo-1,6-glucosidases from *Bacillus mycoides*, *Thermomyces lanuginosus Bifidobacterium*, *Aspergillus niger*, *A. oryzae*, *S. cerevisiae*, were also being expressed and characterized by our group (report elsewhere). Our results have provided a foundation for exploiting oligo-1,6-glucosidases with desired biochemical properties from various microorganisms.

References

1. Murphy C, Powlowski J, Wu M, Butler G, Tsang A (2011) Curation of characterized glycoside hydrolases of fungal origin. Database bar020
2. Deng X, Petitjean M, Teste M-A, Kooli W, Tranier S, François JM, Parrou J-L (2014) Similarities and differences in the biochemical and enzymological properties of the four isomaltases from *Saccharomyces cerevisiae*. FEBS Open Biol 4:200–212
3. Larner J, McNickle C (1955) Gastrointestinal digestion of starch I. The action of oligo-1,6-glucosidase on branched saccharides. J Biol Chem 215:723–736
4. Hauri H-P, Quaroni A, Isselbacher KJ (1979) Biogenesis of intestinal plasma membrane: posttranslational route and cleavage of sucrase–isomaltase. Proc Natl Acad Sci 76:5183–5186

5. Jespersen HM, MacGregor EA, Henrissat B, Sierks MR, Svensson B (1993) Starch- and glycogen-debranching and branching enzymes: prediction of structural features of the catalytic (β/α) 8-barrel domain and evolutionary relationship to other amylolytic enzymes. J Protein Chem 12:791–805
6. Suzuki Y, Yuki T, Kishigami T, Abe S (1976) Purification and properties of extracellular α-glucosidase of a thermophile, *Bacillus thermoglucosidius* KP 1006. Biochim Biophys Acta (BBA)-Enzymol 445:386–397
7. Suzuki Y, Aoki R, Hayashi H (1982) Assignment of a p-nitrophenyl-α-D-glucopyranoside-hydrolyzing α-glucosidase of *Bacillus cereus* ATCC 7064 to an exo-oligo-1,6-glucosidase. Biochim Biophys Acta (BBA)-Protein Struct Mol Enzymol 704:476–483
8. Yamamoto K, Miyake H, Kusunoki M, Osaki S (2010) Crystal structures of isomaltase from *Saccharomyces cerevisiae* and in complex with its competitive inhibitor maltose. FEBS J 277:4205–4214
9. Watanabe K, Kitamura K, Iha H, Suzuki Y (1990) Primary structure of the oligo-l,6-glucosidase of *Bacillus cereus* ATCC7064 deduced from the nucleotide sequence of the cloned gene. Eur J Biochem 192:609–620

Transcriptomic Analysis of *CYP* Genes in *Rhizopus nigricans* and Identification of the Steroid 11α-Hydroxylase Candidate Genes

Jianguo Zhao, Pengcheng Sui, Ruijie Wang, Ziyin Zhang, Fuping Lu, Zhengxiang Wang and Xiaoguang Liu

1 Introduction

Steroids are an important class of drugs widely used as anti-inflammatory, anabolic and anticancer agents, and contraceptives as well [1–3]. Oxyfunctionalization of the steroid core structure at specific positions is required for the appropriate pharmaceutical activities of steroids. However, insertion of an oxygen atom in certain positions of a steroid molecule by synthetic chemistry is inefficient and requires multiple complex protection and deprotection reactions owing to the unique chemical characteristics of steroids [4]. In the steroid industry, microbial biotransformation is routinely used to effect specific oxygenation of steroids because of its high regio- and stereoselectivity and better efficiency of the process [5–7]. *Rhizopus nigricans* is the major fungal species used in China for selective hydroxylation of 16,17α-Epoxyprogesterone (EP) on a industrial scale for the preparation of 11α-hydroxy-16,17α-epoxyprogesterone (HEP) [5], a key steroid intermediate in the synthesis of a variety of glucocorticoid drugs. However, in the industrial setting, the low conversion (40–50% with concentrations of 20–25 g/L of EP) of the current *R. nigrican*s production strains is one of the major factor limiting the process efficiency. It is known that the fungal steroid hydroxylase of *R. nigricans* is a cytochrome P450 protein (CYP) associated with the endoplasmic reticulum [8]. However, the genetic determinant of the responsible hydroxylase in

J. Zhao · P. Sui · R. Wang · Z. Zhang · F. Lu
Key Laboratory of Industrial Fermentation Microbiology, Ministry of Education,
College of Bioengineering, Tianjin University of Science and Technology (TUST),
Tianjin 300457, China

Z. Wang · X. Liu (✉)
Laboratory of Biocatalysis and Biotransformation, The College of Chemical Engineering
and Materials Science, Tianjin University of Science and Technology (TUST),
Tianjin 300457, China
e-mail: liu_xg@tust.edu.cn

H. Liu et al. (eds.), *Advances in Applied Biotechnology*, Lecture Notes
in Electrical Engineering 444, https://doi.org/10.1007/978-981-10-4801-2_3

R. nigricans remains to be identified even though it has been used for commercial steroid hydroxylation reactions for over half of a century since the discovery of its ability to perform selective 11α-hydroxylation of progesterone in 1952 [9]. It is expected that the cloning and identification of the steroid 11α-hydroxylase gene in *R. nigricans* will aid in the engineering of more efficient industrial strains for steroid 11α-hydroxylation reactions.

In the present study, we took advantage of the power of RNA-sequencing technology and also the fact the fungal steroid 11α-hydroxylase is a member of the cytochrome P450 superfamily to establish the list of *CYP* genes expressed under biotransformation conditions, then used quantitative RT-PCR to determine the highly induced *CYP* genes to identify steroid 11α-hydroxylase candidate genes in *R. nigricans*.

2 Materials and Methods

2.1 Materials

16,17α-Epoxyprogesterone (EP) and 11α-hydroxy-16,17α-epoxyprogesterone (HEP) were obtained from Solarbio (Beijing, China). Peptone and Yeast Extract were purchased from Oxoid Ltd. All other chemicals were of analytical grade and purchased from Merck.

2.2 Microorganisms and Cultural Conditions

The filamentous fungus *N. nigricans* TCCC41047 from the microbial strain collection of the applied microbiology lab of Tianjin University of Science and Technology was routinely maintained on potato dextrose agar (PDA) slants.

2.3 Determination of Induction of R. nigricans *Steroid Hydroxylase Activities*

R. nigricans TCCC41047 was cultivated on a PDA agars slant at 28 °C for 5 days, and then conidia were collected using sterile water and adjusted to about 10^6 spores/ml. Conidia suspension (50 μl) was inoculated into 50 ml PDA medium and cultivated at 28 °C on a rotary shaker at 180 r/min. After 12 h, the culturex were treated with 0.05% (W/V) EP in ethanol, while the control was treated by an equal volume of ethanol only. After 6 h induction, mycelia were collected and then filtered, rinsed three times with saline water and resuspended in 20 ml of fresh phosphate buffer (1 mM sodium phosphate, 0.2 mM EDTA, 0.04 mM glutathione, pH 5.5). EP and cycloheximide (100 mg/ml) was then added together to respective final concentrations of 1 and 0.03 mg/ml for each treatment, incubated for 24 h at 28 °C with shaking at 180 rpm, and one milliliter fermentation broth was extracted with ethyl acetate and analyzed by thin layer chromatography (TLC).

2.4 Total RNA Isolation and Transcriptome Sequencing

Total RNAs were isolated using Trizol reagents (Promega, USA) from *R. nigricans* TCCC41047 cultures treated with 0.01% (W/V) EP for 6 h and without EP respectively. After the RNA integrity and purity were assessed by the NanoDrop (NanoDrop Technologies, USA), RNA samples were delivered to BGI-Beijing (Beijing Genomics Institute) for sequencing using Illumina HiSeq™ 2000 (Illumina, Inc. USA) (http://www.genomics.cn/index) for transcriptome sequencing BGI.

2.5 Identification of CYP Genes Expressed Under Steroid Induction

The returned RNA sequencing data were analyzed to identify expressed *CYP* genes under substrate induction. A set of prediction algorithms were used to identify the full length of Open Reading Frames (ORFs) of *CYP* genes, including ORF Finder (http://www.ncbi.nlm.nih.gov/gorf/gorf.html). The catalytic domains of predicted CYPs were identified by InterProScan (http://www.ebi.ac.uk/Tools/InterProScan). The sequences of *CYP* genes were analyzed to identify their similarity to known putative sequences using NCBI BLAST (http://blast.ncbi.nlm.nih.gov/).

2.6 First Strand CDNA Synthesis and Real Time RT-PCR

Reverse transcription of the first cDNA strand was performed using 3 μg of the total RNA with PrimeScript Reverse Transcriptase (TaKaRa, Dalian, PR China) in a 20 μl reaction volume according to the manufacturer's instructions. To quantify expression levels of expressed *CYP* genes under induction and no-induction, real-time quantitative PCR (qRT-PCR) was performed as follows: the total volume of PCR reaction is 20 μl containing 10 μl of Master Mix with SYBR (Solarbio, Beijing, China), 300 nmol/l of both primers (Table 1) and 1 μl of cDNA template. The qRT-PCR amplification parameters comprise denaturation (95 °C, 10 min), 40 cycles of denaturation (95 °C, 30 s) and annealing (60 °C, 30 s) (Applied Biosystems, USA). The transcript level of glyceraldehyde-3-phosphate dehydrogenase gene (GAPDH) was used as an internal control and relative to a calibrator, is given by amount of target = $2^{-\Delta\Delta CT}$ [10].

Primers for real-time PCR reactions were as follows (Table 1):

Table 1 Primers for qRT-PCR of expressed *CYP* genes

Gene name	Upstream (5′–3′)	Downstream (5′–3′)
Unigene4381	CCTGCATATGTTACCACCG	CAATGTCGGCTGCCACTAC
Unigene5826	TGGCCAGATGGTACACATG	CAAGTGCCTCAAGTGTTGC
Unigene2681	GACACGAAGTGAGGATTGC	TGTCCAACCTGCTGTGGTG
Unigene8057	TAGCCATGCATTCCAACC	GGAATCAAGTTCTGGTGC
Unigene11442	TACAGGAGAGGATGGTAG	ACTCCCATTCATCAGACGC
Unigene11598	CGATCCTTTGCGATGGATC	TAATCAGACTTGGGCACACC
Unigene12673	GTGTCTGTGGTTACTGGC	TGGGTACTACATAGGGAC
Unigene9260	CATGGCTTCCTTCACAGAC	AGGTGGTGTGAATACTAGCG
Unigene13341	ACTACTGCTGTTGTCTGTTC	CACACAAGATAATTCACCAC
Unigene3103	ATCAAGACCTACCGTGACGA	GATCAGACACCATCTCAGTC
Unigene14584	CAATGCTTTGCCTTATCTG	AACGCGATGGGTTAAATC
Unigene6658	TTGGTGACGAAACACCTTC	CCAACCTTGTCAGTCTTCC
Unigene2821	ACACGATCCATGTATGCTTC	GCACTGGTCCAACTGTGAC
Unigene612	CAATGGACGAATTCGGTGC	CAAAGAAGGCATGCTTGAC
GAPD	AAA GAC TGG AGA GGT GGT CG	ACC GCT TTC AGA CGC TGT C

3 Results and Discussion

3.1 *EP Induction of* R. nigricans *Steroid 11α-Hydroxylase Activity*

R. nigricans TCCC41047 is a production strain currently used for commercial preparation of 11α-hydroxy-EP. Previously it was reported that steroid 11α-hydroxylase activities in *R. nigricans* were induced by substrate [8, 10, 11, 12]. To facilitate the search for the target steroid 11α-hydroxylase gene, we first evaluated whether 11α-hydroxylase activity was induced by the steroid substrate in strain TCCC41047 and also the extent of induction. Protein synthesis inhibitor cycloheximide (100 mg/ml) was used to examine its effect on EPC 11α-hydroxylase activity. As shown in Fig. 1, treatment of strain TCCC41047 cultures with cyclohexmide inhibited the formation of HEP product, whereas HEP was produced in the control, indicating that the target enzyme 11α-hydroxylase, which converts EP to HEP, was produced only after the addition of substrate EP. The above result demonstrated that the expression of the steroid 11α-hydroxylase gene was strongly induced at the transcriptional level and its expression under non-induction conditions is likely to be at a very low level.

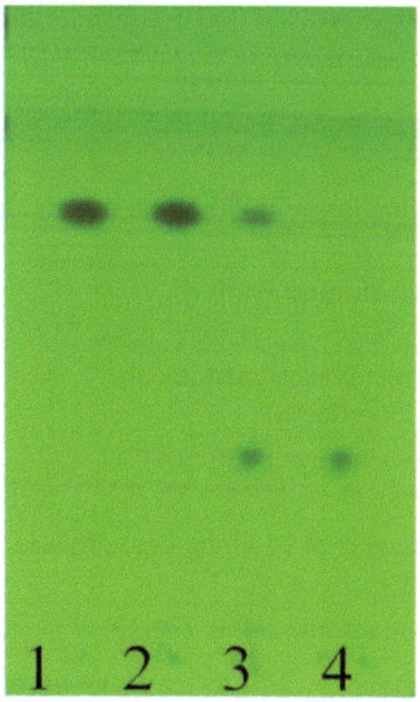

Fig. 1 TLC assay of induction of steroid 11α-hydroxylase activities in *R. nigricans* by EP. The detailed biotransformation procedure was described in materials and methods. *1* Authentic EP, *2* non-induction + EP + cycloheximide, *3* substrate induction for 6 h, followed by EP + cycloheximide, *4* Authentic C11α-OH EP

3.2 Transcriptomic Profiling of CYP *Genes in* R. nigricans *Under Substrate Induction*

Given that the steroid 11α-hydroxylase in *R. nigricans* TCCC41047 is known to be a member of the cytochrome P450 superfamily, we focused on mining *CYP* genes expressed under substrate induction. Analyses of RNA-seq data identified 14 unique transcript sequences capable of encoding CYP enzymes (Table 2), indicating that at least 14 *CYP* genes were expressed under EP induction. The above results also suggest that one of 14 expressed *CYP* genes could encode the target 11α-hydroxylase in *R. nigricans* TCCC4104.

3.3 Identification of Candidate Steroid 11α-Hydroxylase Genes by qRT-PCR

Since we demonstrated that the steroid 11α-hydroxylase was expressed only under EP induction, real-time RT-PCR was employed to determine the degree of expression induction of the 14 *CYP* genes, and a *GAPD* gene was used as an internal control. As shown in Tables 2 and 3 genes (Unigene11442, Unigene3103 and Unigene612) were highly induced after 6 h EP treatment, thus establishing these three *CYP* genes are the 11α-hydroxylase candidate genes in *R. nigricans*.

Table 2 Cytochrome P450 genes expressed under substrate induction in *R. nigricans*

Number	Gene id	Swissprot-annotation	COG-function-description
1	Uni gene4381	Cytochrome P450 3A5	Cytochrome P450
2	Uni gene5826	Cytochrome P450 524A1	Cytochrome P450
3	Uni gene2681	Cytochrome P450 61	Cytochrome P450
4	Uni gene8057	O-methylsterigmatocystin oxidoreductase	Cytochrome P450
5	Uni gene11442	Cytochrome P450 509C	Cytochrome P450
6	Uni gene11598	Lanosterol 14-alpha demethylase	Cytochrome P450
7	Uni gene12673	Cytochrome P450 72A15	Cytochrome P450
8	Uni gene9260	Ent-kaurene oxidase	Cytochrome P450
9	Uni gene13341	Cytochrome P450 503A1	Cytochrome P450
10	Uni gene3103	Paxilline synthesis protein P	Cytochrome P450
11	Uni gene14584	Cytochrome P450 52A2	Cytochrome P450
12	Uni gene6658	Cytochrome P450 52A3	Cytochrome P450
13	Uni gene2821	Phenylacetate 2-hydroxylase	Cytochrome P450
14	Uni gene612	Bifunctional P-450:NADPH-P45	Cytochrome P450

Table 3 *CYP* genes highly induced in *R. nigricans* by substrate as revealed by qRT-PCR

Gene id	Ct/Non-induction	Ct/Induction (6 h)	Fold differences
GAPDH	17.91	18.05	–
Unigene11442	30.54	22.60	270.60
Unigene3103	30.48	22.99	198.09
Unigene612	28.10	21.58	101.13

4 Conclusions

Transcriptomic analyses revealed that an expression repertoire of 14 *CYP* genes under EP induction in the production strain *R. nigricans* TCCC41047 and qRT-PCR identified 3 *CYP* genes that were highly induced by steroid substrate as the 11α-hydroxylase candidate genes.

Acknowledgments This work was financially supported by a grant from the National High Technology Research and Development Program of China (863 Program) (No. 2011AA02A211).

References

1. Ayer DE, Schlage CA, Flynn GL (1976) Anti-inflammatory steroid. US Pat 3980 778
2. Salvador, Jorge AR, et al (2012) Anticancer steroids: linking natural and semi-synthetic compounds. Nat Prod Rep 30(2):324–374
3. Mahat SB, Garai S (1997) Advances in microbial steroid biotransformation. Steroids 62: 332–345
4. Zeelen FJ (1990) Pharmaceutical chemistry of steroids. Elsevier, Amsterdam
5. Žakelj-Mavrič M, Belič I (1987) Hydroxylation of steroids with 11α-hydroxylase of *Rhizopus nigricans*. J Steroid Biochem 28(2):197–201
6. Sonomoto K, Hoq MM, Tanaka A, Sukui S (1983) 11β-Hydroxylation of cortexolone (Reichstein compound S) to hydrocortisone by *Curvularia lunata* entrapped in photo-cross-linked resin gels. Appl Environ Microb 45(2):436–443
7. Donova MV, Egorova OV (2012) Microbial steroid transformations: current state and prospects. Appl Microbiol Biotechnol 94:1423–1447
8. Breskvar K, Hudnik-Plevnik T (1981) Inducibility of cytochrome P-450 and of NADPH-cytochrome c reductase in progesterone treated filamentous fungi *Rhizopus nigricans* and *Rhizopus arrhizus*. J Steroid Biochem 14:395–399
9. Murray HC, Peterson DH (1952) Oxygenation of steroids by Mucorales fungi. US Pat 2602 769
10. Livak KJ, Schmittgen TD (2001) Analysis of relative gene expression data using real-time quantitative PCR and the $2^{-\Delta\Delta CT}$ method. Methods 25:402–408
11. Irrgang S, Schlosser D, Schmauder HP (1992) The steroid 15α-hydroxylase of *Penicillium raistrickii* i477 is inducible. Biotechnol Lett 14:33–38
12. Lin YY, Smith LL (1970) Microbial hydroxylations: VIII. Induction of steroid hydroxylases of *Curvularia lunata* by 19-nortestosterone analogs. Biochim Biophys Acta 218:526–531

Identification of Genes Encoding Receptors for Six *Pseudomonas aeruginosa* Phages

Jiajia You, Xiaoli Cui, Li Sun, Xiaojing Yang and Hongjiang Yang

1 Introduction

Pseudomonas aeruginosa is one of the main pathogens of nosocomial infection, due to the ever increasing occurrence of multidrug resistant clinical isolates, there is an urgent need for development of new types of antimicrobial agents [1]. Phages can specifically lyse host bacterial cells, and can be used as alternatives to antibiotics. Phage therapy has been practiced in the control of *P. aeruginosa* infection for 50 years, and some animal experiments have confirmed the effectiveness of the relevant phage therapies [2–5].

The emergence of phage-resistant strains has become a big obstacle in the practice of phage therapy [6–9]. Phage cocktails are often used in the treatments. Phage cocktails can expand the host range and effectively reduce phage resistant strains in the process of treatment. In Georgia, phage cocktails have been widely applied in the treatments of suppurative skin, wounds, and lung infection causing by some pathogenic bacteria such as *Staphylococcus*, *Streptococcus*, *Proteus*, *Escherichia coli*, and *P. aeruginosa* [10].

Receptor recognition and adsorption is the first step of the phage infection process. A variety of molecules can be used as phage receptors, such as the different components of LPS, outer membrane proteins, teichoic acid, polysaccharide capsule or mucous layer, pilus, and flagella. In *E. coli*, phages K2, SST, T5, and H8 adsorbed to OmpA, OmpC, TonA, and FepA, respectively. Phage K20, Ox2, TuII, T2, and T4 can recognize LPS and some outer membrane proteins as receptors [11]. Phage CP-T1 and φYeO3-12 adsorbed to the side chains of O antigen of *Yersinia*

J. You · X. Cui · L. Sun · X. Yang · H. Yang (✉)
Key Laboratory of Industrial Fermentation Microbiology
Ministry of Education Tianjin Key Laboratory of Industrial
Microbiology College of Biotechnology, Tianjin University
of Science and Technology, Tianjin 300457, China
e-mail: hongjiangyang@tust.edu.cn

H. Liu et al. (eds.), *Advances in Applied Biotechnology*, Lecture Notes
in Electrical Engineering 444, https://doi.org/10.1007/978-981-10-4801-2_4

enterocolitica serotype O:3 [12]. Several *Salmonella* phage receptors are including FhuA, TolC, BtuB, OmpC, Vi capsular antigen, LPS, and flagella [13]. The receptor of *Vibrio cholera* phage CTXΦ is TCP pili. Phage fs1, fs2, VGJ, and KSF recognize MSHA pili as the receptor. In *P. aeruginosa*, phage receptors include LPS and pilus [14–19].

In this study, the receptors of 6 previously isolated *P. aeruginosa* phages were studied by constructing the Tn5G transposon insertion library of *P. aeruginosa* PAK. The phage resistant mutants were characterized and the disrupted genes were identified with inverse PCR. The results will provide the basis of the potential applications of the *P. aeruginosa* phages.

2 Materials and Methods

2.1 Bacterial Strains, Plasmids, and Phages

Pseudomonas aeruginsa strains used in this study were listed in Table 1 *P. aeruginsa* phages O1, O2, K1, K2, K3, and C11 were previously isolated using different *P. aeruginosa* strains as the indicators [11].

2.2 Construction of Tn5G Transposon Library

As described previously, the Tn5G transpon insertion libraries of *P. aeruginosa* PAK were constructed [24]. *P. aeruginosa* phages O1, O2, K1, K2, K3, and C1, were applied to screen the phage-resistant mutants, respectively. Spotting assay and the double-layer plate methods were used to test the phage sensitivity and phage titers. The adsorption rate of the selected phage-resistant mutants was analyzed as previously described [11].

2.3 Inverse PCR Identification of the Tn5G Insertion Sites

Genomic DNA was extracted from the phage-resistant mutants with the phenol-chloroform extraction method [25]. The restriction enzymes *Taq*I enzyme was used to digest the purified genomic DNA at 65 °C for 4 h. After purification, the digested DNA was ligated with T4 DNA ligase at 16 °C for 12 h. The ligation product was subsequently used as templates for amplification using the primers OTn1 and OTn2 listed in Table 2. The sequences obtained from the PCR products were analyzed by searching the database in GenBank of NCBI.

Table 1 Strains and plasmids used in this study

Strain/plasmid/phage	Description	Source
P. aeruginosa		
PAK	Wild-type	[20]
RO1-1, RO1-5	*wbpO*::Gm^R mutant of PAK, resistant to phage O1	This study
RO1-2	*wbpV*::Gm^R mutant of PAK, resistant to phage O1	This study
RO1-3, RO1-6	*wbpR*::Gm^R mutant of PAK, resistant to phage O1	This study
RO1-4	*wbpL*::Gm^R mutant of PAK, resistant to phage O1	This study
RO2-1[a]	*wzy*::Gm^R mutant of PAK, resistant to phage O2	This study
RK1-1	*PA4367*::Gm^R mutant of PAK, resistant to phage K1	This study
RK1-2	*PA0583*::Gm^R mutant of PAK, resistant to phage K1	This study
RK2-2	*PA5004*::Gm^R mutant of PAK, resistant to phage K2	This study
RK3-1	*wbpR*::Gm^R mutant of PAK, resistant to phage K3	This study
RK3-2	*wbpT*::Gm^R mutant of PAK, resistant to phage K3	This study
RK3-4	*PALES_16971*::Gm^R mutant of PAK, resistant to phage K3	This study
RK3-8	*wbpV*::Gm^R mutant of PAK, resistant to phage K3	This study
RC10-1	*PA5001*::Gm^R mutant of PAK, resistant to phage C10	This study
RC10-2	*PALES_16971*::Gm^R mutant of PAK, resistant to phage C10	This study
Plasmid		
pUCP18	Broad host range shuttle vector, Ap^r	[21]
pRK2013Tn5G	used for transposon mutagenesis, Gm^r	[22]
pLY1201	*wzy* gene driven by P_{lac} promoter cloned in pUCP18, Ap^r	[23]
pXW1501	*PA5001* gene driven by P_{lac} promoter cloned in pUCP18, Ap^r	[23]
pZM1502	*wbpT* gene with promoter region cloned in pUCP18, Ap^r	This study
pXW1503	*wbpO* gene driven by P_{lac} promoter cloned in pUCP18, Ap^r	[23]
pXW1502	*wbpL* gene driven by P_{lac} promoter cloned in pUCP18, Ap^r	This study
pFJ1501	*wbpV* gene driven by P_{lac} promoter cloned in pUCP18, Ap^r	[23]
pXL1504	*wbpR* gene with promoter region cloned in pUCP18, Ap^r	[23]
pXL1509	*bifA* gene driven by P_{lac} promoter cloned in pUCP18, Ap^r	This study
pXL1510	*PA0583* gene driven by P_{lac} promoter cloned in pUCP18, Ap^r	This study
pXL1511	*PA5004* gene driven by P_{lac} promoter cloned in pUCP18, Ap^r	This study
Phage		
O1, O2, K1, K2, K3	*P. aeruginosa* lytic phage	[11]
C10	*P. aeruginosa* lytic phage	[11]

[a]The mutants RO2-1, RO2-2, RO2-3, RO2-8, RO2-10, RO2-13, RO2-15, and RO2-25 also had the gene *wzy* disrupted

Table 2 Primers used for PCR

Primer	Primer sequence (5′–3′)	Target region
OTn1	GATCCTGGAAAACGGGAAAG	Genome DNA regions bordering the Tn5G
OTn2	CCATCTCATCAGAGGGTAGT	
bifA-F	CCCAAGCTTCGCATCCCTTCATTCTCTC	Gene *bifA*
bifA-R	CCCAAGCTTATTTCGCTGCCTGAGTCTG	
WbpL-F	CGCAAGCTTATGAAAACCCGCCAAGTG	Gene *wbpL*
WbpL-R	CGCGGATCCGGGCAGTCCATTTCCTTG	
PA0583-F	CCCAAGCTTTGCAGAGCTTCGACGTGT	Gene *PA0583*
PA0583-R	CCCAAGCTTTCGCATTTCCAGTTGGTG	
PA5004-F	CCCAAGCTTACTTCGCTCAATCGCCGTG	Gene *PA5004*
PA5004-R	CCCAAGCTTCCCAACCCAGGTAACGCAC	

2.4 Complementation Experiment

The target genes disrupted in the phage-resistant mutants were amplified using the primers listed in Table 2, the products were cloned the vector pUCP18, and the resulting recombinant plasmids were transformed into the corresponding phage-resistant mutants. With the spotting assay, the phage sensitivity of the transformants was tested as described previously [26].

2.5 Swimming and Swarming Analysis

The ability of swimming and swarming was analyzed according to the method previously described [27]. Briefly, one single colony was inoculated in 5 ml LB medium. The overnight culture was transferred into fresh LB medium and cultured to the logarithmic phase. Take 1 μl culture and drop on the surface of the casein medium with 0.3% agar. After standing at desktop for 30 min, the plates were incubated at 37 °C for 8–12 h.

3 Results

3.1 Host Range Analysis

Host ranges of *P. aeruginosa* phages O1, O2, K1, K2, K3, and C10 were determined by spotting assay. Sixteen clinical *P. aeruginosa* strains were included in the assay. With the tested strains, phage K1 and O2 displayed the same host ranges and the other phages showed different host ranges (Table 3).

Table 3 Host range analysis of the bacteriophages

Strains	Phages[a]					
	O1	O2	K1	K2	K3	C10
PAK	○	○	○	○	○	○
TJC135	○	○	○	○	○	○
TJC196	●	●	●	●	●	●
TJC102	○	○	○	●	●	○
TJC204	○	○	○	●	○	○
TJC124	○	○	○	○	○	○
TJC283	●	○	○	●	○	○
TJC285	●	●	●	●	●	●
TJC1107	●	●	●	●	●	●
TJC729	●	●	●	●	●	○
TJC730	●	●	●	●	●	●
TJC487	○	○	○	○	○	○
TJC422	○	○	○	○	○	○
TJC337	●	●	●	●	●	●
TJC304	●	●	●	●	●	●
TJC415	●	●	●	●	●	○
TJC307	●	●	●	●	●	●

[a]*Open circle* represented transparent zones in spotting assay; *filled circle* represented no transparent zones in spotting assay

3.2 Screening Phage-Resistant Mutants

To further analyze phage receptors, the Tn5G transposon insertion banks of PAK were constructed and used in the screening of the phage-resistant mutants. After verifying with the double-player plate method, totally 23 mutants were selected, including 6 strains resistant to phage O1, 8 strains resistant to phage O2, 2 strains resistant to phage K1, 1 strain resistant to phage K2, 4 strains resistant to phage K3, and 2 strains resistant to phage C10 (Fig. 1).

3.3 Adsorption Rate Determination

The adsorption efficiency was tested between phage-resistant mutants and the corresponding phages. As shown in Fig. 1, all adsorption rates were significantly lower than the parent strain PAK, ranging from 13.84% to 55.41% (Fig. 1). The results implied that all the isolated phage-resistant mutants had their phage receptors impaired.

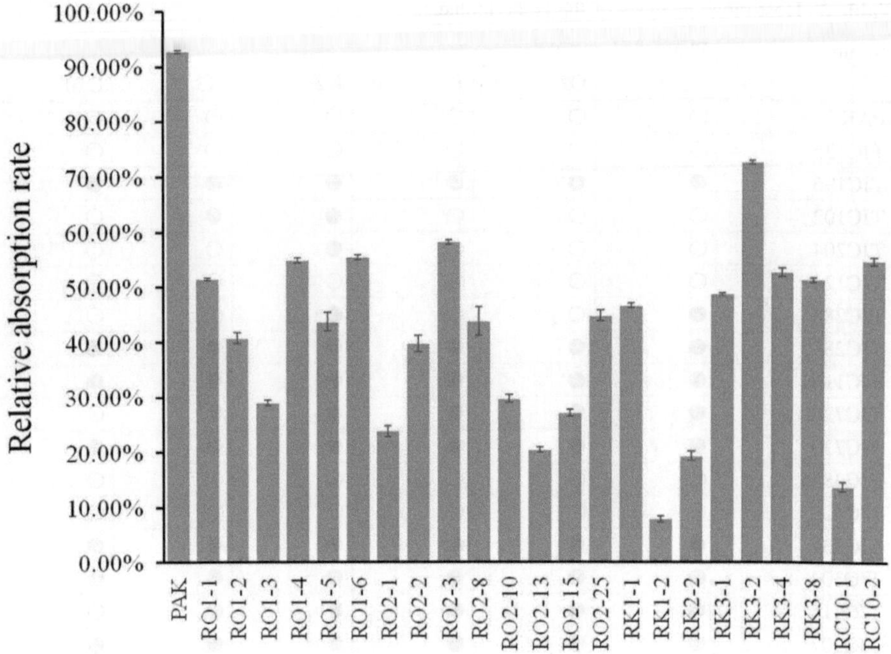

Fig. 1 Adsorption rates of the isolated phage-resistant mutants. PAK was used as control

3.4 Discrimination the Phage-Resistant Mutants

With the spotting assay, the isolated phage-resistant mutants were discriminated by testing the sensitivity to the six phages. The majority of them showed the same patterns as the other members isolated from the same group with a few exceptions, including RO1-4 and RO1-6, RK1-1, and RK3-1 (Table 4). The results suggested the six phages may share common phage receptors.

3.5 Identification of the Tn5G Insertion Sites

With inverse PCR, totally 10 different genes were found disrupted in the 23 phage-resistant mutants (Table 1). In the mutants resistant to phage O1, four genes were found disrupted, including *wbpO* gene encoding UDP-glucose/GDP-mannose dehydrogenase-like protein, *wbpR* gene encoding a glycosyltransferase, *wbpL* gene encoding a glycosyl transferase group 4-like protein, and *wbpV* gene encoding a NAD dependent epimerase/dehydratase-like protein. All mutants resistant to phage O2 had the gene *wzy* disrupted and it encoded an O-antigen polymerase [28]. In the mutants resistant to phage K1, RK1-1 had Tn5G inserted in the gene *PAK_04826*,

Table 4 Sensitivity test of the phage-resistant mutants

Mutants	Phages[a]					
	O1	O2	K1	K2	K3	C10
RO1-1	●	●	●	●	●	●
RO1-2	●	●	●	●	●	●
RO1-3	●	●	●	●	●	●
RO1-4	●	●	●	●	●	○
RO1-5	●	●	●	●	●	●
RO1-6	●	●	●	●	●	○
RO2-1	●	●	●	●	●	●
RO2-2	●	●	●	●	●	●
RO2-3	●	●	●	●	●	●
RO2-8	●	●	●	●	●	●
RO2-10	●	●	●	●	●	●
RO2-13	●	●	●	●	●	●
RO2-15	●	●	●	●	●	●
RO2-25	●	●	●	●	●	●
RK1-1	○	○	●	○	○	○
RK1-2	●	●	●	●	●	●
RK2-2	●	●	●	●	●	●
RK3-1	●	●	●	●	●	○
RK3-2	●	●	●	●	●	●
RK3-4	●	●	●	●	●	●
RK3-8	●	●	●	●	●	●
RC10-1	●	●	●	●	●	●
RC10-2	●	●	●	●	●	●

[a]*Open circle* represented transparent zones in spotting assay; *filled circle* represented no transparent zones in spotting assay

highly similar to gene *bifA* of PAO1, encoding a Cyclic-Di-GMP phosphodiesterase located on the cell membrane, RK1-2 had gene *PAK_00800* disrupted. Its counterpart *PA0583* encodes a protein with the 2-amino-4-hydroxy-6-hydroxymethyldihydropteridine diphosphokinase activity, involved in the biosynthesis of folic acid. The mutant RK2-2 resistant to phage RK2 had gene *PAK_05511* disrupted, which is 99.08% similar to *wapH* gene in PAO1. In the mutants resistant to phage RK3, RK3-1 had gene *wbpR* disrupted, RK3-2 had gene *wbpT* disrupted, RK3-4 had gene *wzy* disrupted, RK3-8 had gene *PAK_02045* disrupted, encoding the NAD-dependent dehydratase WbpV. In the mutants resistant to phage C10, RC10-1 had gene PAK_05508 genes disrupted identical to gene *ssg* in PAO1, RC10-2 had gene *wzy* disrupted.

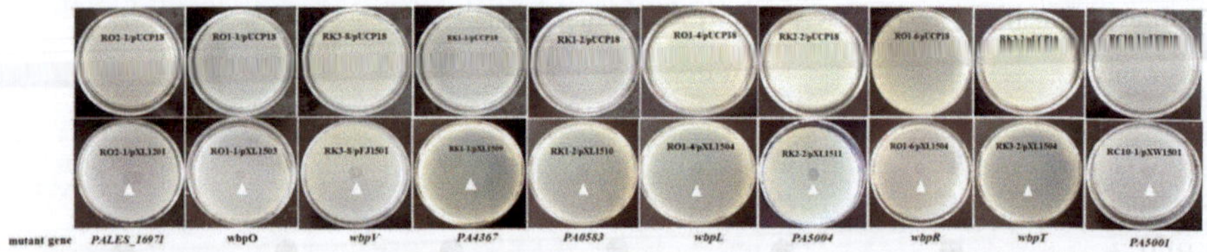

Fig. 2 Confirmation of the phage-resistant mutants. The recombinant plasmids carry the corresponding genes, pXL1201 carrying *wzy* gene, pXL1503 carrying *wbpO* gene; pFJ501 carrying *wbpV* gene, pXL1509 carrying *PA4367* gene, pXL1502 carrying *wbpL* gene, pXL1511 carrying *PA5004* gene; pXL1504 carrying *wbpR* gene, pZM1502 carrying *wbpT* gene, and pXL1511 carrying *PA5001* gene

Fig. 3 Swarming motility test of the *bifA* mutant

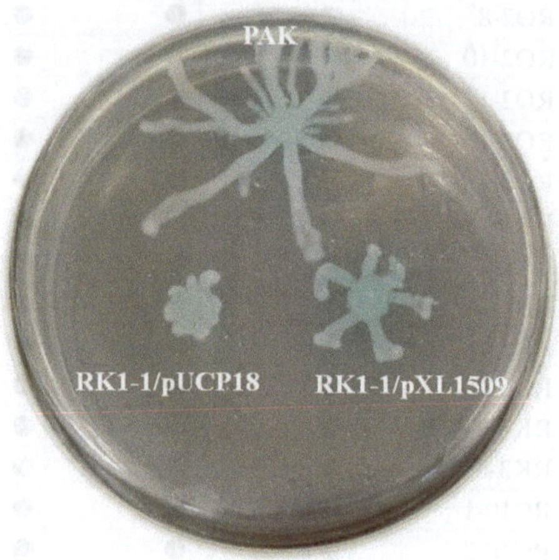

3.6 *Confirmation of the Phage-Resistant Mutants*

The target genes disrupted in the phage-resistant mutants were amplified and cloned into the plasmid pUCP18. The resulting constructs were transformed into the mutants to test the phage sensitivity of the transformants. As shown in Fig. 2, all the cloned target genes can trans-complement the disrupted genes and restore the phage sensitivity individually.

3.7 *Gene* bifA *Affects Swarming Ability of PAK*

Gene *bifA* encodes a phosphodiesterase and can degrade the intracellular second messenger c-di-GMP. In *bifA* mutant, the elevated c-di-GMP level inhibits the swarming motility of bacterial cells [29]. The *bifA* mutant displayed the least mobility ability, and the marginal region of the colony was much smaller than the strain RK1-1/pXL1509 and the parent strain PAK (Fig. 3).

4 Discussion

Phage adsorption is the first step in the phage infection. *P. aeruginosa* phages recognize a variety of molecules on the cellular surface as receptors to initiate the infection process, such as flagella, pili, and LPS. LPS is composed of three parts, including lipid A, core polysaccharide, and O antigen which has two different forms, CPA (common polysaccharide antigen) and OSA (O-specific antigen). OSA is L-heteroglycan, formerly called B-band. Multiple gene clusters are involved in the biosynthesis of LPS. In our work, the disrupted gene *wbpV*, *wbpR*, *wbpP*, *wbpO*, *ssg*, and *wbpL* direct involved in biosynthesis LPS [23]. The gene *wbpL* encoded the enzyme that involved in synthesis of the O-antigen side chain of LPS. Those gene were found related to the synthesis of phage receptors, demonstrating that all the six P. aeruginosa phages recognize LPS as their receptors, though it's possible that the phages may recognize the different structures of LPS during the adsorption step.

Among the ten disrupted genes, *bifA* gene was the only exception not directly involved in the biosynthesis pathway of LPS. BifA is predicted to be a c-di-GMP phosphodiesterase controlling the concentration of the second messenger c-di-GMP which affects cell motility, extracellular polysaccharide production, and biofilm formation [27]. The *bifA* mutant has the increased level of c-di-GMP, which leads the increased synthesis of a polysaccharide produced by the *pel* locus and decreased synthesis of OSA LPS [30]. And that's may be the reason the bifA mutant is resistant phage K1 infection.

In conclusion, all the six *P. aeruginosa* phages recognize LPS as their receptors. The minor difference in the host ranges may imply different phages adsorb to the different structures of LPS.

Acknowledgements This work was partly supported by The National Natural Science Foundation of China (grant 31370205 and 30970114).

References

1. Vaisvila R et al (2001) Discovery and distribution of super-integrons among Pseudomonads. Mol Microbiol 42(3):587–601
2. Soothill J (2013) Use of bacteriophages in the treatment of *Pseudomonas aeruginosa* infections. Appl Microbiol 11(9):909–915
3. Tiwari BR et al (2011) Antibacterial efficacy of lytic Pseudomonas bacteriophage in normal and neutropenic mice models. J Microbiol 49(6):994–999
4. Watanabe R et al (2007) Efficacy of bacteriophage therapy against gut-derived sepsis caused by *Pseudomonas aeruginosa* in mice. Antimicrob Agents Chemother 51(2):446–452
5. Velasquez MCSM, Fralick JA (2007) Phage therapy of *Pseudomonas aeruginosa* infection in a mouse burn wound model. Antimicrob Agents Chemother 51(6):1934–1938
6. Parasion S et al (2014) Bacteriophages as an alternative strategy for fighting biofilm development. Pol J Microbiol 63:137–145

7. Le S et al (2014) Chromosomal DNA deletion confers phage resistance to *Pseudomonas aeruginosa*. Sci Rep 4:4738
8. Friman VP et al (2013) *Pseudomonas aeruginosa* adaptation to lungs of cystic fibrosis patients leads to lowered resistance to phage and protist enemies. PLoS ONE 8(9):e75380
9. Blokhina I et al (1985) Distribution of phage-resistant *Pseudomonas aeruginosa* strains. Zhurnal mikrobiologii, epidemiologii, i immunobiologii 11:14
10. Essoh C et al (2013) The susceptibility of *Pseudomonas aeruginosa* strains from cystic fibrosis patients to bacteriophages. PLoS ONE 8(4):e60575
11. Li L, Yang H, Yue H (2011) Isolation and classification of the bacteriophages of *Pseudomonas aeruginosa* and their application on biofilm control. Chin J Microbiol Immunol 31(4):330–334
12. Pajunen M, Kiljunen S, Skurnik M (2000) Bacteriophage φYeO3-12, Specific for Yersinia enterocolitica Serotype O:3, is related to Coliphages T3 and T7. J Bacteriol 182(18):5114–5120
13. Shin H, Lee J-H, Kim H, Choi Y, Heu S, Ryu S (2012) Receptor diversity and host interaction of bacteriophages infecting *Salmonella enterica* serovar Typhimurium. PLoS ONE 7(8): e43392
14. Yokota S-I, Hayashi T, Matsumoto H (1994) Identification of the lipopolysaccharide core region as the receptor site for a cytotoxin-converting phage, phi CTX, of *Pseudomonas aeruginosa*. J Bacteriol 176(17):5262–5269
15. Patel I, Rao K (1983) Studies on the *Pseudomonas aeruginosa* PAO1 bacteriophage receptors. Arch Microbiol 135(2):155–157
16. Garbe J, Bunk B, Rohde M, Schobert M (2011) Sequencing and characterization of *Pseudomonas aeruginosa* phage JG004. BMC Microbiol 11(1):102, doi:10.1186/1471-2180-11-102
17. Jarrell K, Kropinski A (1981) *Pseudomonas aeruginosa* bacteriophage phi PLS27-lipopolysaccharide interactions. J Virol 40(2):411–420
18. Kropinski AM, Chan L, Jarrell K, Milazzo F (1977) The nature of *Pseudomonas aeruginosa* strain PAO bacteriophage receptors. Canadian J Microbiol 23(6):653–658
19. Bae H-W, Cho Y-H (2013) Complete genome sequence of *Pseudomonas aeruginosa* podophage MPK7, which requires type IV pili for infection. Genome Announce 1(5):e00713–e00744
20. Bradley TK, Han N (1974) The production of extracellular lipids by *Pseudomonas aeruginosa* NCTC 2000 in stationary liquid media containing macrogols. J Pharm Pharmacol 26(11):900–902
21. Schweizer H (1991) Escherichia-pseudomonas shuttle vectors derived from pUC18/19. 97(1):109–112
22. Nunn DN, Lory S (1992) Components of the protein-excretion apparatus of *Pseudomonas aeruginosa* are processed by the type IV prepilin peptidase. Proc Natl Acad Sci 89(1):47–51
23. Pan X, Cui X, Zhang F et al (2016) Genetic evidence for O-specific antigen as receptor of *Pseudomonas aeruginosa* Phage K8 and Its genomic analysis. Front Microbiol 7:252
24. Mueller RS et al (2007) Vibrio cholerae strains possess multiple strategies for abiotic and biotic surface colonization. J Bacteriol 189(14):5348–5360
25. Psifidi ACI, Dovas G Banos (2010) A comparison of six methods for genomic DNA extraction suitable for PCR-based genotyping applications using ovine milk samples. Mol Cell Probes 24(2):93–98
26. Li L et al (2010) Classification of 17 newly isolated virulent bacteriophages of *Pseudomonas aeruginosa*. Can J Microbiol 56(11):925–933
27. Kuchma SL, Brothers KM, Merritt JH et al (2007) BifA, a cyclic-Di-GMP phosphodiesterase, inversely regulates biofilm formation and swarming motility by *Pseudomonas aeruginosa* PA14. J Bacteriol 189(22):8165–8178
28. Be M et al (1991) Functional analysis of genes responsible for the synthesis of the B-band O antigen of Pseudomonas aeruginosa serotype O6 lipopolysaccharide. Microbiology 145:3505–3521

29. Kuchma SL et al (2010) Cyclic-di-GMP-mediated repression of swarming motility by *Pseudomonas aeruginosa*: the pilY1 gene and its impact on surface-associated behaviors. J Bacteriol 192(12):2950–2964
30. Ha D-G, O'Toole GA (2015) c-di-GMP and its effects on biofilm formation and dispersion: a *Pseudomonas aeruginosa* review. Microbiol Spectr 3(2). doi:10.1128/microbiolspec.MB-0003-2014

High Efficiency Expression of Trehalose Synthase in *Escherichia coli* and Its Use in the Production of Trehalose

Hong-Ling Liu, Rui-Ming Wang and Teng-Fei Wang

1 Introduction

Trehalose (also known as α-D-glucopyranosyl-α-D-glucopyranoside, mycose or mushroom sugar) is a kind of white, odorless powder with 45% relative sweetness of sucrose. In nature it can be found in many biological species [1] as carbon energy source [2], structural component of bacterial cell wall [3, 4]. It is a non-reducing sugar formed from two glucose units joined by an α-1,1-glycosidic bond, this special bonding can prevent Maillard reaction with protein or amino [5], and then make it very stable at high temperatures or acidic conditions. Because of those special properties, trehalose have been used as osmoregulation [6], dessication protectant [7, 8], cryoprotection [9], immunogenicity [10], stabilization of drug molecules, vaccines and liposomesgainst sTreSs [11–13], preservation of mammalian cells against dessication [14] and cryoprotection of human fibroblasts and oocytes [15, 16], etc. in various industrial processes including food processing, cosmetics, pharmaceutics. In 2000, trehalose was accepted as a novel food ingredient under the generally recognized as safe (GRAS) terms in the U.S. and EU, which certainly increased the scope of trehalose application. Although trehalose is present quite widely in the nature, the commercial application of this sugar was still limited by their low yields and expensive production costs.

H.-L. Liu
Key Laboratory of Industrial Fermentation Microbiology,
Ministry of Education, Tianjin University of Science and Technology,
Tianjin Economic-Technological Development Area (TEDA),
Tianjin 300457, People's Republic of China

R.-M. Wang · T.-F. Wang (✉)
Key Laboratory of Shandong Microbial Engineering,
QILUUniversity of Technology, Jinan, Shandong, People's Republic of China
e-mail: wangtengfei1981@163.com

H. Liu et al. (eds.), *Advances in Applied Biotechnology*, Lecture Notes in Electrical Engineering 444, https://doi.org/10.1007/978-981-10-4801-2_5

TreS (TreS, EC 2.4.1.245) can catalyzes the trehalose synthesis by a one-step conversion with a low cost substrate maltose. Comparing with other enzyme-based trehalose synthesis methods, the pathway catalyzed by TreS may be more suitable for industrial-scale production. Although many strains have been identified and characterized for industrial production of trehalose, such as *Mycobacterium smegmatis*, *Thermobifida fusca*, *Pseudomonas stutzeri* CJ38, *Pseudomonas putida*, *Enterobacter hormaechei*, *Arthrobacter aurescens* CGMCC 1.1892, *Coryne bacterium glutamicum* ATCC13032 and *Meiothermus ruber* CBS-01. the production of TreS was still restricted because of the growth characteristics of strain itself. Heterologous expression of TreS in *E.coli* has provided a more abundant source of these proteins for industry production of trehalose. Although the gene of *TreS* in different strains have been cloned and expressed in *E.coli* systems [17–20] were reported, the studies on industrial-scale production of trehalose by recombinant TreS in *E.coli* were still not involved.

In this paper, a recombinant *E.coli* BL21(DE3) harboring a *treS* gene from *Pseudomonas putida* ATCC47054 was constructed, the effect of environmental factors such as growth temperature and the concentration of transcription inducer for improving the expression of TreS in *E.coli* BL21(DE3) was investigated and compared. We also established a set of batch fermentation strategy to improve the enzyme activity of unit dry cell weight. And the cell lysis containing TreS was directly used to convert maltose into trehalose.

2 Materials and Methods

2.1 Strains and Plasmids

E.coli BL21(DE3), restriction enzymes, T4 DNA ligase, agarose gel DNA purification kit were purchased from TaKaRa (Dalian, China). DNA sequencing and primer synthesis were performed by Sangon Biotech (Shanghai, China) Co. Ltd. The *Pseudomonas putida*ATCC47054 containing the *treS* gene was obtained from American Type Culture Collection (ATCC; Manassas, VA, USA). All the chemicals used in this study were purchased from Sinopharm Chemical Reagent Co. Ltd. (Shanghai, China) unless otherwise specified. The strains, plasmids, and oligonucleotides used in this study are summarized in Table 1.

2.2 Construction of Expression Vectors

The DNA fragments containing *treS* gene from *Pseudomonas putida* ATCC47054 were prepared with bacteria genome DNA extracting kit. The primer pair (P1/P2, P3/P4 as shown in Table 1) were designed and synthesized for amplification of the

Table 1 Strains, plasmids, and oligonucleotides

Strain, plasmid, ligonucleotide	Relevant genotype or phenotype[1]	Source
E.coli		
BL21(DE3)	F^- *ompT dcm lon* $hsdS_B$ (r_B^-) *gal l*(DE3[*lacI ind 1Sam7 nin5 lacUV5-T7 gene 1*])	TakaRa
DH5α	F^-(*ψ80 lacZ M15*) Δ(*lacZYA-argF*)*U169 deoR recA1 endA1 hsdR17*(r_K^-, m_K^+) *phoA supE44 λ-thi-1 gyrA96 relA1*	Lab stock
Plasmid		
pET15b	Expression vector, P_{T7}::*lacOp*::*6* × *his*, Ori (pBR322), Ap^R	Novagen
pET22b	Expression vector, P_{T7}::*lacOp*::*6* × *his*, Ori (pBR322), Ap^R	Novagen
oligonucleotides		
P1	5′-GGGAATTC*CATATG*ATGACCCAGCCCGACC-3′ (*Nde*I *italic*)	This study
P2	5′-CGC*GGATCC*TCAAACATGCCCGCTGC-3′(*Bam*HI *italic*)	This study
P3	5′-CGC*GGATCC*ATGACCCAGCCCGACC-3′(*Bam*HI *italic*)	This study
P4	5′-CCG*CTCGAG*AACATGCCCGCTGC-3′(*Xho*I *italic*)	This study

[1]Designed restriction sites are underlined and introduced mutations are in *italic*

treS gene (gi = 1042893, NCBI). By using these primers, the *treS* gene was amplified using *pfu* DNA polymerase by polymerase chain reaction (PCR) (2720 Thermal Cycler, Applied Biosystems, Foster, CA) and purified by using a QIA quick polymerase chain reaction purification Kit (Qiagen, Valencia, CA). The PCR products were cloned into pET15b and pET22b and transformed into *E.coli* DH5α and screened. The DNA sequencing was performed to ensure no mutations occurred during PCR. The purified pET15b-*treS* and pET22b-*treS* vector containing *treS* gene fragments were respectively digested by *Nde*I/*Bam*HI and *Bam*HI/*Xho*I and respectively chem-transformed into *E.coli* BL21(DE3) for expression of TreS. The pET15b expression vector containing N-terminal six His-tag used in the experiments was generated for cytoplasmic expression of TreS under control of the T7 lac promoter. The pET22b expression vector containing C-terminal six His-tag was for periplasmic space expression of TreS under control of the T7 lac promoter. The recombinant *E.coli* BL21(DE3)/pET15b-*treS* and *E.coli* BL21(DE3)/pET22b-*treS* were obtained and named BL01 and BL02, respectively.

2.3 Characteristics of TreS

The culture sample was appropriately diluted with sterile water, the cell suspension were centrifuged at 4 °C and 8000 × g for 10 min and washed two times by sterile water. The cell pellet was fastly dried for determining dry cell weight of per unit volume. The cells were harvested by centrifugation at 8000 × g for 10 min at 4 °C. The wet cell pellet was suspended in 10 mmol/L potassium phosphate buffer (pH 7.5) and cells were cyclically disrupted two times by high-pressure homogenizer (APV-2000, Germany) at 880 bar. The insoluble cell debris were removed by centrifugation at 8000 × g for 20 min 4 °C. The recombinant TreS was purified using nickel-nitrilotriacetic acid affinity chromatography (Ni-NTA, Qiagen) as the manufacturer recommended. The purified enzymes were analyzed on 15% SDS-PAGE and protein concentration was determined by the method of Bradford using BSA as a standard.

The optimal pH was determined by measuring the activity of purified TreS at pH 3.0–10.0 (pH 3.0–6.0, 20 mM citrate buffer; pH 6.0–8.0, 20 mM sodium phosphate buffer; pH 7.0–9.0, 20 mM Tris-HCl buffer; and pH 9.0–11.0, 20 mM sodium carbonate buffer). The optimal temperature of purified TreS in 10 mM sodium phosphate buffer (pH 8.0) was measured at 10–65 °C by using maltose substrate. The pure TreS in various pH values of buffer (as shown above) was incubated at 25 °C for 60 min to determine the pH stability. An equal volume of 10 mM sodium phosphate buffer (pH 8.0) was added to maintain the pH at 8.0. The TreS in 10 mM phosphate buffer (pH 8.0) was incubated at 10–65 °C for 60 min and then chilled in ice water immediately for 5 min to determine the thermal stability.

The mixture reaction suspension was diluted to 10 times and centrifuged at 4 °C and 12,000 g for 10 min. After removal of insoluble fraction, the supernatant was filtrated by 0.45 μm filter membrane. The content of trehalose was identified by high pressure liquid chromatography (HPLC, Shimadzu-GL Sciences, Japan). One unit (U) of TreS was defined as the amount of enzyme required to produce 1 μmol trehalose per hour.

2.4 Optimization of Inducing pH and Temperature

Luria-Bertani (LB) medium (10 g/L peptone, 5 g/L yeast extract, 5 g/L NaCL) was used for flask cultures. The recombinant BL01 and BL02, stored at −80 °C, were revived in 250 mL Erlenmeyer flasks containing 50 mL LB medium with 50 μg ampicillin and grown at 37 °C and 200 rpm on rotary shakers for 10 h. Primary seed culture (1 mL) was used to inoculate two 100 mL secondary LB medium and grown for a further 8 h in a rotary shaker at 37 °C and 200 rpm. When the cell density reached approximately 3.0 OD_{600nm}, the culture temperature was slowly cooled down and lactose was added to induce TreS expression for another 7 h. Shake flask cultivation was performed in different culture conditions in order to find optimal induction and growth conditions. To check the effects of the inducer (lactose) concentration, induction temperature, induction pH and induction time on the soluble intracellular expression of TreS, the recombinant BL01 was induced by final concentration 4 g/L lactose at different temperature (22, 24, 27, 30, 34 and

37 °C) and different pH (pH 6.0, pH 6.5, pH 7.0, pH 7.5 and pH 8.0. During shake flask cultivation, samples were taken for analysis of cell density and enzyme activity of unit cell dry weight. All experiments were conducted at least in triplicate.

2.5 *Batch Fermentation*

Aliquots (25 μL) from frozen cell banks of BL01 were used to inoculate 50 mL primary seed culture with 50 μg ampicillin and grown at 37 °C for 10 h. Primary seed culture (10 mL) was used to inoculate 100 mL secondary seed cultures with 100 μg/mL ampicillin and grown for a further 8 h at 37 °C. 200 mL secondary seed cultures were used to inoculate 2.6 L defined mediumin the reactor vessel of a 5-L bioreactor (Shanghai Bailun Co. Ltd). The optimized batch fermentation medium was glucose 35 g/L, NH_4Cl 5 g/L, $K_2HPO_4{\cdot}3HO$ 8 g/L, KH_2PO_4 (3 g/L), peptone (7.5 g/L), yeast extract (5 g/L), $MgSO_4{\cdot}7H_2O$ (1.5 g/L), trace elements (0.5 mL/L), and ampicillin (100 mg/L). Trace elements (400×) was composed of $CoCl_2{\cdot}6H_2O$ (1 mg/mL), $MnCl_2{\cdot}4H_2O$ (6 mg/mL), $CuCl_2{\cdot}2H_2O$ (0.6 mg/mL), H_3BO_3 (1.2 mg/mL), $Na_2MoO_4{\cdot}2H_2O$ (1 mg/mL), Zn(II)acetate·$2H_2O$ (5.2 mg/mL) and Fe (III) citrate(40 mg/mL). Batch fermentation was carried out in fermentation medium with initial glucose concentration of 35 g/L. Agitation was regulated at 800 rpm to keep the dissolved oxygen nature with a constant sterile air flow rate 1 vvm or maintain dissolved oxygen at between 20–30% by stirring and passing into the oxygen. pH was maintained at 7.0 ± 0.2 by automatic pH control with additions of 250 g/L sodium hydroxide solution, and pH was maintained at 8.0 ± 0.2 by 10% H_3PO_4 phosphoric acid solution. The temperature was controlled by a heating sleeve and an integrated cooling system to 37 °C for cell growth, and 27 °C for induction expression of TreS. Sampling was done every 2 h for analysis of residual glucose concentration, acetic acid concentration, dry well weight (DCW) and the enzyme activity of TreS after induction.

2.6 *Production of Trehalose Using High Maltose Syrup*

High maltose syrup (about 90.5 ± 1.2% maltose, 3.3 ± 0.8% glucose, 1.3 ± 0.2% maltotriose, 4.5 ± 0.5% other polysaccharides) was used as substrate for its low cost and easy available. High maltose syrup was mixed reaction with cell disruptionsolution containing TreS (200 ± 10.0 U/g maltose) for 0–24 h with 100 rpm agitation rate at 50 °C in a 10 L bioconversion tank. The weight % (w/v) of trehalose was analyzed by HPLC (Shimadzu-GL Sciences, Japan). The apparatus was equipped by an inertsil-NH_2 column (4.6 × 250 mm, Shimadzu) at 40 °C. Separation was achieved by pumping acetonitrile: water (75:25 v/v) through the column at a flow rate of 1.0 mL/min for 20 min. The trehalose content in the sample was identified by comparing with the HPLC curves produced by the trehalose standard (Sigma Co., St. Louis, USA) of various concentrations. The weight conversion of trehalose (%) is ratio of the weight (g) of trehalose and the weight (g) of maltose in high maltose syrup.

3 Results

3.1 *Expression and Properties of Recombinant TreS*

To evaluate the effect of the plasmid on TreS expression, pET15b-*treS* and pET22b-*treS* containing *treS* gene regulated by the T7 promoter was used. The two recombinant plasmids were successfully transformed into competent *E.coli* BL21 (DE3) by heat shock, and the TreS was successfully expressed, respectively. The growth curves characteristics of two recombinant strains (BL01 and BL02) showed that the change of plasmid did not affect the growth of the two recombinant strains in LB medium, The two recombinant strains come to the logarithmic growth phase at 2 h, and retains this period at 7–8 h. The stability of recombinant plasmid pET15b-*treS* and pET22b-*treS* in *E.coli* BL21(DE3) were determined by the plating test. The consecutive vaccination analysis of 60 times showed that pET15b-*treS* plasmid loss rate maximum was less than 8%, and the pET22b-*treS* plasmid loss rate was less than 14%, indicating that the two recombinant plasmids are relative stability, and can be applied to large scale production.

When using IPTG as inducer, the recombinant TreS protein content can reached 45.5% of total protein in BL01, and reached 20.1% of total protein in BL02, respectively. When using the final concentration 4 g/L lactose as inducer, the recombinant TreS content can reached 34.3% of total protein in BL01, and reached 19.9% of total protein in BL02. The results showed that No TreS protein and enzyme activity could be detected in the extracellular fermentation liquid of two recombinant strains. Although the pET22b-*treS* recombinant plasmid include secretion signal peptide, it also could not secrete TreS into culture medium. By comparing the soluble expression of TreS protein in BL01 and BL02 intracellular using lactose as inducer, the intracellular expression of TreS protein in BL01 was higher than that in BL02, so BL01 was selected for a further study.

For identify the enzymatic characteristics of TreS, TreS purified by nickel nitrilotri acetic affinity chromatography mixed reaction 1 h with maltose (100 g/L final concentration) as a substrate in different pH, temperature. The enzyme activity of TreS at various pH was studied at 25 °C in 100 mM phosphate buffer (pH 3.0–10.0) for 60 min, using 100 g/L maltose as a substrate. To examine the stability of TreS, the TreS were preincubated at various pH values (pH 3.0–10.0) for 60 min at 25 °C. The residual activities were measured at pH 8.0. The results in Fig. 1 showed that the optimal enzyme activity pH was pH 8.0, and was relatively stable in pH 6.5–9.0. The enzyme activity of TreS at various temperature was studied in

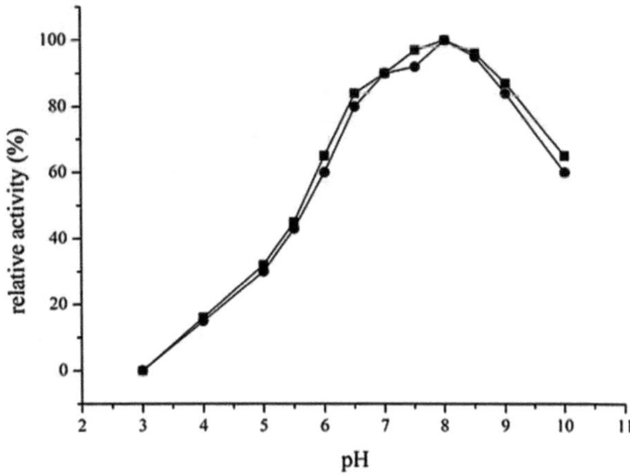

Fig. 1 Effects of pH on the activity and stability of the recombinant TreS. The *filled square* (■) represents effects of pH on the activity of TreS; the *filled circle* (●) represents effects of pH on the stability of TreS. Data are averages of three independent experiments

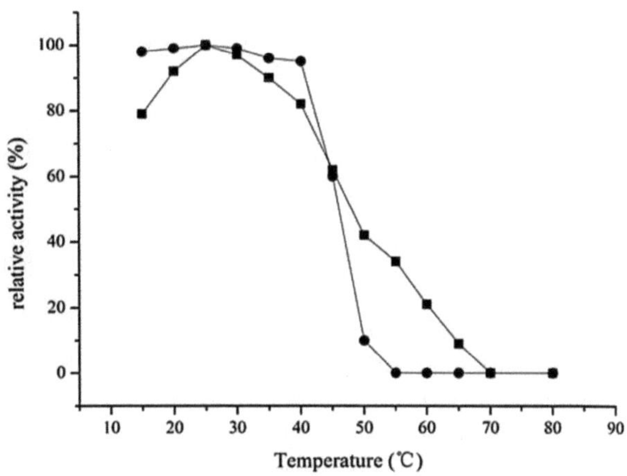

Fig. 2 Effects of temperature on the activity and stability of the recombinant TreS. The *filled square* (■) represents effects of temperature on the activity of TreS; the *filled circle* (●) represents effects of temperature on the stability of TreS. Data are averages of three independent experiments

100 mM phosphate buffers (pH 8.0) for 60 min, using 100 g/L maltose as a substrate. To examine the thermal stability of TreS, the TreS were preincubated at various temperatures (20–60 °C) for 60 min at pH 8.0. the residual activities were measured at 25 °C. The optimal enzyme activity temperature was 25 °C, and was relatively stable from 15 to 40 °C as shown in Fig. 2. But when the reaction temperature was over 40 °C, and with the extension of time, the enzyme activity of TreS decreased rapidly.

In order to better meet the scale-up application of recombinant TreS. The equal amounts of purified TreS was added into phosphate buffer solution (pH 8.0) for conversion reaction. The substrate maltose concentration in conversion system was adjusted to 300, 200, 150, 100, 50 g/L, and the system temperature was controlled at 20, 25, 30, 35, 40, 45, 50, 55 and 60 °C, and the conversion time was 12 h. The yield of trehalose was analyzed by HPLC, and the conversion rate of maltose to trehalose was compared. The results were as shown in Fig. 3. When the substrate

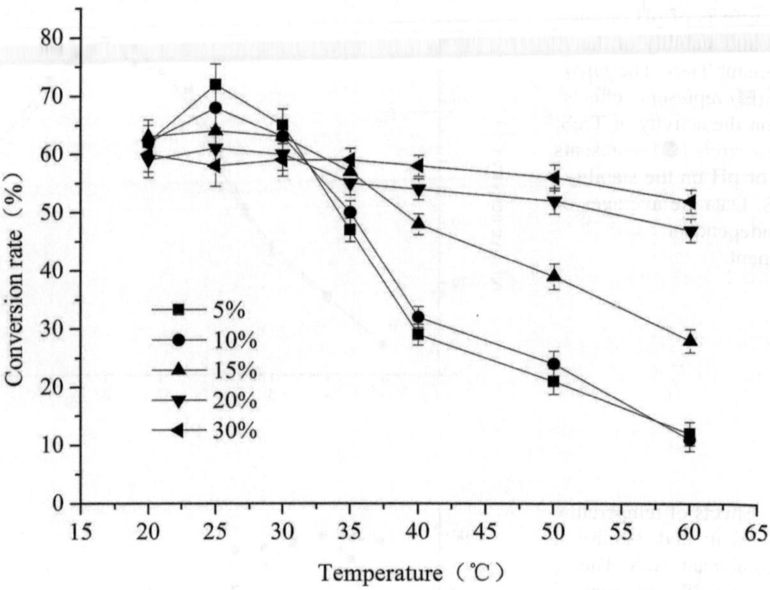

Fig. 3 The maltose conversion rate curves in different substrate high maltose syrup concentration as with different temperature

maltose concentration was 50–100 g/L, the conversion ratio was firstly increased and then decreased with the increase of system temperature, when the system temperature was higher than 40 °C, the conversion rate decreased rapidly. When the substrate maltose concentration was between 150–200 g/L, the conversion rate was higher at 10–30 °C, with the increase of system temperature, the conversion rate was slightly decreased. When the substrate maltose concentration was 300 g/L, with the increase of system temperature, the conversion rate tends to balance, and the conversion rate was over 58% at 50 °C. This reaction temperature is conducive to prevent microbial contamination of conversion system. This phenomenon showed that high substrate concentration of maltose syrup is beneficial to the thermal stability of TreS. Therefore, the enzyme activity unit of TreS was defined as the amount of enzyme required to produce 1 μmol trehalose per hour at 50 °C and pH 8.0, when the substrate maltose concentration was 300 g/L.

3.2 Optimization of Induction pH and Temperature

For increasing the solution expression of TreS, the induction conditions were optimized in LB medium. As shown in Fig. 2, the pH value of the medium was gradually increased with the growth of BL01 in the LB medium of different initial pH values. The occurrence of this phenomenon may be caused by the release of

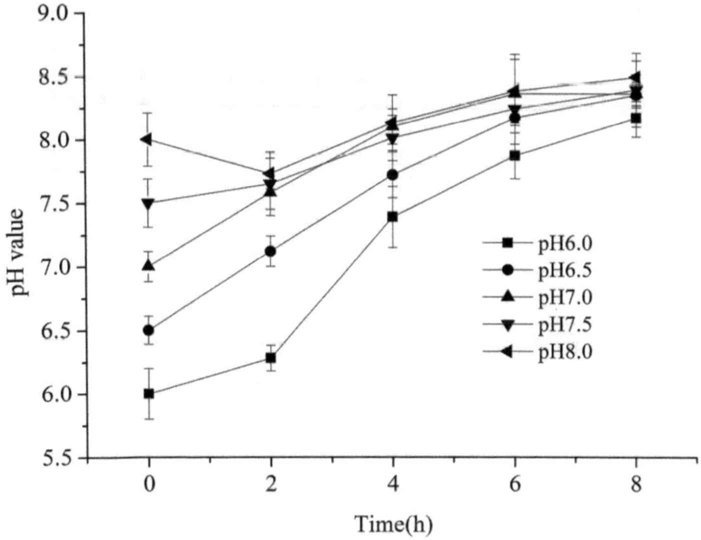

Fig. 4 The pH value change curves of LB medium with different initial pH

ammonium ions in amino acids and so on. This phenomenon is good for the stability of enzyme activities in the later stage of fermentation, because the optimal pH of TreS was 8.0. The recombinant BL01 fermented for 6 h in pH constant LB medium, and the fermentation temperature 37 °C, induction temperature 27 °C and induction time 7 h by final concentration 4 g/L lactose, the growth curves as shown in Fig. 4, the enzyme activity of unit dry cell weight as shown in Fig. 5. The results showed that the growth of recombinant BL01 was optimal in pH 7.0–7.5, and the enzyme activity of unit dry cell weight was optimal in pH 8.0. Under this conditions, the enzyme activity could reach 17,730 ± 673 U per gram dry cell weight. So neutral environment was conducive to the growth of BL01, and the partial alkaline environment (pH 8.0) was conducive to the expression of TreS.

The recombinant BL01 fermented for 6 h at 37 °C in LB medium of constant pH 7.0, and induced for 7 h using final concentration 4 g/L lactose at different temperature, the growth curves as shown in Fig. 6, and the enzyme activity as shown in Fig. 7. The results showed that the induction temperature at 27–30 °C was not only beneficial to the growth of the bacteria, but also was beneficial to increase the enzyme activity of unit dry cell weight. When the induction temperature was at 22 or 24 °C, the biomass of BL01 was decreased firstly and then increased slowly, and the biomass and enzyme activity was lower at the end of induction. When the temperature was at the 34 or 37 °C, the cell growth rate was faster, the final OD_{600} value was higher in the same fermentation time. But with extension of induction time, the enzyme activity of unit dry cell weight was lower than 27–30 °C, and cell autolysis in the late stage of induction. Although the amount of bacteria at

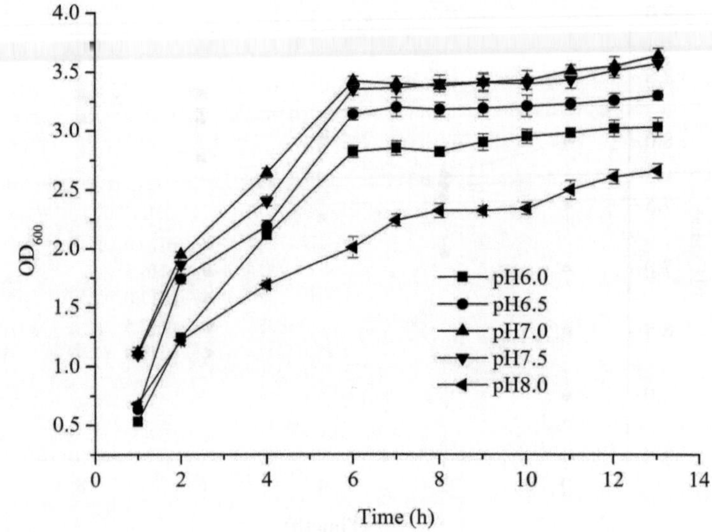

Fig. 5 The OD_{600} curves of BL01 growth with different fermentation pH

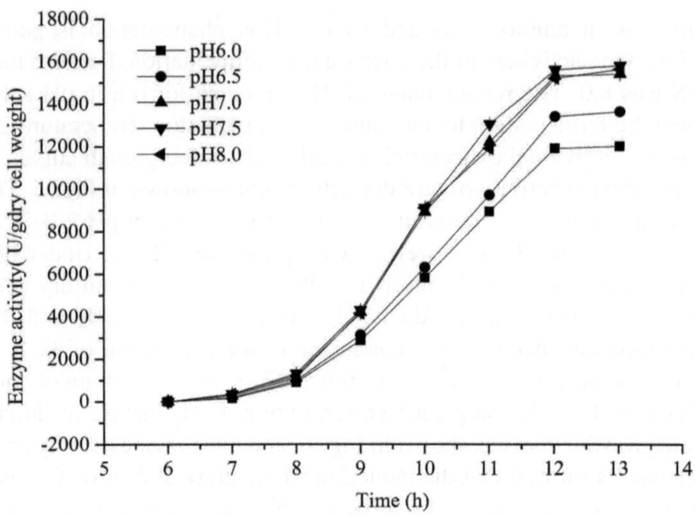

Fig. 6 The enzyme activity curves of BL01 unit dry cell weight with different fermentation pH

27–30 °C induction was slightly lower than that of 34 and 37 °C, but the enzyme activity of unit dry cell weight was relatively higher, and avoided cell autolysis. So the induction temperature 27–30 °C was the optimal choice for high efficiency soluble expression of TreS in BL01.

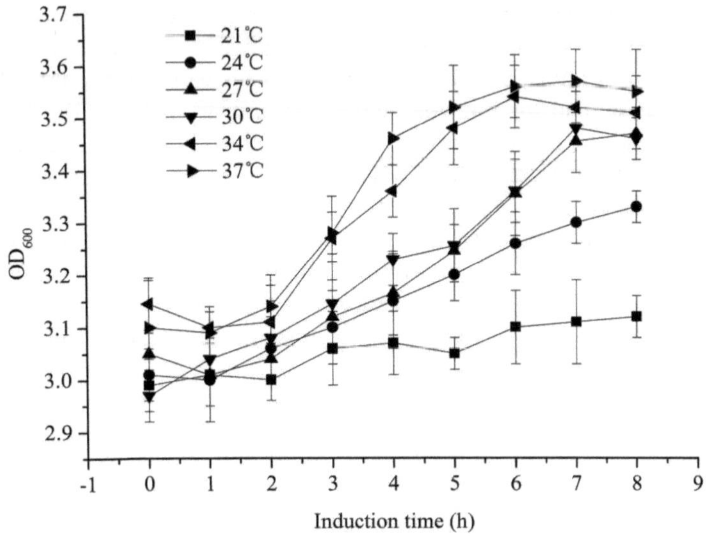

Fig. 7 The OD_{600} curves of BL01 growth with different induction temperature

3.3 Batch Fermentation in 5-L Fermentor

Upscalling of active TreS production in BL01 was performed firstly in batch fermentation. Seed culture (200 mL) was inoculated into a 5 L fermentor with 2.6 L culture medium for fermentation. The pH was automatically controlled at 7.0 by adding NaOH solution (5 M) before induction, and the pH was naturally increasing to pH 8.0 after induction, and was automatically controlled at 8.0 ± 0.5 by adding 10% phosphoric acid solution. The temperature was maintained at 37 °C before induction and 27 °C after induction. The ventilation quantity was controlled at 1 vvm, and stirring speed was controlled at 800 rpm. When the dissolved oxygen increased rapidly from the lowest, adding 4 g/L final concentration lactose into culture medium for 7 h induction and the dissolved oxygen (DO) level was maintained 15–20% of saturation by cascading the agitation speed (100–400 rpm) and aeration rate (0.5 vvm). Figure 8 showed that dissolved oxygen in fermentor decreased close to zero at fermentation 6 h, and dissolved oxygen rised rapidly the continued after 2 h, and the content of reducing sugar in the medium was reduced to 1.67 ± 0.4 g/L by HPLC analysis. At this time, the lactose was added to induce for avoiding the metabolism of glucose repression. As the dry cell weight growth curve shown, the maximum dry cell weight can be reached 10.1 ± 0.4 g/L, and the dry cell weight was 9.64 ± 0.38 g/L after induction. The acetic acid curve in Fig. 9 showed that and the acetic acid content in fermentation broth was rising with recombinant bacteria growth, and reached 8.65 ± 0.11 g/L at 9 h. Subsequently, acetic acid content has a downward trend, because the carbon source reduced in culture medium, the recombinant bacteria use acetic acid as carbon. Acetic acid content was still greater than 7 g/L at the end of fermentation, this is the main cause of cell autolysis in the induction process. Figures 10 and 11 showed that the enzyme activity of unit dry cell weight began to increase rapidly from 2 h in the induction process, and the enzyme activity of unit dry cell weight tended to be stable at 7 h, and the highest enzyme activity reached 39,866 ± 1420 U per gram dry cell weight. the pH value showed an upward trend with extension of induction time during induction.

Oxygen represents an important regulatory stimulus in aerobic bacteria. To obtain the optimal growth of *E.coli* resulting in high yields of active TreS, oxygen saturations of about 10 and 50% and a lack of oxygen were analyzed. Fermentation with an oxygen level of about 20% resulted in a dry cell weight of 15.68 ± 0.33 g/L before induction and 12.6 ± 0.48 g/L after induction. the enzyme activity of about 38,967 ± 1223 U per gram dry cell weight. The 10% oxygen saturation during the fermentation process was responsible for a massive loss of enzyme activity although a dry cell weight of 9.68 ± 0.93 g/L was reached. An increase of the oxygen level to 40% yielded very high dry cell weight of 15.98 ± 0.73 g/L, whereas the enzyme activity reached nearly the same value as fermentation at 20% oxygen saturation. The maximum content of acetic acid in the fermentation broth reached 2.76 ± 0.13 g/L in 20% oxygen saturation, far less than the uncontrolled oxygen fermentation. So using 20% oxygen saturation during batch fermentation higher dry cell weight and higher enzyme yield could be achieved.

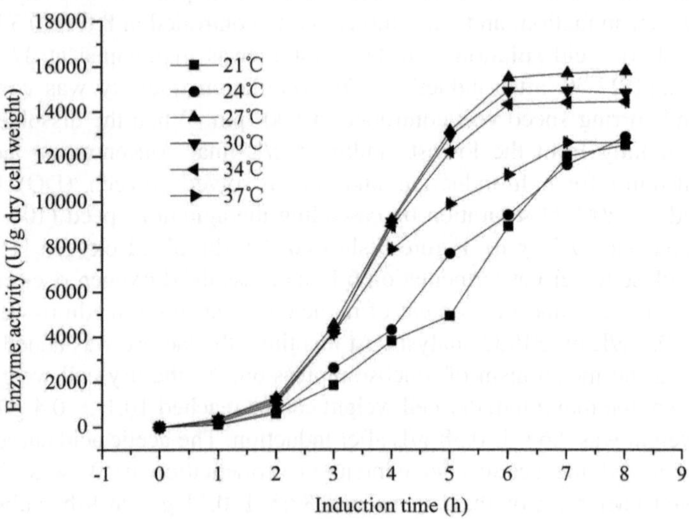

Fig. 8 The enzyme activity curves of BL01 unit dry cell weight with different induction temperature

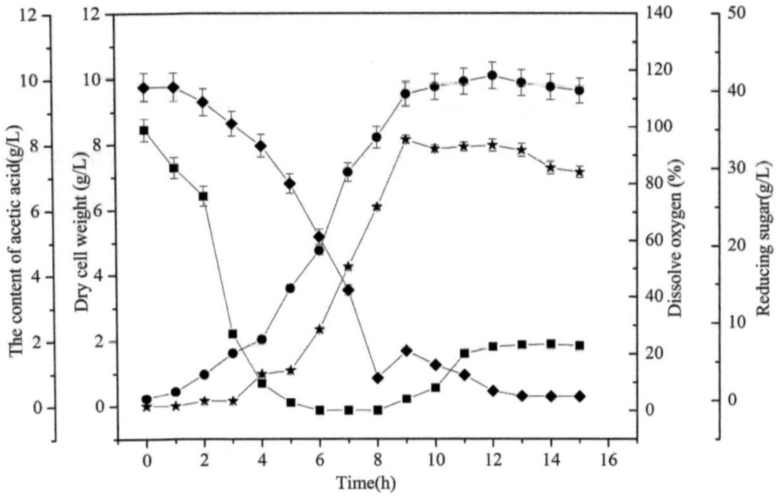

Fig. 9 The parameter variation curves of BL fermentation process in batch fermentation

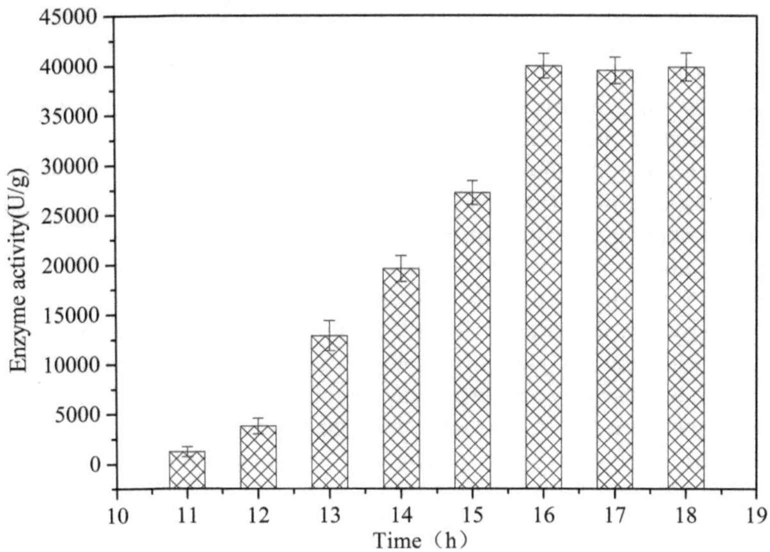

Fig. 10 The enzyme activity of BL01 unit dry cell weight in with different fermentation time

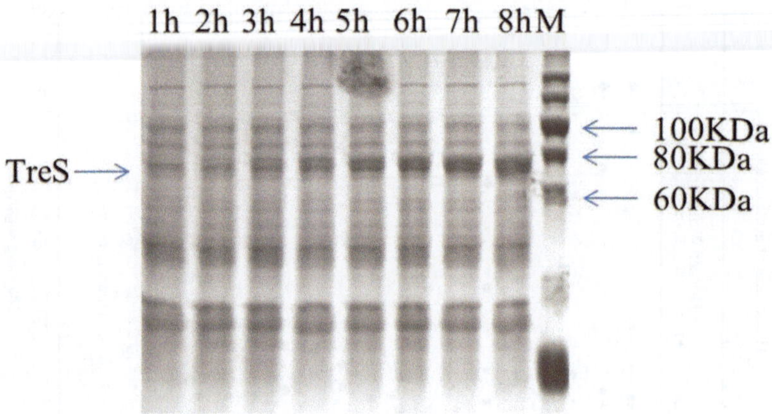

Fig. 11 The TreS SDS-PAGE of BL01 under the same cell mass with different fermentation time

3.4 Preparation of Trehalose

The recombinant cells containing TreS were crushed by high pressure homogeneous machine. Then the cell disruption was directly used to convert high maltose syrup for 24 h with 60 rpm agitation rate at 50 °C and pH 8.0. The content of trehalose was analyzed by HPLC. In order to improve the yield of trehalose in conversion system, the effects of temperature and pH on enzyme activity were studied respectively. As shown in Fig. 12, the highest production of trehalose was 193.5 ± 5.0 g/L at pH 8.0 for 24 h with 300 g/L high maltose syrup as substrate. Under this condition, the trehalose conversion rate can reach 64.5 ± 1.6%. The yield of trehalose was decreased when the temperature was higher than 60 °C or lower than 45 °C. The reason why the trehalose yield reduced below 45 °C may be caused by protease degradation in cell lysate or pH descent for microbial contamination, while the low yield of trehalose when the temperature was higher than 60 °C may be caused by enzyme denaturation. The results showed that higher reaction temperature (50 °C) not only reduced microbial contamination, but also promoted the yield of trehalose. Therefore, this temperature was beneficial for industrial production in our process. As shown in Fig. 13, the highest trehalose yield was reached 193 ± 2.0 g/L at 50 °C for 24 h with 300 g/L maltose as substrate, and 64.3 ± 0.6% maltose was converted into trehalose. Meanwhile, the TreS was highly stable over a broad pH range of 6.5–8.5, retaining more than 92% of its original yield. Above all, the optimal conditions of converting maltose into trehalose was 50 °C, pH 7.5–8.0 and 300 g/L maltose as substrate.

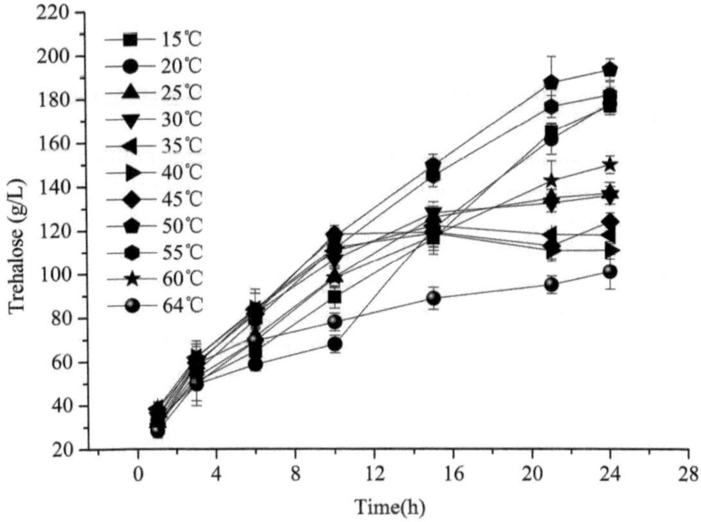

Fig. 12 The yield curves of trehalose under the same substrate high maltose syrup concentration (300 g/L) with different reaction temperatures

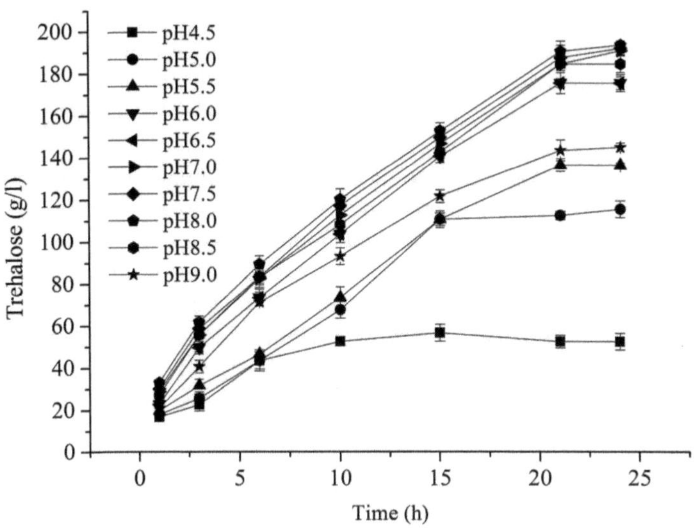

Fig. 13 The yield curves of trehalose under the same substrate high maltose syrup concentration (300 g/L) with different reaction pH values

4 Discussion

In recent years, the production enterprises of high maltose syrup was gradually increasing, the price of high maltose syrup was reduced gradually because of enterprise competition and market oversupply. So this pathway of converting maltose into trehalose has been considered to be an effective way to bring economic benefits for enterprises and a convenient and economical biocatalyst process for industrial production of trehalose. However, scale-up production of TreS is still a main limited factors for large-scale production of trehalose. In this study, we have solved the problem by constructing recombinant *E.coli* that expressed soluble TreS, which promoted the industrialization pace. The high level expression of recombinant TreS in *E.coli*, the production of trehalose by recombinant TreS were the objects of these investigations.

In order to obtain high enzyme activity of unit dry cell weight, induction conditions were optimized. Solubility is a key issue for the production of recombinant protein in heterologous expression systems. Soluble recombinant proteins are often properly folded, functional and easier to purifiy than aggregated proteins from inclusion bodies. We have developed a strategy to drive the expression of TreS in BL01. Our strategy was designed for expression of TreS under the control of the T7 *lac* promoter. Our data show that we have achieved the expression of TreS as an intracellular soluble in the BL01. This results were achieved at a growth temperature of 37 °C and at a growth pH of 7.0, induction temperature of 27 °C, induction pH of 8.0, induction lactose final concentration of 4 g/L, induction time of 7 h. The enzyme activity of unit dry cell weight reached 3.9×10^4 U per gram dry cell weight.

In order to reduce the purification cost of recombinant TreS and the loss of enzyme activity in the purification process, the cells lysis solution of BL01 was directly used to product trehalose, and the application of crude enzyme avoided the tedious purified steps of TreS. More than 64% maltose using substrate high maltose syrup concentration 300 g/L can be converted into trehalose at 50 °C, and pH 8.0 for 24 h in a 10 L reaction system.The research showed that 300 g/L maltose concentration as substrate has a protective effect on TreS, which improves the temperature tolerance and thermal stability of TreS. The high conversion temperature not only avoided microbial contamination, but also save energy and reduce consumption.

In this study, TreS was expressed in BL01 intracellular by lactose induction, which avoided the use of IPTG, so the production process of trehalose was safe without any chemical addition. The production process of TreS using the break fluid of recombinant *E.coli* BL21(DE3) is a new technological breakthrough. It will provide a new direction for the industrial production of trehalose. We expect such high efficient and green process can be widely applied in industrial production.

5 Acknowledgments

This work was supported by the National Nature Science Foundation of China (No. 31501413), Shandong higher education research project (J14LE02), and the Foundation (No. 2016IM005) of Key Laboratory of Industrial Fermentation Microbiology of Ministry of Education and Tianjin Key Lab of Industrial Microbiology (Tianjin University of Science & Technology).

References

1. Elbein AD, Pan YT, Pastuazak I, Carroll D (2003) New insights on trehalose: a multi-functional molecule. Glycobiology 13(4):17–27
2. Whatmore AM, Reed RH (1990) Determination of turgor pressure in *Bacillus subtilis*: a possible role for K^+ in turgor regulation. J Gen Microbiol 136(12):2521–2526
3. Carpinelli J, Kraemer R, Agosin E (2006) Metabolic engineering of *Corynebacterium glutamicum* for trehalose over production: role of the TreYZ trehalose biosynthetic pathway. Appl Environ Microbiol 72(3):1949–1955
4. Murphy HN, Stewart GR, Mischenko VV, Apt AS, Harris R, McAlister MS, Driscoll PC, Young DB, Robertson BD (2005) The OtsAB pathway is essential for trehalose biosynthesis in *Mycobacterium tuberculosis*. J Biol Chem 280(15):14524–14529
5. Higashiyama T (2002) Novel functions and applications of trehalose. Pure Appl Chem 74(7):1263–1269
6. Kempf B, Bremer E (1998) Uptake and synthesis of compatible solutes as microbial sTreSs responses to high-osmolality environments. Arch Microbiol 170(5):319–330
7. Rueda B, Miguelez EM, Hardisson C, Manzanal MB (2001) Changes in glycogen and trehalose content of *Streptomyces brasiliensis* during growth in liquid cultures under sporulating and non-sporulating conditions. FEMS Microbiol Lett 194(2):181–185
8. Crowe LM (2002) Lessons from nature: the role of sugar in anhydrobiosis. Com Biochem Physio A-molecular Integr Physiol 131(3):505–513
9. Duong T, Barrangou R, Russell WM, Klaenhammer TR (2006) Characterisation of the tre locus and analysis of trehalose cryoprotection in *Lactobacillus acidophilus* NCFM. Appl Environ Microbiol 72(2):1218–1225
10. Rao V, Gao F, Chen B, Jacobs WR Jr, Glickman MS (2006) Trans-cyclopropanation of mycolic acids on trehalose dimycolate suppresses *Mycobacterium tuberculosis* induced inflammation and virulence. J Clin Invest 116(6):1660–1667
11. Schwendeman SP, Constantino HR, Gupta RK, Siber GR, Klibanov AM, Langer R (1995) Stabilization of tetanus and diphtheria toxoids against moisture induced aggregation. Proc Natl Acad Sci USA 92(24):11234–11238
12. Jain NK, Roy I (2008) Role of trehalose in moisture induced aggregation of bovine serum albumin. Eur J Pharm Biopharm 69(3):824–834
13. Crowe JH, Leslie SM, Crowe LM (1994) Is vitrification sufficient to preserve liposomes during freeze drying. Cryobiology 31(4):355–366
14. Guo N, Puhlev I, Brown DR, Mansbridge J, Levine F (2000) Trehalose expression confers desiccation tolerance on human cells. Nat Biotechnol 18(2):168–171
15. Eroglu A, Russo MJ, Bieganski R, Fowler A, Cheley S, Bayley H, Toner M (2000) Intracellular trehalose improves the survival of cryopreserved mammalian cells. Nat Biotechnol 18(2):163–167
16. Eroglu A, Toner M, Toth TL (2002) Beneficial effect of microinjected trehalose on the cryosurvival of human oocytes. Fertil Steril 77(1):152–158

17. Lee JH, Lee KH, Kim CG, Lee SY, Kim GJ, Park YH, Chung SO (2005) Cloning and expression of aTreS from *Pseudomonas stutzeri CJ38* in *Escherichia coli* for the production oftrehalose. Appl Microbiol Biotechnol 68:213–219
18. Yue M, Wu XL, Gong WN, Ding HB (2009) Molecular cloning and expression of a novel TreS gene from *Enterobacter hormaechei*. Microb Cell Fact 8:34
19. Wu X, Ding H, Yue M, Qiao Y (2009) Gene cloning, expression, and characterization of a novel TreS from *Arthrobacter aurescens*. Appl Microbiol Biotechnol 83(3):477–482
20. Kim TK, Jang JH, Cho HY, Lee HS, Kim YW (2010) Gene cloning and characterization of a TreS from *Corynebacterium glutamicum* ATCC13032. Food Sci. Biotechnol 19(2):565–569

Functional Modification of the Substrate-Binding Site for Isomaltulose Production Based on Predicted Structure of Sucrose Isomerase from *Pantoea dispersa* UQ68 J

Huijie Liu, Xueyan Xing, Fuping Lu and Yu Li

1 Introduction

Isomaltulose (α-D-glucopyranosyl-1, 6-D-fructofuranose) is a naturally occurring isomer of sucrose commonly referred to as palatinose. As a nutritional sugar found in honey with trace amount [1], it can prevent tooth decay and control insulin levels in the bloodstream. Furthermore, it is healthy for diabetics and non-diabetics, because it is digested more slowly than sucrose in intestine. Therefore, it is a potential functional sugar for human consumption [2]. Compared with chemical synthesis of isomaltulose, the bioconversion of isomaltulose could be carried out at ambient temperature and atmospheric pressure, thus avoiding the use of more extreme conditions causing problems with isomerization, racemization, epimerization and rearrangement [3]. The reaction process of sucrose isomerases (SIs) consists of sucrose binding, sucrose hydrolysis, glucosyl-enzyme intermediate occurrence, fructose release and sucrose isomer formation [4]. The immobilized microbial cells using chitosan embedding of *Serratia plymuthica* can only convert 40% of sugarcane to isomaltulose [5]. SmuA from *Propionibacterium rubrum* produces 85% of isomaltulose, together with a minor amount of trehalulose, monosaccharide and isomaltose [6]. Therefore, improving isomaltulose production while controlling the amounts of side reaction products remains the considerable industrial problem.

Sucrose isomerases are widely used in industry for the production of isomaltulose and trehalulose. However, only a narrow range of bacterial species are

H. Liu · X. Xing · F. Lu · Y. Li
College of Biotechnology, Tianjin University of Science & Technology, Tianjin 300457, China

H. Liu · X. Xing · F. Lu · Y. Li (✉)
Key Laboratory of Industrial Microbiology, Ministry of Education, College of Biotechnology, Tianjin University of Science and Technology, Tianjin 300457, China
e-mail: liyu@tust.edu.cn

H. Liu et al. (eds.), *Advances in Applied Biotechnology*, Lecture Notes in Electrical Engineering 444, https://doi.org/10.1007/978-981-10-4801-2_6

known to be able to convert sucrose into isomaltulose and trehalulose catalyzed by sucrose isomerase or a glucosyltransferase (EC.5.4.99.11), in which they produce trace amount of fructose and glucose as by-products. These bacterial species include *Klebsiella* sp. LX3 [7], *Protaminobacter rubrum* [5], *Serratia plymuthica* [4], *Pantoea dispersa* UQ68 J [8], and *Enterobacter* sp. FMB1 [9]. SIs catalyze the isomer of sucrose to produce isomaltulose and trehalulose with various proportions of glucose and fructose as by-products. Some purified SIs, such as *Erwinia rhapontici* [10], *Klebsiella* sp. LX3 [7] isomaltulose synthase predominantly produce isomaltulose (80%) [11], and others (*Agrobacterium radiobacter* MX-232, *Pseudomonas mesoacidophila* MX-45 trehalulose synthase) produce trehalulose (90%) [12]. However, Sim1 showed the high conversion ratio of isomaltulose in the sucrose isomerase family.

In this paper, the Sim1 was cloned and overexpressed in *Escherichia coli*. The 3D structure of Sim1 was simulated using the crystal structure of *Klebsiella* sp. LX3 isomaltulose synthase (PDB ID: 1M53) as template. Furthermore, we performed the structure-based functional modification, and evaluated the effect of Y296, P297 and E299, which located close to the substrate-binding site, and were considered to affect the conversion rate of isomaltulose by site-directed mutagenesis.

2 Materials and Methods

2.1 Bacterial Strains and Plasmids

The Sim1 was cloned into pET22-b vector, *E. coli* DH5α was used for genetic manipulation, and *E. coli* BL-21 (DE3) was used for the expression of Sim1 and its mutants. Bacterial strains and plasmids used in this study are summarized in Table 1. The mutants were further confirmed by DNA sequencing.

2.2 Plasmid Construction

The sequence from the *P. dispersa* UQ68 J (Gene accession number AY223549) has a 1797-bp of open reading frame (ORF) encoding 598 amino acids residues.

Table 1 Bacterial strains and plasmids used in this study

Strain/plasmids	Description	Source
E. coli BL21 (DE3)	Expression host	Laboratory
E. coli DH5α	Cloning host	This study
Pantoea dispersa UQ68 J	Source for Sim1 gene	This study
pET-22b	Expression vector and PCR sample	Laboratory
PMD-18-T	Cloning vector	Takara Dalian

The amino acids residues of Sim1 include the predicted 33-amino-acid signal peptide, and PCR primers were designed for cloning the Sim1 gene (without noncoding regions and signal sequences) into expression vector of pET-22b. The forward primer (5′-CGC GGA TCC AAT GGC AAC GAA TAT AGA-3′) included a *Bam*H I restriction site and a start codon. The reverse primer (5′-CCC AAG CTT GTT CAG CTT ATA GAT CCC GG-3′) included a *Hin*d III restriction site and a stop codon. The wild-type Sim1 was set up in 5 mL of LB medium with 50 μg/mL of Ampicillin in 30-mL test tubes. Cells were grown at 37 °C with shaking at 250 rev/min. Isopropyl-β-D-thiogalactopyranoside (IPTG) was added to a final concentration of 0.5 mM until an OD600 of 0.6–0.8 is reached, and the cultures was continued to incubate for 12 h at 20 °C. After the incubation, an appropriate volume of culture was used for protein quantification, sodium dodecyl sulfate polyacrylamide gel electrophoresis (SDS-PAGE), and quantification of the conversion efficiency from sucrose to isomaltulose and trehalulose.

2.3 *Protein Purification*

Cells were harvested by centrifugation at 3000*g*, 4 °C for 10 min, re-suspended in 50 mM Tris-HCl (pH 6.0) and then re-centrifuged, and washed twice with 50 mM sodium phosphate lysis buffer (pH 6.0). The cells were then re-suspended in the same lysis buffer containing 200 mM of NaCl, 10 mM β-mercaptoethanol and 10 mM of imidazole, disrupted by sonication, and centrifuged at 10,000*g* for 30 min at 4 °C. The supernatant was pooled. The recombinant protein was purified by nickel-nitrilotriacetic acid (Ni-NTA) affinity column chromatography (Qiagen), and eluted with a linear gradient from 20 to 250 mM of imidazole in 50 mM of sodium phosphate buffer (pH 6.0) after being washed extensively with 20 mM imidazole. The purified protein was concentrated by ultrafiltration after dialysis against 50 mM of sodium acetate buffer (pH 6.0) and used for further investigation. The protein concentration was measured by the Bradford method using the Bradford reagent kit (Solarbio, Dalian China).

2.4 *Activity Assay of Sim1*

Enzyme activity was measured by incubating 5 μL of purified enzyme with 95 μL of a 10% sucrose solution in 0.1 M citrate/phosphate buffer (pH 6.0) at 30 °C. The sugar profiles in the reaction mixture were analyzed at intervals by high-performance liquid chromatography with electrochemical detection as described [13, 14]. Briefly, resolution and quantification of sucrose, isomaltulose, trehalulose, glucose, and fructose were achieved by isocratic high-performance liquid chromatography. The HPLC system (Waters Alliance 2690) was equipped with the refractive index detector (Waters 410). The samples were diluted and

20 μL of samples were injected into the TSK gel Amide-80 (TOSOH) column (4.6 mm × 25 cm) for measurement of the sugar composition in the reaction products. The isocratic mobile phase was acetonitrile: water (90:10), and the temperature was 80 °C [16, 17]. All sugar quantification data presented below were obtained by this method, and calibration against a dilution series of sugar standards was performed for every sample batch. One unit of activity is defined as the amount of enzyme required to catalyze formation of 1 μmol of isomaltulose in 1 min under assay conditions [15].

The effect of pH and temperature to enzyme activity was evaluated between pH 4.0 and 9.0 and 20–60 °C using 0.1 mM of sodium citrate (pH 4.0–6.0), sodium phosphate (pH 6.0–8.0), and Tris-HCl (pH 8.0–9.0) buffers. Heat inactivation of the enzyme was examined at 30, 40, 50, and 60 °C.

2.5 *Conversion of Sucrose*

The conversion of sucrose was carried out in the test tube containing 10 mL of 50% sucrose solution and 1 mL of enzyme at 35 °C in the shaking water bath for 4–6 h, the reactions were stopped by boiling for 10 min [16, 17]. To study the effects of mutants on product formation, the same amount of enzymes of wild type and mutants were incubated separately with the 10 mL of 50% sucrose at 25–45 °C intervals, and the products were subjected to HPLC analysis as described above.

2.6 *Site-Directed Mutagenesis of Sucrose Isomerase*

Site-directed mutagenesis was performed with QuikChange (Stratagene) by PCR and pET-22b-Sim1 plasmid as a template. Each desired amino acid residue replacement was generated using two synthetic oligonucleotide primers. The PCR fragments were amplified using Pfu turbo polymerase (Solarbio, Dalian China). The obtained mutants were expressed and purified according to the method described for wild-type Sim1.

3 Results and Discussion

3.1 *Characterization of the Recombinant Sim1*

To investigate the isomaltulose conversion of recombinant strains, we studied the effects of pH, temperature, and substrate concentration on the production of isomaltulose. As shown in Fig. 1a, the optimal temperature for Sim1 activity was 40 °C.

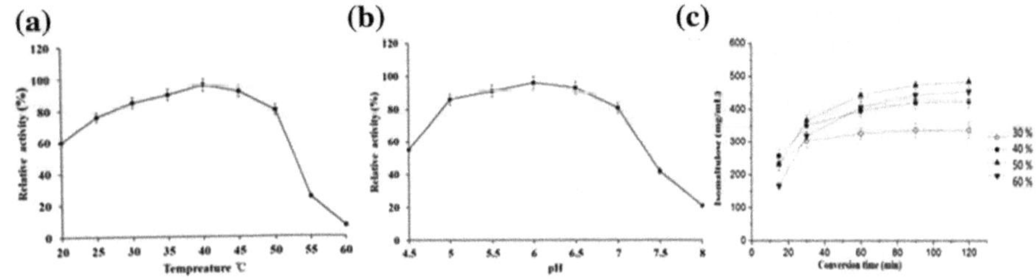

Fig. 1 The effect of the pH (**a**), temperature (**b**) and sucrose concentration (**c**) to Sim1

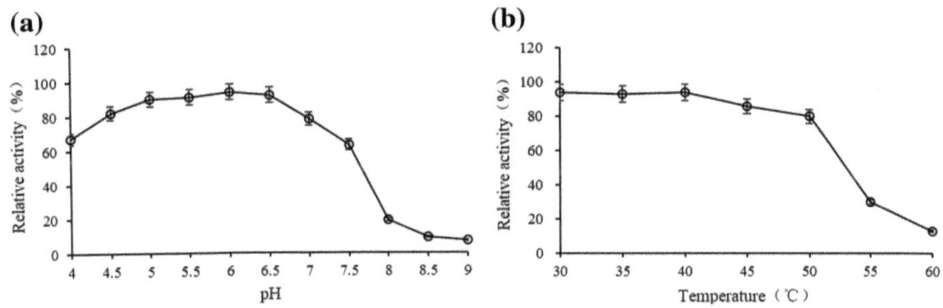

Fig. 2 The Sim1 stability of pH (**a**) and temperature (**b**)

The activity decreased rapidly when the temperature was over 50 °C. Sim1 was active at pH 4.5–8.0, and the high activity was observed at pH 5.5–6.5 (Fig. 1b). Sim1 showed the maximal isomaltulose conversion concentration at 50% substrate sucrose (Fig. 1c). Therefore, the best conversion condition of isomaltulose is 50% of sucrose solution at 40 °C and pH 6.0.

The stability of Sim1 were investigated. It exhibited the pH stability was well at pH 4.5–6.5, and more than 80% of the maximal activity was retained (Fig. 2a). The enzyme activity was examined at different storage temperatures, and 80% of the enzyme activity was retained when the enzyme was incubated at 20–50 °C for 30 min. However, there is only 32% residual activity at 55 °C (Fig. 2b).

3.2 Products Analysis

Four products catalyzed by Sim1 were analyzed with HPLC and saccharides were identified by comparison with standards. Figure 3 showed the HPLC results of the reaction products catalyzed by Sim1. Four products (isomaltulose, trehalulose, glucose, and fructose) were identified by the method described above. We calculated the relative percentage of individual sugar based on its peak area.

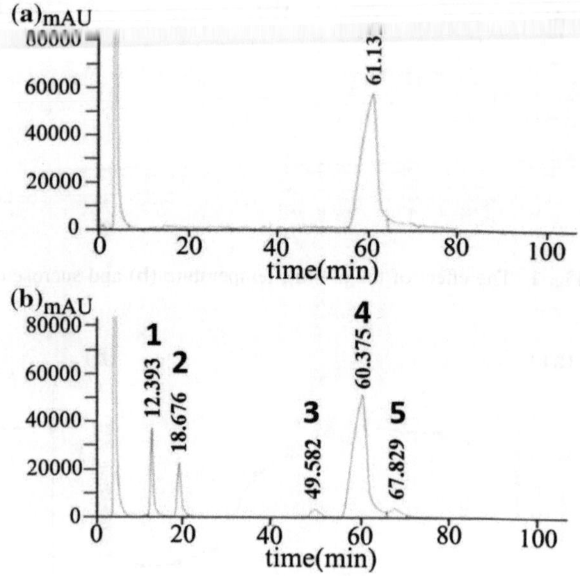

Fig. 3 HPLC analysis of saccharides converted by Sim1 **a** Productions; **b** Sample *1* Fructose; *2* Glucose; *3* Sucrose; *4* Isomaltulose; *5* Trehalulose

3.3 The Structure of the UQ68 J Sucrose Isomerase

Dispersa UQ68 J, which was previously reported to show the high efficiency and specificity for production of isomaltulose from sucrose, has a sucrose isomerase gene that was different from other characterized SI family members previously [18]. The protein sequence alignment of Sim1 exhibited 64–74% with/without the signal peptide to other known SIs encoding sequences.

The sequence of Sim1 revealed that it is different from those of other species. It contains a 1698-bp ORF encoding 565 amino acids, excluding a 33-amino-acid signal sequence (GenBank accession number AY223549). The protein sequence of the Sim1 was submitted in NCBI, it showed that the sucrose isomerase from *Klebsiella* sp. LX3 is the highest similarity sequence as the model structure. The obtained 3D structure of the Sim1 was energy minimized using Smart Minimizer algorithm (Discovery Studio 3.0, default parameters) until the potential energy of the system became constant. The 3D structure of the Sim1 was simulated based on the crystal structure of *Klebsiella* sp. LX3 isomaltulose synthase (PDB ID: 1M53). The structural model of Sim1 obtained by energy minimization showed the relative positions of α-helices and β-sheets in the 3D structure of the proteins (Fig. 4a). The predicted 3D model of Sim1 seemed to share a conserved barrel (α/β)8 domain for sucrose-binding and glycosidase activities with all other SIs and glycosidase [19, 20]. A "RLDRD" motif in proximity to the active site is also well conserved in the predicted 3D model of Sim1. This motif is considered to be responsible for sucrose isomerization. The substrate sucrose was located over the TIM barrel and was bound to the residues D69, R206, D208, E262, R292, R300, H335, D336, E395, and R423 through hydrogen bonds. It also formed hydrophobic interaction with F152, and F264 (Fig. 4b).

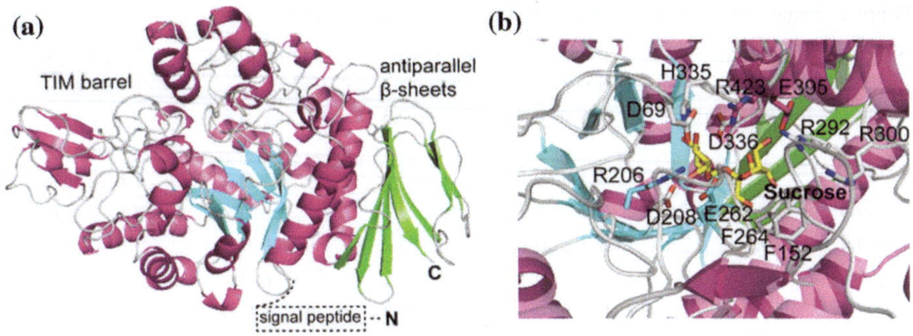

Fig. 4 The simulated 3D structure. **a** The simulated 3D structure of the *P.* dis Sim1; **b** The interaction between sucrose and Sim1. The sucrose was modeled based on the alignment of Sim1 and sucrose isomerase NX-5 (PDB ID: 4HPH)

3.4 Isomaltose Production by Mutated Sim1

It was reported that the specificity for major product formation was regulated not only by the "RLDRD" motif [21], but also by some other region(s) or other key residues surrounding the binding site. Table S3 includes residues in these enzymes that vary from the residues in other SIs particularly close to the "RLDRD" motif. The F296 was located behind "RLDRD" motif in Sim1. However, it was not conserved in the isomerase family. Isomerase enzymes from *Klebsiella* sp., *E. rhapontici*, *P. rubrum* and *P. mesoacidophila* showed aspartic residue. There is tyrosine residue from *P. mesoacidophila* at this position. Therefore, Y296 was replaced with aspartic, glutamic, phenylalanine, and histidine, respectively (Y296D, Y296E, Y296F, and Y296H). The P297 was replaced with aspartic and glutamic. The Q299 was replaced with glutamic, aspartic, asparagine, and knocking down, respectively (Q299E, Q299D, Q299N, ΔQ299). The sim1 mutants kept the similar activity to wild type (Table 2).

The Sim1 was cloned into pET22-b vector, and overexpressed in BL21(DE3). Furthermore, the culture conditions were determined. To investigate Sim1's activity and the importance of residues located behind "RLDRD" motif at active site, the mutants of Sim1 kept the similar activity with the wild type. However, the replacement of Q299 to glutamic acid (Q299E) increased the percentage of isomaltulose and decreased the percentage of trehalulose in the reaction products, while the mutants has minor impact on the formation of productions compared with wild-type. It is interesting that the mutant of deleting Q299 showed the contrary phenomenon to Q299E. It decreased the conversion percentage of isomaltulose, but increased trehalulose in the reaction products. It is suggested that the conserved Q299 is important for the products specificity at the active site. On the other hand, the replacement of Tyrosine 296 to Aspartic enhanced the percentage of glucose and fructose in the reaction production by about 2-fold.

Table 2 Sugar compositions of reaction mixtures of the enzymes

Sim1	Activity U/mL	Glucose (%)	Fructose (%)	Isomaltulose (%)	Trehalulose (%)
Wide type	20.7±0.2	3.6 ± 0.4	3.0 ± 0.5	90.3 ± 0.8	3.1 ± 0.3
Q299E	19.8 ± 0.2	2.3 ± 0.2	2.1 ± 0.2	94.2 ± 0.4	1.6 ± 0.1
Q299D	20.1 ± 0.2	3.2 ± 0.3	2.6 ± 0.1	90.2 ± 0.7	3.8 ± 0.1
Q299N	19.5 ± 0.2	3.3 ± 0.9	3.0 ± 0.2	89.8 ± 0.5	3.5 ± 0.2
ΔQ299	20.6 ± 0.3	4.2 ± 0.1	3.8 ± 0.1	81.0 ± 0.6	11 ± 0.2
Y296D	19.9 ± 0.1	10.6 ± 0.5	9.6 ± 0.3	76.3 ± 0.6	3.6 ± 0.4
Y296E	20.3 ± 0.2	3.9 ± 0.1	3.1 ± 0.1	89.7 ± 0.3	3.3 ± 0.6
Y296F	18.9 ± 0.2	3.7 ± 0.3	2.9 ± 0.2	90.9 ± 0.4	2.4 ± 0.2
Y296H	20.4 ± 0.3	4.7 ± 0.1	3.4 ± 0.2	87.9 ± 0.6	4.1 ± 0.5
P297D	20.0 ± 0.2	3.4 ± 0.3	3.1 ± 0.2	90.3 ± 0.6	3.3 ± 0.5

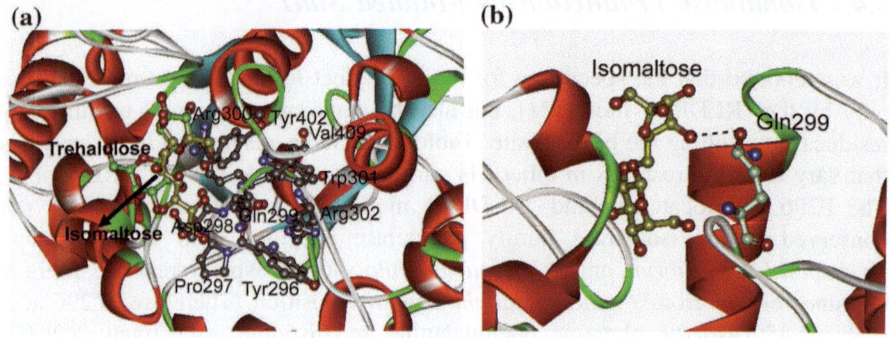

Fig. 5 **a** The docking simulation of isomaltulose (*yellow orange*) and trehalulose (*green*) to the Sim wide-type. **b** The hydrogen bond between isomaltulose and Q299 of Sim1

3.5 *Substrate Docking Simulation*

To explain the product specificity of SIs, isomaltulose and trehalulose were docked into the binding site of Sim1 using the CDOCKER protocol in Accelrys Studio (DS 3.5). The docking results are listed in Table 2. The isomaltulose and trehalulose products were well docked in the active site (Fig. 5). Q299 could form hydrogen bond with the fructofuranose moiety of isomaltulose at substrate-binding site rather than trehalulose, which preferred to bind isomaltulose. Q299E was considered to strengthen the hydrogen bond interaction between isomaltulose and Glutamine. The simulated model is consistent with the experimental data (Table 2). In Table 3, the calculated binding energy of isomaltulose bond to Sim1 is higher than that of trehalulose. In the docking simulation of Sim1 mutants with isomaltulose, Q299E (48.68 kJ/mol) exhibited higher binding energy than wild-type (45.33 kJ/mol), and ΔQ299 (43.64 kJ/mol). Furthermore, ΔQ299 (42.08 kJ/mol) exhibited higher

Table 3 Docker interaction energy and the percentage of product

Number	Name	-cDocker interaction energy (kJ/mol)
1	sim-isomaltulose	45.3
2	sim-trehalulose	39.1
3	Q299E-isomaltulose	48.7
4	Q299E-trehalulose	36.9
5	ΔQ299-isomaltulose	43.6
6	ΔQ299-trehalulose	42.1

binding energy than Sim1 (39.04 kJ/mol), and Q299E (36.87 kJ/mol) in docking simulation of those with trehalulose. Furthermore, sim1 (ΔQ299) broke down the charge balance because the negative charge of glutamine was missing. Therefore, it could affect the binding ability with sugar molecular and change the product specificity. The analysis of binding energy to isomaltulose and trehalulose are also consistent with the experimental data that the isomaltulose is more suitable for binding to the substrate-binding site of Sim1.

Our molecular docking simulation and experimental data studies demonstrated the loop 296–299 of Sim1, as well as the mutants of Q299E, ΔQ299, and Y296D showed significant difference on product specificity. This research will provide a platform for solving the problem of converting sucrose to isomaltulose and trehalulose effectively.

References

1. Lina BA, Jonker D, Kozianowski G (2002) Isomaltulose (palatinose): a review of biological and toxicological studies. Food Chem Oxicol 40:1375–1381
2. Patel RN (2003) Synthesis of chiral pharmaceutical intermediates by biocatalysis. Curr Opin Drug Disc 6:902–920
3. Xu Z, Li S, Li J et al (2013) The structural basis of *Erwinia rhapontici* isomaltulose synthase. PLoS ONE 8:e74788
4. Park SE, Cho MH, Lim JK et al (2007) A new colorimetric method for determining the isomerization activity of sucrose isomerase. Biosci Biotechnol Biochem 71:583–586
5. Duan X, Cheng S, Ai Y et al (2016) Enhancing the thermostability of *Serratia plymuthica* sucrose isomerase using B-factor-directed mutagenesis. PLoS ONE 11:e0149208
6. Lee HC, Kim JH, Kim SY (2008) Isomaltose production by modification of the fructose-binding site on the basis of the predicted structure of sucrose isomerase from "*Protaminobacter rubrum*". Appl Environ 74:5183–5194
7. Ravaud S, Robert X, Watzlawick H et al (2009) Structural determinants of product specificity of sucrose isomerases. FEBS 583:1964–1968
8. Zhang DH, Li XZ, Zhang LH (2002) Isomaltulose Synthase from *Klebsiella* sp. strain LX3: gene cloning and characterization and engineering of thermostability. Appl Environ 68: 2676–2682
9. Wu L, Birch RG (2005) Characterization of the highly efficient sucrose isomerase from *Pantoea dispersa* UQ68 J and cloning of the sucrose isomerase gene. Appl Environ 71:1581–1590

10. Cho MH, Park SE, Lim JK et al (2007) Conversion of sucrose into isomaltulose by *Enterobacter* sp. FMB1, an isomaltulose-producing microorganism isolated from traditional Korean food. Biotechnology 29:453–458
11. Ahn SJ, Yoo JH, Lee HC et al (2005) Enhanced conversion of sucrose to isomaltulose by a mutant of *Erwinia rhapontici*. Biotechnol Lett 25:1179–1183
12. Ren B, Li S, Xu H et al (2011) Purification and characterization of a highly selective sucrose isomerase from *Erwinia rhapontici* NX-5. Bioprocess Biosyst Eng 34:629–637
13. Watzlawick H, Mattes R (2009) Gene cloning protein characterization and alteration of product selectivity for the trehalulose hydrolase and trehalulose synthase from "*Pseudomonas Mesoacidophila*" MX-45. Appl Environ 75:7026–7036
14. Karnaouri AC, Topakas E, Christakopoulos P (2014) Cloning, expression, and characterization of a thermostable GH7 endoglucanase from Myceliophthora thermophila capable of high-consistency enzymatic liquefaction. Appl Micro Biotechnol 98:231–242
15. Holub I, Gostner A, Theis S et al (2010) Novel findings on the metabolic effects of the low glycaemic carbohydrate isomaltulose (Palatinose™). Brit J Nutr 103:1730–1737
16. Li S, Cai H, Qing Y et al (2011) Cloning and characterization of a sucrose isomerase from *Erwinia rhapontici* NX-5 for isomaltulose hyperproduction. Appl Biochem 163:52–63
17. Jonker D, Lina BA, Kozianowski G (2002) 13-Week oral toxicity study with isomaltulose (Palatinose) in rats. Food Chem Toxic 40:1383–1389
18. Kawai K, Okuda Y, Yamashita K (1985) Changes in blood glucose and insulin after an oral palatinose administration in normal subjects. Endocrinol Jpn 32:933–936
19. Uitdehaag JC, Mosi R, Kalk KH (1999) X-ray structures along the reaction pathway of cyclodextrin glycosyltransferase elucidate catalysis in the alpha-amylase family. Nat Struct Mol Biol 6:432–436
20. Lipski A, Watzlawick H, Ravaud S et al (2013) Mutations inducing an active-site aperture in Rhizobium sp. sucrose isomerase confer hydrolytic activity. Acta Crystallogr Sect D Biol Crystallogr 69:298–307
21. Nakayama A, Yamamoto K, Tabata S (2001) Identification of the catalytic residues of bifunctional glycogen debranching enzyme. J Biol Chem 276:28824–28828

Construction of the *Escherichia coli-Bacillus subtilis* Shuttle Vector pBE2R and Identification of the Critical Residues Involved in the Autoprocessing of the Propeptide of the Alkaline Protease

Run Wei, Xiao-mei Liang, Mei-juan Xuan, Hao-yu Yuan, Fu-ping Lu and Ming Li

1 Introduction

The serine alkaline protease, secreted from a wide variety of *Bacillus* species, is an important industrial enzyme and a model system for protein engineering. The alkaline protease can be widely used in washing powders and dehairing hides for its high activity and stability. The alkaline protease generally secreted extracellular for the purpose of scavenging nutrients is specific for aromatic or hydrophobic residues, such as tyrosine, phenylalanine and leucine. However, they are highly sensitive to phenyl methyl sulphonyl fluoride and diisopropyl-fluorophosphate. The *Bacillus*-originated alkaline proteases are mesophilic enzymes with a molecular weight range of 15–30 kDa and an isoelectric point near pI 9. They reflect high activity at 50–70 °C, however, the activity were reduced significantly at low temperature, like 20 °C [1].

Directed evolution has rapidly emerged as a powerful strategy for improving the characteristics of various enzymes in a targeted manner. To generate large gene variant libraries, it is possible to optimize an enzyme for its specific applications through combining error-prone PCR or DNA shuffling with the high-throughput screening which is used to select the specific properities of an enzyme, such as thermostability, catalytic activity and substrate specificity [2]. Therefore, it is available to improve the activity of the alkaline protease at low temperature through

Run Wei and Xiao-mei Liang contributed equally to this work.

R. Wei · X. Liang · M. Xuan · H. Yuan · F. Lu · M. Li (✉)
Key Laboratory of Industrial Fermentation Microbiology, Ministry of Education (TUST), National and Local United Engineering Laboratory of Metabolic Control Fermentation Technology (TUST), College of Biotechnology, Tianjin University of Science and Technology, Tianjin 300457, China
e-mail: liming09@tust.edu.cn

H. Liu et al. (eds.), *Advances in Applied Biotechnology*, Lecture Notes in Electrical Engineering 444, https://doi.org/10.1007/978-981-10-4801-2_7

the directed evolution. The construction of the mutant library was a key step in the process of the alkaline protease directed evolution. In this report, a new *Escherichia coli-Bacillus subtilis* shuttle vector pBE2R was constructed to establish a platform for the high-throughput screening of the alkaline protease in the process of the alkaline protease directed evolution.

The alkaline protease derived from *Bacillus alcalophilus* TCCC11263 can be expressed successfully in *B. subtilis* WB600 by using the pBE2R vector [3]. Apart from this, in the process of construction of the pBE2R vector we found that the correct cleavage of the propeptide after secretion was significant for the formation of the functional protein. As other subtilisins, the *B. alcalophilus* TCCC11263 alkaline protease of 380 residues is synthesized as an inactive precursor that is composed of a 27 residues signal peptide for protein secretion, a 84 residues propeptide for folding and formation of the active protease and a 269 residues mature peptide for the catalytic protein. The signal peptide is cleaved as the protein crosses the inner membrane, and the propeptide remains covalently attached until the protein is secreted from the cell [4]. The propeptides are relatively common in *Bacillus* secretory proteins, and they are divided into two different kinds, long and short [5]. The propeptide region generally functions as a facilitator of folding, stability, and even secretion of the protein. In most cases, the propeptide domain is cleaved from the enzymatic domain autocatalytically to release an active protease. Removal of the propeptide by autocleavage during processing is crucial for the secretion and production of subtilisins [6]. Takahashi et al. [7] showed that the autoprocessing efficiency of a subtilisin E mutant with altered specificity for acid residues was improved by substituting the autoprocessing site Tyr-1 with Asp or Glu. Grande et al. [8] reported that insertion within a 9-amino-acid region in the propeptide caused dramatic reduction in LasA enzymatic activity. All mutant proLasA proteins were still secreted, but extracellular stability was low due to clustered insertions within the propeptide. However, little is known about the critical residues involved in the autoprocessing of the *B. alcalophilus* alkaline protease. Therefore, another primary goal of this study was to identify the critical residues correlated to the autocleavage of the propeptide of the *B. alcalophilus* alkaline protease.

2 Materials and Methods

2.1 Strains and Plasmids

E. coli DH5α, *B. subtilis* WB600, *B. alcalophilu*s TCCC11263 were stored by Tianjin university of science and technology laboratory. The plasmid pWB980 was kindly provided by Sui-Lam Wong (University of Calgary, Canada). The plasmid pBE2 was kindly provided by NanKai University.

2.2 Growth Medium and Conditions

E. coli DH5α, *B. subtilis* WB600 were cultured aerobically in LB medium at 37 °C. For recombinant selection, the kanamycin and ampicillin were applied to the above media. The final concentration of Kanamycin and Ampicillin was 30 and 100 μg/mL respectively. The skim milk plate containing Kanamycin (30 μg/mL) was used for the preliminary screening of the mutants.

2.3 Related Enzymes and Reagents

GeneRuler, restriction endonuclease and T4 DNA ligase were purchased from TaKaRa Biotechnology (Dalian) Co., Ltd. *Taq* DNA polymerases and Unstained Protein Molecular Weight Marker were purchased from Fermentas. PCR primers were synthesized by Shanghai Sangon Biological Engineering Technology & Services Co., Ltd. All other reagents were analytically pure.

2.4 Construction of the pBE2R

P43 promoter is a strong promoter containing two overlapping promoters which are recognized in vitro by σ55- and σ37-containing RNA polymerase holoenzymes from *B. subtilis* [9]. The 317 base pair length fragment P43 promoter was released from the plasmid pWB980 by digestion with *Eco*R I and *Kpn*I. The fragment was isolated by gel purification and inserted into the MCS of the pBE2 vector directionally. Cohesive ends ligation of both fragments resulted in the plasmid pBE2a. Then the DNA fragment containing signal peptide and pro-peptide of alkaline protease derived from *B. alcalophilus* TCCC11263 was amplified by the Polymerase Chain Reaction (PCR). Based on the DNA sequence of the alkaline protease (*aprE*) of *B. alcalophilus* reported on NCBI, a primer pair (Psp1 and Psp2) (Table 1) with indicated engineered restriction sites were designed to amplify the signal peptide and pro-peptide gene (*sp*).

Table 1 Primers used for PCR experiments

Targert gene	Primer sequence
Signal peptide and pro-peptide gene	
Forward	5′-CGGGGTACCATTATAGGTAAGAGAGGAATGTACA CATGAAGAAACCGTTGGGGAAAATTG-3′ (*Kpn*I)
Reverse	5′-CGCGGATCCCGCCATTGTCGTTACTTCTGCATCC-3(*Bam*H I)
Mature peptide encoding gene	
Forward	5′-CGCGGATCCCAATCAGTGCCATGGGGAATTAGCCG-3′ (*Bam*H I)
Reverse	5′-ACGCGTCGACTTAGCGTGTTGCCGCTTCTGCATTGAC-3′ (*Sal*I)

The *Kpn*I-*Bam*H I fragment of the PCR product was inserted into pBE2a directionaly, forming plasmid pBE2R. Then the pBE2R was transformed into *E. coli* DH5α. The pBE2R was sent to Shanghai Sangon Biological Engineering Technology & Services Co., Ltd for DNA sequencing.

The mature peptide encoding gene of alkaline protease was introduced into the pBE2R to detect the function of the new shuttle vector. A primer pair (Pmp1and Pmp2) (Table 1) were designed to amplify the mature peptide encoding gene. The *Bam*H I-*Sal*I fragment of the PCR product of the mature peptide (*mp*) was introduced into pBE2R directionaly to generate pBE2R-*mp*. Then the recombinant plasmid was transformed into *B. subtilis* WB600 to express the alkaline protease.

2.5 Construction of the Mutants with Different Insertional Positions of the Restriction Sites Around the Region Between Propeptide and Mature Peptide

Four pairs primers were designed to amplify the fragment of signal and pro-sequences of alkaline protease with the different insertional positions of restriction site around the region between propeptide and mature peptide. The *Bam*H I-*Sal*I fragments of PCR products were inserted into the pBE2a to construct plasmids containing different insertional positions. According to the insertional position, The recombinant plasmids were named pBE2a::Rs109, pBE2a::Rs110, pBE2a::Rs111, pBE2a::Rs112, respectively. The mature peptide encoding gene was amplified by PCR. Then the PCR products were introduced into the recombinant plasmids described above, generating plasmids pBE2a::Rs109-*mp*′, pBE2a::Rs110-*mp*′, pBE2a::Rs111-*mp*′ and pBE2a::Rs112-*mp*′. Meanwhile, a recombinant plasmid pBE2a-*apr* containing the signal and pro-sequences, as well as the mature sequence of the alkaline protease was constructed without the insertion of restriction site at the junction between propeptide and mature peptide. Then these recombinant plasmids were transformed into the *B. subtilis* WB600 to express the alkaline protease. The skim milk plate containing Kanamycin (30 μg/mL) was used to screen the mutants preliminarily.

2.6 Assay of the AprE Activity

The *B. subtilis* WB600 carrying recombinant plasmids were cultured in 100 mL LB medium containing Kanamycin (30 μg/mL) at 37 °C for 48 h. The culture supernatants were collected by centrifugation at 4 °C and 12,000 r/min for 10 min. The supernatants were used as crude enzyme solutions for the activity determination of the AprE protease. The general methods of determination for industrial enzymes was employed to assay the activity of the alkaline protease quantitatively.

2.7 SDS-PAGE Analysis

The SDS-PAGE was employed to detect the relative content of the extracellular proteins. The hosts were cultured in 100 mL LB medium containing Kanamycin (30 μg/mL) at 37 °C for 48 h. Supernatants and the cells were harvested by centrifugation at 4 °C and 5000×*g* for 10 min, respectively. Supernatants were precipitated with 100% TCA at −20 °C for 5–10 min and then 4 °C for 12 h. Centrifugating at 4 °C and 5000×*g* for 10 min to collect the precipitate. The precipitate was washed by acetone for three times and volatiled thoroughly. Then the precipitates were dissolved with the 1× loading buffer before they were boiled. The cells collected by centrifugation described above were suspended with 40 μL of distilled water and 10 μL of 1× loading buffer. Then the solutions were boiled at 100 °C for 20 min before loading. SDS-PAGE was performed in a 30% polyacrylamide gel.

3 Results

3.1 Expression of the Alkaline Protease (AprE)

The *Escherichia coli-Bacillus subtilis* shuttle vector pBE2R was constructed by fusing P43 promoter and signal as well as pro-sequence of the *aprE* gene to the pBE2 vector (Figs. 1 and 2). The plasmid was verified by double digestion (*Kpn*I/*Bam*H I). The results demonstrated that the sequences inserted into pBE2.

The mature peptide encoding gene of alkaline protease was introduced into pBE2R to generate pBE2R-*mp*, which was transformed into *B. subtilis* WB600 to detect the function of the new shuttle vector. The transformants were cultured on the skim milk plate containing Kanamycin (30 μg/mL) at 37 °C for 16 h. It was observed from the production of the proteolytic ring that the shuttle vector pBE2R can express the alkaline protease successfully in *B. subtilis* WB600 (Fig. 3).

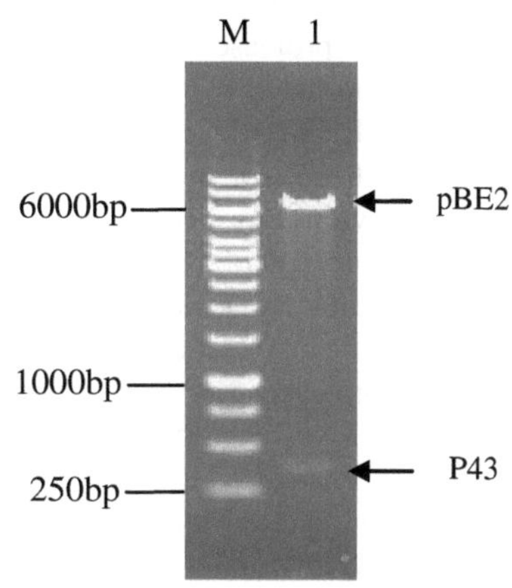

Fig. 1 Verification of the recombinant plasmid pBE2a

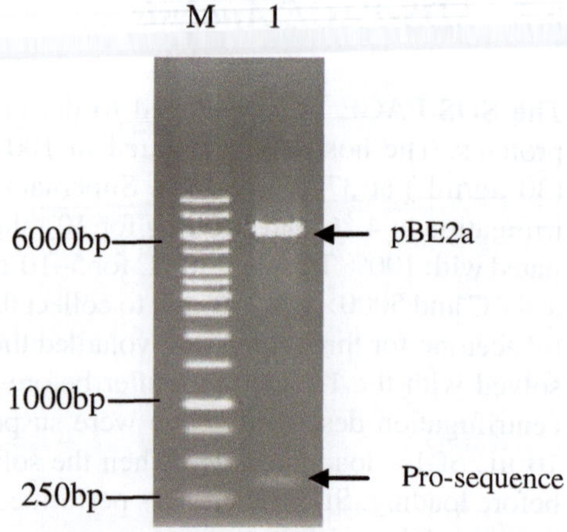

Fig. 2 Verification of the recombinant plasmid pBE2R

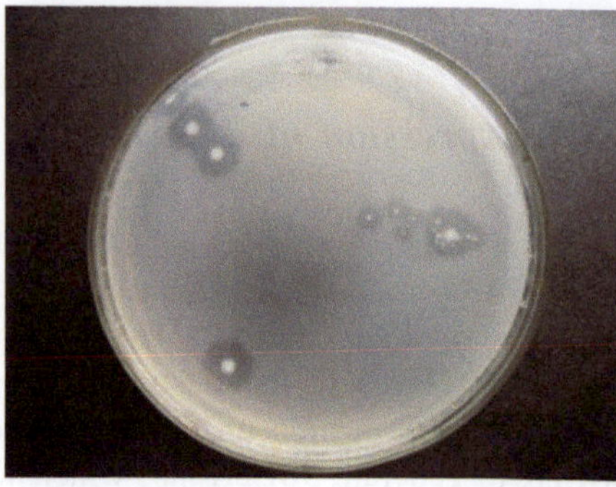

Fig. 3 Expression of the pBE2R-*mp* in the *B. subtilis* WB600

3.2 *Deduction of the Critical Residues Involved in the Autoprocessing of the Propeptide*

The new constructed shuttle vector pBE2R can be used in the directed evolution of the alkaline protease. The mutational gene encoding mature peptide of the alkaline protease produced by error-prone PCR and DNA shuffling can be inserted into the MCS of pBE2R with the connection of *Bam*H I. Therefore, the pBE2R provided a convenient tool for the high-throughput screening of the forward mutation in the directed evolution of the alkaline protease.

In the process of pBE2R construction, we found that the insertional position of *Bam*H I which was located at the junction between the propeptide and mature peptide had a significant effect on the activity of the alkaline protease. It was indicated that the inactive protease was due to the aborting autoprocessing of the propeptide after secretion. Five recombinant plasmids pBE2a::Rs109 ~ pBE2a::Rs112 and pBE2a-*apr* were constructed to detect the critical residues involved in the autoprocessing of the propeptide. Then the mature peptide encoding gene amplified by PCR was ligated to the pBE2a::Rs109 ~ pBE2a::Rs112 to generate pBE2a::Rs-*mp′* respectively.

Finally, the plasmids pBE2a::Rs-*mp′* and pBE2a-*apr* were transformed into *B. subtilis* WB600 to express the alkaline protease.

It was observed that the proteolytic ring could be produced by the hosts carrying pBE2a-*apr* and pBE2a::Rs112 but could not by the others carrying pBE2a::Rs109 ~ pBE2a::Rs111. The results showed that the activity of the alkaline protease produced by the hosts carrying pBE2a-*apr* and pBE2a::Rs112-*mp′* could be detected and were parallel. While the activity of the protease produced by the hosts carrying pBE2a::Rs109-*mp′* ~ pBE2a::Rs111-*mp′* could not be detected (Table 2). The determination of the activity of AprE was consistent with the results of the proteolytic ring.

To determine if there are mutations in the coding domain of the alkaline protease which couldn't produce the proteolytic ring. DNA sequencing was used to detect the mutation in the mature peptide encoding gene which was probably produced by PCR. However, the results of the DNA sequencing revealed that there was no mutation in the coding region of the mature peptide. The protein electrophoresis was employed to determine if the insertions adversely affected the secretion of the alkaline protease. The electrophoresis result of the intracellular protein showed that no increased accumulation of 41-kDa proAprE protein was observed in cell (Fig. 4a). The results of the extracellular protein electrophoresis showed that there was no obvious difference in the secretion level between the two active proteases which were produced by the hosts carrying pBE2a-*apr* and pBE2a::Rs112-*mp′*, respectively, while there was a great difference between the active protease and the inactive protease. The active proteases, a 38-kDa protein bond could be detected in the hosts producing inactive proteases, while the mature peptide which is 29-kDa in molecular mass was not detected (Fig. 4b). The results described above implied that the mutant proAprE protein resulted from the different insertions of *Bam*H I appeared to be secreted normally but abnormally in autoprocessing of the propeptide outside of the cell.

In our experiments, the insertion occured at the residues 109, 110, 111 and 112, respectively (Table 3). The insertion of the restriction sites between propeptide and mature peptide might interfere with the recognition of the autoprocessing site so that the pro-peptide couldn't be cleaved from preproenzyme correctly. It was deduced from the experimental results described above that the amino acids

Table 2 Determine result of the enzyme activity of the samples

Plasmid designation	Average value of OD600	Activity of AprE (U/mL)	Standard deviation
pBE2a (control)	0.000	0	±0.1
pBE2a-*apr*	0.478	18	±0.9
pBE2a::Rs109-*mp′*	0.000	0	±0.2
pBE2a::Rs110-*mp′*	0.000	0	±0.3
pBE2a::Rs111-*mp′*	0.000	0	±0.2
pBE2a::Rs112-*mp′*	0.484	18	±2.0

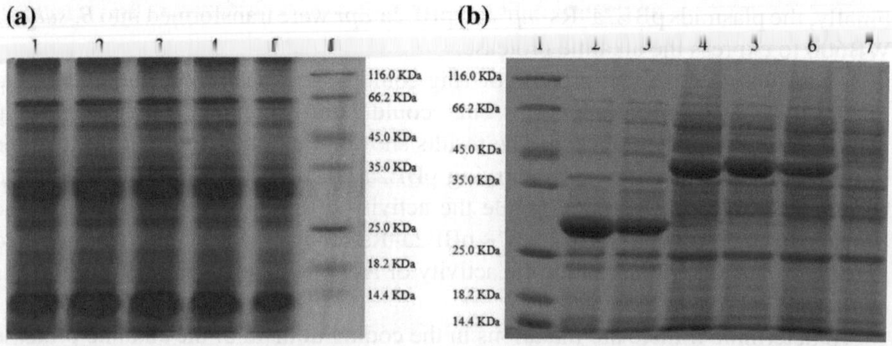

Fig. 4 The SDS-PAGE results of the intracellular and extracellular protein from different Samples. **a** The SDS-PAGE result of the intracellular protein. The strains examined were as follows *1–5* hosts carrying pBE2a::Rs109-*mp'*~pBE2a::Rs112-*mp'* and pBE2a-*apr*, respectively; *6* protein marker. **b** The SDS-PAGE result of the extracellular protein. *1* Protein marker; the strains examined were as follows: *2–3* hosts carrying pBE2a-*apr* and pBE2a::Rs112-*mp'*; *4–6* hosts carrying pBE2a::Rs109-*mp'*~pBE2a::Rs111-*mp'*; *7* host of carrying pBE2 (negative control)

Table 3 The insertional position of the *Bam*H I

Plasmid designation	Insertion position	Amino acids sequence
pBE2a-*apr*	–	TTMAQS
pBE2a::Rs109-*mp'*	109	**TGS**TMAQS
pBE2a::Rs110-*mp'*	110	T**TGS**MAQS
pBE2a::Rs111-*mp'*	111	TTM**GS**AQS
pBE2a::Rs112-*mp'*(pBE2R-*mp*)	**112**	TTMA**GS**QS

The pBE2a-*apr* is the plasmid carrying the intact peptide of alkaline protease without the insertion of the *Bam*H I. pBE2a::Rs109~pBE2a::Rs112 are plasmids containing different insertional positions of *Bam*H I between propeptide and mature peptide. The black body "GS" represents two amino acids introduced by insertion of *Bam*H I

"TTMA" which located at the junction between propeptide and mature peptide, as well as its arranging sequence were critical in autoprocessing of the propeptide. The autoprocessing would be disturbed when this amino acid sequence was destroyed.

4 Discussion

This report described the construction of the *Escherichia coli-Bacillus subtilis* shuttle vector pBE2R and the identification of the critical residues involved in the autoprocessing of the propeptide in alkaline protease. The new shuttle vector contained P43 promoter and the signal as well as pro-peptide encoding gene of alkaline protease. It was demonstrated that the signal peptide of the alkaline protease could guide the foreign protein to secrete effectively. The propeptide acting as an intramolecular chaperone had a significant effects on the folding and stability of the protease, as well as the formation of the active protease [10–15]. In our experiment, It was shown that thc dcletion of the propeptide resulted in the acute instability of the alkaline protease. The alkaline protease could not fold itself into a suitable state without the assistance of the propeptide so that could be recognized and digested by the proteases produced by the host (not data shown). On the other hand, the correct cleavage of the propeptide after the secretion of the protein was also indispensable to an active protease. It was revealed from the mutant experimental results that several crucial amino acids "TTMA" locating at the 3′ terminal of the propeptide, as well as the maintenance of its integrated sequence played a key role in the autoprocessing of the propeptide. The autoprocessing might be disturbed when this amino acid sequence was destroyed. In addition, the inactive protease caused by the aborted autoprocessing of the propeptide suggested that the propeptide was a potent inhibitor of the mature domain although it was needed in the folding of the protein before secreting.

The pBE2R could be used to express the alkaline protease in *B. subtilis* WB600 successfully. The shuttle vector pBE2R not only provided a convenient tool for the expression of the foreign protein in the *B. subtilis*, but also established a platform for the high-throughput screening of the alkaline protease in the process of the directed evolution of the alkaline protease. The mutated encoding genes of the mature peptide which were produced by error-prone PCR and DNA shuffling were introduced into the pBE2R to express the alkaline protease with the different mutation in *B. subtilis* WB600. The desired mutant containing different positive characters was obtained after multiple rounds of error-prone PCR and DNA shuffling as well as the high-throughput screening.

The test for the expression of mesophilic alpha-amylase gene in *B. subtilis* WB600 by using pBE2R revealed that the long propeptide of the alkaline protease could not guide the alpha-amylase which belongs to the short propeptide protein to produce the functional protein.

Acknowledgements This work was supported by the National Natural Science Foundation of China (No. 21176190). We thank professor Sui-Lam Wong (University of Calgary, Canada) for providing the pWB980 plasmid kindly. We also thank NanKai University for providing the pBE2 plasmid kindly.

References

1. Gupta R, Beg Q, Lorenz P (2002) Bacterial alkaline proteases: molecular approaches and industrial applications. Appl Microbiol Biotechnol 59:15–32
2. Bryan PN (2001) Protein engineering of subtilisin. Biochim Biophys Acta 1543:203–222
3. Siezen RJ, Leunissen JAM (1997) Subtilases: the superfamily of subtilisin-like serine proteases. Protein Sci 6:501–523
4. Kessler E, Safrin M, Gustin JD, Ohman DE (1998) Elastase and LasA protease of *Pseudomonas aeruginosa* are secreted with their propeptides. J Biol Chem 273:30225–30231

5. Simonen M, Palva I (1993) Protein secretion in *Bacillus* species. Microbiol Rev 57(1):109–137
6. Hsu CC, Tsai YC, Chang YS, Lisw SH, Mei HC (2002) Mutational analysis of the autoprocessing site of subtilisin YaB-G124A. Biochem Biophys Res Commun 291:165–169
7. Takahashi M, Hasuura Y, Nakamori S et al (2001) Improved autoprocessing efficiency of mutant subtilisins E with altered specificity by engineering of the pro-region. J Biochem 130:99–106
8. Grande KK, Gustin JK, Kessler E et al (2007) Identification of critical residues in the propeptide of LasA protease of *Pseudomonas aeruginsa* involved in the formation of a stable mature protease. J Bacteriol 189(11):3960–3968
9. Wang PZ (1984) Doi RH Overlapping promoters transcribed by *Bacillus subtilis* σ55- and σ37-RNA polymerase holoenzymes during growth and stationary phases. J Biol Chem 259 (13):8619–8625
10. Barequet IS, Ben SGM, Kessler E et al (2004) *Pseudomonas aeruginosa* LasA protease in treatment of experimental staphylococcal keratitis. Antimicrob Agents Chemother 48:1681–1687
11. Cunningham EL, Mau T, Truhlar SME et al (2002) The proregion N-terminal domain provides specific interactions required for catalysis of alpha-lytic protease folding. Biochemistry 41:8860–8867
12. Chang AK, Park JW, Lee EH et al (2007) The N-terminal propeptide of *Vibrio vulnificus* extracellular metalloprotease is both an inhibitor of and a substrate for the enzyme. J Bacteriol 189(19):6832–6838
13. Falzon L, Patel S, Chen YJ et al (2007) Autotomic behavior of the propeptide in propeptide-mediated folding of prosubtilisin E. J Mol Biol 366:494–503
14. Shinde U, Inouye M (2000) Intramolecular chaperones: polypeptide extensions that modulate protein folding. Semin Cell Dev Biol 11:35–44
15. Pulido M, Saito K, Tanaka S et al (2006) Ca^{2+}-dependent maturation of subtilisin from a hyperthermophilic archaeon, *Thermococcus kodakaraensis*: the propeptide is a potent inhibitor of the mature domain but is not required for its folding. Appl Environ Microbiol 72 (6):4154–4162

Effects of *ATF2* Overexpression with *BAT2* Deletion on the Higher Alcohols and Esters in Beer Yeast

Xiaoer Liu, Jia Xu, Li Pi, Cuiying Zhang and Dongguang Xiao

1 Introduction

In the yeast fermentation process, the by-products, carbon dioxide and alcohol form a unique beer flavor. The by-products include higher alcohols, esters, phenolic compounds and so on [1]. Among them, esters and higher alcohols are two of the most important groups of volatile flavor compounds. As we all known, beer alcohols and esters are generated in the main fermentation stage. However, levels of higher alcohols in the fermentation solution are often too high for favorable beer development, while beer yeast strains display low capacity for ester production. Thus, development of methods by which to decrease the generation of higher alcohols and increase the production of aromatic esters in beers, particularly through the use of industrial brewer's yeast, is of great importance.

Alcohol acetyltransferases (AATases), which catalyze the transformation of alcohols and acetyl-coenzyme A into acetate esters, are key enzymes involved in ester synthesis [2–4]. Three types of AATases (namely, AATase, AATase I, and AATase II) have been studied, and they are encoded by the *ATF1*, *Lg-ATF1*, and *ATF2* genes respectively [5–9]. Several researchers have reported that, compared to *ATF1,* overexpression of *ATF2* caused slight increase in the process of synthesis of ester. Similarly, *Lg-ATF1* has a very limited role in the synthesis of volatile esters [10–12]. In the previous study, the ester production was observed to increase significantly when overexpressed *ATF1* in beer yeast, which leaded to the disharmony of beer flavor [13]. Overexpression of *ATF2* in industrial brewer's yeast is observed for appropriately increasing the production of ester, thus beer flavor behave more harmonious.

X. Liu · J. Xu · L. Pi · C. Zhang (✉) · D. Xiao
Tianjin University of Science and Technology,
Tianjin 300457, People's Republic of China
e-mail: cyzhangcy@tust.edu.cn

H. Liu et al. (eds.), *Advances in Applied Biotechnology*, Lecture Notes
in Electrical Engineering 444, https://doi.org/10.1007/978-981-10-4801-2_8

Higher alcohols, also known as fusel alcohols, include propyl alcohol, isoamyl alcohol, isobutyl alcohol, active amyl alcohol and so on. About 80% outputs of higher alcohols are formed during the primary fermentation. Higher alcohols are one of the inherent flavor ingredients of beer, and the coordination between all kinds of higher alcohols can make beer palate soft and unique taste [14]. In the amino acid catabolism pathway, knocking out the amino acid transaminase (BCAT), which is encoded by *BAT1* and *BAT2* gene, can reduce the concentration of isobutanol and isoamyl alcohol [15]. The product of the *BAT2* gene has been previously reported to play an important role in the production of higher alcohols than the *BAT1* gene [16].

In this article, the brewing yeast strain with part *BAT2* allele insteaded by *ATF2* gene was constructed to improve a moderate amount of acetate content and reduce higher alcohol content. Our data show that *ATF2* overexpression and *BAT2* deletion can improve the acetate ester content in beer while significantly reducing its isoamyl alcohol content. The results in this article are useful in future developments in the beer industry.

2 Materials and Methods

2.1 Strains and Media

Escherichia coli DH5α, the parental strain *Saccharomyces cerevisiae* S5, and plasmid pUC-BBAK, pUC-PIA2K were obtained from the Microbiological Culture Collection Center of Tianjin Industrial Microbiology Key Laboratory, Tianjin University of Science and Technology, People's Republic of China.

Plasmid pUC-BBAK contained two homology DNA fragments of upstream and downstream of *BAT2* gene and *loxP-kanMX-loxP* gene disruption cassette, named BA, BB and K, respectively. Plasmid pUC-PIA2K was used in the preparation of the PGK_P-PGK_T expression cassette and the *ATF2* expression genes. The *ATF2* gene was connected between the promoter and the terminator, the direction is consistent with them.

E. coli DH5α used for the preparation and construction of the plasmid, was grown at 37 °C in Luria-Bertani medium (1% Bacto-tryptone, 0.5% yeast extract, and 0.5% NaCl). Ampicillin was added to the medium at a final concentration of 100 μg mL^{-1} to select positive *E. coli* transformants. The parental strain S. cerevisiae S5 was usually cultured in YPD medium (1% yeast extract, 2% Bacto-peptone, and 2% glucose). G418 was added to the medium at a final concentration of 100 μg mL^{-1} to select positive yeast transformants. The optimum growth temperature of *S. cerevisiae* was 30 °C. During fermentation, the yeast cells were cultured in wort medium. Wort medium was prepared from crushed malt and distilled water according to the ratio of material to water 1:4, and saccharification was performed according to certain routes with sugar meter adjusted to 10 °Brix. All of the solid media used in this study contained 2% agar.

2.2 DNA Manipulation

The DNA operation in this study was performed according to the standard procedure described by Ausubel [17]. In-Fusion DNA ligase, TaKaRa LA Taq DNA polymerase, 5000 DNA Marker, 15,000 DNA Marker and restriction enzymes were used for DNA manipulation, and these reagents were purchased from the TaKaRa Biotechnology.

2.3 Plasmid Construction

The polymerase chain reaction (PCR) primers used in this work are listed in Table 1. A 3379-bp PGKp-ATF2-PGKt fragment was amplified via PCR from the plasmid pUC-PIA2K, which contained *ATF2* gene (1608-bp) under the control of phosphoglycerate kinase I gene promoter and terminator (1771-bp). A 5309 bp BB-pUC19-BA-KanMX fragment was amplified via PCR from the plasmid pUC-BBAK, which contained the homology arms BA (488-bp) and BB (522-bp) and *loxP-kanMX-loxP* fragment (1613-bp). The recombinant plasmid pUC-PABBK was obtained from the ligation of PGKp-ATF2-PGKt fragment and BB-pUC19-BA-KanMX fragment reacted for 30 min with the use of In-Fusion DNA ligase under 50 °C.

2.4 Yeast Transformation and Screening

Transformation fragment PABBK was obtained by PCR amplification from the constructed plasmid pUC-PABBK, and then transformed into yeast genome through the method of LiAc transformation. G418-resistant transformants were selected and identified via PCR.

Table 1 Primers used in this study

Primer	Sequence (5′ → 3′)
BB-U	TTCGTTACGGATCCCGGCGTTTTTCTACTGAGTTAAGGGGTC
BK-D	CGGGATCCGGCATAGGCCACTAGTGGATCTG
PGK-U	CCTATGCCGGATCCCGTCTAACTGATCTATCCAAAACTGA
PGK-D	CGGGATCCGTAACGAACGCAGAATTTTC
BA-U	TACCAAACCGCTAACTTCCTTC
BB-D	AGTTTTCATCATCTTTGGGGTC
U-①	ATGTCGCCGCCGTCAATA
D-①	CAAGACTGTCAAGGAGGGTA
U-②	GTTCGGGTTCAGCGTATT
D-②	TGACAAAGGGAGTAGCAT

2.5 Fermentation Method

First of all, the yeast cells were cultured in tube containing 7 mL wort medium and static cultured for 24 h in 30 °C incubator. The culture was added to the 150 mL triangle bottle containing 45 mL wort medium with 10% inoculation quantity, and static cultured for 24 h in 16 °C incubator. Then 15 mL of the culture was transferred into 250 mL triangle bottle containing 135 mL wort medium, and static cultured 7–9 days in 10 °C incubator. Weight lost of Carbon dioxide was detected every 12 h, until the data is less than 0.2 g.

2.6 Gas Chromatography (GC) Analysis

After fermentation, samples were filtered and distilled from the wort medium, then used for GC analysis. Analysis was performed on an Agilent 7890C GC system. Capillary column was 30 m × 320 μm × 0.5 μm and column temperature was 75 °C. The temperature of the flame ionization detector (FID) was adjusted to 230 °C, the injector temperature was 200 °C, and the split ratio was 20:1. Nitrogen was used as the carrier gas, and the injection volume was 1.0 μL. Butyl acetate was used as the internal standard. A specific amount of each of the analytes was measured and used as a standard for machine calibration. Ethyl acetate, amyl acetate, isoamyl acetate, isobutanol and isoamyl alcohol were purchased from Merck.

2.7 Other Basic Performance Test

Weight lost of Carbon dioxide, residual sugar, alcohol degree, appearance fermentation degree and real fermentation degree were referred from People's Republic of China country.

3 Results and Discussions

3.1 Construction of Recombinant Plasmid pUC-PABBK

The construction process of the recombinant plasmid pUC-PABBK was showed in Fig. 1.

Plasmids pUC-BBAK and pUC-PIA2K were used as templates, and BB-pUC19-BA-KanMX (5310-bp) and PGK_P-ATF2-PGK_T (3379-bp) fragments were obtained by PCR amplification with primers BB-U, BK-D and PGK-U, PGK-D, respectively. Data was shown in Fig. 2. The two fragments were connected

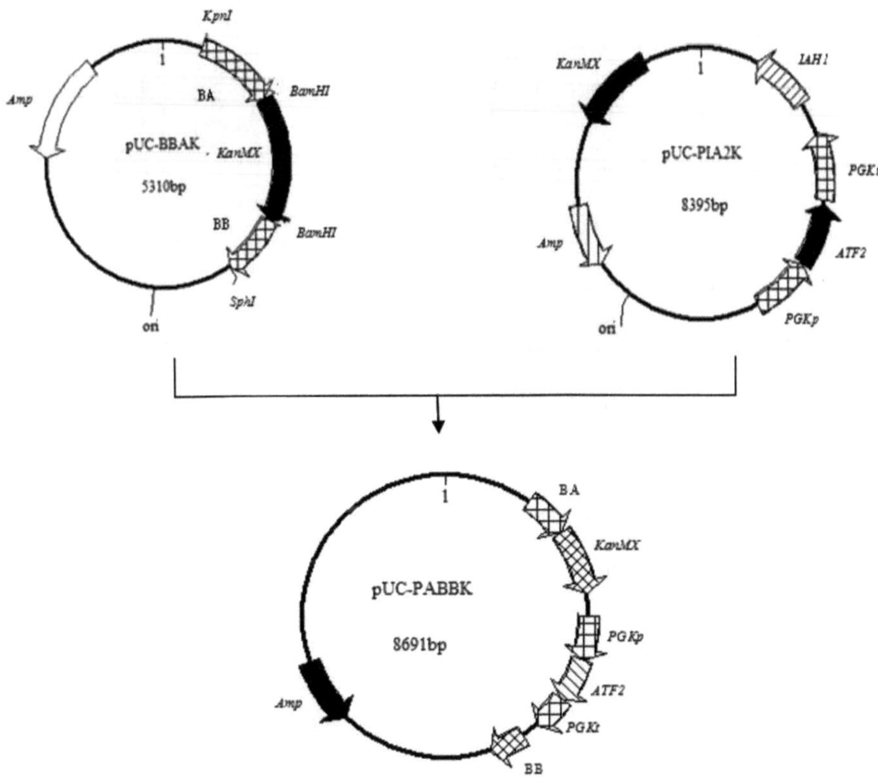

Fig. 1 Construction process of plasmid pUC-PABBK

with In-Fusion DNA ligase. For PCR verification, the 6005-bp size of a fragment (Fig. 3a) would be gotten with the primers BA-U and BB-D. For enzyme digestion verification, the 8691-bp size of a fragment (Fig. 3b) would be gotten using the restriction enzyme *Nco* I.

3.2 Construction of Engineered Brewer's Yeast Strains

The transformation fragment PABBK was amplified via PCR from the plasmid pUC-PABBK and integrated into the homologous genome of the S5 strain via LiAc transformation (Fig. 4). The resulting transformants were screened on YPD plates containing 0.50 mg mL^{-1} G418 [18]. The strain S5-L was selected as the correct recombinant after PCR analysis using the primer pairs U-①, D-① and U-②, D-②, separately (Fig. 5).

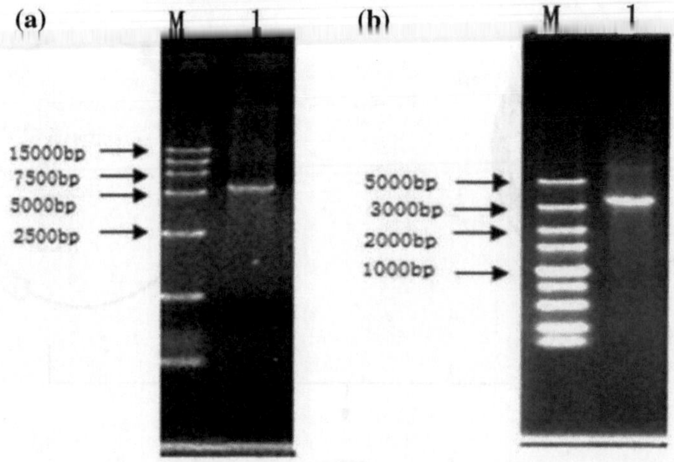

Fig. 2 **a** *M* 15,000 DNA marker; *lane 1* PCR amplification results of BB-pUC19-BA-KanMX fragment (5310-bp). **b** *M* 5000 DNA marker; *lane 1* PCR amplification results of PGK_P-ATF2-PGK_T fragment (3379-bp)

Fig. 3 **a** *M* 15,000 DNA marker; *lane 1* the result of PCR verification using primer BA-U and BB-D. **b** *M* 15,000 DNA marker; *lane 1* the result of enzyme digestion verification using restriction enzyme *Nco* I

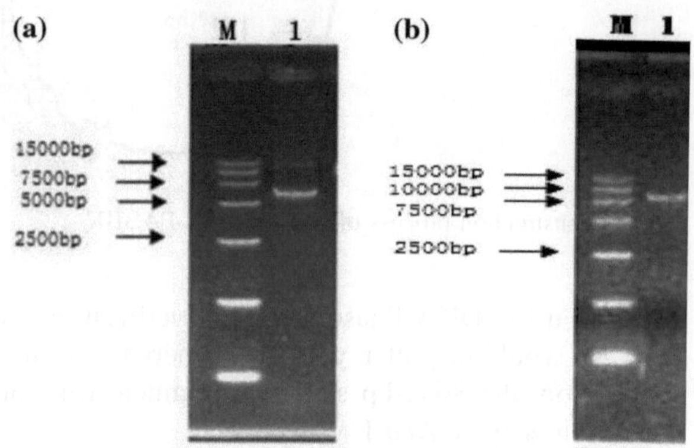

3.3 Basic Fermentation Performance of S5 and S5-L

The fermentation performance of S5 and S5-L were detected. The results (Table 2) showed that there was no significant difference in the basic fermentation performances of the engineered strains S5-L and the parental strain S5.

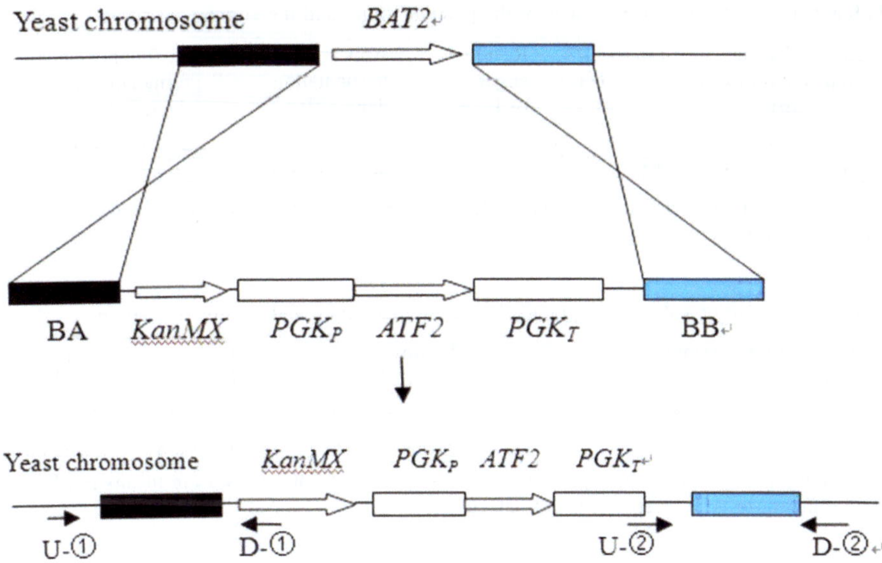

Fig. 4 Construction of engineered brewer's yeast strain

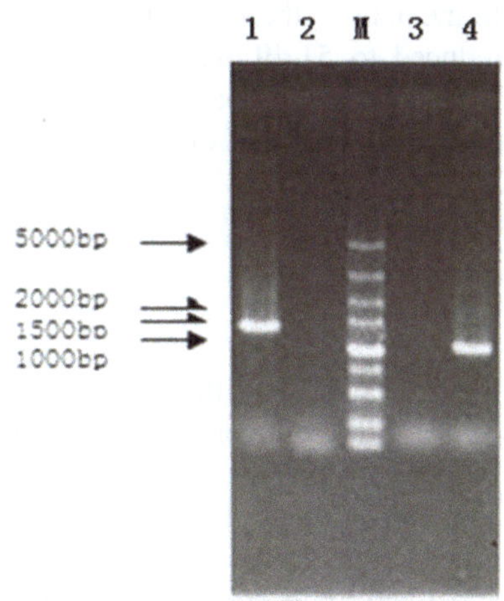

Fig. 5 *lanes 1, 2* were PCR amplification results from the recombinant (S5-L) genome and the parental strain (S5) genome using the primer pairs U-①, D-①; *M* 5000 DNA marker; *lanes 3, 4* were PCR amplification results from the parental strain (S5) genome and recombinant (S5-L) genome using the primer pairs U-②, D-②

3.4 *Production of Volatile Flavor Compounds During Fermentation*

After fermentation, the esters and higher alcohols in parental strain and engineering strain were determined by GC analysis. The results showed that the content of acetate esters in engineering strain S5-L was improved than that of the parental strain S5. The content of ethyl acetate in engineering strain S5-L reached

Table 2 Fermentation performances of the parental strain and the engineered strain

Yeast strains	Weight loss of CO_2 (g)	Ethanol (%, v/v, 20 °C)	Residual sugar (g L^{-1})	Appearance fermentation degree (%)	Real appearance fermentation degree (%)
S5	6.6 ± 0.20	4.27 ± 0.23	8.5 ± 0.32	67.90 ± 0.55	58.27 ± 0.40
S5-L	6.6 ± 0.15	4.12 ± 0.02	9.33 ± 0.80	66.22 ± 0.70	57.62 ± 0.38

Results are averages from three parallel independent experiments. Values are means ± SD from three different tests

Table 3 Content of volatile compounds in the parental strain and engineering strain

Yeast strains	Ethyl acetate (mg L^{-1})	Isoamyl acetate (mg L^{-1})	Isoamyl alcohol (mg L^{-1})	Isobutanol (mg L^{-1})	Propanol (mg L^{-1})
S5	5.92 ± 0.38	–	60.74 ± 0.40	11.35 ± 0.33	12.23 ± 0.16
S5-L	7.60 ± 0.47	–	51.49 ± 0.45	8.30 ± 0.58	9.72 ± 0.37

Results are averages from three parallel independent experiments. Values are means ± SD from three different tests. "–" represents not detected

7.60 mg L^{-1}, which was 1.28-fold of that of the parental strain S5. Isoamyl acetate in engineering strain and parental strain was not detected. It was probably the result of that the content of isoamyl acetate was lower than the GC detection. After fermentation, the content of isoamyl alcohol in the engineering strain S5-L was reduced to 51.49 mg L^{-1}, which was 84.77% of that of the parental strain S5. Isobutanol and propanol were reduced to 8.30 and 9.72 mg L^{-1}, which was 73.13% and 79.47% of that of the parental strain S5 (Table 3), respectively.

4 Conclusion

In this work, with the action of *PGK* promoter, *ATF2* gene, which encodes the AATase II, was overexpressed to increase acetate content. On the other hand, *BAT2* gene, which encodes the amino acid transaminase, was knocked out to decrease the content of higher alcohols. The results of the final fermentation showed that ethyl acetate content was moderate increased while higher alcohols concentrations were effectively reduced. In this paper, the ester content has been improved, but there's still a lot of work to do in order to achieve better results.

References

1. Styger G, Prior B, Bauer FF (2011) Wine flavor and aroma. J Ind Microbiol Biotechnol 38:1145–1159
2. NordstrÖm K (1963) Formation of ethyl acetate in fermentation with brewer's yeast IV: metabolism of acetyl coenzyme A. J Inst Brew 69:142–153

3. NordstrÖm K (1964) Formation of esters from alcohols by brewer's yeast. J Inst Brew 70:328–336
4. NordstrÖm K (1962) Formation of ethyl acetate in fermentation with brewer's yeast III: participation of coenzyme A. J Inst Brew 68:398–407
5. Fujii T, Nagasawa N, Iwamatsu A, Bogaki T, Tamai Y, Hamachi M (1994) Molecular cloning, sequence analysis and expression of the yeast alcohol acetyltransferase gene. Appl Environ Microbiol 60:2786–2792
6. Fujii T, Yoshimoto H, Nagasawa N, Bogaki T, Tamai Y, Hamachi M (1996) Nucleotide sequence of alcohol acetyltransferase genes from lager brewing yeast, *Saccharomyces carlsbergensis*. Yeast 12:593–598
7. Yoshimoto H, Momma T, Fujiwara D, Sone H, Kaneko Y, Tamai T (1998) Characterization of the *ATF1* and *Lg-ATF1* genes encoding alcohol acetyltransferases in the bottom fermenting yeast *Saccharomyces pastorianus*. J Ferment Bioeng 86:15–20
8. Yoshimoto H, Fujiwara D, Momma T, Tanaka K, Sone H, Nagasawa N, Tamai T (1999) Isolation and characterization of the *ATF2* gene encoding alcohol acetyltransferase II in the bottom fermenting yeast *Saccharomyces pastorianus*. Yeast 15:409–417
9. Lilly M, Bauer FF, Lambrechts MG, Swiegers JH, Cozzolino D, Pretorius IS (2006) The effect of increased yeast alcohol acetyltransferase and esterase activity on the flavour profiles of wine and distillates. Yeast 23(9):641–659
10. Nagasawa N, Bogaki T, Iwamatsu A, Hamachi M, Kumagai C (1998) Cloning and nucleotide sequence of the alcohol acetyltransferase II gene (*ATF2*) from *Saccharomyces cerevisiae* Kyokai No. 7. Biosci Biotechnol Biochem 62:1852–1857
11. Lilly M, Lambrechts MG, Pretorius IS (2000) Effect of increased yeast alcohol acetyltransferase activity on flavour profiles of wine and distillates. Appl Environ Microbiol 66:744–753
12. Verstrepen KJ, Van Laere SDM, Vanderhaegen BMP (2003) Expression levels of the yeast alcohol acetyltransferase genes *ATF1*, *Lg-ATF1*, and *ATF2* control the formation of a broad range of volatile esters. Appl environ microbiol 69(9):5228–5237
13. Cui-Ying Z, Yu-Lan L, Ya-Nan Q, Jian-Wei Z, Long-Hai D, Xue L, Dong-Guang X (2013) Increased esters and decreased higher alcohols production by engineered brewer's yeast strains. Eur Food Res Technol 236:1009–1014
14. Stewart GG (2005) Esters: the most important group of flavor-active beer compounds. Proc Congr Eur Brew Conv 30:100–101
15. Yoshimoto H, Fukushige T, Yonezawa T et al (2002) Genetic and physiological analysis of branched-chain alcohols and isoamyl acetate production in *Saccharomyces cerevisiae*. Appl Microbiol Biotechnol 59(4):501–508
16. Eden A, Van Nedervelde L, Drukker M et al (2001) Involvement of branched-chain amino acid aminotransferases in the production of fusel alcohols during fermentation in yeast. Appl Microbiol Biotechnol 55(3):296–300
17. Ausubel FM, Brent R, Kingston RE, Moore DD, Seidman JG, Smith JA, Struhl K (1994) Current protocols in molecular biology. Wiley, New York
18. Schiestl RH, Gietz RD (1989) High efficiency transformation of intact yeast cells using single stranded nucleic acids as a carrier. Curr Genet 16(5–6):339–346

Study on a Staphylococcal Tat Signal Peptide Guided EGFP Translocation in *E. coli*

Qiu-Xiang Zhou, Jun Zhang, Mei-Na Wang, Wen-Hao Yang, Jian Zhang and Qiang Gao

1 Introduction

A variety of protein transport systems have been found in bacteria for decades. Based on their structure, the signal peptide can be mainly divided into the following kinds: Sec (Secretion Translocation Pathway), Tat (Twin Arginine Translocation) [1], LIP (Lipoprotein), Com (type IV pili structure), and ATP-binding cassette (ABC) signal peptide [2]. To date, Sec pathway is widely used for exogenous protein secretion in genetic engineering, but Sec transport system can only secrete unfolded or partially folded proteins [3]. However, Tat signal peptide secretion system can secrete the correctly-folded protein to the outside of the cell [4, 5]. Tat pathway was first found in *E. coli*. Bolhuis et al. found that two integral cytoplasmic membrane proteins, TatB and TatC, consist a structural and functional unit of the twin-arginine translocases in *E. coli*. In the field of genetic engineering, it can be used to secrete heterologous proteins that cannot be transported by Sec translocation pathway [6]. Currently, it has become a focus of protein transportation research in the world due to its feature to help translocate fully folded proteins across the bacterial plasma membrane [7].

Green fluorescent protein (GFP) can be well expressed as a reporter protein in a large number of prokaryotic [8] and eukaryotic cells without affect by the biological, tissue or genotype [9]. In addition, GFP possesses a small molecular weight (*ca.* 27 kDa), and does not affect the property and function of other fused proteins [4]. Thus, the target protein can be fused with GFP for the localization analysis [10]. Moreover, its expression has the advantages of easy detection, high sensitivity, stable fluorescence, non-toxic to cells, among others. Therefore, GFP can be

Q.-X. Zhou · J. Zhang · M.-N. Wang · W.-H. Yang · J. Zhang · Q. Gao (✉)
Key Laboratory of Industrial Fermentation Microbiology,
Ministry of Education, College of Biotechnology, Tianjin University
of Science and Technology, Tianjin 300457, People's Republic of China
e-mail: gaoqiang@tust.edu.cn

H. Liu et al. (eds.), *Advances in Applied Biotechnology*, Lecture Notes
in Electrical Engineering 444, https://doi.org/10.1007/978-981-10-4801-2_9

directly used for the determination of living cells characteristics, and is widely used in various research fields of the life sciences [11].

In this study, the enhanced green fluorescent protein (EGFP) was used as a reporter protein to study the protein secretion properties by Tat pathway in *E. coli* [12]. A Tat signal peptide from *Staphylococcus carnosus* TM300 [7, 13, 14] was ligated with EGFP gene to construct the Tat-EGFP fusion gene. The EGFP and fused Tat-EGFP (EGFPs) were employed to investigate its role in EGFP translocation in *E. coli* host.

2 Materials and Methods

2.1 *Strains and Plasmids*

E. coli DH5α was obtained from the preservation of Tianjin Municipal Industrial Microbiology Key Laboratory, Tianjin, China. *Staphylococcus carnosus* TM300 and plasmid pBT2 (Cm^r, Amp^r) were kindly offered by Professor Dr. Friedrich Goetz of the Department of Microbial Genetics, Eberhard-Karls-Universitaet, Tuebingen, Baden-Wuerttemberg, Germany. The plasmids pBT2-ETG encoding the Tat-EGFP fusion protein with a Tat signal peptide of iron-dependent peroxidase EfeB (GenBank ID: CAL29117.1) in *S. carnosus* TM300 and pBT2-ET-EGFP only encoding EGFP gene without signal peptide were constructed in our previous work [12–14].

2.2 *Media and Culture Condition*

LB broth (10 g/L tryptone, 5 g/L yeast extract, 10 g/L NaCl) was employed for the cultivation of *E. coli*. The solid LB medium was prepared by the addition of 20 g/L agar. The ampicillin resistance was used for the screening of the recombinants at final concentration of 100 μg/mL.

2.3 *Enzymes and Reagents*

The *Nhe*I and *Bam*HI restriction enzymes, T4 DNA ligase and low molecular weight proteins marker were purchased from Fermentas Inc., (Burlington, Ontario, Canada). 2× Taq PCR Mastermix, plasmid extraction kit, 1 kb DNA Ladder and DNA Marker III were bought from Tiangen Biotech (Beijing) Co., Ltd., (Beijing, China). Lysozyme (20,000 U/mL), EGFP antibody, PVDF film and Color Developing Reagent (DAB) kit were bought from Beyotime Biotechnology Research Institute (Nantong, Jiangsu, China). Cell lysis buffer was composed of

0.8 mg/mL lysozyme, 20% sucrose, 50 mmol/L Tris and 1 mmol/L EDTA (pH 8.0). All other reagents were analytical or biological grade.

2.4 *Extraction of EGFPs from* E. coli

For the extraction of EGFPs (EGFP and Tat-EGFP), single recombinant colonies of *E. coli* DH5α/pBT2-ET-EGFP and *E. coli* DH5α/pBT2-ETG were respectively grown in 50 mL LB broth at 180 r/min and 37 °C for 16 h. The cell pellets were harvested by centrifugation at 12,000 r/min and 4 °C for 15 min. Proteins in the culture broth were treated with 10% trichloroacetic acid (TCA) for 12 h at 4 °C. The cells were resuspended in 2 mL PBS buffer, disrupted by sonication, and recovered the supernatant from cell fragments by centrifugation as the cytoplasm fraction.

The fermentation broth of *E. coli* was centrifuged at 12,000 r/min and 4 °C for 10 min, then the cell precipitate was washed twice with PBS (pH 7.4) for collection. Then the cell lysis buffer was added to fully suspend the bacterial cells on ice bath for 30 min. The supernatant after centrifuge was the periplasm protein fraction.

The culture broth of *E. coli* was centrifuged, and the supernatant was extracted and then mixed with 10% TCA precipitation, kept statically at 4 °C overnight. The precipitation was washed once with 100–80% acetone, and then the proteins in the fermentation broth was precipitated by centrifugation, and dried at room temperature as the fermentation broth fraction.

2.5 *Detection of EGFP*

The expression of the EGFPs in the cells and fermentation broth of *E. coli* DH5α/pBT2-ET-EGFP and *E. coli* DH5α/pBT2-ETG was observed by fluorescence microscopy [OLYMPUS (China) Co., Ltd., Beijing, China], and analyzed by SDS-PAGE and Western blot.

3 Result

3.1 *Fluorescence Observation of EGFPs in* E. coli *DH5α Cells*

By dropleting the *E. coli* transformant culture broth on the slide surface with cover glass, fluorescence was observed under a fluorescent microscope. In Fig. 1, *E. coli* DH5α/pBT2-ET-EGFP and *E. coli* DH5α/pBT2-ETG cells emitted green

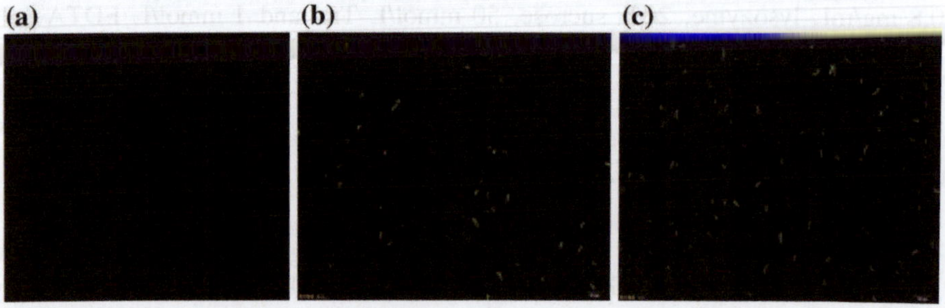

Fig. 1 Result of fluorescence observation of various *E. coli* DH5α recombinants. **a** *E. coli* DH5α/pBT2; **b** *E. coli* DH5α/pBT2-ET-EGFP; **c** *E. coli* DH5α/pBT2-ETG

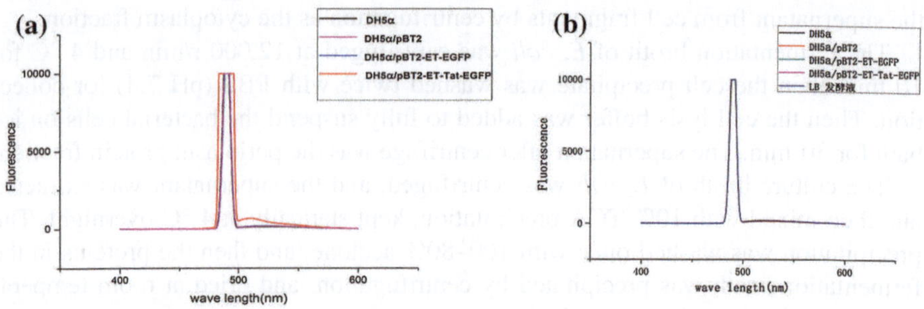

Fig. 2 Fluorescence spectrophotometer determination for EGFPs of various *E. coli* host. **a** The periplasm extract; **b** the fermentation broth

fluorescence, but the control *E. coli* DH5α not. The results indicated that both EGFPs genes were able to be expressed in *E. coli* DH5α host.

3.2 Fluorescence Spectrophotometer Detection

The detection of EGFP and Tat-EGFP in periplasm and fermentation broth of various *E. coli* hosts were performed using an fluorescence spectrophotometer (Hitachi F-7000, Japan). Only Tat-EGFP gave an obvious fluorescence emission peak at 510 nm in the periplasm of *E. coli* DH5α/pBT2-ETG, but not EGFP for *E. coli* DH5α/pBT2-ET-EGFP, indicating that the staphylococcal Tat signal peptide plays a positive role for transmembrane protein transport of Tat-EGFP (Fig. 2).

Additionally, no emission peak was found at 510 nm from the both fermentation broth, suggesting that both kinds of EGFPs could not be secreted into the fermentation broth from their *E. coli* host cells, respectively. This can be referred to the lack of translocation and the outer membrane block for EGFP as well as the later for Tat-EGFP.

3.3 SDS-PAGE Analysis

For the overnight cultures of *E. coli* DH5α, *E. coli* DH5α/pBT2, *E. coli* DH5α/pBT2-ET-EGFP and *E. coli* DH5α/pBT2-ETG, SDS-PAGE was used for the analysis of EGFPs in the cytoplasm, periplasm and fermentation broth to determine whether the EGFPs could be secreted into the culture broth (Fig. 3).

3.4 Western Blot Verification

In this work, whether the expressed EGFPs can be translocated into the periplasm and fermentation broth from various *E. coli* hosts were further detected by Western blot. The Western blot experiment here also exhibited the same result as SDS-PAGE, i.e. both EGFP and Tat-EGFP were correctly expressed in the cytoplasm of *E. coli* hosts, but only Tat-EGFP was expressed in the periplasm (Fig. 4), and none of them could be transported into the fermentation broth (data not shown).

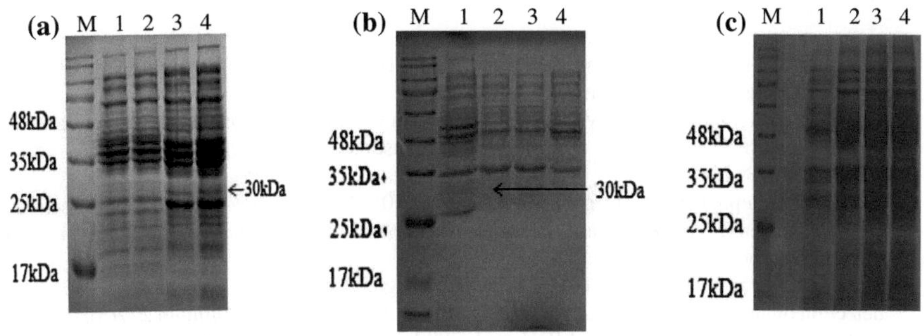

Fig. 3 SDS-PAGE analysis of intracellular proteins in *E. coli*. **a** intracellular; **b** periplasm; **c** fermentation broth. M: protein marker; 1: *E. coli* DH5α/pBT2-ETG; 2: *E. coli* DH5α/pBT2-ET-EGFP; 3: *E. coli* DH5α/pBT2; 4: *E. coli* DH5α

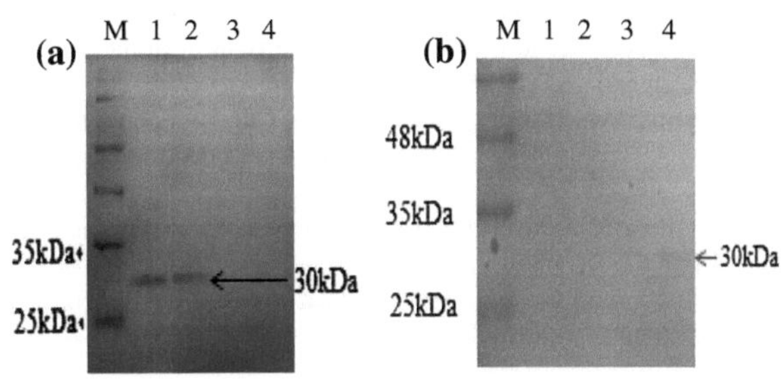

Fig. 4 Western blot analyses of EGFPs in *E. coli*. **a** intracellular. M: protein marker; 1: *E. coli* DH5α/pBT2-ETG; 2: *E. coli* DH5α/pBT2-ET-EGFP; 3: *E. coli* DH5α/pBT2; 4: *E. coli* DH5α; **b** periplasm. M: protein marker; 1: *E. coli* DH5α; 2: *E. coli* DH5α/pBT2; 3: *E. coli* DH5α/pBT2-ET-EGFP; 4: *E. coli* DH5α/pBT2-ETG

4 Conclusion

To our best knowledge, this is the first report that the constructed Tat-EGFP was successfully translocated in the periplasm *E. coli* DH5α host by a Tat signal peptide from *S. carnosus* TM300, but EGFP not. Our results revealed that the heterologous staphylococcal Tat signal peptide is recognized by the *E. coli* host, and exerts a positive role in the transmembrane secretion of Tat-EGFP, which sheds the light as a novel pathway for the secretion of the heterologous proteins in *E. coli* and probably other bacteria.

Acknowledgements This work was financially supported by the National Natural Science Foundation of China (31370075, 31101275 and 61603273), the Undergraduate Laboratory Innovation Fund of Tianjin University of Science and Technology of China (1504A304X) and the Youth Innovation Fund of Tianjin University of Science and Technology of China (2014CXLG28).

References

1. Lee PA, Tullman D, Georgiou G et al (2006) The bacterial twin-arginine translocation pathway. Annu Rev Microbiol 60:373–395
2. Bogsch EG, Sargent F, Stanley NR et al (1998) An essential component of a novel bacterial protein export system with homologues in plastids and mitochondria. J Biol Chem 273(29):18003–18006
3. Jensen CL, Stephenson K, Jørgensen ST et al (2000) Cell associated degradation affects the yield of secreted engineered and heterologous proteins in the *Bacillus subtilis* expression system. Microbiology 146(10):2583–2594
4. Zhang G, Gurtu V, Kain SR (1996) An enhanced green fluorescent protein allows sensitive detection of gene transfer in mammalian cells. Biochem Biophys Res Commun 227(3):707–711
5. Tjalsma H, Bolhuis A, Jongbloed JDH et al (2000) Signal peptide-dependent protein transport in *Bacillus subtilis*: a genome-based survey of the secretome. Microbiol Mol Biol Rev 64(3):515–547
6. Thiemann V, Saake B, Vollstedt A et al (2006) Heterologous expression and characterization of a novel branching enzyme from the thermoalkaliphilic anaerobic bacterium *Anaerobranca gottschalkii*. Appl Microbiol Biotechnol 72(1):60–71
7. Biswas L, Biswas R, Nerz C et al (2009) Role of the twin arginine translocation pathway in *Staphylococcus*. J Bacteriol 191(19):5921–5929
8. Tünnemann G, Martin RM, Haupt S et al (2006) Cargo-dependent mode of uptake and bioavailability of TAT-containing proteins and peptides in living cells. FASEB J 20(11):1775–1784
9. Sheen J, Hwang S, Niwa Y et al (1995) Green-fluorescent protein as a new vital marker in plant cells. Plant J 5(8):777–784
10. Chalfie M, Tu Y, Euskirchen G et al (1994) Green fluorescent protein as a marker for gene expression. Science 263(5148):802–805

11. Chen Y, Müller JD, Ruan QQ et al (2002) Molecular brightness characterization of EGFP in vivo by fluorescence fluctuation spectroscopy. Biophys J 82(1):133–144
12. Yu C, Zheng X, Zhu Y et al (2011) Construction of *tat-gfp* fusion gene and its expression in *Staphylococcus carnosus*. Biotechnol Bull 8:203–207
13. Xu B, Cheng Y, Wang L et al (2015) Construction of *Eschericha coli-Staphylococcus* shuttle vector for EGFP expression and potential secretion via Tat pathway. Lect Notes Electr Eng 333:171–180
14. Gao Q, Xu B, Cheng Y et al (2014) GFP translocation of twin-arginine secretion pathway in *Staphylococcus carnosus*. J Tianjin Univ Sci Technol 5:1–5

A Novel GH10 Xylanase Xyn13-3 from Alkaline Soil: Gene Cloning and Heterogenous Expression

Haiyan Qiu, Zhongyuan Li, Hui Wang, Shuang Li and Tongcun Zhang

1 Introduction

Hemicelluloses are the second most abundant plant polysaccharides in nature. Due to their potential role as sustainable energy sources, hemicelluloses are increasingly becoming an important concern [1]. The main chain of xylan is constituted with xylose unit through the β-1,4-linkage, and it is usually substituted with varying levels of L-arabinofuranosyl, galactosyl, acetyl, glucuronyl, and 4-O-methylglucuronyl groups [2]. Among them, xylanase plays a crucial role in the hydrolysis of the xylan back-bone [3], in which xylanase cleaves the β-1,4-glycosidic bond between xylose residues to release xylooligosaccharides. A great potential for some biotechnological applications in many biotechnological application especially in food, bioconversion, pulp and paper industries [4]. Due to the increasing demand for xylanase production, exploring new enzyme sources has been a hot topic in recent years [5]. Therefore, we tried to explore some novel alkaline xylanase from alkaline soil, in this study, a novel xylanase of family 10 (Xyn13-3) was obtained from the metagenomic DNA of alkaline soil and it was heterologously expressed in *Pichia pastoris* GS115.

Haiyan Qiu and Zhongyuan Li: Co-first authors.

H. Qiu · Z. Li · H. Wang · S. Li · T. Zhang (✉)
College of Biotechnology, Tianjin University of Science and Technology, Tianjin 300457, People's Republic of China
e-mail: lizhongyuan@tust.edu.cn

H. Liu et al. (eds.), *Advances in Applied Biotechnology*, Lecture Notes in Electrical Engineering 444, https://doi.org/10.1007/978-981-10-4801-2_10

2 Materials and Methods

2.1 *Strains, Plasmids, and Culture Conditions*

E. coli DH5α and the vector PMD-19T (TaKaRa, Dalian, China) were used for gene cloning. *Pichia pastoris* GS115 and vector pPIC9 used for gene expression were purchased from Invitrogen (Carlsbad, CA, USA). Reagents and chemicals used for DNA manipulation were purchased from ThermoFisher Scientific (Shanghai, China).Yeast Nitrogen Base (YNB) with amino acids was purchased from Solarbio (Beijing, China). All yeast culture media were prepared according to the Yeast Protocols Handbook.

2.2 *Alkaline Soil Sampling and Metagenomic DNA Extraction*

Alkaline soil was obtained from the coastal saline area in Tianjin (China) and stored in −20 °C. The pH was measured by pH meter (SevenEasy, Shanghai). About 1 g soil sample was firstly grinded by liquid nitrogen, and the metagenomic DNA was extracted by CTAB-SDS method [6]. The crude DNA was purified by DNA purification kit (Solarbio, Beijing, China), and was detected by nucleic acid electrophoresis.

2.3 *Cloning and Sequence Analysis of Xylanase Gene Xyn13-3*

The xylanase gene *Xyn13-3* was obtained by Touch-down PCR and TAIL-PCR methods. Firstly, the conserved sequence of *Xyn13-3* was amplificated by Touch-down PCR with primers GH10F and GH10R according to Wang et al. [7]. The Nucleotide fragment was sequenced by GENEWIZ, Beijing, China. The full gene sequence was amplificated using Tail-PCR method with the six specific primers designed based on the conserved sequence. The primers used in the study were shown in Table 1. The sequence was assembled by Vector NTI7.0 (Invitrogen), and was blasted by BLASTX (https://blast.ncbi.nlm.nih.gov/Blast.cgi).

Table 1 Details of primers used in this investigation

Primers	Primer sequence(5′–3′)
GH10F	CTACGACTGGGAYGTNIBSAAYGA
GH10R	GTGACTCTGGAWRCCIABNCCRT
SP1-U	GTAGTCGTTGATATACAGCTTGGCGGCC
SP1-D	TTGCGTTTGAGGCGGCCCGTGAAG
SP2-U	AGTGCCCGGCGGAGTTTGCGTTAC
SP2-D	GGCCGCCAAGCTGTATATCAACGACTAC
SP3-U	CTAGGAGGCGGGAGAAGACAGAG
SP3-D	CGGTGTGCCCATTGACGGAAATCGG
PIC9F	GGGGAATTCATGGGCACAAAGTACCCGGCGAGCG
PIC9R	GGGGCGGCCGCTCAAAGAATGTTCAGAATAGC

2.4 Heterologous Expression of Xyn13-3

The full-length gene sequence of *Xyn13-3* was digested with *Eco*RI and *Not*I, ligased with pPIC9 vector, transformed into *E. coli* DH5α, verified by restriction digestion, which was then linearized by *Pme*I and transformed into *P. pastoris* GS115 competent cells by electroporation. The transformants were screened and induced by methanol according to the instructions of *Pichia* expression kit (Invitrogen). The recombinant protein Xyn13-3 was secreted into the medium and collected by centrifugation, and subjected to the enzyme activity assay [8]. The transformant with the highest activity was fermented in a 1-L flask for optimization of the induction time. The recombinant protein Xyn13-3 was purified by nickel affinity chromatography (GE Healthcare, Uppsala, Sweden) with 10–500 mM imidazole in buffer [9, 10]. The protein fraction corresponding to different imidazole concentration were collected and subjected to sodium dodecyl sulfate (SDS)-polyacrylamide gel electrophoresis (PAGE; 12% separation gel and 5% spacer gel). The protein concentration was determined using Bradford protein assay kit (BioRad) [11].

3 Results and Discussion

3.1 Cloning and Sequence Analysis of the Xylanse Gene Xyn13-3

A partial gene fragment of 250 bp was amplified by Touch-down PCR, and was cloned into pMD-19T for sequencing. The flanking region was amplified with TAIL-PCR using the nested internal primers which were designed based on the 250 bp core region. Based on the analysis by Vector NTI 7.0 and FGENESH, the complete ORF of Xyn13-3 consisted of 1020 bp (Fig. 1). The mature protein

Fig. 1 The xylanse gene *Xyn13-3* by PCR method *M* DM2000 marker; *1* the sequence of gene *Xyn13-3*

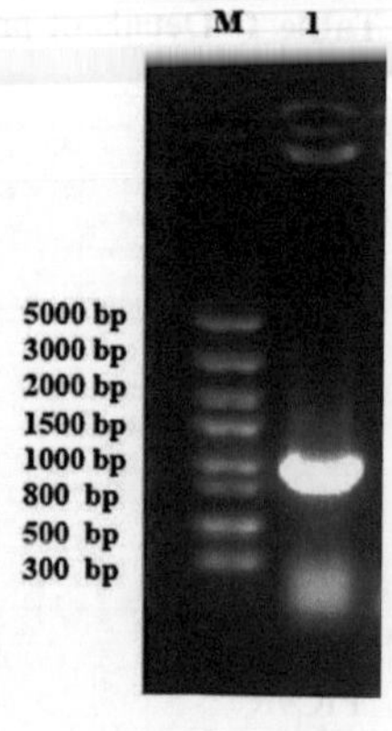

consisted of 339 residues with a calculated molecular mass of 36.97 kDa. *Xyn13-3* had no predicated N-glycosylation site. BlastP search indicates that *Xyn13-3* belongs to GH family 10 and its amino acid sequence showed the highest identity (61%) with the reported GH10 xylanase from *Trichoderma gamsii* (XP_018659952.1).

3.2 Phylogenetic Analysis

MEGA software was used to build *Xyn13-3* and other GH10 family xylanase phylogenetic tree, in order to understand the evolutionary relationship between *Xyn13-3* and other GH10 family xylanase, the results are as shown in Fig. 2. The xylanase has closely related to the genetic relationship with xylanase which originated from *Trichoderma gamsii* (XP_018659952.1).

3.3 Expression and Purification of Recombinant Enzyme

The full-length sequence of gene *Xyn13-3* was cloned into pPIC9 to obtain the recombinant plasmid pPIC9-Xyn13-3 (Fig. 3). The recombinant protein Xyn27-1 was purified to homogeneity at 200 mM imidazole with a molecular weight of 39.97 kDa (Fig. 4), a little larger than the calculated molecular mass, which might be attribute to the His6-tag. The specific activity of purified Xyn13-3 measured by DNS method was 135.24 U/mg.

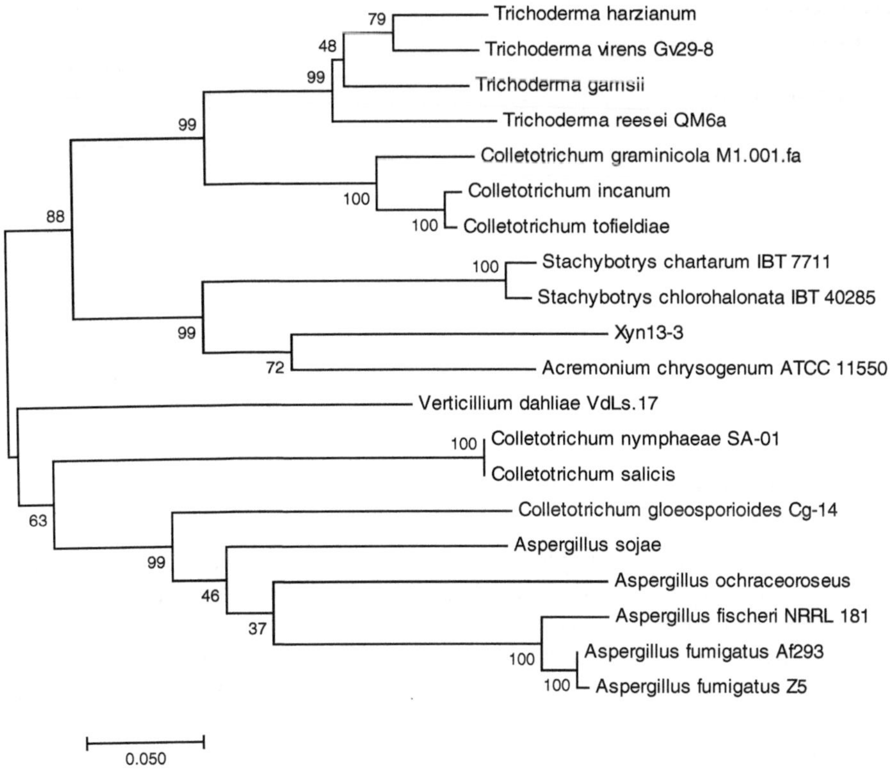

Fig. 2 Phylogenetic tree of *Xyn13-3*

Fig. 3 The structure of recombinant plasmid pPIC9-Xyn13-3

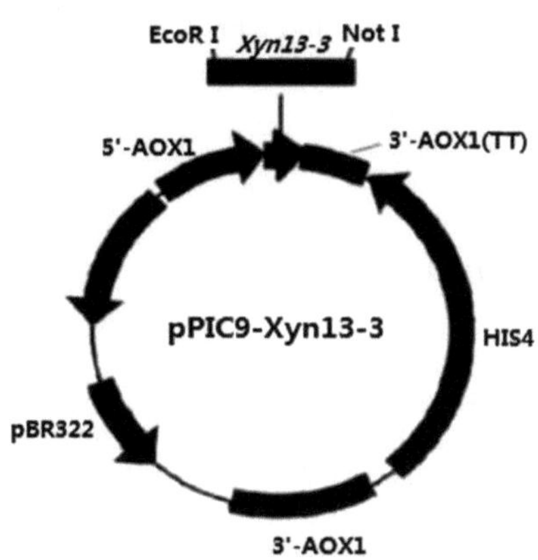

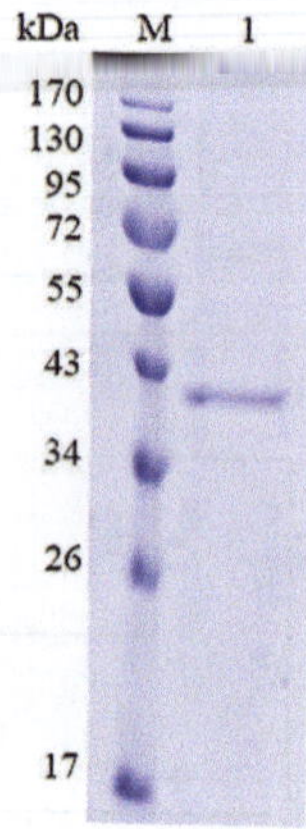

Fig. 4 SDS–PAGE analysis for the expression of Xyn13-3; *M* protein marker; *1* the purified xylanase

4 Conclusion

Xylanase is the most critical xylan degrading glycoside hydrolases, xylan could be degraded into oligosaccharides, which could be further degraded to xylose by glucosidase and other side-chain Hydrolase. In this study, a novel GH family 10 Xyn13-3 from the genomic DNA of Alkaline Soil was cloned and successfully expressed in *Pichia pastoris* GS115. The deduced proteins showed low identities with known fungal xylanases which originated from *Trichoderma gamsii* (61%). The specific activity of purified Xyn13-3 measured by DNS method was 135.24 U/mg. This is a new xylanase that needs further investigation to characterize its use. In order to improve the activity of the recombinant xylanase, severals factors for the optimization of induced culture will be studied, including induction timing, induced time, inducer concentration and inducing temperature, as well as the following study on enzymatic properties which can provide worthful theory data for application.

References

1. Saha BC (2003) Hemicellulose bioconversion. J Ind Microbiol Biotechnol 30:279–291
2. Bastawde KB (1992) Xylan structure, microbial xylanases, and their mode of action. World J Microbiol Biotechnol 8:353–368
3. Kambourova M, Mandeva R, Fiume I (2007) Hydrolysis of xylan at high temperature by co-action of the xylanase from *Anoxybacillus flavithermus* BC and the β-xylosidase/α-rabinosidases from *Sulfolobus solfataricus* Oalpha. J Appl Microbiol 102:1586–1593
4. Lagaert SA, Pollet A, Courtin CM et al (2014) β-Xylosidases and α-l-arabinofuranosidases: accessory enzymes for arabinoxylan degradation. Biotechnol Adv 32:316–332
5. Salmon DNX, Spier MR, Ricardo Carlos (2014) Analysis of inducers of xylanase and cellulose activities production by *Ganoderma applanatum* LPBMR-56. Fungal Biol 118:655–662

6. Brady SF (2007) Construction of soil environmental DNA cosmid libraries and screening for clones that produce biologically active small molecules. Nat Protoc 2:1297–1305
7. Wang G, Luo H, Meng K (2011) High genetic diversity and different distributions of glycosyl hydrolase family 10 and 11 xylanases in the goat rumen. doi:10.1371/journal.pone.0016731
8. Teng C, Jia H, Yan Q et al (2011) High-level expression of extracellular secretion of a beta—xylosidase gene from *Paecilomyces thermophila* in *Escherichia coli*. Bioresour Technol 102:1822–1830
9. Sharma M, Chadha BS (2010) Purification and characterization of two thermostable xylanases from *Malbranchea flava* active under alkaline conditions. Bioresour Technol 101:8834–8842
10. Tseng MJ, Yap MN (2002) Purification and characterization of two cellulase free xylanases from an alkaliphilic *Bacillus firmus*. Enzyme Microb Technol 30:590–595
11. Li Z, Xue X, Zhao H et al (2014) A C-terminal proline-rich sequence simultaneously broadens the optimal temperature and pH ranges and improves the catalytic efficiency of glycosyl hydrolase family 10 ruminal xylanases. Appl Environ Microbiol 80:3426–3432

Exploitation of a *KU70*-Deficient Mutant for Improving Gene Deletion Frequency in *Aspergillus niger*

Liu-hua Yin, Lan Zhang, Ling Liu, Hongfei Zhang, Li Hou and De-pei Wang

1 Introduction

Lack of objective functional gene, often referred as targeted gene knockout, is an essential tool for genetic engineering and reverse genetics [1]. The gene knockout is developed on the basis of homologous recombination between homologous DNA and chromosomal DNA, with site specificity. The target fragment can be used to characterize the genetic stability of chromosomal DNA. The process is mainly to build a plasmid vector into cells after homologous recombination to achieve the target gene knockout. This is of great significance for the transformation of industrial production strains. It is a major obstacle in many industrially important species such as *A. niger*.

The repair mechanism of DNA double-strand breaks is mainly HR and NHEJ [2], both of which are mediated by competitive inhibition [3, 4]. HR repair process mainly uses the homology of the DNA double-strand to carry out specific recognition, the protein responsible for matching and recombinant protein does not include the specificity of the base sequence. NHEJ does not require homologous chromosomes to be a template to repair the broken DNA double strand by connecting the two ends together, which is a powerful repair. Among the different species, the proportion of HR and NHEJ involved in repair is different. In eukaryotes, the repair of DSBs is more of the form of NHEJ. We can improve the efficiency of gene targeting by reducing the NHEJ pathway [5]. Through a large number of studies we affirm that KU protein [6] with NHEJ pathway has an important link, by knock out the KU gene to reduce the NHEJ pathway [7]. Ku protein is a widely located nuclear protein, encoded by the *KU* gene, which is

L. Yin · L. Zhang · L. Liu · H. Zhang · L. Hou · D. Wang (✉)
Key Laboratory of Industrial Fermentation Microbiology, Ministry of Education,
College of Biotechnology, Tianjin University of Science & Technology,
Tianjin 300457, China
e-mail: 15620074987@163.com

H. Liu et al. (eds.), *Advances in Applied Biotechnology*, Lecture Notes in Electrical Engineering 444, https://doi.org/10.1007/978-981-10-4801-2_11

widely existed with a variety of biological functions [8] and involved in a number of important cell metabolism, such as DNA replication, DSBs repair, transcription regulation [9]. When the *kusA* gene is destroyed in *A. niger*, which is a homologous sequence of *ku70*, the gene-targeting frequency of the gene-deleted strain was increased to 80% and the sensitivity to UV and X-Ray increased as well [10].

In this study, we deleted the *ku70* gene fragment to inhibit its non-homologous end of the link and improved the transformation efficiency of *Agrobacterium tumefaciens* in *A. niger*. It can be a convenient way for us to study the filamentous fungi in Genetic Engineering.

2 Materials and Methods

2.1 Material

2.1.1 Strains, Plasmids and Primers

Aspergillus niger CGMCC 10142, a hyper-producer of citric acid, was obtained from Tianjin Key Laboratory of Fermentation Microbiology, Tianjin, China. *Escherichia coli* DH5α was used as the host for the propagation of plasmid DNA, while *Agrobacterium tumefaciens* AGL1 was employed as T-DNA donor for fungal transformation [11] (Table 1). Primers in this study were listed in Table 2.

2.1.2 Reagents, Media and Cultivation

Hygromycin B was purchased from Solarbio institute, Beijing, China. Bleomycin and Geneticin were bought from Sigma institute, USA. All of the chemicals used were of analytical grade.

The bacterial strains were cultivated in LB medium at 37 °C (DH5α) or 28 °C (AGL1), respectively [12]. Potato dextrose agar (PDA) was used for the growth of *A. niger* [13]. Liquefied corn medium, provided by RZBC Co., Ltd., Rizhao, Shandong, China, was used for citric acid fermentation, the content of initial total sugar was 179.8 g L^{-1}, the mineral element: $FeSO_4 \cdot 7H_2O$ 30 mg L^{-1}, $CaCl_2 \cdot 2H_2O$ 10 mg L^{-1}, $MnCl_2 \cdot 4H_2O$ 7 mg L^{-1}, $CuSO_4 \cdot 5H_2O$ 0.3 mg L^{-1}, $MgSO_4 \cdot 7H_2O$ 0.3 g L^{-1}, $ZnSO_4 \cdot 7H_2O$ 10 mg L^{-1}, KH_2PO_4 0.3 g L^{-1}, yeast extract 0.004% (w/v). Fungal culture medium (FCM) was composed of 30 g L^{-1} sucrose, 10 mg L^{-1} $FeSO_4 \cdot 7H_2O$, 0.5 g L^{-1} KCl, 0.5 g L^{-1} $MgSO_4 \cdot 7H_2O$, 1 g L^{-1} K_2HPO_4, and 20 g L^{-1} agar powder (pH 5.0–6.0).

Table 1 Strains and plasmids in this study

Strains and plasmids	Relevant genotype/description	Source
Aspergillus niger CGMCC10142		Tianjin University of Science and Technology
E.coli DH5α		Tianjin University of Science and Technology
Agrobacterium tumefaciens AGL1		Tianjin University of Science and Technology
pGM-T	Amp^r	Purchased from Tiangen, China
pBackZero-T	Amp^r	Purchased from Takara, Japan
p60	$Ble^r HYG^r Kan^r$	Tianjin University of Science and Technology
p80	*E. coli* DH5α-*A. tumefaciens* shuttle vector, $HYG^r Kan^r$, containing the deletion cassette of *ku70*	This work
p10	*E. coli* DH5α-*A. tumefaciens* shuttle vector, $HYG^r Kan^r$, containing the deletion cassette of *cs*	This work
p74	*E. coli* DH5α-*A. tumefaciens* shuttle vector, $HYG^r Kan^r$, containing the deletion cassette of *aox*1	This work
p81	*E. coli* DH5α-*A. tumefaciens* shuttle vector, $G418^r Kan^r$, containing the deletion cassette of *tps*A	This work
p90	*E. coli* DH5α-*A. tumefaciens* shuttle vector, $G418^r Kan^r$, containing the deletion cassette of *pd*	This work

2.2 *Construction of the* ku70 *Deletion Cassette*

Construction of the *ku70* deletion cassette and PCR verification were shown in Figs. 1 and 2, respectively. The deletion cassette containing the hygromycin resistance gene (*hyg*) was designed to replace the *ku70* gene coding region according to a method of double-crossover homologous recombination. Briefly, a 955-bp fragment of the 5′-flanking region (the deletion cassette left arm) and another 874-bp fragment of the 3′-flanking region (the deletion cassette right arm) were amplified from the genomic DNA using primer pairs P3/P4 and P5/P6 (Table 2), respectively, and served as homologous arms for the recombination event. The right arm fragment and plasmid p60 were respectively digested with *Pst*I and *Hin*dIII, the two linear fragments were linked by T_4 DNA ligase to create the intermediate plasmid p80-R. The left arm fragment and plasmid p80-R were

Table 2 The primers used for the experiment

Primers	Sequence (5′ → 3′)	Descriptions
pI	TTAAACGATTCCACAACATTCTACT	Upstream primer of *ku70*
pII	CCTAATGCTCTTCGTTTTGACAT	Downstream primer of *ku70*
p1	GTCGACGTTAACTGATATTG	Upstream primer of *hyg*
p2	TTTGCCCTCGGACGAGTGCT	Downstream primer of *hyg*
p3	GGGGTACCGAGCTCGAGGCCAAACAGGCAGAC	Upstream primer of left deletion cassette
p4	CGGGATCC AGATCTATCGCTGGGCAATGACTTG	Downpstream primer of left deletion cassette
p5	AACTGCAG TCTAGAGCAGCTGCCGAAGGGA	Upstream primer of right deletion cassette
p6	CCCAAGCTT TCTAGATAACTGTACATCGCCT	Downstream primer of right deletion cassette
p7	CCGCCGATGTCCGCATACTG	For verification of the 2684 bp of the *ku70* gene
p8	TCATTGTACGCATACCGTTA	
p9	CTCACCGAGGCAGTTCCATAGG	Upstream primer of *G418*
p10	GCCCGATGCGCCAGAGTTGTTT	Downstream primer of *G418*

digested with *Kpn*I and *Bam*HI, respectively; the two linear fragments were linked by T_4 DNA ligase to generate the deletion plasmid p80 (5′-flanking region-hyg gene-3′-flanking region).

2.3 Electroporation Parameters

The constructed plasmid p80 was transformed into *A. tumefaciens* AGL1 by electric transformation [14]. The electroporational parameters were as follows: plasmid 1.2 μL, *A. tumefaciens* 70 μL, electric shock 0.5 s at 2.5 kV.

2.4 Infection Experiment

The *A. tumefaciens* AGL1 that carried plasmid 80 were cultured untill OD_{600} reached 0.8 in induction medium. Then, 200 μL *A. tumefaciens* AGL1 and 200 μL *A. niger* spores were added on 0.45 μm cellulose acetate membrane which placed in induction medium with 200 μmol/L AS and 100 μg/L Kanamycin. The co-culture was performed in 25 °C for 48 h [15, 16].

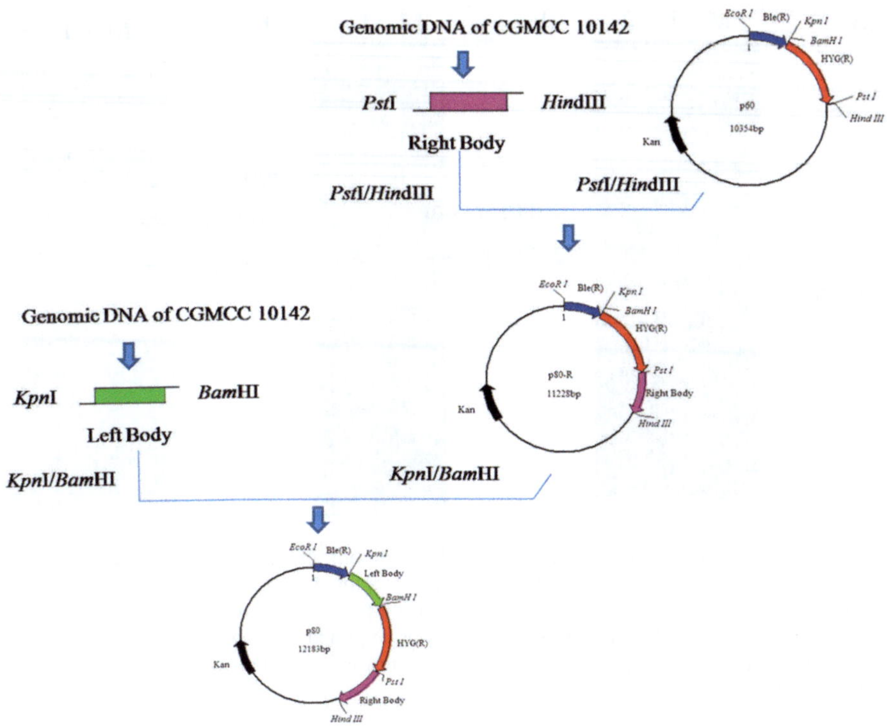

Fig. 1 Construction schematic diagram of plasmid p80

3 Results and Discussions

3.1 Cloning of the ku70 *Gene from* A. niger *CGMCC 10142*

The genomic DNA of *A. niger* was extracted from mycelia that grew on PDA media by the cetyltrimethylammonium bromide method [17]. The BLAST method was used to survey the whole genomic sequences of *A. niger* CBS 513.88 to search for an *ku70*-like gene [18, 19]. PCR was carried out to amplify the homolog of the *ku70*-like gene using a pair of pI and pII primers. The protocol and results were shown in Table 3 and Fig. 2, respectively.

3.2 Sequence Alignment

The result of *ku70* gene purification recovery was shown in Fig. 3. The DNA sequences of *ku70* gene of *A. niger* CGMCC 10142 was assigned with GenBank

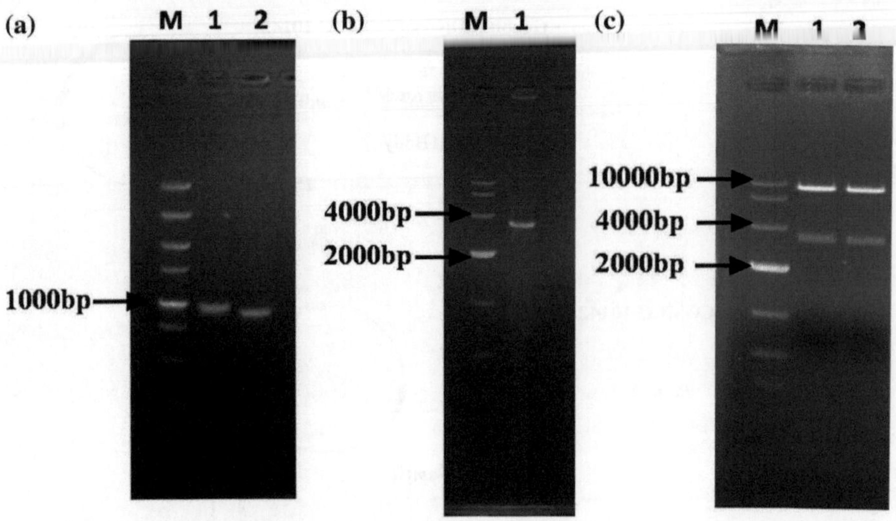

Fig. 2 The PCR test of left arm, right arm, full length and restriction enzyme analysis of p80. **a** *M* 5000 bp DNA marker; *lane 1* PCR product was amplified by using p80 as template with the primers P3/P4; *lane 2* PCR product amplified by using p80 as template with the primers P5/P6; **b** *M* 10,000 bp DNA marker; *lane 1* PCR product was amplified by using p80 as template with the primers P3/P6; **c** *M* 10,000 bp DNA marker; *lanes 1,2* restriction enzyme analysis of p80 amplified with *Kpn*I/*Hind*III

Table 3 The reaction system of PCR using rTaq polymerase

Reaction system		Condition		
Total volume	50 μL	Step 1 (1 cycle)	95 °C	5 min
10 ×buffer	5 μL			
d NTPs (2.5 mmol/L)	4 μL		95 °C	30 s
Template	1 μL	Step 2 (30 cycles)	60 °C	30 s
Primer I (10 μmol/L)	2 μL		72 °C	3 min
Primer II (10 μmol/L)	2 μL			
rTaq (5 U/μL)	0.5 μL	Step 3 (1 cycle)	72 °C	10 min
ddH$_2$O	33 μL			
		Step 4 (1 cycle)	4 °C	30 min

accession numbers XM001396771 and the homeology was 99.75% between the augmented sequence after identified by sequencing appraisal and nucleotide sequence reported in Genbank.

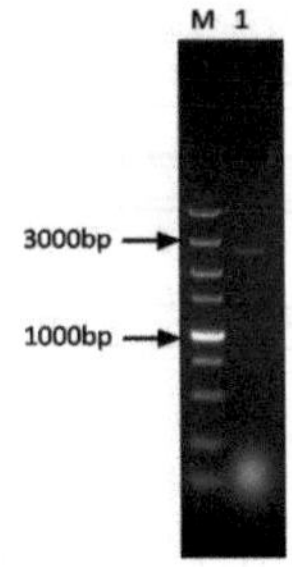

Fig. 3 The PCR product of *ku70* gene

3.3 Analysis of Phylogenetic Tree of ku70 *Gene*

A Neighbor-Joining tree was constructed from the amino acids sequences encoding *ku70* gene of the same 22 fungi of filamentous fungi and sacchromyces classes which were used to construct phylogenetic tree of *ku70* gene (Fig. 4). According to the phylogenetic tree, *Aspergillus niger* strain *CGMCC10142* have the closest relation compare to the other strains incorporated filamentous fungi and *saccharomyces*. The distantly regarding *ku70* gene from *saccharomyces* was showed the farthest relation are *Candida ablicans* strain *SC 5314* and *Pichia farinosa* strain *CBS 7064*.

3.4 Screening of the ku70 *Deletion Mutant*

The plasmid p80 was transformed into *A. niger* CGMCC 10142 by AMAT, and 35 transformants were obtained on CM plates containing 250 μg/mL hygromycin. All of transformants were verified by PCR of deletion cassette gene and hygromycin resistant in Fig. 5. The results demonstrated that transformant was the *ku70* gene knockout mutant therefore named as *Δku70.*

3.5 The Genetic Stability of Transformant

15th passage culture from the transformant was amplified in Fig. 6. The result indicated that *Δku70* was genetically stable.

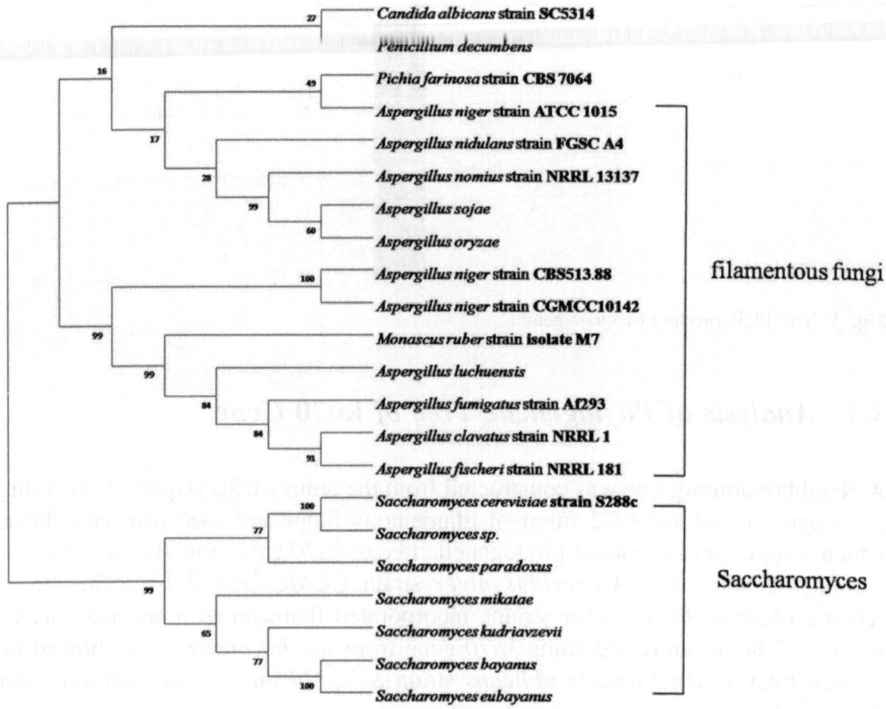

Fig. 4 Phylogenetic tree constructed from *ku70* gene amino acid sequence

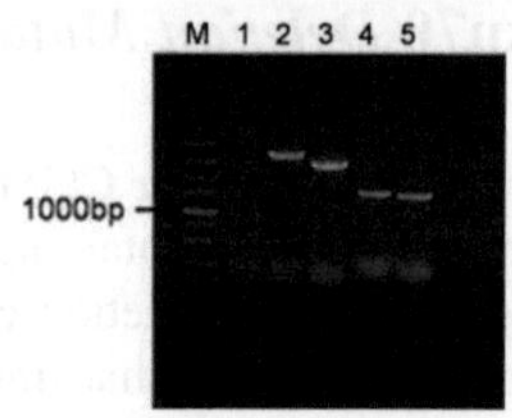

Fig. 5 The PCR test of *A. niger* transformation of *ku70* deletion. *M* 5000 bp DNA marker; *lane 1* negative control; *lane 2* 3240 bp PCR product amplified by using *Δku70* as template wit, the primers P1/P7; *lane 3* 2375 bp PCR product amplified by using *Δku70* as template with the primers P2/P8; *lane 4* 1400 bp fragment of *hyg* amplified by using p80 as template with the primers P1/P2; *lane 5* 1400 bp fragment of *hyg* amplified using transformant as template with the primers P1/P2

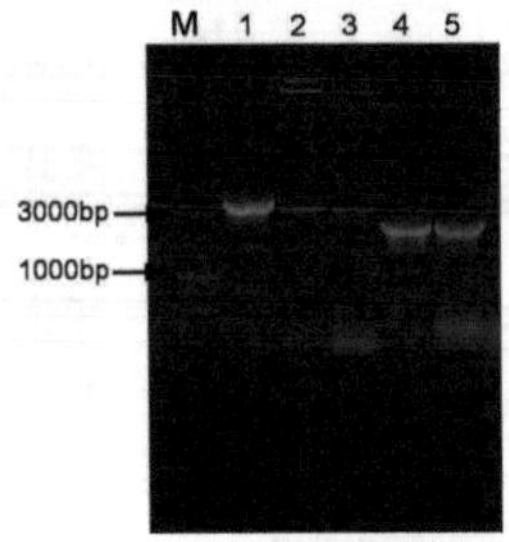

Fig. 6 The PCR test of *A. niger* transformation. *M* 5000 bp DNA marker; *lane 1* PCR product amplified by using the genome of 1st Δ*ku70* subculture as template with the primers P1/P7; *lane 2* PCR product amplified by using the genome of 10th Δ*ku70* subculture as template with the primers P1/P7; *lane 3* PCR product amplified by using the genome of 15th Δ*ku70* subculture as template with the primers P1/P7; *lane 4* PCR product amplified by using the genome of 10th Δ*ku70* subculture as template with the primers P2/P8; *lane 5* PCR product amplified by using the genome of 15th Δ*ku70* subculture as template with the primers P2/P8

Table 4 The number of transformants by different plasmids infect *A. niger*

	p10	p74	p81	p90
CGMCC 10142	8	18	39	31
Δku70	12	31	56	47
Efficiency (%)	50	72.2	43.5	51.6

3.6 *Reduce-Representation Genome Sequencing*

The result of Reduce-Representation Genome Sequencing showed that the *hyg* had a unique insertion site in 68,064 at An15. It is consistent within the locus of CBS513.88 *ku70* gene (GenBank No: ANI_1_420134) in 679,864–682,385.

3.7 *The Efficiency of Different Plasmids Affected by* ku70 *Gene Insertion*

The strain of *Δku70* and original strain CGMCC 10142 were infected by plasmids p10, p74, p81 and p90 which preserved in our laboratory, respectively. The transformants were selected in the plate form containing 150 μL/mL *hyg* after incubating 72 h at 35 °C. The numbers of transformants were counted in Table 4, and partial transformants verification as shown in Fig. 7. The result indicated that strain *Δku70* had a higher efficiency comparing with original strain CGMCC 10142.

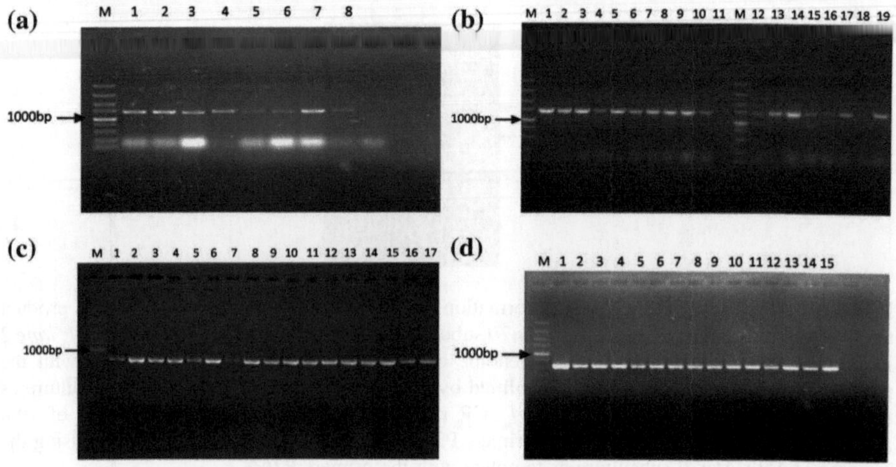

Fig. 7 The PCR test of *A. niger* transformation. **a** *M* 5,000 bp DNA marker; *lane 1–8* 1400 bp fragment of *hyg* amplified by using the genome of CGMCC Δ*cs* as template with the primers P1/P2; ***b*** *M* 5000 bp DNA marker; *lane 1–19* 1400 bp fragment of *hyg* amplified by using the genome of CGMCC Δ*aox* 1 as template with the primers P1/P2; **c** *M* 5000 bp DNA marker; *lane 1–17* 550 bp fragment of *G418* amplified by using the genome of CGMCC Δ*tps* A as template with the primers P9/P10; **d** *M* 5000 bp DNA marker; *lane 1–15* 550 bp fragment of *G418* amplified by using the genome of CGMCC Δ*pd* as template with the primers P9/P10

4 Conclusion

Aspergillus niger CGMCC10142 was successfully transformed by *Agrobacterium tumefaciens* with plasmid p80 and the hereditary stability of the transformant *Δku70* was confirmed by subculture verification. *A. niger* CGMCC10142 and *Δku70* were transformed by different plasmids respectively. It is indicated that all of genes segment insertion efficiency were dramatically improved by 40–80% compared with the original strain. The plasmids p10 and p74 (*hyg*) have a higher deletion efficiency of 50 and 72.2% respectively in *Δku70* compared with *A. niger* CGMCC 10142. Meanwhile, the deletion efficiency conducted with plasmids p81 and p90 (*G418*) was improved 43.5 and 51.6% respectively. The results showed that *G418* was a beneficial marker for inserting target fragment in *A. niger*.

References

1. Meyer V (2008) Genetic engineering of filamentous fungi-progress, obstacles and future trends. Biotechnol Adv 26:177–185
2. Shrivastav M, De Haro LP, Nickoloff JA (2008) Regulation of DNA double-strand break repair pathway choice. Cell Res 18:134–147

3. Krappmann S (2007) Gene targeting in filamentous fungi: the benefits of impaired repair. Fungal Biol Rev 21:25–29
4. Kück U, Hoff B (2010) New tools for the genetic manipulation of filamentous fungi. Appl Microbiol Biotechnol 86:51–62
5. Weld RJ, Plummer KM, Carpenter MA, Ridgway HJ (2006) Approaches to functional genomics in filamentous fungi. Cell Res 16:31–44
6. Walker JR, Corpina RA, Goldberg J (2001) Structure of the Ku heterodimer bound to DNA and its implications for double-strand break repair. Nature 412:607–614
7. Lieber MR (2010) The mechanism of double-strand DNA break repair by the nonhomologous DNA end-joining pathway. Ann Rev Biochem 79:181–211
8. Daley JM, Palmbos PL, Wu D et al (2005) Nonhomologous end joining in yeast. Ann Rev Genet 39:431–451
9. Ninomiya Y, Suzuki K, Ishii C et al (2004) Highly efficient gene replacements in *Neurospora* strains deficient for nonhomologous end-joining. Proc Natl Acad Sci U S A 101:12248–12253
10. Meyer V, Arentshorst M, El-Ghezal A et al (2007) Highly efficient gene targeting in the *Aspergillus niger kus*A mutant. J Biotechnol 128:770–775
11. Lazo GR, Stein PA, Ludwig RA (1991) A DNA transformation-competent *Arabidopsis* genomic library in *Agrobacterium*. Nat Biotech 9:963–967
12. Sambrook J, Fritsch EF, Maniatis T (1989) Molecular cloning: a laboratory manual, 2nd edn. Cold Spring Harbor Laboratory Press, Cold Spring Harbor
13. Atlas RM (1997) Handbook of microbiological media, 2nd edn. CRC Pr I Llc, pp 283–284
14. Haq I, Ali S, Iqbal J (2003) Direct production of citric acid from raw starch by *Aspergillus niger*. Process Biochem 38:921–924
15. de Groot MJA, Bundock P, Hooykaas PJJ et al (1998) *Agrobacterium tumefaciens*-mediated transformation of filamentous fungi. Nat Biotechnol 16:839–842
16. Sugui JA, Chang YC, Kwon-Chung KJ (2005) *Agrobacterium tumefaciens*-mediated transformation of *Aspergillus fumigatus*: an efficient tool for insertional mutagenesis and target gene disruption. Appl Environ Microbiol 71:1798–1802
17. Shao YC, Ding YD, Zhao Y et al (2009) Characteristic analysis of transformants in T-DNA mutation library of *Monascus ruber*. World J Microbiol Biotechnol 25(6):989–995
18. Yuan LX, van der Kaaij RM, van den Hondel CA et al (2008) *Aspergillus niger* genome-wide analysis reveals a large number of novel alpha-glucan acting enzymes with unexpected expression profiles. Mol Genet Genomic 279:545–561
19. Cheng Z, Xue YF, Zhang YL et al (2009) Recombinant expression and characterization of *Thermoanaerobacter tengcongensis* thermostable α-glucosidase with regioselectivity for high-yield isomaltooligosaccharides synthesis. J Microbiol Biotechnol 19:1547–1556

Biotransformation to Produce Boldenone by *Pichia pastoris* GS115 Engineered Recombinant Strains

Rui Tang, Peilin Ji, Ying Yu, Xu Yang, Mengjiao Liu, Kaiyuan Chen, Yanbing Shen and Min Wang

1 Introduction

Steroid medications are used widely in clinical applications and form an important and large category in the pharmaceutical industry. Boldenone (BD) is androgenic anabolic steroid; which increases nitrogen retention, protein synthesis, appetite and stimulates the release of erythropoietin in the kidneys. Boldenone undecylenate is often prescribed for patients that have lost muscle mass due to extended periods of bed rest or bouts with significantly diminished muscle mass. In addition, Boldenone is a popular drug for administration in various animals and extensively used in the cattle and meat production industry.

Traditionally, boldenone is obtained from 4-AD by chemical synthesis [1, 2] with many by-products, low quality and low yield. In contrast to chemical synthesis, biotransformation provides an alternative method for the production of steroid medicine intermediates and has been used extensively as a common and economical process in the pharmaceutical industry [1, 3, 4]. The methods of BD biosynthesis usually use AD as the substrate, through the Δ1-dehydrogenation and the reduction reaction of C17-one to C17-alcohol. While most of natural microorganisms do not have the capability of high expressing 17Hsd and 3-ketosteroid-Δ1-dehydrogenase at the same time. This bottlenecks seriously limit the wide industrial application of microbial transformation to produce BD. While recently *P. pastoris* GS115 was found to be capable of transforming the C17-one to

Rui Tang and Peilin Ji contributed equally to this work.

R. Tang · P. Ji · Y. Yu · X. Yang · M. Liu · K. Chen · Y. Shen · M. Wang (✉)
Key Laboratory of Industrial Fermentation Microbiology, Ministry of Education,
Tianjin Key Laboratory of Industrial Microbiology, College of Biotechnology,
Tianjin University of Science and Technology, No. 29, 13th Avenue, TEDA, Tianjin 300457,
People's Republic of China
e-mail: minw@tust.edu.cn

H. Liu et al. (eds.), *Advances in Applied Biotechnology*, Lecture Notes
in Electrical Engineering 444, https://doi.org/10.1007/978-981-10-4801-2_12

C17-alcohol of steroids, indicating that *P. pastoris* GS115 contain the gene for 17Hsd, but its 3-ketosteroid-Δ1-dehydrogenase has low activity. The construction of genetically engineered recombinant strains for bacterial KstDs is usually insoluble [5–10]. The expression of $KstD_F$ from *Aspergillus fumigatus* in *P. pastoris* was found to be the only case that was soluble [6]. And the KsdD2 from *R. rhodochrous* DSM43269 was found to have strong Δ1-dehydrogenation [9].

In this work, a 3-ketosteroid-Δ1-dehydrogenase KsdD2 from *R. rhodochrous* DSM43269 was successfully expressed intracellularly in *P. pastoris* GS115. The engineered recombinant strains *P. pastoris* GS115 pPIC3.5K-*ksdd2* was found to be capable of transforming AD to both ADD and BD. The transformation of AD to ADD and BD is certainly the result of synergism between exogenous KsdD2 and endogenous 17Hsd. In this way, BD could be obtained by biotransformation from AD. Thus we have established a green BD biosynthetic pathway, and provided theory basis for the wide industrial application of microbial transformation to produce BD.

2 Materials and Methods

2.1 Strains and Plasmids Construction

P. pastoris GS115 was obtained from Tianjin University of Science and Technology Culture Collection Center (TCCC), Tianjin, China. *R. rhodochrous* DSM43269 was purchased from China General Microbiological Culture Collection Center (CGMCC) and was used as the template for cloning the *ksdd2* gene. The plasmids pPIC3.5K was used to express the *ksdd2* gene. Primers were designed based on the published sequence of the *ksdd2* gene (Accession No. KU257604), and the sequences for the primers were: Sense: 5-CCG<u>GAATTC</u>ATGGCCAAG-ACCCCTGTACC-3. Antisense: 5-AAGGAAAAAA<u>GCGGCCGC</u>TCAGGCGA-CCGGTGCGT-3. The underlined characters indicate the sites of *Eco*R I and *Not* I, respectively. *P. pastoris* GS115 integration vector pPIC3.5K carrying the bacterial kanamycin gene (kan from Tn903) that confers resistance to Geneticin® and Ampicillin (Amp) resistance were used. The plasmids pPIC3.5K was the common plasmids used in *P. pastoris* GS115 for gene expression [11]. The purified DNA was digested with *Eco*R I and *Not* I and ligated into the *Eco*R I—*Not* I double-digested pPIC3.5K to construct pPIC3.5K-*ksdd2*. The pPIC3.5K-*ksdd2* vector was imported into *P. pastoris* GS115 by electroporation. The transformants were screened by MD medium. The selected mutants were designated as *P. pastoris* GS115-pPIC3.5K-*ksdd2*, for further characterization.

2.2 Media and Culture Conditions

Substrate androst-4-ene-3,17-dione (AD) was obtained from Sigma Aldrich Co. (USA). All chemical solvents and salts used were of analytical grade or higher. The cultivation and bioconversion of microorganisms, as well as the preparation and analysis of transformation products, were performed following the procedures described by Shao et al. [12]. The AD concentration added in the fermentation system were 2 g/L.

2.3 Analytical Methods

The method used for TLC and HPLC analysis of androst-4-ene-3,17-dione (AD), androst-1,4-ene-3,17-dione (ADD) and boldenone (BD) concentrations was in accordance with our previous study [13]. Samples extracted with ethyl acetate were detected by TLC and HPLC. TLC was performed on a TLC plate (Merck, KGaA; 20 × 20, 0.25 mm, Germany), developed by petroleum ether/ethyl acetate (4:6v/v), and visualized by spraying with 20% H_2SO_4 and heating at 100 °C for about 5 min until the colors developed. HPLC was carried out with an Agilent 1260 HPLC instrument equipped with a C18 column (ZORBAX Eclipse Plus C18, 3.5-Micron, 150 mm × 4.6 mm) and a UV/visible detector. AD, ADD and BD were monitored at 254 nm and the mobile phase composed of methanol and water (70:30, v/v). The flow rate was 1 mL/min and the column temperature was 30 °C.

3 Results and Discussion

3.1 Cloning of the Gene Encoding KsdD2

The complete sequence of the *ksdd2* gene from *R. rhodochrous* DSM43269 was obtained by PCR. The resulting fragment containing 1713 bp nucleotides is a complete open reading frame. Sequence analysis revealed that the sequence of the *ksdd2* ORF uses ATG as the start codon and the G+C content (%) of the *ksdD2* is 69.06 mol%. Also, a possible ribosomal binding site (GAAAGG) sixteen nucleotides upstream was found. The *ksdd2* gene consists of 1713 nucleotides and encodes a deduced protein of 571 amino acids. The molecular weight of KsdD2 was estimated to be 60.63 kDa, and the pI value was calculated to be 5.54 by the ExPASy compute pI/Mw program algorithm (http://www.expasy.org/cgi-bin/protparam).

The putative N-terminal FAD-binding motifs in the KsdD2 was consistent with the sequence as previously described [14], GSG(A/G)(A/G)(A/G)X17E (Fig. 1). Online analysis software showed that KsdD2 has two transmembrane segments,

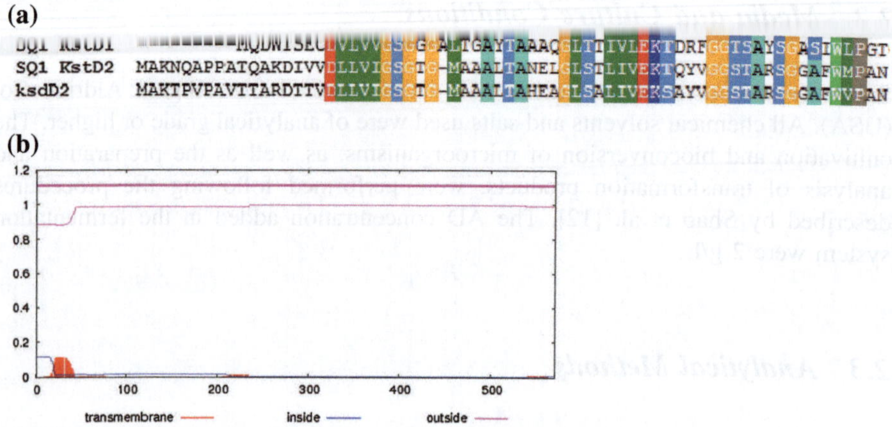

Fig. 1 Conserved sequence and the analysis of transmembrane region in KsdD2; **a** N-terminal FAD-binding motif of the putative KsdD2; **b** the analysis diagram of transmembrane region by TMHMM

four transmembrane helices, and a strong transmembrane domain near the N-terminal, indicating that KsdD2 may be a membrane protein.

Then the amplified gene was subcloned into the *Eco*R I/*Not* I—digested sites of the *E. coli/P. pastoris* shuttle vector pPIC3.5K to form pPIC3.5K-*ksdd2*. The resultant plasmid pPIC3.5K-*ksdd2* was verified by PCR and double endonuclease digestion (Fig. 2). The sequence of the cloned *ksdd2* gene was analyzed and the results indicated that the sequence was consistent with the reported data.

3.2 *Expression and Characterization of KstD2 in* P. pastoris

Generally, biotransformation from ADD to BD is usually carried out by 17β-hydroxysteroid-dehydrogenases (17Hsd), which are widespread in microorganisms such as yeasts. Previous research implied that *Saccharomyces cerevisiae* can transform the C17-one to C17-alcohol of steroids [15, 16], but its 3-ketosteroid-Δ1-dehydrogenase has low activity. To enhance its Δ1-dehydrogenation and the proportion of BD in the end products of AD transformation, eukaryotic host *P. pastoris* strain GS115 was then chosen to overexpress the *ksdD2* from *R. rhodochrous* DSM43269. pPIC3.5K, a powerful intracellular expression plasmid, was firstly selected. The recombinant *ksdD2* strain was designated as *P. pastoris* GS115 pPIC3.5K-*ksdd2*. The expression profile of *P. pastoris* GS115 pPIC3.5K-*ksdd2* showed that the expressed KsdD2 was larger than 66 kDa (Fig. 3).

As shown in Fig. 4, *P. pastoris* GS115 pPIC3.5K-*ksdd2* transformed AD to different products. This may be attributed to the original AD transformation

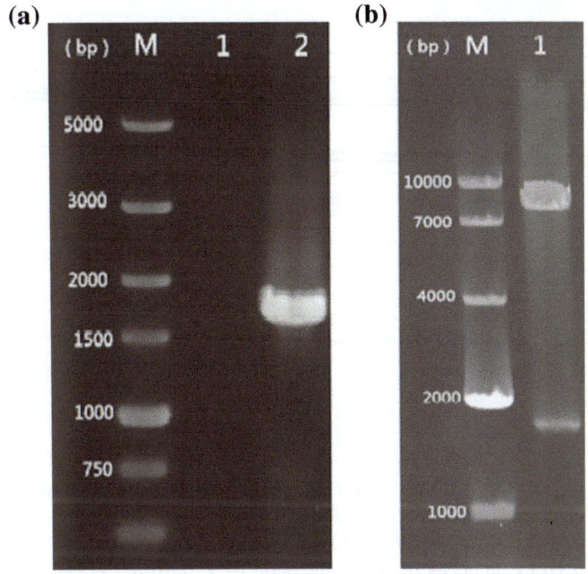

Fig. 2 Identification of recombinant plasmid pPIC3.5K-*ksdd2*. *M* DNA marker; **a** PCR analysis of the recombinant plasmid: *lane 1* PCR analysis of *E. coli* DH5a containing the empty-vector, pPIC3.5K; *lane 2* PCR analysis of *E. coli* DH5a containing the recombinant plasmid pPIC3.5K-*ksdd2*; **b** double endonuclease digestion analysis of the recombinant plasmid: *lane 1* double endonuclease digestion analysis of *E. coli* DH5a containing the recombinant plasmid pPIC3.5K-*ksdd2*

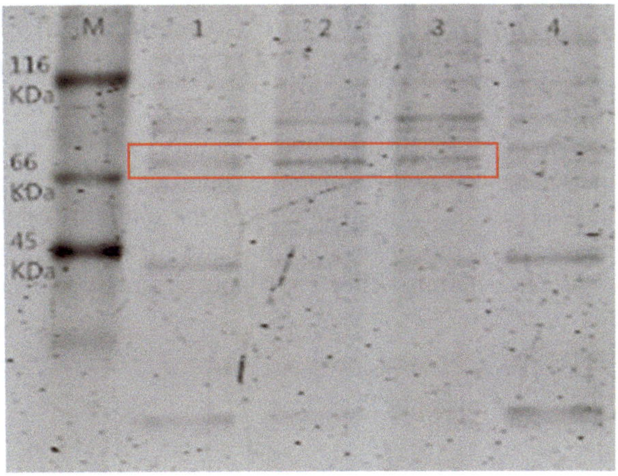

Fig. 3 SDS-PAGE analysis of cell extract of different *MNR* M3 strains. *M* Molecular weight standards; *lane 1–3* the whole protein expression of *P. pastoris* GS115 containing the plasmid, pPIC3.5K-*ksdd2*; *lane 4* protein expression profiles of *P. pastoris* GS115 containing the empty-vector, pPIC3.5K

capacity of *P. pastoris* GS115 (Fig. 4a). One AD product produced by *P. pastoris* GS115 pPIC3.5K-*ksdd2* developed as jacinth spots on TLC and was later identified as ADD by an authentic sample. Another product, produced by *P. pastoris* GS115 pPIC3.5K-*ksdd2* developed as brick-red spots on TLC in AD transformation and was identified as boldenone (BD). The AD transformed by *P. pastoris* GS115 pPIC3.5K-*ksdd2* was further analyzed by HPLC. The HPLC profiles of the samples transformed by *P. pastoris* GS115 pPIC3.5K-*ksdd2* showed four chemical peaks: ADD (t = 4.528 min), BD (t = 4.949 min), and AD (t = 5.632 min) (Fig. 4b).

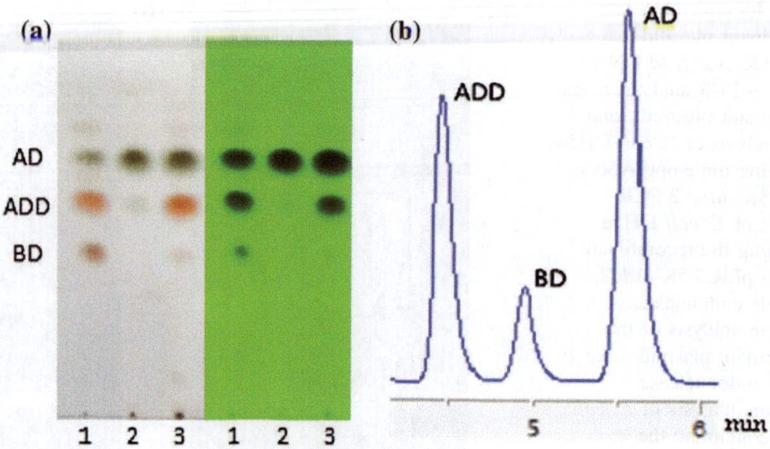

Fig. 4 KsdD2 activity for AD transformation by *P. pastoris* GS115 pPIC3.5K-*ksdd2*. **a** TLC chromatogram of AD transformation for 2d. *Lane 1* The standard of AD, ADD and BD; *lane 2* AD transformed by wild-type GS115; *lane 3* AD transformed by *P. pastoris* GS115 pPIC3.5K-*ksdd2*. **b** HPLC analysis of AD conversion by *P. pastoris* GS115 pPIC3.5K-*ksdd2*

3.3 Improved Substrate Concentration by Screening the Cosolvents on AD Biotransformation

As steroids' water-insoluble seriously restricts its biotransformation efficiency, measures need to be explored to improve their dispersity and solubility in reaction media [1]. Cosolvents can improve the solubility of steroid substrates which further increases its bioconversion efficiency. However, high organic solvent concentrations have harmful effects on biocatalyst. By increasing the water solubility and

Fig. 5 Effects of different HP-β-CD:AD molar ratios on AD conversion rate

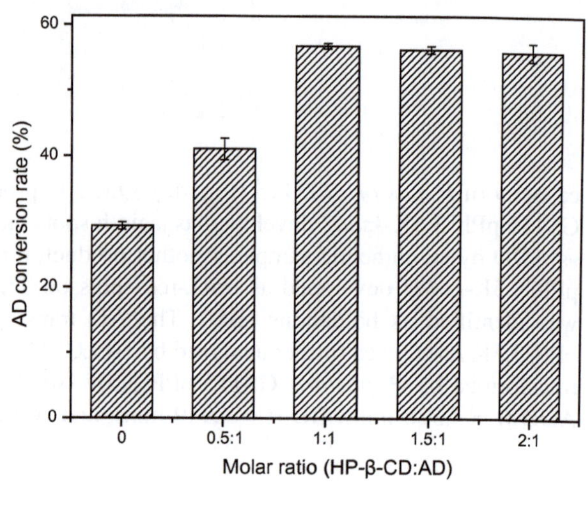

reducing the biotoxicity of substrates, CDs have been widely used to obtain a higher bioconversion efficiency [1, 17, 18]. In this study, the biotransformation of AD was conducted with HP-β-CD, its solubility and conversion rate in *P. pastoris* GS115 pPIC3.5K-ksdd2 were improved, the conversion efficiency of AD on ADD and BD with different molar ratios of HP-β-CD were investigated as below.

As shown in Fig. 5, AD conversion rate increased obviously with the increase of HP-β-CD concentration to some extent. When its molar ratio increased to 1:1, the AD conversion rate reached the maximum 56.76%. So we selected 1:1 as the optimized molar ratio in AD conversion.

4 Conclusions

In conclusion, we analysed the molecular property of a 3-ketosteroid-Δ1-dehydrogenase KsdD2 from *R. rhodochrous* DSM43269 by TMpred program prediction and supposed KsdD2 to be a membrane-bound protein. It may therefore be required for them to bind to the cellular membrane tightly by hydrophobic transmembrane domains to make it easier to contact with steroids in cellular membrane. Then KsdD2 was successfully expressed intracellularly in *P. pastoris* GS115. The engineered recombinant strains *P. pastoris* GS115 pPIC3.5K-*ksdd2* was found to be capable of transforming AD to both ADD and BD. The transformation of AD to ADD and BD is certainly the result of synergism between exogenous KsdD2 and endogenous 17Hsd. In this way, BD could be obtained by biotransformation from AD, but how to enhance the 17Hsd activity still needs further study. Although the BD was still at a low conversion, the research developed in this study could provide theory basis for the wide industrial application of microbial transformation to produce BD.

Acknowledgements This work was supported by National Natural Science Foundation of China (No. 21276196, 21406167), Key Project of Chinese Ministry of Education (213004A) and the Tianjin College students' innovative entrepreneurial training plan (201410057106).

References

1. Malaviya A, Gomes J (2008) Androstenedione production by biotransformation of phytosterols. Bioresour Technol 99(15):6725–6737
2. Brabander HF, Poelmans S, Schilt R et al (2004) Presence and metabolism of the anabolic steroid boldenone in various animal species: a review. Food Addit Contam 21(6):515–525
3. Fernandes P, Cabral JMS (2007) Phytosterols: applications and recovery methods. Bioresour Technol 98:2335–2350
4. Bhatti HN, Khera RA (2012) Biological transformations of steroidal compounds: a review. Steroids 77(12):1267–1290

5. Choi KP, Molnar IJ, Yamashita M (1995) Purification and characterization of the 3-ketosteroid Δ1 dehydrogenase of *Arthrobacter simplex* produced in *Streptomyces lividans*. J Biochem 117(5):1043–1049
6. Chen MM, Wang FQ, Lin LC et al (2012) Characterization and application of fusidane antibiotic biosynethsis enzyme 3-ketosteroid-Δ1-dehydrogenase in steroid transformation. Appl Microbiol Biotechnol 96:133–142
7. Molnar I, Choi KP, Yamashita M et al (1995) Molecular cloning, expression in *Streptomyces lividans*, and analysis of a gene cluster from *Arthrobacter simplex* encoding 3-ketosteroid-Δ1-dehydrogenase, 3-ketosteroid-Δ5-isomerase and a hypothetical regulatory protein. Mol Microbiol 15:895–905
8. Morii S, Fujii C, Miyoshi T (1998) 3-Ketosteroid-Δ1-dehydrogenase of *Rhodococcus rhodochrous* sequencing of the genomic DNA and hyperexpression purification, and characterization of the recombinant enzyme. J Biochem 124:1026–1032
9. Liu Y, Shen YB, Qiao YQ et al (2016) The effect of 3-ketosteroid-Δ1-dehydrogenase isoenzymes on the transformation of AD to 9α-OH-AD by *Rhodococcus rhodochrous* DSM43269. J Ind Microbiol Biotechnol 43:1303–1311
10. Wagner M, Atrat PG, Wagner B et al (1992) Overexpression of a *Rhodococcus erythropolis* protein in *Escherichia coli* with immunological identity to the *Rhodococcus* steroid-1-dehydrogenase. Immunoelectron microscopic localization and electrophoretic studies. J Basic Microbiol 32:269–277
11. Wei W, Fan SY, Wang FQ et al (2014) Accumulation of androstadiene-dione by overexpression of heterologous 3-ketosteroid D1-dehydrogenase in *Mycobacterium neoaurum* NwIB-01. World J Microbiol Biotechnol 30:1947–1954
12. Shao ML, Zhang X, Rao ZM et al (2015) Enhanced production of androst-1, 4-Diene-3, 17-Dione by *Mycobacterium neoaurum* JC-12 using three-stage fermentation strategy. PLoS ONE 10(9):1–13
13. Xie RL, Shen YB, Qin N et al (2015) Genetic differences in *ksdD* influence on the ADD/AD ratio of *Mycobacterium neoaurum*. J Ind Microbiol Biotechnol 42:507–513
14. van der Geize R, Hessels GI, van Gerwen R et al (2002) Molecular and functional characterization of the kstD2 gene of *Rhodococcus erythropolis* SQ1 encoding a second 3-ketosteroid-Δ1-dehydrogenase isoenzyme. Microbiol 148:3285–3292
15. Singer Y, Shity H, Bar R (1991) Microbial transformation in a cyclodextrin medium. Part 2. reduction of androstenedione to testosterone by *Saccharomyces cerevisiae*. Appl Microbiol Biotechnol 35:731–737
16. Marina VD, Olga VE, Vera MN (2005) Steroid 17β-reduction by microorganisms–a review. Process Biochem 40(7):2253–2262
17. Wang M, Zhang LT, Shen YB et al (2009) Effects of hydroxypropyl-β-cyclodextrin on steroids 1-en-dehydrogenation biotransformation by *Arthrobacter simplex* TCCC 11037. J Mol Catal B Enzym 59:58–63
18. Zhang LT, Wang M, Shen YB et al (2009) Improvement of steroid biotransformation with hydroxypropyl-β-cyclodextrin induced complexation. Appl Biochem Biotechnol 159(3):642–654

The Prokaryotic Expression of Cyclin-Dependent Kinase 2, and the Establish of Its Inhibitor Screening System

Yuan Yuan, Meile Gao, Huan Liu, Tingting Ruan, Weiran Xie, Meng Wu, Xin Qu, Zhen Liu, Peng Yu and Yuou Teng

1 Introduction

Cyclin dependent kinases (CDKs) are serine/threonine protein kinases and play essential roles in the intracellular control of the cell division [1, 2], including CDK1, CDK2, CDK4, CDK6 and different Cyclins (A, B, D and E). At least there are twenty different CDKs and Cyclins have been reported in mammalian cells [3]. CDKs-Cyclins complex can boost the interphase to mitosis of cell cycle in a continuous and orderly manner [4, 5]. Meanwhile the abnormal regulated of CDKs-Cyclins is indispensible ingredient in biological characteristics of tumour an important. The increased expression of Cyclin or CDK have been observed in various cancers such as stomach and lung cancer [6]. As the star of CDKs family, CDK2 causes much attention for its close relationship with cancer [7–9]. The activation of CDK2 can accelerate the cell cycle progression. Therefore, CDK2 as an effective target for anti-cancer has attracted the attention of many researchers. The monomer of CDK2 does not show any activity, but the complex of CDK2 and Cyclin is the activation type. In addition, a huge number of work committed to develop small molecule inhibitors of CDK2-Cyclins complex [10]. In this study, we mainly constituted the prokaryotic expression and screening system of CDK2-Cyclin A.

There are over 300 crystal structure of CDK2 monomers or binding with its various inhibitor in PDB-database, more than the sum of other members of CDKs.

Y. Yuan · M. Gao · H. Liu · T. Ruan · W. Xie · M. Wu ·
X. Qu · Z. Liu · P. Yu · Y. Teng (✉)
Sino-French Joint Lab of Food Nutrition/Safety and Medicinal Chemistry,
Key Lab of Industrial Fermentation Microbiology,
Tianjin Key Lab of Industrial Microbiology,
College of Biotechnology, Tianjin University of Science and Technology,
No. 29, 13th Street, Tianjin 300457, People's Republic of China
e-mail: tyo201485@tust.edu.cn

H. Liu et al. (eds.), *Advances in Applied Biotechnology*, Lecture Notes in Electrical Engineering 444, https://doi.org/10.1007/978-981-10-4801-2_13

CDK2 monomer consists of 298 amino acids, which has the similar structure with the classical protein kinase- a typical dual-stack synthetic lobed structure [11]. CDK2 has four cavities: one competitive site called ATP binding site, two non-competitive sites, and ATP non-competitive sites (ANS) or allosteric binding sites, which appear when conformational changed after Cyclin binds CDK2, also called cell cycle protein binding site (CBG), which can be used as a small molecule inhibitor docking target [12].

Small molecule inhibitors of CDK2 have been discovered with the development of the structure of CDK2 and Cyclin A. The chemical structures of the CDK2 inhibitors include purine, pyrimidine, flavonoids, indole and its derivatives, imidazole and pyridine, pyrazine and pyridine, pyrazine and thiazine ketones, piperacillin, pyrazine, butyric acid lactone [13]. The first stage of the CDK2 inhibitors are flavopiridol, purine ((R)-roscovitine), SNS-032 [6] and PHA-793887. However, the adverse pharmacological properties and low specificity made them interrupt in phase II and phase III clinical. Indolocarbazole analogy, indirubin and its derivatives, pyrazine and other 13 kinds of small molecule inhibitors are undergoing clinical trials [14]. Roscovitine, a classic CDK inhibitor, can work by directly binding the ATP binding domain of CDK. As reported, Roscovitine potently inhibits CDK2/Cyclin A with IC_{50} of 0.7–1.5 μM [15]. In this study, Roscovitine was used as a positive control to identify the successful establishment of CDK2 inhibitor screening system.

2 Materials and Methods

2.1 Construction of Recombinant Plasmids

CDK2 full-length cDNA was amplified by the cell of K562, which was used as a template for PCR amplification of the region in the open reading frame encoding the human CDK2. PCR amplification was performed using 25 ng of template DNA, 0.1 ng of the appropriate primers [5′-GCCGAATTCATGGAGAACTTCCAA-3′ (5′-primer) and 5′-GCCTCGAGTCAGAGTCGAA GATG-3′(3′-primer)], 1 μL 1.0 U/μL KOD DNA polymerase, 5 μL 10 × KOD-plus Buffer, 5 μL 2 mM dNTP and 2 μL 25 mM-$MgSO_4$ in a final reaction volume of 50 μL. The mixture reaction was subjected to amplification for 35 cycles (95 °C, 0.5 min; 58 °C, 1 min; 72 °C, 1.5 min) with a thermal cycler. After amplification, PCR products were cleaved with *Eco* RI and *Xho* I and ligated between the corresponding restriction sites of the vector pET-28a. This vector has a His-tag coding sequence and can produce a C-terminal 6 His-tag fusion protein. Then the ligation product was transformed into *E. coli.* Top 10. Through further bacteria liquid polymerase chain reaction, we confirmed positive transformants and then utilized EndoFree Plasmid Mini Kit (CW Bio) to extract CDK2 plasmids for gene sequencing. Correct transformants were used to induce the expression of CDK2 and Cyclin A protein.

The method of constructing Cyclin A recombinant plasmids is the same as that of CDK2 recombinant plasmids. The appropriate primers for performing PCR amplification of the region in the open reading frame encoding the human Cyclin A is [5′-GCGAATTCATGTTGGGCAACTCT-3′ (5′-primer) and 5′-GCCCTCGAGT TACAGATTTAGTGT-3′ (3′-primer)].

2.2 Expression and Purification of Cyclin A and CDK2

To express the recombinant human CDK2/Cyclin A, *E. coli* BL21 (DE3) Plyss cells were grown in solid LB medium (1.5% agar, 1% bactotryptone, 0.5% yeast extract, 1% NaCl) containing 10 mg/mL Kanamycin at 37 °C overnight to obtain single colonies. Then the single colonie was used to inoculate a 5 mL LB medium pre-culture, containing the appropriate antibiotic and was then grown at 37 °C overnight. The culture was used to inoculate 50 mL of LB medium at 37 °C to OD_{600} at 0.6–0.8. Isopropyl β-D-1-thiogalactopyranoside (IPTG) was then added to a final concentration of 0.1 mM and cells were further incubated for 12 h at 25 °C.

Following induction, cells were collected by centrifugation (5,000 g for 5 min), the supernatant was decanted, and the cell pellet was suspended in lysis buffer containing 50 mM Tris-HCl (pH 8.0), 100 mM NaCl, 5 mM DTT and 10 mM imidazole and broken with a noise isolating tamber (Ningbo Scienfz Biotechnology Co, LTD). The lysate was then centrifuged at 5000 g for 5 min at 4 °C to remove cell debris. The CDK2/Cyclin A proteins supernatants were loaded on a Ni-Sepharose TM 6 Fast Flow at 4 °C overnight. After that the Ni-beads were eluted three times by an elution liquid of 250 mM imidazole. The fractions containing CDK2/Cyclin A protein was then pooled and concentrated by dialysis (46 mM $Na_2HPO_4 \cdot 12H_2O$, 3 mM $NaH_2PO_4 \cdot 2H_2O$, 0.1 mM NaCl and 10% ml Glycerin) at 4 °C overnight.

2.3 Construction of CDK2 Kinase Inhibitors Screening Platform In Vitro

In this study, Kinase-Glo Luminescent Kinase Assay (Kinase-Glo luminescent kinase detection) kit was used to establish CDK2 kinase inhibitors screening platform in vitro. According to Kinase-Glo® Luminescent Kinase Assay, the content of ATP in the system after the kinase reaction is detected to quantitatively determine the activity of the purified kinase [15]. The content of ATP was determined by fluorescence intensity which was produced by the oxidation of Mg^{2+}, ATP and beetle luciferin. Adding a certain amount of ATP into the reaction system, ATP was consumed and the remaining ATP reacted with Oxyluciferin as well as emitted fluorescence, which could quantitatively detect the remaining amount of ATP in kinase reaction. Luminescence is inversely related to kinase activity.

The reagents used in the screening system are as followed: 2.5 μM ATP, 0.004 μg/mL substrate Histone H1, 5 μg/mL CDK2 and Cyclin A proteins, Kinase Glo plus, Assay Buffer (25 mM Hepes, 10 mM $MgCl_2$, 0.01% Trition X-100,100 μg/mL BSA, 2.5 mM DTT, pH 7.4) and Roscovitine. The test would be divided into 3 groups including the blank control group, kinase control group, and the Roscovitine test group. Then add 5 μL Assay Buffer into blank control group; add 2.5 μL Assay Buffer, 1 μL CDK2 kinase, 1 μL Cyclin A kinase and 0.5 μL DMSO into kinase control group and add 2.5 μL Assay Buffer, 1μL CDK2 kinase, 1 μL Cyclin A kinase and 0.5 μL Roscovitine of different concentrations into Roscovitine test group. 384 well was centrifuged at 26 g for 2 min and incubated at 30 °C for 30 min. Followed by 5 μL 2 × ATP-substrate mixtures in the same well, which was centrifuged at 26 g of 2 min, and incubated at 30 °C for 3 h. 10 μL Kinase-Glo plus was finally added, centrifuged at 26 g for 2 min and incubated for 10 min. Relative light unit (RLU) was concluded from microplate reader.

Inhibition rate (%) of Roscovitine = (RLU Roscovitine test group − RLU kinase control group)/(RLU blank control group − RLU kinase control group) 100%. The IC_{50} value of each concentration point half is calculated by linear regression equation which was obtained according to RLU results.

2.4 *Statistical Analysis*

All data were expressed as mean ± S.D. Results were analyzed by one-way analysis of variance (ANOVA), and significant differences were determined by post hoc Tukey test using SPSS 21.0 software. Both $^{*}P < 0.05$ and $^{**}P < 0.01$ are compared to blank group in Fig. 3a, and both $^{*}P < 0.05$ and $^{**}P < 0.01$ are compared to DMSO control group in Fig. 3b. $P < 0.05$ was considered as significant.

3 Results and Discussion

3.1 *Construction of CDK2 and Cyclin A Recombinant Plasmids*

To obtain recombinant plasmids of CDK2 and Cyclin A, mRNA was extracted from human leukemia cells and reversed transcription into cDNA as a template for PCR amplification with designed primers. The results of cell PCR products from agarose gel electrophoresis were in good agreement with the size of CDK2 and Cyclin A protein (897 and 1299 bp, respectively), and the sizes of fragments obtaining from double digestion were consistent with target fragments too.

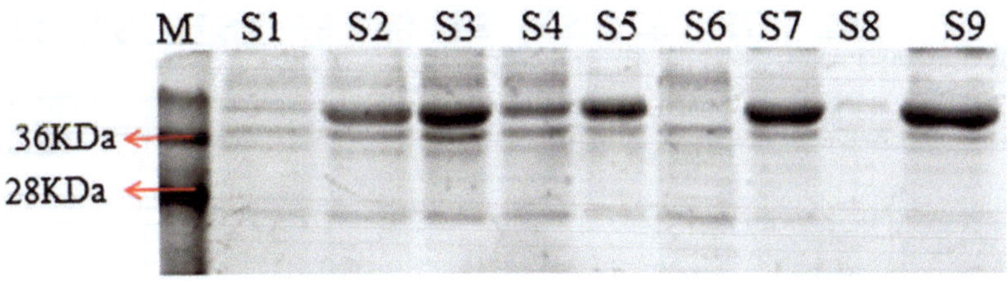

Fig. 1 Western blot result of CDK2 protein. Lane M is multicolor protein marker; *Lane S1* is cell sample before adding 0.1 mM IPTG; *Lane S2* is the cell sample after adding 0.1 mM IPTG; *Lane S3* is the cell sample after adding lysis buffer; *Lane S4* is the supernatant sample after sonication; *Lane S5* is the precipitate sample after sonication; *Lane S6* is the supernatant sample after binding Ni-Beads; *Lane S7* is the protein sample after elution; *Lane S8* is the beads sample after elution; *Lane S9* is the purified protein sample after dialysis

3.2 Expression and Purification of Cyclin A and CDK2

3.2.1 Western Blot Analysis of CDK2 Kinase

The size of CDK2 gene fragment is 897 bp and its protein size is 39 KDa. The prokaryotic expression system of BL21 was applied to induce expression of the target proteins. When the OD_{600} value reached 0.6–0.8, 0.1 mM IPTG was added as an inducer at 25 °C for 12 h. The result of western blot analysis was used to identify the molecular weight and purity of CDK2 (Fig. 1).

3.2.2 Western Blot Analysis of Cyclin A Protein

The size of Cyclin A gene fragment is 1299 bp and its protein size is 55 KDa. The expression and purification procedure of Cyclin A protein was the same as CDK2. The result of western blot analysis was used to identify the molecular weight and purity of Cyclin A (Fig. 2).

As is shown in Figs. 1 and 2, the expression of CDK2 and Cyclin A increased significantly after IPTG induction, and a part of soluble CDK2 and Cyclin A protein appeared in supernatant by cell disruption and centrifugation. Then the supernatant was combined with the Ni-Beads, imidazole elution and dialysis for protein purification. It was found that 250 mM imidazole could be capable of eluting target protein for the first time and the eluted Ni-Beads had little residues of CDK2 and Cyclin A protein.

3.3 Construction of CDK2 Kinase Inhibitors Screening Platform In Vitro

In order to confirm whether the preliminary purified CDK2 and Cyclin A protein was activated or not and eliminate the possibility of only one enzyme worked which

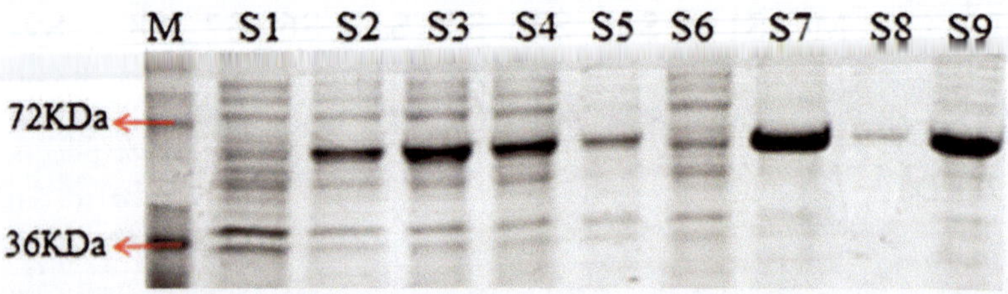

Fig. 2 Western blot result of Cyclin A protein. *Lane M* is multicolor protein marker; *Lane S1* is cell sample before adding 0.1 mM IPTG; *Lane S2* is the cell sample after adding 0.1 mM IPTG; *Lane S3* is the cell sample after adding lysis buffer; *Lane S4* is the supernatant sample after sonication; *Lane S5* is the precipitate sample after sonication; *Lane S6* is the supernatant sample after binding Ni-Beads; *Lane S7* is the protein sample after elution; *Lane S8* is the beads sample after elution; *Lane S9* is the purified protein sample after dialysis

lead to the inhibitory effects in kinase test system, this study would be divided into 4 groups totally, including the blank, CDK2 group, Cyclin A group and CDK2 and Cyclin A group (Fig. 3a).

The results in Fig. 3a showed that CDK and Cyclin A protein were purified by the above method. The light values did not reveal significant decrease in any single kinase control group compared with kinase test group, which means a single kinase cannot exert function and consume ATP. When different concentrations of Roscovitine was added into the system (Fig. 3b), light values rose with the increase of the inhibitor concentration, proving that CDK2 kinase must be activated by Cyclin A to play a role. IC_{50} value of Roscovitine was 1.87 μM, calculating through linear regression equation which was obtained according to RLU results. The IC_{50} value of Roscovitine in our system is in agreement with the reported [15].

The above results show that the CDK2 kinase inhibitors screening platform in vitro was successfully constructed.

4 Conclusions and Perspectives

CDKs play significant roles not only in regulation of cell cycle progression but also in a wide variety of vital physiological processes including neuronal and transcription functions [16]. Their deregulation associated with overexpression, amplification or mutation of the specific CDK or Cyclin proteins have been reported in various human cancers [17–20]. This suggests that developing novel CDKs inhibitors is conducive to the development of pharmacological agents for the treatment of cancer.

In this study, the prokaryotic expression system of CDK2 and Cyclin A was inaugurated. The soluble expression protein was inaugurated by plasmid pET-28a/CDK2 and pET-28a/Cyclin A in *Escherichia* coli BL21 with 0.1 mM IPTG at 25 °C for 12 h. The protein was purified by Ni-beads and eluted by

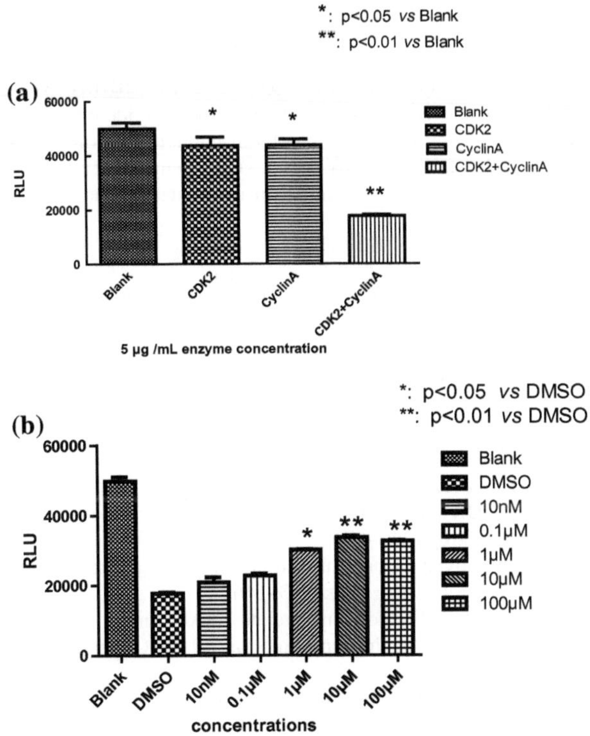

Fig. 3 The results of RUL between four test groups. **a** The reagents used in the screening system are as followed: 2.5 μM ATP, 0.004 μg/mL Histone H1, 5 μg/mL CDK2 and Cyclin A proteins in kinase test group while only one of kinases were added into the system in CDK or Cyclin A control group, and kinases were replaced by assay buffer in blank control group. **b** The results of RUL between different concentrations of Roscovitine. Five different concentrations of inhibitors were designed to confirm whether there are concentration-dependent between inhibitors and kinases or not, and DMSO, the solvent of inhibitors were existed as control group

imidazole to obtain recombinant CDK2 and Cyclin A. Then IC_{50} value of Roscovitine within 4 h was successfully tested in the system in agreement with the reported [15]. We established a rapid and stable CDK2 kinase inhibitor screening platform, which laid a durable foundation for the study of new CDK inhibitors in the future.

Acknowledgements The authors sincerely thank the financial support from the financial support from National Natural Science Foundation of China (31301142), Tianjin Natural Science Foundation of China (12JCYBJC31600 and 11JGYBJC14300) and International Science & Technology Cooperation Program of China (2013DFA31160).

References

1. Malumbres M (2014) Cyclin-dependent kinases. Chem Rev 101(8):2511–2526
2. Mascarenhas NM, Ghoshal N (2008) An efficient tool for identifying inhibitors based on 3D-QSAR and docking using feature-shape pharmacophore of biologically active conformation—a case study with CDK2/CyclinA. Eur J Med Chem 43(12):2807–2818
3. Malumbres M, Barbacid M (2005) Mammalian cyclin-dependent kinases. Trends Biochem Sci 30(11):630–641
4. Camalier CE et al (2013) An integrated understanding of the physiological response to elevated extracellular phosphate. J Cell Physiol 228(7):1536–1550
5. Takuwa N, Takuwa Y (2001) Regulation of cell cycle molecules by the Ras effector system. Mol Cell Endocrinol 177(1–2):25–33
6. Roskoski R (2016) Cyclin-dependent protein kinase inhibitors including palbociclib as anticancer drugs. Pharmacol Res 107:249–275
7. Cicenas J, Valius M (2011) The CDK inhibitors in cancer research and therapy. J Cancer Res Clin Oncol 137(10):1409–1418
8. Cheng PH et al (2013) Molecular basis for viral selective replication in cancer cells: Activation of CDK2 by adenovirus-induced cyclin E. PLoS ONE 8(2):e57340
9. Hongo F et al (2014) CDK1 and CDK2 activity is a strong predictor of renal cell carcinoma recurrence. Urol Oncol 32(8):1240–1246
10. Duca JS (2009) Recent advances on structure-informed drug discovery of cyclin-dependent kinase-2 inhibitors. Fut Med Chem 1(8):1453–1466
11. Dölker N et al (2014) The SH2 domain regulates c-Abl kinase activation by a cyclin-like mechanism and remodulation of the hinge motion. PLoS Comput Biol 10(10):e1003863–e1003863
12. Schulman BA, Harlow E (1998) Substrate recruitment to cyclin-dependent kinase 2 by a multipurpose docking site on Cyclin A. Proc Natl Acad Sci USA 95(95):10453–10458
13. Wang Lixing et al (2014) Mini review for the inhibitors of cyclin dependent kinase(CDK). Chin Med Sci J 13:44–47
14. Shwrr CJ, Beach D (2016) Targeting CDK4 and CDK6: from discovery to therapy. Cancer Discov 6(4):353–367
15. Meijer L, Meijer L, Borgne A, Mulner O, Chong JP, Blow JJ, Inagaki N et al (1997) Biochemical and cellular effects of roscovitine, a potent and selective inhibitor of the cyclin-dependent kinases CDC2, CDK2 and CDK5. Eur J Biochem 243(1–2):527–536
16. Koresawa M, Okabe T (2004) High-throughput screening with quantitation of ATP consumption: a universal non-radioisotope, homogeneous assay for protein kinase. Assay Drug Dev Technol 2(2):153–160
17. Peyressatre M et al (2015) Targeting cyclin-dependent kinases in human cancers: from small molecules to peptide inhibitors. Cancers 7(1):179–237
18. Lim S, Kaldis P (2013) Cdks, cyclins and CKIs: roles beyond cell cycle regulation. Development 140(15):3079–3093
19. Malumbres M, Barbacid M (2009) Cell cycle, CDKs and cancer: a changing paradigm. Nat Rev Cancer 9(3):153–166
20. Malumbres M (2011) Physiological relevance of cell cycle kinases. Physiol Rev 91(3):973–1007

Effects of Heterologous Pyruvate Carboxylase Expression on Synthesis of L-Threonine in *Escherichia coli*

Junpeng Wang, Yan Zhao, Tao Liu, Ting Wang, Chao Han, Qian Mao, Ning Chen and Yanjun Li

1 Introduction

L-Threonine is an essential amino acid that is widely used in animal feeds, human foods, and the pharmaceutical and cosmetics industries [4]. The global demand for L-threonine is around 0.7 million tons annually and China has become the biggest L-threonine producer with a proportion of 60% of the world's production. Currently, L-threonine is produced mainly through fermentation by *Escherichia coli.*

L-threonine, which belongs to the aspartic family of amino acids, is synthesized from the direct precursor oxaloacetate. There are three ways for supplying oxaloacetate in biological systems: namely the carbon dioxide fixation through anaplerotic reaction, the TCA cycle and the glyoxylate shunt. In bacteria, the anaplerotic reaction includes the carboxylation of phosphoenolpyruvate (PEP) catalyzed by PEP carboxylase (Ppc) and/or the carboxylation of pyruvate catalyzed by pyruvate carboxylase (Pyc). In *Rhodocacter capsulatus*, Pyc is the only enzyme involves

J. Wang · Y. Zhao · T. Liu · T. Wang · C. Han · Q. Mao · N. Chen (✉) · Y. Li (✉)
College of Biotechnology, Tianjin University of Science and Technology, Tianjin 300457, China
e-mail: ningch@tust.edu.cn

Y. Li
e-mail: yli@tust.edu.cn

N. Chen · Y. Li
National and Local United Engineering Lab of Metabolic Control Fermentation Technology, Tianjin 300457, China

N. Chen · Y. Li
Key Laboratory of Industrial Fermentation Microbiology, Ministry of Education, Tianjin University of Science and Technology, Tianjin 300457, China

N. Chen · Y. Li
Tianjin Economic and Technological Development Area (TEDA), No. 29, 13th. Avenue, Tianjin 300457, China

H. Liu et al. (eds.), *Advances in Applied Biotechnology*, Lecture Notes in Electrical Engineering 444, https://doi.org/10.1007/978-981-10-4801-2_14

the anaplerotic reaction [6], while the carbon dioxide is only fixed with pyruvate catalyzed by Ppc in *E. coli* [2]. In a few bacteria such as *Corynebacterium glutamicum*, both Pyc and Ppc function in the carboxylation reactions, although Pyc contributes to more than 90% of the oxaloacetate supply [7].

In *E. coli*, the oxaloacetate is mainly synthesized through TCA cycle, which implies that strengthening of the anaplerotic reaction might be a feasible approach for increasing oxaloacetate supply and consequently enhancing L-threonine synthesis. Zelle et al. [11] investigated the effects of Pyc and Ppc overexpression on succinate production, and found that overexpression of Pyc not only increased the flux towards succinate production, but decreased the accumulation of the by-product lactate. The experiments conducted by Chao and Liao [1] indicated that overexpression of Ppc in *E. coli* reduced glucose uptake rate by 30% in aerobic condition. Similarly, Gokarn et al. [3] found that the glucose consumption rate decreased by 14% when Ppc was overexpressed in anaerobically cultured *E. coli*. From these results, it can be conclude that the enhancement of Ppc activity is not a reasonable way, since its substrate PEP is also needed during the process of glucose uptake via PTS system. Therefore, this study expressed the heterologous Pyc in an L-threonine producer, *E. coli* THRD, to investigate the influence on L-threonine fermentation.

2 Materials and Methods

2.1 *Strains and Plasmids*

The *E. coli* THRD is the producer of L-threonine and stored at the Culture Collection Center of Tianjin University of Science and Technology. Strains and plasmids used in this work are listed in Table 1.

2.2 *Plasmid Construction and Chromosomal Integration of PycA$_{bsu}$*

The genomic DNA of *B. subtilis* 168 and *C. glutamicum* 13032 was used as template to amplify the gene *pycA*, respectively. Primers *pycA*-pTrc99a-S and *pycA*-pTrc99a-A were used to clone the gene fragment of *pycA* from *B. subtilis*. The amplified fragment was purified from agarose gel, digested with *Kpn* I and *Sal* I and subsequently inserted into the *Kpn* I-*Sal* I site of pTrc99a vector to yield pTrc99a-*pycA*$_{bsu}$. The plasmids pTrc99a-*pycA*$_{cgl}$ and pWSK29-*pycA* were constructed similarly with the corresponding primers and restriction sites indicated in Table 2. There plasmids were then transformed into the competent cells of *E. coli* THRD treated with $CaCl_2$.

Table 1 Strains and plasmids used in this work

Name	Characteristics	Source
Strains		
E. coli DH5α	Wild type	Lab stock
E. coli THRD	L-threonine producer (ILE^L, AHV^r)	Lab stock
E. coli THRD Δ*pykF*	*E. coli* THRD, *pykF*$^-$	Lab stock
E. coli THRD *pykF*::*pycA*	*E. coli* THRD, *pykF* replaced with *pycA*	This study
Plasmids		
pKD46	Amp^R, λ Red-expressing vector, temperature-sensitive origin	Lab stock
pKD3	Cm^R, template vector	Lab stock
pCP20	Amp^R, Cm^R, FLP recombinase expression vector, temperature-sensitive origin	Lab stock
pTrc99a	Amp^R, cloning vector	Lab stock
pWSK29	Amp^R, cloning vector	Lab stock
pTrc99a-*pycA*	pTrc99a carrying *pycA* gene from *Bacillus subtilis*	This study
pTrc99a-$pycA_{cgl}$	pTrc99a carrying *pycA* gene from *Corynebacterium glutamicum*	This study
pWSK29-*pycA*	pWSK29 carrying *pycA* gene from *Bacillus subtilis*	This study

The genomic integration of $pycA_{bsu}$ in *E. coli* THRD was performed using Red and FLP-mediated recombination method. Upstream and downstream regions of *pykF* fragments were obtained by PCR using the primer pair *pykF*-up-S and *pykF*-up-A, and *pykF*-down-S and *pykF*-down-A, respectively. The Cm^r gene (about 1000 bp in length) was obtained using primers pKD3-S and pKD3-A with helper plasmid pKD3 as template. The $pycA_{bsu}$ fragment was amplified using primers *pycA*-S and *pycA*-A with genomic DNA of *B. subtilis* 168 as template. Splicing by overlap extension was used to fuse the fragments up- and downstream of the *pykF*, Cm^r genes and $pycA_{bsu}$ with the primers *pykF*-up-S and *pykF*-down-A. In order to remove the Cm^r gene from the integrated locus, cells were transformed with plasmid pCP20 carrying the FLP recombinant gene. Mutant sequences were verified by PCR and the subsequent sequencing.

2.3 Media and Shake-Flask Cultivation

The seed medium contains (per liter): glucose 40 g, $(NH_4)_2SO_4$, 20 g; KH_2PO_4, 1 g; $MgSO_4{\cdot}7H_2O$, 1 g; yeast extract, 1 g; $FeSO_4{\cdot}7H_2O$, 10 mg; $MnSO_4{\cdot}H_2O$, 10 mg; L-isoleucine, 1 mg; pH 7.0–7.2.

Table 2 Primers used in this work

Name	Sequences (5′–3′)	Restriction site
pycA-pTrc99a-S	GCGCGGTACCAAAAGTTTGTCTCAGCAATCGATAC	*Kpn* I
pycA-pTrc99a-A	CGCCGTCGACCTCTTCTAACCATCTGTTTCACTCC	*Sal* I
pycA$_{cgl}$-pTrc99a-S	CATGCCATGGGCGGTCTCTTGTTGAAAGGAATA	*Nco* I
pycA$_{cgl}$-pTrc99a-A	GCGCAAGCTTCCAACTCACCCATCTCCCAT	*Hind* III
pycA-pWSK29-S	GCGCTCTAGAAAGAAGGAGATATACAATGTCTCAGCAATCGATACAAAAAG	*Xba* I
pycA-pWSK29-A	GCGCGGTACCTCTTCTAACCATCTGTTTCACTCCA	*Kpn* 1
pykF-up-S	GGCTGAGACACAAGCACACATT	
pykF-up-A	CCAGCCTACACAGTGAAAGTAAAGGTCTGACCAGCT	
pykF-down-S	GGTTAGAAGAG TGATGAACAGCCGTCTCGAGT	
pykF-down-A	CGTGAACAGATGCGGTGTTAGT CCTTTACTTTCACTGTGTAGGCTGGAGCTGCTTC	
pKD3-S	GCACAGCCAT GCCATGGTCCATATGAATATCCT	
pKD3-A	GGACCATGGCATGGCTGTGCAGGTCGTAAAT	
pycA-S		
pycA-A	CGGCTGTTCATCACTCTTCTAACCATCTGTTTCACTCC	
pycA-qPCR-S	TTCAGGGATGCCCACCAA	
pycA-qPCR-A	AGGGAAGCAACATCTGGAATAAG	

Fermentation medium contains (per liter): glucose, 40 g; $(NH_4)_2SO_4$, 18 g; KH_2PO_4, 2 g; $MgSO_4 \cdot 7H_2O$, 1.5 g; yeast extract, 1 g; $FeSO_4 \cdot 7H_2O$, 10 mg; $MnSO_4 \cdot H_2O$,10 mg; L-isoleucine, 1 mg; pH 7.0–7.2.

Shake-flask cultivation was carried out by adding 30 ml of seed culture to a 500-ml shake-flask, incubated at 37 °C, and rotated at 200 rpm. Then 3 ml of seed medium at an OD_{600} of 4–6 was inoculated into 27 ml fermentation medium in a 500-ml shake-flask. As the inducer, 0.1 mmol/L IPTG (final concentration) was added four hours after the fermentation began. During fermentation, the pH was maintained at about 7.0 by addition of NH_4OH with a micro-injector, and 1 ml of 60% glucose stock was added with a transfer pipette each time when the residual glucose was low. The phenol red was used to show changes in the pH and the status of residual glucose. All the shake-flask fermentations were conducted in triplicate.

2.4 Protein Preparation and SDS-PAGE

The strain *E. coli* THRD carrying pTrc99a, pTrc99a-*pycA* and pTrc99a-$pycA_{cgl}$ was cultured in LB media at 37 °C, 200 rpm until the OD_{600} reached about 0.6, when a final concentration of 0.1 mmol/L IPTG was added to induce the protein synthesis. After 8 h's induction, the cells were harvested and washed twice with deionized water. The suspension cells were then ultrasonically lysed and centrifuged. Both the supernatants and precipitates were mixed with the sample buffer, boiled for 3 min, and then separated on 12% SDS/PAGE gels.

2.5 RNA Extraction and RT-QPCR Analysis

The strain *E. coli* THRD, THRD/pTrc99a-*pycA* and THRD/pWSK29-*pycA* were cultivated according to the shake-flask method described above. The samples were harvested at 12 h and the total RNA was extracted using RNAiso plus (Takara Bio Inc., Dalian, China). To avoid DNA contamination, the total RNA samples were treated with gDNA Eraser (Takara Bio Inc., Dalian, China) at 42 °C for 5 min, following the manufacturer's instructions. The RNA was reverse transcribed into cDNA using the PrimeScriptTM RT reagent Kit (Perfect Real Time) (Takara Bio Inc., Dalian, China) following the manufacturer's protocol. The qPCR was performed with the StepOnePlusTM Real-Time PCR System (Thermo Fisher Scientific, USA) using SYBR Premix Ex TaqTM II kit (Takara Bio Inc., Dalian, China) with the primers $pycA_{bsu}$-qPCR-S and $pycA_{bsu}$-qPCR-A for amplifying a 212 bp amplicon. The data were analyzed using the $2^{-\Delta\Delta C_T}$ method [5]. The relative abundance of 16S RNA was used as the internal standard. All assays were performed in triplicate.

2.6 Analytical Procedures

Cell growth was measured by monitoring the optical density at 600 nm (OD_{600}) and dry cellular weight (DCW) was calculated from previously determined calibration curve. One unit of OD_{600} was considered to be equal to 0.37 g of DCW.

L-threonine was quantified by HPLC involving pre-column derivatization. The samples were diluted and derivatized with 2,4-dinitrofluorobenzene, and measured by HPLC (1200 series, Agilent Technologies, USA) equipped with an Agilent ZORBAX Eclipse AAA column (4.6 mm × 150 mm, 5 μm) and an Agilent 1200 G1329A auto sampler. Elution was performed using a gradient of reagent A (50% acetonitrile v/v) and reagent B (0.05 M CH_3COONa, pH 6.4) at a flow rate of 1.0 mL/min. UV absorption was measured at 360 nm and the column temperature was maintained at 33 °C.

Glucose concentration was measured using a biosensor analyzer (SBA-40E, Institute of Biology, Shandong Academy of Sciences, Jinan, China). The yield (%) of L-threonine was calculated based on the glucose consumed.

3 Results

3.1 SDS-PAGE Analysis of Pyruvate Carboxylase

The heterologous pyruvate carboxylase (Pyc) from *B. subtilis* and *C. glutamicum* was expressed in *E. coli* THRD, respectively, and the results of SDS-PAGE analysis were shown in Fig. 1. It can be seen that the target bands appear in lanes 3–6 with the molecular weight of about 120 kDa, which is well corresponded to the expected value of $PycA_{bsu}$ (127.96 kDa) and $PycA_{cgl}$ (123.12 kDa). In control group (*E. coli* THRD/pTrc99a), neither the supernatant (lane 1) nor the precipitate (lane 2) contains target band, which confirms that the Pyc does not exist in *E. coli* THRD. The supernatant of THRD/pTrc99a-*pycA* (lane 3) contains more Pyc than the precipitate (lane 4) does, which implies that most of the Pyc from *B. subtilis* is in its active state. However, vast majority of the Pyc from *C. glutamicum* appears in the form of inclusion body, since the band in lane 6 is much thicker than that in lane 5. This may be due to the incorrect folding of Pyc derived from *C. glutamicum* in *E. coli* THRD.

In conclusion, the Pyc from *B. subtilis* was successfully expressed in *E. coli* THRD, which was used for the following study.

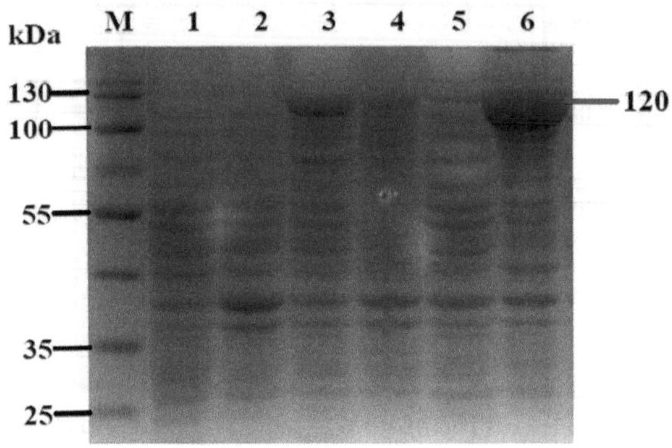

Fig. 1 SDS-PAGE analyses of pyruvate carboxylases from *B. subtilis* and *C. glutamicum.* M: Marker; *1 E. coli* THRD/pTrc99a, supernatant; *2* THRD/pTrc99a, precipitate; *3* THRD/pTrc99a-*pycA*, supernatant; *4* THRD/pTrc99a-*pycA*, precipitate; *5* THRD/pTrc99a-$pycA_{cgl}$, supernatant; *6* THRD/pTrc99a-$pycA_{cgl}$, precipitate

3.2 *Relative Expression of PycA in Different Copy Number Plasmid*

Although $pycA_{bsu}$ was successfully expressed in *E. coli* THRD within plasmid pTrc99a-*pycA*, we wondered whether the activity of the heterologous Pyc was too high for L-threonine synthesis, because pTrc99a is a well-known high copy number plasmid. We then constructed a new expression vector of *pycA* from pWSK29a, a generally regarded low copy number plasmid. The RT-qPCR analyses were performed to determine relative expression levels of *pycA* in these two plasmids, and the results were presented in Fig. 2. We hardly detect expression of *pycA* in *E. coli* THRD at 12 h of shaking flask fermentation, which was consistent with the result of SDS-PAGE (Fig. 1). In *E. coli* THRD harboring recombinant plasmids pTrc99a-*pycA* and pWSK29-*pycA*, the gene was significantly expressed, and the relative expression level of the former compared with THRD was about 3.8 times of that of the latter (Fig. 2).

3.3 *Effect of PycA Expression on L-Threonine Fermentation*

To study the effects of plasmid-borne *pycA* expression with different intensities on L-threonine production by *E. coli* THRD, shake-flask fermentations were carried out for 28 h. The bacterial biomass, L-threonine titer and yield were determined (or calculated) according to the methods stated in Materials and Methods section, and the results are shown in Fig. 3.

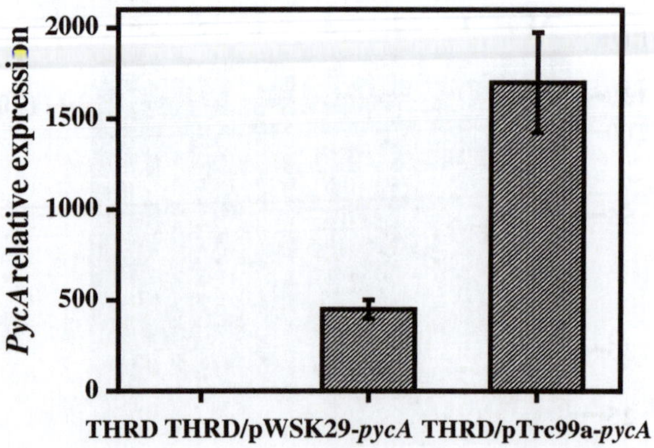

Fig. 2 Relative expression levels of $pycA_{bsu}$ in pWSK29-*pycA* and pTrc99a-*pycA*

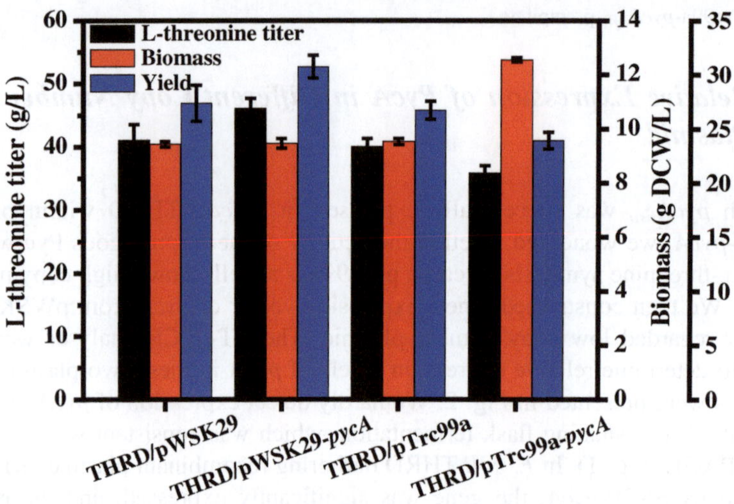

Fig. 3 Effect of *pycA* expression on L-threonine fermentation (biomass, L-threonine titer and yield)

The biomass of *E. coli* THRD/pWSK29-*pycA* was equivalent to that of THRD/pWSK29, which indicated that the heterologous expression of *pycA* had no influence on the growth of *E. coli* THRD. At the end of fermentation, the L-threonine titer and yield of THRD/pWSK29 was 41.05 g/L and 27.37%, respectively. The values of THRD/pWSK29-*pycA* were 12.3% higher compared with those of the control (THRD/pWSK29). At the end of fermentation, the biomass of THRD/pTrc99a-*pycA* was 31.69% higher compared with that of THRD/pTrc99a (9.53 g DCW/L); however, the titer (35.91 g/L) and yield (23.94%) of L-threonine were 10.4% lower than those of the control. Taken together, the results of shake-flask fermentation indicated that the heterologous expression of Pyc with low copy number plasmid pWSK29 was beneficial to the accumulation of L-threonine and that whit high copy number plasmid pTrc99a had a role in promoting bacterial growth.

3.4 *Effect of Chromosol Integration of PycA on L-Threonine Fermentation*

In previous study, we found that deletion of *pykF* that encodes pyruvate kinase significantly improved L-threonine titer and yield of *E. coli* THRD [10]. Since the expression of Pyc with low copies was beneficial for L-threonine accumulation, we proposed that the integration of *pycA* into the chromosome might be also useful for L-threonine fermentation. Therefore, we integrated *pycA* in the locus of *pykF*, and investigate the influence on L-threonine fermentation. The biomass, L-threonine titer and yield of THRD, THRDΔ*pykF* and THRD *pykF*::*pycA* at the end of fermentation (28 h) are shown in Fig. 4.

As you can see from Fig. 4, the biomass, L-threonine titer and yield of THRDΔ*pykF* was significantly higher than that of THRD, respectively. The biomass of THRD *pykF*::*pycA* was slightly less than that of THRD (10.03 g DCW/L), while the titer (37.64 g/L) and yield (28.95) of L-threonine was significantly higher, respectively. The titer and yield of L-threonine of THRD *pykF*::*pycA* was comparable to that of THRDΔ*pykF*, respectively, although the biomass of the former was obviously lower. The results revealed that the heterologous expression of *pycA* on the basis of knockout of *pykF* did not further improved L-threonine synthesis.

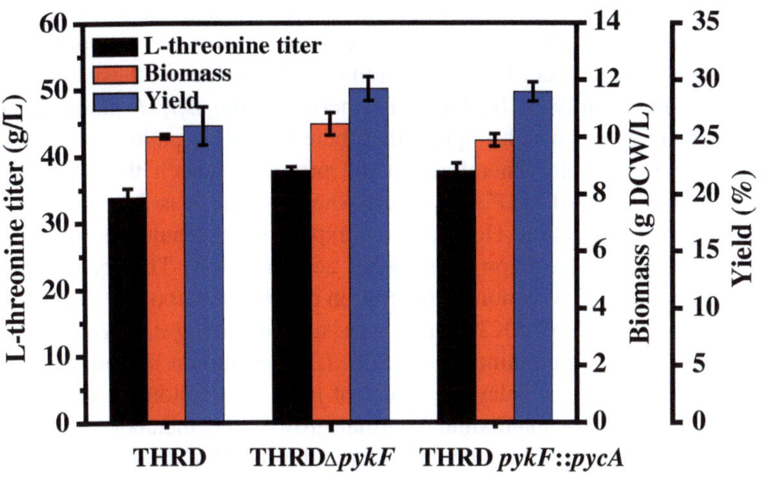

Fig. 4 Effect of *pykF* deletion and *pycA* integration on L-threonine fermentation (biomass, L-threonine titer and yield)

4 Discussion

In bacteria, metabolite interconversion at the PEP-pyruvate-oxaloacetate node involves a set of reactions that interconnects the major pathways of carbon metabolism and thus, is responsible for the distribution of the carbon flux among catabolism, anabolism and energy supply of the cell. In contrast to the relatively conserved central metabolic pathways, the reactions catalyzed by the enzymes at this node are very diverse [8]. Peters-Wendisch et al. [7] confirmed that Pyc was an important enzyme for the molecular construction of amino acid hyperproducing *C. glutamicum* strains. The results of their study indicated that overexpression of the *pyc* gene in a lysine producer resulted in about 50% higher lysine accumulation in the fermentation broth, while the lysine production of *pyc* knockout strain decreased by 60%. In *E. coli*, only Ppc is responsible for the anaplerotic reaction (CO_2 fixation), which cannot be readily overexpressed because its substrate, PEP, is also involves in the process of sugar translocation through the membrane mediated by the PTS system. Therefore, this study focused on the effects of increasing enzyme activity of Pyc via heterologous gene overexpression on L-threonine production by the L-threonine-producing strain *E. coli* THRD.

The results of shake-flask fermentation showed that the biomass of THRD/pTrc99a-*pycA* increased by 31.69% compared with that of THRD/pTrc99a. The reason might be that in addition to participating in the reducing reaction of TCA, Pyc also played an important role in the gluconeogenesis pathway. The oxaloacetate produced, on one hand, flew into the reducing reaction of TCA, and on the other hand poured into the TCA cycle, which provided more ATP and intermediate metabolites and thus, promoted the bacterial growth. The more oxaloacetate flew into TCA cycle, the less flux distributed for synthesis of L-asparagine and finally the accumulation of L-threonine. Through metabolic flux analysis based on ^{13}C-labeling experiments, Siddiaquee et al. [9] found that the pentose phosphate pathway increased in the mutant of *pykF* deletion. Similarly, our previous study revealed the positive effect of *pykF* deletion on L-threonine synthesis, probably due to the increasing supply of NADPH. Furthermore, the Δ*pykF* strain showed no obvious growth defect, which implied that there was other pathway to compensate the decrease of pyruvate. Therefore, in the present research the *pycA* gene was integrated in the locus of *pykF* to evaluate whether there was further improvement for L-threonine fermentation. However, the experimental results showed no further impact of *pycA* integration on L-threonine accumulation. The reason might be: (1) the expression level of genomic integration of *pycA* was too low, despite the low copy number plasmid pWSK29 was more suitable for *pycA* overexpression in consideration of the L-threonine production; (2) the amount of substrate pyruvate was not high enough to display the effect of *pycA* expression, as *pykF* was inactivated; (3) the ATP for maintaining normal growth and metabolisms of the cells was inadequate, which hampered the accumulation of L-threonine, because the carboxylation reaction of pyruvate catalyzed by Pyc was ATP-dependent.

References

1. Chao YP, Liao JC (1993) Alteration of growth yield by overexpression of phosphoenolpyruvate carboxylase and phosphoenolpyruvate carboxykinase in *Escherichia coli*. Appl Environ Microbiol 59(12):4261–4265
2. Gottschalk G (2009) Bacterial metabolism. Springer, Berlin
3. Gokarn RR, Eiteman MA, Altman E (2000) Metabolic analysis of *Escherichia coli* in the presence and absence of carboxylating enzymes phosphoenolpyruvate carboxylase and pyruvate carboxylase. Appl Environ Microbiol 66(5):1844–1850
4. Liu YF, Li FR, Zhang XR et al (2014) A fast and sensitive coupled enzyme assay for the measurement of L-threonine and application to high-throughput screening of threonine-overproducing strains. Enzyme Microb Technol 67:1–7
5. Livak KJ, Schmittgen TD (2001) Analysis of relative gene expression data using real-time quantitative PCR and the $2_T^{-\Delta\Delta C}$ method. Methods 25(4):402–408
6. Modak HV, Kelly DJ (1995) Acetyl-CoA-dependent pyruvate carboxylase from the photosynthetic bacterium *Rhodobacter capsulatus*: rapid and efficient purification using dye-ligand affinity chromatography. Microbiology 141(10):2619–2628
7. Peters-Wendisch PG, Schiel B, Wendisch VF et al (2001) Pyruvate carboxylase is a major bottleneck for glutamate and lysine production by *Corynebacterium glutamicum*. J Mol Microbiol Biotechnol 3(2):295–300
8. Sauer U, Eikmanns BJ (2005) The PEP-pyruvate-oxaloacetate node as the switch point for carbon flux distribution in bacteria. FEMS Microbiol Rev 29(4):765–794
9. Siddiquee KAZ, Arauzo-Bravo MJ, Shimizu K (2004) Metabolic flux analysis of *pykF* gene knockout *Escherichia coli* based on ^{13}C-labeling experiments together with measurements of enzyme activities and intracellular metabolite concentrations. Appl Microbiol Biotechnol 63 (4):407–417
10. Xie XX, Liang Y, Liu HL et al (2014) Modification of glycolysis and its effect on the production of L-threonine in *Escherichia coli*. J Ind Microbiol Biotechnol 41(6):1007–1015
11. Zelle RM, de Hulster E, van Winden WA et al (2008) Malic acid production by *Saccharomyces cerevisiae*: engineering of pyruvate carboxylation, oxaloacetate reduction, and malate export. Appl Environ Microbiol 74(9):2766–2777

A Crystal Structure of the *R*-Amine Transaminase from *Arthrobacter* for Asymmetric Catalysis of Chiral Amines

Lijun Guan, Nina Ji, Qinghong Meng, Xuejun Xie, Ye Zhou, Yijuan Cui and Shuwen Lu

1 Introduction

Synthesis of chiral compounds is significant due to the side effects of undesired enantiomers. Pure chiral amines are also crucial building blocks in the synthesis of some pharmaceutical drugs, for example, antidiabetic (*S*)-Repaglinide [1], HIV-protease inhibitor (S)-Lopinavir [2], antihistamine (R)-Levocetirzine [3] and (R)-Cinacalcet [4] which is used to treat secondary hyperparathyroidism. During the synthesis of the chiral compounds, the generation of chirality is most often the actual challenge. Thus, optically pure amines became increasingly significant as building blocks in pharmaceutical industry.

Chemical transformations used to synthesize chiral amines have shortcomings such as harsh reaction conditions, effluent of toxic catalytic metal, and even insufficient enantiopurity. Several approaches to produce active chiral amines using biocatalysts, such as hydrolases, oxidoreductases and transferases [5] has been developed to overcome the shortcomings of chemical synthesis. However, these enzymes have their own drawback such as low reactivity and reaction equilibrium issue [6]. As a solution to these problems, amine transaminases (ATAs), trivially referring to amine-pyruvate transaminase have been paid increased attention due to their unique ability for the asymmetric synthesis of pure chiral amines with high turnover rate, excellent stereoselectivity and stable catalytic activity [7]. Thus they are being seen as useful catalytic enzymes in industry production of chiral pharmaceutical drugs.

L. Guan (✉) · N. Ji · Q. Meng · X. Xie · Y. Zhou · Y. Cui · S. Lu (✉)
Food Processing Institute, Heilongjiang Academy of Agricultural Science,
368# Xuefu Road, Nangang District, Harbin 150086, China
e-mail: qqaipphh@sina.com

S. Lu
e-mail: shuwenl@sina.com

H. Liu et al. (eds.), *Advances in Applied Biotechnology*, Lecture Notes
in Electrical Engineering 444, https://doi.org/10.1007/978-981-10-4801-2_15

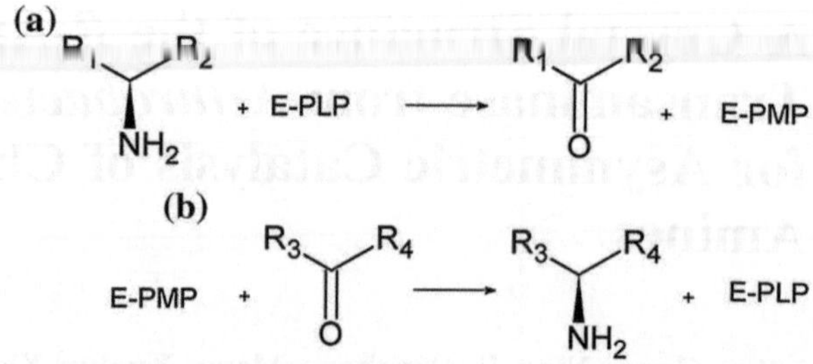

Fig. 1 Two steps transamination reaction

ATAs reversibly catalyze the transfer of an amino group from an amino compound to a ketone compound using the cofactor pyridoxal 5′-phosphate (PLP) two half reactions following a typical ping pong bi bi kinetic. In the first half reaction, the amino group of the amino donor is transferred to the PLP to form pyridoxamine-5′-phosphate (PMP) and the corresponding ketone product is released. In the second half reaction, the amino group is transferred from the PMP to the amino acceptor, while the cofactor is recycled to PLP and the product is released (Fig. 1).

ATAs are divided into two groups, (R)- and (S)-ATAs, according to the chirality at the amino carbon atom of the substrate. In particular, R-stereospecific ATA (R-ATA) is quite valuable, and the first (R)-stereoselective ATA identified from *Arthrobacter* sp. KNK168 (FERM-BP-5228) [R-ATA] by enrichment screening shows activity on secondary amine and methyl ketones or small cyclic ketons [8] and was successfully applied to the asymmetric synthesis of several kinds of chiral amines, such as (R)-dimethoxyamphetamine (DMA) and so on. Codexis and Merck & Co. created a variant of an R-ATA allowing industrial-scale synthesis of sitagliptin, which is an active ingredient for type-2 diabetes. The improved enzyme, designated as ATA-117-Rd11, gave a 13% increase in yield, a 53% increase in productivity and a 19% reduction in total cost, as compared to the chemical approach conventionally used to synthesize sitagliptin [9]. Recently, a research group determined the crystal structures of R-ATA form *Arthrobacter* sp. KNK168 (Ab-R-ATA) and demonstrated the mechanism of the substrate recognition and substrate specificity of R-ATAs. The active-site residue Arg138 functions in the dual substrate recognition that is a unique characteristic of ATAs. The recognition mechanism means to recognize both hydrophilic and hydrophobic substrates in the same active site. Moreover, the structures of ATA-117-Rd11 and the G136F mutant of Ab-R-ATA revealed the mechanism that the mutations in both enzymes resulted in a change of substrate specificity and showed that a loop near the active catalytic site was a new target region for the rational design, allowing to change the substrate specificity of R-ATAs. These findings would lead to the rational design of R-ATAs based on the structural information to develop novel biocatalysts useful for the production of a diverse range of chiral amine compounds, which would accelerate the industrial uses of chiral-amine synthesis using R-ATAs.

Gabaculine (5-amino-1,3-cyclohexadienylcarboxylicacid) is a neurotoxic natural product from *Streptomyces toyocaenis* and is known to be a covalent inhibitor for transaminases. During the transamination of gabaculine, a cyclohexatrienyl system, which is bound to the cofactor PLP, is formed. After spontaneous aromatization, m-carboxyphenylpyridoxamine phosphate (mCPP, Scheme 1) is formed and this finally inhibits the enzyme irreversibly [10]. Here, we report the crystal structure of the R-ATA from *Arthrobacter* (Ab-R-ATA) with the bound inhibitor gabaculine to clarify the substrate specificity, enantioselectivity for R-ATAs.

2 Materials and Methods

The Ab-R-ATA expression plasmid constructed by inserting the gene (GenBank accession number AB638718) under the lac promoter of the pUCN vector was used for transformation of *Escherichia coli* Rosetta (DE3). The transformants were cultivated at 310 K in LB medium containing 100 μg ml^{-1} ampicillin. Overexpression was induced by adding 1 mM isopropyl-β-D-thiogalactopyranoside (IPTG) at 293 K.

To purify Ab-R-ATA, the cells were resuspended in 50 mL lysis-buffer containing 20 mM potassium phosphate buffer (pH 6.8), 0.01% 2-mercaptoethanol, 1 mM PLP and 10% glycerol (v/v). The cells were disrupted by sonication for 6 min using BRANSON Digital Sonifier at power 7. After centrifugation at 40,000 g at 277 K for 30 min, the supernatant was applied onto DEAE-Sepharose, and the Ab-R-ATA protein was eluted with 400 mM NaCl dissolved in 20 mM potassium phosphate buffer (pH 6.8) containing 1 mM DTT and 10% glycerol (v/v). The eluted protein with 1 mM PLP additive was dialyzed against 20 mM potassium phosphate buffer (pH 6.8) containing 1 mM DTT and 10% glycerol (v/v). The dialyzate was applied to a ResourceQ column 6 ml (GE Healthcare) equilibrated with 20 mM potassium phosphate buffer (pH 6.8) containing 1 mM DTT and 10% glycerol (v/v), and the protein was eluted with a NaCl gradient ranging from 0 to 1 M. The active fractions obtained by ResourceQ chromatography were concentrated and loaded on a Superdex 200 h 10/30 column (GE Healthcare) equilibrated with 20 mM potassium phosphate buffer containing 1 mM DTT and 5% glycerol. The active fractions were concentrated to 8 mg ml^{-1} by centrifugation at 4000 g using Vivaspin-20 (10,000 molecular weight cutoff) prior to crystallization trials.

Cocrystallization was performed to obtain the complex of Ab-R-ATA and gabaculine. Initial crystallization screening was carried out with the sparse-matrix screening kits Crystal Screen HT (Hampton Research), Index HT (Hampton Research) and Wizard I and II (Emerald BioSystems) in 96-well plates using the sitting-drop vapour-diffusion method. For refinement of the crystallization conditions, 1 μL of protein solution was mixed with an equal volume of reservoir solution and equilibrated against 0.5 ml of reservoir solution at 293 K in 24-well plates (Hampton Research).

The dataset was collected with a wavelength of 1.0000 Å, a rotation angle of 0.5°, an exposure time of 1 s for each image. Data set was composed of 360 images.

The diffraction data was indexed, integrated and scaled with XDSme [11]. Initial phasing of the two diffraction data set was conducted by molecular replacement using MOLREP [12] in the CCP4 [13]. The search model was the crystal structure of Ab-R-ATA. Molecular replacement was performed using a resolution range from 20 to 3.0 Å. The initial phase was refined using Refmac5 [14] then examined using Coot [15]. mCPP was added on the basis of 2*Fo*-*Fc* and *Fo*-*Fc* electron density maps. The final manual fitting and structure refinement was also completed in Coot and validated with the Ramachandran plot drawn by the PROCHECK program.

3 Results and Discussion

After incubation with the gabaculine, sample changed from yellow to colourless. It was supposed that Ab-R-ATA had been inhibited by gabaculine in this condition. By streak seeding, crystals grew in 4 days. The crystallization drops were prepared by mixing 1.5 μL sample (15 mg/ml) and 1.5 μL reservoir solution (0.2 M magnesium chloride, 0.1 M Bis-Tris pH 6.5, 18% PEG 3350).

Crystal structure analysis of the ATA from *Arthrobacter* with bound gabaculine was carried out to obtain detailed insight into substrate recognition and enantioselectivity. The crystals of the complex of Ab-R-ATA and gabaculine belonged to space group $P4_22_12$. The unit cell and the other crystallographic data collection statistics are shown in Table 1. Matthew coefficient value indicted that there is one molecule per asymmetric unit. The initial phase was also determined by molecular replacement using the crystal structure of Ab-R-ATA as search model. The electron density of mCCP was observed (Fig. 2), which is consistent with the results of UV-visible spectroscopy. Thus, mCCP was added into the corresponding electron density. Manual structure construction and refinement were finished using Refmac5 and Coot. The refinement statistics are given in Table 1. The Ramachandran plot showed that 96.3 and 100% of φ-ψ pairs lie in the most favored and additionally allowed regions, respectively.

The resulting crystal structure could be solved and refined to a resolution of 1.7 Å and the asymmetric unit contains one monomer like the wild type. A comparison of the complex and Ab-R-ATA structures showed that they were very similar, with an r.m.s.d. of 0.38 Å. However, the N-terminal region (residues 1–28) of the complex was not possible to build due to the absence of continuous electron density. In addition, there were much fewer water molecules in the complex than Ab-R-ATA structure. It is indicated that the N-terminal loop may adopt an extended conformation and partially shield the active site from the bulk solvent.

The mCPP adduct formed by gabaculine and PLP was observed in the active site of Ab-R-ATA, which will contribute to understanding the binding pockets. The mCPP adduct is located in the center of the active site of the enzyme and directs to the S pocket constituted by the side chains of residues Val69, Thr283 and Ala284

Table 1 Data collection and refinement statistics

Complex of Ab-R-ATA and gabaculine	
Data collection	
Wavelength (Å)	1.0000
Space group	$P4_22_12$
Cell dimensions (Å)	74.83, 74.83, 89.5
Cell angles (Å)	90.00, 90.00, 90.00
Resolution (Å)	45.0–1.77
Number of observations	297,565
Number of unique reflections	25,302
Completeness (%)	99.1 (95.5)
R_{merge} (%)	6.0 (57.4)
Redundancy	11.76
I/σ (I)	23.99 (2.49)
Refinement	
Resolution range (Å)	33.49–1.86
R_{work}/R_{free} (%)	19.8/23.0
Number of non-hydrogen atoms	
Protein	2315
Water	30
Ligand	mCCP
	25
RMSD bond length (Å)	0.018
RMSD bond angle (°)	1.827
Ramachandran plot (%)	
Preferred	96.3
Allowed	3.7
Outliers	0.0

The values in parentheses are for the highest resolution shell $R_{work} = \sum||Fo| - |Fc||/\sum|Fo|$, where $|Fo|$ and $|Fc|$ are the observed and calculated structure factor amplitudes of a particular reflection. R_{free} was calculated from 5% of reflections omitted from the refinement

with a close distance (<4.5 Å) (Fig. 3), which probably interact with the methyl group of MBA. The L pocket constituted by the side chains of Tyr67, His62 and Trp192 from the adjacent subunit locates on the opposite side of the S pocket. This L pocket of Ab-R-ATA can accommodate the phenyl group of substrates (R)-MBA, which is the model substrate used for the research of ATAs. Based on the information obtained from the crystal structure of this complex, it is indicated that the shape of the two pockets and the relative position of the L pocket and the small pocket determines the enantioselectivity of the ATAs.

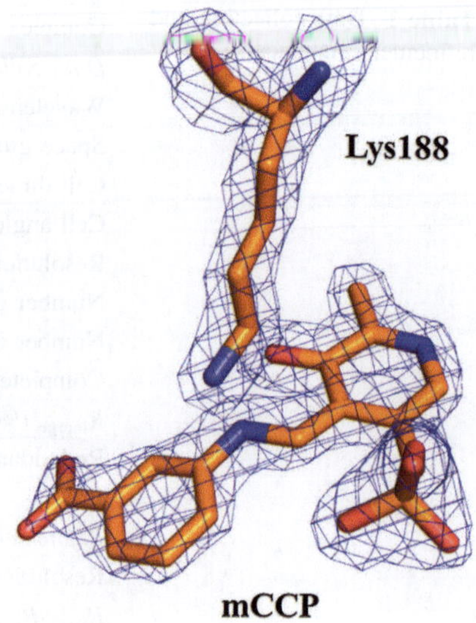

Fig. 2 2*Fo*-*Fc* omit density map of Lys188 and the mCCP formed by PLP and gabaculine, contoured at 2.0 σ

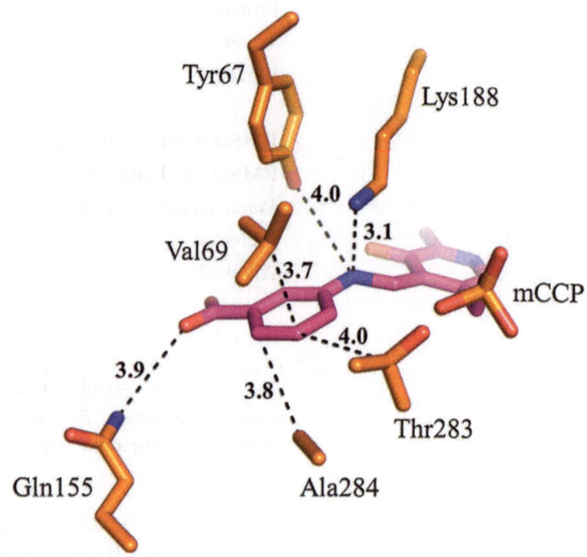

Fig. 3 Binding mode of mCCP in Ab-R-ATA. mCCP is shown in *purple* and the residues of Ab-R-ATA are shown in *orange*

4 Conclusion

The crystal structural analysis could provide lots of information for the rational design of proteins. Thus, it is a powerful tool in the protein-engineering field. The crystal structure of the complex of mCCP and Ab-R-ATA reported here provided essential information and insight into the substrate recognition and revealed the mechanism of high stereoselective catalysis of ATAs. These results will probably provide structural basis for the molecular modification of the ATAs and theoretical basis for designing the novel chiral amine compounds.

References

1. Plosker GL, Figgitt DP (2004) Repaglinide. Pharmacoeconomics 22:389–411
2. Eron J et al (2006) The klean study of fosamprenavir-ritonavir versus lopinavir-ritonavir, each in combination with abacavir-lamivudine, for initial treatment of HIV infection over 48 weeks: a randomised non-inferiority trial. Lancet 368:476–482
3. Chen C (2008) Physicochemical, pharmacological and pharmacokinetic properties of the zwitterionic antihistamines cetirizine and levocetirizine. Curr Med Chem 15:2173–2191
4. Block GA et al (2004) Cinacalcet for secondary hyperparathyroidism in patients receiving hemodialysis. N Engl J Med 350:1516–1525
5. Höhne M, Bornscheuer UT (2009) Biocatalytic routes to optically active amines. ChemCatChem 1:42–51
6. Shin JS, Kim BG (1999) Asymmetric synthesis of chiral amines with ω-transaminase. Biotechnol Bioeng 65:206–211
7. Taylor PP, Pantaleone DP, Senkpeil RF, Fotheringham IG (1998) Novel biosynthetic approaches to the production of unnatural amino acids using transaminases. Trends Biotechno. 16:412–418
8. Iwasaki A, Yamada Y, Ikenaka Y, Hasegawa J (2003) Microbial synthesis of (R)- and (S)-3,4-dimethoxyamphetamines through stereoselective transamination. Biotechnol Lett 25:1843–1846
9. Savile CK et al (2010) Biocatalytic asymmetric synthesis of chiral amines from ketones applied to sitagliptin manufacture. Science 329:305–309
10. Rando RR (1977) Mechanism of the irreversible inhibition of GABA transaminase by the neurotoxin gabaculine. Biochemistry 16:4604–4610
11. Kabsch W (2010) XDS. Acta Crystallogr D 66:125–132
12. Vagin A, Teplyakov A (2010) Molecular replacement with MOLREP. Acta Crystallogr D 66:22–25
13. Winn MD et al (2011) Overview of the CCP4 suite and current developments. Acta Crystallogr D 67:235–242
14. Murshudov GN et al (2011) REFMAC5 for the refinement of macromolecular crystal structures. Acta Crystallogr D 67:355–367
15. Emsley P, Lohkamp B, Scott WG, Cowtan K (2010) Features and development of Coot. Acta Crystallogr D 66:486–501

Part II
Optimization and Control of Biological Process

Optimization of Culture Medium for Rifamycin SV Production by *Amycolatopsis kentuckyensis* 22-187

Mingyu Xu, Meng Li, Ying Zhang and Huitu Zhang

Nomenclature

AHBA	3-amino-5-hydroxy benzoic acid
RF	Rifamycins
RSM	Response surface methodology
CCD	Central composite design
ANOVA	Analysis of variance
PBD	Plackett-Burman design

1 Introduction

Rifamycins (RF), a group of antiobiotics of the ansamycin family [1], are clinically important antibacterial agents active against gram-positive bacteria. Several semisynthetic rifamycins variants (for example, rifampin and refapentine) have been used clinically for the treatment of tuberculosis and other bacterial infections.

Rifamycins A, B, C, D, E, G, L, O, S, SV, W, X and Y are known to be produced by the soil actinobacterial species Amycolatopsis mediterranei ATCC13685 [2], while S.tolypophorous produces O and S [3], S.albovinaceous and Nocardia asiatica produce B, Micromonospora chalcea produces SV, Nocardia mediterranei produces R and production of RF- P and Q by a mutant of Nocardia mediterranei has been reported [4]. 22-187 strain as a new isolate was obtained in our ansamycins antibiotic producers screening program by targeting the conserved regions of AHBA (3-amino-5-hydroxybenzoic acid) synthase gene from streptomyces [5]. On the basis of chemotaxonomic and phylogenetic analysis of 16S rDNA sequences, the isolate is related most closely to Amycolatopsis kentuckyensis.RF-SV and B (Fig. 1) were chemically identified from 22-187 fermentation broth [5].

M. Xu · M. Li · Y. Zhang · H. Zhang (✉)
Key Laboratory of Industrial Fermentation Microbiology,
Ministry of Education, College of Biotechnology, Tianjin University
of Science and Technology, No. 29, 13 Main Street, Tianjin, China
e-mail: hzhang@tust.edu.cn

H. Liu et al. (eds.), *Advances in Applied Biotechnology*, Lecture Notes
in Electrical Engineering 444, https://doi.org/10.1007/978-981-10-4801-2_16

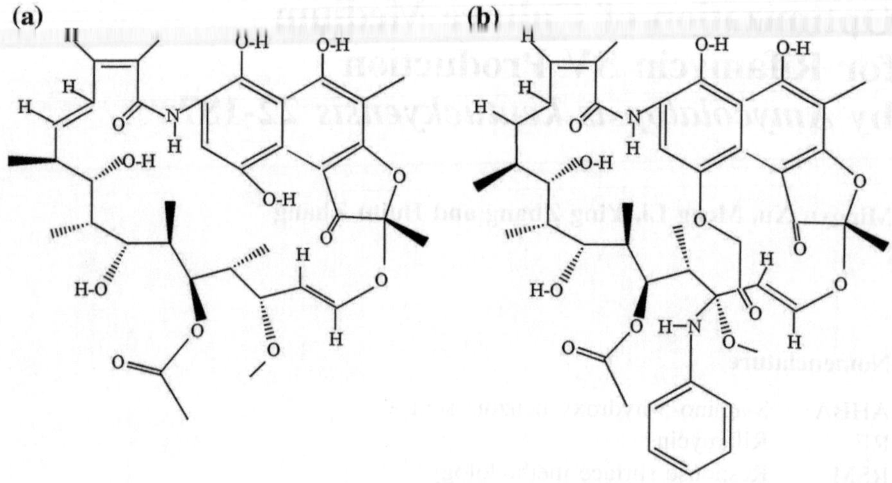

Fig. 1 Chemical structures of rifamycin SV (**a**) and rifamycin B (**b**)

RF-B, as one of components discovered in the first naturally occurring RFs complex, is moderately active against gram-positive bacteria and mycabacteria. RF-SV which is very active against gram-positive bacteria and mycobacteria initially was obtained by removing the glycolic chain in position 4 of RF-B by oxidative cyclization, hydrolysis and reduction of the quinone chemically. RF-SV today can be considered a natural RF after programmed research guided by biosynthetic considerations. To our knowledge, this is the first report to find RF producer inAmycolatopsis kentuckyensis. The aim of the present work was to acquire an optimized fermentation medium for favoring RF-SV production in Amycolatopsis kentuckyensi. Response surface methodology (RSM) was employed here to optimize the medium components. RSM are important tools to study the effect of both the primary factors and their interactions [6]. First Plackett-Burman design was applied to identify the most important components in the media makeup. Then central composite design was used to examine and optimize the fermentation media for RF-SV production by Amycolatopsis mediterranei 22-187 in shake flask experiments.

2 Materials and Methods

2.1 Microorganism and Culture Conditions

Amycolatopsis mediterranei 22-187 was grown on the YMG medium [5]. The seed culture preparation and fermentation was carried out as described [5].

2.2 Analytical Methods

After 120 h of fermentation, samples were centrifuged, the supernatant was extracted twice with an equal volume of ethyl acetate. The mycelia were extracted twice with an equal volume of acetone. The acetone extracts were concentrated and further extracted with ethyl acetate. The combined ethyl acetate extracts were evaporated under vacuum to dryness to obtain crude materials. The crude materials were weighted then dissolved in an appropiate volume of methanol to make a final concentration of 10 μg/mL solution. The methanol solute was clarified by centrifugation and 10 μl of the supernatant was injected and analyzed by HPLC. HPLC analysis of RF-SV was carried out with Diamonsil C18 column (5 μm particle size, length × i.d.150 × 4.6 mm) using linear gradient of 20–100% acetonitrile. UV detection was performed at 304 nm wave length.

2.3 Experimental Design and Data Analysis

Plackett-Burman design was chosen to study the effects of the most important components in the media. Then a central composite design (CCD) was performed to optimize the critical components and maximize the RF-SV productivity.

Based on a Plackett-Buramn factorial design, each factor was examined at 2 levels: low and high factors were coded as −1 and +1, and the center point was coded as 0. Table 1 illustrates the factors investigated, as well as the levels of each factor used in the experimental design. The response chosen was the production of RF-SV as peak area in HPLC analysis under RT 14.471 (supplementary Fig. 2a).

In order to fit empiric second-order polynomial model, a central composition design with five coded levels was performed (Table 2). The quadratic model for predicting the optimal point was expressed according to following equation:

$$Y = b0 + \sum bixi + \sum biix + \sum bijxixj$$

Table 1 Media components and levels for the Plackett-Burman experiment

Media# component	Component	Low level (−1) (g/l)	High level (+1) (g/l)
Ammonium citrate	X1	0	5
KH_2PO_4	X2	0.05	0.1
$MgSO_4$	X3	0.05	0.1
$CoCl_2$	X4	0	0.005
Glucose	X5	50	100
Peptone	X6	0	5
KNO_3	X7	0	5
Starch	X8	0	20

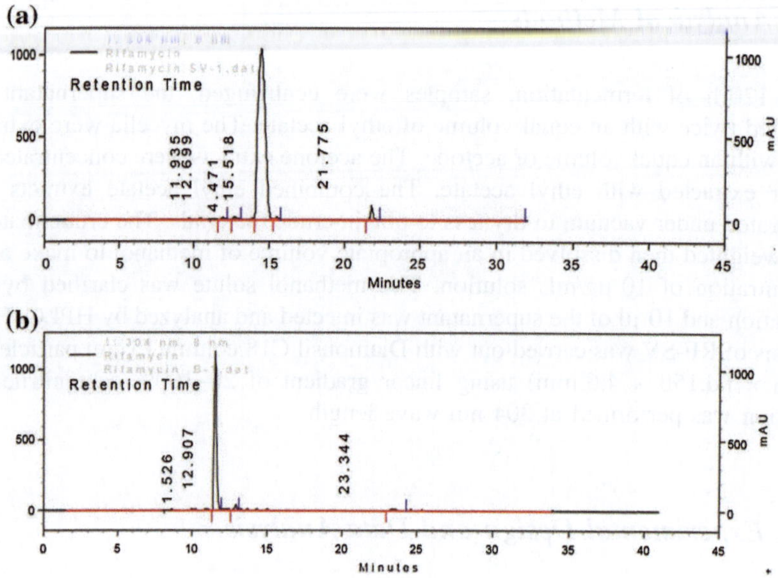

Fig. 2 HPLC analysis of rifamycin. **a** Rifamycin SV and **b** rifamycin B

Table 2 Factors and levels of CCD experimental design

Factor	Level				
	−1.414	−1	0	+1	+1.414
Glucose (g/l)	95.9	110.0	144.0	178.0	192.1
peptone (g/l)	6.5	7.5	10.0	12.5	13.5

where Y is the response variable, b0 is the value of the fixed response at the central point of the experiment, which is the point (0,0); bi and bii the linear and quadratic coefficients of a factor Xi, respectively; bij is the interaction coefficient between two factors Xi and Xj.

The Design Expert software (version 6.0.10, stat-ease, inc., Minneapolis, USA) was used for regression and graphical analysis of the experimental data. The statistical and analysis of the model was performed in the form of analysis of variance (ANOVA). The determination coefficient R2 measures the goodness of fit of regression model. It also includes the *t*-value for the estimated coefficients and associated probabilities.

3 Results and Discussion

3.1 Detection of Main Medium Components for RF-SV Production

Eight medium components were tested using the Plackett-Burman Design (PBD). Fifteen experiments were carried out that were the result of the combination of the variables at the high (+1) and low (−1) level,and three experiments in the center-point conditions (experiments 3, 6 and 10). The experiment design matrix and the results are shown in Table 3. The results from the PBD were analyzed using analysis of variance (ANOVA) (Table 4). The non-significant coefficients were eliminated on the basis of *p*-values after examining the coefficients. It was observed from Table 4, that positively effect RF-SV production by Amycolatopsis mediterranei 22-187 were KH_2PO_4, glucose and peptone concentrations in media, whereas KNO_3 presence showed negative effect [*p*-values ($p > 0.05$)]. The interaction between ammonium citrate and $MgSO_4$ could have a positive effect on RF-SV productrion. A coefficient of determination (Adj R2) value of 0.9868 showed that the obtained data is highly reliable. Hence, in the subsequent experiment, glucose and peptone were selected and optimized using CCD. In contrast, the content of KNO_3 was eliminated. KH_2PO_4, Ammonium citrate and $MgSO_4$ concentrations were kept at high value.

Table 3 Experimental matrix factorial design

Experiment	Variable								RF-SV (peak area × 105)
	X1	X2	X3	X4	X5	X6	X7	X8	
1	−1	−1	−1	−1	−1	−1	−1	−1	19.99
2	1	1	−1	1	1	1	−1	−1	27.61
3	0	0	0	0	0	0	0	0	26.01
4	−1	−1	1	−1	1	1	−1	1	25.71
5	1	−1	−1	−1	1	−1	1	1	12.87
6	0	0	0	0	0	0	0	0	26.08
7	1	−1	1	1	−1	1	1	1	25.17
8	−1	−1	−1	1	−1	1	1	−1	23.97
9	1	1	1	−1	−1	−1	1	−1	15.68
10	0	0	0	0	0	0	0	0	25.23
11	−1	1	1	1	−1	−1	−1	1	19.86
12	−1	1	1	−1	1	1	1	−1	24.6
13	−1	1	−1	1	1	−1	1	1	22.28
14	1	−1	1	1	1	−1	−1	−1	25.3
15	1	1	−1	−1	−1	1	−1	1	25.44

Table 4 Analysis of variance by ANOVA

Variables	β	F value	*p*-value
X1	−0.16	0.529181	0.5426
X2	1.31	22.94838	0.0409[a]
X3	−0.02	0.014373	0.9155
X4	0.56	4.200666	0.1769
X5	1.79	43.09511	0.0224[a]
X6	4.10	604.0425	0.0017[a]
X7	−2.67	255.8468	0.0039[a]
X8	0.78	13.0053	0.0690
X1, X3	2.69	19.22	0.0114[a]

[a] $P < 0.05$

Table 5 Experimental designs and the results of central composite design (CCD)

Run	X5	X6	Y (peak area × 105)	
			Observed	Predicted
1	−1	−1	15	14.08
2	1	−1	12.7	12.79
3	0	0	12.9	12.82
4	1.414	0	12	11.08
5	0	0	12.9	12.82
6	0	1.414	6.9	7.1
7	0	−1.414	12.2	12.82
8	−1.414	0	15.6	16.87
9	1	1	5.7	6.82
10	0	0	11.8	12.82
11	0	0	11.8	12.82
12	0	0	14.7	12.82
13	−1	1	14.7	13.72

3.2 *Optimization of Medium Components for RF-SV Production*

Further optimization of RF-SV production was carried out by using a Box-Wilson central composite design with four star points and five replicates at center point for each of two factors. Table 5 showed the design of this experiment and the results. Regression analysis was performed to fit the response function with the experimental data and results were shown in Table 6. The statistical significance of the second-order model equation was checked by an F-test (ANOVA) and data shown in Table 7.

Table 6 Results of parameter estimate for RF-SV productivity of CCD

Parameters	Parameter estimate	df	Standard error	p-value (Prob > F)
Intercept	12.82	1	0.55	0.0000
X1	−2.05	1	0.44	0.0022
X2	−1.85	1	0.44	0.0039
X1X2	−1.67	1	0.62	0.0302
X12	0.58	1	0.47	0.2577
X22	−1.55	1	0.47	0.0131

Table 7 Analysis of variance (ANOVA) for the model

Regression	Sum of square	DF	F-ratio	p value
Linear	60.94653	2	7.107211	0.0120[a]
interaction	11.2225	1	3.190824	0.1077
Quadratic	20.95758	2	6.857546	0.0224[a]

[a]$P < 0.05$

Three-dimensional response surface plot of X_5 and X_6 against RF-SV productivity can explain the results of the statistical analysis. The response surface open its mouth downwards demonstrated there must be a maximum in the stable range (Fig. 3). In light of multi-regressive-analysis of the central composite experiments showed in Table 6, the second-order polynomial prediction model was obtained as Eq. (1). Table 6 and Eq. (1) show positive effects of X_5^2, and negative effects of X_5, X_6, X_5, X_6 and $X_6{}^2$.

The statistical significance of the second-order model was checked by an F-test (ANOVA) and data showed in Table 7. The fit value, termed R^2 (determinant coefficient), of the polynomial model was calculated to be 0.8970, indicating that 89.7% of the variability in the response could be explained by the second-order polynomial prediction equation given below (Eq. 1).

$$\text{RF-SV} = 12.82 - 2.05 \times 1 - 1.85 \times 2 - 1.67 \times 1 \times 2 + 0.58 \times 12 - 1.55 \times 22 \quad (1)$$

The ANOVA results showed that this model is appropriate. It also suggested that RF-SV production was primarily determined by the linear and quadratic terms of peptone and glucose of the model and no significant interaction existed between the two factors. The Eq. 1 for RF-SV production showed negative linear effect and positive quadratic effect. The three-dimensional graph obtained from the calculated response surface was showed in Fig. 3a and the contour plot in Fig. 3b. Three-dimensional response surface plot and the contour plot of peptone and glucose concentrations against RF-SV production can further explain the results of the statistical and mathematical analyses. It is evident from the three-demension plot that RF-SV production reached its maximum at a combination of coded level of X_1 and X_2 at −0.99 and −0.082, respectively. The contour plot predicted a maximum response of RF-SV production at the point when the concentrations of glucose and

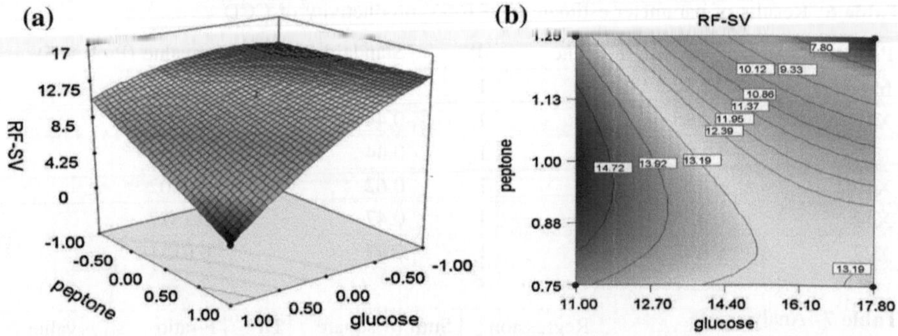

Fig. 3 **a** Three-demension plot for RF-SV production as a function of peptone and glucose concentrations. **b** the contour plot of peptone and glucose (g/100 ml) against RF-SV productivity

peptone were 11.02 and 0.98% (w/v). The maximum RF-SV production should be reached to 14.72 × 10^5 (HPLC peak area) by prediction from RSM analysis. To confirm these results, optimal medium was verified and compared with the initial medium. The average RF-SV production was 13.96 × 10^5 (HPLC peak area) from triple-duplicated experiments and was two folds higher than in the initial medium. But the production of RF-B (detected by HPLC analysis under RT 11.526 as shown in supplementary Fig. 2(B)) was remained the same as in initial medium. The optimal medium composition for RF-SV production by *Amycolatopsis mediterranei* 22-187 was finalized as follows (%): citrate NH_4(0.5%), K_2HPO_4 (0.1%), $MgSO_4$ (0.1%), $CaCO_3$(0.5%), glucose (11.02%), peptone (0.89%).

4 Conclusion

Batch culture of Amycolatopsis kentuckyensis 22-187 was investigated and optimized for the production of the RF-SV. Statistical optimization method for fermentation overcome the limitations of classic empirical methods and was proved to be a powerful tool for the optimization of RF-SV production. In this study the experimental results clearly showed that the RF-SV production was dependent mainly on glucose and peptone and a final concentration of 11.02% (w/v) glucose and 0.98% (w/v) of peptone was found to be best for rifamycin production. Through the statistically designed optimization the title of fermentation of rifamycin SV was increased two folds comparing with that in the initial medium by HPLC analysis. But the production of RF-B was remained the same as in initial medium.

Acknowledgements This study was financially supported by the Municipal Science and technology program of Tianjin (Grant number: 12JCYBJC31800). It was also supported by the National High Technology Research and Development Program (2012AA022108) and the program for Changjiang Scholars and Innovative Research Team in University(IRT1166).

References

1. Dewick PM (2009) Medicinal natural products: a biosynthetic approach, 3rd edn. Wiley, Chichester, p 656
2. Schupp T, Taxier P, Auden JAL (1981) New rifamycins produced by a recombinant strain of *Nocardia mediterranei*. J Antiobiot 34(8):965–970
3. Martinelle E, Antonini P, Crichio R, Lancini G, White RJ (1978) Rifamycin R, a novel metabolite from a mutant of Nocardia meditarranea. J Antibiot 31(10):949–951
4. Crichio R, Antonini P, Sartori G (1980) Biosynthesis of rifamycins P, Q and verde novel metabolites from a mutant of *Nocardia mediterranea*. J Antibiot 33(8):842–856
5. Zhang H, Linzhuan W, Aiming L, Guizhi S, Feng H, Qiuping L, Yuzhen W, Huanzhang X, Qunjie G, Yiguang W (2009) PCR screening of 3-amino- 5-hydroxybenzoic acid synthase gene leads to identification of ansamycins and AHBA-related antibiotic producers in *Actinomycetes*. J Appl Microbiol 106:755–763
6. Montgomery DC (1997) Design and analysis of experiments. Wiley, New York

Optimization of Processing Parameters of Stabilizers After Enzymes Hydrolysis for Cloudy Ginkgo Juice

Haifeng Yu, Junyan Liu and Jingxi Yang

1 Introduction

Ginkgo *biloba*, dating back 300 million years, is a living fossil. Ginkgo seeds are nutritious, contain starch, protein, lipid, pectin, amino acid, vitamins, trace elements and abundant phenolic compounds and have been used as food and herbal medication in china for several thousand years [1]. Ginkgo *biloba* is wildly grown in China and the yields of ginkgo seeds are rich. However, they are not made full use. In the daily life, fruit juice is regular consumed, especially cloudy juice. The sensory qualities of cloudy juice are important factors for consumer acceptance [2, 3]. On processing juice, enzymatic hydrolysis is used to obtain higher yield and improved stability [4–6]. Ginkgo juice is studied recently, but there are many sediments in it because of its abundant starch. The major sensory problem is the generation of a large of sediments in ginkgo juice storage. In order to produce stable ginkgo juice, enzymes were used to hydrolyze starch, after that the juice was centrifuged to remove sediment [7]. On the processing, ginkgo juice only has little solids. In other way, stabilizers were added into the juice to improved stability, but the ginkgo juice produced could not stay stability for long time.

So the experiment is studied to improve the ability of ginkgo cloudy juice. We find that if stabilizers are added into the juice after enzymes hydrolysis, ginkgo juice would be more stable(data unpublished). Therefore, the aim of this study was to investigate the effect of stabilizers (CMC, pectin and SA) dosage after enzymes hydrolysis on ginkgo cloudy juice and optimize the process conditions by RSM.

H. Yu (✉) · J. Liu · J. Yang
Shandong Provincial Key Laboratory of Microbial Engineeringy,
Qilu University of Technology, No. 3501 Daxue Road, Changqing District 250353
Jinan Shandong Province, People's Republic of China
e-mail: yhfdzz@126.com

H. Liu et al. (eds.), *Advances in Applied Biotechnology*, Lecture Notes
in Electrical Engineering 444, https://doi.org/10.1007/978-981-10-4801-2_17

2 Materials and Methods

Ginkgo seeds were purchased from Tancheng city Shandong province. From the pre-experiment we know that the water content is 55.88% (w/w), the protein content is 5.85% (w/w) and the starch content is 28.30% (w/w).

2.1 *Enzyme and Stabilizer Source*

Medium temperature α-amylase and neutral protease were purchased from Ruiyang Biotechnology Company, Jiangsu, China. CMC, pectin and SA were purchased from Quankang Food Ingredients Company, Jinan, China.

2.2 *Preparation of Ginkgo Juice*

Based on pre-experiment: a ratio of 1:20(kernel:water,w/v) was used in the comminuting process; the dosage of enzymes were determined that α-amylase was 0.015 g/100 mL ginkgo juice, neutral protease was 0.02 g/100 mL ginkgo juice, enzyme temperature was 80 ± 1 °C and enzyme treatment time was 90 min; the range of the variables for stabilizer conditions was selected, there were CMC dosage, X_1(0.01 − 0.13 g/100 mL ginkgo juice), pectin dosage, X_2(0.02 − 0.18 g/100 mL ginkgo juice) and SA dosage, X_3(0.01 − 0.11 g/100 mL ginkgo juice). The processing of ginkgo juice is showed in Fig. 1.

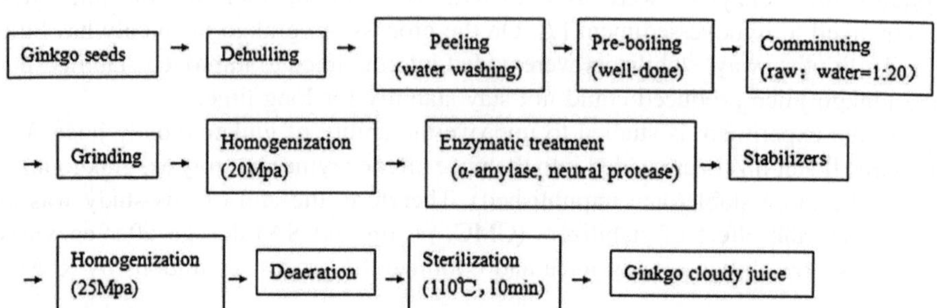

Fig. 1 Flow chart of ginkgo cloudy juice production

2.3 Centrifugal Sedimentation Rate

Ginkgo juice was shaken and certain quality juice was removed to centrifuge tube, the juice was centrifuged at 3000 rpm for 30 min to measure the quality of sediments. The ratio of sediments quality and juice quality was considered a measure of centrifugal sedimentation rate.

2.4 Experimental Design and Statistical Analysis

RSM was used in designing this experiment. A Design-Expert V8.0 was used to generate the experimental designs, statistical analysis and regression model. The independent variables were the dosage of CMC (x_1).the dosage of pectin (x_2), the dosage of SA (x_3). Each independent variable had coded levels of −1, 0 and 1. The experimental designs of the coded (x) and actual (X) levels of variables are shown in Table 1. The response (y) is centrifugal sedimentation rate (%).

The response function (y) was related to the coded variables (x_i, i = 1, 2, 3) by a second-degree polynomial (1) using the method of least squares.

Table 1 The RSM experimental design (in coded level of three variables) employed for processing ginkgo cloudy juice with CMC, pectin, SA

Experim-ent number	CMC (g/100 mL ginkgo juice)	Pectin (g/100 mL ginkgo juice)	SA (g/100 mL ginkgo juice)
	$X_1(x_1)$	$X_2(x_2)$	$X_3(x_3)$
1	0.13(1)	0.1(0)	0.11(1)
2	0.07(0)	0.18(1)	0.01(−1)
3	0.13(1)	0.18(1)	0.06(0)
4	0.01(−1)	0.1(0)	0.01(−1)
5	0.13(1)	0.1(0)	0.01(−1)
6	0.01(−1)	0.18(1)	0.06(0)
7	0.07(0)	0.1(0)	0.06(0)
8	0.07(0)	0.1(0)	0.06(0)
9	0.07(0)	0.02(−1)	0.11(1)
10	0.13(1)	0.02(−1)	0.06(0)
11	0.01(−1)	0.02(−1)	0.06(0)
12	0.07(0)	0.1(0)	0.06(0)
13	0.07(0)	0.1(0)	0.06(0)
14	0.01(−1)	0.1(0)	0.11(1)
15	0.07(0)	0.18(1)	0.11(1)
16	0.07(0)	0.02(−1)	0.01(−1)
17	0.07(0)	0.1(0)	0.06(0)

$$y = b_0 + b_1x_1 + b_2x_2 + b_3x_3 + b_{11}x_1^2 + b_{22}x_2^2 + b_{33}x_3^2 + b_{12}x_1x_2 + b_{13}x_1x_3 + b_{23}x_2x_3 \quad (1)$$

Analysis of variance (ANOVA) was performed. The significances of all terms were judged statistically by computing the *F*-value at a probability (*p*) of 0.001, 0.01 or 0.05.The regression coefficients were used to make statistical calculations to generate contour maps from the regression models.

3 Results and Discussion

3.1 Statistical Analysis

The experimental results on the effect of the dependent variables namely CMC dosage, pectin dosage and SA dosage on the response functions are shown in Table 2.

The corresponding R^2 and coefficients of the variables in the models are shown in Table 3. The closer the value R^2 is to unity, the better the empirical model fits the actual data. The value R^2 for centrifugal sedimentation rate (after enzymes hydrolysis) and for centrifugal sedimentation rate (no enzymes hydrolysis) were 0.9472 and 0.9574, indicating that the regression models explained the reaction well. The probability (*p*) values of all regression models were less than 0.05.

Table 2 Effect of stabilizers on two dependent variables

Experiment number	Centrifugal sedimentation rate (%) (after enzymes hydrolysis)	Centrifugal sedimentation rate (%) (no enzymes hydrolysis)
	Y_1	Y_2
1	9.94	35.03
2	9.85	37.23
3	10.33	36.6
4	9.31	35.9
5	12.42	40.02
6	11.22	40.19
7	5.32	22.44
8	5.33	23.48
9	11.32	40.59
10	10.23	36.19
11	9.96	35.11
12	5.36	23.6
13	7.25	28.19
14	12.17	44.01
15	11.12	39.78
16	8.11	31.06
17	5.58	25.68

3.2 Centrifugal Sedimentation Rate (After Enzymes Hydrolysis)

From Table 3, it is observed that the quadratic terms of CMC dosage ($p \leq 0.001$), pectin dosage ($p \leq 0.01$) and SA dosage ($p \leq 0.001$) had a significant effect on centrifugal sedimentation rate. CMC dosage and SA dosage had a negative interaction effect ($p \leq 0.05$).

Figure 2a describes that the dependence of centrifugal sedimentation rate with CMC dosage and pectin dosage at fixed SA dosage. From Fig. 2a, it is clear that at constant pectin dosage and SA dosage, centrifugal sedimentation rate decreased with CMC dosage at the beginning then increased gradually. Likewise, with the increase of pectin dosage, the centrifugal sedimentation rate decreased gradually first, then increased.

Figure 2b presents the variation of centrifugal sedimentation rate with CMC dosage and SA dosage at constant pectin dosage. It is evident that at a fixed CMC dosage and pectin dosage, the centrifugal sedimentation rate of ginkgo juice decreased with SA dosage at the beginning and then increased.

Table 3 Regression coefficients, R^2 and p value for the response function

Coefficient of the regression equation	Centrifugal sedimentation rate (%)	Centrifugal sedimentation rate (%)
	After enzymes hydrolysis	No enzymes hydrolysis
b0 (intercept)	11.76123	40.98628
b1	−74.18611	−196.18819
b2	−43.44896	−108.11927
b3	−61.143	−182.86633*
b11	767.63889***	1933.26389***
b22	297.42187**	841.36719**
b33	971.4***	2840.9***
b12	−60.41667	−243.22917
b13	−445*	−1091.66667*
b23	−121.25	−436.25
R2	0.9472	0.9574
p	0.0011	0.0005

b was the coefficient of polynomial
b0 (constant); b1, b2 and b3 (linear effects); b11, b22 and b33 (quadratic effects); and b12, b13 and b23 (interaction effects)
1 CMC dosage, *2* pectin dosage, *3* SA dosage
*Significant at $p \leq 0.05$. **Significant at $p \leq 0.01$. ***Significant at $p \leq 0.001$

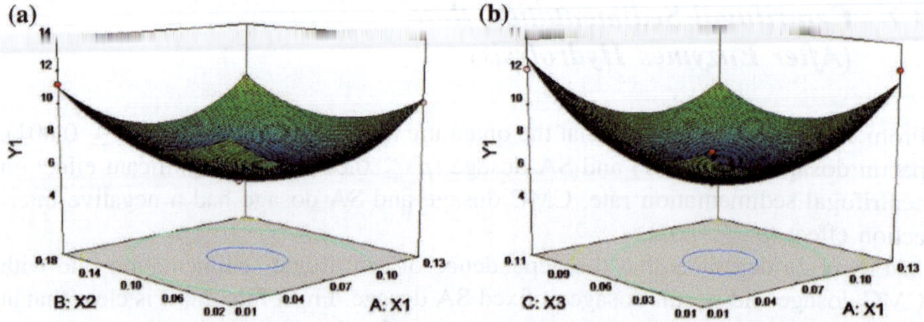

Fig. 2 Response surface diagram for centrifugal sedimentation rate (after enzymes hydrolysis) of ginkgo juice as a function of **a** CMC dosage and pectin dosage (the SA dosage was kept constant at the central point which was 0.06 g/100 mL ginkgo juice) and **b** CMC dosage and SA dosage (the pectin dosage was kept constant at the central point which was 0.10 g/100 mL ginkgo juice). X_1 was CMC dosage (g/100 mL ginkgo juice). X_2 was pectin dosage (g/100 mL ginkgo juice). X_3 was SA dosage (g/100 mL ginkgo juice)

3.3 *Optimization (After Enzymes Hydrolysis)*

The optimum processing conditions to centrifugal sedimentation rate were investigated.

Figure 3 shows the superimposed contour plot for optimization of centrifugal sedimentation rate (after enzymes hydrolysis) keeping the SA dosage constant at the central point and keeping pectin dosage constant at the central point. The zone of optimization, as shown in the superimposed contour plot, depicts CMC dosage to be in the range of 0.05–0.09 g/100 mL ginkgo juice, pectin dosage in the range of 0.06–0.12 g/100 mL ginkgo juice and SA dosage between 0.03 and 0.08 g/100 mL ginkgo juice.

3.4 *Centrifugal Sedimentation Rate (No Enzymes Hydrolysis)*

It is clear from Table 3 that the linear term of SA dosage ($p \leq 0.05$) had a negative effect on centrifugal sedimentation rate. The quadratic terms of CMC dosage (p 0.001), pectin dosage ($p \leq 0.01$) and SA dosage ($p \leq 0.001$) had a significant effect on centrifugal sedimentation rate. CMC dosage and SA dosage had a negative interaction effect ($p \leq 0.05$).

Figure 4a describes that the dependence of centrifugal sedimentation rate with CMC dosage and pectin dosage at determined SA dosage. From Fig. 4a, it is shown that at constant pectin dosage and SA dosage, centrifugal sedimentation rate

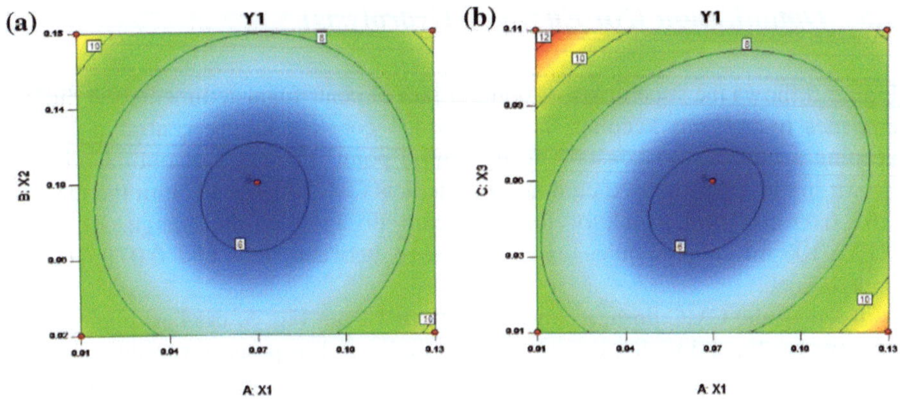

Fig. 3 Superimposed contour plots for optimization of centrifugal sedimentation rate (after enzymes hydrolysis) when SA dosage was kept constant at central point (0.06 g/100 mL ginkgo juice) (**a**) and CMC dosage and SA dosage (the pectin dosage was kept constant at the central point which was 0.10 g/100 mL ginkgo juice) (**b**). X_1 was CMC dosage (g/100 mL ginkgo juice). X_2 was pectin dosage (g/100 mL ginkgo juice). X_3 was SA dosage (g/100 mL ginkgo juice)

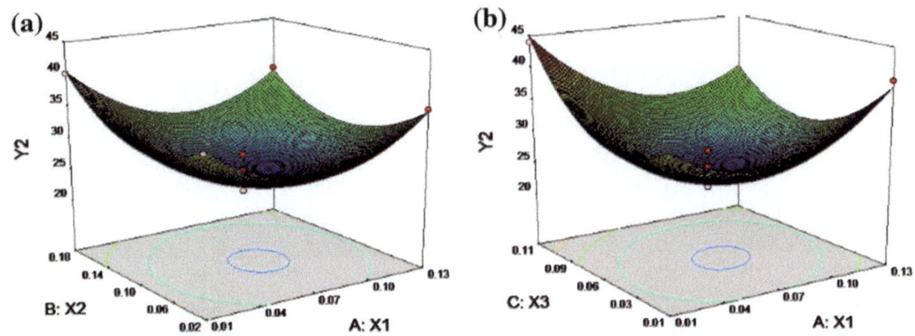

Fig. 4 Response surface diagram for centrifugal sedimentation rate (no enzymes hydrolysis) of ginkgo juice as a function of **a** CMC dosage and pectin dosage (the SA dosage was kept constant at the central point which was 0.06 g/100 mL ginkgo juice) and **b** CMC dosage and SA dosage (the pectin dosage was kept constant at the central point which was 0.10 g/100 mL ginkgo juice). X_1 was CMC dosage (g/100 mL ginkgo juice). X_2 was pectin dosage (g/100 mL ginkgo juice). X_3 was SA dosage (g/100 mL ginkgo juice)

decreased with CMC dosage at the beginning then increased gradually. With the increase of pectin dosage, the centrifugal sedimentation rate decreased gradually first, then increased.

Figure 4b presents the variation of centrifugal sedimentation rate with CMC dosage and SA dosage at constant pectin dosage. It is evident that at a certain CMC dosage and pectin dosage, the centrifugal sedimentation rate of ginkgo juice decreased with SA dosage at the beginning and then increased.

3.5 Optimization (No Enzymes Hydrolysis)

Figure 5 shows the superimposed contour plot of centrifugal sedimentation rate (no enzymes hydrolysis) keeping the SA dosage constant at the central point and keeping pectin dosage constant at the central point. The zone of optimization, as shown in the superimposed contour plot, depicts CMC dosage to be in the range of 0.05–0.09 g/100 mL ginkgo juice, pectin dosage in the range of 0.06–0.12 g/100 mL ginkgo juice and SA dosage between 0.03 and 0.07 g/100 mL ginkgo juice.

Keeping the CMC dosage constant as firmed from Fig. 5, the best combination of response functions, SA dosage was determined. The process variables for best combination of response functions were CMC dosage 0.07 g/100 mL ginkgo juice, pectin dosage 0.09 g/100 mL ginkgo juice and SA dosage 0.05 g/100 mL ginkgo juice.

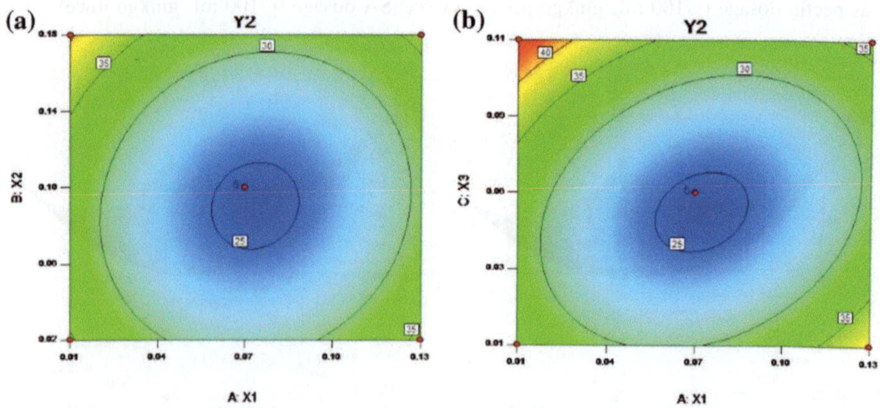

Fig. 5 Superimposed contour plots for optimization of centrifugal sedimentation rate (no enzymes hydrolysis) when SA dosage was kept constant at central point (0.06 g/100 mL ginkgo juice) (**a**) and CMC dosage and SA dosage (the pectin dosage was kept constant at the central point which was 0.10 g/100 mL ginkgo juice) (**b**). X_1 was CMC dosage (g/100 mL ginkgo juice). X_2 was pectin dosage (g/100 mL ginkgo juice). X_3 was SA dosage (g/100 mL ginkgo juice)

4 Conclusion

Using RSM, the optimum condition was obtained. These were CMC dosage 0.07 g/100 mL ginkgo juice, pectin dosage 0.09 g/100 mL ginkgo juice and SA dosage 0.05 g/100 mL ginkgo juice (the optimum condition was the same between "after enzymes hydrolysis" and "no enzymes hydrolysis" processing for ginkgo juice).

Enzymes complete the hydrolysis of starch into low-molecular substances relative to starch. After that, put stabilizers into ginkgo juice, these low-molecular substances in juice are more stable. By this experiment, it is known that the lowest centrifugal sedimentation rate is 5.70% for "after enzymes hydrolysis" and 24.44% for "no enzymes hydrolysis" processing. From these two figures, it is clear that if ginkgo juice is enzymes hydrolyzed first and then put into stabilizers, we can get much more stable ginkgo juice. For this, you can store ginkgo juice for longer time with little sedimentation.

Acknowledgements This work was supported by the Excellent Middle Aged and Young Scientist Award Foundation of Shandong Province (No. BS2011SW029) and a Project of Shandong Province Higher Educational Science and Technology Program (No. J13LE01) and Science and technology development project of Shandong Province (No. 2014GSF121039).

References

1. Fu XH, Li F, Xu CQ, Wei X (1997) Studies on growth of fruits and components of endosperm in seeds of Ginkgo *biloba* cv. Jiangsu DafoShou. GuangXi Plant 17(3):263–269
2. Cameron R (1997) Citrus tissue extracts affect juice cloudy stability. J Food Sci 62:242–245
3. Rubico SM, Resurreccion AVA, Frank JF, Beuchat L (1987) Suspension stability, texture, and color of high temperature treated peanut beverage. J Food Sci 52(6):1676–1679
4. Abastasakis M, Lindamood JB, Chism GW, Hansen PMT (1987) Enzymatic hydrolysis of carrot for extraction of a cloud-stable juice. Food Hydrocolloids 1:247–261
5. Pilnik W, Voragen GJ (1993) Pectic enzyme in food and vegetable juice manufacture. In: Enzymes in food processing. Academic Press, New York, pp 367–371
6. Reiter M, Stuparic M, Neidhart S, Carle R (2003) The role of process technology in carrot juice cloudy stability. Lebensmittle-Wissenschaft Und Technologie 36:165–172
7. Zhang H, Wang Z, Xu S-Y (2007) Optimization of processing parameters for cloudy ginkgo (Ginkgo biloba Linn.) juice. J Food Eng 80:1226–1232

Characterization of Recombinant Dioxygenase from *Bacillus thuringiensis*

Ning Xue, Zhixiang Li, Lei Zhao, Jie Ma, Qingyang Xu,
Chenglin Zhang and Ning Chen

1 Introduction

4-hydroxyisoleucine (4-HIL) possesses glucose-dependent insulinotropic activity. Up till now, the main method for 4-HIL production was extracting from fenugreek seeds (*Trigonella foenum-graecum*) [1–4]. However, its 4-HIL yield is rather low (only about 150 mg of 4-HIL can be extracted from 1 kg of fenugreek seeds) [5], moreover, 4-HIL has several stereo configurations and only (2S,3R,4S)-4-HIL showed biological activity [6]. Besides seed extraction method, chemical synthesis was also developed [7–10]. But the two methods was considered to be of low-efficiency, high-cost and heavy pollution [11].

L-isoleucine dioxygenase (IDO) from *Bacillus thuringiensis* 2e2 was found to specifically convert L-isoleucine to (2S,3R,4S)-4-HIL. Further study showed that IDO was a member of α-KG-dependent hydroxylase family and need Fe^{2+}, and α-KG as cofactors for the hydroxylating reaction [12, 13]. So (2S,3R,4S)-4-HIL can be produce by IDO via biotransformation.

In previous study, we cloned an *ido* from isolated *B. thuringiensis* TCCC 11826, of which the affinity to L-isoleucine and catalytic rate was higher that IDO from *B. thuringiensis* 2e2 [14]. Gaining IDO with excellent enzymatic characteristic is essential for (2S,3R,4S)-4-HIL synthesis.

N. Xue · Z. Li · L. Zhao · J. Ma · Q. Xu · C. Zhang (✉) · N. Chen (✉)
College of Biotechnology, Tianjin University of Science and Technology,
Tianjin 300457, China
e-mail: zcl@tust.edu.cn

N. Chen
e-mail: ningch@tust.edu.cn

Q. Xu · C. Zhang · N. Chen
National and Local United Engineering Lab of Metabolic
Control Fermentation Technology, Tianjin 300457, China

H. Liu et al. (eds.), *Advances in Applied Biotechnology*, Lecture Notes
in Electrical Engineering 444, https://doi.org/10.1007/978-981-10-4801-2_18

In this study, IDO encoding gene *tido* from *B. thuringiensis* TUST-1 was cloned and expressed in *Escherichia coli* BL21(DE3). Then enzymatic characteristics of the recombined IDO including K_m and V_{max}, optimum temperature and pH as well as thermostability was studied.

2 Materials and Methods

2.1 Strains and Plasmids

E. coli BL21(DE3) and *Bacillus thuringiensis* TUST-1 were stored in our lab. Plasmid pET-His (laboratory stock) were used for gene cloning and expression, respectively.

2.2 Media and Culture Condition

E. coli was grown at 37 °C in Luria-Bertani medium [LB, 1% (w/v) tryptone, 0.5% (w/v) yeast extract and 1% (w/v) NaCl]. Kanamycin (50 μg/mL) was added for strain selection. *B. thuringiensis* TUST-1 was grown at 32 °C in the medium containing 0.5% (w/v) peptone and 0.3% (w/v) beef extract.

2.3 Amplification of L-isoleucine Dioxygenase Encoding Gene tido from Bacillus thuringiensis TUST-1

Genome DNA of *B. thuringiensis* TUST-1 was extracted according to the manufacturer instructions of the kit [TIANGEN BIOTECH(Beijing) Co., Ltd.]. *Tido* was PCR amplified using the primers Bt-1(5′-GGATCCATGAAAATGAGTGGCTTTAGCATAG-3’) and Bt-2(5′-GAATTCTTATTTGTCTCCTTATAAGAAAATGTTACTAA-3′) and *B. thuringiensis* TUST-1 genome DNA as template. The PCR condition was used as follows, 94 °C 5 min 1 cycle, 94 °C 30 s, 56 °C 30 s, 72 °C 1 min 30 cycles, 72 °C 10 min 1 cycle, the system volume was 100 μL. The PCR product was detected by agarose gel electrophoresis.

2.4 Expression of tido Gene

PCR production of *tido* was extracted with gel extraction kit [TAKARA Biotech (Dalian) Co., Ltd.]. The fragment was digested by *Bam*H I and *Eco*R I and was cloned into the corresponding restriction site of pET-His, generating in pET-tido. Then the recombinant plasmids were transformed to *E. coli* BL21(DE3), resulting EB-tido.

IPTG (0.1 mmol/L, final concentration) was added when EB-tido was grown in LB medium[with kanamycin (50 μg/mL)] to the midexponential stage. The cells continued to be cultured for another 4 h. The cells were harvested by centrifuging and broken with sonication. Cell lysates were separated into soluble and insoluble fractions by centrifugation at 10,000×*g* for 30 min. The expression of recombinat TIDO was confirmed by sodium dodecyl sulfate polyacrylamide gel electrophoresis (SDS-PAGE). Protein bands were visualized in gels by staining with coomassie brilliant blue R 250.

The 6×His tagged TIDO was purified using a Ni^{2+}-NTA affinity column. And the enzyme concentration was determined by the BCA assay kit (Bio-Rad, USA).

2.5 Assay for Recombinant TIDO Activity

For catalytic characterization of TIDO, the reaction mixture was composed of 10 mmol/L α-ketoglutarate and L-isoleucine, 5 mmol/L $FeSO_4$, 10 mmol/L ascorbic acid and 100 mmol/L Tris-HCl (pH 7.0), and 0.5 mg/mL purified recombinant IDO, and the reaction was performed at 30 °C for 30 min. Enzymatic activity was determined as the production of 4-HIL measured by HPLC. In order to determine K_m values, L-isoleucine was used at concentrations of 0.02–5 mmol/L.

In order to examine the pH dependency of the reaction, citric acid sodium-citrate buffer (pH 3.0–5.0), phosphate buffered saline (pH 6.0–8.0), tris-HCl (pH 9.0) and sodium bicarbonate-sodium carbonate buffer (pH 10.0–11.0) were used. Reactions were performed at 10–70 °C to examine the temperature dependency of the reaction.

2.6 Analytical Procedure of 4-HIL Concentration by High-Performance Liquid Chromatography

4-HIL was analyzed by precolumn derivatization with 2, 4-fluoro-dinitrobenzene and detected by high-performance liquid chromatography (HPLC) with an Agilent C18 column (150 mm × 4.6 mm, 3.5 μm). Elution was performed using a gradient of 50% acetonitrile (v/v)/50 mmol/L (CH_3COONa), and fed at a constant flow rate of 1.0 mL/min. UV absorption was measured at 360 nm and the column temperature was maintained at 33 °C.

3 Results and Discussions

3.1 Cloning and Sequence Analysis of Tido

Tido was successfully amplified from *B. thuringiensis* TUST-1. The sequence of the gene and its encoding products was analyzed, which showed that the gene was 723 bp in length and encoded a 240-amino acid protein. The gene exhibited 99.31% (718/723) and 98.33% (236/240, L28F, V38L, L157W, A224F) similarity with nucleotides and amino acids of *ido* that we cloned before [14]. The different amino acid sequence might endow the enzyme distinctive characteristics, compared to the *ido* from *B. thuringiensis* TCCC 11826.

3.2 Expression of Tido in E. Coli BL21(DE3)

Transmembrane helices were predicted by the program of TMHMM Server v.2.0 (http://www.cbs.dtu.dk/services/TMHMM-2.0), which showed that no transmembrane helices existed in the TIDO. So the total gene was cloned into expression vector pET-His after digested with *Bam*H I and *Eco*R I.

Expression of *tido* was analyzed by SDS-PAGE. As Fig. 1 showed, an additional band was detected at approximately 29 kDa, corresponding to the expected size of the 240-amino acid protein in lysate of EL-ido, indicating that *tido* was successfully expressed. Notably, the band also appeared in soluble fraction of EB-ido, indicating that the recombinant TIDO could exist as soluble form.

Cells induced by IPTG were collected and cell lysate was applied to a Ni^{2+}-NTA affinity column. Proteins binding to the column were washed with imidazole solution of different concentrations. As Fig. 2 showed, increased concentrations of imidazole was helpful to elution of recombinant TIDO. The optimum imidazole concentration was 60 mmol/L and concentration of recombinant TIDO was 3.6 mg/L.

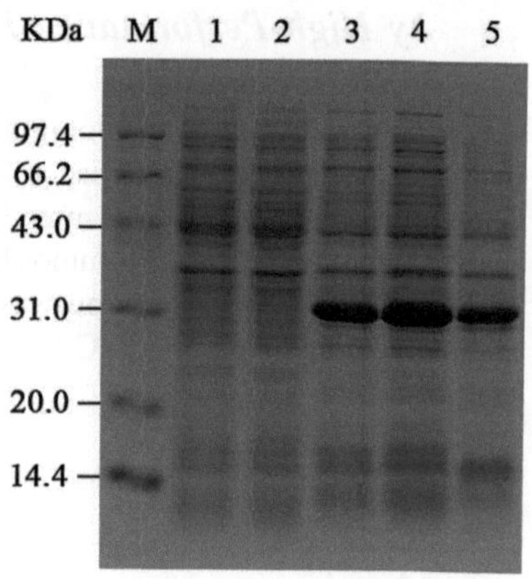

Fig. 1 SDS-PAGE analysis of recombinant TIDO expressed in *E. coli* BL21 (DE3). *M* Marker; *1* total cell lysate of *E. coli* BL21DE harboring pET-His; *2* total cell lysate of EB-ido without IPTG induction; *3* total cell lysate of EB-ido induced by IPTG; *4* soluble fraction of EB-ido induced by IPTG; *5* insoluble fraction of EB-ido induced by IPTG

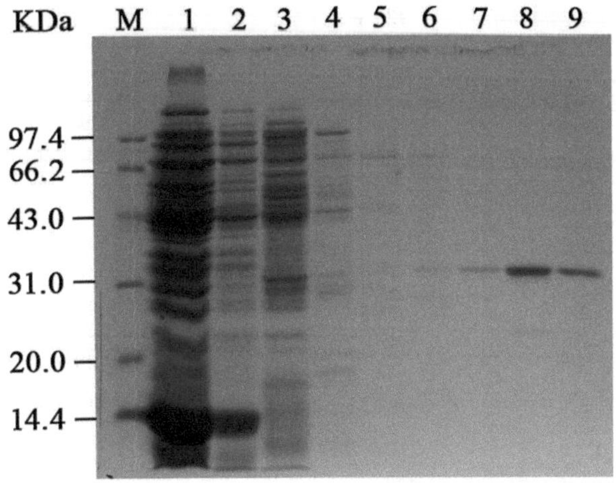

Fig. 2 SDS-PAGE analysis recombinant TIDO purification. *M* marker; *1* flowthrough effusing from Ni^{2+}-NTA affinity column; *2* flowthrough of binding proteins washed with NTA-0 buffer (20 mmol/L Tris-HCl pH7.9, 0.5 mol/L NaCl, 10% Glycerol); *3* soluble fraction of EB-ido induced by IPTG; *4–9* flowthrough of bing proteins washed with 5, 10, 20, 40, 60 and 100 mmol/L imidazole solution

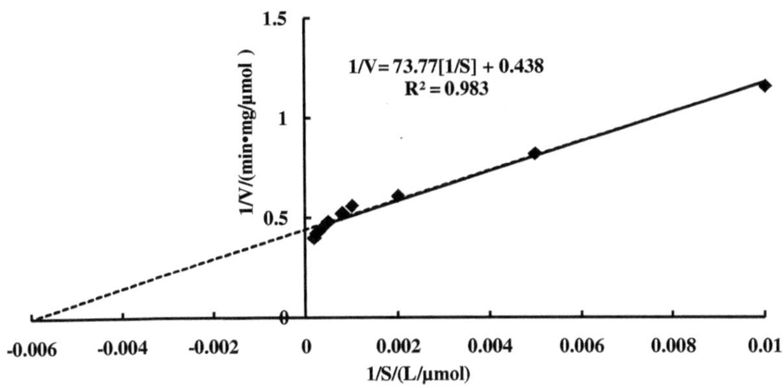

Fig. 3 Hyperbolic kinetics of recombinant TIDO

3.3 *Enzymatic Characterization of TIDO*

Recombinant TIDO was purified and kinetics of the enzyme activity was examined. As shown in Fig. 3, the Michaelis-Menten equation for the enzyme was 1/V = 73.770 [1/S] + 0.438, in which the K_m and V_{max} value for L-isoleucine was 0.168 mmol/L and 2.283 μmol/min/mg, respectively. The result indicated that TIDO exhibited higher affinity to L-isoleucine and catalytic rate than IDO from *B. thuringiensis* TCCC 11826 (0.18 mmol/L and 2.10 μmol/min/mg, respectively) [14].

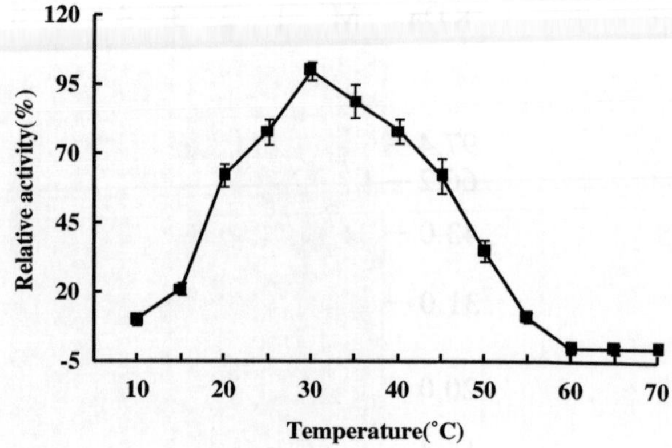

Fig. 4 Effect of temperature on recombinant enzyme activity

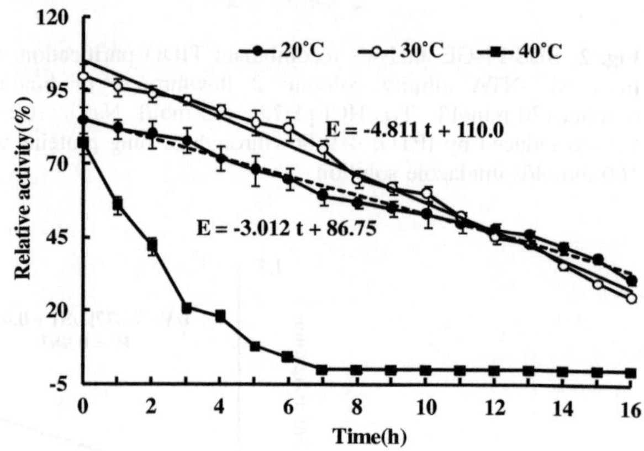

Fig. 5 Thermal stability of recombinant IDO

The effect of temperature (10–70 °C) on recombinant TIDO activity was examined. As shown in Fig. 4, the enzyme activity became higher with the increasing temperature from 10 to 30 °C but lower from 40 to 70 °C. So it could be concluded that the optimal temperature for recombinant TIDO was 30 °C, which was lower than that of IDO from *B. thuringiensis* TCCC 11826 (35 °C) [14].

The thermostability of recombinant TIDO was detected under 20, 30 and 40 °C. As shown in Fig. 5, the enzyme activity decreased with the extension of reaction time, while, it should be notable that the recombinant TIDO at 30 °C exhibited highest activity before 11 h but at 20 °C after 11 h and it almost has non-activity after 4 h at 40 °C. However, it showed that recombinant TIDO at 20 °C was more thermostable than that at 30 °C since the slope of the curve at 20 °C is lower than that at 30 °C (−3.012 vs. −4.811) So it could be explained as that TIDO optimal temperature of TIDO was 30 °C and thus exhibited high activity but the activity decreased after 11 h due to the thermal lability.

The effect of pH (3.0–11.0) on recombinant TIDO activity was determined. As shown in Fig. 6, the enzyme activity rose with the increased pH till pH 8.0 and then

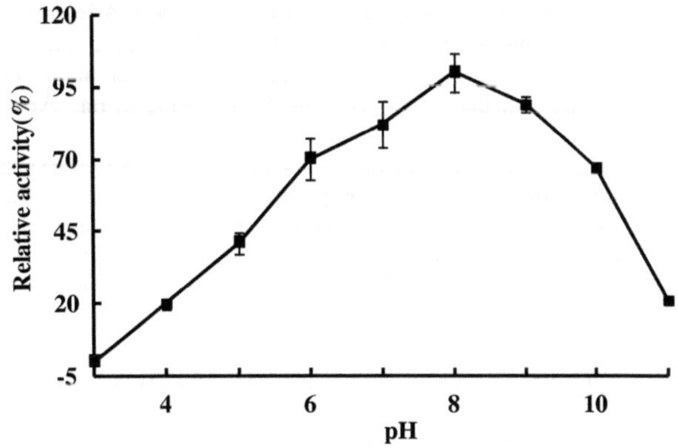

Fig. 6 Effect of pH on recombinant enzyme activity

decreased. So the optimal pH for recombinant TIDO was 8.0, which was higher than that of IDO from *B. thuringiensis* TCCC 11826 (7.0 °C) [14].

4 Conclusion

The gene *tido* from *B. thuringiensis* TUST-1 was successfully cloned and expressed. Homology alignment of *tido* and *ido* from *B. thuringiensis* TCCC 11826 showed that *tido* exhibited higher similarity to the latter gene. Enzyme kinetic analysis result showed that K_m and V_{max} value of for *tido* L-isoleucine was 0.168 mmol/L and 2.283 μmol/min/mg, respectively, indicating that TIDO exhibited higher affinity to L-isoleucine and catalytic rate than IDO from *B. thuringiensis* TCCC 11826. The optimal temperature and pH for TIDO was 30 °C and 8.0, respectively. This work will lay theoretical foundation and practical basis on the microbial manufacture technology of 4-HIL and other amino acid derivatives.

Acknowledgements This work was supported by National High Technology Research and Development Program 2013AA102106), by the National Natural Science Foundation of China (31300069); Tianjin Municipal Science and Technology Commission (grant No. 15JCTPJC62800); Tianjin Undergraduate Training Program for Innovation and Entrepreneurship (201510057063). China Postdoctoral Science Foundation Funded Project (2017M61170) and Innovation training program for college students in Tianjin(201710057039) need to be added as the last two supported project.

References

1. Fowden L, Pratt HM, Smith A (1973) 4-Hydroxyisoleucine from seed of *Trigonella foenum-graecum*. Phytochemistry 12(7):1707–1711
2. Al-Habori M, Raman A (1998) Antidiabetic and hypocholesterolaemic effects of fenugreek. Phytother Res 12(4):233–242

3. Broca C, Manteghetti M, Gross R et al (2000) 4-Hydroxyisoleucine: effects of synthetic and natural analogues on insulin secretion. Eur J Pharmacol 390(3):339–345
4. Broca C, Breil V, Cruciani-Guglielmacci C et al (2004) Insulinotropic agent ID-1101 (4-hydroxyisoleucine) activates insulin signaling in rat. Am J Physiol Endocrinol Metab 287 (3):463–471
5. Narender T, Puri A, Shweta KT et al (2006) 4-hydroxyisoleucine an unusual amino acid as antidyslipidemic and antihyperglycemic agent. Bioorg Med Chem Lett 16:293–296
6. Alcock N, Crout D, Gregorio M et al (1989) Stereochemistry of the 4-hydroxyisoleucine from *Trigonella foenum-graecum*. Phytochemistry 28:1835–1841
7. Wang Q, QuazzaniJ Sasaki NA et al (2002) A practical synthesis of (2S, 3R, 4S)-4-hydroxyisoleucine, a potent insulinotropic alpha-amino acid from fenugreek. Eur J Org Chem 5:834–839
8. Aouadi K, Jeanneau E, Msaddek M et al (2012) 1,3-Dipolar cycloaddition of a chiral nitrone to (*E*)-1,4-dichloro-2-butene: a new efficient synthesis of (2S,3S,4R)-4-hydroxyisoleucine. Tetrahedron Lett 53(23):2817–2821
9. Smirnov SV, Samsonova NN, Novikova AE et al (2007) A novel strategy for enzymatic synthesis of 4-hydroxyisoleucine: identification of an enzyme possessing HMKP (4-hydroxy-3-methyl-2-keto-pentanoate) aldolase activity. FEMS Microbiol Lett 273(1):70–77
10. Ogawa J, Yamanaka H, Mano J et al (2007) Synthesis of 4-hydroxyisoleucine by the aldolase-transaminase coupling reaction and basic characterization of the aldolase from *Arthrobacter simplex* AKU 626. Biosci Biotech and Bioch 71(7):1607–1615
11. Smirnov SV, Kodera T, Samsonova NN et al (2010) Metabolic engineering of *Escherichia coli* to produce (2S,3R,4S)-4-hydroxyisoleucine. Appl Microbiol Biotechnol 88(3):719–726
12. Ogawa J, Kodera T, Smirnov SV et al (2011) A novel L-isoleucine metabolism in *Bacillus thuringiensis* generating (2S,3R,4S)-4-hydroxyisoleucine, a potential insulinotropic and anti-obesity amino acid. Appl Microbiol Biotechnol 89(6):1929–1938
13. Hibi M, Kawashima T, Kodera T et al (2011) Characterization of *Bacillus thuringiensis* L-isoleucine dioxygenase for production of useful amino acids. Appl Microbiol Biotechnol 77 (19):6926–6930
14. Zhang CL, Liu Y, Xue N et al (2014) Characterization of recombinant L-isoleucine-4-hydroxylase from *Bacillus thuringiensis* and its application in 4-hydroxyisoleucine biosynthesis. Acta Microbiol Sin 54(8):889–896

Optimization of Condition for Recombinant L-Isoleucine Dioxygenase Expression in *Escherich coli* BL21(DE3)

Jie Ma, Zhixiang Li, Lei Zhao, Qingyang Xu, Chenlin Zhang, Xixian Xie and Chen Ning

1 Introduction

The enzymatic characteristics of IDO has been determined with an industrially important biocatalyst, α-KG-dependent hydroxylase L-isoleucine dioxygenase (IDO), which was found to be the enzyme responsible for the generation of (2S,3R,4S)-4-hydroxyisoleucine in *Bacillus thuringiensis* 2e2. Depending on the amino acid used as the substrate, IDO could catalyze three different types of oxidation reactions: hydroxylation, dehydrogenation, and sulfoxidation [1]. Importantly, IDO with the ability of converting L-isoleucine to 4-Hydroxyisoleucine [2, 3]. Moreover, IDO need α-ketoglutarate, Fe^{2+}, and ascorbic acid as cofactors for the hydroxylating reaction.

In particular, 4-Hydroxyisoleucine is a natural nonproteinogenic amino acid that first isolated from the seeds of *Trigonella foenum-graecum* [4] with absolute stereo configuration of (2S,3R,4S)-4-Hydroxyisoleucine [5], which demonstrated insulinotropic biological activity [6, 7] in rat models of T2DM [8–10]. Furthermore, 4-Hydroxyisoleucine also possesses effective antidiabetic activity in a model of type-1 diabetes mellitus (T1DM) without insulin [11].

To gain enough IDO is the precondition for studying the characteristic of the enzyme or producing 4-Hydroxyisoleucine with the enzyme. So in this study, conditions for recombinant IDO expression, such as IPTG concentration, induction

J. Ma · Z. Li · L. Zhao · Q. Xu · C. Zhang (✉) · X. Xie · C. Ning (✉)
College of Biotechnology, Tianjin University of Science and Technology, Tianjin 300457, China
e-mail: zcl@tust.edu.cn

C. Ning
e-mail: ningch@tust.edu.cn

Q. Xu · C. Zhang · X. Xie · C. Ning
National and Local United Engineering Lab of Metabolic Control Fermentation Technology, Tianjin 300457, China

H. Liu et al. (eds.), *Advances in Applied Biotechnology*, Lecture Notes in Electrical Engineering 444, https://doi.org/10.1007/978-981-10-4801-2_19

time, expression time and induction temperature was investigated and optimized. These results may be useful for the industrial production of optically active amino acids, as well as 4-HIL

2 Materials and Methods

2.1 Materials and Strain Construction

The 6×His-tagged IDO-expressing strain *E. coli* BL21(DE3) carrying pEThis-*ido* was constructed. All restriction enzymes, Primer STAR HS DNA polymerase and T4 DNA ligase were purchased from Takara Bio Inc. (Dalian China). All oligonucleotides used in this study were synthesized by GENEWIZ Bio Inc. (Suzhou China).

In order to construct plasmid, the 723 bp-sized *ido* fragment was amplified by PCR with the primers HIL-F(TTAGGATCCATGAAAATGAGCGGTTTTA-GCAT) and HIL-R(GCA AAGCTTTTATTTGGTTTCTTTATAGCTAAAGGTC) using the genomic DNA of *B. thuringiensis* TCCC11826 as template. The following PCR protocol was used: 95 °C-5 min; [94 °C-30 s; 55 °C-30 s; 72 °C-60 s] × 25cycles. The PCR product was digested with *Bam*H I and *Hin*d III, ligated into the *Bam*H I and *Hin*d III restriction sites of the pET-his vector to yield pEThis-*ido*, and then transformed into *E. coli* BL21(DE3).

2.2 Preparation for Expression of Recombinant IDO

Expression experiments were prepared by transferring appropriate number of strains from agar slant cultured into 500-mL Erlenmeyer flask containing 30 mL of LB medium (1% tryptone, 1% NaCl, 0.5% yeast, pH 7.0), and was cultured at 37 °C. The culture was inoculated (1%, V/V) into 30 mL of LB medium, at different temperature and induction time points, with different IPTG concentration and induction time. Cells were collected by centrifugation and was broken at 4 °C.

2.3 Detection of Recombinant IDO Expression

Recombinant IDO expression amount was detected by SDS-PAGE through monitoring the coloring degree of the protein bands. Total protein concentration was detected by BCA Protein Assay Kit (Sigma-Aldrich, Germany). Samples with the same concentration of total protein were loaded.

3 Results and Discussion

3.1 Construction of Recombinant Plasmid and Strain

The plasmid pEThis-*ido* which contains the endogenous *ido* of *B. thuringiensis* TCCC11826 was constructed. For the construction of pEThis-*ido*, the 723 bp-sized *ido* fragment was amplified by PCR (Fig. 1), which was same as the expected molecular weight (723 bp)

The PCR product was digested with *Bam*H I and *Hin*d III, and ligated with the *Eco*R I-*Bam*H I-digested DNA fragment of pET-his to construct the plasmid pEThis-*ido*, and then transformed into *E. coli* DH5α. The obtained pEThis-*ido* was digested with *Bam*H I and *Hin*d III, as shown in Fig. 2, the recombinant plasmid pEThis-*ido* was been constructed successfully. Finally, pEThis-*ido* was transformed into *E. coli* BL21(DE3) and screening positive strains (Fig. 3).

In order to detect the recombinant strain *E. coli* BL21/pEThis-*ido* with ability to express protein, cultivation experiments were implemented in 500 mL Erlenmeyer flask containing 30 mL LB medium with 0.1 mmol/L IPTG at 37 °C. Then, cells were collected by centrifugation and was broken, as shown in Fig. 4, protein of lysate supernate and sediment were detected by SDS-PAGE through monitoring the coloring degree of the protein bands. Luckly, IDO (28.92 kDa) could been expresses by recombinant strains in the lysate supernate.

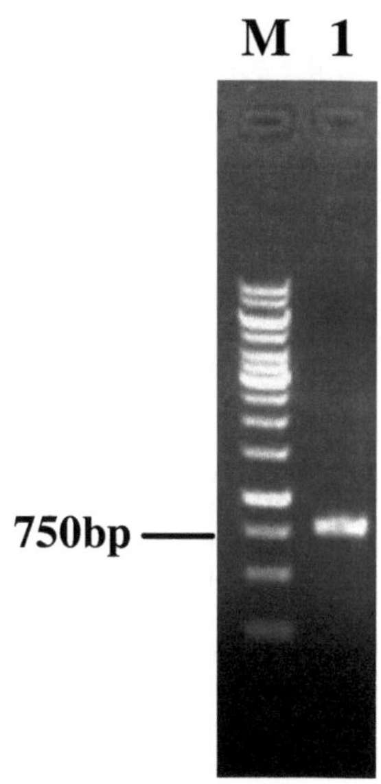

Fig. 1 Map of *ido* PCR product. *M* marker; *lane 1 ido* PCR product

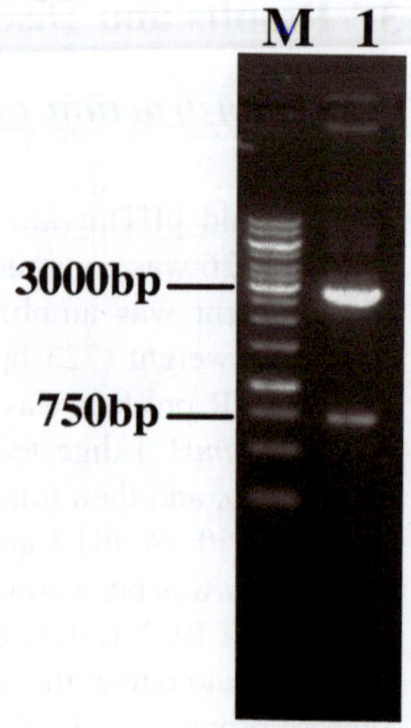

Fig. 2 Identification of recombinant plasmids digested by restricted endonuclease. *M* marker; *lane 1* pEThis-*ido* digested by *BamH* I and *Hind* III

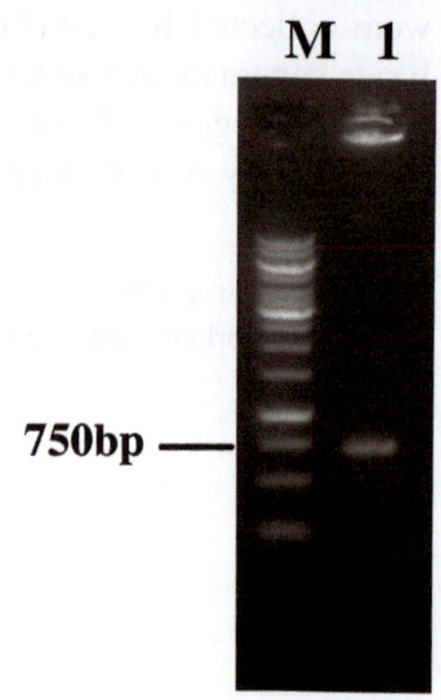

Fig. 3 Identification of recombinant strains. *M* marker; *lane 1 ido* PCR product

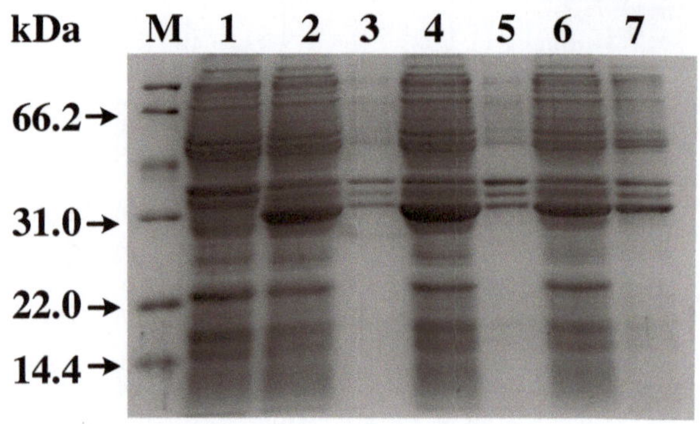

Fig. 4 Expression of *E. coli* BL21(DE3)/pEThis-*ido* protein production. *M* protein standard marker; *lane 1* bacteria induced for 0 h; *lane 2, 4, 6* lysate supernate of *E. coli* BL21(DE3)/pEThis-*ido*; *lane 3, 5, 7* lysate sediment of *E. coli* BL21 (DE3)/pEThis-*ido*

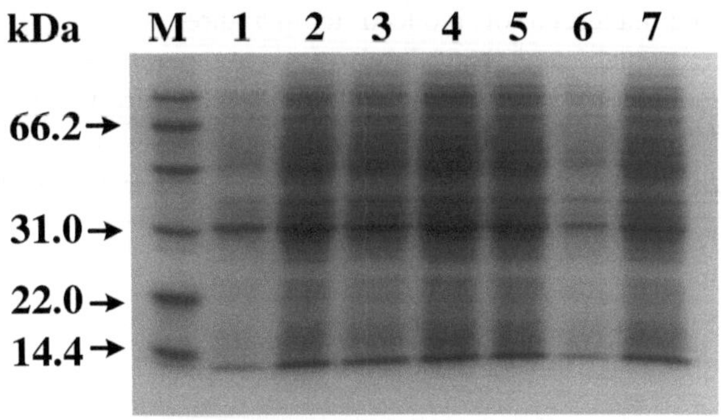

Fig. 5 Expression of *E. coli* BL21(DE3)/pEThis-ido protein production. *M* protein standard marker; *lane 1–7* different concentration of IPTG (0.05, 0.1, 0.15, 0.2, 0.25, 0.3, 0.35 mmol/L)

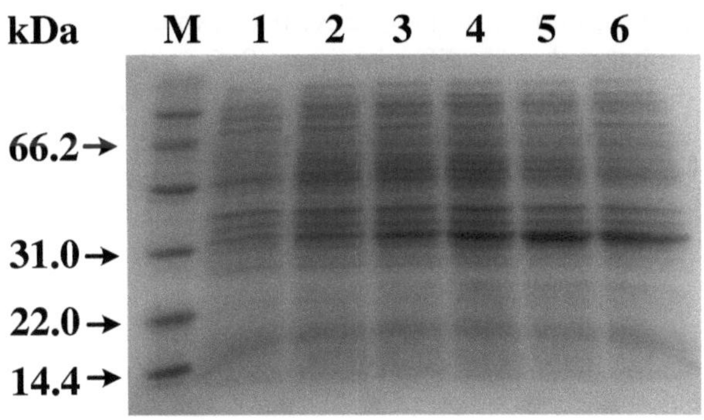

Fig. 6 Expression of *E. coli* BL21(DE3)/pEThis-*ido* protein production. *M* Protein standard marker; lane 1-6: different temperature of induction (20, 25, 28, 32, 37, 42 °C)

3.2 *Influence of IPTG for Protein Expression*

Due to the influence of differernt concentration of IPTG, IDO-expressing strain *E. coli* BL21(DE3)/pEThis-*ido* could express different amount of protein. Then, checking the condition of expression by SDS-PAGE to find out the biggest protein band, which is the most appropriate concentration of IPTG for protein expression. In this study, different concentration of IPTG (0.05, 0.1, 0.15, 0.2, 0.25, 0.3, 0.35 mmol/L) was added into the medium. As shown in Fig. 5, with the increasing concentration of IPTG, protein production did not follow the trend declined rather than increased. As an inducer, too high concentration of IPTG can affect the growth of bacteria. Therefore, 0.15 mmol/L IPTG is the most appropriate concentration for protein expression, which was used in subsequent expression experiments.

3.3 *Expression Temperature for IDO*

Temperature is an important factor for the growth of bacteria and protein expression. In this study, 0.15 mmol/L IPTG was added into the medium, then expressing in different temperature (20, 25, 28, 32, 37, 42 °C). As shown in Fig. 6, along with the increased of induced temperature, first of all the protein bands increased extremely and then slowly. Low temperature could inhibit the activity of intracellular enzyme activity and influence the growth of bacteria, which is an important reason for reducing protein expression. However, high temperature could inactivate intracellular coenzyme. Therefore, 37 °C is the most appropriate temperature for protein expression, which was used in subsequent expression experiments.

3.4 *Effect of Induction Time*

Concentration of bacterial cells is also an important factor that affects expression level of protein. In this study, IDO-expressing strain was cultured at 37 °C, on an orbital shaker for different times (0–6 h). Then, 0.15 mmol/L IPTG was added into the medium and expressing the same time. As shown in Fig. 7, Recombinant IDO was hardly detected when the culture time was short. Therefore, the strain should been cultivated at least 4 h, suggesting that 4 h is the most appropriate induction time for protein expression, which was used in subsequent expression experiments.

3.5 *Influence of Expression Time*

As a limiting factor for protein expression, expression time plays an important role in protein production. In order to increase the protein production of IDO-expressing strain *E. coli* BL21(DE3)/pEThis-*ido*, 0.15 mmol/LIPTG was added into the medium at 4 h and expression times were adjusted different gradient(1–7 h). As shown in Fig. 8, at the early stage, protein production increased with expression time. However, after 6 h it did not increase with the enhancive expression time. The results indicated that 6 h is the most appropriate expression time for protein expression.

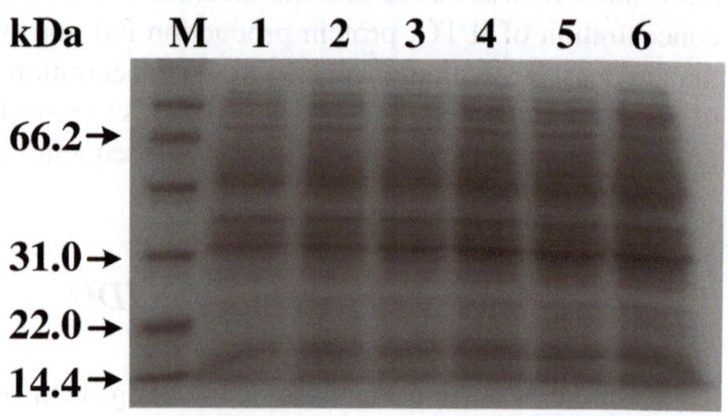

Fig. 7 Expression of *E. coli* BL21(DE3)/pEThis-*ido* protein production. *M* Protein standard marker; *lane 1–6* different time of induction (0–6 h)

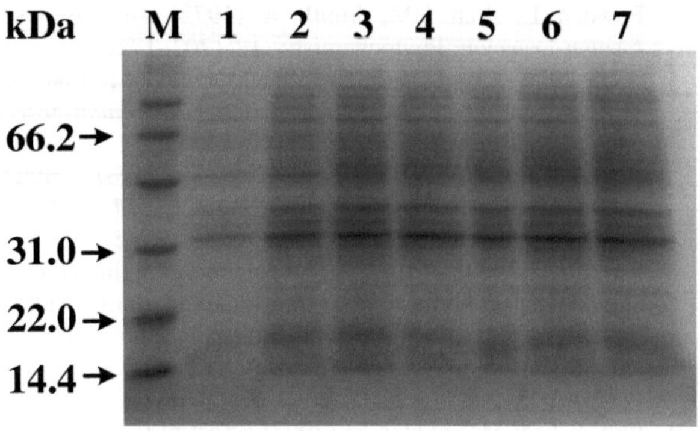

Fig. 8 Expression of *E. coli* BL21(DE3)/pEThis-*ido* protein production. *M* protein standard marker; *lane 1–7* different time of expression (1–7 h)

4 Conclusions

As an inducer, suitable concentration of IPTG is an important factor for cell growth and protein expression. We found that high concentration of IPTG could affect the growth of bacterial cells and 0.15 mmol/L IPTG is the most appropriate concentration for protein expression. Moreover, low temperature could inhibit growth of cells and thus decrease protein expression level, however, high temperature could inactivate intracellular coenzyme. Therefore, 37 °C is selected for the most appropriate temperature for protein expression. Furthermore cells cultivated for 4 h and induced for at least 6 h exhibited excellent protein expression level.

Acknowledgements This work was supported by National High Technology Research and Development Program 2013AA102106), by the National Natural Science Foundation of China (31300069); Tianjin Municipal Science and Technology Commission (grant No. 15JCTPJC62800). China Postdoctoral Science Foundation Funded Project (2017M61170) and Innovation training program for college students in Tianjin (201710057039) need to be added as the last two supported project.

References

1. Hibi M, Kawashima T, Kodera T, Smirnov SV, Sokolov PM, Sugiyama M, Shimizu S, Yokozeki K, Ogawa J (2011) Characterization of *Bacillus thuringiensis* L-isoleucine dioxygenase for production of useful amino acids. Appl Environ Microbiol 77:6926–6930
2. Kodera T, Smirnov SV, Samsonova NN, Kozlov YI, Koyama R, Hibi M, Ogawa J, Yokozeki K, Shimizu S (2009) A novel L-isoleucine hydroxylating enzyme, L-isoleucine dioxygenases from *Bacillus thuringiensis*, produces (2S,3R,4S)-4-hydroxyisoleucine. Biochem Biophys Res Commun 390:506–510
3. Ogawa J, Kodera T, Smirnov SV, Hibi M, Samsonova NN, Koyama R, Yamanaka H, Mano J, Kawashima T, YokozekiK Shimizu S (2011) A novel L-isoleucine metabolism in *Bacillus thuringiensis* generating (2S,3R,4S)-4-hydroxyisoleucine, a potential insulinotropic and anti-obesity amino acid. Appl Microbiol Biotechnol 89:1929–1938

4. Fowden L, Pratt HM, Smith A (1973) 4-Hydroxyisoleucine from seed of *Trigonella foenum graecum*. Phytochemistry 12:1707–1711
5. Alcock NW, Crout DHG, Gregorio MVM, Lee E, Pike G, Samuel CJ (1989) Stereochemistry of the 4-hydroxyisoleucine from *Trigonella foenum-graecum*. Phytochemistry 28:1835–1841
6. Sauvaire Y, Petit P, Broca C, Manteghetti M, Baissac Y, Fernandez-Alvarez J, Gross R, Roye M, Leconte A, Gomis R, Ribes G (1998) 4-Hydroxyisoleucine: a novel amino acid potentiator of insulin secretion. Diabetes 47:206–210
7. Broca C, Gross R, Petit P, Sauvaire Y, Manteghetti M, Turnier M, Masiello P, Gomis R, Ribes G (1999) 4-Hydroxyisoleucine: experimental evidence of its insulinotropic and antidiabetic properties. Am J Physiol Endocrinol Metab 277:617–623
8. Narender T, Puri A, Shweta KT, Saxena R, Bhatia G, Chandra R (2006) 4-Hydroxyisoleucine an unusual amino acid as antidyslipidemic and antihyperglycemic agent. Bioorg Med Chem Lett 16:293–296
9. Eidi A, Eidi M, Sokhteh M (2007) Effect of fenugreek (*Trigonella foenum-graecum* L) seeds on serum parameters in normal andstreptozotocin-induced diabetic rats. Nutr Res 27:728–733
10. Singh AB, Tamarkar AK, Narender T, Srivastava AK (2010) Antihyperglycaemic effect of an unusual amino acid (4-Hydroxyisoleucine) in C57BL/KsJ-db/db mice. Nat Prod Res 24: 258–265
11. Haeri MR, Limaki HK, White CJ, White KN (2012) Non-insulin dependent anti-diabet activity of (2S,3R,4S) 4-Hydroxyisoleucine of fenugreek (*Trigonella foenum graecum*) in streptozotocin-induced type I diabetic rats. Phytomedicine 19:571–574

Protein Sparing Effect of Carbohydrate on Growth Performance, Digestion Ability of Common Carp (*Cyprinus carpio*) at Different Feeding Frequencies

Ze Fan, Xiuting Qiao, Jinhui Sun, Pei Cui, Zhenzhen Fang, Dongqing Bai and Zhenyan Cheng

Over the last few decades, finding best practices on how to decrease fish meal consumption and improve the protein utilization by fish has became an environmental and economical goal for the sustainable development of aquaculture feed industry [1, 2]. Accordingly, many studies have been designed to evaluate and promote the protein-sparing feasibility of lipid and carbohydrates in fish diets [3, 4]. However, it is worth noting that fish, in their complex living environment, create a "glucose intolerant" predisposition, resulting in poorer ability to utilize carbohydrate as energy sources than protein and lipids [5, 6]. Consequently, finding the most efficient way to improve access to carbohydrate has became a vital subject of fish nutrition research.

Of the cultivation process, besides focusing on feed quality, feeding strategy is one of the key factors influencing growth and feed conversion, such as feeding rhythm, feeding regime, feeding rate, feeding frequency, feeding methods and so on [7, 8]. Among those, feeding frequency has close correlation with the ability to use carbohydrates [8, 9]. Previous study has indicated that weight growth rate (WGR), feed efficiency (FE), and protein efficiency ratio (PER) of tilapia (*Oreochromis niloticus* × *O.aureus)* were significantly higher in fish fed 44% starch at a feed frequency of 6 meal/d in comparison to those fed at 2 meal/d [10].

Common carp (*Cyprinus carpio*), an omnivorous fish, are native to Asia. The recent study mainly focused on the aspect of its requirement of protein [11, 12]. Although the protein-sparing of dietary carbohydrate [13] or feed frequency [14] has been well demonstrated in common carp, most studies concerned mainly on one or other of two factors, and the interaction effect of two factors on growth and feed

Z. Fan · X. Qiao · J. Sun · P. Cui · Z. Fang · D. Bai · Z. Cheng (✉)
Tianjin Key Lab of Aqua-Ecology and Aquaculture, College of Fisheries Tianjin Agricultural University, Tianjin 300384, P.R. China
e-mail: chengzhenyan2005@126.com

D. Bai
e-mail: dankey1534417945@163.com

H. Liu et al. (eds.), *Advances in Applied Biotechnology*, Lecture Notes in Electrical Engineering 444, https://doi.org/10.1007/978-981-10-4801-2_20

utilization of common carp have received insufficient attention currently. Based on this, the present study was conducted to investigate the interaction effect of dietary carbohydrate or feed frequency on growth performance and digestion ability of common carp to evaluate the impact of feed frequency on the protein-sparing of dietary carbohydrate.

1 Materials and Methods

1.1 Experimental Diet

The formulation and proximate composition of experimental diets are specified in Table 1. The experimental diets were isocaloric (16.20 MJ/kg DM) and isolipidic (6.7% DM) with three levels of carbohydrate (5, 10 and 20%). Fish meal, soybean meal, peanut meal and cotton meal were used as protein source. Soybean oil and wheat starch served as single lipid source and carbohydrate source, respectively. The amount of wheat starch in the diets was increased at the cost of fish meal and micronized cellulose. The three diets were, respectively, recorded as C5P32 (C5 = 5% carbohydrate, P32 = 32% protein), C10P30 (C10 = 10% carbohydrate, P30 = 30% protein), and C20P28 (C20 = 20% carbohydrate, P28 = 28% protein). All ingredients were provided by Tianjin Tianxiang Aquatic Co. Ltd., and were pulverized by 40 mesh sieve, thoroughly mixed, and made into 2 mm pellets using a pellet machine (MUZLMV4, Jiangsu Muyang Group Co., Ltd., Yangzhou, China). The experimental diets were stored at −20 °C until used.

1.2 Fish and Experimental Design

Common carp with an initial weight of (55.37 ± 3.35) g were obtained from Tianjin Huanxin Aquatic Breeding Farm. Fish were acclimated to the experimental conditions and fed a commercial formulated feed (32% protein, supplied by Tianxiang Aquatic Co. Ltd.) for 2 weeks in the 3 m × 3 m × 1.5 m cages before the beginning of feeding trial. 50 common carp per cage randomly selected and assigned into the 18 cages (1 m × 1 m × 1.5 m). Water quality parameters during the experimental period were pH 7.8 ± 0.2, temperature 26.5–30.5 °C, dissolved oxygen 6.0–8.0 mg/L and total ammonia nitrogen <0.2 mg/L. Feeding frequencies were two meals a day at 08:00 and 17:00, and four meals a day at 08:00, 11:00, 14:00 and 17:00. Feeding was conducted manually at a daily ration size 4% of the initial body weight by the same person during the 8-week feeding trial.

Table 1 Formulation and proximate composition of experimental diets (% dry matter)

Ingredients (%)	Diet		
	C5P32	C10P30	C20P28
Fish meal[1]	7	5	3
Soybean meal[1]	22	22	22
Peanut meal[1]	15	15	15
Cotton meal[1]	12	12	12
Rapeseed meal[1]	13	13	13
Microcrystalline cellulose	12	8.8	0.6
DDGS[2]	3	3	3
Soybean oil	5	5.2	5.4
Premix[3]	2	2	2
Wheat starch	5	10	20
CMC-Na	2	2	2
Monocalcium phosphate	2	2	2
Total	100	100	100
Nutrient levels			
Crude protein (% DM)	31.94	30.46	28.38
Crude lipid (% DM)	6.66	6.69	6.67
Energy (MJ/kg DM)	16.21	16.23	16.25

[1]Fish meal: 62.3% crude protein, 12% crude lipid; soybean meal: 43.7% crude protein; peanut meal: 46.2% crude protein; cotton meal: 43.3% crude protein; rapeseed meal: 36.2% crude protein

[2]DDGS: dried distillers grains with solubles, 26% crude protein, 8% crude lipid

[3]Per kg diet: vitamin A 1500 IU, vitamin B_1 4.5 mg, vitamin B_2 5 mg, vitamin B_6 3.75 mg, vitamin B_{12} 5 mg, vitamin D_3 500 IU, vitamin E 25 mg, vitamin K_3 2.5 mg, biotin 0.03 mg, folic acid 1 mg, D-pantothenic acid 5 mg, nicotinic acid 25 mg, inositol 20 mg, VC 30 mg, antioxidant 1 mg, Cu (as copper sulfate) 2.25 mg, Fe (as ferrous sulfate) 22.5 mg, Mn (as manganese sulfate) 0.5 mg, Zn (as zinc sulfate) 12.5 mg, I (as potassium iodide) 0.1 mg, Se (as sodium selenite) 0.005 mg, Co (as cobalt chloride) 0.075 mg

[4]Calculated as protein: 24 kJ/g; fat: 38 kJ/g; and starch: 17 kJ/g

1.3 Sample Collection

At the end of the trial, the common carp were counted and weighted under moderate anesthesia (MS-222 at 100 μg/kg) after a 24-h fasting period. Total length (0.01 cm) of 45 fish per diet (15 fish per cage) were measured randomly, and whole fish (0.01 g) of them were weighed for determination of condition factor (CF). Blood samples were extracted from the caudal vein, and immediately centrifuged at 4500 rpm for 15 min at 4 °C, and the obtained serum was stored at −80 °C until the biochemical criterion determinations were made. After blood collection, the individual intestines and hepatopancreas were quickly excised, and hepatopancreas

(0.01 g) were weighed for the calculation of hepatosomatic index (HSI). After washed thoroughly with chilled physiological saline (0.85% w/v NaCl), tissue samples were stored at −80 °C for further analysis.

1.4 Analytical Method

1.4.1 Diets Nutrient Composition Analysis

Nutrient composition of experimental diets was determined using the following methods: crude protein (N × 6.25) was estimated by the Dumas combustion methods using an Auto Dumas combustion System (Buchi, Flawil, Switzerland); crude lipid was measured by the Soxhlet extraction method with petroleum ether extraction; gross energy was analyzed in an Oxygen Bomb Calorimeter Model 6100 (PARR, Moline, IL, USA).

1.4.2 Tissue Processing

Tissues were quickly thawed, and weighed samples were automatically homogenized applying an electric homogenizer with a metal pestle on ice. Hepatopancreas for the glycogen assays was rapidly thawed, and promptly hydrolyzed in alkaline liquor at the ratio of 1:3 (w:v) with no centrifugation step. For the digestive enzymes and carbohydrate metabolic enzymes, tissue samples were homogenized in 9 volumes of ice-cold physiological saline (0.85% w/v NaCl). Homogenates were centrifuged at 3500 rpm for 15 min at 4 °C (Thermo Fisher Scientific ST 16R centrifuge), and the resulting supernatants were aliquoted and stored at −80 °C until subsequent analysis.

1.4.3 Determination of Intestinal Digestive Enzymes

Protease activity was assayed by Folin-reagent method. One unit of protease activity was defined as 1 μg tyrosine released 1 g tissue in 1 min at 37 °C Amylase activity was measured using iodine solution to expose non-hydrolyzed starch. One unit was defined as the amount of enzyme catalyzing 10 mg starch hydrolyzed in 30 min at 37 °C. Lipase activity was determined based on the measurement of the release of fatty acids due to enzymatic hydrolysis of triglycerides in a stabilized olive oil emulsion. One unit of lipase activity was defined as 1 g of substrate released per min per gram of tissue at 40 °C. Enzyme activities were expressed as U/g tissue.

1.5 Calculation and Statistical Analysis

$$\text{Survival rate} = 100\% \times N_t/N_0; \tag{1.1}$$

$$\text{Weight growth rate (WGR, \%)} = 100\% \times (W_t - W_0)/W_0 \tag{1.2}$$

$$\text{Specific growth rate (SGR, \%/d)} = 100\% \times [\ln(W_t) - \ln(W_0)]/t \tag{1.3}$$

$$\text{Feed conversion ratio (FCR)} = F/(W_t - W_0) \tag{1.4}$$

$$\text{Protein efficiency ratio (PER, \%)} = 100\% \times (W_t - W_0)/(F \times P) \tag{1.5}$$

$$\text{HIS (\%)} = 100\% \times W_g/\ W_t \tag{1.6}$$

$$\text{CF (\%)} = 100\% \times W_t/L_t^3 \tag{1.7}$$

where N_t and N_0 are the final and initial body number, W_t and W0 are the final and initial body weight (g), t-time of rearing (days), L-fish body length (cm), F-total feed intake (g), P-content of dietary protein (g) and W_g-liver weight (g).

Results were subjected to two-way analysis of variance (ANOVA), followed by a Duncan's test to delineate significance among fish groups using SPSS 16.0 software (SPSS Inc., Chicago, IL, USA). For all analyzes, significant levels were set at $P < 0.05$ unless otherwise stated. All data are represented as means ± SD.

2 Results

2.1 Growth Parameters and Feed Utilization

Growth parameters and feed utilization were presented in Table 2. Weight growth rate (WGR), specific growth rate (SGR), and protein efficiency ratio (PER) were generally decreased with increasing dietary carbohydrate levels, but no significant difference was observed ($P > 0.05$), whereas WGR, SGR, and PER of fish fed 4 meal/d were significantly higher than those of fish fed 2 meal/d ($P < 0.05$). Feed conversion ratio (FCR) tended to increase with increasing dietary carbohydrate levels, whereas FCR tended to decrease with increasing feeding frequencies although no significant difference was observed ($P > 0.05$).

2.2 Morphology Index

As can be seen from Table 3, hepatosomatic index (HSI) of fish fed 4 meal/d was significantly lower than that of fish fed 2 meal/d ($P < 0.05$). Besides, HIS of fish fed diet C5P32 was significantly lower than that of fish fed diet C20P28 in term of diet

Table 2 Growth parameters and feed utilization of common carp reared at 2 feeding frequencies (2 and 4 meal/d) and fed with three diets (C5P32, C10P30 and C20P28) (means ± SD)

Diet	Feeding frequencies	WGR (%)	SGR (%/d)	FCR (g/g)	PER (g/g)
C5P32	4	258.10 ± 13.61[a]	2.97 ± 0.11[a]	1.67 ± 0.09[c]	1.87 ± 0.10[a]
C10P30	4	243.92 ± 42.23[ab]	2.85 ± 0.34[ab]	1.80 ± 0.33[bc]	1.77 ± 0.31[ab]
C20P28	4	239.06 ± 43.33[ab]	2.81 ± 0.35[ab]	1.84 ± 0.33[bc]	1.73 ± 0.31[ab]
C5P32	2	193.55 ± 3.90[bc]	2.42 ± 0.04[abc]	2.22 ± 0.05[abc]	1.40 ± 0.03[bc]
C10P30	2	168.33 ± 30.88[c]	2.18 ± 0.30[c]	2.62 ± 0.51[a]	1.22 ± 0.21[c]
C20P28	2	187.04 ± 42.82[bc]	2.32 ± 0.39[bc]	2.38 ± 0.51[ab]	1.36 ± 0.31[bc]
Main effects					
C5P32		225.83	2.70	1.95	1.64
C10P30		206.12	2.51	2.21	1.50
C20P28		213.05	2.58	2.11	1.54
	4	247.03[a]	2.87[a]	1.77	1.79[a]
	2	182.97[b]	2.32[b]	2.41	1.33[b]
P value					
Feeding frequency		0.00	0.00	0.70	0.00
Diet		0.60	0.56	0.15	0.59
Interaction		0.83	0.80	0.98	0.84

All values are expressed as mean values (n = 3) ± SD, and different alphabets in the same column denote significant difference (*P < 0.05*)

composition (*P < 0.05*). Condition factor (CF) was significantly higher in fish fed 4 meal/d compared with that of fish fed 2 meal/d, but unaffected by diet composition (*P > 0.05*). It is worth noting that HIS was significantly affected by the interaction of diet composition and feed frequency. The highest HIS was found in fish fed with diet C10P30 at 2 meal/d group.

2.3 Intestinal Digestive Enzyme Activity

2.3.1 Protease Activity

Protease activities in foregut of fish generally reduced as dietary carbohydrate levels increased, but no significant difference was observed (*P > 0.05*), whereas protease activities in foregut of fish maintained at a feed frequency of 4 meal/d were significantly higher than those of fish at 2 meal/d (*P* < 0.05). Protease activities in midgut of fish (CF) was significantly lower in fish offered diet C20P28, compared with those of fish offered diet C5P32 and C10P30 (*P* < 0.05), but unaffected by feed frequency (*P > 0.05*). In addition, protease activities in hindgut were not affected by either diet composition or feed frequency (*P > 0.05*) (Table 4).

Table 3 Hepatosomatic index (HSI, %), condition factor (CF, % g/cm^3) of common carp reared at 2 feeding frequencies (2 and 4 meal/d) and fed with three diets (C5P32, C10P30, and C20P28) (means ± SD)

Diet	Feeding frequencies	HSI	CF
C5P32	4	0.80 ± 0.08^{c}	2.50 ± 0.18^{ab}
C10P30	4	0.99 ± 0.04^{bc}	2.63 ± 0.07^{a}
C20P28	4	1.18 ± 0.15^{ab}	2.80 ± 0.08^{a}
C5P32	2	1.22 ± 0.17^{a}	2.14 ± 0.08^{b}
C10P30	2	1.29 ± 0.10^{a}	2.20 ± 0.20^{b}
C20P28	2	1.21 ± 0.06^{a}	2.48 ± 0.17^{ab}
Main effects			
C5P32		1.01^{b}	2.32
C10P30		1.14^{ab}	2.42
C20P28		1.19^{a}	2.65
	4	0.99^{b}	2.65^{a}
	2	1.24^{a}	2.27^{b}
P value			
Feeding frequency		0.00	0.00
Diet		0.00	0.07
Interaction		0.03	0.91

All values are expressed as mean values (n = 3) ± SD, and different alphabets in the same column denote significant difference ($P < 0.05$)

Table 4 Protease activities (μg/ml) in intestine of common carp reared at 2 feeding frequencies (2 and 4 meal/d) and fed with three diets (C5P32, C10P30, and C20P28) (means ± SD)

Diet	Feeding frequencies	Foegut	Midgut	Hindgut
C5P32	4	147.18 ± 7.13	172.19 ± 52.22^{abc}	86.56 ± 7.24
C10P30	4	168.26 ± 34.11	185.78 ± 20.84^{a}	134.08 ± 9.69
C20P28	4	157.88 ± 28.42	123.04 ± 5.55^{c}	105.71 ± 7.81
C5P32	2	105.37 ± 44.59	158.99 ± 16.32^{abc}	85.80 ± 3.41
C10P30	2	128.53 ± 46.33	183.12 ± 6.58^{bc}	92.71 ± 7.73
C20P28	2	107.40 ± 48.58	134.40 ± 21.70^{bc}	89.74 ± 6.34
Main effects				
C5P32		137.86	165.59^{a}	81.41
C10P30		136.82	184.44^{a}	111.91
C20P28		132.64	128.72^{b}	99.22
	4	157.78^{a}	160.34	108.78
	2	113.76^{b}	158.84	89.42
P value				
Feeding frequency		0.04	0.90	0.58
Diet		0.98	0.01	0.52
Interaction		0.66	0.71	0.41

All values are expressed as mean values (n = 3) ± SD, and different alphabets in the same column denote significant difference ($P < 0.05$)

2.3.2 Amylase Activities

As can be seen from Table 5, amylase activities in foregut and hindgut of fish were not affected by either diet composition or feed frequency ($P > 0.05$). Amylase activities in midgut of fish maintained at a feed frequency of 4 meal/d were significantly lower than those of fish at 2 meal/d ($P < 0.05$). Besides, amylase activities in midgut of fish were significantly affected by diet composition with the lowest observed in fish fed with diet C20P28 ($P < 0.05$).

2.3.3 Lipase Activities

As can be seen from Table 6, fish fed at a feed frequency of 2 meal/d exhibited significantly higher lipase activities in midgut when compared with fed at 4 meal/d ($P < 0.05$). However, there was no significant difference in lipase activities in midgut among fish fed three different carbohydrate levels ($P > 0.05$). Lipase activities in foregut and hindgut of fish increased first and then decreased as dietary carbohydrate levels increased, but the difference was not significant ($P > 0.05$). Besides, lipase activities in foregut and hindgut were unaffected by feed frequency ($P > 0.05$).

Table 5 Amylase activities (μg/ml) in intestine of common carp reared at 2 feeding frequencies (2 and 4 meal/d) and fed with three diets (C5P32, C10P30, and C20P28) (means ± SD)

Diet	Feeding frequencies	Foegut	Midgut	Hindgut
C5P32	4	100.82 ± 31.90	101.36 ± 50.90[bc]	68.95 ± 23.64
C10P30	4	124.28 ± 40.70	130.16 ± 35.90[ab]	83.66 ± 63.16
C20P28	4	119.72 ± 57.60	60.54 ± 21.91[c]	75.97 ± 48.57
C5P32	2	128.83 ± 48.87	149.17 ± 16.57[ab]	121.72 ± 16.59
C10P30	2	157.81 ± 64.91	175.15 ± 26.79[a]	124.79 ± 26.15
C20P28	2	148.85 ± 39.94	90.63 ± 49.44[bc]	107.11 ± 49.70
Main effects				
C5P32		115.88	139.66[a]	105.34
C10P30		146.26	138.25[a]	104.26
C20P28		134.29	63.08[b]	91.54
	4	114.94	89.02[a]	86.19
	2	145.16	138.32[b]	127.87
Feeding frequency				
Diet		0.84	0.02	0.06
Interaction		0.87	0.01	0.90
Feeding frequency		0.34	0.46	0.87

All values are expressed as mean values (n = 3) ± SD, and different alphabets in the same column denote significant difference ($P < 0.05$)

Table 6 Lipase activities (μg/ml) in intestine of common carp reared at 2 feeding frequencies (2 meal/d and 4 meal/d) and fed with three diets (C5P32, C10P30, and C20P28) (means ± SD)

Diet	Feeding Frequencies	Foegut	Midgut	Hindgut
C5P32	4	332.64 ± 11.19	63.48 ± 19.02	27.57 ± 11.86
C10P30	4	435.60 ± 18.08	90.82 ± 12.23	85.00 ± 11.47
C20P28	4	98.83 ± 16.41	69.69 ± 11.33	64.37 ± 11.95
C5P32	2	89.51 ± 12.41	175.97 ± 10.74	88.13 ± 16.31
C10P30	2	172.90 ± 19.06	184.03 ± 11.79	89.60 ± 54.63
C20P28	2	339.79 ± 11.24	135.27 ± 12.38	84.21 ± 16.62
Main effects				
C5P32		211.08	119.72	57.85
C10P30		304.252	135.43	87.30
C20P28		219.31	102.48	72.94
	4	289.03	74.67^{a}	58.98
	2	200.74	131.76^{b}	86.41
Feeding frequency				
Diet		0.34	0.04	0.32
Interaction		0.65	0.58	0.67
Feeding frequency		0.07	0.19	0.67

All values are expressed as mean values (n = 3) ± SD, and different alphabets in the same column denote significant difference ($P < 0.05$)

3 Discussion

Irrespective of feeding to satiation or feeding on fixed rations, previous research have demonstrated that appropriate increasing feeding frequency can improve growth performance and feed conversion of fish [15, 16]. However, it is necessary to note that our study adopts feeding on fixed rations, which aims to eliminate any consideration of the change of feeding rate. On the premise of feeding on fixed ration, the present study demonstrated that regardless of dietary composition, WGR, SGR and PER of common carp were significantly affected by feeding frequency. WGR and PER of fish fed 4 meal/d were significantly higher than those of fish at 2 meal/d except fish fed with diet C20P28, suggesting that increasing feeding frequency can promote diet protein utilization and boost the protein-sparing of dietary carbohydrate. Unlike the case of WGR, SGR, PER, FCR was not affected by feeding frequency. However, it is worth noting that FCR showed a reducing trendwith increased feeding frequency, indicating that improvement of feed efficience was gained with appropriate increasing feeding frequency. Different results were observed in *Pseudosciaena crocea R.* [17], *Salmo galrdnera*. Two explanations could be able to explain this difference. Firstly, different from the former two, as a typical non-stomach, the digestive organ of common carp is intestine, which would inevitably reduce the length of time feed remains in fish body, consequently, appropriate increasing feeding frequency can increase the length of time feed remains in fish body, thus enhancing the digestion ability and the assimilation efficiency of food. Secondly, the previous study reported that for fish, the key to absorbing the dietary carbohydrate was the absorption percent within 2 h. Thus, increasing feeding frequency, under the case of feeding on fixed ration, might be advanced to effectively regulate the carbohydrate absorption rate to boost the utilization ability of fish for high-carbohydrate diets. In the aspect of diet composition, WGR, SGR, PER of common carp tended to decrease with increasing dietary carbohydrate levels, but no significant difference was observed. This may suggest that increasing dietary carbohydrate levels inhibit growth and feed conversion of common carp to some extent. Similar results on *Acipenser gueldenstaedtii* were reported by Cui [18]. Further studies concerned how to improve the utilization ability of fish for dietary carbohydrate in common carp ought to be designed.

The liver occupies a very important position in the metabolism of carbohydrate, lipid, protein and other nutrients, and is also an important nutrient storage sites, which is why the hepatosomatic indexes (HSI) of fish is treated as very sensitive indicators for the long-term and the short-term nutrition way [19]. In the present study, HSI of common carp was significantly affected by the interaction of dietary carbohydrate levels and feeding frequencies. Firstly, HSI of fish fed 4 meal/d were significantly lower than those of fish at 2 meal/d except fish fed with diet C20P28. Secondly, when analyzed by feeding frequency, improvements in HIS of common carp were observed. At 4 meal/d group, HIS of fish fed with diet C5P32 was significantly lower than that of fish fed with diet C20P28. Previous study on *Pelteobagrus fulvidraco* have also reported that HSI tended to decrease with increasing feeding frequency [20]. In fish biological study, condition factor (CF) is regarded as a measure of evaluating growth and feeding intensity of fish [21]. CF of common carp, in the present study, was significantly affected by feeding frequency. Irrespective of dietary composition, improvement in CF was found in fish fed 4 meal/d compared with those fed 2 meal/d. Whilst independent of feeding frequency, CF of common carp tended to increase with increasing dietary carbohydrate levels. Similar findings were also found in *Acipenser gueldenstaedtii* [18]. This coupled with HIS obtained may demonstrate that, under feeding on fixed ration, appropriate increasing feeding frequency can reduce the proportion of the body organs and promote the increase of the body length. This may testify that the protein-sparing effect of carbohydrate can be enhanced by appropriate increasing feeding frequency on the other side. However, it is a remarkable fact that improvement in HSI and CF of common carp was observed in fish offered high-carbohydrate diet. This may suggest that more dietary protein or carbohydrate might be transformed into lipid or glycogen which is accumulated in the liver or viscera.

Digestion is the first limiting factor, which have an effect on utilizing dietary carbohydrate for growth. In the present study, intestinal digestive enzyme activities were significantly affected by dietary carbohydrate levels and feeding frequencies to some extent. Firstly, independent of dietary composition, improvement in protease activities was found in fish fed 4 meal/d in comparison to those fed 2 meal/d, whereas amylase and lipase activities tended to decrease with increasing feeding frequency. Among these, protease activity of foregut, amylase and lipase activities of midgut were significantly affected by feeding frequency. The present study may suggest that under feeding fixed ration, appropriate increasing feeding frequency would produce the dietary restriction; whilst fish are able to increase the secretion of digestive enzymes or improve the digestive enzyme activities to enhance nutrient absorption under the case of the dietary restriction. This was also certificated by increasing protease activity of midgut with increasing feeding frequency in the present study. Similar findings on *Acipenser gueldenstaedtii* were observed [18]. Secondly, irrespective of feeding frequency, protease and amylase activities of fish fed with diet C5P32 and C10P30 were significantly higher than those of fish fed with diet C20P28. Different results were observed in *Larimichthys crocea* [22], and *Silurus meridionalis* [23]. That might be attributed to the difference of feeding habits, growth phase, dietary carbohydrate sources and so on. However, further studies should be needed to reveal this.

4 Conclusion

The present study suggests that the growth and the digestive abilities of midgut of common carp are significantly affected by dietary carbohydrate levels and feeding frequencies. An increase of feeding frequency from 2 to 4 meal/d can improve the growth and intestinal digestive enzyme activity of common carp and enhance the carbohydrate utilization and boost the protein-sparing effect of dietary carbohydrate. Meanwhile, increasing carbohydrate level from 5 to 10% leads to the improvements of protease and amylase activities of midgut. Therefore, it is recommended that common carp should be fed at 4 meal/d, and the carbohydrate content of the dietary can be appropriately increased to 5–10%.

Acknowledgements This study was supported by the Tianjin Research Program of Application Foundation and Advanced Technology (No. 14JCQNJC15100), the Key Technology R&D Program of Tianjin (No. 13ZCZDNC00900), and the National Natural Science Foundation of China (No. 31402313). We would also like to acknowledge Tianjin Tianxiang Aquatic Co. Ltd. for provide the farming equipment and supplements, and LI Jinghui YIN Shuai, ZHANG Zhiyuan, ZHANG Tengxian, LIN Chengli and WANG Anqi from Tianjin Agriculture University for their assistance with sampling and index determine.

References

1. Ai CX, Tao QY (2013) The replacement of fish meal—the technical strategy of development of aquatic compound feed in the case of the high price of fish meal. Feed Ind 34(10):1–7
2. Dong SL(2011) High efficiency with low carbon: the only way for China aquaculture to develop. J Fish China 35(10): 1595–1600
3. Fan Z, Ling JH, Cheng ZY et al (2015) Protein sparing effect of lipid diets for common carp (*Cyprinus carpio*). Adv Eng Res 45:357–368

4. SunJ H, Fan Z, Cheng ZY et al (2016) Effects of dietary corn starch supplemental level on growth performance, digestive enzyme activities and serum biochemical indices of common carp. Chin J Anim Nutr 28(4):1152–1159
5. Stone DAJ (2003) Dietary carbohydrate utilization by fish. Rev Fish Sci 11(4):337–369
6. Wilson RP (1994) Utilization of dietary carbohydrate by fish. Aquaculture 124:67–80
7. Biswas G, Thirunavukkarasu AR, Sundaray JK, Kailasam M (2010) Optimization of feeding frequency of Asian seabass (*Lates calcarifer)* fry reared in net cages under brackishwater environment. Aquaculture 305:26–31
8. Wang H, Li Y, Chen K et al (2008) Research progress of feeding rhythm and feeding regime for aquatic animal. Feed Ind 29(24):17–20
9. Tung PH, Shiau SY (1991) Effects of meal frequency on growth performance of hybrid tilapia *Oreochromis niloticus* × *O. Aureus*, fed different carbohydrate diets. Aquaculture 92: 343–350
10. Shiau SY (1997) Utilization of carbohydrates in warm water fish-with particular reference to tilapia, *Oreochromis niloticus* × *O.aureus*. Aquaculture 151:79–96
11. Guo L, Jing RZ, Cheng ZY et al (2013) Preliminary study on the effects of reducing dietary protein of common Carp (*Cyprinus carpio*). Feed Ind 34(8):41–45
12. Zhang BL, Gao MZ, Cheng ZY et al (2015) Effects of reducing dietary protein level on growth performance, body composition and immunity of common carp (*Cyprinus carpio*). Feed Res 8:49–55
13. Li JN, Xu QY, Wang CA et al (2015) Effects of different dietary carbohydrates and carbohydrate levels on GH/IGF-I mRNA expression and the fish body composition of juvenile mirror carp (*Cyprinus carpio*). J Shanghai Ocean Univ 24(4):489–495
14. Deng ZW (2015) Effects of feeding frequency on feeding and growth of juvenile FFRC. J Fujian Fish 37(1):68–72
15. Dwyer KS, Brown JA, Parrish C, Lall SP (2002) Feeding frequency affects food consumption, feeding pattern and growth of juvenile yellowtail flounder (*Limanda ferruginea*). Aquaculture 213:279–292
16. Silva CR, Gomes LC, Brandao FR (2007) Effect of feeding rate and frequency on tarnbaqui (*Colossoma macropomum*) growth, production and feeding costs during the first growth phase in cages. Aquaculture 264:135–139
17. Sun RJ, Zhang WB, Xu W et al (2013) Effects of dietary protein level and feeding frequency on the growth performance, body composition and protein metabolism of juvenile large yellow croakers *Pseudosciaena crocea* R. Acta Hydrobiol Sin 37(2):281–289
18. Cui C (2014) Effects of light, dietary carbohydrate and feeding strategy on growth of Russian sturgeon. East China Normal University, Shanghai
19. Peres H, Oliva-Teles A (1999) Effect of dietary lipid level on growth performance and feed utilization by European seabass juveniles *Dicentrarchus labrax*. Aquaculture 179(1):325–334
20. Fang w (2010) Food intake and feeding strategy of yellow catfish *Pelteobagrus fulvidraco*. Huazhong Agricultural University, Wuhan
21. Froese R (2006) Cube law condition factor and weight-length relationships: history, meta-analysis and recommendations. J Appl Ichthyol 22(4):241–253
22. Wang MQ, Huang WW, Zhou PP et al (2015) Effects of dietary protein and wheat starch levels on growth performance, hepatic glycolysis and gluconeogenic keyn enzymes activities in large yellow croaker (*Larimichthys crocea* Richardson). J Fish China 39(11):1691–1701
23. Gao M, Luo YP, Cao ZD (2006) Effect of dietary carbohydrate on digestive enzyme activities in southern catfish (*Silurus meridionalis* Chen) juveniles. J Southwest China Norm Univ (Nat Sci) 31(2):120–123

Engineering of Industrial *Aspergillus ochraceus* Strains for Improved Steroid 11α-Hydroxylation Efficiency via Overexpression of the 11α-Hydroxylase Gene *CYP68J5*

Xingwei Yang, Fan Wu, Xiangjiang Hou, Benfeng Lin, Ruijie Wang, Fuping Lu, Zhengxiang Wang, Bin Zeng and Xiaoguang Liu

1 Introduction

Selective hydroxylation of steroid compounds is crucial for their appropriate pharmacological activities and 11α-hydroxylation of steroids is one of the most important reactions for steroid functionalization, which is mainly accomplished in the industrial setting by using filamentous fungi-mediated biotransformation due to their high regio- and steroselectivity of the steroid hydroxylation reactions [1–4]. For instance, *Aspergillus ochraceus* TCCC41060 is commercially used for the preparation of 11α-hydroxylation of 16,17α-epoxyprogesterone (EP), a key intermediate for the production of a series of valuable glucocorticoids (see Fig. 1). However, the productivity of the 11α-hydroxylation process is limited by the moderate conversion rates and/or the lengthy process required to obtain desired yields. Previous efforts to enhance conversion efficiency have mainly focused on improving solubility of steroid substrates by using biocompatible cosolvents, adoption of biphasic reaction systems and ionic liquid, [5, 6] and by employing immobilized cell systems [7, 8]. Much less effort has been devoted to the rational

X. Yang · F. Wu · X. Hou · B. Lin · R. Wang · F. Lu
Key Laboratory of Industrial Fermentation Microbiology, Ministry of Education, College of Bioengineering, Tianjin University of Science and Technology (TUST), Tianjin 300457, China

Z. Wang · X. Liu (✉)
Laboratory of Biocatalysis and Biotransformation, College of Chemical Engineering and Materials Science, Tianjin University of Science and Technology (TUST), Tianjin 300457, China
e-mail: liu_xg@tust.edu.cn

B. Zeng
College of Life Science, Jiangxi Science and Technology Normal University, Nanchang 330013, China

H. Liu et al. (eds.), *Advances in Applied Biotechnology*, Lecture Notes in Electrical Engineering 444, https://doi.org/10.1007/978-981-10-4801-2_21

breeding of more efficient fungal strains for selective steroid hydroxylation. Enzymes of filamentous fungi involved in the steroid hydroxylation are members of the cytochrome P450 superfamily. A cytochrome P450 gene encoding a steroid 11α-hydoxylase (CYP68J5) was previously cloned from *A. ochraceus* and functionally expressed in the insect Sf-9 cells [9]. Recently, we showed that the *CYP68J5* gene is solely responsible for the hydroxylating activities of the production strain *A. ochraceus* TCCC41060 on EP since disruption of *CYP68J5* resulted in a complete loss of hydroxylation activities toward EP in the mutant (unpublished data).

In the present study, we tested the idea that 11α-hydroxylation efficiency of *A. ochraceus* TCCC41060 is limited by the expression levels of the steroid 11α-hydoxylase and thus overexpression of the CYP68J5 gene should enhance 11α-hydroxylation efficiency on EP.

2 Materials and Methods

2.1 Bacterial Strains, Plasmids and Culture Conditions

Aspergillus ochraceus TCCC41060 was obtained from the microbial strain collection of the applied microbiology lab of Tianjin University of Science and Technology (TUST). TCCC41060 is an industrial strain used for the commercial preparation of 11α-hydroxy-16,17α-epoxyprogesterone, a key intermediate for the production of a series of glucocorticoids. The fungus was routinely maintained on potato dextrose agar (PDA) or in steroid transformation medium (STM) (20 g/l glucose, 20 g/l yeast extract, 20 g/l tryptone, pH 5.8). The plasmids pB-TrpC and pBluescript II KS+ vector were obtained from the microbial strain collection of the applied microbiology lab of TUST and plasmid pAg1-h3-ble was kindly provided by Prof. Hao Liu (College of Biotechnology, TUST). Unless noted, all enzymes for DNA molecular manipulations in this study were purchased from the TaKaRa (Dalian, China) and all primers were synthesized by AuGCT (Beijing, China).

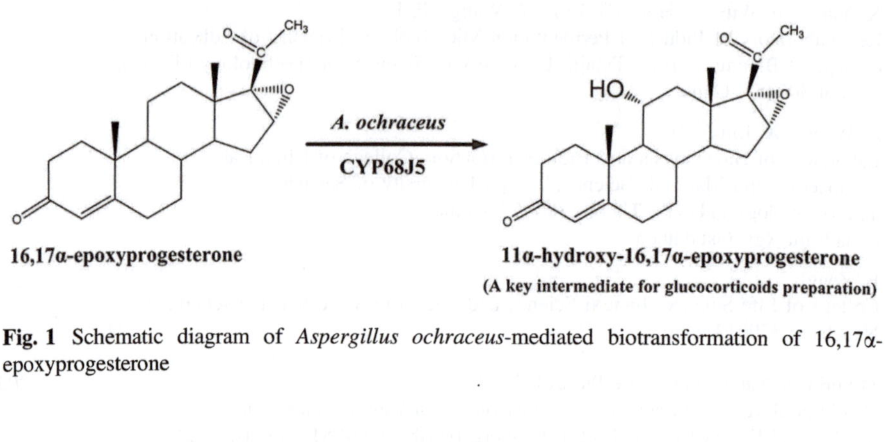

Fig. 1 Schematic diagram of *Aspergillus ochraceus*-mediated biotransformation of 16,17α-epoxyprogesterone

2.2 *Construction of Expression Constructs*

To create *A. ochraceus* recombinant strains over-expressing the steroid 11α-hydroxylase, pB-CYP68J5-TrpC was constructed. To construct expression vector pB-CYP68J5-TrpC, a ~2.2 kb DNA fragment encompassing the full ORF of *CYP68J5* plus 334 bp of the 3′ terminator region was PCR-amplified using the primer set (Forward: 5′-GAATTCATGCCCTTCTTCACTG-3′; reverse: 5′-CTCGAGGACAGAAGGAAACCTG-3′) and then inserted into the *Sma*I site of pBluescript II KS+ to produce plasmid vector pB-T-CYP68J5. The *ble* gene was released from pB-ble by digestion with *HindIII* and *Xho*I. The *CYP68J5*-T fragment was obtained from pB-T-CYP68J5 by digestion with *Xho*I and *Eco*RI. Then the *Xho*I/*Eco*RI CYP68J5-T and *Xho*I/*HindIII* ble fragments were ligated into *HindIII* and *Eco*RI sites of pB-TrpC to produce CYP68J5-expressing construct pB-CYP68J5-TrpC (Fig. 2).

2.3 *Preparation of* A. ochraceus *Protoplasts*

Conidia of *A. ochraceus* were inoculated onto PDA slants and incubated at 28 °C for 4–6 days. Conidia were harvested from slants using 1 M sorbitol and adjusted to 1×10^7 conidia/ml. Conidia of 100 μl were inoculated into 100 ml YPD in a 500 ml flask and incubated for 20–28 h at 28 °C with shaking, followed by still incubation for 10–15 h at 28 °C. The fungal mycelia were then collected and washed thoroughly with 1 M sorbitol and resuspended in enzyme mixture solutions prepared in osmotic stabilizer of 1 M sorbitol and incubated at 30 °C for 2 h with shaking. The lysing enzyme mixture was composed of 10 g/l cellulase (Solarbio, Beijing, China), 1 g/l lysing enzymes (Solarbio, Beijing, China) and 10 g/l snailase (Solarbio, Beijing, China). Protoplasts were collected by centrifugation at $3000\times g$ for 15 min at 4 °C and then washed twice using 0.6 M KCl + STC (1 M sorbitol, 10 mM Tris-HCl and 50 mM $CaCl_2$; pH 7.5) and resuspended in STC to $>5 \times 10^6$ protoplasts/ml.

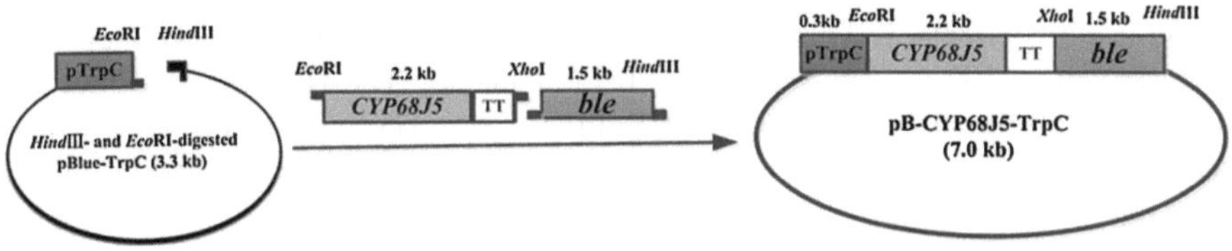

Fig. 2 The schematic diagram for the construction of *CYP68J5* exprssion vectors: pB-CYP68J5-TrpC

2.4 Protoplasts Transformation

Transformation was carried out as follows: (1) an aliquot of 200 μl of the protoplasts suspension was added to a 10 ml microcentrifuge tube, followed by gently mixing with 5–10 μg of linearized plasmid and 100 μl of PEG solution (300 g/l PEG 8000; 50 mM $CaCl_2$; 10 mM Tris-HCl; pH7.5; Sangon Biotech) and incubating for 30 min on ice; (2) PEG solution (2 ml) was added, gently mixed with the protoplasts and then kept at room temperature for 5 min before the addition of STC (4 ml) and (3) the transformation mixture was then added to 50 ml liquid regeneration medium (PDA containing 0.6 M KCl and 30 μg/ml bleomycin; Solarbio) and plated on PDA plates. Transformants started to appear after incubation at 28 °C for 3–5 days and were transferred to PDA slant containing bleomycin (30 μg/ml). Selected transformants were propagated for 3–5 generations to assess their mitotic stability before being verified by PCR amplification of the *ble* gene with the primer set (forward: 5′-CCAATGGCTCATAGTACCAG-3′, reverse: 5′-GACCTAGAC TTCAGGTTGTC-3′).

2.5 Biotransformation Experiments

A. ochraceus was grown for 6 days in PDA slants at 28 °C, conidia were collected and inoculated (10^6/ml) into 50 ml STM in a 250 ml flask and cultivated for 22–24 h at 28 °C with shaking at 180 rpm. Then predetermined amounts of substrate 16,17α-epoxyprogesterone (EP) was added and allowed transformations to proceed at 28 °C on the rotary shaker. The samples were extracted with ethyl acetate, dried and dissolved in acetonitrile and analyzed by HPLC.

2.6 HPLC Analyses

To determine the transformation products, 0.5 ml of fungal transformation broth was extracted with 400 μl of ethyl acetate. The product was analyzed by HPLC as follows: the extracts (200 μl) were allowed to dry completely and then were re-dissolved in 200 μl acetonitrile for HPLC analysis. The HypersilTM ODS C18 columns (250 mm L * 4.6 mm I.D Column, 5 μm; Thermo Scientific) was used and the mobile phase consists of 80% acetonitrile (v/v) and 20% H_2O (v/v) with a flow rate of 0.8 ml/min. 11α-hydroxy-16,17α-epoxyprogesterone (HEP) standard was used to confirm the authenticity of the product and 11α-hydroxylation product was quantified by comparison to peak areas of known amounts of HEP standards. Conversion rate of EP to HEP was estimated using the following equation:

$$\text{Conversion rate }(\%) = \frac{(mHEP \div MWHEP)}{mEP \div MWEP} \times 100\%$$

where mEP and mHEP are the weights of EP and HEP, respectively; MWEP and MWHEP are the molecular weights of EP and HEP, respectively [10].

2.7 RNA Isolation and CDNA Synthesis

A. ochraceus was cultured for 24 h at 28 °C in liquid STM medium, followed by adding 0.05 mg/ml EP for induction for 3 h. Total RNAs were isolated using Trizol reagents (Promega, USA) from the resultant mycelia. The synthesis of the first cDNA strand was performed using 5 μg total RNA with PrimeScript Reverse Transcriptase in a 20 μl reaction volume according to the manufacturer's instructions.

2.8 Real-Time Quantitative PCR

To determine the transcript levels of *CYP68J5* in recombinant strains, real-time quantitative PCR (qRT-PCR) was conducted as follows: the total volume of PCR reaction is 20 μl containing 10 μl of MasterMix with SYBR (Solarbio, Beijing, China), 300 nmol/l of both primers (forward: 5′-CCAATCGTCGTCGTGTTC-3′, reverse: 5′-CTTTAGTCAGGTTCGGGTCT-3′) and 1 μl of cDNA template. The PCR amplification parameters comprise denaturation (95 °C, 10 min), 40 cycles of denaturation (95 °C, 30 s) and annealing (60 °C, 30 s) (Applied Biosystems, USA). The transcript levels of the glyceraldehyde-3-phosphate dehydrogenase gene (*GAPDH*) (forward: 5′-CCAAGAAGGTTGTCATCTCC-3′; reverse: 5′-CATGA GACCCTCAACGATAC-3′) were used respectively as an internal control for determining the expression levels of *CYP68J5*. The $2^{-\Delta\Delta CT}$ method was used for determining the transcript level of the target gene [11].

2.9 Statistical Analysis

The statistical analyses were performed using GraphPad Prism 5.0 (GraphPad, San Diego, CA). All quantitative data were expressed as mean ± SD for each condition.

3 Results and Discussion

3.1 Construction of Recombinant Strains

To increase the steroid 11α-hydroxylation activity of the current production strain *A. ochraceus* TCCC41060, *CYP68J5* expression vectors (pB-CYP68J5-TrpC) was constructed according to Materials and methods (Fig. 2). The expression of the *CYP68J5* gene in expression vector pB-CYP68J5-TrpC was under the control of the constitutive TrpC promoter of *Aspergillus nidulans*. The linearized pB-CYP68J5-TrpC vector was introduced into *A. ochraceus* TCCC41060 by protoplast transformations and subsequent selection on bleomycin-containing plates (30 μg/ml) and positive transformants were verified by PCR amplification (data not shown). In total, two transformants were obtained and designated as CYP68J5-TrpC-1 and CYP68J5-TrpC-2.

3.2 Recombinant Strains Show Elevated Levels of CYP68J5 Expression

To assess transcript levels of *CYP68J5* in two recombinant strains, a quantitative reverse transcription PCR was performed to compare the expression levels of *CYP68J5* between the recombinant strains and the recipient strain TCCC41060. As shown in Fig. 3, after induction for 3 h by 16,17α-epoxyprogesterone (EP),

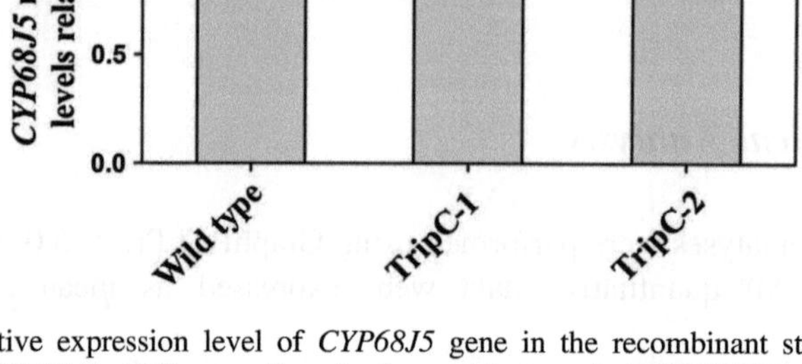

Fig. 3 The relative expression level of *CYP68J5* gene in the recombinant strains (CYP68J5-TrpC-1 and CYP68J5-TrpC-2) as well as in the recipient strain TCCC41060. Total RNA was extracted from *Aspergillus ochraceus* mycelia treated with 16,17α-epoxyprogesterone (EP) for 3 h induction and quantitative RT-PCR was used to quantify the relative expression levels of *CYP68J5*

expression levels of *CYP68J5* in recombinant strains CYP68J5-TrpC-1, CYP68J5-TrpC-2 were 35 and 62% higher than that of the start strain TCCC41060, indicating that increasing the genomic copy number of *CYP68J5* in the production strain resulted in a significant enhancement of the transcript levels of *CYP68J5*. However, two recombinant strains showed quite different transcript levels, which could be due to differences in the inserted copy number of *CYP68J5* or position effects of the inserted *CYP68J5*.

3.3 Recombinant Strains Show Increased Conversion of EP

The EP conversion rates of recombinant strains CYP68J5-TrpC-1 and CYP68J5-TrpC-2 were compared with the recipient strain TCCC41060. As shown in Fig. 4a, with 2 g/l of EP as substrate, conversion rates of the recombinant strains were significantly higher than the control strain. Strain CYP68J5-TrpC-2 exhibited a better performance than strain CYP68J5-TrpC-1. For the first 12 h, the conversion rate for the control strain was less than 5%, while recombinant strains CYP68J5-TrpC-1 and CYP68J5-TrpC-2 had conversion rates of 12.5, 17.8% respectively. For the entire course of biotransformation, the highest conversion rates for the recombinant strains were observed at 60 h, which is about 12 h shorter than that for recipient strain TCCC41060 which reached its highest conversion rates at 72 h. These results demonstrate that increasing expression of the steroid 11α-hydroxylase gene *CYP68J5* is an effective strategy to breed more efficient strains for 11α-hydroxylation of EP. It is well established that in steroid biotransformation substrate concentrations have a great effect on the whole-cell transformation efficiency. To explore the potential of the recombinant strains for EP conversion, the effect of EP concentrations on the 11α-hydroxylation of EP was investigated. As shown in Fig. 4b, after 48 h transformation, recombinant strains maintained a high EP conversion rate (>75%) when EP concentrations were ≤ 5 g/l, whereas strain TCCC41060 showed lower EP conversion rates (<70%) even when the EP concentrations were 2 g/l. It is also apparent that both recombinant strains showed a dramatic decline in EP conversion rates when the EP concentrations exceeded 5 g/l (Fig. 4b), although they still exhibited much higher conversion rates compared with strain TCCC41060, suggesting that conversion rates were either limited by the level of the steroid hydroxylase activity or by the entry of EP into the fungal cells or both.

3.4 Tween-80 Enhances 11α-Hydroxylation of EP

To improve the entry of EP into the fungal cells, Tween-80 was used to optimize 11α-hydroxylation of 16,17α-epoxyprogesterone to better evaluate the transformation potential of the recombinant strains. As shown in Fig. 4c, sharp increases in

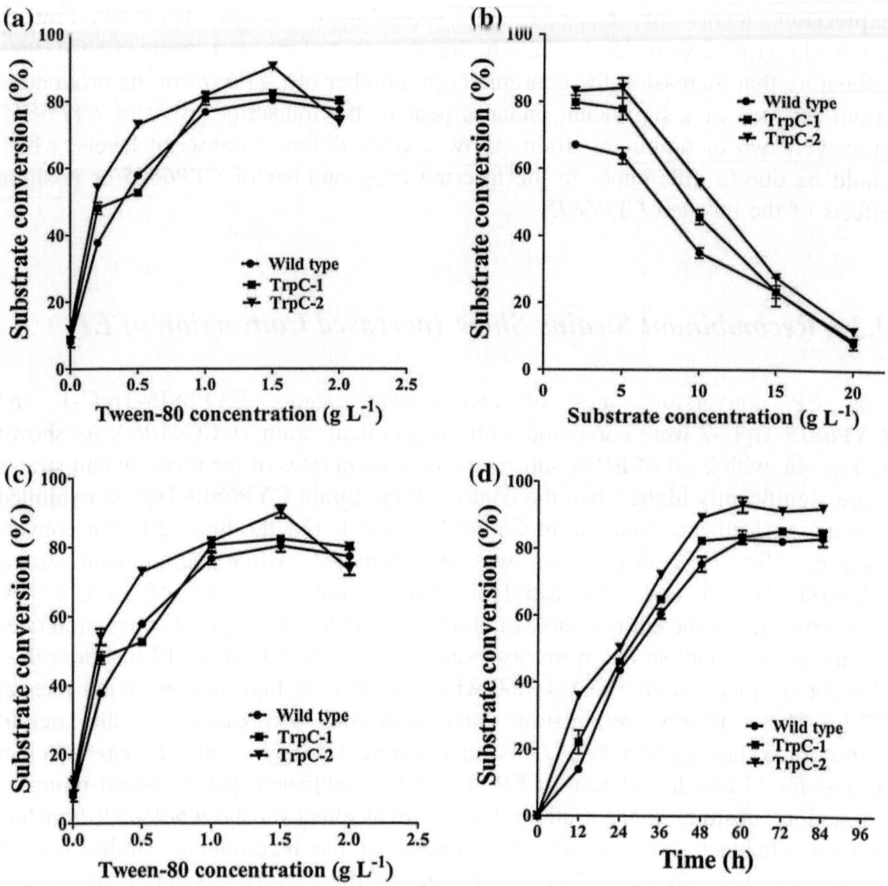

Fig. 4 The time-course of substrate conversion profiles of the recombinant strains (CYP68J5-TrpC-1 and CYP68J5-TrpC-2) and the control strain TCCC41060 under: **a** 2 g/l of EP; **b** conversion of at various EP concentrations for 48 h; **c** 20 g/l of EP with various amounts of Tween 80 supplement for conversion of 48 h; and **d** 20 g/l of EP supplemented with 1.5 g/l of Tween 80

EP conversion rates for all strains tested were observed with increased Tween-80 concentrations from 0.2 to 1.5 g/l with constant EP concentrations of 20 g/l. However, a further increase in the concentration of Tween-80 resulted in declined conversion rates. As shown in Fig. 4d, with 20 g/l of EP as substrate supplemented with 1.5 g/l of Tween 80, the current production strain TCCC41060 achieved the highest conversion rates at 60 h (80%), whereas recombinant strain CYP68J5-TrpC-2 showed the highest conversion rates (92%) at 48 h, a 12% increase compared with TCCC41060. Importantly, the time required for the two recombinant strains to achieve the highest conversion rates is 12 h shorter than that of the control strain, demonstrating that the two recombinant strains CYP68J5-TrpC have much higher hydroxylation efficiency than the production strain TCCC41060.

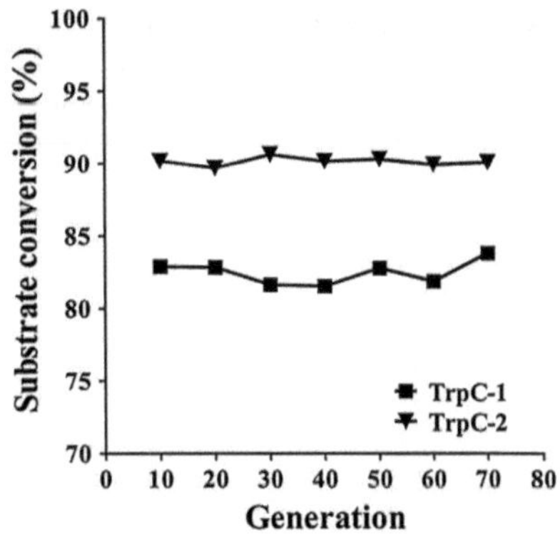

Fig. 5 Substrate conversion assessment of recombinant strains CYP68J5-TrpC-1 and CYP68J5-TrpC-2 of different generations. Transformations were performed for 48 h with 20 g/l of EP and 1.5 g/l of Tween 80

Interestingly, a higher catalytic efficiency was observed in CYP68J5-TrpC-2 than in CYP68J5-TrpC-1, consistent with the different *CYP68J5* transcript levels observed in these two recombinant strains. Further study is required to address the underlying genetic basis for the observed differences in transcript levels in the two recombinant strains.

3.5 *The Genetic Stability of the Recombinant Strain*

To investigate the genetic stability of the recombinant strains during long-term cultivation, the recombinant strains (CYP68J5-TrpC-1 and CYP68J5-TrpC-2) were grown by continuous subculture without antibiotic selection through 70 generations, and the conversion rates of each 10-generation were determined. The results revealed that the two recombinant strains retained the higher levels of conversion rates after 70 generations (Fig. 5).

4 Conclusion

Recombinant strain CYP68J5-TrpC-2 over-expressing the steroid 11α-hydroxylase gene *CYP68J5* not only showed a 12% increase in 16,17α-epoxyprogesterone (EP) conversion rates but also shortened the transformation time by 12 h compared with the recipient production strain TCCC41060, indicating that the transcript level of 11α-hydroxylase gene *CYP68J5* is a limiting factor for 11α-hydroxylation efficiency of strain TCCC41060 on EP. Thus, increasing steroid 11α-hydroxylase expression is an effective strategy for engineering highly efficient commercial strains for steroid 11α-hydroxylation in *A. ochraceus* TCCC41060.

Acknowledgements This work was supported by National High-tech Research and Development Program of China (863 Program) (no. 2011AA02A211) and Natural Science Foundation of China (no. 21206127).

References

1. Carballeira JD, Quezada MA, Hoyos P et al (2009) Microbial cells as catalysts for stereoselective red–ox reactions. Biotechnol Adv 27:686–714
2. Črešnar B, Petrič Š (2011) Cytochrome P450 enzymes in the fungal kingdom. Biochim Biophys Acta 1814:29–35
3. Donova MV, Egorova OV (2012) Microbial steroid transformations: current state and prospects. Appl Microbiol Biot 94:1423–1447
4. Girvan HM, Munro AW (2016) Applications of microbial cytochrome P450 enzymes in biotechnology and synthetic biology. Curr Opin Chem Biol 31:136–145
5. Ceen EG, Dunnili P, Herrmann JPR (1988) Two-liquid phase reactor studies of 11α-hydroxylation of progesterone by *Aspergillus ochraceus*. Biotechnol Bioeng 31:743–746
6. Wu DX, Guan YX, Wang HQ et al (2011) 11α-Hydroxylation of EP by *Rhizopus nigricans* in a biphasic ionic liquid aqueous system. Bioresour Techn 102:9368–9373
7. Bihari V, Goswami PP, Rizvi SHM et al (1985) Studies on immobilized fungal spores for microbial transformation of steroids: 11α-hydroxylation of progesterone with immobilized spores of *Aspergillus ochraceus* G8 on polyacrylamide gel and other matrices. Biotechnol Bioeng 27:1392
8. Ohlson S, Larsson PO, Mosbach K (1979) Steroid transformation by living cells immobilized in calcium alginate. Appl. Microbiol Biot 7:103–110
9. Suzanne BL, Clayton RA, Easton AM et al (2003) *Aspergillus ochraceus* 11 alpha hydroxylase and oxidoreductase. US Patent 20030148420A1
10. Sripalakit P, Wichai U, Saraphanchotiwitthaya A (2006) Biotransformation of various natural sterols to androstenones by *Mycobacterium* sp. and some steroid-converting microbial strains. J Mol Catal B-enzym 41:49–54
11. Livak KJ, Schmittgen TD (2001) Analysis of relative gene expression data using real-time quantitative PCR and the $2^{-\Delta\Delta}$CT method. Methods 25:402–408

Response Surface Methodology Optimization Extraction of Polysaccharides from Maca (*Lepidium meyenii*) Leaves and Primary Characterization

Caicai Kang, Liming Zhang, Limin Hao, Huanhuan Ge, Meng Xu, Jie Cao, Jianyong Yu and Yongwu Yang

1 Introduction

Response surface methodology (RSM) has proved quite an effective method for the expected purpose. Recently, RSM has been widely used to optimize processes in many studies, including the extraction of polysaccharides. Box-Behnken design (BBD), the most impotent and common design of RSM, is helpful to design experiments, build models, evaluate the effects of factors and searching optimum condition of factors for desirable responses [1]. At present, the RSM has been successfully applied to extract Bletillastriata polysaccharides [2].

Maca (*Lepidium meyenii*), a medicine food homology plant, belongs to the Brassicaceae family [2]. Maca roots (hypocotyls), the subterranean part, have been a traditional medicinal agent and dietary staple since pre-Columbian times [3]. Maca roots are consumed worldwide as liquor, pills, soft drinks and candies for its high nutritional value and multi-functional properties [4]. Reported properties

C. Kang · L. Zhang (✉) · L. Hao (✉) · H. Ge · M. Xu · J. Cao
Key Laboratory of Industrial Fermentation Microbiology, Ministry of Education, Tianjin University of Science and Technology, Tianjin 300457, People's Republic of China
e-mail: zhanglmd@126.com

L. Hao
e-mail: hlm2005@163.com

L. Hao · J. Yu
The Quartermaster Equipment Institute of General Logistics, Department of People's Liberation Army, Beijing 100010, People's Republic of China

Y. Yang
Lijiang Maca Biological Technology Co., Ltd., Kunming 674119, Yunnan, People's Republic of China

H. Liu et al. (eds.), *Advances in Applied Biotechnology*, Lecture Notes in Electrical Engineering 444, https://doi.org/10.1007/978-981-10-4801-2_22

include energizing, fertility-enhancing effects, antifatigue [5], anti-inflammation [6], antiosteoporosis [7], antidepression [8], antioxidant activity [9], and so on. However, little information has been published on maca leaves, such as polysaccharides. Polysaccharides are macromolecules composed of more than ten monosaccharide units. And the monosaccharides are linked by glycosidic bonds. Plant polysaccharides have been used in the food industry and in medicine for a long time. They have attracted excessive attention in recent years [10] because of their exhibited biocompatibility, low toxicity and pharmaceutical biomedical activity, such as antioxidant, immunomodulation, anti-tumor, anti-virus, anti-complementary, and anti-inflammatory activities [11].

Therefore, this paper optimized the extraction of maca leaves polysaccharide by RSM. In addition, the primary characterization of the crude MLPs were described, which may improve some theory value for the further study and comprehensive utilization of maca leaves.

2 Materials and Methods

2.1 *Material*

Fresh maca leaves were collected from Yunnan Province during the months of November-December, 2013, and identified by Professor Wenbin Hou, Tianjin Institute of Pharmaceutical Research, China. The leaves were shade dried, powdered, sieved through 800 μm and stored at room temperature. 1-phenyl-3-methyl-4-benzene formyl pyrazolone (PMP) was from Solarbio Co. Ltd (Beijing, China). All of the other chemicals were analytical grade.

2.2 *Pretreatment of Maca Leaves*

All the raw materials were immersed in petroleum ether (b. p. 60–90 °C) to remove the liposoluble constituents using Soxhlet apparatus. The crude polysaccharides were extracted from the pretreated maca leaves under extraction conditions and determined by the phenol-sulfuric acid method [12].

$$\text{Extraction yield (\%)} = (\mathrm{C} \times \mathrm{N} \times \mathrm{V}) / \mathrm{W} \times 100 \qquad (1)$$

where C is the concentration of polysaccharide as calculated from the calibrated regression equation (mg/mL); N is the dilution factor; V is the total volume of extraction solution (mL); and W is the weight of samples (mg).

2.3 Extraction of the Crude MLPs

The crude MLPs was obtained under the extraction conditions. The extraction solution were precipitated by the addition of anhydrous ethanol to a final concentration of 80% (v/v). The precipitate was collected by centrifugation at 4000 rpm for 20 min, and dissolved with water. After that, proteins were removed by the Sevag reagent (chloroform: butyl alcohol, 4:1, v/v). The precipitated were washed with anhydrous ethanol and dissolved with water. After lyophilization, the crude MLPs were obtained.

2.4 Single-Factor Design for the Crude MLPs Extraction

A single-factor design was used to determine the preliminary range of extraction factors including A (water to raw material ratio: 10, 20, 30, 40, 50 and 60 mL/g), B (extraction time: 30, 60, 90, 120, 150 and 180 min), C (extraction temperature: 55, 65, 75, 85, 95 and 100 °C) and D (extraction times: 1, 2, 3, 4 and 5). The extraction yield of the crude MLPs were the dependent variable.

2.5 BBD for the Crude MLPs Extraction

The software Design-Expert 8.0 was used to design the experiment, analyze the data and build the model. Box-Behnken design (BBD) with four independent variables at three levels was applied to determine the optimal extraction variables. Liquid-solid ratio (X_1), extraction time (X_2), extraction temperature (X_3) and extraction number (X_4) were selected to design the experimental project using BBD. Y means the extraction yield of polysaccharides from maca leaves and was taken as the response of the design experiments. Finally, Y was fitted a second-order model in order to correlate the response variable to the independent variables. The non-linear computer-generated quadratic model was given as below:

$$Y = \beta_0 + \sum_{j=1}^{k} \beta_j X_j + \sum_{j=1}^{k} \beta_{jj} X_j^2 + \sum\sum_{i<j} \beta_{ji} X_i X_j \tag{2}$$

where Y was the estimated response, and β_0, β_j, β_{jj} and β_{ji} were the regression coefficients for intercept, linearity, square and interaction. X_i and X_j were the independent coded variables (i $\neq$ j, i and j range from 1 to k), and k was the number of independent parameters (k = 3 in this study).

2.6 *Analysis of the Crude MLPs*

The crude MLPs were obtained under the finally optimum extraction conditions. Protein was removed by Sevag method [13] and the protein content was determined by the method of Bradford test [14]. All measurements were conducted in triplicate. Then, the crude MLPs were further depigmented by using D101 type macroporous adsorption resin [15].

The IR spectrum of the crude MLPs was determined using a fourier transform IR spectrophotometer (FT-IR) (Bruker, Germany). The freeze-dried crude MLPs powder was grinded with potassium bromide power and pressed into 1 mm pellets in the frequency range of 4000–400 cm^{-1}.

2.7 *Statistical Analysis*

Values from analysis are expressed as mean ± standard deviation (S.D.). Data sets with multiple comparisons were evaluated by one-way analysis of variance (ANOVA) with Duncan's test. Differences were considered significant at $P < 0.05$.

3 Results and Discussion

3.1 *Single-Factor Experiments Relating to the Crude MLPs Extraction*

3.1.1 Effect of Water to Raw Material Ratio on the Crude MLPs Yield

The ratio of water to raw material was an important parameter of the polysaccharides. As the Fig. 1a shown that the yield of the crude MLPs were influenced under the ratio from 10 to 60. The other extraction conditions were set as follows: extraction time 60 min, extraction temperature 75 °C and extraction times 1. The yield of the crude MLPs was increased rapidly with the increase of the ratio from 10 to 30 mL/g, increased gradually at ratios from 30 to 50 mL/g, but, decreased at ratios above 50 mL/g. Thus, to minimize electricity and time costs, the water to raw material range of 40–60 mL/g was considered for use in further BBD experiments.

3.1.2 Effect of Extraction Time on the Crude MLPs Yield

Extraction time was another factor that would influence the extraction efficiency. As the Fig. 1b shown that the yield of the crude MLPs were influenced under the time

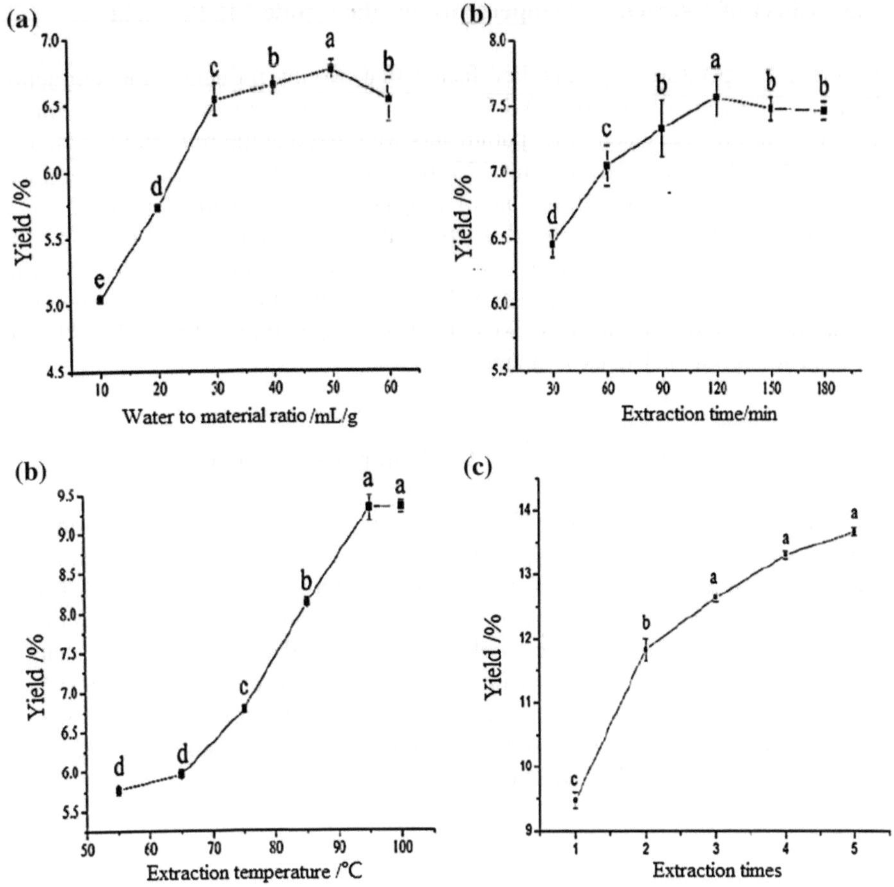

Fig. 1 Effects of water to raw material ratio (**a**), extraction time (**b**), extraction temperature (**c**), extraction times (**d**) on extraction yield of the crude MLPs. Different *lower case letters* in the same curve indicate significant differences ($p < 0.05$). Date shown in mean ± standard deviation (n = 3)

from 30 to 180 min. Other extraction conditions were set as follows: the ratio of water to raw material (40 mL/g), extraction temperature 75 °C and extraction times 1. The Fig. 1b illustrated the yield of the crude MLPs increased from 6.46 ± 0.10 to 7.57 ± 0.15% with the extraction time varied from 30 to 120 min and decreased thereafter, indicating that longer extraction time can lead to thermal instability and degradation of the polysaccharides. Consequently, extraction time of from 90 to 150 min was used for further BBD experiments.

3.1.3 Effect of Extraction Temperature on the Crude MLPs Yield

Extraction temperature was another factor that would influence the extraction efficiency. The yield of the crude MLPs affected by different extraction temperature was shown in Fig. 1c, when other parameters were fixed at the ratio of water to raw material (40 mL/g), extraction time 120 min and extraction times 1. As evident from Fig. 1c, the maximum amount of polysaccharides obtained was 9.32% at 95 °C, the extraction yield increased sharply with extraction temperature from 45 to 95 °C (5.58 ± 0.07 to 9.33 ± 0.07%). This can be explained by the fact that polysaccharides solubility increases with temperature, thus facilitating polysaccharides diffusion out of the cells. So, extraction temperature of from 85 to 100 °C was used for further BBD experiments.

3.1.4 Effect of Extraction Time on the Crude MLPs Yield

The extraction time was another factor that would influence the extraction efficiency. As shown in Fig. 1d that the yield of the crude MLPs were influenced under the extraction times from 1 to 5 when the other conditions were 40 (mL/g), 120 min and 90 °C (the ratio of water to raw material, extraction time and extraction temperature). Figure 1d shows that the yield of the crude MLPs was increased sharply with the increase of the extraction times from 1 to 2, reaching to 11.8 ± 0.18% under the extraction times 2, increased gradually at extraction times from 3 to 5. Thus, to minimize electricity and time costs, the extraction times of from 1 to 3 was considered for use in further BBD experiments.

3.2 The Result of BBD Design

On the basis of single factor experiment, using the method of Box-Behnken Design (BBD) designed four factors three levels center test. Then using the Design-Expert 8.0.6 software based on the yield of the crude MLPs as response value table, test Design scheme and test results were shown in Table 1.

3.3 Response Surface Results of the Crude MLPs Yield

Response surface plots and contour plots can be obtained according to the regression model combined with the actual value of the response surface optimization analysis, as shown in Figs. 2 and 3. And the comprehensive analysis can be concluded that the interaction was between the various factors on the yield of the crude MLPs.

Table 1 BBD with the experimental values and predicted values for extraction yield of the crude MLPs

Run	Actual (Code)				Yield (%)	
	X_1 Water/solid	X_2 Time	X_3 Temperature	X_4 Number	Observed	Predicted
1	30 (−1)	120 (0)	100 (1)	2 (0)	1.0529	1.0998
2	40 (0)	120 (0)	80 (−1)	3 (1)	1.1037	1.1039
3	40 (0)	150 (1)	100 (1)	2 (0)	1.1737	1.2568
4	30 (−1)	120 (0)	80 (−1)	2 (0)	0.9793	0.8578
5	50 (1)	90 (−1)	90 (0)	2 (0)	1.0442	1.0371
6	30 (−1)	150 (1)	90 (0)	2 (0)	1.0899	1.0496
7	40 (0)	120 (0)	90 (0)	2 (0)	1.1724	1.1491
8	50 (1)	120 (0)	100 (1)	2 (0)	1.1559	1.1718
9	40 (0)	120 (0)	80 (−1)	1 (−1)	0.7295	0.7218
10	40 (0)	120 (0)	90 (0)	2 (0)	1.1625	1.1491
11	50 (1)	120 (0)	90 (0)	1 (−1)	0.7402	0.7628
12	50 (1)	150 (1)	90 (0)	2 (0)	1.0902	1.0271
13	40 (0)	150 (1)	80 (−1)	2 (0)	0.9606	1.0588
14	40 (0)	120 (0)	90 (0)	2 (0)	1.2054	1.1493
15	50 (1)	120 (0)	80 (−1)	2 (0)	1.0073	0.9698
16	30 (−1)	120 (0)	90 (0)	1 (−1)	0.6225	0.6286
17	40 (0)	150 (1)	90 (0)	3 (1)	1.3861	1.3046
18	40 (0)	120 (0)	90 (0)	2 (0)	1.0254	1.1491
19	50 (1)	120 (0)	90 (0)	3 (1)	1.0113	1.1384
20	30 (−1)	90 (−1)	90 (0)	2 (0)	0.8024	0.8275
21	40 (0)	90 (−1)	100 (1)	2 (0)	1.1909	1.1708
22	40 (0)	150 (1)	90 (0)	1 (−1)	0.7931	0.7789
23	30 (−1)	120 (0)	90 (0)	3 (1)	1.0282	1.0967
24	40 (0)	90 (−1)	90 (0)	1 (−1)	0.7189	0.7742
25	40 (0)	120 (0)	100 (1)	3 (1)	1.4358	1.3638
26	40 (0)	90 (−1)	90 (0)	3 (1)	1.1149	1.0833
27	40 (0)	90 (−1)	80 (−1)	2 (0)	0.9306	0.9248
28	40 (0)	120 (0)	90 (0)	2 (0)	1.1824	1.1493
29	40 (0)	120 (0)	100 (1)	1 (−1)	0.9606	0.9058

3.4 *Model Fitting and Optimization of the Crude MLPs Extraction*

3.4.1 Model Fitting

By applying multiple regression analyses on the experimental data, the response variables and test variables were related by the following second-order polynomial equation:

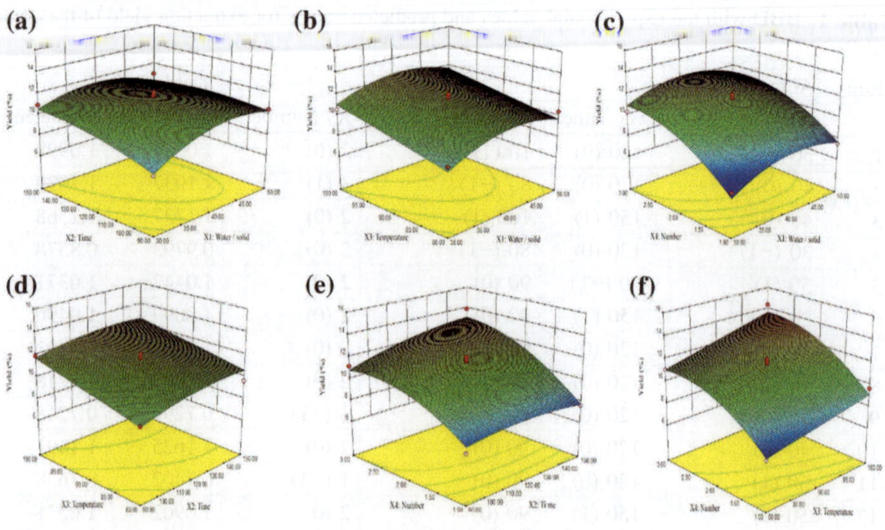

Fig. 2 Response surface plots (3D) showing the effects of variables on response yield of the crude MLPs

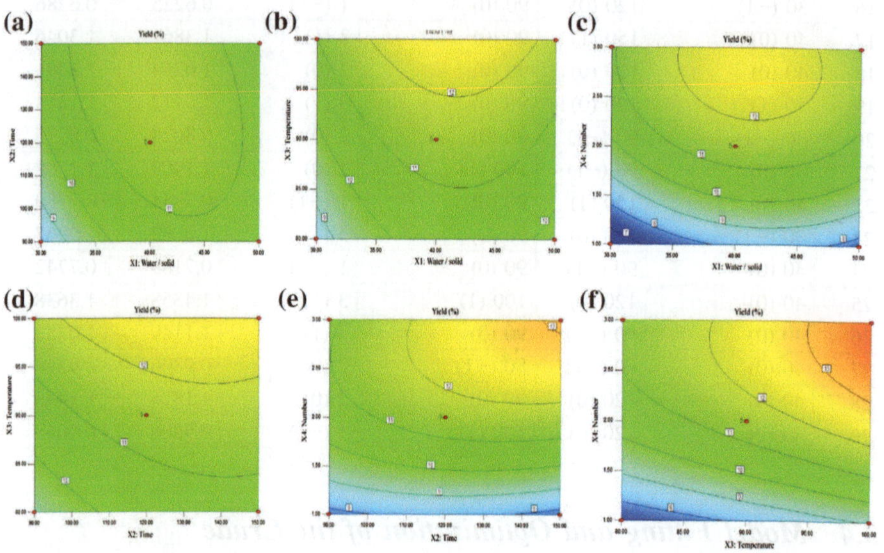

Fig. 3 *Contour plots* showing the effects of variables on response yield of the crude MLPs

$$\begin{aligned} Y = {} & 11.49 + 0.46X_1 + 0.55X_2 + 1.11X_3 + 2.10X_4 \\ & - 0.57X_1X_2 - 0.10X_1X_3 - 0.22X_1X_4 - 0.12X_2X_3 \\ & + 0.57X_2X_4 + 0.19X_3X_4 - 1.22X_1^2 - 0.44X_2^2 - 0.022X_3^2 - 1.23X_4^2 \end{aligned} \quad (3)$$

where *Y* represents the yield of MLPs, and X_1, X_2, X_3, X_4 are water to raw material ratio, extraction time, extraction temperature and extraction times, respectively.

The ANOVA was shown in Table 2. *P*-values were used to check the significance of each coefficient, and the smaller *P*-value means the more significance relevant coefficient. As shown in Table 2: a model *P*-value of $P < 0.0001$ indicated very high significance of the regression, and the lack of Fit *F*-value of 1.51 implies the lack of Fit is not significantly relative to the pure error. Meanwhile, the determination coefficient R^2 (0.9081) and the adjusted R^2 (0.8162) were enough approximate to the model. It also can be found that X_3, X_4, X_1^2 and X_4^2 were high significant ($P < 0.01$), X_2 was significant to explain the effect of variables on the yield of MLPs. And others were not significant to explain the yield of the crude MLPs.

Table 2 ANOVA for response surface quadratic model analysis of variance table

Source	Sum of squares	df	Mean square	*F*-value	*P*-value	Significant levels
Model	94.46	14	6.75	9.88	<0.0001	***
A	2.49	1	2.49	3.65	0.0769	
B	3.62	1	3.62	5.31	0.0371	*
C	14.8	1	14.8	21.68	0.0004	***
D	53.07	1	53.07	77.73	<0.0001	***
AB	1.3	1	1.3	1.9	0.1895	
AC	0.042	1	0.042	0.062	0.8077	
AD	0.2	1	0.2	0.3	0.5949	
BC	0.057	1	0.057	0.083	0.7774	
BD	1.31	1	1.31	1.92	0.1875	
CD	0.14	1	0.14	0.2	0.6597	
A^2	9.71	1	9.71	14.22	0.0021	**
B^2	1.23	1	1.23	1.81	0.2004	
C^2	3.10E−03	1	3.10E−03	4.54E−03	0.9473	
D^2	9.83	1	9.83	14.4	0.002	**
Residual	9.56	14	0.68			
Lack of fit	7.56	10	0.76	1.51	0.3675	
Pure error	2	4	0.5			
Correlation Total	104.02	28				
R^2	0.9081		Adj-R^2	0.8162		

Note *$P < 0.05$, significant; **$P < 0.01$, high significant, ***$P < 0.001$, very high significance

3.4.2 Optimization for the Extraction Conditions of the Crude MLPs

In order to get a profound understanding of the results as shown in Table 2 and the interactions of the variables, as shown in Figs. 2 and 3, the 3D plots and their corresponding contour plots depicted the interactions between two variables when the others were kept at their zero levels. Also the 3D plots and their corresponding contour plots provided a vivid interpretation of the interaction for two variables and facilitated the location of optimum experimental conditions. The optimal extractions of MLPs from maca leaves were as follows: ratio of water to raw material (38 mL/g), extraction time 150 min, extraction temperature 100 °C and extraction times 3. Under the above conditions, the maximum predicted yield was 14.24%, which corresponded fairly well to that of experimental value 14.20 ± 0.25% (n = 3).

3.5 Primary Characterization of the Crude MLPs

The protein content of the crude MLPs obtained by the finally optimum extraction conditions was 5.63 ± 0.26%. After deproteinization and depigment, the protein level of the crude MLPs were 0.09% by the Bradford test. The deproteinization rate was 95.7%.

In order to characterize primarily the crude MLPs, FT-IR analysis was performed. The FT-IR results was shown in Fig. 4. According to the FT-IR analysis,

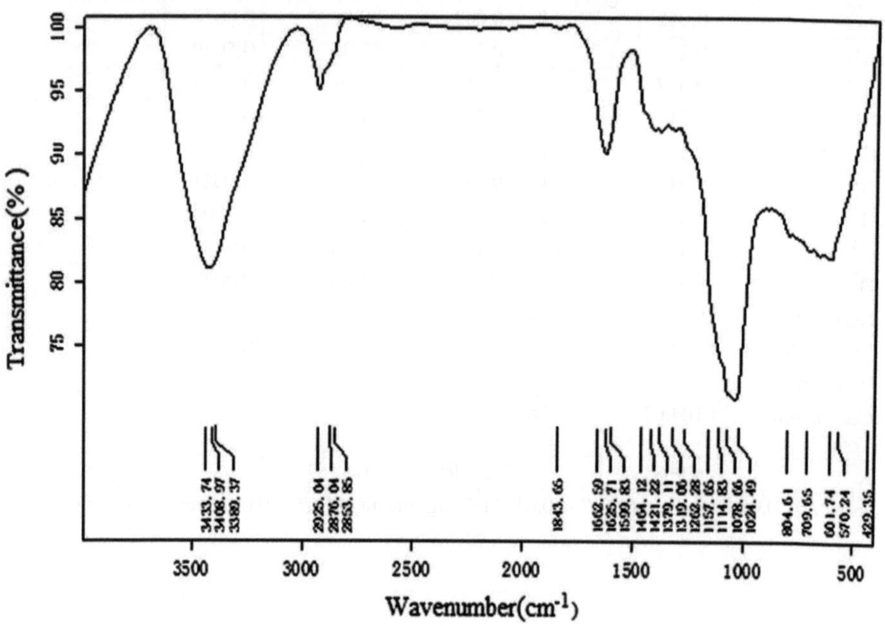

Fig. 4 FT-IR spectrum analysis of the crude MLPs

the strong band at 3445 cm^{-1} represented the stretching of the hydroxyl group (O–H), and the band at 2929 cm^{-1} attributed to the C–H stretching of the CH^2 and CH^3 group. The asymmetrical peak at around 1562 cm^{-1} and symmetrical peak at 1443 cm^{-1} were due to the carboxyl group in the crude MLPs. The peak at around 1100 cm^{-1} indicated C–O stretching vibration [16]. Three absorption peaks at around 1100–1010 cm^{-1} indicated a pyranose form, which was performance (1114, 1078 and 1024 cm^{-1}). The crude MLPs was similar to sugar characteristic absorption peaks in the IR spectrum.

4 Conclusions

In this study, response surface methodology (RSM) was used to optimize extraction process of polysaccharides from maca leaves. The optimal conditions for MLPs extraction were as follows: ratio of water to raw material (38 mL/g); extraction time, 150 min; extraction temperature, 100 °C and extraction times, 3. Under the optimal extraction condition, the maximum predicted yield of the crude MLPs was 14.4% which was close to the experimental yield 14.2 ± 0.25% (n = 3) and had a good reproducibility. The FT-IR analysis revealed the general characteristic absorption peaks of the crude MLPs. These results indicated that the MLPs have enormous potential value as a novel natural polysaccharides in functional food. Also the MLPs need further works on polysaccharides structures, pharmacological activity evaluation, and application studies.

Acknowledgements This research was financially supported by the Research Project of People's Liberation Army (No. AX110C002, BX115C007 and HX-04-13-022).

References

1. Wu SH, Gong GL, Wang YY, Li F, Jia SY, Qin FX et al (2013) Response surface optimization of enzyme-assisted extraction polysaccharides from Dictyophoraindusiata. Int J Biol Macromol 61:63–68
2. Qu Y, Li CX, Zhang C, Zeng R, Fu CM (2016) Optimization of infrared-assisted extraction of Bletillastriata polysaccharides based on response surface methodology and their antioxidant activities. Carbohyd Polym 148:345–353
3. Zha SH, Zhao QS, Chen JJ, Wang LW, Zhang GF, Zhang H, Zhao B (2014) Extraction, purification and antioxidant activities of the polysaccharides from maca (*Lepidium meyenii*). Carbohyd Polym 111:584–587
4. Esparza E, Hadzich A, Kofer W, Mithöfer A, Cosio Eric G (2015) Bioactive maca (*Lepidium meyenii*) alkamides are a result of traditional Andean postharvest drying practices. Phytochem 116:138–148
5. Chen JJ, Zhao QS, Wang LW, Zha SH, Zhang LJ, Zhao B (2015) Physicochemical and functional properties of dietary fiber from maca (*Lepidium meyenii Walp.*) liquor residue. Carbohyd Polym 132:509–512

6. Uchiyama F, Jikyo T, Takeda R, Ogata M (2014) *Lepidium meyenii* (Maca) enhances the serum levels of luteinising hormone in female rats. J Ethnopharmacol 151:897–902
7. Bai N, He K, Roller M, Lai CS, Bai L, Pan MH (2015) Flavonolignans and other constituents from *Lepidium meyenii* with activities in anti-inflammation and human cancer cell lines. J Agric Food Chem 63:2458–2463
8. Liu H, Jin W, Fu C, Dai P, Yu Y, Huo Q, Yu L (2015) Discovering anti-osteoporosis constituents of maca (*Lepidium meyenii*) by combined virtual screening and activity verification. Food Res 77:215–220
9. Stojanovska L, Law C, Lai B, Chung T, Nelson K, Day S et al (2015) Maca reduces blood pressure and depression, in a pilot study in postmenopausal women. Climacteric 18:69–78
10. Sandoval M, Okuhama NN, Angeles FM, Melchor VV, Condezo LA, Lao J et al (2002) Antioxidant activity of the cruciferous vegetable Maca (*Lepidium meyenii*). Food Chem 79:207–213
11. Li E, Nie SP, Yang C, Qiu ZH, Xie MY (2011) Extraction optimization, characterization and bioactivity of the crude polysaccharides from Herba Moslae. Carbohyd Polym 83:1201–1206
12. Wang QH, Shu ZP, Xu BQ, Xing N, Jiao WJ, Yang BY, Kuang HX (2014) Structural characterization and antioxidant activities of polysaccharides from *Citrus aurantium* L. Int J Biol Macromol 67:112–123
13. Dubois M, Gilles KA, Hamilton JK, Rebers PT, Smith F (1956) Colorimetric method for determination of sugars and related substances. Anal Chem 28:350–356
14. Vilkas E, Radjabi NF (1986) The glucomannan system from Aloe vahombe (liliaceae). III. Comparative studies on the glucomannan components isolated from the leaves. Biochimie 68 (9):1123–1127
15. Bradford MM (1976) A rapid and sensitive method for the quantitation of microgram quantities of protein utilizing the principle of protein-dye binding. Anal Biochem 72:248–254
16. Li C, Huang Q, Fu X, Yue XJ, Liu RH, You LJ (2015) Characterization, antioxidant and immunomodulatory activities of polysaccharides from *Prunella vulgaris* Linn. Int J Biol Macromol 75:298–305

Metabolomic Analysis of *Dunaliella salina* upon Concurrent Deprivation of Nitrogen and Phosphor

Hexin Lv, Xianggan Cui, Shilei Wang and Shiru Jia

1 Introduction

Dunaliella salina is a halotolerant microalga and could accumulate 14% beta-carotene of dry weight under stress conditions such as nutrients deprivation, and thus it is a model alga for osmoregulation and carotenoids biosynthesis researches. We have reported that the concurrent deprivation of nitrogen and phosphor has sever effects on many aspects such as decreased growth rates and photosynthetic efficiencies, altered activities of superoxide dismutase and catalase, accumulation of carotenoids and increased expression levels of methylerythritol 4-phosphate pathway and carotenoids biosynthesis pathway genes [1].

Metabolomics could detect global metabolites changes of cells. We have used metabolomics strategy to interrogate the global metabolites changes when cells were subjected to nitrogen deprivation and many differential small metabolites were identified [2]. Herein, GC-MS were used to detect the metabolites changes when *D. salina* cells shifting culture conditions to concurrent deprivation of nitrogen and phosphor. Differential metabolites were identified and statistical analyses by principal component analysis (PCA) and partial least-squares (PLS) indicated that cells metabolites changed significantly upon concurrent deprivation of nitrogen and phosphor.

H. Lv · X. Cui · S. Wang · S. Jia (✉)
Key Laboratory of Industrial Fermentation Microbiology, Ministry of Education,
Tianjin Key Lab of Industrial Microbiology, Tianjin University of Science and Technology,
13th. 29, Tianjin 300457, People's Republic of China
e-mail: jiashiru@tust.edu.cn

H. Lv
e-mail: lvhx@tust.edu.cn

H. Liu et al. (eds.), *Advances in Applied Biotechnology*, Lecture Notes
in Electrical Engineering 444, https://doi.org/10.1007/978-981-10-4801-2_23

2 Materials and Methods

2.1 Algae and Cultivation Conditions

D. salina was cultured in the modified Johnson's medium [1]. Cells were cultured in 500 mL Erlenmeyer flasks containing 250 mL of medium under 60 μmol m^{-2} s^{-1} continuous illumination under 30.0 °C. Cells were inoculated at 2.0 × 10^5 cell/ml. Cultures were manually shaken three times per day. All the cells were cultured for two 16/8 h light/dark cycles to synchronize the growth phases before inoculation and thereafter transfer to continuous light conditions. Concurrent nitrogen and phosphor deprivation was performed according to our previous reports [1].

2.2 Extraction of Intracellular Metabolites and Fatty Acid Methyl Esters

Cells were quenched with −40 °C 60% (v/v, methanol/water) methanol solution for 5 min. Cells were collected by 4000 rpm centrifugation for 5 min at 4 °C and washed with 0.5 mol/L NaCl for three times. Cell pellets were put into in liquid nitrogen and ground into powder. The intracellular metabolites were extracted according to previously reported methods [3]. Metabolites quenching and extraction were performed according to previous report [4] and with some modifications. The fatty acids were extracted according to a modified method [5]. Briefly, 10 mg cell pellets was suspended in 0.8 mL fatty acid extract (Trichloromethane:methanol, 2:1, v:v) and 10 μL 80 μg/ml internal standard solution nonadecanoic acid was added, and 1.2 mL FAMES reagent (10% HCl-methanol, m:m) was added into the samples, and incubated at 80 °C for 1 h. 1 mL hexane was used to extract fatty acid methyl esters.

2.3 Derivatization and GC-MS Analysis

Metabolites were derivatized according to our previous report [2]. A GC-MS system (Agilent 7890 GC-5975 MSC, USA) equipped with a HP-5 capillary column (60 m × 320 μm i.d., 0.25 μm film thickness; Agilent J&W Scientific, Folsom, CA, USA) was selected in our study. 1 μL of sample was injected without a split ratio. Helium with flow rate of 1 mL/min was used as the carrier gas. Full scan mode with (m/z: 50-800) electron impact ionization (70 eV) was used. The oven temperature was set to 70 °C, after 2 min raised to 290 °C at 5 °C/min, and maintained for 6 min. The oven temperature for measuring fatty acid methyl esters was set to 70 °C for 2 min, followed by 8 °C/min to 200 °C for 2 min, then increased at a rate 3 °C/min to 245 °C and maintained at 245 °C for 6 min.

2.4 Data Analysis and Statistical Analyses of GC-MS

The MSD Productivity ChemStation software (version E.02.01.1177 Agilent) was used for typical total ion chromatograms analyses. Metabolite were identified by NIST mass spectral library 2011 (version 2.0 g). PCA was used for initial analysis. PLS analysis was used to further verify the differences and identify the metabolites responsible for distinguishing the three different growth phases. Loading plots were used for finding biomarkers. Score plots were used to previewing the clustering effect. The relative contents of metabolites and fatty acids were calculated according to a previous method [6]. HCE 3.5 software (Human-computer interaction lab, University of Maryland) was used for unsupervised hierarchical cluster analysis (HCA), the metabolite levels of three pigmentation phase were normalized by default firstly, then average linkage method were adopted for hierarchical clustering.

3 Results and Discussion

3.1 Metabolites Identified by GC-MS and Clustering

To investigate the global metabolites changes between nutrients replete and depleted conditions, metabolites from *D. salina* cells cultured under concurrent nitrogen and phosphor deprivation conditions for seven days were derivatized by methoxamine hydrochloride and N-methyl-N-(trimethylsilyl) trifluoroacetamide and then analyzed by GC-MS. By this method, only thermally stable and volatile compounds with molecular weight generally less than 1000 Daltons could be detected. By comparing with the NIST library and filtering for match degrees greater than 700, 81 kinds of metabolites were identified at the initial threshold between 0.1 and 15.0, including 16 amino acids, 13 fatty acids, 14 organic acids, 18 sugars, 12 alcohols, 5 amines and 3 others which is unable to be recognized by NIST library (Table 1). In order to know their changing trends, hierarchical clustering was used to categorize the detected metabolites according to the relative contents in comparison with inner control (Fig. 1). Results showed that most sugars and part of amino acids were subcategorized into one group, whose contents were increased upon concurrent deprivation of nitrogen and phosphor, indicted the activities of EMP pathway or starch metabolism pathway were changed. The accumulation of glucose, sucrose, fructose and other sugars implicated that these metabolites provided more substrates for triacylglycerol biosynthesis, because we investigated that lipid droplets were substantially increased when concurrent deprivation of nitrogen and phosphor. In addition, as expected, the contents of protective metabolites such as tagatose, sorbose, talose, myo-inosito and meso-erythrotol were increased too, suggested that cells triggered that protective system to resist the unfavorable conditions. Conversely, metabolites from another group mainly included organic acids and alcohols, their contents were decreased.

Table 1 Metabolites detected by GC-MS of cells cultured under complete media and that deprived of nitrogen and phosphor[a]

Class	Components	Ratio (-N-P/CM)
Amino acids	N,N-Dimethylglycine	0.75
	L-Valine	0.78
	l-Alanine	0.63
	l-Leucine	1.82
	l-Isoleucine	0.85
	L-Norvaline	0.56
	Serine	0.83
	l-Threonine	1.29
	L-Serine	0.80
	N,O,O-Tris(trimethylsilyl)-L-threonine	1.09
	DL-Ornithine	1.32
	N-α-Acetyl-L-Lysine	2.46
	L-Proline	2.90
	Glycyl-l-glutamic acid	1.12
	Glutamic acid	1.90
	L-Lysine	ND in -N-P
Sugars	D-(-)-Erythrose	ND in CM
	L-(-)-Trehalose	1.30
	D-Ribose	2.96
	D-(-)-Tagatose	ND in CM
	L-(-)-Sorbose	ND in CM
	d-Galactose	1.31
	D-(+)-Talose	2.66
	D-Mannose	ND in CM
	D-Glucose	5.75
	2-Deoxy-D-ribose	ND in CM
	D-Erythro-Pentose	1.15
	D-erythro-2-Pentulose	ND in CM
	D-Psicofuranose	ND in CM
	Lactose	1.00
	D-Fructose	ND in CM
	D-Mannose	3.75
	Sucrose	6.61
	D-(+)-Cellobiose	22.16
Fatty acid	Nonanoic acid	ND in -N-P
	Dodecanoic acid	ND in -N-P
	Hexadecanoic acid	1.81
	trans-9-Octadecenoic acid	0.14
	Myristate	1.33
	Tetradecanoic acid	ND in CM

(continued)

Table 1 (continued)

Class	Components	Ratio (-N-P/CM)
	cis-11-Eicosenoic acid	ND in -N-P
	cis-13-Eicosenoic acid	ND in -N-P
	11-Eicosenoic acid	ND in -N-P
	Decanoic acid	ND in -N-P
	cis-13-Docosenoic acid	ND in -N-P
	Docosanoic acid	ND in -N-P
	Octadecanoic acid	1.17
Oganic acid	L-(+)-Lactic acid	2.99
	Acetic acid	1.10
	Ethanedioic acid	1.22
	2,3,4-Trihydroxybutyric acid	1.14
	Pentanedioic acid	1.00
	Heptanedioic acid	ND in CM
	Phosphoric acid	2.77
	Azelaic acid	1.16
	Propanoic acid	0.64
	2-Ketoglutaric acid	2.62
	Sebacic acid	2.62
	1,2,3-Propanetricarboxylic acid	1.38
	Deoxycholic acid	0.84
	L-Ascorbic acid	2.17
Alcohols	Glycerol	0.98
	Pentitol	1.06
	D-Erythro-Pentitol	1.27
	D-(+)-Arabitol	0.62
	L-(-)-Arabitol	3.07
	Xylitol	1.58
	Adonitol	1.34
	L-Fucitol	ND in CM
	d-Mannitol	ND in -N-P
	D-Sorbitol	ND in -N-P
	Myo-Inositol	1.51
	meso-Erythritol	1.18
Amines	Acetamide	0.64
	Cadaverine	1.37
	3-methylol-methylamine	1.29
	Urea	0.96
	N-[4-[Bis-amino]butyl] acetamide	1.36
Others	Unknown 3	1.36
	Unknown 5	1.38
	Unknown 6	ND in CM

[a]*ND* not detected

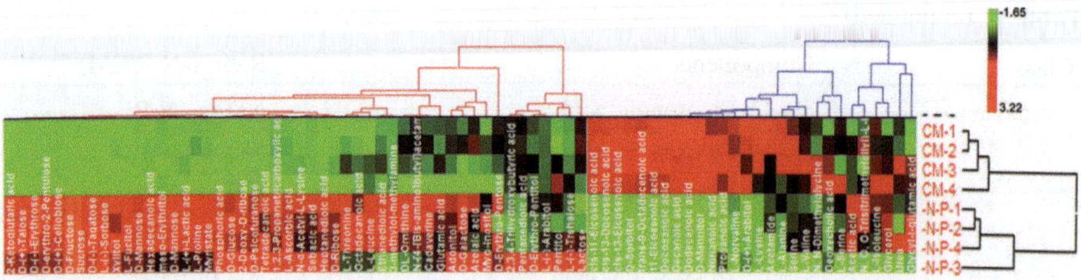

Fig. 1 Clustering of metabolites from *D. salina* cells cultured under complete media and that deprived of nitrogen and phosphor

Table 2 Statistical data from PCA and PLS at different culture conditions

Sample	PCA		PLS		
D. salina (-N-P)	R^2X	Q^2	R^2X	R^2Y	Q^2
	0.991	0.820	0.991	1	0.999

3.2 *Multivariate Data Analysis*

Multivariate statistical analysis PCA and PLS were used to study changes in the intracellular metabolites between nutrients replete and deplete condition. Results showed that both PCA and PLS have satisfactory predictive and fitting abilities based on R2X and Q2 which are more than 0.8 (Table 2). The unsupervised clustering method PCA was used to identify and rank major sources of variance within the two data sets. Based on similarities and differences in the measured parameters, PCA was able to cluster biological samples into both expected and unexpected groups. Samples in different phases were separated clearly on the PCA score plot (Fig. 2). The first principal component (PC1) accounted for 92.6% of the total variance between the two experimental groups. The PC1 vs. PC2 of PCA plot (these two components account for 98.1% of the variance) clearly separates the two kinds of cultures. These results show that the intracellular metabolic profiles of concurrent nitrogen and phosphor deprived cells differ significantly. PCA loading plots were used to analyse the contributions of each metabolite to the principal components. The contributions of data points were evaluated based on their distances from the origin point [7]. The potential biomarkers identified by PCA loading plots were sucrose, glycerol, L-(+)-Lactic acid, DL-Ornithine, hexadecanoic acid, trans-9-Octadecenoic acid, cis-13-Eicosenoic acid, cis-11-Eicosenoic acid, L-Proline, suggested that the activities of the related pathways were changes significantly (Fig. 3).

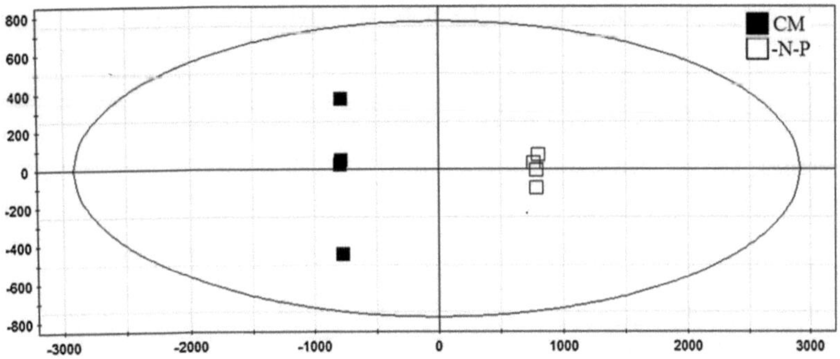

Fig. 2 PCA score plot of metabolites from *D. salina* cells cultured under complete media and that deprived of nitrogen and phosphor

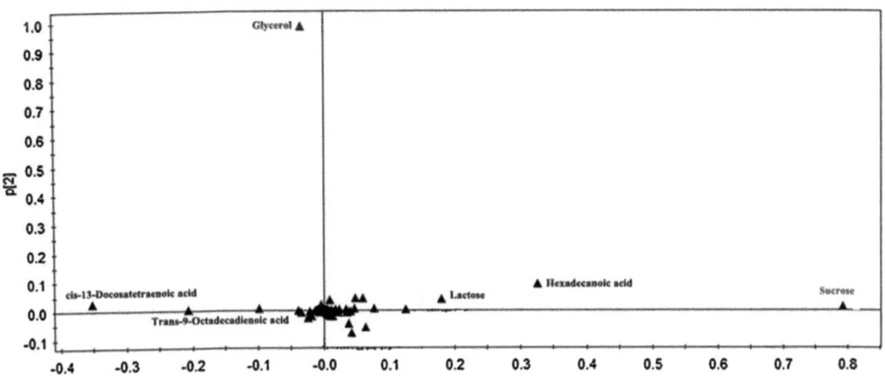

Fig. 3 PCA loading plot of metabolites from *D. salina* cells cultured under complete media and that deprived of nitrogen and phosphor

3.3 *Profiles of Fatty Acids Metabolism*

To further investigate the fatty acids profiles, fatty acids were extracted and esterified by FAMES reagent, and analyzed by GC-MS. A total of 15 kinds of fatty acids were detected (Table 3). The contents of total fatty acids were increased significantly and the saturation of fatty acids also increased. These results were similar to that of single nitrogen deprivation [1]. Among the detected fatty acids, the most abundant are hexadecanoic acid and 9,12,15-Octadecatrienoic acid. 4,7,10,13,16,19-Docosahexaenoic acid which have special pharmaceutical value was detected and its content was decreased upon concurrent deprivation of nitrogen and phosphor.

Table 3 Statistical data from PCA and PLS at different culture conditions

FA	CX:Y	Relative content (%)	
		-N-P	CM
Tridecanoic acid	(13:0)	0.55 ± 0.02	0.45 ± 0.02
Hexadecanoic acid	(16:0)	28.5 ± 0.7	23.0 ± 0.4
Heptadecanoic acid	(17:0)	1.38 ± 0.17	0.81 ± 0.09
∑SFA		30.5 ± 0.9	24.2 ± 0.5
9-Hexadecenoic acid	(16:1)	1.70 ± 0.40	2.28 ± 0.04
10-Octadecenoic acid	(18:1)	ND	0.17 ± ND
7,10-Hexadecadienoic acid	(16:2)	3.24 ± 0.11	3.45 ± 0.09
7,10-Octadecadienoic acid	(18:2)	0.91 ± 0.04	0.85 ± ND
9,12-Octadecadienoic acid	(18:2)	12.2 ± 0.0	11.5 ± 0.1
10,13-Eicosadienoic acid	(20:2)	ND	0.22 ± 0.01
γ-Linolenic acid	(16:3)	1.14 ± 0.04	1.22 ± ND
9,12,15-Octadecatrienoic acid	(18:3)	38.1 ± 0.7	35.4 ± 0.4
cis-5,8,11-Eicosatrienoic acid	(20:3)	2.99 ± 0.06	3.87 ± 0.06
5,8,11,14-Eicosatetraenoic acid	(20:4)	ND	3.65 ± 0.14
6,9,12,15-Docosatetraenoic acid	(22:4)	0.72 ± 0.11	0.75 ± 0.04
4,7,10,13,16,19-Docosahexaenoic acid	(22:6)	8.5 ± 0.1	11.7 ± 0.2
∑UFA		69.5 ± 1.8	75.1 ± 1.0
Total FA		711.0 ± 24.0	546.0 ± 13.0

[a]*ND* not detected

4 Conclusion

In this study, the metabolome of *D. salina* upon concurrent deprivation of nitrogen and phosphor was profiled. Results showed the contents of most sugars and protective metabolites increased but the contents of fatty acids and part of amino acids decreased upon nutrients deprivation. Statistical analyses suggested metabolites changes significantly and biomarkers which could distinguish the replete and depleted nutrients conditions were identified.

Acknowledgements This project was supported by the National Natural Science Foundation of China (No. 31401029). There are no conflicts of interest to declare.

References

1. Lv H, Cui X, Wang S et al (2016) Metabolic profiling of *Dunaliella salina* shifting cultivation conditions to nitrogen deprivation. Metabolomics 6:170
2. Lv H, Cui X, Wahid F et al (2016) Analysis of the physiological and molecular responses of *Dunaliella salina* to macronutrient deprivation. PLoS ONE 11:e0152226

3. Weckwerth W, Wenzel K, Fiehn O (2004) Process for the integrated extraction, identification and quantification of metabolites, proteins and RNA to reveal their co-regulation in biochemical networks. Proteomics 4:78–83
4. Villas-Boas SG, Hojer-Pedersen J, Akesson M et al (2005) Global metabolite analysis of yeast: evaluation of sample preparation methods. Yeast (Chichester, England) 22:1155–1169
5. Lee S-Y, Kim S-H, Hyun S-H et al (2014) Fatty acids and global metabolites profiling of Dunaliella tertiolecta by shifting culture conditions to nitrate deficiency and high light at different growth phases. Process Biochem 49:996–1004
6. Liu M, Zhong C, Wu X-Y et al (2015) Metabolomic profiling coupled with metabolic network reveals differences in Gluconacetobacter xylinus from static and agitated cultures. Biochem Eng J 101:85–98
7. Jalali-Heravi M, Masoum S, Shahbazikhah P (2004) Simulation of 13C nuclear magnetic resonance spectra of lignin compounds using principal component analysis and artificial neural networks. J Magn Reson (San Diego, California 1997) 171:176–185

Effect of Yeast Extract on Production of ε-poly-L-lysine by *Streptomyces diastatochromogenes*

Fengzhu Guo, Haoran Zheng, Xue Zhang, Yawen Cheng, Zhilei Tan and Shiru Jia

1 Introduction

As a natural antimicrobial peptide, ε-poly-L-lysine (ε-PL) has been widely used as a food preservative and granted as Generally Recognized As Safe (GRAS) status by the FDA for application as a preservative in certain foods [1, 2]. In April 2014, ε-PL and ε-PL·HCl has been approved as new varieties of food additive in China. Currently, it can be used in fruits and vegetables, rice and rice products, grains products, meat and meat products, spices and other products in seven categories. In order to reduce production costs and to improve its application rate for ε-PL, there is need for further study of the fermentation process for ε-PL.

Nitrogen sources, as a nutrient, is closely related to the growth of microbes which can be divided into organic sources (e.g. yeast extract, peptone, soybean cake powder) and inorganic nitrogen sources (e.g. ammonium sulfate, nitrates). The categories and quantities of organic nitrogen sources are essential to the microorganism's growth and the secretion of the products.

Yeast extract is one of the commonly used organic nitrogen sources, which is very complex as it contains many kinds of nutrients including polypeptide, nucleotide, amino acid, vitamin B and trace elements. The yeast extract from different manufacturers show different fermentation results due to the complex of substrate for yeast extracts and the unmanageable of the processing technology.

Shima et al. have reported that the L-lysine molecule is directly utilized in ε-PL biosynthesis [3]. In most bacteria, L-lysine is biosynthesized by the amino acid biosynthetic pathway from L-aspartic acid (aspartate pathway). The first enzyme in this pathway is aspartokinase (Ask; EC.2.7.2.4), which is subject to complex

F. Guo · H. Zheng · X. Zhang · Y. Cheng · Z. Tan · S. Jia (✉)
Key Laboratory of Industrial Fermentation Microbiology, Ministry of Education,
College of Biotechnology, Tianjin University of Science and Technology, Tianjin 300457,
People's Republic of China
e-mail: jiashiru@tust.edu.cn

H. Liu et al. (eds.), *Advances in Applied Biotechnology*, Lecture Notes
in Electrical Engineering 444, https://doi.org/10.1007/978-981-10-4801-2_24

regulation by L-lysine and/or L-threonine to provide sufficient amounts of L-lysine for ε-PL biosynthesis [4].

Hamano et al. [4] have reported the purification of an ε-PL synthetase (Pls; 130KDa) and the cloning of its gene from an ε-PL producing strain of *Streptomyces albulus*. Pls was found to be a membrane protein with adenylation and thiolation domains characteristic of the non-ribosomal peptide synthetases (NRPSs). It had no traditional condensation or thioesterase domain; instead, it had six transmembrane domains surrounding three tandem soluble domains [5]. Pls play an important role as a key enzyme in the synthetic of ε-PL. At present, studies of the relationship between the activity of Pls and the synthetic of ε-PL has rarely been discussed.

In Japan, ε-PL has been manufactured at the commercial scale by a fermentation process using the mutant of No. 346 [6], while domestic study still rest on the lab level and pilot-plant scale. Most recent advances focuses on screening of high producing strains, ferment process control, separating and extracting method of ε-PL and industrial applications of these polymers. Few research on the relationships among the activities of key enzyme, production of ε-PL and the concentration of yeast extract has been reported. Therefore in this paper, we focus on the effect of yeast extract which from different manufacturers or concentrations on production of ε-PL and provide some elementary theories for the industrially produced of ε-PL.

2 Materials and Methods

2.1 *Microorganisms*

Streptomyces diastatochromogenes 6#-7 was used for ε-PL fermentation and preserved in our laboratory.

2.2 *Culture Medium*

The strain was maintained on modified Bennett's agar slant, which contained (per liter): 10 g glucose, 1 g beef meat extract, 2 g polypepton, 1 g yeast extract, and 20 g agar. The pH was adjusted to 7.7 with 2 M NaOH solution.

Medium M3G was used for both seed culture and production culture throughout the study, which contained (per liter): 50 g glucose, 5 g yeast extract (Oxoid Ltd., England or Angel yeast Co., Ltd.), 10 g $(NH_4)_2SO_4$, 0.8 g $K_2HPO_4 \cdot 3H_2O$, 1.36 g KH_2PO_4, 0.5 g $MgSO_4 \cdot 7H_2O$, 0.04 g $ZnSO_4 \cdot 7H_2O$, 0.03 g $FeSO_4 \cdot 7H_2O$, and the pH was adjusted to 7.2 with NH_4OH solution (24–28%, w/v).

The above media were autoclaved at 121 °C for 20 min, and in each case, glucose and yeast extract were autoclaved separately.

2.3 Culture Conditions

For seed culture, a loopful of the spore from the slant cultures was inoculated into a 500-mL flask containing 100 mL M3G medium and cultured at 30 °C, 180 rpm on a rotary shaker for 30 h. For shake-flask production with two-stage fermentation, 100 mL M3G in 500-mL Erlenmeyer flask was inoculated with 6% v/v of pre-cultured seed culture and then cultured at 30 °C, 180 rpm for 96 h. Bead rings were added in seed medium to control the initial OD_{600} 1.9–2.1.

The fed-batch fermentation was carried out in a 5-L fermentor (B. Braun Biotech, Germany) with a 3L work volume using a two-stage pH control strategy at 30 °C and agitation 300–1000 r/min, dissolved oxygen (DO) was controlled at about 30% monitored with a DO electrode. The pH monitored with a pH electrode was kept 4.0 automatically with aqueous ammonia when pH dropped to below 4.0. When the consumption rate of glucose decreased significantly, the ε-PL fed-batch fermentation was finished.

2.4 Analytical Methods

The samples were intermittently obtained from the Erlenmeyer flask or 5-L fermentor for analysis. The cells were collected by centrifuged (6000 × g, 5 min), and washed twice with ultra-pure water. The supernatant was used to determine the ε-PL concentration according to the Itzhaki's method improved by our laboratory [7, 8]. The concentrations of residual glucose were determined using a biosensor SBA-40E (Shandong academy of sciences). The pH was determined using an acidometer (FE20, Mettler Toledo) and biomass accumulation was determined using dry cell weight analysis.

2.5 Enzyme Activity Assays

The aspartokinase (Ask) activity was assayed as described by Hamano et al. [4]. One unit of enzyme activity is defined as the amount of enzyme catalyzing the formation of 1 μmol aspartyl-β-hydroxamate per minute per g of cell at 30 °C. The standard assay mixture was consisted of 100 mM Tris–H_2SO_4 (pH 7.0), 600 mM $(NH_4)_2SO_4$, 600 mM hydroxylamine–KOH (pH 7.0), 10 mM ATP, 10 mM $MgSO_4$, 10 mM aspartic acid–KOH (pH 7.0), and crude enzyme solution in a 0.5 mL volume. The reaction mixtures were incubated for 10 min at 30 °C, and 0.75 mL ferric chloride solution (10% $FeCl_3 \cdot 6H_2O$ and 3.4% trichloroacetic acid in 0.7 N HCl) was added. After centrifugation, the A_{540} was measured in the supernatant meanwhile background activity was measured in the absence of aspartic acid.

The ε-PL synthetase (Pls) activity was determined as described by Yamanaka et al. [5]. The assay mixtures contained 100 mM N-Tris(hydroxymethyl) methyl-3-aminopropanesulfonic acid (TAPS)-NaOH (pH 8.5), 1000 mM L-lysine, 5 mM $MgCl_2$, 5 mM ATP, 1 mM dithiothreitol (DTT), 20% glycerol (v/v), 0.2% (w/v) NP-40 and 4.0 μL enzyme solution in a final volume of 40 μL. The reaction mixtures were incubated at 30 °C for 10 min. All assays were carried out under linear conditions, and one unit was defined as the amount of enzyme catalyzing the incorporation of 1 pmol of L-lysine into ε-PL per second at 30 °C.

The activity of homoserine dehydrogenase (HSD) was measured as described by YÂlmaz et al. [9]. The activity of HSD was expressed as micromoles of NADPH formed per min per g of cell. All the enzyme assays were performed in triplicate.

3 Results and Discussion

3.1 *Production Behavior In* Streptomyces diastatochromogenes *6#-7*

The fermentation parameters of strain 6#-7 in 500 mL Erlenmeyer flask were shown in Fig. 1a. The biomass accumulation and the concentration of ε-PL reached 3.47 and 0.64 g/L at 72 h respectively.

The activities of Ask and Pls showed a trend to gradually increase as the fermentation continued. Finally, the total enzyme activity of two enzymes (Ask and Pls) reached 78.6 U/g wet cells and 1287 U/g wet cells at 72 h respectively (shown in Fig. 1b). So we speculate that the metabolism of cell had obviously positive correlation with enzyme activities.

To further confirm the fermentation parameters and the trend of enzyme activities variation of strain 6#-7, the fed-batch fermentation method was studied by 5-L fermentor to measure the fermentation parameters every 24 h.

The enzyme activity of Ask increased firstly and then decreased with the increase of time, and the activity was up to 59 U/g at 96 h. The change of Pls

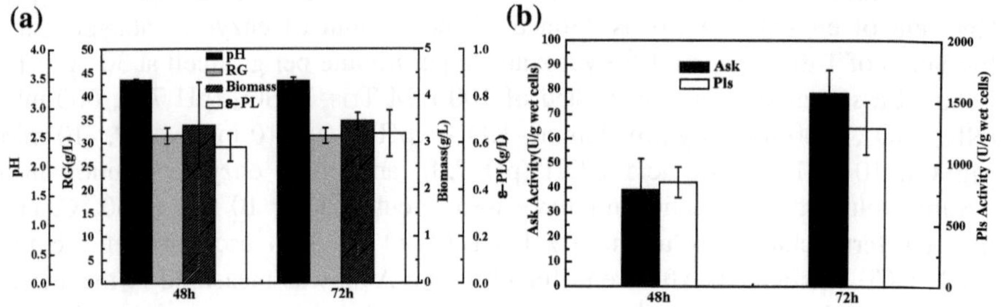

Fig. 1 The results of fermentation and enzyme activity in strain 6#-7

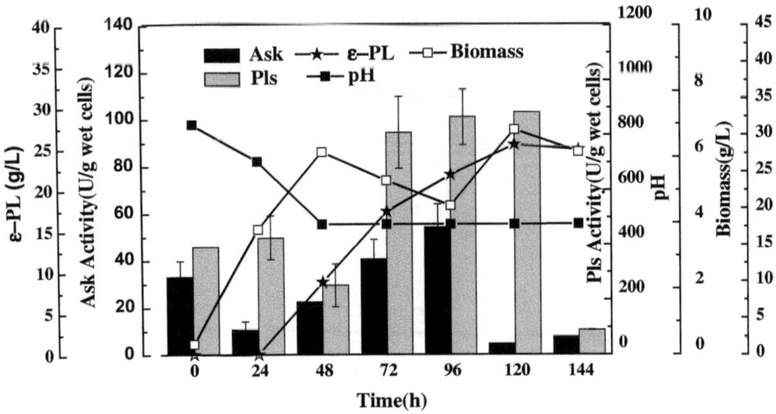

Fig. 2 The changes of ε-PL, biomass, pH and key enzyme activities in the process fed-batch fermentation

activity has the same trend as Ask (shown in Fig. 2). It is not difficult to find that the activities of Ask and Pls have obviously positive correlation with the concentration of ε-PL and biomass.

3.2 *Effect of Yeast Extract from Different Manufacturers and Batches on ε-PL Production*

Li et al. [10] find that the addition of yeast extract greatly inhibited pyruvate accumulation, while peptone was shown to be the most favorable nitrogen source and indicates that nitrogen level plays an important role in the production of pyruvate. Gorret et al. [11] demonstrated that yeast extract plays an important role during the whole process of fermentation. It can not only accelerate the cell's growth but also directly affects the production and the accumulation of by product.

Because yeast extracts play an important effect during the fermentation and constituent which from different manufacturers have a slightly different. This article investigated the effect of yeast extract from different production batches during the ε-PL fermentation to investigate whether it have noticeable influence or not.

These two kinds of yeast extract from the same manufacturer (Oxoid Ltd., England) with different production date were defined as group A and group B, respectively. The yeast extract from another manufacturer (Angel yeast Co., Ltd.) was named group C, as control group. In order to guarantee the validity and reliability, each sample was assayed in triplicate wells. The T tests were performed with SPSS 20.0 software to analysis the experimental results.

According to Table 1, the result shows that there is a significant difference when adding yeast extract which are from different factories ($p < 0.05$), yet no significant

Table 1 The results of fermentation in flask and SPSS analysis

Group	pH	ε-PL (g/L)	Biomass (g/L)	RG (g/L)	Ask (U/g wet cells)	Pls (U/g wet cells)	HSD (U/g wet cells)
A	3.27 ± 0.05	0.64 ± 0.02	5.79 ± 0.44[a]	29.2 ± 0.9	30.87 ± 3.04[a]	1411.4 ± 418	66.52 ± 10.86
B	3.25 ± 0.04	0.72 ± 0.04[b]	5.45 ± 0.04[b]	29.0 ± 0.6	26.91 ± 8.15	1902.2 ± 405.6[b]	70.1 ± 7.1
C	3.38 ± 0.04	0.53 ± 0.04	4.43 ± 0.21	30.5 ± 1.3	6.81 ± 4.84	513.1 ± 94.4	47.03 ± 4.22

[a]There was significant difference between group A and group C ($p < 0.05$)
[b]There was significant difference between group B and group C ($p < 0.05$)

difference in yeast extract which are from different production date of same factory ($p > 0.05$). Furthermore, the production of ε-PL was affected by the enzyme activities.

3.3 *Effect of Yeast Extract Concentration on the Fermentation of Strain 6#-7*

Aeschlimann et al. [12] discovered that the yeast extract concentration had significant effect on the accumulation of ε-PL. Therefore, research about the effect of yeast extract concentration on the fermentation parameters of strain 6#-7 was carried out.

According to the data in Fig. 3a, the effect of yeast extract in ε-PL production performed in a concentration-dependent manner. Strain 6#-7 possessed clearly higher activities of production of ε-PL with 15 g/L yeast extract, with a 6-fold enhancement compared to group without yeast extract. The results indicated that the higher concentration of yeast extract, the more favorable to increase the content of ε-PL. However, as shown in Fig. 3b, pH was always maintained at the range of

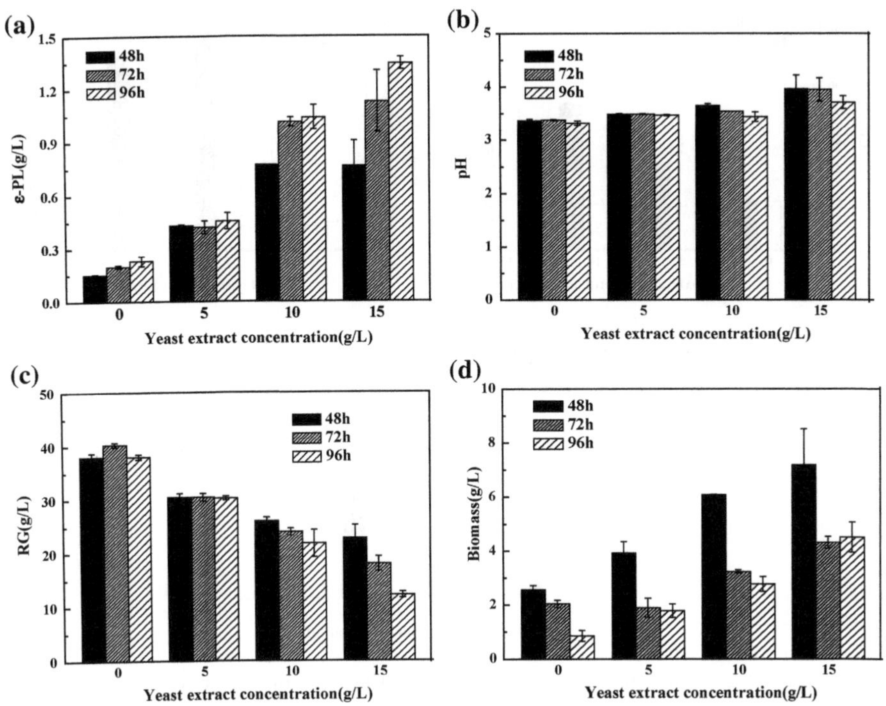

Fig. 3 The influence of different concentrations of yeast extract of strain 6#-7

3.5–3.7. When the concentration of yeast extract was 0 and 5 g/L, the residual glucose (RG) barely changed. At the level of 10 and 15 g/L, the RG significantly decreased to 21 and 11 g/L, respectively (shown in Fig. 3c). In Fig. 3d, the biomass decreased at different degrees with the increase of yeast extract, especially when the yeast extract was 15 g/L, the biomass decreased the most, which was about 4.4 g/L. Hence, it can be inferred that cell autolysis may occur during the fermentation.

To further explore the improved ε-PL synthetic ability of different concentration of yeast extract, key enzyme activities (e.g. Ask, Pls, and HSD) were investigated. Compared with the control group, the levels of Ask had significantly upward tendency in response to the concentration of yeast extract and reached maximum at 125 U/g (shown in Fig. 4a). According to the data in Fig. 4b, it seems that there were no evident law between the activity of Pls and yeast extract.

According to the date in Fig. 4c, the activity of HSD present a trend from increase to decrease and the activity reached its maximum in 72 h when the yeast extract concentration was 10 g/L.

Comprehensive analysis of Figs. 3 and 4, we can find that the production of ε-PL and enzyme activities increased significantly with the increase of yeast extract concentration.

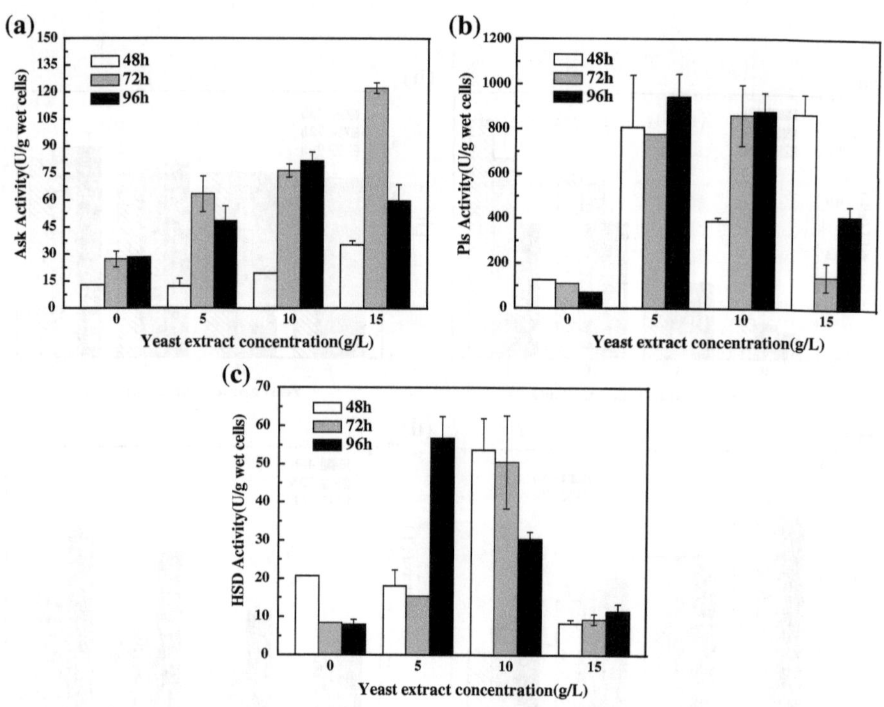

Fig. 4 The influence of different concentrations of yeast extract on key activities of strain 6#-7

4 Conclusions

In this study, the results shown the following conclusion:

(1) The results presented here suggest that there is a significant difference when using yeast extract which are from different manufactures. By contrast, there is no significant difference in yeast extract which are from different production date of the same factory.
(2) Strain 6#-7 possessed clearly higher activities of production of ε-PL with 15 g/L yeast extract, with a 6-fold enhancement compared to group without yeast extract.
(3) The concentration of yeast extract has a significant influence on the activity of key enzyme and further lead to the improvement of the yield of ε-PL.

In recent years, special attention has also been devoted to amino acid, which play important roles during the fermentation of ε-PL. Further study in detail will focus on the influence of different amino acids for fermentation. The obtained results would be useful to large-scale ε-PL production and provide new information on ε-PL biosynthesis mechanism.

Acknowledgements We are grateful for financial support from the National Natural Science Foundation of China (Project 21276197), the National High-tech R&D Program (863 Program) (2013AA102106), the Tianjin science and technology commissioner project (15JCTPJC59700) and the National Key Technology Support Program (No. 2015BAD16B04).

References

1. Hiraki J, Ichikawa T, Ninomiya S, Seki H, Uohama K, Seki H, Kimura S, Yanagimoto Y, Barnett JW (2003) Use of ADME studies to confirm the safety of epsilon-polylysine as a preservative in food. Regul Toxicol Pharmacol 37(2):328–340
2. Shih IL, Shen MH, Van YT (2006) Microbial synthesis of poly (epsilon-lysine) and its various applications. Biores Technol 97(9):1148–1159
3. Shima S, Oshima S, Sakai H (1983) Biosynthesis of ε-poly-L-lysine by washed mycelium of *Streptomyces albulus* No. 346. Nippon Nogei Kagaku Kaishi 57:221–226
4. Hamano Y, Nicchu I, Shimizu T et al (2007) ε-poly-L-lysine producer, *Streptomyces albulus*, has feedback-inhibition resistant aspartokinase. Appl Microbiol Biotechnol 76(4):873–882
5. Yamanaka K, Maruyama C, Takagi H et al (2008) ε-poly-L-lysine dispersity is controlled by a highly unusual nonribosomal peptide synthetase. Nat Chem Biol 4(12):766–772
6. Yoshida T, Nagasawa T (2003) e-poly-L-lysine: microbial production, biodegradation and application potential. Appl Microbiol Biotechnol 62(1):21–26
7. Itzhaki RF (1972) Colorimetric method for estimating poly-lysine and poly-arginine. Anal Biochem 50(2):569–574
8. Cao W, Tan Z, Yuan G et al (2007) Improvement of the assay method for ε-poly-lysine. Tianjin Univ Sci Technol 22:9–11 (in Chinese)
9. Yılmaz EI, Çaydasi AK, Özcengiz G (2008) Targeted disruption of homoserine dehydrogenase gene and its eVect on cephamycin C production in Streptomyces clavuligerus. J Ind Microbiol Biotechnol 35:1–7

10. Li Y, Chen J, Liang DF et al (2000) Effect of nitrogen source and nitrogen concentration on the production of pyruvate by Torulopsisglabrata. J Biotechnol 81:27–34
11. Gorret N, Maubois JL, Engasser JM et al (2001) Study of the effects of temperature, pH and yeast extract on growth and exopolysaccharides production by Propionibacterium acidi-propionici on milk microfiltrate using a response surface methodology. J Appl Microbiol 90:788–796
12. Aeschlimann A, von Stockar U (1990) The effect of yeast extract supplementationion the production of lactic acid from whey permeate by Lactobacillus helveticus. Appl Microbiol Biotechnol 32:398–402

Comparison of Aroma Compounds in Sauce-Flavor and Sesame-Flavor "Shan Zhuang Lao Jiu" Liquors by Headspace Solid-Phase Microextraction Coupled with Gas Chromatography-Mass Spectrometry

Fei Liu, Wanqiang Yin, Liping Du and Dongguang Xiao

1 Introduction

Chinese liquor, one of the oldest distillates, and Whisky, Brandy, Vodka, Rum and Gin are known as the six main distilled liquors in the world. Meanwhile, Chinese liquor is a traditional alcoholic beverage and well received by consumers in China. It is usually fermented from grains, mainly including sorghum, wheat, rice, sticky rice and corn, etc. After fermentation, the fresh spirit is distilled off and then aged under controlled conditions. The aged distillate is adjusted to the designated ethanol concentration and blended to ensure the quality of the final product [1]. Chinese liquor contains quite a number of volatile compounds which can greatly influence its flavor and aroma, including acids, esters, alcohols, aldehydes, ketones, hydrocarbons, pyrazines, phenolic compounds and others. According to their aroma characteristics, Chinese liquor can be classified into many types [2]. Among them, sauce-flavor liquor and sesame-flavor liquor are two typical Chinese liquors.

In order to achieve a practical and reliable method for the analysis of various volatiles in alcoholic beverages, several extraction methods have been developed and used, such as liquid–liquid extraction (LLE) [2], liquid–liquid micro-extraction extraction (LLME) [3], solid-phase extraction (SPE) [4], stir bar sorptive extraction (SBSE) [5], and solid-phase micro-extraction (SPME) [6]. Among, the SPME method was widely used in the analysis of aroma compounds with its solvent-free, simple operation, high sensitivity and low sample loading. Due to the difference of

---

F. Liu · W. Yin · L. Du (✉) · D. Xiao
Key Laboratory of Industrial Fermentation Microbiology, Ministry of Education, Tianjin Industrial Microbiology Key Laboratory, Tianjin Food Safety and Low Carbon Manufacturing Collaborative Innovation Center, College of Bioengineering, Tianjin University of Science and Technology, Tianjin 300457, People's Republic of China
e-mail: dlp123@tust.edu.cn

H. Liu et al. (eds.), *Advances in Applied Biotechnology*, Lecture Notes in Electrical Engineering 444, https://doi.org/10.1007/978-981-10-4801-2_25

the structure and volatility of the analytes, the different SPME fibers have been used, including polyacrylate (PA), carboxen/polydimethylsiloxane (CAR/PDMS), polydimethylsiloxane (PDMS), divinylbenzene/carboxen/polydimethylsiloxane (DVB/CAR/PDMS), and polydimethylsiloxane/divinylbenzene (PDMS/DVB), etc. The DVB/CAR/PDMS fiber was more suitable for the analysis of aroma compounds in alcoholic beverages according to the relevant research [7]. So far, there were some studies on volatile aroma compounds by SPME in alcoholic beverages, such as wine [6], Chinese liquor [8], and Chinese rice wine [9].

As a result of the multiple variations, such as raw materials, environment and manufacturing practices, the wines from the different regions or the same region but different categories all have unique aroma profiles. "Shan Zhuang Lao Jiu" liquor is a China liquor produced in Chengde, Hebei, China, and is welcomed by consumers in the north of China. The objective of this study was to identify aroma compounds, and to quantitate the key aroma compounds in sauce-flavor and sesame-flavor "Shan Zhuang Lao Jiu" liquors by HS-SPME-GC-MS to find out the aroma similarities and differences of the two liquors. The findings in this study will help to the further understanding of Chinese liquor.

2 Materials and Methods

2.1 *Materials and Chemicals*

Chinese liquor samples: sauce-flavor and sesame-flavor "Shan Zhuang Lao Jiu" liquor samples were supplied by Summer Resort Industrial Group in Chengde city.

Chemicals: acetic acid, butyric acid, hexanoic acid, ethyl acetate, isobutyl acetate, butyl acetate, isoamyl acetate, ethyl hexanoate, ethyl heptanoate, ethyl lactate, ethyl caprylate, ethyl butyrate, n-propanol, isobutanol, n-butanol, isoamylol, phenethyl alcohol and hydrocarbon mixture (C8–C40) were purchased from Sigma-Aldrich (St. Louis, MO).

2.2 *GC-MS Conditions and Analysis*

Identification was carried out using an Agilent 7890A GC coupled with an Agilent 5975C mass selective detector (MSD). The sample was analyzed on a CP-Wax column (50 m × 250 μm inner diameter, 0.2 μm film thickness). The injector temperature was 250 °C and the split mode was used (ratio 15:1). The oven temperature was held at 50 °C for 3 min, raised to 70 °C at a rate of 3 °C min^{-1}, increased to 170 °C at a rate of 3 °C min^{-1}, then increased to 240 °C at a rate of 8 °C min^{-1} and held at 240 °C for 3 min. The column carrier gas was helium at a constant flow rate of 1 mL min^{-1}. The mass spectrometer was operated in electron-impact (EI) mode at 70 eV. The mass scan range was 30–500 amu. The temperatures of the interface, ion source and quadrupole were 280, 230 and 150 °C, respectively. The aroma compounds were determined by comparing the MS fragments detected with the mass spectra present in the NIST08 MS spectral database. The compounds identified by MS were further confirmed by comparing the retention times generated for each reference compound analyzed, using a commercial hydrocarbon mixture (C8–C40) for determination of the retention indices (RI). The relative content of aroma compounds was calculated by peak area normalization method. The quantitative analysis of the key aroma compounds was performed by calculation from the approximated curve using the linear least-squares method. The respective quantitative values of the key aroma compounds were determined by averaging the triplicate experiments.

2.3 SPME Sampling Conditions

The sample was diluted with ultrapure water to a final concentration of 10% (v/v) ethanol for analysis. A total of 8 mL diluted sample was put into a 20 mL vial and spiked with 3 g NaCl, and a small magnetic stirrer was added. The sample was equilibrated for 10 min and extracted for 50 min at 60 °C with continuous stirring. The SPME fiber holder equipped with DVB/CAR/PDMS fibre (Supelco, Inc., Bellefonte, PA, USA) was used for aroma compounds extraction in this study. After extraction, the fiber was inserted into the injection port of a GC-MS system (at 250 °C for 5 min).

3 Results and Discussion

3.1 Identification of Aroma Compounds in "Shan Zhuang Lao Jiu" Liquor

A total of 120 aroma compounds were identified in the two liquors. Among them, 103 aroma compounds were identified in sauce-flavor liquor and 115 in sesame-flavor liquor. The total ion chromatograms (TIC) were shown in Fig. 1.

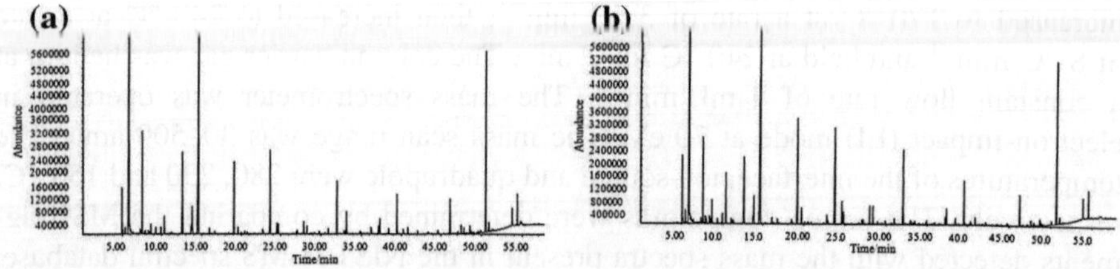

Fig. 1 Total ion chromatograms (TIC) of the sauce-flavor liquor (**a**) and the sesame-flavor liquor (**b**)

As can be seen in Table 1, aroma compounds were abundant in sauce-flavor and sesame-flavor "Shan Zhuang Lao Jiu" liquors. According to their chemical structures, these compounds were grouped as esters, alcohols, acids, aldehydes, ketones, alkanes, furans, phenols, pyrazines, sulphur-containing compounds and aromatic compounds. Obviously, there were some unique aroma compounds in the two liquors. As can be seen in Fig. 2, the relative content of esters (87.63%) and alkanes (0.75%) were both same in the two liquors. And the relative content of acids, ketones and others were similar. However, the relative content of alcohols (5.47 and 4.47%) and aldehydes (3.43 and 4.95%) had larger difference. Simultaneously, it also showed that the main aroma compounds were esters, alcohols and aldehydes in the two liquors, however, the content of acids (1.27 and 0.84%) was less.

The relative content of ethyl caproate (23.977 and 26.7196%) was highest in the two liquors. Ethyl acetate, ethyl butyrate and ethyl lactate are the four main esters of Chinese liquor with ethyl caproate, and the total relative content of them was 8.7776 and 8.1837%, respectively. In addition, the content of ethyl oenanthate, ethyl caprylate, ethyl decanoate, ethyl laurate, ethyl tetradecanoate, ethyl palmitate, ethyl linoleate and isoamylol were also higher. The content of some aroma components was less. However, they have contributed much to the Chinese liquor flavor. Among, carbonyl complexes were the key aroma compounds to the soft characteristic of sauce-flavor, such as furfural, ethanol, acetal, isovaleraldehyde diethyl acetal, nonanal diethyl acetal and phenylglyoxylic acid diethyl acetal. Sulphur-containing compound, such as dimethyltrisulfide, was also detected in the two "Shan Zhuang Lao Jiu" liquors, and its relative content was 0.2178 and 0.1428%, respectively. Sulphur-containing compounds could mix with other aroma substances to produce amplificatory and combined action, though their concentrations were lower [10].

Table 1 Aroma compounds in sauce-flavor and sesame-flavor "Shan Zhuang Lao Jiu" liquors

No	Compounds[a]	RI[b]	Relative content[c] (%)		No	Compounds[a]	RI[b]	Relative content[c] (%)	
			Sauce-flavor	Sesame-flavor				Sauce-flavor	Sesame-flavor
Esters					26	Ethyl lactate	1340.1	0.5141	0.9667
1	Ethyl acetate	854.5	2.5189	3.1493	27	Isobutyl caproate	1349.2	0.1587	0.1373
2	Ethyl propionate	926.5	0.807	0.6619	28	Isoamyl valerate	1357.3	ND[d]	0.1287
3	Ethyl isobutyrate	937	0.6187	0.3831	29	Heptyl acetate	1367.6	0.028	0.0435
4	Isobutyl acetate	1000.5	0.1902	0.1709	30	Butyl hexanoate	1408.7	0.3571	0.5244
5	Ethyl butyrate	1023.7	5.7446	4.0677	31	Hexyl butyrate	1411.1	0.1988	0.2045
6	Ethyl 2-methylbutyrate	1040.6	0.29	0.3116	32	Propyl heptanoate	1413.6	0.0296	0.0747
7	Ethyl 3-methylbutyrate	1057	0.3305	0.3469	33	Ethyl octanoate	1431.4	6.4805	6.4978
8	Butyl acetate	1060.4	ND[d]	0.0825	34	Isoamyl hexanoate	1456.3	0.7297	0.7198
9	Isoamyl acetate	1114.5	.0.6443	0.6777	35	7-Ethyl caprylate	1478.8	ND[d]	0.0644
10	Sec-butyl butyrate	1122.7	ND[d]	0.0298	36	Hexanoic acid, pentyl ester	1508.4	0.2167	0.3589
11	Ethyl valerate	1126.3	1.4337	2.7055	37	Propyl octanoate	1515.6	0.0414	0.054
12	Butyl propionate	1130.8	ND[d]	0.0301	38	Ethyl pelargonate	1532.1	0.901	1.0457
13	Methyl caproate	1178.3	0.0449	0.0397	39	2-Hydroxy-4-methyl ethyl valerate	1542.1	0.6017	1.0534
14	Isoamyl isobutyrate	1190	0.0377	0.0506	40	Isobutyl n-octanoate	1550.3	0.1195	0.0686
15	Ethyl hexanoate	1230.1	23.977	26.7196	41	Heptanoic acid-3-methyl butyl ester	1556.1	0.1406	0.1763
16	Isoamyl butyrate	1262	0.4947	0.4057	42	Isoamyl lactate	1570.3	0.138	0.0836
17	Hexyl acetate	1267	0.1073	0.1133	43	3-Nonene acid ethyl ester	1579.5	ND[d]	0.2256
18	Isoamyl 2-methyl butyrate	1273.4	0.0275	ND[d]	44	Hexyl hexanoate	1609	0.426	0.365
19	Ethyl 5-methyl caproate	1281.8	ND[d]	0.1338	45	Butyl caprylate	1611.3	0.0729	0.0808
20	Ethyl cis-3-hexenoate	1292.2	ND[d]	0.0308	47	Ethyl caprate	1635.4	3.7062	4.5134

(continued)

Table 1 (continued)

No	Compounds[a]	RI[b]	Relative content[c] (%)		No	Compounds[a]	RI[b]	Relative content[c] (%)	
			Sauce-flavor	Sesame-flavor				Sauce-flavor	Sesame-flavor
21	Butyl pentanoate	1310.4	0.0374	0.0696	48	Ethyl benzoate	1659.2	0.3374	0.3277
22	Amyl butyrate	1311.7	0.0510	0.0822	49	Ethyl trans-4-decenoate	1660.8	0.2762	0.2154
23	Propyl hexanoate	1313.8	0.1438	0.263	50	Diethyl succinate	1672.9	0.1182	0.2232
24	Ethyl oenanthate	1329.2	5.5348	7.3111	51	Hexyl oenanthate	1709.5	0.1469	0.1449
25	Hexyl propionate	1333.8	ND[d]	0.0531	52	Ethyl dodecanoate	1738.6	0.1101	0.129
53	Ethyl phenylacetate	1780.8	1.14	0.9674	7	1-Pentanol	1256.7	0.0462	0.0561
54	Hexyl octanoate	1810.5	ND[d]	0.4726	8	Hexyl alcohol	1359.3	0.2582	0.1273
55	Phenethyl acetate	1812.3	0.5586	ND[d]	9	n-Heptanol	1462.2	0.0778	0.0547
56	Ethyl laurate	1842.8	2.7239	2.564	10	1-Octanol	1565.2	0.1487	0.0579
57	Furfuryl caproate	1861.6	0.0362	ND[d]	11	Nonanol	1667.9	ND[d]	0.188
58	Isoamyl decanoate	1863.4	0.0515	0.0775	12	1-Decanol	1770.8	0.0518	ND[d]
59	Ethyl 3-phenylpropionate	1883.5	0.5923	0.4931	13	Phenethyl alcohol	1922	0.3474	0.2247
60	3-phenyl-1-propanol acetate	1943.1	0.0871	0.0292	Acids				
61	Ethyl tridecanoate	1946.1	0.0597	0.0537	1	Acetic acid	1465	0.0512	0.0513
62	Phenethyl butyrate	1964.4	0.1778	0.1073	2	Butyric acid	1643.1	0.1163	0.1068
63	Ethyl myristate	2054	2.1095	1.5393	3	Isovaleric acid	1685.1	0.1794	0.1661
64	Ethyl pentadecanoate	2156.6	0.3832	0.2545	4	Hexanoic acid	1863	0.4734	0.2531
65	Phenylethyl caproate	2178.4	0.1166	0.0559	5	Heptoic acid	1976.8	0.0914	0.0638
66	Ethyl palmitate	2260.8	10.262	7.2524	6	Octanoic acid	2087.7	0.1737	0.086
67	Ethyl 9-Hexadecenoate	2286.5	0.3864	0.2288	7	Decanoic acid	2307.8	0.0799	0.0571
68	Ethyl octadecanoate	2468.7	0.039	0.0583	Aldehydes and Ketones				
69	Ethyl oleate	2490.4	0.7528	0.6568	1	Acetaldehyde	645.5	0.1051	0.1096
70	Ethyl linoleate	2540.4	1.3786	1.011	2	Isobutyraldehyde	809.7	ND[d]	0.1328

(continued)

Table 1 (continued)

No	Compounds[a]	RI[b]	Relative content[c] (%)		No	Compounds[a]	RI[b]	Relative content[c] (%)	
			Sauce-flavor	Sesame-flavor				Sauce-flavor	Sesame-flavor
71	Diisobutyl phthalate	2570.6	0.3596	0.0743	3	Isovaleraldehyde	884.8	0.6684	1.084
Alcohols					4	2-Pentanone	956.3	0.2832	0.5143
1	Acetal	861.7	0.3038	0.3302	5	Isovaleraldehyde diethyl acetal	1069.9	0.6022	1.0331
2	Propyl alcohol	1036.8	0.2211	0.4261	6	3-Octanone	1250.6	0.0746	0.0681
3	Isobutanol	1096	0.4131	0.2458	7	2-Nonanone	1385.4	0.1098	0.0275
4	n-Butyl alcohol	1145.2	0.1189	0.1342	8	Furfural	1451	0.992	1.5976
5	2-methyl-1-butanol	1210.3	0.7377	0.504	9	Benzaldehyde	1511.9	0.398	0.3564
6	Isoamylol	1212.9	2.2771	1.843	10	Nonanal diethyl acetal	1526.7	0.1139	0.0596
11	2-Undecanone	1596.7	0.1343	0.0678	7	Heptadecane	1699.1	0.0949	0.0713
12	Acetophenone	1646	0.0955	0.065	Others				
13	Phenylglyoxylic acid diethyl acetal	1712.3	0.2523	0.2719	1	2-Amyl furan	1219.8	0.0378	0.0449
14	cis-Geranylacetone	1854	0.0979	0.054	2	Phenyl ethylene	1243.7	ND[d]	0.1216
Alkanes					3	Trimethylbenzene	1324.1	ND[d]	0.0186
1	1,1-diethoxy-3-methylButane	1065.1	0.1341	0.2151	4	Dimethyltrisulfide	1365.6	0.2178	0.1428
2	Dodecane	1200.7	0.0706	0.0478	5	2-ethyl-6-methyl pyrazine	1387.8	ND[d]	0.0343
3	Tridecane	1300	ND[d]	0.0199	6	Naphthalene	1730	0.1157	0.1049
4	Tetradecane	1398.9	0.1303	0.1171	7	1,7-Dimethyl naphthalene	1997.8	0.1197	ND[d]
5	Pentadecane	1499.3	0.1216	0.1202	8	Metacresol	2094.6	0.0334	0.0178
6	Hexadecane	1599.2	0.1354	0.1126					

[a]All compounds detected by GC-MS were identified by MS spectra and RI comparisons to pure standards
[b]Retention index
[c]The relative content of each compound was calculated by peak area normalization method
[d]Not detected

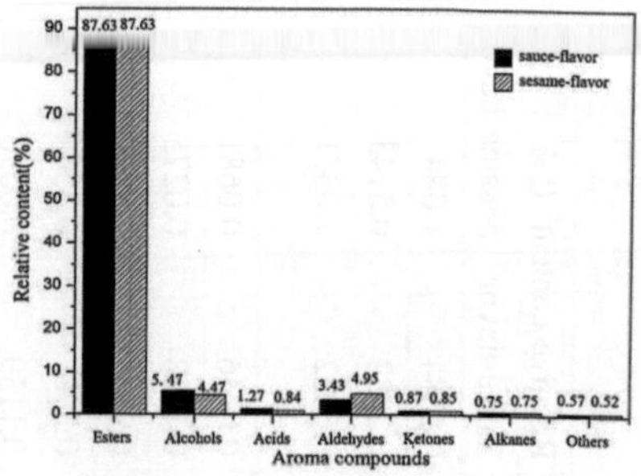

Fig. 2 Relative content of various types of aroma compounds in two kinds of "Shan Zhuang Lao Jiu" liquors

3.2 Quantitative Analysis of the Key Aroma Compounds

The result of quantitative analysis of the key aroma compounds in sauce-flavor and sesame-flavor "Shan Zhuang Lao Jiu" liquors can be seen in Table 2.

The most important aroma compounds seemed to be esters in Chinese liquor. The relative content of esters was 87.63% in the two liquors. Among them, ethyl hexanoate showed the highest content (398.96 and 667.09 mg/L). As we all know, the quantity relative ratio relationship of esters is crucial to Chinese liquor flavor. The content ratio of the four main esters (ethyl hexanoate, ethyl acetate, ethyl butyrate and ethyl lactate) of Chinese liquor was 1: 0.11: 0.24: 0.26 and 1: 0.12: 0.15: 0.38 in the two liquors, respectively. Furthermore, another five important esters were also quantified. Butyl acetate (2.06 mg/L) was a unique component of the sesame-flavor liquor. Alcohols were as the main source of auxiliary aroma agents in Chinese liquor. Isoamyl alcohol showed the highest content (468.59 and 504.82 mg/L) in the two liquors (except for ethanol). Meanwhile, propyl alcohol, n-butyl alcohol, isobutanol and phenethyl alcohol were also quantified. However, the content of propyl alcohol (45.50 and 116.74 mg/L) had larger difference, which could be related to the distillation process. Acids, other some important aroma components, were the key factors affecting the mouthfeel and after taste of Chinese liquor. In this study, the content of acids is relative less in the both liquors. Acetic acid (10.18 and 13.37 mg/L), butyric acid (23.16 and 27.88 mg/L) and hexanoic acid (94.23 and 66.06 mg/L) were the main acids.

4 Conclusions

The study has revealed the aroma similarities and differences of the two Chinese liquors. The results showed that a total of 120 aroma compounds were identified in the two liquors, and the unique components of sauce-flavor and sesame-flavor liquors were 5 and 17 kinds, respectively. According to the chemical structure, aroma compounds were grouped as esters, alcohols, acids, aldehydes, ketones, alkanes, furans, phenols, pyrazines, sulphur-containing compounds and aromatic

Table 2 Quantitative analysis of the key aroma compounds in sauce-flavor and sesame-flavor "Shan Zhuang Lao Jiu" liquors

No	Compounds	Concentration[a] (mg/L)		No	Compounds	Concentration[a] (mg/L)		No	Compounds	Concentration[a] (mg/L)	
		Sauce-flavor	Sesame-flavor			Sauce-flavor	Sesame-flavor			Sauce-flavor	Sesame-flavor
1	Ethyl acetate	41.91	78.62	7	Ethyl heptanoate	92.10	182.53	13	Isoamylol	468.59	504.82
2	Isobutyl acetate	3.16	4.27	8	Ethyl caprylate	107.83	162.23	14	Phenethyl alcohol	69.14	58.66
3	Ethyl butyrate	95.59	101.55	9	Ethyl lactate	102.31	252.33	15	Acetic acid	10.18	13.37
4	Butyl acetate	ND[b]	2.06	10	n-Propanol	45.50	116.74	16	Butyric acid	23.16	27.88
5	Isoamyl acetate	10.72	16.92	11	Isobutanol	85.01	67.3	17	Hexanoic acid	94.23	66.06
6	Ethyl hexanoate	398.96	667.09	12	n-Butanol	24.47	36.77				

[a]The concentration of each compound was calculated on the basis of the approximated curve using the linear least-squares method. Data are the mean value of triplicate measurements

[b]Not detected

compound. The main aroma compounds were esters, alcohols and aldehydes. However, the content of acids was relatively less in the two liquors. Meanwhile, the content of some key aroma compounds had great difference. Among them, isoamyl alcohol showed the highest content in sauce-flavor liquor, while ethyl hexanoate showed the highest content in sesame-flavor liquor. And butyl acetate was a unique compound of the sesame-flavor liquor. Therefore, this study demonstrated that aroma profile was the key factor for flavor and taste of Chinese liquors produced by the different brewing processes and raw materials.

References

1. Fan W, Qian MC (2006) Identification of aroma compounds in Chinese 'Yanghe Daqu' liquor by normal phase chromatography fractionation followed by gas chromatography/olfactometry. Flavour Fragr J 21(2):333–342
2. Zhu S, Lu X, Ji K et al (2007) Characterization of flavor compounds in Chinese liquor Moutai by comprehensive two-dimensional gas chromatography/time-of-flight mass spectrometry. Anal Chim Acta 597(2):340–348
3. Ferreira V, López R, Cacho JF (2000) Quantitative determination of the odorants of young red wines from different grape varieties. J Sci Food Agric 80:1659–1667
4. López R, Aznar M, Cacho J et al (2002) Determination of minor and trace volatile compounds in wine by solid-phase extraction and gas chromatography with mass spectrometric detection. J Chromatogr A 966:167–177
5. Fan W, Shen H, Xu Y (2011) Quantification of volatile compounds in Chinese soy sauce aroma type liquor by stir bar sorptive extraction and gas chromatography–mass spectrometry. J Sci Food Agric 91(7):1187–1198
6. Boutou S, Chatonnet P (2007) Rapid headspace solid-phase microextraction/gas chromatographic/mass spectrometric assay for the quantitative determination of some of the main odorants causing off-flavours in wine. J Chromatogr A 1141:1–9
7. Du LP, He TT, Li W (2015) Analysis of volatile compounds in Chinese Laobaigan liquor using headspace solid-phase microextraction coupled with GC-MS. Anal Methods 7:1906–1913
8. Fan W, Qian MC (2005) Headspace solid phase micro-extraction (HS–SPME) and gas chromatography-olfactometry dilution analysis of young and aged Chinese "Yanghe Daqu" liquors. J Agric Food Chem 53(20):7931–7938
9. Luo T, Fan W, Xu Y (2008) Characterization of volatile and semi-volatile compounds in Chinese rice wines by headspace solid phase microextraction followed by gas chromatography-mass spectrometry. J Inst Brew 114(2):172–179
10. Chin HW, Lindsay RC (1994) Mechanisms of formation of volatile sulfur compounds following the action of cysteine sulfoxide lyases. J Agric Food Chem 42:1529–1536

Optimization the Fermentation Conditions of *Marasmius androsaceus* by Desirability Function Method

Jia Song, Le Cui, Xiaobo Ma, Yan Su, Zhengmei Huang and Min Wang

1 Introduction

Marasmius androsaceus is one of the most valuable and rare fungus which is known as Chinese traditional medicinal fungus. *M. androsaceusis* is also called "Gui Mao Zhen" in folk and widely distribute in the dense forest in China. It has many pharmacological effects such as tendon relaxation, pain alleviation, and antihypertension [1]. Although it possesses many health benefits, only the pharmacological activities of *M. androsaceus* which related to analgesic and antioxidant effects were preliminarily reported [2, 3].

M. androsaceus has been used as a Chinese medicine known as "An Luo Tong" which has analgesia function and possesses pertinent market demand. Besides, the fermentation metabolites of *M. androsaceus* contain abundant active compounds, such as adenosine, polysaccharide, which contributed to various pharmacological activities and have gained great attention recently. Therefore, Submerged fermentation, is an efficient way to produce mycelia and bioactive metabolites of fungus, that has been studied for years in numerous research groups [4–6]. Recently, the studies on this important fungus are mainly focused on the following aspects, extraction and isolation of bioactives [7, 8] and some work about optimization

J. Song · L. Cui · Y. Su · M. Wang (✉)
Key Laboratory of Industrial Fermentation Microbiology, Ministry of Education,
National Engineering Laboratory for Industrial Enzymes, The College of Biotechnology,
Tianjin University of Science and Technology, Tianjin 300457, People's Republic of China
e-mail: minw@tust.edu.cn

Z. Huang
Onlystar Biotechnology Co. Ltd., Cross of Shuangfu dr and Hongdu dr,
Economic Development Zone, Dezhou 253000, Shandong, People's Republic of China

X. Ma
The International College, Tianjin University of Science and Technology,
Tianjin 300457, People's Republic of China

H. Liu et al. (eds.), *Advances in Applied Biotechnology*, Lecture Notes
in Electrical Engineering 444, https://doi.org/10.1007/978-981-10-4801-2_26

M. androsaceus submerged fermentation conditions [9]. However, there have been few investigations on the optimization fermentation condition of *M. androsaceus* in a flask [10, 11].

Our previous work reported that desirability function combining Plackett-Burman (PB) design and response surface methodology (RSM) were successfully applied to optimize submerge fermentation medium of *Paecilomyces tenuipes* N45 and *Saccharomyces cerevisiae* [12, 13]. Similarity, in this work, desirability function designed to combine multi-response values into one response value [14], was also used to unite four indexes (mycelium yield, the contents of adenosine, cordycepic acid and polysaccharide) together.

The objective of the present work is to optimize submerged fermentation condition in a 500 mL shake flask for *M. androsaceus* products and its active metabolites (adenosine, cordycepic acid, and polysaccharide) by using statistical and mathematical techniques including PB design and Box-Behnken (BB) design.

2 Materials and Methods

2.1 *Strains and Materials*

Marasmius androsaceus (CCTCC M2013175, obtained from China Center for Type Culture Collection, China) was cultivated in 250 mL conical flask with a defined liquid medium containing: 24.1 g/L sucrose, 10 g/L peptone, 11.3 g/L yeast extract powder, 1 g/L $MgSO_4 \cdot 7H_2O$, 0.955 g/L $KH_2PO_4 \cdot 3H_2O$, and 0.1 g/L Vitamin B_1.

2.2 *Desirability Function Establishment*

During the optimization process, the mycelium yield, and the concentration of adenosine, mannitol and intracellular polysaccharide were uniformed into one index —*Da* (Ranged: 0–1) based on desirability function [15]. The response values (y) were transfer to d values (Ranged: 0–1) according to Eq. (1), and then uniformed into *Da* value following as Eq. (2).

$$d_i = \left| \frac{\hat{y}_i - y_{\min}}{y_{\max} - y_{\min}} \right| \quad \hat{y} > y_{\max}, \quad d_i = 1; \quad \hat{y} < y_{\min}, \quad d_i = 0 \qquad (1)$$

where, $\hat{y}_i$ is the response value of an i analyzed factor.

Table 1 Parameters of the desirability function

Parameter	Mycelium yield (g/L)	Adenosine (g/L)	*Cordycepic* acid (g/L)	Polysaccharide (g/L)
y_{il}	2.00	0.01	0.00	0.10
y_{ih}	15.00	0.15	0.80	2.50
w_i	0.25	0.30	0.15	0.30

$$Da = d_1^{w_1} \times d_2^{w_2} \times d_3^{w_3} \times \cdots \times d_i^{w_i} \quad w_1 + w_2 + w_3 + \cdots w_i = 1 \tag{2}$$

where, W_i is weighted value of index i.

The parameters of desirability function in our study were displayed in Table 1.

2.3 The Concentration of Effective Constituents Determination

Measurement of biomass. After fermentation, the fermentation products were centrifuged at 4000 × g for 10 min. The precipitate of mycelium was dried by a freeze dryer. Then, the biomass, *M. androsaceus* mycelium was measured by a electronic balance.

Measurement of polysaccharide. The amount of total polysaccharide was determined by anthrone–sulfuric acid method [16].

Measurement of cordycepic acid (mannitol). The concentration of cordycepic acid in *M. androsaceus* mycelium was measured using the colorimetric method as described previously [17].

Measurement of adenosine. As it is recommended in the Chinese Pharmacopoeia that the adenosine concentration in *Marasmius androsaceus* is detected via high performance liquid chromatography (HPLC) [18].

2.4 Plackett-Burman Design

Plackett-Burman design was performed to evaluate the contribution of the following factors during *M. androsaceus* submerged fermentation in a 500 mL shake flask [19]. Each independent variable is tested at a high (+1) and a low (−1) level separately. All the experiments were carried out according to the design matrix (Table 2). The significance of each variable was determined by one-way variance analysis (ANOVA) and defined statistical significance as $P < 0.05$.

Table 2 The design matrix and the results of Plackett-Burman design

Num	X_1	X_2	X_3	X_4	X_5	X_6	X_7	X_8	*Da*
	Culture temperature (°C)	Rotate speed (rpm)	Initial pH	Inoculum size (%)	Inoculum age (day)	Culture time (day)	Loading volume (mL * 500 mL^{-1})		
1	1 (28)	−1 (100)	1 (7)	−1 (3)	−1 (3)	−1 (5)	1 (240)	1	0.3332
2	1 (28)	1 (200)	−1 (5)	1 (7)	−1 (3)	−1 (5)	−1 (160)	1	0.4255
3	−1 (24)	1 (200)	1 (7)	−1 (3)	1 (5)	−1 (5)	−1 (160)	−1	0.2727
4	1 (28)	−1 (100)	1 (7)	1 (7)	−1 (3)	1 (7)	−1 (160)	−1	0.3168
5	1 (28)	1 (200)	−1 (5)	1 (7)	1 (5)	−1 (5)	1 (240)	−1	0.3635
6	1 (28)	1 (200)	1 (7)	−1 (3)	1 (5)	1 (7)	−1 (160)	1	0.3828
7	−1 (24)	1 (200)	1 (7)	1 (7)	−1 (3)	1 (7)	1 (240)	−1	0.2763
8	−1 (24)	−1 (100)	1 (7)	1 (7)	1 (5)	−1 (5)	1 (240)	1	0.2681
9	−1 (24)	−1 (100)	−1 (5)	1 (7)	1 (5)	1 (7)	−1 (160)	1	0.2221
10	1 (28)	−1 (100)	−1 (5)	−1 (3)	1 (5)	1 (7)	1 (240)	−1	0.2634
11	−1 (24)	1 (200)	−1 (5)	−1 (3)	−1 (3)	1 (7)	1 (240)	1	0.2305
12	−1 (24)	−1 (100)	−1 (5)	−1 (3)	−1 (3)	−1 (5)	−1 (160)	−1	0.2129

2.5 Response Surface Methodology

In order to optimize three selected factors in a 500 mL shake flask of *M. androsaceus*, RSM was applied. The factors which were applied in BB design were investigated at three levels (−1, 0, 1) based on the results obtained from PB design, the experiments were performed according to the design matrix (Table 3). The statistical significance of the coefficients in quadratic model were identified by One-way ANOVA. Statistical significance was defined as $P < 0.05$.

The results obtained from BB design via Artificial neural network-genetic algorithm (ANN-GA) was analyzed by Matlab2012a software [20]. In our present study, 11 data from BB design was randomly chosen as calibration set. The rest 4 data was randomly divided into 2 groups, test set and prediction set. The network was operated via Levenberg–Marquardt back propagation algorithm. Sigmoid function and linear function were used for the hidden layer and output layer respectively. Approaching degree (Dv), defined as Eq. (3), was used to search the optimal number of hidden nodes which strongly influence the quality and predictive ability of the developing network. In this study, we should defined the c into 0.005. The best ANN model was selected according to related Dv values.

$$Dv = \frac{c}{\frac{nc}{n} \times MSEc + \frac{nt}{n} \times MSEt + |MSEc - MSEt|} \quad (3)$$

where, nc and nt were the number of calibration and test sets; n was the total number of the calibration and test sets; MSEc and MSEt were the root mean square error of the calibration and test set, respectively.

GA was further applied to search optimal culture conditions in test regions. After a 50 generation evaluation by GA, based on the given range of input parameters, optimal culture conditions were achieved [21].

Table 3 Regression analysis of Plackett-Burman design experiment

Source	DF	SS	MS	*F* value	*Pr* > *F*
X_1	1	0.0302	0.0302	132.6515	0.0014
X_2	1	0.0094	0.0094	41.3235	0.0076
X_3	1	0.0014	0.0014	6.3777	0.0858
X_4	1	0.0026	0.0026	11.3382	0.0435
X_5	1	<0.0001	<0.0001	0.1772	0.7022
X_6	1	0.0028	0.0028	12.3924	0.0389
X_7	1	0.0008	0.0008	3.5154	0.1575
X_8	1	0.0020	0.0020	8.9078	0.0584
Model	9	0.0493	0.0062	27.0855	0.0102
Error	2	0.0007	0.0002		
Total	11	0.0500		R^2	0.9863

2.6 Independent Verification Test

Based on the analysis via software SAS V 9.0 and Matlab2012a software, three parallel experiments for each analysis result were performed to test the selected fermentation conditions obtained from RSM and ANN-GA respectively. After comparing, the optimum submerged fermentation conditions of *M. androsaceus* in a 500 mL shake flask were achieved.

3 Results and Discussion

3.1 Results of PB Design

PB design was used to filtrate the significant variables via a 500 mL shake flask during *Marasmius androsaceus* submerge fermentation. Experiments performed the designed matrix (Table 2). *Da* value was influenced by culture temperature, rotate speed, initial pH, inoculum size, inoculum age, culture time, loading volume. After comparing F-value and *P*-value, according to the Eq. (4), culture temperature, rotate speed, inoculum size and culture time played significant influence on *M. androsaceus* submerge fermentation ($P < 0.05$) (Table 3). Furthermore, the order of each investigating factor was $X_1 > X_2 > X_6 > X_4 > X_8 > X_3 > X_7 > X_5$. Considering the *P*-value of inoculum size was 0.0435 and the next RSM designed, we only chose culture temperature, rotate speed and culture time for the next experiment.

$$Y = 0.297 + 0.050X_1 + 0.028X_2 + 0.011X_3 + 0.015X_4 - 0.002X_5 - 0.015X_6 - 0.008X_7 + 0.013X_8 \quad (4)$$

3.2 Results of RSM Design and ANN-GA

The selected variables including culture temperature, rotate speed and culture time were optimized by RSM design. The concentrations used in Plackett–Burman design served as central points, and the experiments were performed following with the design matrix (Table 4).

Significant F-value (22.3623) and *P*-value (0.0016), with non-significant lack of fit ($P = 0.2196$) suggest that the established model was well adapted to the response (Table 5). And, the value of coefficient (R^2) was 97.58% indicating the well fitness of the quadratic regression model and nearly 97.58% changes on *Da* value can be explained by Eq. (5). Furthermore, the once migration (U_1, U_2 and U_3), interaction migration ($U_1 * U_2$, $U_1 * U_3$ and $U_2 * U_3$), and quadratic migration (${U_1}^2$, ${U_2}^2$ and ${U_3}^2$) showed strong influence on Eq. (5) indicating that the relationship between the

Table 4 The design matrix and the results of Box-Behnken design

Num	U_1 culture temperature (°C)	U_2 rotate speed (rpm)	U_3 culture time (day)	*Da*	NUM	U_1 culture temperature (°C)	U_2 rotate speed (rpm)	U_3 culture time (day)	*Da*
1	−1 (24)	0 (150)	−1 (3)	0.3941	9	−1 (24)	−1 (100)	0 (6)	0.4348
2	−1 (24)	0 (150)	1 (7)	0.4591	10	1 (28)	−1 (100)	0 (6)	0.4301
3	1 (28)	0 (150)	−1 (5)	0.4578	11	−1 (24)	1 (200)	0 (6)	0.4156
4	1 (28)	0 (150)	1 (7)	0.4359	12	1 (28)	1 (200)	0 (6)	0.4706
5	0 (26)	−1 (100)	−1 (5)	0.4091	13	0 (26)	0 (150)	0 (6)	0.4700
6	0 (26)	1 (200)	−1 (5)	0.4085	14	0 (26)	0 (150)	0 (6)	0.4733
7	0 (26)	−1 (100)	1 (7)	0.4253	15	0 (26)	0 (150)	0 (6)	0.4648
8	0 (26)	1 (200)	1 (7)	0.4685					

Table 5 Regression analysis of Box–Behnken design experiment

Sourc	DF	SS	MS	F value	P value
U_1	1	0.0010	0.0010	21.0816	0.0059
U_2	1	0.0018	0.0018	36.5944	0.0018
U_3	1	0.0005	0.0005	10.1890	0.0242
U_1^2	1	0.0005	0.0005	9.9225	0.0254
$U_1 * U_2$	1	0.0019	0.0019	39.2637	0.0015
$U_1 * U_3$	1	0.0009	0.0009	18.1900	0.0080
U_2^2	1	0.0017	0.0017	34.1939	0.0021
$U_2 * U_3$	1	0.0005	0.0005	9.8992	0.0255
U_3^3	1	0.0015	0.0015	30.9868	0.0026
Model	9	0.0098	0.0011	22.3623	0.0016
Lack of Fit	3	0.0002	0.0001	3.7092	0.2196
Pure Error	2	<0.0001	<0.0001		
Cor Total	14	0.0100		R^2	0.9758

selected variables and Da values was rather a simple linear relation. After analyzing via RSM, the fermentation conditions of *M. androsaceus* in a 500 mL shake flask was as follows: initial pH of 6.0, rotating speed of 155 rpm, culture time 6.48 days, culture temperature 27.4 °C, inoculum size of 5%, inoculum age of 4 days and a loading volume of 200/500 mL, and the predictive *Da* value was 0.4762.

$$\begin{aligned} Y = &\ 0.469 + 0.011U_1 + 0.015U_2 + 0.007U_3 - 0.011U_1^2 \\ &- 0.022U_1U_2 + 0.015U_1U_3 - 0.021U_2^2 + 0.011U_2U_3 - 0.020U_3^2 \end{aligned} \tag{5}$$

During the establishment of the ANN model, *Dv* value supplied as evaluation index to optimize the number of hidden layer nodes which was finally chosen as 10 (Fig. 1a). The satisfied determination coefficient (R^2) (0.9945) demonstrated the well fitness of ANN model. The values of RMSEc, RMSEt, and RMSEp were 0.0027, 0.0025 and 0.0068. Furthermore, GA was performed to search for the best conditions with maximum *Dv* based on experimental results. The best fitness plot revealed successive generations toward the final optimum value (Fig. 1b). After analyzing via ANN-GA, the fermentation conditions of *M. androsaceus* in a 500 mL shake flask was as follows: initial pH of 6, rotating speed of 169 rpm, culture time 6.57 days, culture temperature 26.7° C, inoculum size of 5%, inoculum age of 4 days and a loading volume of 200 mL/500 mL, and the predictive *Da* value was 0.5059.

3.3 Independent Verification Test

Based on the fermentation conditions obtained from RSM and ANN-GA, six independent verification tests (three for each result) were applied to investigate the

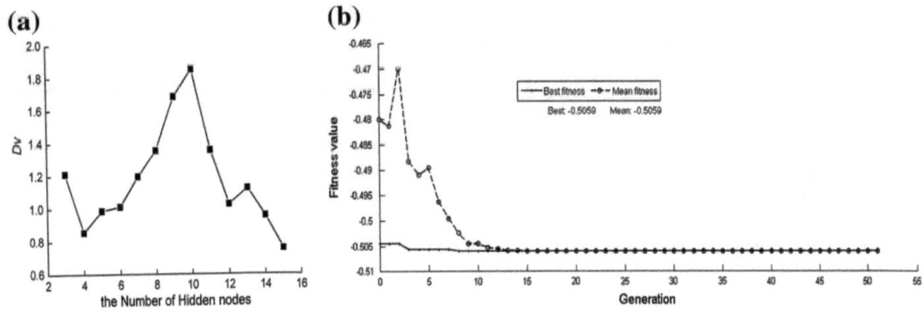

Fig. 1 **a** The effects of the number of hidden nodes on *Dv*. **b** The effects of genetic algebra on model fitness

Table 6 The results of ANN-GA independent verification test

Num	Mycelium yield (g/L)	Adenosine (g/L)	*Cordycepic* acid (g/L)	Polysaccharide (g/L)	*Da*
1	16.01	0.043	0.28	1.61	0.4940
2	15.99	0.044	0.29	1.55	0.4948
3	16.17	0.043	0.28	1.59	0.4936

similarity between experimental data and model predictive values. The predictive *Da* value from RSM via SAS V 9.0 software was 0.4762, and the average experimental value was 0.4427. Comparatively, the predictive *Da* value from ANN-GA via Matlab2012a software was 0.5059, and the average experimental value was 0.4941(Table 6). Finally, the optimum submerged fermentation condition of *M. androsaceus* in a 500 mL shake flask was obtained after ANN-GA optimization.

4 Conclusions

As a medicine traditionally used for pain alleviation, the large-scale application of *M. androsaceus* is limited by immature artificial cultivation technology. The fermentation conditions of *M. androsaceus* in a 500 mL shake flask have not been yet determined. In general, the classical optimal methods including a one-factor-at-a-time approach are laborious and time consuming, and ignores the explanation of interaction among selected variables [22]. In our present study, submerged fermentation conditions of *M. androsaceus* in a 500 mL shake flask were optimized by the application of Plackett–Burman design and Box–Behnken design. Both RSM and ANN-GA were applied to analyze the data obtained from Box–Behnken design. After comparison, the optimum fermentation conditions were obtained as follows: initial pH of 6.0, rotating speed of 169 rpm, culture time 6.57 days, culture temperature 26.7 °C, inoculum size of 5%, inoculum age of

4 days and a loading volume of 200 mL/500 mL, and the predictive *Da* value was 0.5059. Our finding reveals that Plackett-Burman design and Box-Behnken design are effective tools for mathematical modeling and factor analysis of the fermentation optimization process.

Acknowledgements This work is supported by the Natural Science Foundation of Tianjin (No. 16JCQNJC09100).

References

1. Zhang L, Yang M, Song Y et al (2009) Antihypertensive effect of 3,3,5,5-tetramethyl-4-piperidone, a new compound extracted from *Marasmius androsaceus*. J Ethnopharmacol 123(1):34–39
2. Wang X, Liang QM, Ting T et al (2006) Study on extraction and anti-oxidation of *Marasmius androsaceus* mycelium polysaccharides. Food Sci Technol 12:80–83
3. Ye WB, Yang XT, Chen Y et al (2002) Long-time analgesic effect of *Marasmius androsaceus* in rats. Pharm Clin Chin Mat Med 18(4):19–21
4. Fang QH, Zhong JJ (2002) Submerged fermentation of higher fungus *Ganoderma lucidum* for production of valuable bioactive metabolites—ganoderic acid and polysaccharide. Biochem Eng J 10(1):61–65
5. Saha T, Sasmal S, Alam S et al (2014) Tamarind kernel powder: a novel agro-residue for the production of cellobiose dehydrogenase under submerged fermentation by *Termitomyces clypeatus*. J Agric Food Chem 62(15):3438–3445
6. Zhou ZX, Yin Z, Hu XQ (2014) Corncob hydrolysate, an efficient substrate for Monascus pigment production through submerged fermentation. Biotechnol Appl Biochem 61(6): 716–723
7. Gao Y, Liu D, Yang X et al (2014) Isolation and characterization of glycopeptides with analgesic activities from *Marasmius androsaceus*. Eur J Integr Med 6(6):728
8. Song J, Xing GY, Cao JM et al (2016) Investigation of the antidepressant effects of exopolysaccharides obtained from *Marasmius androsaceus* fermentation in a mouse model. Mol Med Rep 13(1):939–946
9. Meng FX, Xing GY, Li, YT et al (2016) The optimization of *Marasmius androsaceus* submerged fermentation conditions in five-liter fermentor. Saudi J Biol Sci 23(Suppl 1):S99–S105
10. Ahmed A, Mukhtar H, Gohar UF et al (2013) Production of lovastati from aspergillus terreus through sub-merged fermentation. Pak J Bot 45(5):1795–1800
11. Gupta G, Sahai V, Gupta R (2013) Optimization of xylanase production from Melanocarpus albomyces using wheat straw extract and its scale up in stirred tank bioreactor. Indian J Chem Technol 20(4):282–289
12. Dong Y, Zhang N, Lu JH et al (2012) Improvement and optimization of the media of Saccharomyces cerevisiae strain for high tolerance and high yield of ethanol. Afr J Microbiol Res 6(10):2357–2366
13. Du LN, Song J, Wang HB et al (2012) Optimization of the fermentation medium for Paecilomyces tenuipes N45 using statistical approach. Afr J Microbiol Res 6(32):6130–6141
14. Heidari H, Razmi H, Jouyban A (2014) Desirability function approach for the optimization of an in-syringe ultra-sound-assisted emulsification-microextraction method for the simultaneous determination of amlodipine and nifedi-pine in plasma samples. J Sep Sci 37(12):1467–1474
15. Kleijnen JPC, Sargent RG (2000) A methodology for fitting and validating metamodels in simulation. Eur J Oper Res 120(1):14–29

16. Leyva A, Quintana A, Sanchez M et al (2008) Rapid and sensitive anthrone-sulfuric acid assay in microplate format to quantify carbohydrate in biopharmaceutical products: Method development and validation. Biologicals 36(2):134–141
17. Dong CH, Yao YJ (2008) In vitro evaluation of antioxidant activities of aqueous extracts from natural and cultured mycelia of Cordyceps sinensis. Lwt-Food Sci Technol 41(4):669–677
18. Chinese Pharmacopoeia Commission (2010) Pharmacopoeia of the People's Republic of China, 2010th edn. Chemical Industry Press, Beijing, China
19. Giordano PC, Beccaria AJ, Goicoechea HC (2011) Significant factors selection in the chemical and enzymatic hydrolysis of lignocellulosic residues by a genetic algorithm analysis and comparison with the standard Plackett-Burman methodology. Biores Technol 102 (22):10602–10610
20. Fatiha B, Sameh B, Youcef S et al (2013) Comparison of artificial neural network (ANN) and response surface methodology (RSM) in optimization conditions for lipase from Candida rugosa on amberjet (R) 4200-cl. Prep Biochem Biotechnol 43(1):33–47
21. Butt MA, Ahmad M, Fatima et al (2015) Ethnomedicinal uses of plants for the treatment of snake and scorpion bite in Northern Pakistan. J Ethnopharmacol 168:164–181
22. Din MI, Hussain Z, Mirza ML et al (2014) Adsorption optimization of lead (II) using sacharum bengalense as a non-conventional low cost biosorsent: isothrem anf thermodynamics modeling. Int J Phytorem 16(9):889–908

Heterologous Expression and Enzyme Properties of β-mannanase from *Trichoderma reesei* in *Pichia pastoris*

Qing Ma, Lijuan Ma, Rui Cai, Fengchao Jiang, Pan Song and Dongguang Xiao

1 Introduction

Hemicellulose, one of the most abundant biomass in nature, is considered to be a promising material for biorefinery [1, 2]. Mannan is one of the major constituents of hemicellulose in the cell wall of higher plants [3]. Hydrolysis of mannan requires synergy within several glycoside hydrolases (GH), including β-mannanase, β-glucosidase and β-mannosidase [4].

β-mannanase, which is the second abundant enzymes to hydrolysis hemicellulose, catalyzes the random breaking of β-D-1,4-mannopyranosyl linkages in the main chain of mannans and various heteromannans [5, 6]. The source of β-mannanase is extensive, including bacterial, fungi, animals and plants, and microorganism is the main source of β-mannanase [7, 8]. Currently, with the discovery of the new function of β-mannanase and mannan, as well as with the exploitation of hemicellulose resource, studies focus on β-mannanase have caused more and more concern [9]. As a growing research interest, they have been widely applied in the food and feed, paper and pulp, and detergent industries [10]. At present, researches on β-mannanase was focus on gene cloned and expression from different sources to investigate their characteristics and structures. Hu et al. cloned the gene of β-mannanase from *Paenibacillus* CH-3 and expressed it in *E. coli* BL21, obtained a high activity of 967 IU/mL [11]. Chen et al. successfully constructed a *Pichia pastoris* strain to highly expresse the recombinant β-mannanase with an activity of 470 IU/mL [7].

Trichoderma reesei Rut-C30, which was used as a gene source in this study, is a kind of fungi widely used in the production of cellulase and hemicellulase [12].

Q. Ma · L. Ma · R. Cai · F. Jiang · P. Song · D. Xiao (✉)
Key Laboratory of Industrial Fermentation Microbiology, Ministry of Education,
Tianjin Industry Microbiology Key Laboratory, College of Biotechnology,
Tianjin University of Science and Technology, Tianjin 300457, People's Republic of China
e-mail: xdg@tust.edu.cn

H. Liu et al. (eds.), *Advances in Applied Biotechnology*, Lecture Notes in Electrical Engineering 444, https://doi.org/10.1007/978-981-10-4801-2_27

β-mannanase form *T. reesei* was first purified and characterized by Arisan-Atac et al. in 1993 [13]. Stålbrand et al. successfully expressed the β-mannanase from *T. reesei* in *Saccharomyces cerevisiae* in 1995 [14]. However, there were little reports about the characteristic of the *T. reesei* β-mannanase expressed in *Pichia pastoris* which is the most popular expression system for fungal enzymes.

Here, the β-mannanase gene from *Trichoderma reesei* Rut-C30 was cloned and expressed in *P. pastoris* GS115. The purified protein was then used to be investigated the enzyme properties of recombinant β-mannanase, including the optimal pH and temperature, influence of metal ions and catalytic kinetics constants. The thermal and pH stabilities were also investigated.

2 Method

2.1 Strains and Vectors

T. reesei Rut C30 (ATCC 56765) was maintained on potato dextrose agar (PDA) at 4 °C and sub-cultured once in every three months. *Escherichia coli* DH5α was used for the construction and propagation of recombinant vectors. *P. pastoris* GS115 and pPIC9 K (Invitrogen) were used as a host-vector system for protein expression.

2.2 Total RNA Isolation

The spores of *T. reesei* Rut C30 were prepared into a suspension with the spore concentration of 10^7–10^8 spores/mL. 1 mL spore suspension was transferred into 250 mL Erlenmeyer flasks with 50 mL basal medium containing Mandel' s salt solution supplemented with 0.3% (w/v) glucose as the carbon source and incubated at 28 °C, 180 rpm for 24 h [15].

The mycelia harvested at 24 h was immediately frozen and grounded to a fine powder in liquid nitrogen. Total RNA was extracted using RNA prep pure plant kit (TIANGEN, China). The quality and concentration of the extracted RNA were determined at 260 nm using a nucleic acid spectrometer (Implen, Germany).

2.3 Construction of the Expression Plasmid

Approximately 500 ng total RNA was treated with gDNA Eraser to remove genomic DNA and subjected to reverse transcription using a Prime ScriptTMRT reagent kit (TaKara, China).

The gene of β-mannanase was amplified from reverse transcription cDNA by PCR. The forward primer and downstream primer used in the PCR were 5′-CGGAATTCATGATGATGCTCTCAAAG-3′ and 5′-ATGCGGCCGCTCAGTGGTGGTGGTGGTGGTGTGTATTCAGGCATTGCG-3′, respectively. The PCR products of the target genes were purified with DNA purification kit (Omega), digested with *Eco*RI and *Not*I and ligated into the pPIC9 K vector digested with the same enzymes. The recombinant expression vectors were transformed into *E. coli* DH5α to obtain single colonies and the positive recombinant plasmids were confirmed by restriction digestion and DNA sequencing.

2.4 Transformation and Expression in P. pastoris GS115

The recombinant plasmids were linearized and transformed into *P. pastoris* GS115 cells. Transformants were screened on MD plate containing 0.04 mg/L biotin and YPD plates containing Zeocin. The *P. pastoris* recombinants were identified by PCR.

The recombinant *P. pastoris* colonies were inoculated into BMGY medium at 30 °C, 240 rpm for 24 h. The yeast cells were aseptically harvested by centrifugation, resuspended in 25 ml of BMMY medium at 30 °C, 240 rpm for 168 h. At 24 h intervals methanol was added to the culture to maintain a final concentration of 0.5%. The culture supernatant was collected by centrifugation and subjected to β-mannanase activity assay.

2.5 Determination of Enzyme Activity

Mannanse activity was assayed using the 3,5-dinitrosalicylic acid (DNS)-method with 0.5% locust bean gum (LBG, Sigma G-0753) as substrate as described by Stålbrand with minor modifications [14]. One unit of enzyme activity was defined as the amount of enzyme liberating 1 μmol of mannose equivalent per minute.

2.6 Purification and Gel Electrophoresis Analysis

The fermentation broth was centrifuged (5000 × g, 5 min, 4 °C) to remove yeast cells. Then the supernatant was filtered through a 0.22 μm filter, and loaded onto the Ni-Agarose Resin for 6×His-Tagged Proteins (CWBIO, China) according to the manufacturer's protocol. The loaded sample was eluted with a step gradient of 40–300 mM imidazole containing pH 7.9 Tris-HCl and 0.5 mM NaCl, which was followed by eluting with 500 mM imidazole. Finally the purified sample was dialyzed using a dialysis membrane (Solarbio, 8–14 kDa molecular-weight cut off) with 50 mM citrate buffer as the dialysate to remove imidazole.

Sodium dodecyl sulfate-polyacrylamide gel electrophoresis (SDS-PAGE) was performed using a 12% (w/v) polyacrylamide gel according to the method of Laemmli [16]. Protein bands were visualized by staining with Coomassie brilliant blue R-250.

2.7 *Enzyme Properties*

The optimal pH and temperature were determined using the standard enzyme assay conditions by changing the pH (3.0–8.0) or temperature (50–90 °C) respectively. For pH stability assay, the purified enzymes were pre-incubated at 40 °C in citrate buffer (50 mM, pH 5.0) over the pH range of 3.0–8.0 for 30 min without substrate, and then the residual enzyme activities were measured. For the determination of thermal stability, aliquots of purified enzymes were incubated in citrate buffer (50 mM, pH 5.0) at 60 °C for 24, 48, 72, 96, and 120 h without substrate respectively. Thermal inactivation was stopped by cooling the samples on ice, and the residual activities were determined as described above. Each reaction was run in triplicate.

Kinetic parameters, K_m and V_{max}, were determined in citrate buffer (50 mM, pH 5.0) containing 0–5 mg/mL LBG or galactomannan at 60 °C for 5 min. The data were calculated according to the Lineweaver-Burk Method. All experiments were carried out with three replicates.

3 Result and Discussion

3.1 *Cloning and Expression of Recombinant β-mannanase in* P. pastoris

The positive *P. pastoris* recombinant was induced by methanol with a concentration of 0.5%. The biomass and enzyme production curves were shown in Fig. 1. According to the figure, with the increase of time, the biomass of strain and the enzyme activity of supernatant liquid increased gradually. The maximum activity of β-mannanase was 34.5 IU/mL after 7 days induction. Therefore, the recombinant showed high efficiency in expressing and secreting the soluble recombinant protein after 168 h methanol induction.

After the centrifugation of the fermentation broth, the supernatant was loaded onto Ni-Agarose Resin to purify the β-mannanase. The optimal concentration of imidazole to elute the target protein was 200 mM. The result of SDS-PAGE was shown in Fig. 2. The apparent molecular weight was 65 kD, which was bigger than

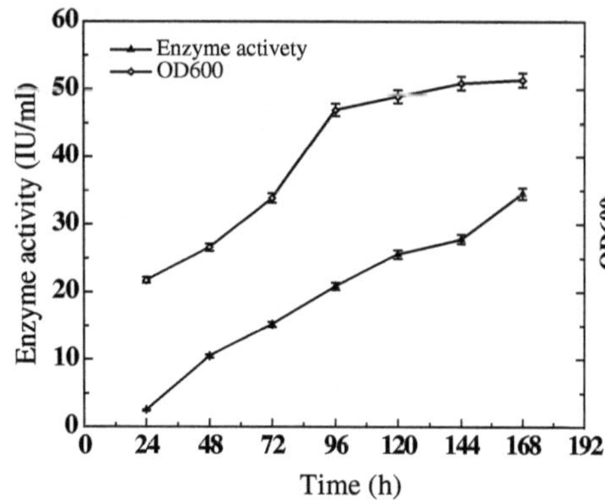

Fig. 1 Secretion courses of mannanase and OD600 of the recombinant *P. pastoris*

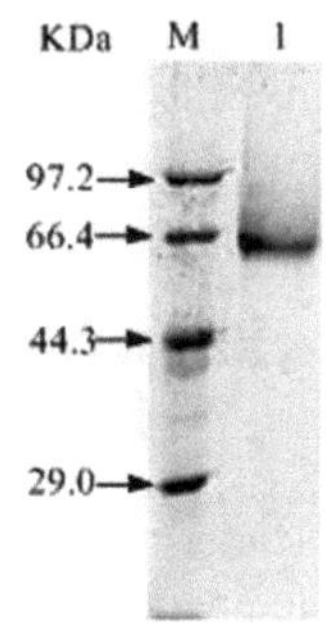

Fig. 2 SDS-PAGE analysis of recombinant mannanase

the theoretical molecular weight. This is probably due to the glycosylation which is widely existed in the expression system of *P. pastoris*, as the literatures mentioned [17–21].

3.2 The Enzyme Properties of the Recombinant β-mannanase

3.2.1 Influence of pH on the Activity of Recombinant β-mannanase

The activities of recombinant β-mannanase at different pH were shown in Fig. 3a. According to the results, the activity of recombinant β-mannanase was obtained at pH 6.0. The purified enzyme was pre-incubated at 40 °C in citrate buffer (50 mM, pH 5.0) over the pH range of 3.0–8.0 for 30 min without substrate, and then the residual enzyme activities were measured.

From Fig. 3b, when the pH was at 4.5–8.0, the relative avtivities of the recombinant β-mannanase were moer than 85%. While when the pH was lower

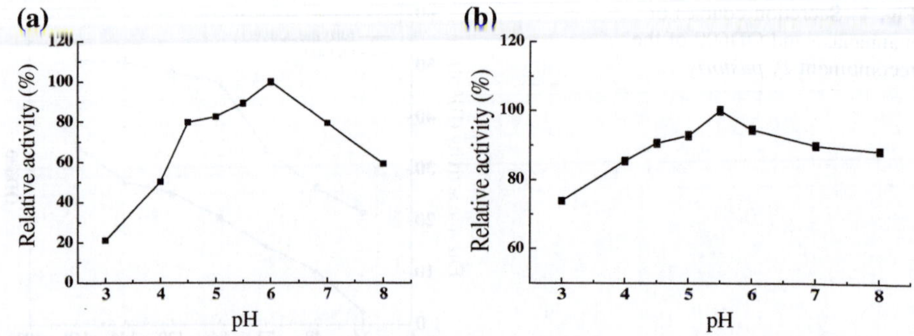

Fig. 3 Effect of pH on recombinant mannanase

than 4.5, the relative activity decreased rapidly. It could be inferred that the recombinant β-mannanase has a good alkali resistance and a relatively poor acid resistance.

3.2.2 Influence of Temperature on the Activity of Recombinant β-mannanase

Under the optimal pH, the activities of recombinant β-mannanase at different temperatures with a range of 50–90 °C were shown in Fig. 4a. According to the figure, the optimal temperature for the action of recombinant β-mannanase was 80 °C. The recombinant enzyme was processed for different time at 60 and 70 °C to determine the thermal stability of β-mannanase, and the results were shown in Fig. 4b, c. As the figures shown, the relative activity of the recombinant enzyme was stable after 120 h incubation at 60 °C, and the relative activity was 72.67% after 120 h, while the relative activity of the recombinant β-mannanase decreased drastically after being processed for 30 min at 70 °C and the relative activity was lower than 20% after 60 min processing, which suggested that the recombinant β-mannanase has a higher thermal stability at 60 °C than 70 °C.

3.2.3 Influence of Metal Ions on the Activity of Recombinant β-mannanase

Different metal ions (5 mM) and EDTA were added to enzyme reaction liquid to investigate the influence of metal ions on the activity of the recombinant enzyme. The results were presented in Fig. 5. According to the data, Fe^{2+}, Cu^{2+}, Li^{2+} and EDTA have a strong inhibitory effect on the activity of the recombinant β-mannanase. The influence of EDTA was the most obviously, which made the activity declined to 38.4%. K^{+} and Zn^{2+} promoted the activity of the recombinant β-mannanase obviously. Fe^{3+}, Ca^{2+} and Mg^{2+} also promoted the activity of the

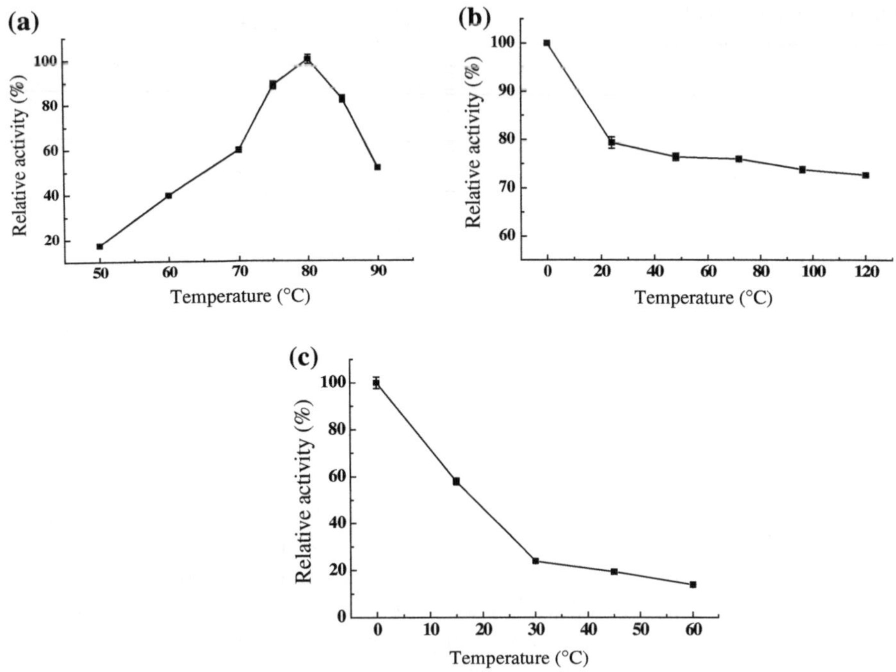

Fig. 4 Effects of temperature on the recombinant β-mannanase

Fig. 5 Influence of metal ions on the activity of recombinant β-mannanase

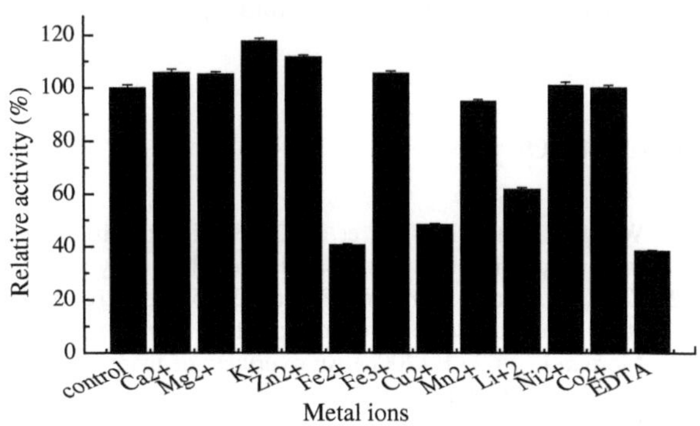

enzyme in different degree. In addition, Ni^{2+} and Co^{2+} have little influence on the activity of the recombinant enzyme.

3.2.4 Specific Activity and Kinetic Constants

The kinetic constant of the curve was calculated by linewaver-burk double counting method. The result was shown in Fig. 6. Through calculation, the K_m was 0.5 mg/mL, V_{max} was 253.8 IU/mg.

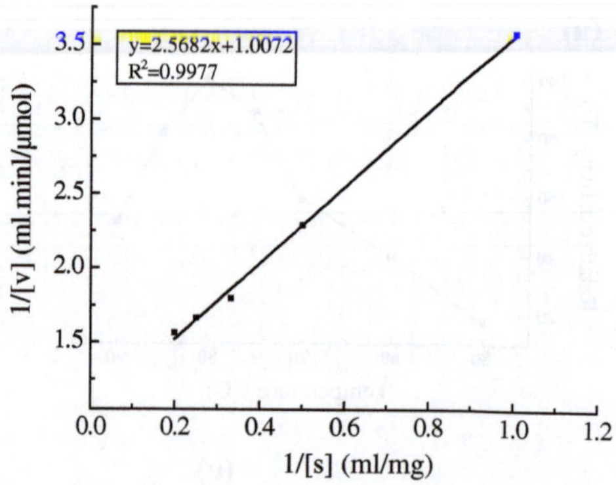

Fig. 6 Linewaver-Burk plot for recombinant mannanase

4 Conclusion

The gene of β-mannanase from *T. reesei* was successfully cloned and expressed in *P. pastoris.* The enzyme was used to investigate its enzyme properties. The optimal pH and temperature of the purified recombinant β-mannanase were 6.0 and 80 °C, respectively. The enzyme had a high alkali resistance and a relatively poor acid resistance. It showed higher thermostability at 60 °C than 70 °C. K^+, Zn^{2+}, Fe^{3+}, Ca^{2+} and Mg^{2+} promoted the activity of the recombinant β-mannanase in different degree, while Fe^{2+}, Cu^{2+}, Li^{2+} and EDTA have strong inhibitory effects on the activity of the recombinant β-mannanase. The K_m and V_{max} of the recombinant β-mannanase were 0.5 mg/mL and 253.8 IU/mg, respectively.

References

1. Wang Y, Yuan B, Ji Y et al (2013) Hydrolysis of hemicellulose to produce fermentable monosaccharides by plasma acid. Carbohyd Polym 97:518–522
2. Zhong C, Wang C, Huang F et al (2015) Selective hydrolysis of hemicellulose from wheat straw by a nanoscale solid acid catalyst. Carbohyd Polym 131:384–391
3. Moreira L, Filho E (2008) An overview of mannan structure and mannan-degrading enzyme systems. Appl Microbiol Biotechnol 79:165–178
4. Malgas S, Dyk J, Pletschke B (2015) A review of the enzymatic hydrolysis of mannans and synergistic interactions between β-mannanase, β-mannosidase and α-galactosidase. World J Microbiol Biotechnol 31:1167–1175
5. Sabini E, Schubert H, Murshudov G et al (2000) The three-dimensional structure of a *Trichoderma reesei* β-mannanase from glycoside hydrolase family 5. Acta Crystallogr D 56:3–13
6. Stålbrand H, Saloheimo A, Vehmaanpera J et al (1995) Cloning and expression in Saccharomyces cerevisiae of a *Trichoderma reesei* β-mannanase gene containing a cellulose binding domain. Appl Environ Microbiol 61:1090–1097
7. Chen B, Qiao Y, Ding HB et al (2007) Expression of Endo-β-mannanase Gene from *Trichoderma reesei* in *Pichia pastoris.* J Agric Biotechnol 15(1):142–143
8. Gao HB (2003) Clone and expression of endo-β-mannanase. Graduation thesis

9. Xu HR (2008) Domestic and foreign research progress of β-mannanase. Feed Hunan 2:30–32
10. Daskiran M, Teeter R, Fodge D et al (2004) An evaluation of endo-β-D-mannanase (Hemicellulose) effects on broiler performance and energy use in diets varying in β-mannan content. Poult Sci 83:662–668
11. Hu WB (2012) Clone, expression and enzyme properties of β-mannanase. Graduation thesis
12. Peterson R, Nevalainen H (2012) *Trichoderma reesei* RUT-C30-thirty years of strain improvement. Microbiology 158(1):58–68
13. Arisan-Atac I, Hodits R, Kristufek D et al (1993) Purification, and characterization of a β-mannanase of *Trichoderma reesei* C-30. Appl Microbiol Biotechnol 39:58–62
14. Stålbrand H, Siika-aho M, Viikari L (1993) Purification and characterization of two β-mannanases from *Trichoderma reesei*. J Biotechnol 29:229–242
15. Mandels M, Reese E (1960) Induction of cellulase in fungi by cellobiose. J Bacteriol 79:816–826
16. Laemmli U (1970) Cleavage of structural proteins during the assembly of the head of bacteriophage T4. Nature 227:680–685
17. Trimble RB, Lubowski C, Hauer CR et al (2004) Characterization of N- and O-linked glycosylation of recombinant human bile salt-stimulated lipase secreted by *Pichia pastoris*. Glycobiology 14(3):265–274
18. Contreras R, Callewaert NLM, Geysens SCJ (2004) Protein glycosylation modification in *Pichia pastoris*. Europe PMC, 2002 506172
19. Zhao L, Geng J, Guo Y et al (2015) Expression of the Thermobifida fusca xylanase Xyn11A in *Pichia pastoris* and its characterization. BMC Biotechnol 15(1):1–12
20. Montesino R, Cremata J, Rodríguez M, Besada V, Falcón V, de la Fuente J (1996) Biochemical characterization of the recombinant Boophilus microplus Bm86 antigen expressed by transformed *Pichia pastoris* cells. Biotechnol Appl Biochem 23(Pt 1):23–28
21. Mochizuki S, Hamato N, Hirose M, Miyano K, Ohtani W, Kameyama S, Kuwae S, Tokuyama T, Ohi H (2001) Expression and characterization of recombinant human antithrombin III in *Pichia pastoris*. Protein Expr Purif 23(1):55–65

Reduction of Characteristic Biogenic Amines Production by Synergistic Fermentation of Salt-Tolerant Yeast in Soy Sauce

Wei Qi, Wen-Tao Zhang and Fu-Ping Lu

1 Introduction

Biogenic amines, basic nitrogenous compounds, are considered as the indicators of food quality and freshness [1, 2]. They have been widely detected in raw and processed foods, especially in fermented foods due to amino acid decarboxylation activities of microbes [3]. Several biogenic amines, such as tyramine, histamine, cadaverine and spermidine were found in foods.

Study on formation, toxicity and detection of biogenic amines had been basically completed [4]. Based on analysis of biogenic amines formation process, three scenarios have been described for controlling biogenic amine in foodstuffs:

(1) Control the level of biogenic amines and their amino acids precursors

To a certain extent, biogenic amines could be controlled by decreasing the level of amino acids, however, there might have some unfavorable influence on the product quality.

(2) Control the level of biogenic amines by inhibiting or decreasing decarboxylase activity

Biogenic amine formation can be controlled through inhibiting or decreasing the level of decarboxylase activity of microbes, or inhibiting microbial growth [5]. Limit microbial growth through chilling and freezing, high hydrostatic pressure, irradiation [6, 7], cold chain coupled with the use of a modified atmosphere packaging [8], vacuum packaging, oxygen scavenger, microbial modeling or use of food additives and preservatives [9].

W. Qi · W.-T. Zhang · F.-P. Lu (✉)
Key Laboratory of Industrial Fermentation Microbiology, Ministry of Education, Tianjin University of Science & Technology, Tianjin 300457, People's Republic of China
e-mail: qiweismiling@126.com

H. Liu et al. (eds.), *Advances in Applied Biotechnology*, Lecture Notes in Electrical Engineering 444, https://doi.org/10.1007/978-981-10-4801-2_28

(3) Biogenic amines degradation by microorganism

Biogenic amines can be degraded by special microbial, such as *Natrinema gari* [7], methanogenic bacteria [10], lactic acid bacteria (*Lactobacillus sakei, Staphylococcus equorum*, and *Lactobacillus casei* subsp. *Case*) [11–13] and vineyard ecosystem fungi [14]. Bacteria with amine oxidase activity have become a particular interest to reduce biogenic amines concentration in food products such as meat and fish sausages [15].

PCR and DNA hybridization, allowed a faster detection identification of food-borne biogenic amine-producing bacteria and introduction of early control measures to avoid development of these bacteria. However, further research on reduction or breakdown of biogenic amines is required.

The flavour and aroma of soy sauce rely extensively on strain improvement for commercialization. The primary importance of the salt-tolerant yeast (such as *C. versatilis* and *Z. rouxii*) is the ability to accelerate and enhance the flavour in a high salt concentration present in moromi [16]. However, a certain level of biogenic amines has been found in soy sauce, which may be associated with *C. versatilis* and *Z. rouxii* [17–19]. Although there is no specific legislation regarding BA content in many fermented products, their toxicity has led to the general agreement that they should not be allowed to accumulate in food [20]. Moreover, there have been several researches focused on co-culture not only enhancing the production of volatile flavor compounds [21] but also decreasing the formation of biogenic amines [22]. Therefore, in the present work, biogenic amines, such as histamine, tyramine, cadaverine and spermidine were determined by the treatment of *C. versatilis*, *Z. rouxii*, *C. versatilis* S3-5 yeast strain, initially isolated for improved tolerance to salt [23], and the synergistic fermentation of salt-tolerant yeast strains during the whole high-salt liquid-state soy sauce fermentation.

2 Materials and Methods

2.1 Raw Materials, Chemicals and Substrate Pretreatment

All chemicals and solvents were in analytical and chromatographic grade. Histamine, tyramine, cadaverine, spermidine and dansyl chloride were purchased from Sigma (St. Lous, MO). Sodium hydroxide, sodium hydrogen carbonate, ammonium hydroxide and perchloric acid were purchased from Tianjin Chemical Reagent Research Institute (Tianjin, China). HPLC grade acetonitrile and ammonium acetate (0.1 M) were obtained from Merk (Whitehouse Station, NJ). The water for the dilution of samples was purified using a Milli-Q system (Millipore, Bedford, MA).

High-salt liquid-state mash was made with steam-cooked soybeans and wheat (1:1, w/w) and supplemented with 17% (w/v) NaCl and water as previously described [24, 25].

2.2 Yeast Strains, Culture Maintenance and Inoculum Preparation

Yeast strains *Candida versatilis*, *Zygosaccharomyces rouxii* and *Aspergillus oryzae* HN3.042 used in this work were kept in our laboratory. *Candida versatilis* S3-5 was derived from the yeast strain *C. versatilis* wild type by genome shuffling [23].

For inoculum preparation, yeast cells were grown in yeast extract peptone dextrose (YPD) medium with shaking (180 rpm) at room temperature (30 ± 1 °C) for 24 h. Five milliliter of YPD medium was transferred to 50 ml of stage 1 inoculum culture containing 8% (w/v) salt of YPD medium with shaking for 48 h. Ten milliliter of the stage 1 inoculum culture was transferred to 100 ml of the stage 2 inoculum culture containing 12% (w/v) salt of YPD medium with shaking for 16 h.

Cells of stage 2 inoculum culture were harvested by centrifugation (5000×*g*, 5 min), suspended with 500 ml soy mash at a final concentration of $\sim 10^6$ cells per ml in anaerobic bottle for 15 days. The synergistic fermentation of yeast strains was assessed by adding mixed *Z. rouxii* and *C. versatilis* or mixed *Z. rouxii* and S3-5 to a cell density of $\sim 10^6$ cells per ml in 500 mL soy mash. The control soy mash sample was treated without yeast. Fermentation was carried out at 30 ± 1 °C and samples were periodically withdrawn. Fermentation experiments were carried out in triplicate using the same batch of soy mash. Statistical significance was determined using the SAS statistical analysis program, version 8.01.

2.3 Analytical Methods

2.3.1 Preparation of Standard Amine Solution

Stock solutions of 1,7-diaminoheptane (internal standard) and all 4 standard biogenic amines (HIS, TYR, CAD and SPR) were separately prepared at 1 g L^{-1} concentration in distilled water. Working solution was prepared by diluting 1 mL of each stock solution in distilled water to yield a final volume of 10 mL. These solutions were stored at 4 °C until used.

2.3.2 Preparation of Sample Extracts

Extraction of fermented mash samples and HPLC determination of biogenic amines were detected as previously described [17, 26]. To extract biogenic amines, 10 mL of 0.4 M perchloric acid containing a known amount of 1,7-diaminoheptane (internal standard) was added to 5 g of fermented soy mash samples, and the mixture was homogenized for 3 min. The homogenate was centrifuged at 3000×*g* for 10 min at 4 °C. The supernatant was collected and the residue was extracted twice with the same volume of 0.4 M perchloric acid solution. All supernatants were combined and the final volume was adjusted to 25 mL using 0.4 M perchloric acid. The extract was filtered through Fisher paper 09-804-142J (Beijing Dingguo Biotechnology Co. Ltd., Beijing, China). One milliliter of each extract was used for the HPLC analysis followed by derivatization with dansyl chloride.

2.3.3 Derivatization of Extracts and Standards

Derivatization of biogenic amines was carried out as previously described [17, 26]. One milliliter of each extracted sample or standard amine solution was mixed with 200 μL of 2 M sodium hydroxide and 300 μL of sodium hydrogen carbonate. Two milliliters of a dansyl chloride solution (10 mg mL^{-1} in acetone) were added to the mixture and then incubated in water bath at 40 °C for 45 min. One hundred micro-liters of 250 g L^{-1} ammonium hydroxide were added to stop the reaction and to remove residual dansyl chloride. After 30 min incubation at room temperature, the final volume was adjusted to 5 mL by adding acetonitrile. Finally, the mixture was centrifuged at 2500×*g* for 5 min and the supernatant was filtered through 0.45 μm-pore-size filter (13 mm, 0.45 μm, Tianjin Jinteng Co., Tianjin, China). The filtered supernatant was kept at −25 °C until HPLC analysis. All samples were subjected to HPLC injection for two times.

2.3.4 Chromatographic Conditions

Biogenic amines in the samples were measured by HPLC using a Shimadzu 20AB with ultraviolet detector at 254 nm. HPLC column was a Spursil C18 column (250 mm × 5 μm × 4.6 mm) and was set to 30 °C during all running procedures, and the injection volume was 20 μL. Different gradient programs were evaluated to achieve good resolution of all biogenic amines in the shortest time. The mobile phase was 0.1 M ammonium acetate (solvent A) and acetonitrile (solvent B) at the flow rate of 1 mL min^{-1} with gradient elution. The gradient elution procedure was 65% A + 35% B at 0.01 min, 55% A + 45% B at 5 min, 35% A + 63% B at 10.05 min, 20% A + 80% B at 17.05 min, 10% A + 90% B at 26.25 min and 65% A + 35% B at 35.00 min. The peak areas of biogenic amines standard solutions were identified to determine the amine concentrations in mash samples.

2.3.5 Statistical Analysis

For the chromatographic data obtained from HPLC was performed with package SPSS Statistics 18.0 software.

3 Results and Discussion

3.1 Total Content of Biogenic Amines in Soy Mash

Recent research has focused on biogenic amines because of their detrimental effects on human beings such as migraine, hypertension, hypotension, rash and digestive problems, which may lead to death in severe cases. Moreover, biogenic amines are considered potential precursors of carcinogenic N-nitroso compounds [27]. The content of total biogenic amines in soy mash was analyzed in this study.

The results showed that the level of total biogenic amines in sauce mash was gradually increased until 120 days and then decreased during the fermentation process (Fig. 1). The total biogenic amines reached their maximum levels at 120 days, in which S3-5 produced 264.57 mg/kg, while the *Z. rouxii* and *C. versatilis* produced 238.96 mg/kg and 176.65 mg/kg biogenic amines, respectively. The synergetic fermentation of *Z. rouxii* + *C. versatilis* generated 117.12 mg/kg biogenic amines, while *Z. rouxii* + S3-5 treatment yielded 201.65 mg/kg biogenic amines. In the control, the total content of biogenic amines increased slowly and reached its maximum level (60.46 mg/kg) at 150 days. At the end of the fermentation, the total concentration of biogenic amines in soy mash of S3-5, *Z. rouxii*, *C. versatilis*, *Z. rouxii* + S3-5, *Z. rouxii* + *C. versatilis*, and control was 185.12, 122.71, 69.96, 151.14, 65.51 and 45.74 mg/kg, respectively. The level of biogenic amines began to decline from 5 month [25].

The biogenic amines in the soy mash was higher after yeast strain treatment, which might be due to the action of yeast decarboxylase [19, 28]. The strongest capability to produce biogenic amines was S3-5, followed by *Z. rouxii* + S3-5, *Z. rouxii*, *C. versatilis*, *Z. rouxii* + *C. versatilis*. The content of biogenic amines produced by S3-5 was higher than *C. versatilis*, which might because genome shuffling technology not only improves strain tolerance to salt but changes the metabolic ability of strain. However, the concentration of biogenic amines yielded by two strains, especially *Z. rouxii* and *C. versatilis*, was lower than any one strain. This suggests that two yeast strains might have opposite effects on generating biogenic amines. The result was similar to the previous research that biogenic amine was reduced via co-culture of *Lactobacillus plantarum* and *Zygosaccharomyces rouxii* during the manufacture of Chinese sauerkraut [22].

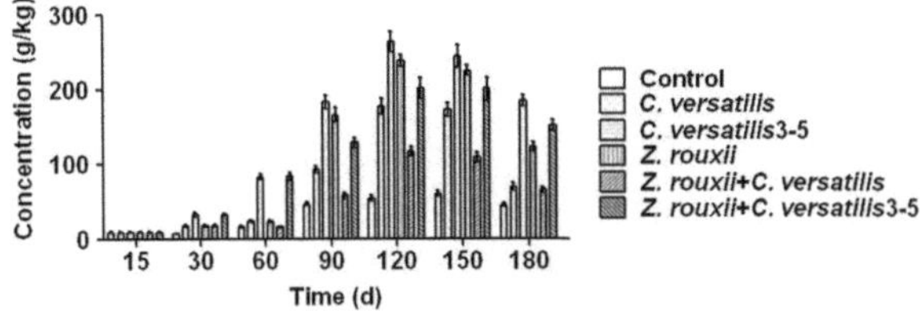

Fig. 1 Total content of biogenic amines in soy mash

3.2 Biogenic Amines Produced by Different Strains in High Salt Soy Mash

3.2.1 Tyramin Produced by Different Strains in High Salt Soy Mash

High levels of tyramine (>100 mg kg^{-1}) can cause adverse effects on human health [17, 29]. Compared with the control, the content of tyramine after yeast treatment was significantly higher. The level of tyramine firstly increased and then decreased until the completion of fermentation (Fig. 2a). The strongest ability to produce tyramine was *Z. rouxii*, followed by S3-5, *Z. rouxii* + S3-5, *C. versatilis*, *Z. rouxii* + *C. versatilis*, and the concentration of tyramine in soy mash was 64.24, 30.64, 25.02, 20.87 and 19.08 mg/kg at the end of fermentation [25]. The trends of tyramine production by S3-5 and synergistic fermentation (*Z. rouxii* + S3-5) were different from other groups, the content of tyramine increased dramatically at the beginning of fermentation, and then reached its maximum level. The metabolism process of tyrosine by S3-5 might be faster than other yeast strains, and thus is easy to adapt to the high salt environment quickly. In addition, tyramine from the synergistic fermentation of two yeast strains was lower than one strain at the mid-, late-fermentation process. It indicates that the synergistic fermentation of yeast strains could control the content of tyramine in the soy mash.

3.2.2 Histamine Produced by Different Strains in High Salt Soy Mash

Histamine poisoning is a worldwide problem that occurs after the consumption of food containing BAs, particularly histamine at concentration higher than 500 ppm [30]. The production of histamine in soy mash was significantly elevated, the trend

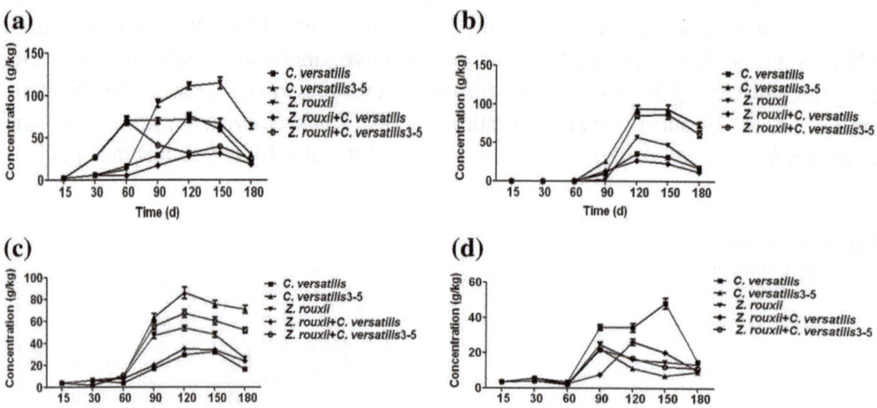

Fig. 2 Biogenic amines produced by different strains in soy mash. **a** Tyramine, **b** Histamine, **c** Cadaverine, **d** Spermidine

was abruptly increased from 60 to 120 days and then gradually decreased until 180 days (Fig. 2b). The strongest ability to produce histamine was S3-5, followed by *Z. rouxii* + S3-5, *Z. rouxii*, *C. versatilis*, *Z. rouxii* + *C. versatilis*. At the end of the fermentation, the level of histamine in soy mash after the treatments of S3-5, *Z. rouxii* + S3-5, *Z. rouxii*, *C. versatilis* and *Z. rouxii* + *C. versatilis* was 73.99, 62.61, 18.46, 17.56 and 12.30 mg/kg, respectively [25]. Moreover, the content of histamine produced by S3-5 was stronger than that by *C. versatilis* and *Z. rouxii*, that is because genome shuffling technology not only improves the S3-5 strain to tolerance salt stress, but improves the S3-5 strain to synthesize and excrete histidine decarboxylase or aromatic L-amino acid decarboxylase.

3.2.3 Cadaverine Produced by Different Strains in High Salt Soy Mash

The level of cadaverine in soy mash was significantly increased from 60 to 120 days after the treatment of yeasts, but gradually decreased until the end of fermentation (Fig. 2c). The strongest ability to produce cadaverine was S3-5, followed by *Z. rouxii* + S3-5, *Z. rouxii*, *Z. rouxii* + *C. versatilis*, *C. versatilis*. At the end of the fermentation, the content of histamine in soy mash of S3-5, *Z. rouxii* + S3-5, *Z. rouxii*, *Z. rouxii* + *C. versatilis* and *C. versatilis* was 71.01, 52.02, 26.35, 24.30 and 16.50 mg/kg, respectively [25].

3.2.4 Spermidine Produced by Different Strains in High Salt Soy Mash

The level of spermidine was remarkably elevated after yeast treatment (Fig. 2d). The strongest ability to produce histamine was *C. versatilis*, followed by *Z. rouxii* + *C. versatilis*, *Z. rouxii*, *Z. rouxi* + S3-5 and S3-5. At the end of the fermentation, the content of spermidine in soy mash of *C. versatilis*, *Z. rouxii* + *C. versatilis*, *Z. rouxii*, *Z. rouxii* + S3-5 and S3-5 was 15.02, 9.82, 13.67, 11.49 and 9.47 mg/kg, respectively [25]. Moreover, the concentration of histamine produced by S3-5 was lower than that by *C. versatilis* and *Z. rouxii*, this suggests genome shuffling technology not only improves S3-5 strain to tolerate to salt, but reduces the strain to synthesize and excrete arginine decarboxylase and spermidine synthase.

3.3 *Content of Biogenic Amines in Soy Mash After Synergistic Fermentation of Yeast Strains*

The highest contents of biogenic amines in the synergistic fermentation mash of *Z. rouxii* + *C. versatilis* (Fig. 3b) was cadaverine, followed by tyramine, histamine and spermidine; while the most important biogenic amines in the synergistic fermentation mash of *Z. rouxii* + S3-5 (Fig. 3c) was histamine, followed by cadaverine, tyramine and spermidine. These data suggest that the synergistic fermentation of yeast strains can control the content of BAs, especially tyramine and histamine in the soy mash.

The level of biogenic amines in soy mash after yeast treatment was higher than that of the control group. The changes in biogenic amines in high salt liquid soy mash during fermentation process indicate that a variety of biogenic amines are increased in the fermentation aging period, which may be due to yeast amino acid decarboxylation to form biogenic amines by yeast decarboxylase. The tyramine produced by *Z. rouxii* was higher than other strains. The histamine and cadaverine produced by S3-5 yeast strain was the highest, while spermidine produced by yeast *C. versatilis* was higher than other strains. At the end of the fermentation, the total level of biogenic amines was ranged from 45.74 to 185.12 mg/kg in the soy mash samples.

In the sauce mash, *Zygosaccharomyces rouxii* appeared in the early fermentation stage, these strains are the dominant moromi yeast which produce the flavours ethanol and higher alcohols. While *Candida versatilis* were detected at the middle and last stage of brine fermentation, these strains produce phenolic compounds i.e. 4-ethylguaiacol and 4-ethylphenol, which contribute to soy sauce some aroma [31, 32]. These might explain why biogenic amines produced by *C. versatilis* was lower than that by *Z. rouxii*. Moreover, The probable reason for reduction of biogenic amines by the treatment of yeast strains synergistic fermentation is that the species of *Candida* are not active in the moromi at the same time as *Zygosaccharomyces rouxii* [33, 34].

The potential formation of biogenic amines is a concern in fermented foods because of the intense microbial activity [21]. In traditionally manufactured soy sauce and soybean paste, the added salt limits protease activity, prolonging

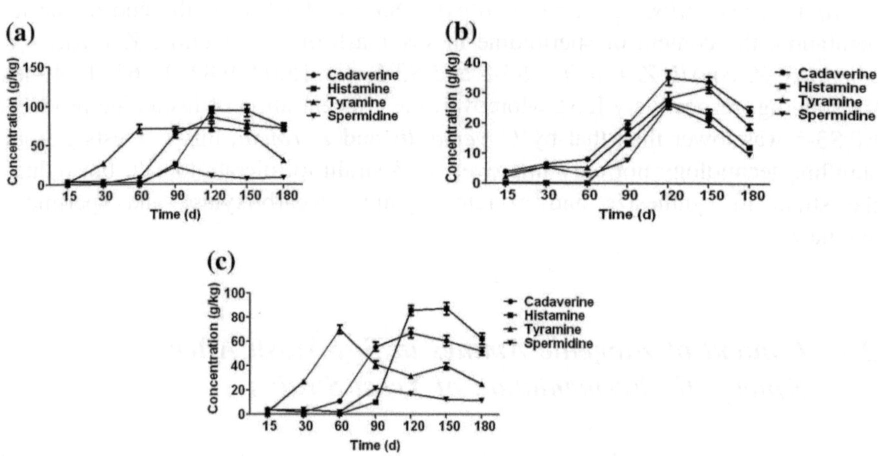

Fig. 3 Content of biogenic amines in soy mash after synergistic fermentation of yeast strains. **a** S3-5, **b** *Z. rouxii* + *C. versatilis*, **c** *Z. rouxii* + S3-5

fermentation time, therefore, minimizes amine formation [6, 18]. In our study, the total level of biogenic amines generated by the synergistic fermentation of *Z. rouxii* + *C. versatilis* was lower than that from other strains, which suggest that the content of biogenic amines could be controlled by the synergistic fermentation of yeast strains, especially tyramine and histamine. These results may revolutionize the manufacturing process of sauce soy manufacture by producing a product of improved quality.

Acknowledgements The project was supported by grants from National Natural Science Foundation of China (31501449) and the Foundation (No. 2015IM102) of Key Laboratory of Industrial Fermentation Microbiology of Ministry of Education and Tianjin Key Lab of Industrial Microbiology (Tianjin University of Science & Technology).

References

1. Sirocchi V, Caprioli G, Cecchini C, Coman MM, Cresci A, Maggi F, Papa F, Ricciutelli M, Vittori S, Sagratini G (2013) Biogenic amines as freshness index of meat wrapped in a new active packaging system formulated with essential oils of *Rosmarinus officinalis*. Int J Food Sci Nutr 64(8):921–928. doi:10.3109/09637486.2013.809706
2. Lázaro CA, Conte-Júnior CA, Canto AC, Monteiro MLG, Costa-Lima B, Cruz AGd, Mársico ET, Franco RM (2015) Biogenic amines as bacterial quality indicators in different poultry meat species. LWT—Food Sci Technol 60(1):15–21. doi:10.1016/j.lwt.2014.09.025
3. Isabel Rodriguez-Naranjo M, Ordonez JL, Callejon RM, Cantos-Villar E, Carmen Garcia-Parrilla M (2013) Melatonin is formed during winemaking at safe levels of biogenic amines. Food Chem Toxicol 57:140–146. doi:10.1016/j.fct.2013.03.014
4. Mohedano ML, López P, Spano G, Russo P (2015) Controlling the formation of biogenic amines in fermented foods. Adv Fermented Foods Beverages 12:273–310
5. Naila A, Flint S, Fletcher G, Bremer P, Meerdink G (2010) Control of biogenic amines in food-existing and emerging approaches. J Food Sci 75(7):R139–R150. doi:10.1111/j.1750-3841.2010.01774.x
6. Kim JH, Kim DH, Ahn HJ, Park HJ, Byun MW (2005) Reduction of the biogenic amine contents in low salt-fermented soybean paste by gamma irradiation. Food Control 16(1):43–49. doi:10.1016/j.foodcont.2003.11.004
7. Tapingkae W, Parkin KL, Tanasupawat S, Kruenate J, Benjakul S, Visessanguan W (2010) Whole cell immobilisation of Natrinema gari BCC 24369 for histamine degradation. Food Chem 120(3):842–849. doi:10.1016/j.foodchem.2009.11.025
8. Chong C, Abu Bakar F, Russly A, Jamilah B, Mahyudin N (2011) The effects of food processing on biogenic amines formation. Int Food Res J 18(3):867–876
9. Mah J-H, Hwang H-J (2009) Effects of food additives on biogenic amine formation in Myeolchi-jeot, a salted and fermented anchovy (*Engraulis japonicus*). Food Chem 114 (1):168–173. doi:10.1016/j.foodchem.2008.09.035
10. Roeder J, Schink B (2009) Syntrophic degradation of cadaverine by a defined methanogenic coculture. Appl Environ Microb 75(14):4821–4828. doi:10.1128/aem.00342-09
11. Capozzi V, Russo P, Ladero V, Fernandez M, Fiocco D, Alvarez MA, Grieco F, Spano G (2012) Biogenic amines degradation by *Lactobacillus plantarum*: toward a potential application in wine. Front Microbiol 3:122. doi:10.3389/fmicb.2012.00122
12. Garcia-Ruiz A, Gonzalez-Rompinelli EM, Bartolome B, Moreno-Arribas MV (2011) Potential of wine-associated lactic acid bacteria to degrade biogenic amines. Int J Food Microbiol 148(2):115–120. doi:10.1016/j.ijfoodmicro.2011.05.009

13. Rabie MA, Siliha H, el-Saidy S, el-Badawy AA, Xavier Malcata F (2011) Reduced biogenic amine contents in sauerkraut via addition of selected lactic acid bacteria. Food Chem 129 (4):1778–1782. doi:10.1016/j.foodchem.2011.05.106
14. Cueva C, Garcia-Ruiz A, Gonzalez-Rompinelli E, Bartolome B, Martin-Alvarez PJ, Salazar O, Vicente MF, Bills GF, Moreno-Arribas MV (2012) Degradation of biogenic amines by vineyard ecosystem fungi. Potential use in winemaking. J Appl Microbiol 112 (4):672–682. doi:10.1111/j.1365-2672.2012.05243.x
15. Zaman MZ, Abu Bakar F, Jinap S, Bakar J (2011) Novel starter cultures to inhibit biogenic amines accumulation during fish sauce fermentation. Int J Food Microbiol 145(1):84–91. doi:10.1016/j.ijfoodmicro.2010.11.031
16. van der Sluis C, Tramper J, Wijffels RH (2001) Enhancing and accelerating flavour formation by salt-tolerant yeasts in Japanese soy-sauce processes. Trends Food Sci Tech 12(9):322–327. doi:10.1016/s0924-2244(01)00094-2
17. Lu Y, Chen X, Jiang M, Lv X, Rahman N, Dong M, Gujun Y (2009) Biogenic amines in Chinese soy sauce. Food Control 20(6):593–597. doi:10.1016/j.foodcont.2008.08.020
18. Guidi LR, Abreu Gloria MB (2012) Bioactive amines in soy sauce: validation of method, occurrence and potential health effects. Food Chem 133(2):323–328. doi:10.1016/j.foodchem.2012.01.033
19. Caruso M, Fiore C, Contursi M, Salzano G, Paparella A, Romano P (2002) Formation of biogenic amines as criteria for the selection of wine yeasts. World J Microb Biot 18(2):159–163. doi:10.1023/a:1014451728868
20. Spano G, Russo P, Lonvaud-Funel A, Lucas P, Alexandre H, Grandvalet C, Coton E, Coton M, Barnavon L, Bach B, Rattray F, Bunte A, Magni C, Ladero V, Alvarez M, Fernandez M, Lopez P, de Palencia PF, Corbi A, Trip H, Lolkema JS (2010) Biogenic amines in fermented foods. Eur J Clin Nutr 64:S95–S100. doi:10.1038/ejcn.2010.218
21. Wah TT, Walaisri S, Assavanig A, Niamsiri N, Lertsiri S (2013) Co-culturing of *Pichia guilliermondii* enhanced volatile flavor compound formation by *Zygosaccharomyces rouxii* in the model system of Thai soy sauce fermentation. Int J Food Microbiol 160(3):282–289. doi:10.1016/j.ijfoodmicro.2012.10.022
22. Wu C, Zheng J, Huang J, Zhou R (2014) Reduced nitrite and biogenic amine concentrations and improved flavor components of Chinese sauerkraut via co-culture of *Lactobacillus plantarum* and *Zygosaccharomyces rouxii*. Ann Microbiol 64(2):847–857. doi:10.1007/s13213-013-0724-8
23. Cao X, Hou L, Lu M, Wang C (2010) Improvement of soy-sauce flavour by genome shuffling in *Candida versatilis* to improve salt stress resistance. Int J Food Sci Tech 45(1):17–22. doi:10.1111/j.1365-2621.2009.02085.x
24. Gao XL, Cui C, Zhao HF, Zhao MM, Yang L, Ren JY (2010) Changes in volatile aroma compounds of traditional Chinese-type soy sauce during moromi fermentation and heat treatment. Food Sci Biotechnol 19(4):889–898. doi:10.1007/s10068-010-0126-7
25. Qi W, Hou L-H, Guo H-L, Wang C-L, Fan Z-C, Liu J-F, Cao X-H (2014) Effect of salt-tolerant yeast of *Candida versatilis* and *Zygosaccharomyces rouxii* on the production of biogenic amines during soy sauce fermentation. J Sci Food Agr 94(8):1537–1542. doi:10.1002/jsfa.6454
26. Ben-Gigirey B, De Sousa J, Villa TG, Barros-Velazquez J (1998) Changes in biogenic amines and microbiological analysis in albacore (*Thunnus alalunga*) muscle during frozen storage. J Food Protect 61(5):608–615
27. Russo P, Spano G, Arena M, Capozzi V, Grieco F, Beneduece L (2010) Are consumers aware of the risks related to biogenic amines in food. Curr Res Technol Edu Top Appl Microbiol Microb Biotechnol 1087–1095
28. Goni DT, Azpilicueta CA (2001) Influence of yeast strain on biogenic amines content in wines: relationship with the utilization of amino acids during fermentation. Am J Enol Viticult 52(3):185–190

29. Rauscher-Gabernig E, Grossgut R, Bauer F, Paulsen P (2009) Assessment of alimentary histamine exposure of consumers in Austria and development of tolerable levels in typical foods. Food Control 20(4):423–429. doi:10.1016/j.foodcont.2008.07.011
30. Brillantes S, Paknoi S, Totakien A (2002) Histamine formation in fish sauce production. J Food Sci 67(6):2090–2094. doi:10.1111/j.1365-2621.2002.tb09506.x
31. Aidoo KE, Rob Nout M, Sarkar PK (2006) Occurrence and function of yeasts in Asian indigenous fermented foods. FEMS Yeast Res 6(1):30–39
32. Y-z Yan, Y-l Qian, F-d Ji, J-y Chen, B-z Han (2013) Microbial composition during Chinese soy sauce koji-making based on culture dependent and independent methods. Food Microbiol 34(1):189–195. doi:10.1016/j.fm.2012.12.009
33. S-i Sugiyama (1984) Selection of micro-organisms for use in the fermentation of soy sauce. Food Microbiol 1(4):339–347. doi:10.1016/0740-0020(84)90067-4
34. Tanaka Y, Watanabe J, Mogi Y (2012) Monitoring of the microbial communities involved in the soy sauce manufacturing process by PCR-denaturing gradient gel electrophoresis. Food Microbiol 31(1):100–106. doi:10.1016/j.fm.2012.02.005

Effect of Temperature, NaCl and Ferulic Acid Concentration on Bioconversion of Ferulic Acid to 4-Vinylguaiacol and 4-Ethylguaiacol by Halotolerant Yeasts *Candida versatilis*

Wei Qi, Wen-Tao Zhang and Fu-Ping Lu

1 Introduction

4-EG, more often described as smoky and sauce flavour, is appreciated in Japanese style and Chinese traditional fermented soy sauce [1, 2]. The threshold of 4-EG is reported to be 0.5 mg/L [3], whereas its precursor has a much higher threshold value of ~600 mg/L [4]. Moreover, there have been reported that 1–2 mg/L 4-EG could promote aroma and flavor of soy product [5, 6]. Therefore, formation of 4-EG has extremely close relationship with the quality of soy sauce [7].

Chinese soy sauce can be classified into high-salt liquid-state soy sauce and low-salt solid-state soy sauce base on the raw materials, brine, fermentation temperature and fermentation period. The conventional fermentation process of high-salt liquid-state soy sauce starts with *koji* culturing which a solid-state fermentation of *Aspergillus* species on a mixture of soybeans and wheat, after that, moromi fermentation which the *koji* is mixed with saline water (22–23% salt), and resulting brine solution about 17%, furthermore, about 6 month process time for high-salt liquid-state soy sauce at 30 °C [8, 9]. The raw material of low-salt solid-state soy sauce are soy beans and wheat bran. Followed the *koji* is mixed with saline water (13–14% salt), the resulting brine solution has a lower salt concentration about 11%, about one month fermentation for low-salt solid-state soy sauce at 40 °C [10].

In the cell walls of cereals, ferulic acid monomers as well as dimers are esterified to arabinose residues of arabinoxylans or etherified to lignin surfaces [4]. In the soy sauce fermentation process, of which raw materials, such as soybeans and wheat or

W. Qi · W.-T. Zhang · F.-P. Lu (✉)
Key Laboratory of Industrial Fermentation Microbiology, Tianjin University of Science & Technology, Ministry of Education, Tianjin 300457, People's Republic of China
e-mail: qiweismiling@126.com

H. Liu et al. (eds.), *Advances in Applied Biotechnology*, Lecture Notes in Electrical Engineering 444, https://doi.org/10.1007/978-981-10-4801-2_29

wheat bran [11], cell wall of cereal was degraded to small molecular weight substances by degrading enzymes (such as xylanases) secreted by *Aspergillus*, then feruloyl esterases destroy ester bond between FA and arabinose residues leading to release the FA [12]. FA can be decarboxylated to 4-VG followed metabolized to 4-EG though 4-VG reductase produced by yeast [13, 14]. It has been reported that 4-EG could be yielded by *Candida versatilis* but not *Zygosaccharomyces rouxii* [15].

Phenolic compounds are appreciated for their antioxidant activity preventing human from many diseases associated with oxidative stress [16]. Moreover, the flavour and quality of soy sauce are attracting consumer attention, because the condiment has a closely relation with their daily life. Thus, the ability of *Candida versatilis* biotransform of FA to 4-VG and of 4-VG to 4-EG by several treatments, such as initial FA concentration, NaCl concentration and temperature, have been investigated.

2 Materials and Methods

2.1 Chemicals

All chemicals and solvents in analytical grade were purchased from Tianjin Chemical Reagent Research Institute (Tianjin, China). HPLC grade acetonitrile and methanol were obtained from Merk (Darmstadt, Germany). Doubly distilled water for the dilution of samples was purified using a Milli-Q system (Millipore, Bedford, MA).

Sixteen sauce samples were purchased from local supermarkets. Twelve samples were brewed by high-salt liquid-state fermentation (HLS) and four samples were brewed by low-salt solid-state fermentation (LSS). The samples were carried to the laboratory and stored in a refrigerator at −4 °C until analysis.

2.2 Yeast Strains, Culture Maintenance and Inoculum Preparation

Yeast strain *Candida versatilis* used in this work was kept in our laboratory. Yeast were inoculated into a resting tube containing a 10 mL YPD medium. After incubation with shaking at room temperature (30 ± 1 °C) for 24 h, yeast cells were transferred to 150 ml YPD medium and incubated on a rotary shaker at room temperature for 20 h. Yeast cells were harvested by centrifugation (5000 g, 5 min, 4 °C) and re-suspended in physiological water (0.9% NaCl). Yeast cell densities were determined microscopically with a Thoma counting chamber.

Table 1 Levels of NaCl concentration, FA concentration and temperature used for fermentation

Representation	NaCl concentration ($g\ L^{-1}$)	FA concentration ($g\ L^{-1}$)	Temperature (°C)
N0F100T30	0	100	30
N9F100T30	9	100	30
N18F100T30	18	100	30
N0F10T30	0	10	30
N0F100T25	0	100	25
N0F200T30	0	200	30
N0F100T40	0	100	40

2.3 *Fermentation Experiments*

250 mL medium containing (per liter) 10 g glucose, 10 g peptone, 5.0 g $(NH4)_2SO_4$, 0.5 g $MgSO_4 \cdot 7H_2O$, 1.0 g KH_2PO_4, 0.1 g $CaCl_2$ (pH = 7) was pitched with $\sim 10^7$ propagated yeast cells per milliliter in 250 ml anaerobic bottle. Different levels of NaCl, FA and temperature were used for fermentation (Table 1). FA were sterilized by filtration (0.2 μm). During fermentation, temperature was set at 30 ± 1 °C for 12 days and samples were periodically taken.

2.4 *Analytical Methods*

Analysis of FA, 4-VG and 4-EG was done according to Max [17] with slight modifications. 100 μL of sample of culture was added to 900 μL of deionized water and the mixture was filtered through a 0.22 μm micron filter. The samples were injected into Shimadzu 20AB high performance liquid chromatography with ultraviolet detector at 280 nm operating at the following conditions: Spursil C18 column (250 mm × 5 μm × 4.6 mm), column temperature 30 °C, flow rate 1.0 mL/min, injection volume 20 μL and mobile phase A (1% formic acid in distilled water, v/v) and B (methanol) from 0 to 52% of solvent B for 30 min at a flow rate of 1 ml/min. The peak areas of standard solutions were identified to determine the amine concentrations in mash samples. Statistical significance was determined using the SAS statistical analysis program, version 8.01.

3 Results and Discussion

3.1 *The Influence of NaCl Concentration on Production of 4-Vinylguaiacol and 4-Ethyl Guaiacol*

Soy sauce brewed with wheat or wheat bran contains considerable level of FA which could form 4-EG and 4-VG by yeast decarboxylation. This study aimed to investigate the influence of fermentation parameter on the level of 4-EG and 4-VG. The content of ferulic acid, 4-vinylguaiacol and 4-ethyl guaiacol were detected by HPLC (Fig. 1).

The content of FA gradually decreased with increasing of cultivated time (Fig. 2), the concentration of FA reached its minimum level at different time: $N_0F_{100}T_{30}$ at 6 days with 14.77 mg L^{-1}, while $N_9F_{100}T_{30}$ and $N_{18}F_{100}T_{30}$ at 9 days with 13.24 and 13.34 mg L^{-1}, respectively.

The level of 4-VG increased with decreasing of FA (Fig. 2a). The 4-VG after $N_0F_{100}T_{30}$ treatment reached its maximum at 9 days with 14.40 mg L^{-1}, while that after treated with $N_9F_{100}T_{30}$ and $N_{18}F_{100}T_{30}$ reached their maximum at 12 days with 15.37 and 16.13 mg L^{-1}, respectively.

The level of 4-EG increased with decreasing of FA concentration and increasing of 4-VG concentration (Fig. 2b). 4-EG in $N_0F_{100}T_{30}$ increased from 3 days and maintained a value then reached its maximum after 9 days with 5.39 mg L^{-1}. 4-EG in $N_9F_{100}T_{30}$ increased from 3 days, and the content was 6.33 mg L^{-1} at the end of fermentation. The concentration of 4-EG in $N_{18}F_{100}T_{30}$ increased from 6 days and the final content was 7.72 mg L^{-1}.

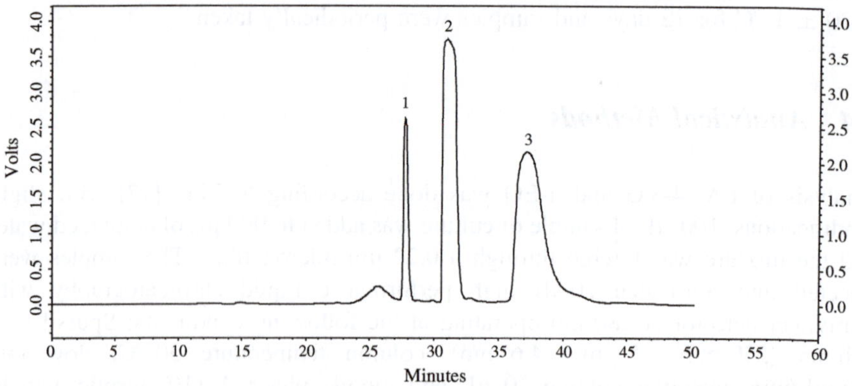

Fig. 1 HPLC chromatogram: ferulic acid (*1*), 4-vinylguaiacol (*2*) and 4-ethyl guaiacol (*3*) have isolated peaks respectively

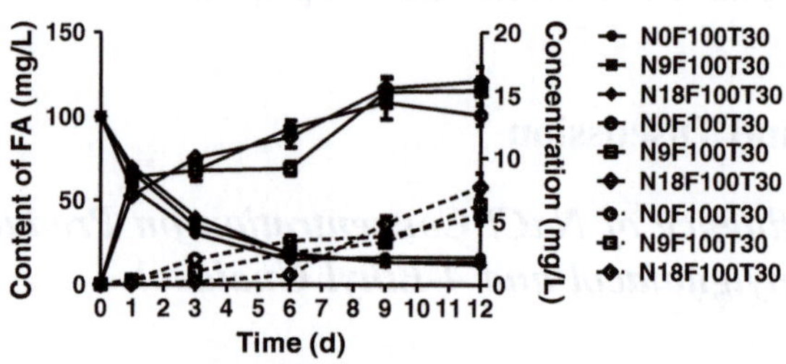

Fig. 2 The influence of NaCl concentration on production of 4VG and 4EG: the level of FA (0%NaCl 9%NaCl 18%NaCl); the level of 4-VG (0%NaCl 9%NaCl 18%NaCl); the level of 4-EG (0%NaCl 9%NaCl 18%NaCl)

The content level of 4-VG and 4-EG produced by *C. Versatilis* cultivated under both 9 and 18% NaCl was higher than cultivated under salt-free. Phenolic metabolism might be enhanced by appropriate salinity in *C. Versatilis*. The results of the research pointed towards the NaCl could affected consumption rate of FA, but not affected its final content. However, the salt concentration not only influenced the yield rate of 4-VG and 4-EG but also affected their final content, which was agree with previous research [15].

3.2 *The Influence of Ferulic Acid Concentration on Production of 4-Vinylguaiacol and 4-Ethyl Guaiacol*

The concentration of FA decreased with increasing of cultivated time (Fig. 3). 4-VG increased at the initial period and decreased rapidly at the later period (Fig. 3a), which may be attributed to form 4-EG.

With the lower level of precursor, the level of 4-VG production was lower. On the basis of the higher levels of FA, it was expected that more metabolite is formed during the fermentation process, nevertheless, the results shown in Fig. 3a do not indicate higher levels of 4-VG. On the contrary, the level of 4-VG and consumption rate of FA has no difference between $N_0F_{100}T_{30}$ and $N_0F_{200}T_{30}$, and the residual content of FA was 16.04 and 31.48 mg L^{-1} in $N_0F_{100}T_{30}$ and $N_0F_{200}T_{30}$, respectively. Concentration of 4-VG was increased significantly from initial stage to 6 days and maintained a value thereafter. The final content of 4-VG was 13.4 and 11.9 mg L^{-1} in $N_0F_{100}T_{30}$ and $N_0F_{200}T_{30}$, respectively.

The content level of 4-EG increased with decreasing of FA and 4-VG (Fig. 3b). 4-EG in samples began to increase from 3 days, the final concentration of 4-EG was 4.3, 5.3 and 6.1 mg L^{-1} in $N_0F_{10}T_{30}$, $N_0F_{100}T_{30}$ and $N_0F_{200}T_{30}$, separately.

In the fermentation, higher precursor concentration slightly affected residual content of FA and its consumption rate, moreover, higher precursor concentration influenced generation rate of 4-VG but not significantly influenced its final content.

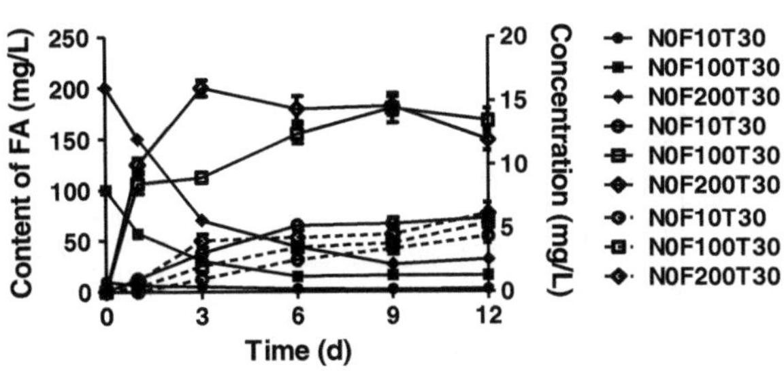

Fig. 3 The influence of FA concentration on production of 4VG and 4EG: the level of FA (-•- 10 mg/L -■- 100 mg/L -◆- 200 mg/L); the level of 4-VG (-⊖- 10 mg/L -⊟- 100 mg/L -◈- 200 mg/L); the level of 4-EG (-⊖· 10 mg/L -⊟· 100 mg/L -◈· 200 mg/L)

The initial concentration of FA not affected the final generation of 4-EG particularly [15]. There was slightly impact of precursor that reached a certain concentration on the final content of 4-EG.

3.3 The Influence of Temperature on Production of 4-Vinylguaiacol and 4-Ethyl Guaiacol

The concentration of 4-VG increased with decreasing content of FA cultivated under different temperature (Fig. 4a). The level of FA decreased in $N_0F_{100}T_{30}$ was fastest, followed $N_0F_{100}T_{40}$ and $N_0F_{100}T_{25}$, however, the content of FA in all samples reached their minimum after 9 days. The content of 4-VG increased in $N_0F_{100}T_{30}$ was fastest, followed $N_0F_{100}T_{25}$ and $N_0F_{100}T_{40}$, which indicated that the FA in $N_0F_{100}T_{40}$ was not bioconversion to 4-VG entirely under high temperature environment. The concentration of 4-VG began to decline after reached the maximum which might because the 4-VG biotransform to 4-EG.

The impact of temperature on the generation of 4-EG was similar on formation of 4-VG. (Figure 4b). The content of 4-EG increased in $N_0F_{100}T_{30}$ was fastest, followed $N_0F_{100}T_{25}$ and $N_0F_{100}T_{40}$. The highest level of 4-EG in $N_0F_{100}T_{30}$ was 5.39 mg L^{-1}, followed $N_0F_{100}T_{25}$ with 3.62 mg L^{-1} and $N_0F_{100}T_{40}$ with 2.96 mg L^{-1}. 4-EG in $N_0F_{100}T_{40}$ decreased in the later period which may be caused by volatilizing or converting to other metabolites.

The one reason of temperature affect the generation of 4-EG is yeast required temperature, the suitable growth temperature of yeast was 28–32 °C, lower temperature would affect its normal growth and metabolic activity, while higher temperature would suppress the growth of the yeast strain, even lead to its autolysis. In our research, although the growth of yeast was slightly slower under 25 °C, but the strain could maintain normal fermentation ability, therefore, the FA can be still metabolized to 4-EG, however the conversion rate was lower than that under 30 °C. The content of FA declined in $N_0F_{100}T_{40}$ was faster than $N_0F_{100}T_{25}$, whereas the

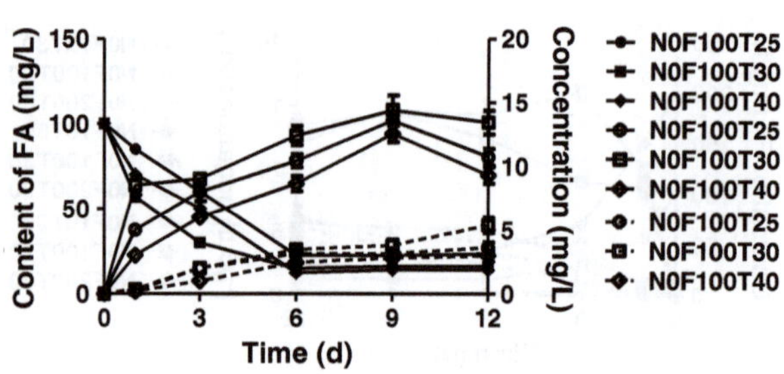

Fig. 4 The influence of temperature on production of 4VG and 4EG: the level of FA (-•- 25 °C -■- 30 °C -◆- 40 °C); the level of 4-VG (-●- 25 °C -■- 30 °C -◆- 40 °C); the level of 4-EG (-●· 25 °C -■· 30 °C -◆· 40 °C)

generation of EG in $N_0F_{100}T_{40}$ was lower than $N_0F_{100}T_{25}$, which may be due to volatilization or the metabolic ability of yeast decreased cultivated under 40 °C. The temperature influenced the reducing rate of FA but not influenced its final content. Moreover, the temperature influenced 4-VG and 4-EG generation rate and their final content.

3.4 Ferulic Acid, 4-Vinylguaiacol and 4-Ethyl Guaiacol in Commercial Chinese Soy Sauce

The levels of the phenolic compound 4-EG, 4-VG and FA were analyzed in several brands of two types of soy sauce: high-salt liquid-state soy sauce (HLS) and low-salt solid-state soy sauce (LSS) (Table 2). In the analyzed soy sauce, phenolic compounds level of HLS was higher compared to LSS. The soy sauces were found to contain the following compounds in concentrations: 0.23–4.08 mg L^{-1} for FA, 0–2.03 mg L^{-1} for 4-VG, 0–1.12 mg L^{-1} for 4-EG. The higher levels of FA possibly arise from the raw material or *Aspergillus* used. The higher level of 4-VG and 4-EG possibly arise from the initially brine solution, fermentation temperatures or yeast strains used.

In summary, 4-VG and 4-EG could be formed from FA metabolized by halo-tolerant yeasts *Candida versatilis*, moreover, production of 4-EG and 4-VG depends on the fermentation conditions. The conversion of FA to 4-EG and 4-VG to 4-EG by yeast was not very high, the yield of 4-ethylguaiacol and

Table 2 Ferulic acid (FA), 4-vinylguaiacol (4-VG) and 4-ethyl guaiacol (4-EG) levels in commercial soy sauce: high salt liquid state soy sauce (HLS) and low salt solid state soy sauce (LSS)

Soy sauce	FA (mg L^{-1})	4-VG (mg L^{-1})	4-EG (mg L^{-1})
HLS1	2.96 ± 0.12	0.24 ± 0.02	0.45 ± 0.11
HLS2	0.23 ± 0.07	0.11 ± 0.01	ND
HLS3	0.93 ± 0.05	0.13 ± 0.03	ND
HLS4	4.08 ± 0.25	2.03 ± 0.14	0.35 ± 0.02
HLS5	1.31 ± 0.21	0.85 ± 0.10	0.11 ± 0.06
HLS6	1.38 ± 0.11	0.89 ± 0.06	1.12 ± 0.17
HLS7	1.85 ± 0.12	0.74 ± 0.03	0.01 ± 0.01
HLS8	0.72 ± 0.13	0.94 ± 0.08	0.08 ± 0.01
HLS9	3.32 ± 0.31	0.86 ± 0.09	0.10 ± 0.05
HLS10	0.65 ± 0.03	ND	0.15 ± 0.06
HLS11	1.67 ± 0.11	0.45 ± 0.01	0.72 ± 0.03
HLS12	1.10 ± 0.06	0.84 ± 0.14	0.61 ± 0.12
LSS1	0.31 ± 0.05	0.40 ± 0.06	ND
LSS2	0.39 ± 0.09	0.32 ± 0.07	0.14 ± 0.05
LSS3	0.79 ± 0.03	ND	0.33 ± 0.04
LSS4	0.52 ± 0.07	0.12 ± 0.04	0.20 ± 0.02

4-vinylguaiacol on ferulic acid was 20% (w/w), which may because the other metabolic intermediates generated during this process. Prior study showed that 4-VG can convert to vanillin and vanillic acid [17, 18]. Suggestions for further research, NMR technology could be used to figure out and identify the metabolites from 4-EG metabolism.

Acknowledgements The project was supported by grants from National Natural Science Foundation of China (31501449) and the Foundation (No. 2015IM102) of Key Laboratory of Industrial Fermentation Microbiology of Ministry of Education and Tianjin Key Lab of Industrial Microbiology (Tianjin University of Science & Technology).

References

1. Yoshikawa S, Kurihara H, Kawai Y, Yamazaki K, Tanaka A, Nishikiori T, Ohta T (2010) Effect of halotolerant starter microorganisms on chemical characteristics of fermented chum salmon (oncorhynchus keta) sauce. J Agr Food Chem 58:6410–6417
2. Feng J, Zhan X-B, Wang D, Zhang L-M, Lin C-C (2012) Identification and analysis of the metabolic functions of a high-salt-tolerant halophilic aromatic yeast *Candida etchellsii* for soy sauce production. World J Microb Biot 28:1451–1458
3. Hamada T, Sugishita M, Motai H (1990) Contributions of immobilized and free cells of salt-tolerant *Zygosaccharomyces rouxii* and *Candida versatilis* to the production of ethanol and 4-ethylguaiacol. Appl Microbiol Biot 11:624–628
4. Coghe S, Benoot K, Delvaux F, Vanderhaegen B, Delvaux FR (2004) Ferulic acid release and 4-vinylguaiacol formation during brewing and fermentation: indications for feruloyl esterase activity in *Saccharomyces cerevisiae*. J Agric Food Chem 52:602–608
5. Yokotsuka T, Asao Y, Sakasaki T (1967) Studies of the flavorous substances in shoyu part XXVII. The production of 4-ethylguaiacol during shoyu fermentation, and its role for shoyu flavor. Nippon Nogeikagaku Kaishi 41:442–447 (in Japanese)
6. Nussinovitch A (2010) Polymer macro- and micro-gel beads: fundamentals and applications. Springer, Berlin
7. Querol A, Fleet G (2006) Yeasts in food and beverages. Springer, Berlin
8. van der Sluis C, Tramper J, Wijffels RH (2001) Enhancing and accelerating flavour formation by salt-tolerant yeasts in Japanese soy-sauce processes. Trends Food Sci Tech 12:322–327
9. Xian-Li G, Chun C, Hai-Feng Z, Mou-Ming Z, Lan Y, Jiao-Yan R (2010) Changes in volatile aroma compounds of traditional Chinese-type soy sauce during moromi fermentation and heat treatment. Food Sci Biotechnol 19:889–898
10. Zhang Y, Tao W (2009) Flavor and taste compounds analysis in Chinese solid fermented soy sauce. Afr J Biotechnol 8:673–681
11. Wanakhachornkrai P, Lertsiri S (2003) Comparison of determination method for volatile compounds in Thai soy sauce. Food Chem 83:619–629
12. Koseki T, Hori A, Seki S, Murayama T, Shiono Y (2009) Characterization of two distinct feruloyl esterases, AoFaeB and AoFaeC, from *Aspergillus oryzae*. Appl Microbiol Biotechnol 83:689–696
13. Suezawa Y, Suzuki M (2007) Bioconversion of ferulic acid to 4-vinylguaiacol and 4-ethylguaiacol and of 4-vinylguaiacol to 4-ethylguaiacol by halotolerant yeasts belonging to the genus *Candida*. Biosci Biotech Biochem 71:1058–1062
14. Suezawa Y (1995) Bioconversion of ferulic acid and p-coumaric acid to volatile phenols by halotolerant yeasts. Nippon Nôgeikagaku Kaishi 69:1587–1596 (in Japanese)

15. van der Sluis C, Stoffelen CJ, Castelein SJ, Engbers GH, ter Schure EG, Tramper J, Wijffels RH (2001) Immobilized salt-tolerant yeasts: application of a new polyethylene-oxide support in a continuous stirred tank reactor for flavour production. J Biotechnol 88:129–139
16. Piazzon A, Forte M, Nardini M (2010) Characterization of phenolics content and antioxidant activity of different beer types. J Agr Food Chem 58:10677–10683
17. Max B, Carballo J, Cortes S, Dominguez JM (2012) Decarboxylation of ferulic acid to 4-vinyl guaiacol by *Streptomyces setonii*. Appl Biochem Biotech 166:289–299
18. Baqueiro-Pena I, Rodriguez-Serrano G, Gonzalez-Zamora E, Augur C, Loera O, Saucedo-Castaneda G (2010) Biotransformation of ferulic acid to 4-vinylguaiacol by a wild and a diploid strain of *Aspergillus niger*. Bioresour Technol 101:4721–4724

Near Infrared Spectroscopic (NIRS) Analysis of Polysaccharides and Ergosterol Contents in *Tricholoma matsutake* Mycelium by Improved Chemometric Model

Qiubo Chu, Quan Li, Shuang Hu, Xue Jiang, Yanzhen Wang, Hao Zeng, Lesheng Teng and Di Wang

1 Introduction

Tricholoma matsutake, an ectomycorrhizal fungus, is an important commercial mushroom, which is highly valued as a delicacy and medicine in Asia [1]. *T. matsutake* not only exhibits a delicate flavor, but also shows various pharmacological activities including immunomodulatory, anti-tumor and anti-oxidant effects [2–4]. Lots of bioactive substances obtained from *T. matsutake* such as polysaccharides and ergosterol show bioactivities. Various polysaccharides with antioxidant activity [5], anti-tumor effect [6] and immune regulation property [7] were purified from *T. matsutake.* However, the traditional detection methods of polysaccharides (Anthrone sulfuric acid method) and ergosterol (High performance liquid chromatography) are time-consuming and high cost. Due to large application of *T. matsutake,* its identification is especially important, and new analytical techniques providing simple and fast measurement are required.

Infrared Spectroscopy Near (NIRS) technology, with fast, simple, and nondestructive characteristics, are widely used in qualitative and quantitative analysis of food and/or Traditional Chinese Medicines (TCM) [8, 9]. A single spectrum allows the simultaneous determination of various properties in one subject. However, the spectrum is complex, and easy to be affected by the background of subjects. Moreover, the effective information rate of NIRS is slightly lower. Least Squares Partial (PLS) has good selectivity and prediction accuracy, and is suitable for

Q. Chu · Q. Li · S. Hu · X. Jiang · Y. Wang · H. Zeng · L. Teng · D. Wang (✉)
School of Life Sciences, Jilin University, No. 2699, Qianjin Avenue,
Changchun 130012, Jilin, People's Republic of China
e-mail: jluwangdi@gmail.com

Y. Wang
Zhuhai College, Jilin University, Zhuhai 519041, China

H. Liu et al. (eds.), *Advances in Applied Biotechnology*, Lecture Notes
in Electrical Engineering 444, https://doi.org/10.1007/978-981-10-4801-2_30

complex multicomponent spectral, which is one of the most widely used multivariate calibration methods [10]. Monte Carlo partial least square (MCPLS) is commonly applied to eliminate abnormal samples and select the number of calibration set [11], which helps to achieve a good improvement model robustness and improve the model generalization capability [12].

In the present study, NIR spectra combined with PLS were applied to analyze the contents of polysaccharides and ergosterol in *T. matsutake* mycelium obtained from submerged fermentation. During model development, MCPLS was used to identify the outliers and select the number of calibration set. Various pro-processing methods including fast Fourier transform (FFT), Savitzky-Golay smoothing, First derivative and Second derivative were performed on the raw spectra to remove signal noise.

2 Methods and Materials

2.1 *Sample Preparation*

Based on original *T. matsutake* stain (CCTCC M2013175 T-08) (obtained from China Center for Type Culture Collection), lots of mutant strains were obtained via chemical mutagenesis, which contain different contents of polysaccharides and ergosterol. *T. matsutake* mycelium of different strains was obtained by submerged fermentation in a defined medium as follows: 20 g/L of sucrose, 10 g/L of peptone, 10 g/L of yeast extract powder, 1 g/L of $MgSO_4$, 1 g/L of KH_2PO_4, and 0.1 g/L of vitamin B1. Fermentation conditions were as follows: 200 ml medium in a 500 ml conical flask, 7% (V/V) inoculum, 120–180 rpm/min shake speed and 25 °C fermentation temperature. All agents used in fermentation were obtained from Sigma, USA.

150 NIR spectra of *T. matsutake* mycelium were got in a random order via UV-3150 UV-Vis-NIR spectrophotometer (Shimadzu Corporation, Japan). Spectrum of each sample was displayed as an average of five parallel scanning from 800 to 2500 nm. The scanned interval was 1 nm and the entrance slit width was 12 nm.

2.2 *Content Measurement*

2.2.1 Ergosterol Measurement

The content of ergosterol in *T. matsutake* mycelium was detected via high performance liquid chromatography (HPLC) as reported previously [13].

2.2.2 Polysaccharides Measurement

The content of polysaccharides in *T. matsutake* mycelium was detected via anthrone sulfuric acid method similar as previous study [14].

2.3 Establishment of PLS Quantitative Analysis Models

2.3.1 Abnormal Samples Remove via MCPLS

Using leave-one-out cross validation, the root mean square errors of the cross-validation (*RMSECV*) was served as evaluation index shown in Eq. (1.1), and the initial hidden variables (n_{LV}) was chosen. For calibration set, Eq. (1.1) is used to calculate root mean square errors of calibration (*RMSEC*). For prediction set, Eq. (1.1) is used to calculate root mean square errors of prediction (*RMSECP*).

$$RMSE = \sqrt{\frac{\sum (y_p - y_c)^2}{n}} \tag{1.1}$$

In the formula, n is the sample quantity, the y_p is the experimental value, and the y_r is calculated or predicted value.

Outliers were identified using MCPLS. 50% samples were randomly selected as calibration set to establish the PLS model, and the rest of the sample as a prediction set. The calculation was repeated for 10,000 times to make sure each sample has served as a predictor set. The predictive residual error (*PRE*) of each sample was calculated and recorded. The mean of the *PRE* (*MPRE*) and the standard deviation of *PRE* (*SDPRE*) were calculated, based on which, samples with high *MPRE* and *SDPRE* were identified as outliers.

2.3.2 Numbers of Calibration Set Chosen

In order to prevent over fitting phenomenon, randomly selected 10, 20, 30, 40, 50, 60, 70, 80 and 90% of whole samples as the calibration set for PLS model establishment. The remained samples serve as prediction set. Following as Eq. (1.2), the degree of fitting (D_f) was calculated, and the process was repeated 5000 times. When the mean of D_f reached to the highest, the number of calibration set was chosen for further experiments.

$$D_f = \frac{c}{\frac{RMSEC}{n_c} + \frac{RMSEP}{n_p} + |RMSEC - RMSEP|} \tag{1.2}$$

In the equation, n_c is the number of samples in the calibration set; n_p is the number of samples in the prediction set, the constant c is chosen 10.

2.3.3 Establishment of Quantitative Analysis Models of PLS

Fast Fourier transform (FFT), Savitzky-Golay smoothing, First derivative and Second derivative was applied to preprocess raw spectra to remove the background or useless noise. The window chosen in the experiments was 5, 7, 9, 11, 13 and 15. During the analysis model establishment via move window partial least square (MWPLS) with various n_{LV} (from 1 to 20, interval of 1), W wavelengths were served as one window (W is defined as a window size), and W values were investigated from 25 to 370 (1–15% of the overall wavelengths) with interval of 12. The models were established repeatedly using wavelengths in a defined window, which shifted one wavelength each time, until the last variable (2000 nm) was enrolled. The degree of approach (Da), Defined as Eq. (1.3), was served as indexes, when it reached to the highest, all wavelengths in the front of n_w were applied to develop analysis model. n_w was chosen from 5 to 75 (Interval of 5).

$$Da = \frac{c}{\frac{n_c}{n_t} \times RMSEC + \frac{n_p}{n_t} RMSEP + |RMSEC - RMSEP|} \tag{1.3}$$

In the equation, n_t is the numbers of all samples; n_c is the numbers of calibration set; n_p is numbers of prediction set; c value is 10.

2.4 *Analysis Procedure*

All of the above procedures were applied by Matlab 2008a software (Mathwork, USA).

3 Results and Discussion

3.1 *Spectra Characterization*

NIRS has been widely applied in qualitative and quantitative analyses due to its fasting detection and nondestructive [15, 16]. The chemical properties of subjects are reflected in NIR spectra. As shown in Fig. 1, the significant spectral variations appear around 1100–1250 nm, 1650–1800 nm, 2100–2200 nm and 2250–2350 nm, which indicate the complex of spectrum.

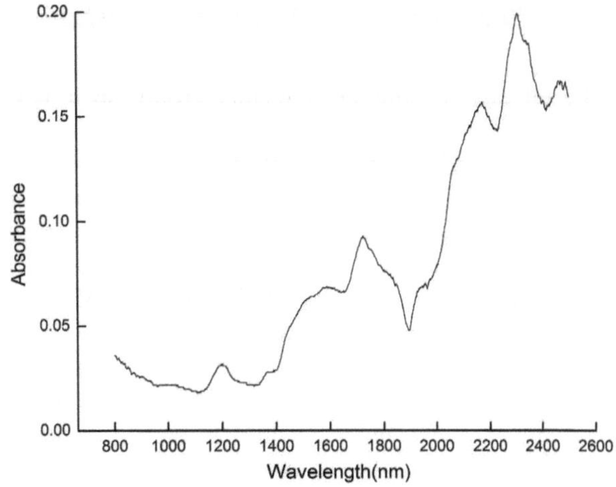

Fig. 1 The average of NIRs of *T. matsutake* mycelium samples

Table 1 The statistical polysaccharides and ergosterol contents of *T. matsutake*

Items		Sample number	Max (g/L)	Min (g/L)	Mean ± S.D. (g/L)
Polysaccharide	Calibration	129	2.4402	0.3043	1.0224 ± 0.4075
	Validation	14	2.2689	0.3299	1.1142 ± 0.5041
Ergosterol	Calibration	129	0.3579	0.0380	0.1493 ± 0.0659
	Validation	14	0.3318	0.0423	0.1639 ± 0.0799

3.2 Outliers Identification and Number of Calibration Set Selection

The number of variables existing in NIR spectra is far more than sample numbers. Outliers among all samples with higher values of MPRE and SDPRE were identified via MCPLS. During the process of PLS-NIRS quantitative analysis models of polysaccharides and ergosterol development, 7 outliers were excluded.

MCPLS was applied to select the numbers of calibration set. 90% samples were served as calibration set for polysaccharides and ergosterol in *T. matsutake* mycelium analysis model developing.

Furthermore, among all using samples, a large variability of polysaccharides and ergosterol contents in *T. matsutake* mycelium were noted. Statistical analysis for the calibration and validation sets, data ranges, number of samples and means ± standard deviations (S.D.) of polysaccharides and ergosterol contents were displayed in Table 1

3.3 PLS Prediction Model Establishment

The interactions among molecular groups such as C–H and O–H in *T. matsutake* mycelium increased the difficulty for polysaccharides and ergosterol determination. In order to eliminate the interference of spectra, four pre-processing methods including FFT, Savitzky-Golay smoothing, First derivative and Second derivative were performed to remove the background, noise, overlapping bands and baseline drift [16]. During PLS predictive model developing, according to *Da,* the values of n_{LV}, W and n_w were optimized. Furthermore, MWPLS was applied to determine the relationship between the wavelength and Da values (Table 2; Fig. 2).

Based on Da value, 800–2500 nm wavelength variables after Second derivative pre-processing were selected for developing polysaccharides predictive model. The optimum n_{LV}, W and n_w values were 8, 136 and 5, respectively (Table 2). The correlation coefficients of calibration set (Rc) 0.9999 and prediction set (Rp) were 0.9863 (Fig. 3) which demonstrated the well fitting and predictive ability of polysaccharides analysis model. 800–2500 nm wavelength variables after Second derivative pre-processing were selected for developing ergosterol predictive model. The optimum n_{LV}, W and n_w values were 15, 187 and 30, respectively (Table 2). The correlation coefficients of calibration set (Rc) and prediction set (Rp) were 0.9999 and 0.9684 (Fig. 4).

Table 2 Results of suitable indexes in PLS models for polysaccharides and ergosterol detection

Components	Preprocessing methods	Window	W_n	n_w	RMSEC (g/L)	RMSEP (g/L)	Da	n_{LV}
Ergosterol	Original spectra		102	5	0.0246	0.02768	3.574	13
	First order derivative	7	119	5	0.0165	0.0251	3.864	10
	Second order derivative	7	187	30	0.0164	0.0230	4.229	15
	FFT	9	34	5	0.0426	0.0423	2.338	13
	Savizky-Golay smoothing	5	221	5	0.1311	0.1441	0.6405	11
Polysaccharide	Original spectra		85	5	0.1193	0.1313	7.547	15
	First order derivative	9	136	5	0.1095	0.1091	9.103	15
	Second order derivative	5	136	5	0.0894	0.1009	9.796	8
	FFT	11	85	15	0.2389	0.2384	4.179	15
	Savizky-Golay smoothing	9	102	15	0.1332	0.1340	7.458	14

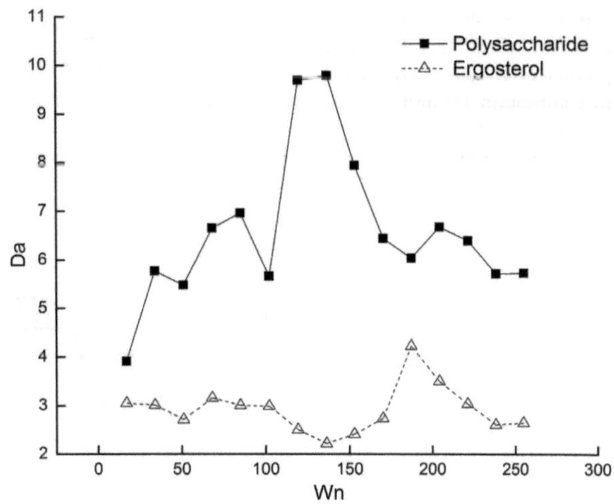

Fig. 2 The influences of Wn on Da value during the PLS models developing for analyzing the contents of polysaccharides and ergosterol

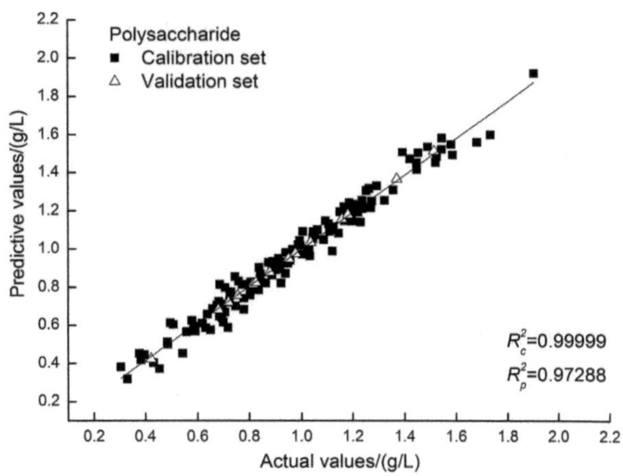

Fig. 3 The correlation between reference values and predictive values of polysaccharides in Calibration set and Validation set of optimum PLS-NIRS models

4 Conclusion

The present study successfully developed predictive models via PLS combined with NIRS to analyze the contents of polysaccharides and ergosterol in *T. matsutake* mycelium. Optimum polysaccharides content analysis model displayed RMSECV of 0.0894 g/L and Rc of 0.9999, and a good prediction ability with RMSEP of 0.1009 g/L and Rp of 0.9863. For ergosterol contents analysis, the optimum PLS model also displayed well fitness and predictive ability indicating by higher Rc (0.9999) and Rp (0.9684), and lower RMSEP value (0.0230 g/L). Our present study provide a fast, accurate and non-destructive quantitative analysis method to detect the contents of polysaccharides and ergosterol in *T. matsutake* mycelium, which may also performed for quality control of other large fungi.

Fig. 4 The correlation between reference values and predictive values of ergosterol in Calibration set and Validation set of optimum PLS-NIRS models

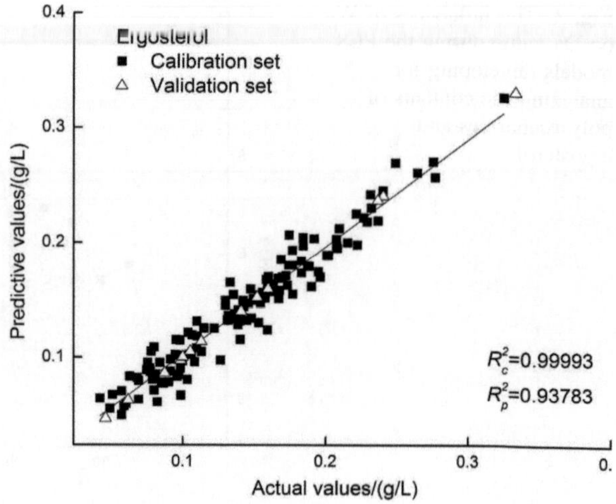

Acknowledgements This work was supported by Science and Technology Development Program of Jilin Province in China (Grant No. 20160520036JH), and Key Scientific and Technological Project of Jilin Province in China (Grant No. 20160204029YY and 20140311072YY).

Conflict of Interest The authors have declared that there is no conflict of interest.

References

1. Murata H, Yamada A, Maruyama T et al (2015) Ectomycorrhizas in vitro between *Tricholoma matsutake*, a basidiomycete that associates with Pinaceae, and Betula platyphylla var. japonica, an early-successional birch species, in cool-temperate forests. Mycorrhiza 25(3):237–241
2. Hoshi H, Yagi Y, Iijima H et al (2005) Isolation and characterization of a novel immunomodulatory alpha-glucan-protein complex from the mycelium of *Tricholoma matsutake* in basidiomycetes. J Agric Food Chem 53(23):8948–8956
3. Ikekawa T, Uehara N, Maeda Y et al (1969) Antitumor activity of aqueous extracts of edible mushrooms. Can Res 29(3):734–735
4. Mau JL, Lin HC, Song SF (2002) Antioxidant properties of several specialty mushrooms. Food Res Int 35(6):519–526
5. Ding X, Tang J, Cao M et al (2010) Structure elucidation and antioxidant activity of a novel polysaccharide isolated from *Tricholoma matsutake*. Int J Biol Macromol 47(2):271–275
6. You L, Gao Q, Feng M et al (2013) Structural characterisation of polysaccharides from *Tricholoma matsutake* and their antioxidant and antitumour activities. Food Chem 138(4):2242–2249
7. Hoshi H, Iijima B, Ishihara Y et al (2008) Absorption and tissue distribution of an immunomodulatory alpha-D-glucan after oral administration of *Tricholoma matsutake*. J Agric Food Chem 56(17):7715–7720

8. Rohman A, Nugroho A, Lukitaningsiha E, Sudjadia (2014) Application of vibrational spectroscopy in combination with chemometrics techniques for authentication of herbal medicine. Appl Spectrosc Rev 49(8):603–613
9. Wang P, Yu Z (2015) Species authentication and geographical origin discrimination of herbal medicines by near infrared spectroscopy: a review. J Pharm Anal 5(5):277–284
10. Ding CC, Fang XJ, Zhao YL et al (2013) Rapid identification of Dendrobium plants based on near infrared diffuse reflection spectroscopy. Spectrosc Spectral Anal 34(1):82–86
11. Guo WL, Zhang ZY, Lu JH (2010) Application of near infrared spectroscopy in screening Cordyceps militaris mutation strains and optimizing their fermentation process. Spectrosc Spectral Anal 8(30):2077–2082
12. Du YP, Kasemsumran S, Maruo K et al (2006) Ascertainment of the number of samples in the validation set in Monte Carlo cross validation and the selection of model dimension with Monte Carlo cross validation. Chemometr Intell Lab Syst 82(1–2):83–89
13. Muniroh MS, Sariah M, Zainal Abidin MA et al (2014) Paterson RRM. Rapid detection of Ganoderma-infected oil palms by microwave ergosterol extraction with HPLC and TLC. J Microbiol Methods 100:143–147
14. Wang Y, Shao S, Xu P et al (2012) Fermentation process enhanced production and bioactivities of oolong tea polysaccharides. Food Res Int 46(1):158–166
15. Zhang X, Li W, Yin B et al (2013) Improvement of near infrared spectroscopic (NIRS) analysis of caffeine in roasted Arabica coffee by variable selection method of stability competitive adaptive reweighted sampling (SCARS). Spectrochim Acta A Mol Biomol Spectrosc 114:350–356
16. Guo WL, Du YP, Zhou YC et al (2012) At-line monitoring of key parameters of nisin fermentation by near infrared spectroscopy, chemometric modeling and model improvement. World J Microbiol Biotechnol 28(3):993–1002

Antioxidant Activity and Hepatoprotective Activity of Shanxi Aged Vinegar in Hydrogen Peroxide-Treated HepG-2 Cells

Ting Xia, Jiahui Yao, Jiankang Wang, Jin Zhang and Min Wang

1 Introduction

Traditional vinegars, typical fermented foods, produced by spontaneous fermentation technology. It can produce complex composition, including sugars, organic acids and various bioactive components. These bioactive components can provide health and therapeutic effects [1]. Shanxi aged vinegar (SAV), a popular condiment in China, is not only rich of pleasant aromas but also valued for its health benefits [2]. It is produced by spontaneous solid-state fermentation including the preparation of daqu, starch saccharification (SS), alcoholic fermentation (AF), and acetic acid fermentation (AAF) [3]. During the process, it can produce rich nutrients and functional factors, such as polyphenols, flavonoids and melanoidins, are the main antioxidant activity ingredients of SAV [4]. It has been reported that these antioxidants played a crucial role in the prevention of free radical-induced disease such as cardiovascular disease, hepatic damage, neurodegenerative diseases, cancer and ageing [1, 5, 6]. Recently, the antioxidant activity and healthy properties of vinegar have become the focus of extensive studies.

T. Xia · J. Yao · J. Zhang · M. Wang (✉)
Key Laboratory of Industrial Fermentation Microbiology, Ministry of Education,
Tianjin Key Lab of Industrial Microbiology, College of Biotechnology,
Tianjin University of Science and Technology, Tianjin 300457,
People's Republic of China
e-mail: minw@tust.edu.cn

J. Wang
Tianjin Key Laboratory of Integrated Design and On-line Monitoring for Light Industry & Food Machinery and Equipment, College of Mechanical Engineering,
Tianjin University of Science and Technology, Tianjin 300222,
People's Republic of China

J. Wang
Department of Mechanical and Automotive Engineering, Zhuhai College of Jilin University,
Zhuhai, Guangdong 519041, People's Republic of China

H. Liu et al. (eds.), *Advances in Applied Biotechnology*, Lecture Notes in Electrical Engineering 444, https://doi.org/10.1007/978-981-10-4801-2_31

Liver, a vital organ, plays a major role in metabolism and excretion of xenobiotic in the human body [7]. Liver damage is a major health problem, which is caused by some toxic chemicals, drugs, excessive alcohol consumption and microbes [8, 9]. Oxidative stress is associated with liver damage and plays an important role in the pathogenesis of liver damage [10]. Oxidative stress, a phenomenon related to the aerobic nature of cellular metabolism, is a redox imbalance between the production of reactive oxygen species (ROS) and antioxidant defense [11]. Excessive ROS results in many patho-physiological conditions in the body, and leads to the damage of cell structures [12]. Prevention or impairment of oxidative stress reduces liver damage and constitutes a therapeutic target to hepatoprotection. In this study, we are interested to investigate the antioxidant activity and protective effects of SAV on damaged hepatocytes.

In the present study, total phenolic and flavonoid content and antioxidant activity of SAV with different manufacturers were determined. Then oxidative damage model induced by H_2O_2 in HepG2 cell was established. The protective effect of SAV was investigated in H_2O_2-treated HepG2 cells. These findings would elucidate the antioxidant properties of traditional Chinese vinegars, and provide a strategy for prevention and treatment of liver injury.

2 Materials and Methods

2.1 Chemicals and Samples

Folin-Ciocalteu reagent, 2,2-diphenyl-2-picrylhydrazyl (DPPH), gallic acid and rutin were purchased from Sinopharm Chemical Reagent Co., Ltd. (Shanghai, China). 2,2′-azino-bis(3-ethylbenzthiazoline-6-sulfonate) (ABTS) method and ferric reducing ability of plasma (FRAP) method were obtained from Beyotime Institute of Biotechnology (Shanghai, China). RPMI-1640, fetal bovine serum (FBS) and penicillin-streptomycin B were obtained from HyClone (Beijing, China). H_2O_2 was purchased from Alfa Aesar (Shanghai, China). Cell Counting Kit-8 (CCK-8) was purchased from Dojindo (Tokyo, Japan).

SAV samples were obtained from local supermarkets. The total acidity and aging time were showed in Table 1.

Table 1 The samples of SAV

Manufacturer	Total acidity (g/100 mL)	Aging time (Year)	Manufacturer	Total acidity (g/100 mL)	Aging time (Year)
A-1	4.5	2	C-1	5.0	2
A-2	6.0	5	C-2	6.0	5
A-3	6.5	8	C-3	6.0	8
B-1	6.0	2	D-1	4.0	2
B-2	8.0	5	D-2	6.0	5
B-3	8.0	8	D-3	6.0	8

2.2 Detection of Total Phenolic and Flavonoid Contents

Total phenolic contents of SAV samples were estimated using Folin-Ciocalteu reagent according to the enclosed guidelines. Absorbance was measured at 765 nm in an ELISA Plate Reader. A standard calibration curve was plotted using gallic acid. Total phenolic content was calculated as equivalent of per mg gallic acid per gram (mg GAE/g). Total flavonoid content was estimated by a colorimetric assay using commercial kits according to the enclosed guidelines. The absorbance was measured at 510 nm. Rutin was used to plot the calibration curve. Total flavonoids were calculated as mg rutin equivalents per gram of dried sample (mg RE/g).

2.3 Antioxidant Activity of SAV in Vitro

DPPH Radical-Scavenging Activity Assay: DPPH (3.9 mg) was dissolved in 100 mL anhydrous ethanol and stored at 4 °C. Samples (20 μL) and DPPH (180 μL) were mixed in a 96-well plates and incubated at room temperature in the dark for 30 min. The absorbance was measured at 517 nm.

ABTS Redical-Scavenging Activity Assay: The ABTS radical cation ($ABTS^{+}$) solution was prepared from the reaction of 7 mM ABTS and 2.45 mM potassium persulfate after incubation and kept at room temperature for 16 h in the dark. Then the solution was diluted with 80% ethanol to obtain an absorbance of 0.70 ± 0.05 at 734 nm. The $ABTS^{+}$ solution (200 μL) was mixed with vinegar sample (10 μL). The reaction mixture was allowed to stand for 6 min in the dark at room temperature. The absorbance was measured at 734 nm.

Ferric Reducing Antioxidant Power (FRAP) Assay: A working solution was prepared freshly by mixing TPTZ dilution (2,4,6-tripyridyl-s-triazine), detective buffer and TPTZ solution in a ratio of 10:1:1 (v/v/v), respectively. A vinegar sample (5 μL) was mixed with FRAP (180 μL) working solution and incubated at 37 °C for 5 min. The absorbance was measured at 593 nm.

2.4 Cell Culture

Human liver cell line HepG2 was obtained from were cultured in RPMI-1640, 10% FBS, and 1% penicillin-streptomycin B solution, and incubated at 37 °C in a humidified atmosphere containing 5% CO_2. Cells were treated with SAV at different concentration. An equal volume of phosphate buffered saline (PBS) was used in the control experiments. Under all conditions we added 20 mM HEPES buffer. Culture medium was changed every two days.

2.5 *Cell Viability Assay*

Cell viability was monitored by CCK-8 assay. Cells were seeded in 96-well microplate and exposed to H_2O_2 or vinegar. After treatment, cells were added with 10 μL CCK-8 solution, and incubated for 4 h at 37 °C. Absorbance in each well was quantified at 450 nm using an ELISA reader.

2.6 *Statistical Analysis*

All data were expressed as the mean ± standard deviation (SD). All data were analyzed by the Student t test or one-way ANOVA procedures using the GraphPad Prism 5.01 statistical software (GraphPad Software Inc., La Jolla, CA, USA). A P value less than 0.05 was considered statistically significant.

3 Results and Discussion

3.1 *Total Phenolic and Flavonoid Contents*

Phenolic and flavonoid compounds are considered to the most potent and therapeutically useful biocompounds since they have antioxidant activity and therapeutic properties [13]. In this study, total phenolic and flavonoid contents of SAV from different manufacturers were determined. The results showed that total phenolic contents of SAV samples ranged between 1.810 mg GAE/g and 5.836 mg GAE/g (Fig. 1a). The concentration of flavonoids in different vinegar samples was in the range of 0.332 mg RE/g to 4.554 RE/g (Fig. 1b). The concentration of phenolics and flavonoids followed the order of B > A > C > D. Total phenolic and flavonoid contents of B-3 were 5.836 mg GAE/g and 4.554 RE/g, respectively, which were the highest among all the samples. Therefore B-3 was selected for the subsequent experiments.

3.2 *Antioxidant Activity in Vitro*

DPPH, ABTS and FRAP are used as the main radicals for testing the antioxidant activities [14]. In this study, the antioxidant capacities of SAV samples were evaluated by these three assays. In DPPH assay, the radical scavenging ability of B-3 was 65.50 ± 1.11%, which has the highest DPPH radical scavenging activity among all the samples (Fig. 2a). In ABTS assay, the radical scavenging ability of B-3 was 76.84 ± 2.40%, which showed an effect similar to that in DPPH assay

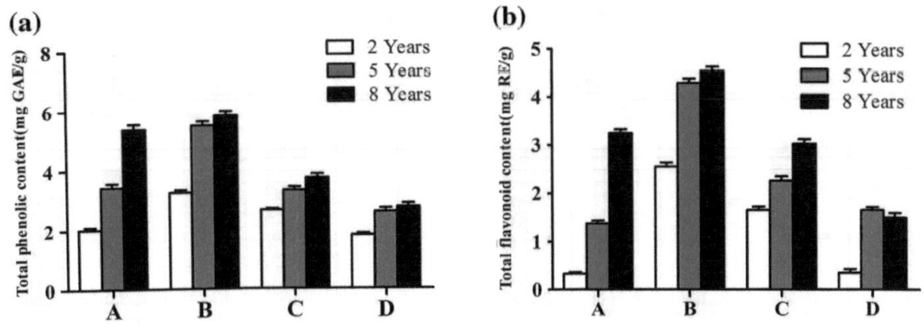

Fig. 1 Total phenolic and flavonoid contents of SAV samples. **a** Total phenolic content was expressed in terms of gallic acid equivalent (mg GAE/g). **b** Total flavonoid content was expressed in terms of rutin equivalent (mg RE/g). Data are means ± SD (n = 3)

(Fig. 2b). Another index for evaluating the antioxidant activity is the reducing power. Reductive capability is the power to transform Fe^{3+} to Fe^{2+}, which can be reflected through the absorbance. As shown in Fig. 2c, The $FeSO_4$ values of vinegar samples with the same aging time were followed the order: B-3 > A-3 > C-3 > D-3. Taken together, these results indicated that B-3 showed the highest antioxidant capability in these vinegar samples. Then, we chose B-3 to evaluate hepatoprotective effects on HepG-2 cells.

3.3 *Effect of B-3 on Cell Viability*

We examined the effect of B-3 on cell viability in human liver cells by using CCK-8 assay. HepG2 cells were exposed to increasing concentrations of B-3 for 24 h. As shown in Fig. 3, cell viability was not significantly changed when HepG2 cells were treated with 2.5–10 μL/mL vinegar. These results indicated that B-3 had no toxicity at the tested concentrations and could be studied for potential protective effects.

3.4 *Protective Effect of B-3 Against H_2O_2-Induced Hepatocyte Damage In Vivo*

H_2O_2 was used to cause liver injury and accelerate free radical derivatives to measure the effect of antioxidants and liver protection in vivo [15]. Firstly, H_2O_2 was used to establish an oxidative injury cell model. HepG2 cells were treated with different concentrations (0–800 μM) for 6 h. As shown in Fig. 4a, Cell viability was significantly decreased in a dose-dependent manner. Treatment of 400 μM H_2O_2 to hepatocytes resulted in a dramatic ($p < 0.05$) decrease in the cell viability

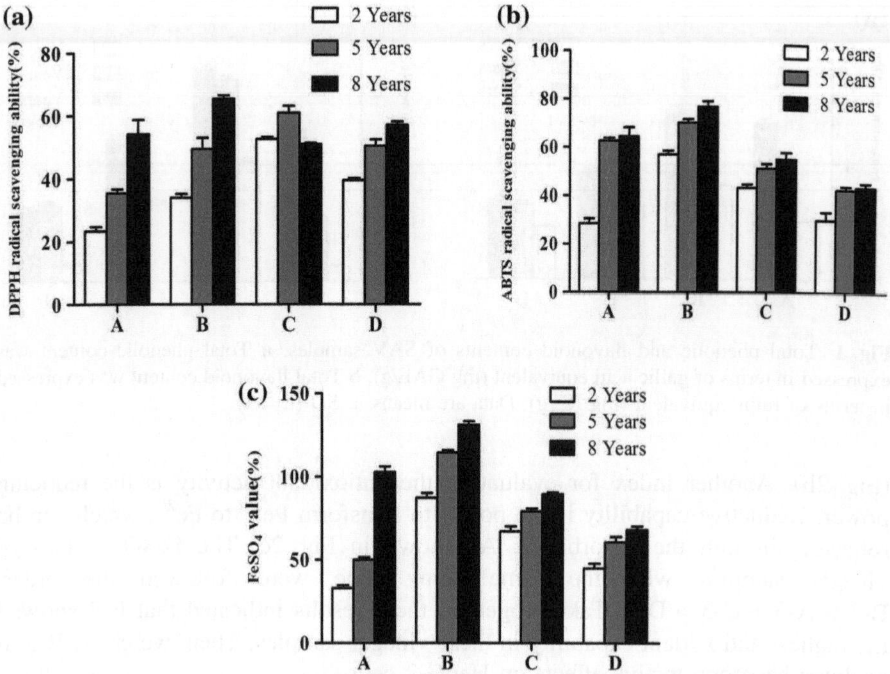

Fig. 2 Total antioxidant capacity of vinegar samples measured by different methods. Data represent mean ± SD (n = 3). **a** DPPH radical scavenging activity. **b** ABTS radical scavenging activity. **c** Ferric reducing antioxidant power

Fig. 3 Effect of B-3 on cell viability in human liver cells

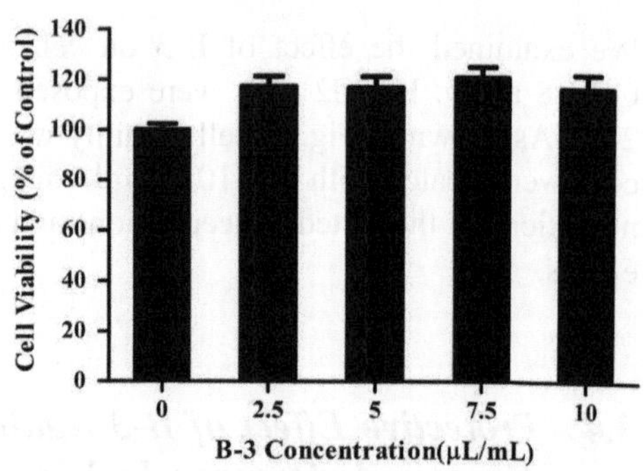

up to about 55%. Then, 400 μM H_2O_2 treatment for 6 h was chosen to examine the protective effect of B-3.

Next, Protective effect of B-3 against H_2O_2-induced hepatotoxicity was detected by CCK-8 assay. HepG2 cells were pretreated with various concentrations (2.5, 5, 7.5, 10 μL/mL) of vinegar for 24 h, and treated with 400 μM H_2O_2 for 6 h. As shown in Fig. 4b, pretreatment with B-3 of prior to H_2O_2 protected cell death in a

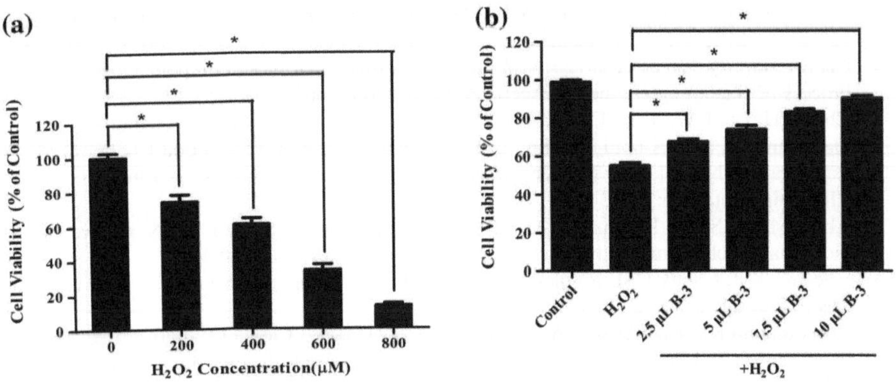

Fig. 4 Effect of B-3 on cell viability in H_2O_2 -treated HepG2 cells. **a** Cells were treated with increasing concentrations of H_2O_2 for 6 h. **b** Cells were pretreated with various concentrations of vinegar for 24 h, and treated with 400 μM H_2O_2 for 6 h. Values are means ± SD (n = 3)

dose-dependent manner ($p < 0.05$). Cell viability was restored up to 90.49% by B-3 pretreatment (10 μL/mL). This result indicates that exposure of hepatocytes to B-3 confers significant protective effects against H_2O_2.

4 Conclusions

In this study, antioxidant activity and hepatoprotective activity of SAV was assessed. SAV was rich in phenolic and flavonoid compounds, and B-3 had the highest phenolic and flavonoid contents among all the samples. In addition, SAV exhibited protective effect against H_2O_2-induced oxidative damage in HepG2 cells. These findings would provide fermented food products with antioxidant and hepatoprotective potentials.

Acknowledgements This work was supported by the National Natural Science Foundation of China (No. 81600126, No. 31471722 and 31671851), National High Technology Research and Development Program of China (2013AA102106), Program for Changjiang Scholars and Innovative Research Team in University (IRT15R49), National Ministry of science and technology (2016YFD0400505).

References

1. Samd A, Azlan A, Ismail A (2016) Therapeutic effects of vinegar: a review. Curr Opin Food Sci 8:56–61
2. Chen FS, Li L, Qu J et al (2009) Cereal vinegars made by solid-state fermentation in China. Vinegars of the World. Springer, Milan, pp 243–259

3. Nie Z, Zheng Y, Du H et al (2015) Dynamics and diversity of microbial community succession in traditional fermentation of Shanxi aged vinegar. Food Microbiol 47:62–68
4. Chen T, Gui Q, Shi JJ et al (2013) Analysis of variation of main components during aging process of Shanxi aged vinegar. Acetic Acid Bacteria 2:e6
5. Chou CH, Liu CW, Yang DJ et al (2015) Amino acid, mineral, and polyphenolic profiles of black vinegar, and its lipid lowering and antioxidant effects in vivo. Food Chem 168:63–69
6. Rajendran P, Nandakumar N, Rengarajan T et al (2014) Antioxidants and human diseases. Clin Chim Acta 436:332–347
7. Abdel-Misih SRZ, Bloomston M (2010) Liver anatomy. Surg Clin N (However, the molecular mechanisms of the interaction bet America) 90(4):643–653
8. Saleem TSM, Chetty CM, Ramkanth S et al (2010) Hepatoprotective herbs—a review. Int J Res Pharm Sci 1(1):1–5
9. Arulkumaran KS, Rajasekaran A, Ramasamy A et al (2009) Cassia rox-burghii seeds protect liver against toxic effects of ethanol and carbontetrachloride in rats. Int J PharmTech Res 1(2):273–276
10. Cubero FJ, Trautwein C (2010) Oxidative stress and liver injury, vol 5. Springer US, pp 427–435
11. Borut P, Dusan S, Irina M (2013) Achieving the balance between ROS and antioxidants: when to use the synthetic antioxidants. Oxidative Med Cell Longevity 1:956792
12. Valko M, Leibfritz D, Moncol J et al (2007) Free radicals and antioxidants in normal physiological functions and human disease. Int J Biochem Cell Biol 39(1):44–84
13. Lamport DJ (2012) The effects of flavonoid and other polyphenol consumption on cognitive function: a systematic review of human experimental and epidemiological studies. Nutr Aging 1:5–25
14 Prior RL, Wu X, Schaich K (2005) Standardized methods for the determination of antioxidant capacity and phenolics in foods and dietary supplements. J Agric Food Chem 53(10): 4290–4302
15 Barry H, Okezie A (1991) DNA damage by oxygen-derived species its mechanism and measurement in mammalian systems. FEBS Lett 281:9–19

Analysis and Application of the Promotion Action of Methanol on *Sorangium cellulosum* Produce Epothilones

Lin-hao Zhang, Lin Zhao, Xin Sun and Xin-li Liu

1 Introduction

Epothilones are a class of secondary metabolites mainly generated by *Sorangium cellulosum* of the suborder *myxobacteria*, belongs to the natural macrolides [1, 2]. By the end of 2005, there are five kinds of epothilone compounds in different stages of clinical trials as anticancer drugs, including epothilone B, epothilone D, ixabepilone (BMS-247550), ZK-EPO and BMS-310705 [3]. Epothilone A and epothilone B are the two main products which produced by *Sorangium cellulosum*, and epothilone B has been approved by the United States FDA to be first-line agents for the treatment of advanced breast cancer [4, 5]. Although they could be synthesised by chemical process, it can not be industrially produced due to the too many steps of chemical synthesis, high cost, low yield and high reaction conditions of chemical synthesis of epothilones, biological fermentation is the best way for produce epothilones.

A variety of methods to improve the yield of epothilones has been reported, which included strains selection, fermentation medium and process optimization, the application of fermentation adsorbent and biological process optimization, etc., but these methods to enhance the production of epothilone has reached a bottleneck [7–9]. The yield of epothilones are still low due to the toxic effects of itself to the host strain,

Lin-hao Zhang and Lin Zhao have contribute equally to this paper.

L. Zhang · L. Zhao · X. Sun · X. Liu (✉)
School of Bioengineering, QiLu University of Technology, No. 3501 Daxue Road, Jinan 250353, People's Republic of China
e-mail: vip.lxl@163.com

L. Zhang · L. Zhao · X. Sun · X. Liu
Shandong Provincial Key Laboratory of Microbial Engineering, QiLu University of Technology, No. 3501 Daxue Road, Jinan 250353, People's Republic of China

H. Liu et al. (eds.), *Advances in Applied Biotechnology*, Lecture Notes in Electrical Engineering 444, https://doi.org/10.1007/978-981-10-4801-2_32

though it can be successfully heterologous expression by receptor cells, it is imperative to improve the epothilones production through new methods [10, 11]. It has been reported that a lot of small molecule inducers can affect DNA binding protein directly or indirectly to promote the transcriptional expression of the gene, such as cellobiose can induce many kinds of fungi to produce cellulose, the phenolic compounds, ethanol and some metal ions have an effect on the production and secretion of laccase, methanol can be used as a strong promoter of mRNA transcription in *Pichia pastoris* [12–15]. A variety of small molecular organic molecules were screened in this laboratory, we found that added small traces of methanol to fermentation broth of *Sorangium cellulosum* can significantly improve the yield of epothilone. It can be excluded the probability of methanol to be precursor substance, nutrient carbon source or change the permeability of cell membranes as the amount of methanol added was only 0.05%, it is more likely to be some specific enzyme activators or signaling substances to influence primary or secondary metabolism of *Sorangium cellulosum.* Firstly, the specificity of the methanol promoting effect was screened and validated in this study, then the effect of methanol on the primary and secondary metabolism of *Sorangium cellulosum* were studied and evaluated, finally, the further increase in production of epothilones by biological fermentation method according to the optimization of adding amount and time of methanol became possible.

2 Materials and Methods

2.1 Materials

2.1.1 Strain

Sorangium cellulosum (*Sorangium cellulosum So ce*2161) is provided by Shandong Provincial Key Laboratory of Microbial Engineering and The Preservation Center of Shandong provincial industrial microorganisms.

2.1.2 Culture Medium

(1) CNST solid medium (g/L): $MgSO_4 \cdot 7H_2O$ 1.0, KNO_3 0.5, $Na_2HPO_4 \cdot 12H_2O$ 0.65, TE 1 mL, pH adjusted to 7.2. Add 2% agar. Sterilized at 121° for 30 min.
(2) M26 medium (g/L): soybean peptone 2, potato starch 8, glucose 2, yeast extract 2, $MgSO_4 \cdot 7H_2O$ 1.0, $CaCl_2$ 1.0, EDTA-Fe 1 mL, TE 1 mL, pH adjusted to 7.2. Sterilized at 121° for 30 min.
(3) EMP fermentation medium (g/L): potato starch 2, skim milk powder 1, glucose 2, soybean flour 2, $MgSO_4 \cdot 7H_2O$ 1.0, $CaCl_2$ 1.0, EDTA-Fe 1 mL, TE 1 mL, pH adjusted to 7.2. Add 2% XAD-16 macroporous adsorption resin. Sterilized at 121° for 30 min.

2.1.3 Instrument and Equipment

STIC-E temperature oscillation incubator (Ji'nan zhanchuang Technology Co. Ltd); SPX-intelligent biochemical incubator (Ningbo Jiangnan instrument factory); DHG-9240A electric constant temperature drying oven (Shanghai Jinghong Experimental Equipment Co. Ltd.); adjustable pipettes (Thermo instruments Ltd); LC-20AT high performance liquid chromatograph (Shimadzu); SGD-IV automatic reducing sugar analyzer (Shandong Provincial Academy of Sciences Institute of biological research); PLG-203 precision electronic balance (Mettler Toledo instruments); 320 PerHect pH (Thermo Orporation); HP-02 filtration apparatus (Beijing Jingyi Baifang Technology Co. Ltd.) etc.

2.2 *Experimental Methods*

2.2.1 Determination Method for Epothilones Production

Column specifications (Shimadzu C18 column): 250 mm * 4.6 mm * 5 μm.

Chromatographic condition: detection wavelength 249 nm; column temperature 40 °C; mobile phase (methanol: water = 65:35), flow rate of 1 mL/min; sample volume of 20 μL.

2.2.2 Dry Weight Method

Suction filtration the zymotic fluid by Buchner funnel to interception all resin and cells, then measure total mass when it was thorough drying in the air dry oven. Calculate the biomass dry weight according to the formula $DCW = W_{total\ mass} - W_{resin} - W_{filter\ paper}$.

2.2.3 Determination Method for Reducing Sugar

According to the instructions of the instrument, using the 1% standard sugar solution as the reference, centrifugal the fermentation liquid at 8000 rpm for 2 min to absorb a certain amount of supernatant and add into the sample hole, conduct automatic titration determination and print the result.

3 Results and Discussion

3.1 *Screening of Small Molecule Organic Inducer*

Small molecules can play a role as inducers in low concentration while at higher concentration it may have a toxic effect on microorganisms. We add small organic molecules which selected at the volume ratio of 0.2% concentration, and add them before inoculate by filtration sterilization. The small organic molecules which we selected as the inducer of *Sorangium cellulosum* produce epothilones are relatively simple and easily soluble in water, the types of small organic molecules we selected and the promoting effect for the produce of epothilones are showed in Table 1 (+express have stimulative effect for produce epothilones, −express inhibited the growth of *Sorangium cellulosum* or decrease the yield of epothilones, 0 express almost have no effect on the growth and metabolism of *Sorangium cellulosum*).

3.2 *Effect and Evaluation of Methanol Addition on Primary Metabolism of* Sorangium cellulosum

3.2.1 Effect of Methanol on the Growth of *Sorangium cellulosum*

DCW were determined at the first 2–6 days of the fermentation culture (Fig. 1).

3.2.2 Effect of Methanol on pH Value of Fermentation Broth

The value of pH of the fermentation culture was determined at the first 2–6 days (Fig. 2).

Table 1 Screening of small molecule inducers and their effects

Small molecule inducer	Effect
Ethanol	0
Formaldehyde	−
Ammonia water	0
Methanol	+
Glycerin	0
Glacial acetic acid	−
Acetone	−
Normal alcohol	0
Butyl alcohol	0
DMSO	0
Ethylene glycol	0

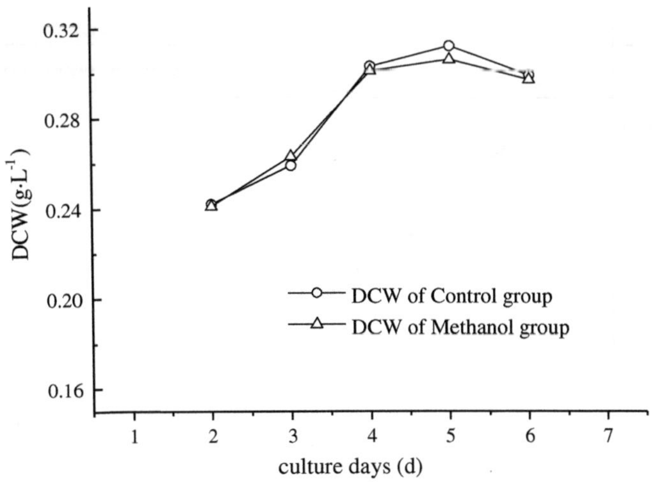

Fig. 1 Changes of DCW in different culture days

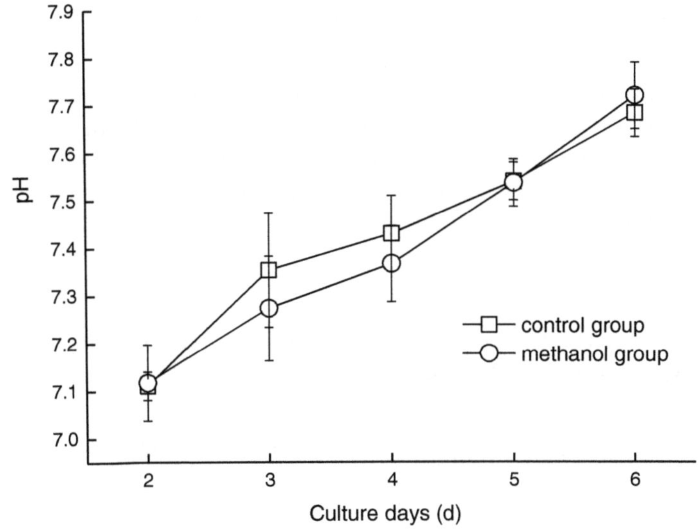

Fig. 2 Changes of pH value in different culture days

3.2.3 Effect of Methanol on the Change of Reducing Sugar in Fermentation Broth

The value of reducing sugar was determined at the first 2–6 days (Fig. 3).

The results of DCW, pH value and reducing sugar content at the first 2–6 days of fermentation culture showed that the DCW of the methanol group and control

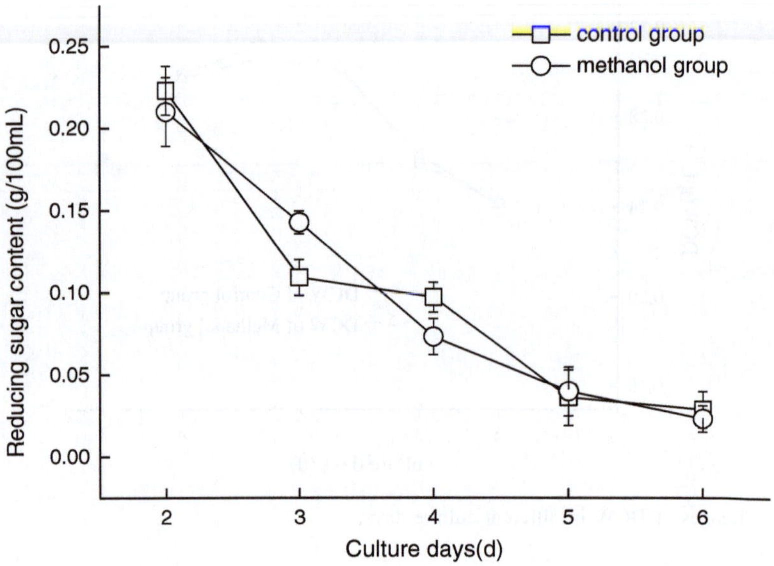

Fig. 3 Changes of reducing sugar content in different culture days

group gradually increased, the value of pH increased continuously while the reducing sugar was declining with the continuous culture at the first 4–6 days. The most important is that there is almost no difference between the control group and the methanol group, it can be preliminarily judged that the addition of methanol has little effect on the growth and primary metabolism of *Sorangium cellulosum.*

3.3 *Effect of Methanol on the Secondary Metabolism of* Sorangium cellulosum

3.3.1 The Experiment to Find Out Which Day Is the Maximum Promotion of Methanol

The yield of epothilones was determined at the first 3–7 days (Figs. 4 and 5).

The above results show that the larger D-value are appeared on the first 5 day and first 6 day which is the stable period of *Sorangium cellulosum*, in other words, the maximum facilitation of methanol for produce epothilones is in the stable period, this is consistent with the prediction of the methanol effect on the secondary metabolism of the *Sorangium cellulosum*. One possibility is that the methanol promote more metabolic pathway of *Sorangium cellulosum* for the synthesis of secondary metabolites epothilones through the overexpression of certain genes or improve the activity of some key enzymes.

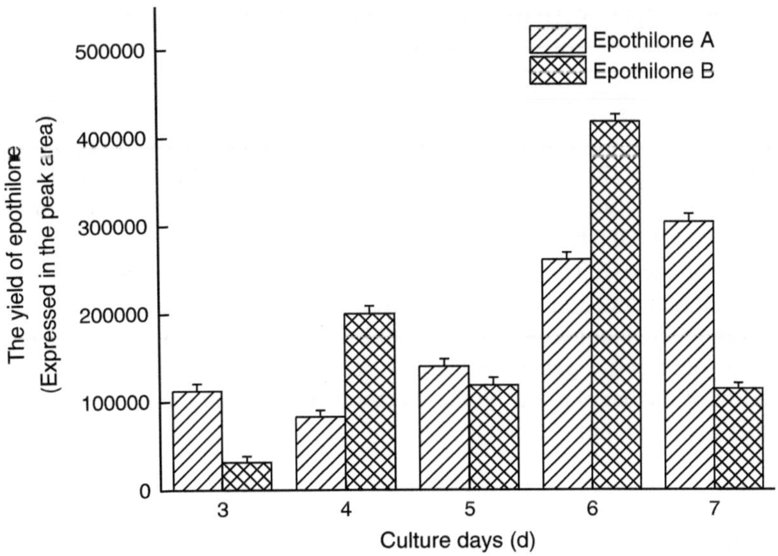

Fig. 4 The D-value of different days between methanol group to control group (a)

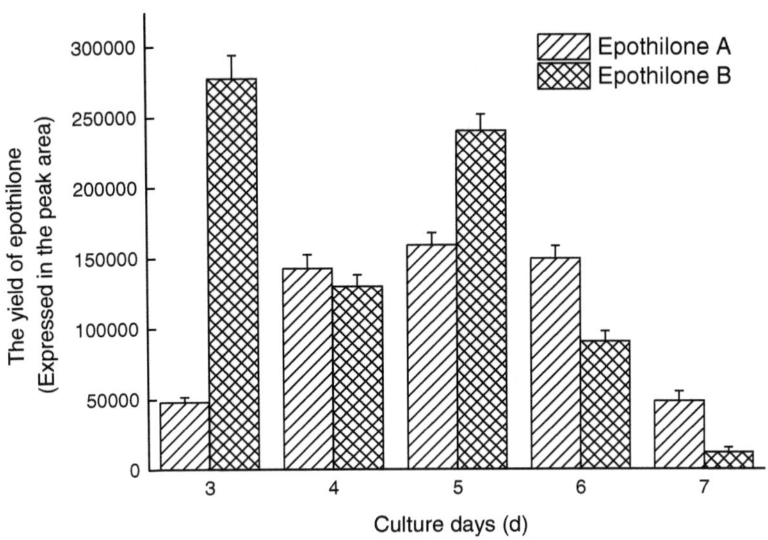

Fig. 5 The D-value of different days between methanol group to control group (b)

3.3.2 The RT-QPCR Analysis of the Effect of Methanol on *Sorangium cellulosum* Produce Epothilones

After the extraction of the RNA from *Sorangium cellulosum* which culture in the fifth day and the reverse transcription, we analyzed the relative expression of the key enzymes of epothilones synthesis by RT-qPCR, including polyketide synthase PK1 and PK2, S-adenosylmethionine synthetase (SAMs), alpha ketoglutarate dehydrogenase (OGDHC) and thioesterase (TE, can be used to measure the yield of epothilone synthesis). The results showed that the relative expression of TE and SAMs are larger and improved to a certain extent, increased by 0.71 and 0.67 respectively. As a consequence, the addition of methanol could increase the expression of specific genes in the two key enzymes or downstream, and the mechanism of methanol effect on the metabolism of *Sorangium cellulosum* need to be further explored and verified.

3.4 *The Optimize of Methanol Addition and Analysis of the Promotion for Epothilones*

3.4.1 The Broad Experiment on Methanol Addition

The addition amount of methanol is in the range of 0–300 μL (the volume of fermentation medium is 50 mL), set a gradient of 25 μL, the relationship between the yield of epothilones and methanol dosage was following (Fig. 6).

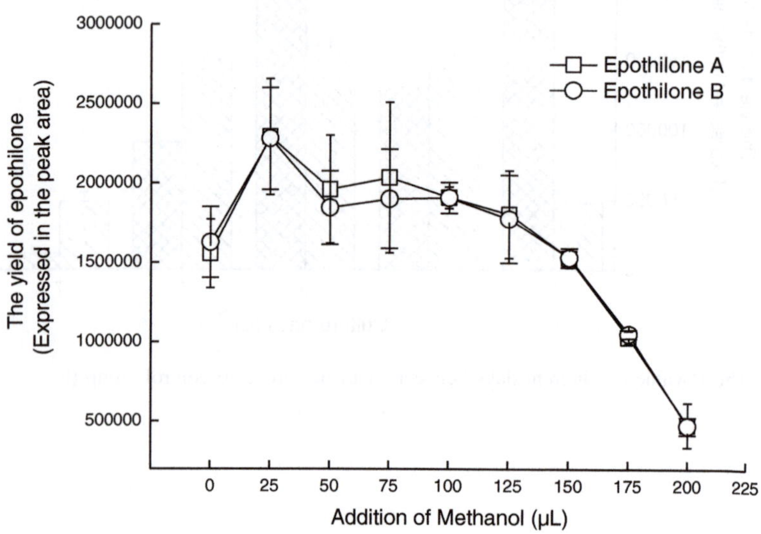

Fig. 6 The relationship between the yield of epothilones and methanol dosage

The results showed that the adding amount of methanol in the range of 0–150 μL has a different promote effect on the yield of epothilones, while the adding amount of methanol is more than 150 μL, the yield of epothilones decreased gradually with the increase of the concentration of methanol in fermentation liquid as methanol toxicity was greater than the effect. Notably, the yield of epothilones contributes close to maximum value with an increase of more than 10% while the adding amount of methanol in the range of 0–50 μL.

3.4.2 The Accurate Experiment on Methanol Addition

In order to verify the results of the broad experiment on methanol addition, and considering the adding amount of methanol in range of 0–150 μL (the volume of fermentation medium is 50 mL) has certain promoting effect, set a gradient of 25 μL while add an addition experiment of 10 μL. The results was shown (Fig. 7).

The experimental results showed that the promoting effect of methanol addition in range of 10–100 μL is quite reasonable, as a result the amount should be in the range of intermediate role in value, the most reasonable dosage should be 25 μL (the volume ratio is 0.05%) addition according to the industrialized production as well as the sake of security.

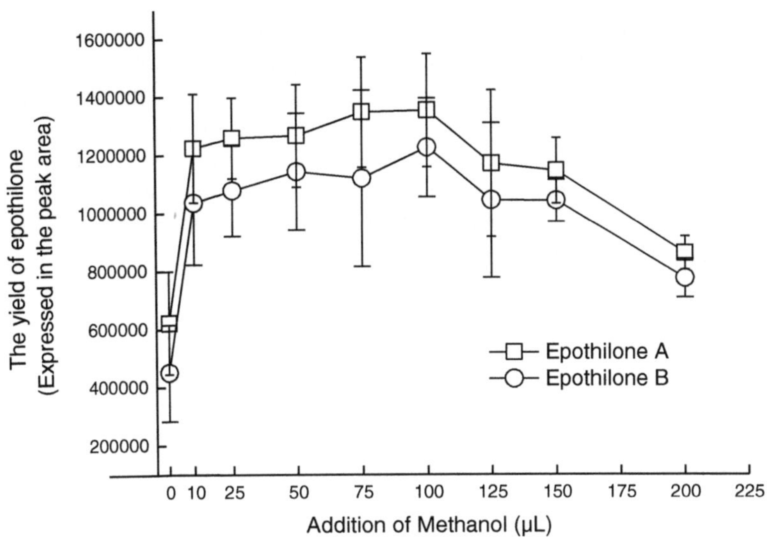

Fig. 7 The relationship between the yield of epothilones and methanol dosage

3.4.3 The Optimize of the Time for Methanol Addition and the Experiment for Fed-Batch of Methanol

We found that the earlier added methanol to the medium the more obvious promoting the yield of epothilones by adding the volume ratio of 0.05% methanol to the fermentation medium in different days. Considering simplicity of operation, we add the volume ratio of 0.05% methanol into the fermentation culture medium after sterilization.

We tried to pass in the way of adding a certain amount of methanol into the medium in order to verify whether the methanol added to the fermentation medium will losing its promotion as it may be utilized by *Sorangium cellulosum*. The results are listed in Table 2.

The results showed that the yield of epothilone A and B have no conspicuous difference between B group and C group, in a word the way of adding methanol by group B and C have the same effect on epothilones synthetization. In other words, the methanol in fermentation broth is stabilize enough for persistent promotion on *Sorangium cellulosum* produce epothilones, the best way for methanol addition is adding before inoculate while adding methanol for fed-batch is meaningless.

4 Conclusion and Prospect

The yield of epothilone A and B in the fermentation broth can be stably increased by 15–20% through add methanol as per the proportion of 0.05% volume into the fermentation medium of *Sorangium cellulosum*. The higher of the yield means the better of the metabolism of the fermentation flask, and the production rate will be higher than 20%. This study gives us a new revelation, the most common methanol as a simple organic molecule has a larger role in the effects of microbial metabolites, and provide a probability to explore new methods to promote the accumulation of microbial metabolites.

Table 2 The result of the experiment for fed-batch of methanol

Group (Each set of four parallel)	Control group (no methanol addition)	The B group (add 50 μL methanol before inoculate)	The C group (add 50 μL methanol in the first, third and fifth days respectively)
The yield of epothilone A (expressed in the Peak area)	2693379	2854270	2926007
The yield of epothilone B (expressed in the Peak area)	1970804	2114215	2127012

Many small molecules can play a role in the induction of the primary and secondary metabolism of microorganism, which can regulate the metabolic flux of microbial metabolites through the expression level of activity of some enzymes or certain genes, and there are more small molecules have a similar regulating effect to different microorganisms. When the other conventional methods to improve the microbial metabolism of the target product yield reaches a certain limit, we can found the corresponding small molecule inducers through the screening of small molecule inducers, which have a positive promote role to obtain a higher yield of target product.

Acknowledgements The work was financially supported by Shandong Provincial Key Research and Development Program (No. 2015ZDZX05001, 2015ZDXX0502B01, 2015ZDXX0403B01, 2016GGB01409), National Natural Science Foundation (No. 31501396), Shandong Provincial Natural Science Foundation (No. ZR2012CQ027).

References

1. Meng FX, Li YX, Guo WL et al (2010) Optimization of fermentation medium for epothilones production with sequential statistical approach. Chem Res Chin Univ 26(1):86–91
2. Park SW, Han SJ, Kimb D-S et al (2007) Improvement of epothilone B production by in situ removal of ammonium using cation exchange resin in *Sorangium cellulosum* culture. Biochem Eng J 37:328–331
3. Kolman A (2005) Activity of epothilones. Curr Opin Invest Drugs 6(6):616–622
4. Regentin R, Frykman S, Lau J et al (2003) Nutrient regulation of epothilone biosynthesis in heterologous and native production strains. Appl Microbiol Biotechnol 61:451–455
5. Wang DH, Yuan JF, Guo WJ (2012) Mutant and optimization of epothilone B produced by *Sorangium cellulosum*. Biotechnology 22(3):70–73
6. Frykman S, Tsuruta H, Lau J et al (2002) Modulation of epothilone analog production through media design. J Ind Microbiol Biotechnol 28(1):17–20
7. Wang XY, Chen SX (2011) Breeding of high yield strain of epothilone and optimization of fermentation medium. Chin J Pharm Ind 42(4):258–261
8. Gerth K, Bedorf N, Höfle G et al (1996) Epothilons A and B: antifungal and cytotoxic compounds from *Sorangium cellulosum* (*Myxobacteria*). J Antibiot
9. Reichenbach H (2001) *Myxobacteria* producers of novel bioactive substances. J Ind Microbiol Biotechnol 27(3):149–156
10. Tang L, Shah S (2000) Cloning and heterologous expression of the epothilone gene cluster. Science 287:640–642
11. Julien B, Shah S (2002) Heterologous expression of epothilone biosynthetic genes in *Myxococcus xanthus*. Antimicrob Agents Chemother 46:2772–2778
12. Kubicek CP, Messner R, Gruber F et al (1993) The Trichoderma cellulase regulatory puzzle: from the interior life of a secretory fugus. Enzyme Microb Technol 15(2):90–99
13. Mandels M, Reese ET (1960) Induction of cellulase in fungal by cellobiose. J Bacteriol 79 (6):816–826
14. Li HR (2005) Biology and biotechnology of white rot fungi. Chemical Industry Press, Beijing
15. Sreekrishna K, Brankamp RG, Kropp KE et al (1997) Strategies for optimal synthesis and secretion of heterologous proteins in the methylotrophic yeast Pichia pastoris, ☆. Gene 190 (1):55–62

Optimization of Solid Fermentation Process of *Bacillus megaterium* and Its Application in Crop Growth

Jinzhao Liu, Lin Zhao, Dong Ma, Xin Sun and Xin-li Liu

1 Introduction

Bacillus megaterium is a kind of aerobic, thermophilic gram positive bacteria, which is able to produce spores. With the characteristics of decomposition of inorganic phosphorus and improving soil environment, increasing soil fertility, improving crop quality in agricultural production and improving the environment, reducing the role of livestock disease in animal husbandry, *B. megaterium* have a great potential as a kind of functional bacteria [3]. It is widely used in the preparation of microbial fertilizer, Because of *B. megaterium* has the characteristics of suitable for industrialized production, such as low nutritional requirements, simple culture conditions, fast growth, able to produce resistant spores and so on [4].

At present, the way of industrialized production of *B. megaterium* is mainly liquid submerged aerobic fermentation. Compared to submerged aeration culture's shortcomings of high energy consumption, high aseptic conditions, higher equipment requirements, solid fermentation using simple equipment, consume low energy, and low probability of a large area of contamination due to poor transmission of solid materials [5]. Therefore, this paper takes large industrial and agricultural low value waste such as cassava residue, furfural residue, mushroom residue as matrix, by optimizing the ratio of carbon sources, nitrogen sources and

---

Authors Jinzhao Liu, Lin Zhao and Dong Ma contributed equally to this paper.

J. Liu · L. Zhao · D. Ma · X. Sun · X. Liu
School of Bioengineering, QiLu University of Technology, No.3501, Daxue Road, Jinan 250353, People's Republic of China

J. Liu · L. Zhao · D. Ma · X. Sun · X. Liu (✉)
Shandong Provincial Key Laboratory of Microbial Engineering, No.3501, Daxue Road, Jinan 250353, People's Republic of China
e-mail: vip.lxl@163.com

H. Liu et al. (eds.), *Advances in Applied Biotechnology*, Lecture Notes in Electrical Engineering 444, https://doi.org/10.1007/978-981-10-4801-2_33

fermentation process to making solid fermentation of *B. megaterium*, preparation of agricultural probiotics, and evaluate its effect.

2 Materials and Methods

The materials and methods are listed as follows in this study.

2.1 *Microorganisms and Medium*

B. megaterium was screened by Shandong Provincial Key Laboratory of Microbial Engineering, and was preserved by the Shandong provincial Industrial Microbial Culture Preservation Center. LB culture medium used as the culture medium and tablet counting medium. Liquid culture medium contains glucose 15.0 g, potato starch 8.0 g, sodium chloride 5.0 g, calcium carbonate 1.5 g, distilled water 1 L, pH: 7.5–8.5. The main raw materials such as furfural residue, cassava residue, mushroom residue, corn flour, amino acid and corn syrup and molasses were come from manufacturers and farmland nearby, urea, peptone, glucose and sucrose were purchased from Sinopharm Chemical Reagent Co., Ltd.

2.2 *The Cultivation of Liquid Strain*

Take one of the *B. megaterium* glycerol tube preserved in the low temperature refrigerator of −80 °C, and activated at 37 °C on LB nutrient medium for 24 h. Then vaccinate it into the seed culture medium (per 1 L flask medium 150 mL), culture 14 h at 37 °C and 200 rpm to acquired shake flask seed liquid.

2.3 *Optimization of Solid Fermentation Conditions*

Crushing the cassava residue, furfural residue, mushroom residue to obtained the raw material of 50–80 mesh. After mixing the material, adjust the moisture, pH, 121 °C sterilized 20 min. Culture the fermentation material under the appropriate conditions after inoculation. Regular sampling examination growth conditions, measured solid matrix of viable cells until spores rate rise to 85% or more.

The experimental design was shown in Table 1 to optimize solid matrix, carbon, nitrogen, water content and temperature of fermentation [6].

With cassava residue as substrate, 40% moisture, 2×10^7 CFU g^{-1} inoculation amount, orthogonal test was used to test the effect of different contents of glucose,

Table 1 Optimization of solid fermentation conditions

Number	Category	Condition	Optimization project
1	Substrate	5% glucose, 5% peptone, 50% moisture, pH 7.0, 2×10^7 CFU g^{-1} inoculum, 37 °C	Furfural residue, cassava residue, mushroom residue
2	Carbon source	Cassava residue, 5% peptone, 50% moisture, pH 7.0, 2×10^7 CFU g^{-1} inoculum, 37 °C	5% content of glucose, corn flour, sugar, molasses
3	Nitrogen source	Cassava residue, 5% glucose, 50% moisture, pH 7.0, 2×10^7 CFU g^{-1} inoculum, 37 °C	5% content of peptone, urea, amino acid and corn syrup
4	Moisture	Cassava residue, 5% glucose, 5% corn syrup, pH 7.0, 2×10^7 CFU g^{-1} inoculum, 37 °C	35, 40, 45, 50, 55%
5	Temperature	Cassava residue, 5% glucose, 5% Corn syrup, pH 7.0, 2×10^7 CFUg^{-1} inoculum, 40% moisture	25, 30, 37, 42 °C

corn syrup and different initial pH on the fermentation level and fermentation process.

Cell growth curve: The above optimized fermentation process was used for solid fermentation, pH and water content of the samples were measured every 2 h and the number of viable bacteria was determined by the plate count in 85% and more mature spore rate.

Test method: Weigh 10 g sample into a clean glass plates, 105 °C bake 2 h to measure the moisture content. Measure the pH after 5 g samples and 45 mL distilled water fully stirred [7]. The mixed material was placed into an oven of 60 °C after fermentation until dry enough, weigh 10 g of dry material into 90 mL sterile saline, shaking 30 min in shaker and radient dilution coating for plate count [8].

2.4 *Pepper Field Experiment*

Measured and statistics of the experimental group and control group of pepper yield per plant, fruit weight, flesh thickness, fruit length and fruit diameter [9] (Table 2).

Table 2 Pepper field test conditions

Experimental group	Category agents	Consumption
Control group	×	×
Group A	0.5×10^8 cfu/g agent	30 kg/ha
Group B	1.0×10^8 cfu/g agent	30 kg/ha
Group C	2.0×10^8 cfu/g agent	30 kg/ha

3 Results and Discussion

We analyse the effect of the solid substrate, carbon sources, nitrogen sources, moisture content, temperature to obtained the optimal fermentation conditions and analysis the results of the field test.

3.1 *Effect of Solid Substrate*

As Fig. 1 shows, after fermentation with furfural residue, mushroom residue, cassava residue as the matrix, the bacteria content was 13.1×10^8 CFU g^{-1}, 16.5×10^8 CFU g^{-1}, 21.6×10^8 CFU g^{-1} respectively. After fermentation, the content of the viable in cassava residue was 30.9% higher than that of furfural residue, which was 64.9% higher than that of the mushroom residue.

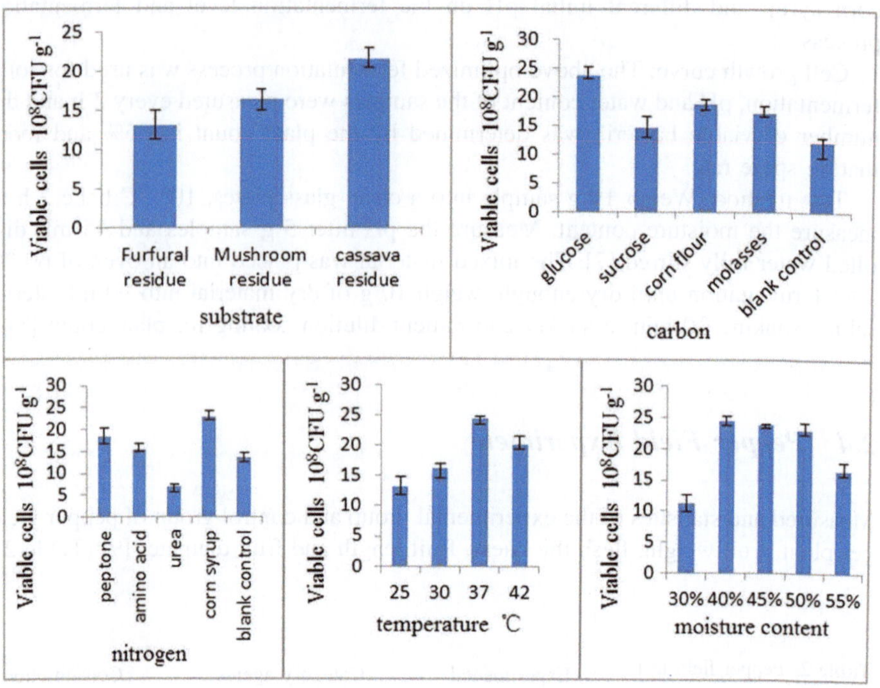

Fig. 1 Effect of solid substrate, carbon source, nitrogen, moisture content, temperature on fermentation

3.2 *Effects of Carbon Sources*

Comparing the fermentation results of glucose, sucrose, corn flour, molasses as carbon source shows that glucose can be used as the best carbon sources with the fermentation level reached 23.7×10^8 CFU g^{-1}, compared with the control (12.1×10^8 CFU g^{-1}), the fermentation level improved 95.9%.

3.3 *Effects of Nitrogen Sources*

Comparing the fermentation results of peptone and amino acid, urea, corn syrup as nitrogen source shows that corn syrup can be used as the best nitrogen sources with the fermentation level reached 22.9×10^8 CFU g^{-1}, compared with the control, the fermentation level improved 49.6%. Compared with the blank control, the fermentation level of the experimental group was 49.6% lower than that of the control group, only reached 6.9×10^8 CFU g^{-1}, which shows that the urea had a certain inhibitory effect on the solid fermentation of *B. megaterium*.

3.4 *Effects of Moisture Content*

Comparing the fermentation results of 30, 40, 45, 50 and 55% water content, the highest level of fermentation of *B. megaterium* is 24.6×10^8 CFU g^{-1} when the water content is 40%. Under the same conditions, the fermentation level increased by 3.4% than 45% moisture content (23.8×10^8 CFU g^{-1}) and 117.7% than 30% moisture content (11.3×10^8 CFU g^{-1}).

3.5 *Effects of Moisture Content*

Compared the fermentation results at different incubation temperatures, the fermentation results reached the highest level of 24.5×10^8 CFU g^{-1} at incubation temperatures was 37 °C, which was increased by 85.6% than the fermentation level of incubation temperature at 25 °C. And under the same conditions, the fermentation level reached 16.3×10^8 CFU g^{-1} at 30 °C and 20.1×10^8 CFU g^{-1} at 42 °C.

3.6 Orthogonal Optimization Experiment

In order to further clarify the fermentation conditions and achieve a higher level of fermentation, the three factors (carbon source, nitrogen source, pH) and levels of orthogonal test was carried out. We obtained the optimal fermentation process by evaluated the fermentation results with viable cells in fermented materials.

The effect of initial pH on the fermentation results was maximum with the range reached 7600 and the effect of corn syrup was minimum as Table 3 shows. Therefore,the optimized fermentation density reached 30.9×10^8 CFU g^{-1}, Table 4 with the optimum fermentation conditions of 5% of glucose, 10% of corn syrup and 6.5–7.0 of the initial pH, which increased 26.1% compared with the initial condition (Figs. 2 and 3).

3.7 Temperature and pH Trends

The whole process of *B. megaterium* solid fermentation which was designed in this paper period about 24 h, the cell in 4–8 h propagate fast for the long chain in the logarithmic growth phase, 8–14 h long chain began to collapse into a single bacterium, the pH value increased from 7 to 7.5 due to the rapid growth in 4–14 h of *B. megaterium*, after 14 h, the growth come into stable period and began to form spores, the fermentation ended after 24 h (Figs. 4 and 5).

3.8 Field Experiments

As Table 5 shows, the microbial fertilizer can significantly improve the pepper quality of all aspects such as yield of per plant, fruit weight, flesh thickness, fruit length, fruit diameter, etc. The quality of pepper increases along with the increased

Table 3 Orthogonal optimization of fermentation conditions

Number	Glucose (%)	Corn syrup (%)	Initial pH	Viable cells (10^8 CFU g^{-1})
1	2.5	5.0	6.5	24.2
2	2.5	10.0	7.0	27.4
3	2.5	15.0	7.5	19.7
4	5.0	5.0	7.0	29.8
5	5.0	10.0	7.5	22.1
6	5.0	15.0	6.5	30.9
7	7.5	5.0	7.5	22.7
8	7.5	10.0	6.5	30.8
9	7.5	15.0	7.0	30.1

Table 4 Orthogonal experiment optimization of fermentation conditions analysis

Fermentation conditions		Glucose	Corn syrup	Initial pH
Viable cells 10^8 CFU g^{-1}	K1	23,767	25,567	28,633
	K2	27,600	26,767	29,100
	K3	27,867	26,900	21,500
	Range R	4100	1333	7600

K1, K2, K3, respectively, represent the average number of viable bacteria of fermentation by different levels of glucose, corn syrup and initial pH

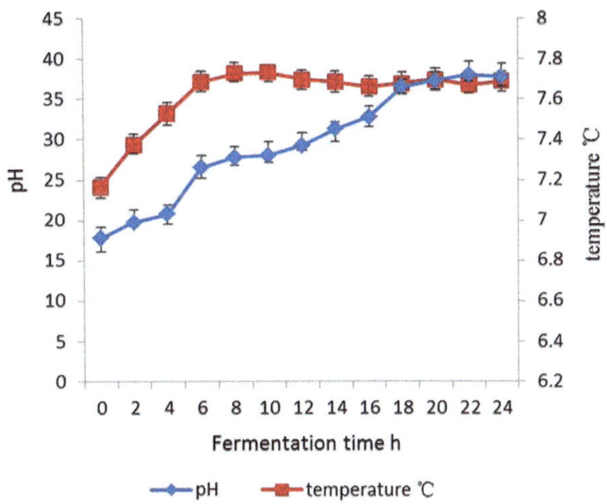

Fig. 2 The change of pH value and temperature in the fermentation process

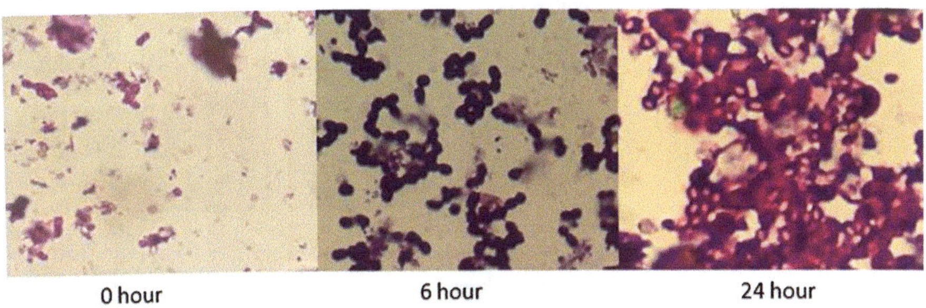

Fig. 3 Microscopic examination of the fermentation

concentration of viable in fertilizer. Pepper diameter increased highest of about 25.91% and the Longitudinal diameter improved the lowest percentage is about 3.40% after the application of microbial fertilizer as Table 3 shows. The pepper fruit become more rounded appearance cause above reason changed the shape index of pepper fruit from 1.26 into 1.01.

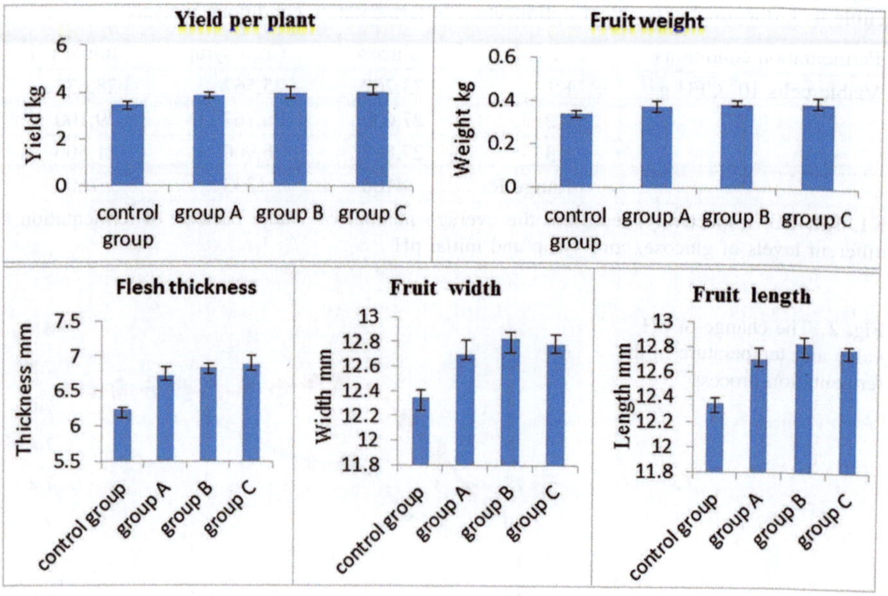

Fig. 4 Effect of different content of fertilizer on fruit quality

Fig. 5 Effect of different content of fertilizer on fruit quality

4 Conclusion

In this study, the cassava residue was identified as the solid fermentation medium of *B. megaterium* by the culture medium optimization experiment. Determined cassava residue as substrate, 5% glucose, 10% corn syrup, initial pH 6.5–7.0, 40% initial moisture, 37 °C was the finally optimum condition of solid fermentation of

Table 5 Effect of fertilizer on pepper quality improvement

Catalogue	Control group	Experience group	Lift %
Fruit weight (g)	0.34 ± 0.03	0.39 ± 0.04 ns	14.71
Flesh thickness (mm)	6.23 ± 0.11	6.82 ± 0.47*	9.47
Yield per plant (kg)	3.48 ± 0.15	3.94 ± 0.26*	13.22
Fruit length (cm)	12.35 ± 0.19	12.77 ± 0.76 ns	3.40
Fruit diameter (cm)	9.84 ± 0.63	12.39 ± 0.49**	25.91
Fruit type index	1.26 ± 0.06	1.01 ± 0.12**	

**, *, ns, respectively, represent the significant differences at $P < 0.01$, $P < 0.05$ and non-significant differences compared to the control group

B. megaterium through single factor experiment and orthogonal experiment. The pepper field experiment shows that the bacterial fertilizer improved pepper yield per plant, fruit weight, flesh thickness, fruit length, fruit width by 14.71, 9.47, 13.22, 3.40, 25.91% respectively and has a good application value.

This article uses different organic as raw materials to develop solid fermentation of *B. megaterium* technology, The viable content of developed microbial agents reached 30.9×10^8 CFU g^{-1}, and validate the effect of fertilizer products through the field to further. The microbial fertilizer developed in this article have good market prospect cause it will significant improved fertilizer efficiency and crop yield and quality [10].

Acknowledgements The work was financially supported by Shandong Provincial Key Research and Development Program (No. 2015ZDZX05001, 2015ZDXX0502B01, 2015ZDXX0403B01, 2016GGX107003), National Natural Science Foundation (No. 31501396), Shandong Provincial Natural Science Foundation (No. ZR2012CQ027).

References

1. Ahmed SA (2008) Optimization of production and extraction parameters of *Bacillus megaterium* levansucrase using solid-state fermentation. J Appl Sci Res 4(10):1199–1204
2. Liu H, Gao S, Zhou Y et al (2014) Optimization of solid state fermentation conditions and culture medium of *Bacillus coagulans*. Cereal Feed Ind
3. Alkotaini B, Koo H, Kim BS (2016) Production of polyhydroxyalkanoates by batch and fed-batch cultivations of *Bacillus megaterium* from acid-treated red algae. Korean J Chem Eng 33(5):1–5
4. Alkotaini B (2015) Potential of *Bacillus megaterium* for production of polyhydroxyalkanoates using the red algae *Gelidium amansii*. Biotechnol Bioprocess Eng 20(5):856–860
5. Hölker U, Lenz J (2005) Solid-state fermentation—are there any biotechnological advantages? Curr Opin Microbiol 8(3):301–306
6. Kanjanachumpol P, Kulpreecha S, Tolieng V et al (2013) Enhancing polyhydroxybutyrate production from high cell density fed-batch fermentation of *Bacillus megaterium* BA-019. Bioprocess Biosyst Eng 36(10):1463–1474
7. Ogbo FC (2010) Conversion of cassava wastes for biofertilizer production using phosphate solubilizing fungi. Bioresour Technol 101(11):4120–4124

8. Casillas-Buenrostro RM, Heredia NL, Benesh DAL et al (2012) Efficacy of 3m™ petrifilm™ aerobic count plates for enumerating *Bacillus sporothermodurans*, and *Geobacillus stearothermophilus*, in UHT milk. Int Dairy J 25(2):147–149
9. Cavalcante ÍHL, Cavalcante LF, Dos Santos GD et al (2012) Impact of biofertilizers on mineral status and fruit quality of yellow passion fruit in brazil. Commun Soil Sci Plant Anal 43(15):2027–2042
10. Cheng L (2013) Short-term effects of organic and inorganic fertilizers on soil microbial community structure and function. Biol Fertil Soils 49(6):723–733

Comparative Quantitative Analysis of Probiotic Bacteria During Puer Tea Pile Fermentation

Shuang Li, Zhongyuan Li, Cuixia Feng and Tongcun Zhang

1 Introduction

Puer tea (puer shucha) is a very popular post-fermented tea in China because of the distinctive color, savoury flavours and aromas [1]. Different to non-fermentation tea, puer tea chooses large leafs of *Camellia sinensis* as raw material, and has a special manufacturing process called "pile fermentation", which is a solid-state fermentation (SSF) with turned over once several days until the color of the tea was changed into rufous [1]. The pH of tea pile is acidic and the center temperature could reach to 60 °C [2]. Complex biological transformations are implemented by a symbiosis of many fermenting microbes (including fungi, yeast and bacteria) and their kinds of extracellular enzymes, and ultimately the unique flavours and aromas of puer tea are formed at the end of pile fermentation [1].

Modern clinical medicine has shown that puer tea displays a variety of health care functions for human body, such as lowering blood pressure [3], preventing cardiovascular disease [4] and moderating the risk of cancer [5]. These benefits are not only in relation to the secondary metabolites themselves in puer tea, but also associated with these fermenting microbes. The specific identity of the microbial population present in puer tea has been the focus on attention, but to date, the majority of studies have relied on culture-based analysis. Previous study have shown fungi, yeast and bacteria were all found in puer tea, fungi *Candida* and *Aspergillus niger* are the dominant microbe in fermentation process [6, 7].

Shuang Li and Zhongyuan Li: Co-first authors.

S. Li · Z. Li · C. Feng · T. Zhang (✉)
College of Biotechnology, Tianjin University of Science and Technology, Tianjin 300457, People's Republic of China
e-mail: lizhongyuan@tust.edu.cn

H. Liu et al. (eds.), *Advances in Applied Biotechnology*, Lecture Notes in Electrical Engineering 444, https://doi.org/10.1007/978-981-10-4801-2_34

However, the culture-based analysis is limited, because only about 1% of the environmental microbes can be uncultured and reliance on the phenotypic traits could lead to misidentification [8, 9]. Recently, a preliminary metagenomic study of puer tea reveals that bacteria are the dominant microbes in pile fermentation [2]. Especially, some probiotic bacteria are also found in puer tea pile fermentation process [10, 11]. Probiotic bacteria are reported to confer many health benefits such as preventing food allergy, gut inflammation and immunomodulation, and also potentially provide anticarcinogenic activity, which plays important role in many fermented food like kimchi [12] and cheese [13]. However, the knowledge of the species, amount and function analysis of probiotic bacteria in puer tea fermentation is limited. To gain a more comprehensive insight into the probiotic bacteria in puer tea fermentation, in this study, specific real-time quantitative PCR assays (qPCR) were developed to accurately determine the distribution and population sizes of the main probiotic bacteria including *Enterococcus* spp., *Bacillus* spp., *Lactococcus* spp., *Bifidobacterium* spp. and *Lactobacillus* group, in addition, the change patterns of these probiotic bacteria during puer tea pile fermentation were also investigated.

2 Materials and Methods

2.1 Puer Tea Sample Collection and Genomic DNA Extraction

Puer tea samples used in this study were provided by Xinghai Tea Factory in Yunnan province, China. At this factory, puer leaf samples were turned over once a week in pile fermentation process to ensure the homogeneity of the fermented tea. About 100 g puer tea samples of 0, 15, 30, 45 fermentation days were separately obtained from the same tea pile, all samples were frozen in liquid nitrogen within 5 min, transported to the laboratory on dry ice and then stored at −70 °C until used. To measure the pH of different tea samples, 1 g of each sample was firstly suspended in 9 ml distilled water and then mixed continuously for 30 min at room temperature (about 25 °C) [2], and pH of the supernatant for each tea sample was measured using a glass pH electrode (Starter 2000, Ohaus, USA). To normalize for different tea samples, 1 mg DNA of plasmid pEGFP-1 harboring enhanced green fluorescent protein (EGFP) encoding gene (*egfp*) which is used as reference gene were firstly added to each tea sample (5 g) and grinded completely with tea by liquid nitrogen before DNA extraction. To confirm the absence of *egfp* in puer tea, *egfp* was amplified by PCR using the gene specific primers EGFP-F/EGFP-R (Table 1) and genomic DNA of puer tea without adding plasmid pEGFP-1. The grinded tea samples were washed by 20 ml washing buffer according to Lyu et al. [2], and the supernatant was incubated for 5 min in 65 °C, and the strain deposits were collected by centrifugation at 6000×*g* for 10 min, and genomic DNA of tea were extracted following a protocol specific for high-molecular-weight environmental DNA [14]. The genomic DNA was analyzed by 1% (w/v) agarose gel electrophoresis and the purity of DNA were determined from the absorbance ratios of A_{260}/A_{280} by Ultrospec 2100 pro UV/visible spectrophotometer (Amersham Biosciences, Uppsala, Sweden).

2.2 Primers Design and qPCR Condition

The specific qPCR primers of universal bacteria and specific-genus qPCR primers of *Bifidobacterium* spp., *Enterococcus* spp., *Bacillus* spp., *Lactococcus* spp. and *Lactobacillus* group (including *Leuconostoc* spp., *Pediococcus* spp., *Aerococcus* spp. and *Weissella* spp.) were summarized in Table 1. The specific qPCR primers for gene *egfp* were designed by the Primer3 program [15]. Strains *Bifidobacterium bifidum*, *Enterococcus faecium* A1, *Bacillus subtillis*, *Lactococcus lactis* NZ9000 and *Lactobacillus plantarum* CGMCC No. 8198 obtained in our laboratory were used as the reference strain for each specific-genus qPCR primer, and *Escherichia coli* harboring pEGFP-1 was used as the reference strain for *egfp*. The specificity of each qPCR primer was further verified by product sequencing and melting curve analysis. To avoid the mutual contamination with other primers, each pair primer was assayed with the other genus reference strains by PCR method and analyzed by agarose gel electrophoresis. The optimal qPCR reaction mixture consisted of 10 μl SYBR Green qPCR mastermix (DBI Bioscience, Shanghai, China), 1 μl template, 0.4 μl 50 × ROX Reference Dye, 0.5 μl PCR forward/reverse primer (10 $\mu mol\ l^{-1}$), and 7.6 μl double distilled H_2O using ABI 7500 Fast Real-Time PCR System (Applied Biosystems, Carlsbad, CA), and the optimal qPCR condition for each qPCR primer was 95 °C for 2 min, followed by 40 cycles of 95 °C for 10 s, 60 °C for 30 s, and 72 °C for 30 s.

2.3 Standard Curves Generation and Estimation of the Copy Numbers of Universal Bacteria and Probiotic Bacteria

The PCR products of each primer with the responding reference strain were purified by DNA product purification kit (TIANGEN Biotech, Beijing, China), ligated with PMD19-T vector (TaKaRa, Dalian, China) and sequenced by Genewiz (Beijing, China), the correct recombinant plasmids were used as the standard plasmid for the following standard curves generation. The standard plasmids for each qPCR primer were extracted by Plasmid Mini Kit (Omega Bio-tek, USA) and quantified by A_{260} measurements. 10-fold dilution series of about 10^1 to 10^7 copies for each target standard plasmid DNA were applied for qPCR analysis. The copy numbers of *Bifidobacterium* spp., *Enterococcus* spp., *Lactobacillus* group, *Bacillus* spp., *Lactococcus* spp. and universal bacteria during tea fermentation were quantified by

Table 1 qPCR primers for Universal bacteria and probiotic bacteria used in this study

Target	Name of primers	Sequences for primers (5′-3′)	Product size (bp)	Efficiency (%)	Reference
Universal bacteria	Bacteria-F	CCTACGGGAGGCAGCAG	193	104.94	[18]
	Bacteria-R	ATTACCGCGGCTGCTGG			
Enterococcus spp.	Ent-F	CCCTTATTGTTAGTTGCCATCATT	144	101.83	[19]
	Ent-R	ACTCGTTGTACTTCCCATTGT			
Lactobacillus group[a]	Lacb-F	AGCAGTAGGGAATCTTCCA	431	94.28	[20, 21]
	Lacb-R	CACCGCTACACATGGAG			
Lactococcus spp.	Lacc-F	GCGGCGTGGCTAATACATGC	305	95.22	[22]
	Lacc-R	CTGCTGCGTCCCGTAGGAGT			
Bacillus spp.	Bac-F	ACGCCGTAAACGATGAGT	424	96.89	[23]
	Bac-R	GTGTGTAGCCCAGGTCATAA			
EGFP	EGFP-F	ACAAGACCCGCGCCGAGGTGAA	253	94.83	This study
	EGFP-R	TCGCCGATGGGGGTGTTCTGCT			

[a] *Lactobacillus* group including *Lactobacillus*, *Leuconostoc*, *Pediococcus*, *Aerococcus* and *Weissella* but not *Enterococcus* or *Streptococcus*

qPCR with each target qPCR primer. The PCR condition and reaction mixture were the same as mentioned above. Fluorescent products were detected at the last step of each cycle. Each sample was assayed in triplicate. The populations of universal bacteria and other genera at different time points were calculated by threshold cycle values based on each standard curves after normalization against gene *egfp*.

2.4 Statistical Analysis

Correlations between/among pH and amount of universal bacteria and *Enterococcus* spp., *Bacillus* spp., *Lactococcus* spp., *Bifidobacterium* spp. and *Lactobacillus* group were tested with Pearson's correlation analysis using the SAS software package, version 8.01 (SAS Institute Inc., Cary, NC). Data were considered significantly correlated for P values of <0.05 and very significantly correlated for P values of <0.01.

3 Results and Discussion

In order to investigate the species and abundance of probiotic bacteria, qPCR is a very precise method which could detect not only the microbe with very low amount but also those uncultured microbe [16]. The template DNA with high quality and concentration is an important factor for the accuracy and validity of qPCR. In this study, puer tea were firstly washed by the washing buffer to reduce the interference of impurity substance in tea for genomic DNA extraction, and the results showed these genomic DNA of puer tea at 0, 15, 30 and 45 days all have high quality with A_{260}/A_{280} ratio of approximately 1.8 (Fig. 1a), and the DNA concentrations of them were 660, 990, 2250, and 2758 μg ml^{-1}, respectively. According to Fig. 1f, Gene *egfp* was verified as an exogenous gene to puer tea, it was added before genomic DNA extraction and used as the reference gene for quantitative analysis, which could usefully normalize the genomic DNA extraction efficiency and reduce technical variations between each tea sample of four time points, this method was also mentioned in our previous study [17]. The high specificity of qPCR primers for each genus and *egfp* was also essential for qPCR assay. Agarose gel electrophoresis of the PCR products in Fig. 1 showed that only one single DNA band for each corresponding genus's reference strain with the expected lengths (Table 1), and no DNA bands for each primer with any other reference strains as DNA template (Fig. 1). The melt curves analysis of these primers showed their high specificity without interference (data not shown) and DNA product sequences were verified belong to the responding genus or *egfp*. With the standard plasmid as template, standard curve equation of cycle threshold values versus plasmid DNA concentrations for each genus primer were generated as shown in Table 2, the resulting

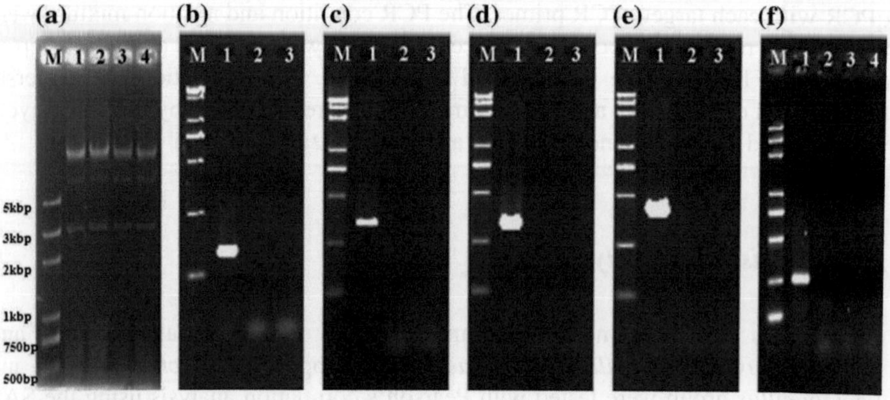

Fig. 1 Verification by agar gel electrophoresis of genomic DNA of puer tea (**a**) at 0 day (*lane 1*), 15 days (*lane 2*), 30 days (*lane 3*), 45 days (*lane 4*) and the specificity of qPCR primers for *Enterococcus* spp. (**b**), *Lactobacillus* spp. (**c**), *Lactococcus* spp. (**d**), *Bacillus* spp. (**e**) and gene *egfp* (**f**). *Lane M* standard molecular size marker; **b–f** *lane 1* PCR product with each reference strain as template; *lane 2* PCR product with sterile water as template; *lane 3* PCR product with other reference strains as template. *Lane 4* PCR product with genome DNA of tea samples without pEGFP-1 as template to evaluate the feasibility of gene *egfp* as the internal reference gene

linear regression analysis suggested the PCR efficiencies ranged from 94.83 to 104.94% (Table 1).

The copy numbers for universal bacteria and other probiotic bacteria were obtained from the threshold cycle values based on standard curves after normalization to that of gene *egfp*. The results showed that the population of universal bacteria was 7.56×10^9–3.45×10^{10} copy g^{-1} in the puer tea fermentation. Among the five genera detected, *Enterococcus* spp. showed 3.43×10^4–6.25×10^5 copy g^{-1}, and *Lactococcus* spp. showed 1.68×10^9–2.78×10^9 copy g^{-1}, which has much more copy numbers than the other two genera *Lactobacillus* spp. (4.26×10^3–5.17×10^4 copy g^{-1}) and *Bacillus* spp. (2.21×10^3–2.31×10^5 copy g^{-1}) according to Fig. 2. These four genera probiotic bacteria all belong to the phylum Firmicutes, which is 20.23% of the total microbe in puer tea fermentation by metagenomic sequencing analysis and played the major role in pile fermentation [2]. Genera *Enterocooccus*, *Lactobacillus*, *Lactococcus*, *Leuconostoc*, *Oenococcus*, *Pediococcus*, *Streptococcus*, *Weissella* and some of *Bacillus* like *Bacillus coagulans* are the main genera belong to probiotic bacteria [18]. These results suggested that probiotic bacteria are involved in puer tea fermentation, which could produce variety of polyphenol oxidase and peroxidase, which could change catechin polyphenols into benzene quinone, and then further changed into polypeptide with red color, and benefit for improving the quality of puer tea and reduce the fermentation time [10]. Only *Bifidobacterium* spp. was not detected during the whole fermentation process, and also there were no reports about the *Bifidobacterium* spp. in puer tea, although *Bifidobacterium* spp. belonged to the phylum Actinobacteria which is up to 30.08% of the microbe

Table 2 Standard curve equation of qPCR primers of universal bacteria and probiotic bacteria (*Enterococcus* spp., *Lactobacillus* spp., *Lactococcus* spp., *Bacillus* spp. and gene *egfp*) used in this study

Strains	Standard curve equation	Variance (R^2)
Bacteria	Y = −3.1072X + 38.145	0.9968
EGFP	Y = −3.523X + 40.147	0.9955
Entq	Y = −3.2788X + 39.549	0.996
lacq	Y = −3.4946X + 39.844	0.995
laccoq	Y = −3.4214X + 44.329	0.997
Bacq	Y = −3.3987X + 41.503	0.9959

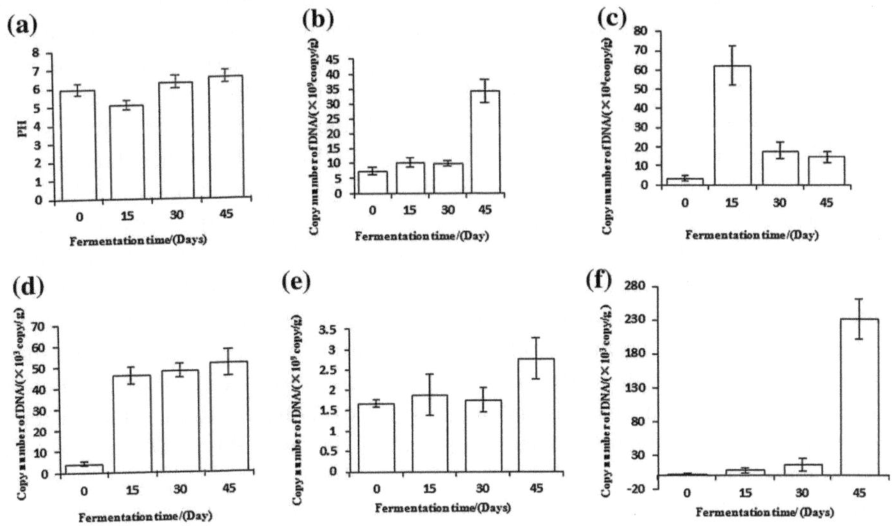

Fig. 2 pH and copy numbers of universal bacteria and bacteria and four genera at different fermentation time. pH (**a**), universal bacteria (**b**), *Enterococcus* spp. (**c**), *Lactobacillus* spp. (**d**), *Lactococcus* spp. (**e**), and *Bacillus* spp. (**f**). *Error bars* represent standard deviations of triplicate measurements

in puer tea fermentation [2]. It might because *Bifidobacterium* spp. as an obligate anaerobe bacterium, could not survive in the SSF because the pile fermentation environment is aerobic.

Among the four genera, family *Bacillaceae* was previously reported to be the main predominant bacteria in the last stage of puer tea fermentation by 16S rRNA clone library analysis [19]. *Lactobacillus* was made up to 30% in the black tea [11]. *Enterococcus* spp. was identified in puer tea by 16S rRNA clone library [19] and *Enterococcus camelliae* sp. nov. was isolated from another fermentation tea from Tailand [20]. But there was no report about *Lactococcus* in puer tea to date. However, based on populations of these four genera in present study, we interestingly found that *Lactococcus* spp. and *Enterococcus* spp. owned larger population (about 100 folds) than *Lactobacillus* spp. and *Bacillus* spp., which suggested *Lactococcus* spp. and *Enterococcus* spp. were the main probiotic bacteria and play important role in this puer tea fermentation. This inconsistent result with previous studies might be due to the use of different puer tea material and production condition.

The pH profiles of puer tea sample were acidic during the tea fermentation (Fig. 2a), the initial pH of the raw material was 6.17, and decreased to pH 5.07 after 15 days' pile fermentation, and increased to pH 6.28 at 30 days and finally reached to pH 6.58 at the end of the pile fermentation. During the four different time points, universal bacteria and four genera also showed different change patterns throughout fermentation process (Fig. 2). Universal bacteria had a trend of slowly, continually and simply increasing from 7.56×10^9 to 3.45×10^{10} copy g^{-1} during 45 days fermentation (Fig. 2b). Different to the universal bacteria, *Bacillus* spp. increased slightly from 2.21×10^3 to 1.55×10^4 copy g^{-1} at the first 30 days, but was found to significantly increase to the maximum in the last 15 days with 2.31×10^5 copy g^{-1}, which are 104 folds of the abundance of initial stage (Fig. 2f) which is consistent to the previous reported [19], The other three genera displayed much more varied change in the 45 days fermentation. *Lactobacillus* spp. increased quickly to the maximum copy numbers (4.62×10^4 copy g^{-1}) in the first 15 days, and then continue increased to 5.17×10^4 copy g^{-1} at the end of fermentation which is still about 12 folds of that of the initial stage (Fig. 2d). *Enterococcus* spp. showed a similar increased trend with *Enterococcus* spp. in the first 15 days by about 18 folds from 3.43×10^4 to 6.25×10^5 copy g^{-1}, but then decreased sharply to 1.49×10^5 copy g^{-1} at the end of the fermentation which is almost the same to that of the initial stage (Fig. 2c). Different to the other genera, *Lactococcus* spp. firstly maintained from 1.68×10^9 to 1.90×10^9 copy g^{-1} in first 30 days, and then increased rapidly to 2.78×10^9 copy g^{-1} (Fig. 2e). Previous study has also shown that the bacteria, yeast, mold were all changeable during the fermentation process [2], which suggests that complex changes are happened in fermenting microbe community during the puer tea fermentation.

Although Pearson's correlation analysis suggested that there are no significantly correlation between pH values and the amount changes of universal bacteria and *Enterococcus* spp., *Bacillus* spp., *Lactococcus* spp. and *Lactobacillus* group (data not shown) during the fermentation, but at the end of the fermentation with pH 6.58, *Enterococcus* spp., *Lactobacillus* spp., *Lactococcus* spp. were all decreased and only *Bacillus* spp. reached to the high abundance, which might be due to *Bacillus* spp. is much adaptable to neutral pH, but *Enterococcus* spp., *Lactobacillus* spp. and *Lactococcus* spp. prefer to the acidic environment. In addition, the correlation among the abundance changes of universal bacteria and *Enterococcus* spp., *Bacillus* spp., *Lactococcus* spp. and *Lactobacillus* group were also not significantly. The synergism and antagonism among the different groups of microbes and even among different genera within the same group the microbe interactions may be another affect factor for complex change patterns.

References

1. Chen Y, Liu B, Chang Y (2010) Bioactivities and sensory evaluation of Pu-erh teas made from three tea leaves in an improved pile fermentation process. J Biosci Bioeng 109:557–563
2. Lyu C, Chen C, Ge F, Liu D, Zhao S, Chen D (2013) A preliminary metagenomic study of puer tea during pile fermentation. J Sci Food Agr 93:165–3174
3. Anderson R, Polansky M (2002) Tea enhances insulin activity. J Agric Food Chem 50:7182–7186
4. Yang D, Hwang L (2006) Study on the conversion of three natural statins from lactone forms to their corresponding hydroxy acid forms and their determination in Pu-Erh tea. J Chromatogr A 30:1–2
5. Hayakawa S, Kimura T, Saeki K, Koyama Y, Aoyagi Y, Noro T, Nakamura Y, Isemura M (2001) Apoptosis-inducing activity of high molecular weight fractions of tea extracts. Biosci Biotech Biochem 65:459–462
6. Wen Q, Liu S (1991) Variation of the microorganism groups during the pile-fermentation of dark green tea. J Tea Sci 11:10–16
7. Zhou H, Li J, Zhao L, Han J, Yang X, Yang W, Wu X (2004) Study on main microbes on quality formation of Yunnan puer tea during pile-fermentation process. J Tea Sci 24:212–218
8. Pace NR (1997) A molecular view of microbial diversity and the biosphere. Science 276: 734–740
9. Raspor P, Goranovic D (2008) Biotechnological applications of acetic acid bacteria. Crit Rev Biotechnol 28:101–124
10. Monhammad FG, Alireza T (2007) Isolation and characterization of polyphenol oxidase and peroxidase-producing *Bacillus* strains from fully fermented tea (*Camellia sinensis*). World J Microbiol Biotechnol 23:1327–1332
11. Klayraung S, Okonogi S (2009) Antibacterial and antioxidant activities of acid and bile resistant strains of *Lactobacillus fermentum* isolated from miang. Braz J Microbiol 40: 757–766
12. Cho K, Math RK, Islam SM, Lim WJ, Hong SY, Kim JM, Yun MG, Cho JJ, Yun H (2009) Novel multiplex PCR for the detection of lactic acid bacteria during kimchi fermentation. Mol Cell Probes 232:90–94
13. Ganesan B, Weimer B, Pinzon J, Dao KN, Rompato G, Brothersen C, McMahon D (2014) Probiotic bacteria survive in Cheddar cheese and modify populations of other lactic acid bacteria. J Appl Microbiol 116:1642–1656
14. Brady SF (2007) Construction of soil environmental DNA cosmid libraries and screening for clones that produce biologically active small molecules. Nat Protoc 2:1297–1305
15. Rozen S, Skaletsky H (2000) Primer3 on the WWW for general users and for biologist programmers. Methods Mol Biol 132:365–386
16. Matijašić B, Obermajer T, Rogelj I (2010) Quantification of *Lactobacillus gasseri*, *Enterococcus faecium* and *Bifidobacterium infantis* in a probiotic OTC drug by real-time PCR. Food Control 21:419–425
17. Li Z, Zhao H, Yang P, Zhao J, Huang H, Xue X, Zhang X, Diao Q, Yao B (2013) Comparative quantitative analysis of gene expression profiles of glycoside hydrolase family 10 xylanases in the sheep rumen during a feeding cycle. Appl Environ Microbiol 79:1212–1220
18. Lubbs DC, Vester BM, Fastinger ND, Swanson KS (2009) Dietary protein concentration affects intestinal microbiota of adult cats: a study using DGGE and qPCR to evaluate differences in microbial populations in the feline gastrointestinal tract. J Anim Physiol Anim Nutr (Berl) 93:113–121
19. Rinttila T, Kassinen A, Malinen E, Krogius L, Palva A (2004) Development of an extensive set of 16S rDNA-targeted primers for quantification of pathogenic and indigenous bacteria in faecal samples by real-time PCR. J Appl Microbiol 97:1166–1177

20. Walter J, Hertel C, Tannock GW, Lis CM, Munro K, Hammes WP (2001) Detection of *Lactobacillus*, *Pediococcus*, *Leuconostoc*, and *Weissella* species in human feces by using group-specific PCR primers and denaturing gradient gel electrophoresis. Appl Environ Microbiol 67:2578–2585
21. Heilig HG, Zoetendal EG, Vaughan EE, Marteau P, Akkermans AD, de Vos WM (2002) Molecular diversity of *Lactobacillus* spp. and other lactic acid bacteria in the human intestine as determined by specific amplification of 16S ribosomal DNA. Appl Environ Microbiol 68:114–123
22. Klijn N, Weerkamp AH, de Vos WM (1995) Detection and characterization of lactose-utilizing *Lactococcus* spp. in natural ecosystems. Appl Environ Microbiol 61:788–792
23. Han GQ, Xiang ZT, Yu B, Chen DW, Qi HW, Mao XB, Chen H, Mao Q, Huang ZQ (2012) Effects of different starch sources on *Bacillus* spp. in intestinal tract and expression of intestinal development related genes of weanling piglets. Mol Biol Rep 39:1869–1876

Prediction of Hot Spots in Dimer Interface of Green Fluorescent Protein

Wenjuan Zhang, Lin Wang, Zhiwei Sun, Bianqiang Zhang, Qiaoqiao Tang and Qiang Gao

1 Introduction

Green fluorescent protein (GFP) is a kind of bioluminescent protein that exists in the bodies of coelenterate such as jellyfish, hydra and coral. GFP protein, which consists of 238 amino acid residues, was first isolated from Aequorea victoria jellyfish in 1962 by Shimomura et al. [1]. GFP can emit visible green fluorescence upon exposure to ultraviolet or blue light. The chromophore of GFP structure is generated by three amino acid residues (glycine, tyrosine and threonine) in its protein primary sequence, and the creation of green fluorescence does not need substrate or cofactor [2]. GFP has been expressed in many species as an ideal reporter, but its usage is limited by sensitivity to temperature and weak fluorescence in some cells [3]. Thus many different mutants of GFP have been available, for example, the enhanced green fluorescent protein (EGFP), which was engineered with two independent residue mutations (F64L, S65T), increased fluorescence well as photostability, and allowed the practical use in mammalian cells [4].

Wild-type GFP has a weak dimerization tendency at higher concentrations, which could lead to artefacts in imaging experiments, and its structure was reported by Yang et al. [5]. This dimer structure sheds light on the neighbouring residue

W. Zhang
Faculty of Fundamental Courses, Tianjin Foreign Studies University, Tianjin 300204, People's Republic of China

L. Wang (✉) · Z. Sun
School of Computer Science and Information Engineering, Tianjin University of Science and Technology, Tianjin 300457, People's Republic of China
e-mail: linwang@tust.edu.cn

B. Zhang · Q. Tang · Q. Gao
Key Laboratory of Industrial Fermentation Microbiology, Ministry of Education, College of Biotechnology, Tianjin University of Science and Technology, Tianjin 300457, People's Republic of China

H. Liu et al. (eds.), *Advances in Applied Biotechnology*, Lecture Notes in Electrical Engineering 444, https://doi.org/10.1007/978-981-10-4801-2_35

interactions. Here we focused on critical residue analysis on dimer interface of wild-type GFP. The study of residues at protein-protein interfaces has shown that only a small portion of all interface residues is actually essential for binding. These residues are termed as hot spots which contribute most binding free energy in protein interfaces [6]. The identification of hot spots in dimer interface of GFP can help to determine site-directed mutagenesis for GFP optimization for stability and fluorescent intensity.

Experimental alanine scanning mutagenesis is a technique used to uncover hot spots by determining the binding free energy change caused by substituting a specific interface residue with alanine. However, this experimental approach requires substantial efforts and is often expensive. Therefore, only a limited number of complexes were conducted to identify hot spots, and deposited in two consortiums, the Alanine Scanning Energetics Database (ASEdb) [7] and the Binding Interface Database (BID) [8].

Based on the existing studies on the characteristics of hot spots, several computational methods have been developed for the prediction of hot spots in protein-protein interface. These methods included energy-based methods and knowledge-based methods [9, 10]. In this article, we adopted several well-established methods to predict hot spots in dimer interface of GFP. According to the interface definition in Tuncbag et al. [11], 17 residues appeared on the dimer interface of GFP. Among them, 10 residues were predicted as hot spots by at least one method.

2 Materials and Methods

The dimer structure of GFP was obtained from the Protein Data Bank (PDB) with PDB ID: 1GFL. The protein was in the shape of a cylinder, comprising 11 strands of beta-sheet with an alpha-helix inside and short helical segments on the ends of the cylinder. The methods that were adopted to predict hot spots in dimer interface of GFP included HotPoint [11], KFC2 [12], PredHS [13] and Robetta [6]. HotPoint, KFC2 and PredHS were knowledge-based methods which try to learn the complex relationship between hot spots and various residue features in training data and predict new hot spots. Among them, HotPoint was an intuitive efficient method to determine computational hot spots based on conservation, solvent accessibility and statistical pairwise residue potentials of the interface residues. KFC2 was a support vector machine-based method to predict hot spots with features of both sequence and structure, and included two hot spot predictors KFC2-A and KFC2-B. PredHS was an effective structure-based hot spot prediction method with focus on using structural neighbourhood properties, and included two predictors PredHS-SVM and PredHS-Ensemble. Robetta used a free energy function to calculate the binding free energy of alanine mutation in a protein-protein complex. The interface residues with estimated binding free energy ($\Delta\Delta G$) $\geq$ 1.0 kcal/mol were predicted as hot spots.

3 Results

3.1 Interface Residues in the Dimer Structure of GFP

A protein interface was defined as a set of amino acids which represents a region that links two protein chains by non-covalent interactions. As to two residues from different chains, if the distance between any two atoms belonging to the two residues, respectively, is less than the sum of their van der Waals radii plus a 0.5 Å tolerance, these two residues are defined as interface residues [11]. According to the interface residue definition, there are 16 interface residues in each chain of GFP, which are illustrated in Fig. 1 in 3D using the PDB file of 1GFL. Notably the interface residues of the two chains overlapped by 15 residues, and in total 17 unique residues were appeared on the dimer interface of GFP. These residues were used as candidates for hot spot prediction.

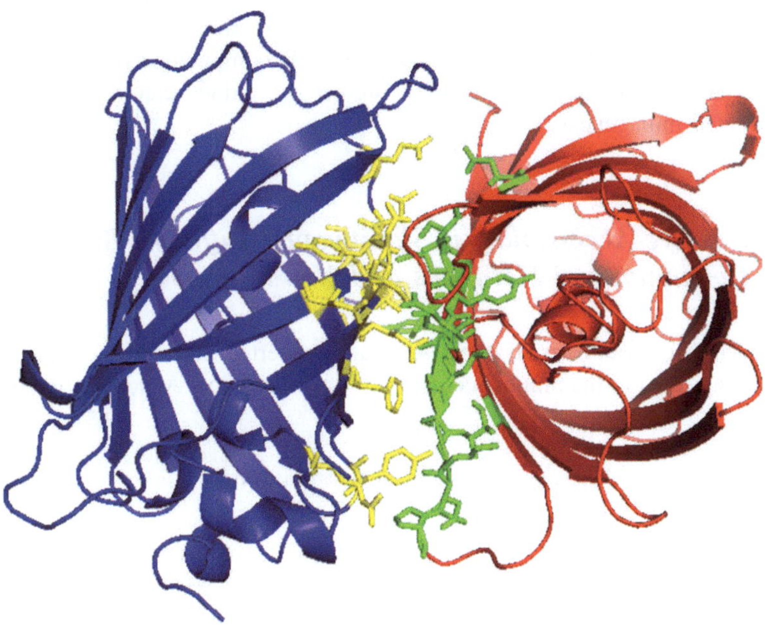

Fig. 1 The protein-protein interface residues in the dimer structure of GFP. The two protein chains are colored by *blue* and *red*, respectively. Their interface residues shown as sticks are colored by *yellow* and *green*, respectively

3.2 Computational Hot Spots of GFP Identified by Well-Established Methods

We adopted well-established hot spot prediction methods to predict hot spots in dimer interface of GFP, i.e. HotPoint, KFC2, PredHS and Robetta. KFC2 contains two predictors KFC2-A and KFC2-B. PredHS contains two predictors PredHS-SVM and PredHS-Ensemble. Three-dimensional structure of 1GFL shows that it is an asymmetric homodimer. We identified computational hot spots in both chains of 1GFL, and the prediction results are shown in Table 1. Among 17 interface residues, 10 residues were identified as computational hot spots by at least one method. Furthermore, we sorted these hot spots according to the number of prediction methods (Table 2). Figure 2 shows the computational hot spots in the dimer interface of GFP. The residues Q204, F223, N146 and A206 were predicted as hot spots by at least four methods. Notably, by examining the structure of dimer GFP, Zacharias et al. identified a hydrophobic patch centred on A206, P221 and F223 of each monomer. Through mutating one or several of these residues to positively charged amino-acid side chains, the interaction between monomers was efficiently removed [14]. These results verified the computational hot spot residues in part, and suggested the remaining predicted residues can be candidates for site-directed mutagenesis.

3.3 Relative Accessibility is Significantly Different Between Computational Hot Spots and Non-hot Spots

The residue accessibilities in dimer state are converted into relative accessibilities by dividing them to maximum accessibility of that residue. As to 10 computational

Table 1 Hot spot prediction by various state-of-the-art methods

Method	Hot spots	Non-hot spots
HotPoint	Y145, Q204, A206, L207, F223	Y39, R73, E142, N144, N146, S147, N149, S208, K209, D210, P211, V219
KFC2-A	N146, S147, Q204, A206, F223	Y39, R73, E142, N144, Y145, N149, L207, S208, K209, D210, P211, V219
KFC2-B	Y145, Q204, L207, F223	Y39, R73, E142, N144, N146, S147, N149, A206, S208, K209, D210, P211, V219
PredHS-SVM	Y39, N146, S147, Q204, A206, F223	R73, E142, N144, Y145, N149, L207, S208, K209, D210, P211, V219
PredHS-Ensemble	Y39, N144, N146, S147, Q204, A206, S208, F223	R73, E142, Y145, N149, L207, K209, D210, P211, V219
Robetta	Y39, N146, Q204, F223	R73, E142, N144, Y145, S147, N149, L207, S208, K209, V219

Table 2 Computational hot spots sorted by the number of prediction methods

Hot spot	Number of methods that predict the residue as a hot spot
Q204	6
F223	6
N146	4
A206	4
Y39	3
S147	3
Y145	2
L207	2
N144	1
S208	1

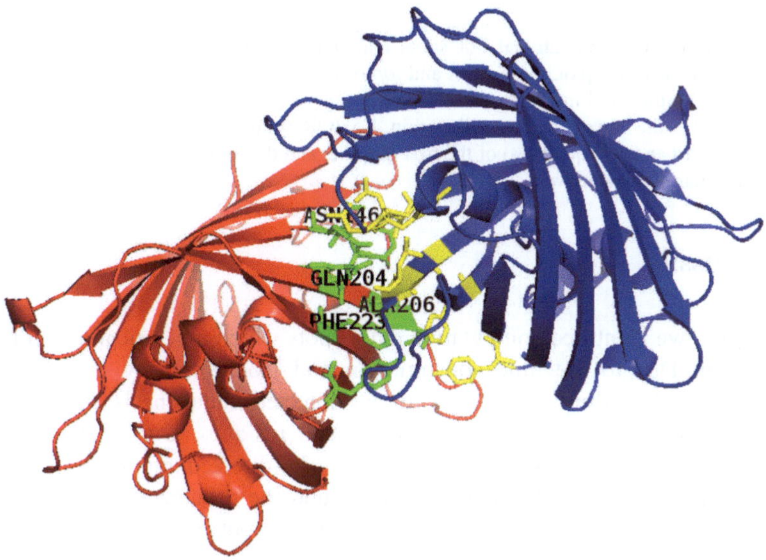

Fig. 2 The computational hot spot residues in the dimer structure of GFP. The two protein chains are colored by *blue* and *red*, respectively. Their computational hot spot residues shown as sticks are colored by *yellow* and *green*, respectively. Furthermore, we labelled the top four computational hot spots (Q204, F223, N146 and A206) based on the number of prediction methods

hot spots of 16 interface residues in chain A, their relative accessibilities are significantly different from those of computational non-hot spots (one-sided Mann–Whitney U test, *P*-value = 1.50×10^{-3}). Figure 3 shows the box plots of relative accessibilities in different residues.

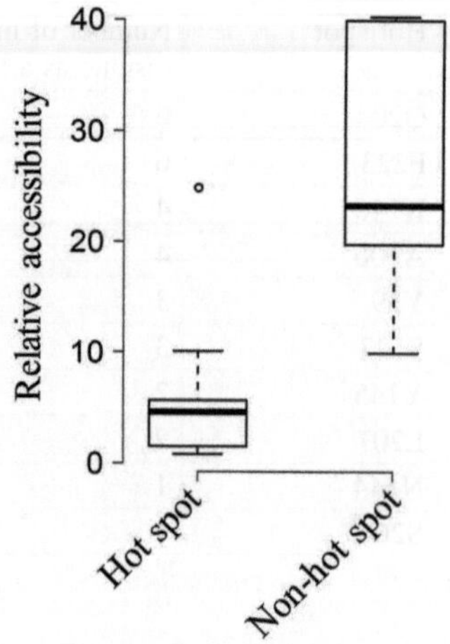

Fig. 3 *Box* plots of computational hot spots and non-hot spots with respect to their relative accessibility. In each *box* plot, the *bottom* and *top* of the *box* represent the first and third quartiles individually with the band inside the *box* being the second quartile. The *lower* and *upper ends* of the whiskers are the lowest datum still within 1.5 IQR (interquartile range, i.e. the difference between the third and first quartiles) of the first quartile, and the highest datum still within 1.5 IQR of the third quartile, respectively

4 Discussion

In this work, we identified computational hot spots which are critical for GFP dimer stabilization. Due to the time consumption and labor intensity in experimental determination of binding free energy for site-directed mutagenesis, there are still a limited number of known site-directed mutated residues. Our prediction results provide candidate residues in the studies of site-directed mutagenesis for GFP optimization for stability. Furthermore, we found the difference of relative accessibility between computational hot spots and non-hot spots is statistically significant.

Acknowledgements This work was sponsored by the Natural Science Foundation of China (31370075 and 61603273), Tianjin Research Program of Application Foundation and Advanced Technology (14JCQNJC00300 and 16JCYBJC18500), Tianjin University of Science and Technology (2014CXLG28), and Tianjin Foreign Studies University (13QN15).

References

1. Shimomura O, Johnson FH, Saiga Y (1962) Extraction, purification and properties of aequorin, a bioluminescent protein from the luminous hydromedusan, *Aequorea*. J Cell Comp Physiol 59:223–239
2. Chalfie M, Tu Y, Euskirchen G et al (1994) Green fluorescent protein as a marker for gene expression. Science 263(5148):802–805

3. Pédelacq JD, Cabantous S, Tran T et al (2006) Engineering and characterization of a superfolder green fluorescent protein. Nat Biotechnol 24(9):79–88
4. Cormack BP, Valdivia RH, Falkow S (1996) FACS-optimized mutants of the green fluorescent protein (GFP). Gene 173(1):33–38
5. Yang F, Moss LG, Phillips GN (1996) The molecular structure of green fluorescent protein. Nat Biotechnol 14(10):1246–1251
6. Kortemme T, Baker D (2002) A simple physical model for binding energy hot spots in protein-protein complexes. Proc Natl Acad Sci 99(22):14116–14121
7. Thorn KS, Bogan AA (2001) ASEdb: a database of alanine mutations and their effects on the free energy of binding in protein interactions. Bioinformatics 17(3):284–285
8. Fischer TB, Arunachalam KV, Bailey D et al (2003) The binding interface database (BID): a compilation of amino acid hot spots in protein interfaces. Bioinformatics 19(11):1453–1454
9. Wang L, Liu ZP, Zhang XS et al (2012) Prediction of hot spots in protein interfaces using a random forest model with hybrid features. Protein Eng Des Sel 25(3):119–126
10. Wang L, Zhang W, Gao Q et al (2014) Prediction of hot spots in protein interfaces using extreme learning machines with the information of spatial neighbour residues. IET Syst Biol 8 (4):184–190
11. Tuncbag N, Keskin O, Gursoy A (2010) HotPoint: hot spot prediction server for protein interfaces. Nucl Acids Res 38(Web Server issue):W402–W406
12. Zhu X, Mitchell J (2011) KFC2: a knowledge-based hot spot prediction method based on interface solvation, atomic density, and plasticity features. Proteins 79(9):2671–2683
13. Deng L, Zhang QC, Chen Z et al (2014) PredHS: a web server for predicting protein-protein interaction hot spots by using structural neighborhood properties. Nucl Acids Res 42(Web Server issue):W290–W295
14. Zacharias DA, Violin JD, Newton AC et al (2002) Partitioning of lipid-modified monomeric GFPs into membrane microdomains of live cells. Science 296(5569):913–916

Regulation of NAD (H) Pool by Overexpression of Nicotinic Acid Phosphoribosyltransferase for AD (D) Production in *Mycobacterium neoaurum*

Liqiu Su, Yanbing Shen, Tian Gao, Le Cui, Jianmei Luo and Min Wang

1 Introduction

Androst-4-ene-3,17-dione (AD) and androst-1 4-diene-3 17-dione (ADD) are two versatile biological intermediates, kinds of strains have been reported capable of converting phytosterol (PS) to AD (D) via the side chain cleavage [1]. *Mycobacterium* is one of the most efficient AD (D) producer still [2]. It was reported by many studies that manipulation of the enzyme activities which included in steroid nucleus oxidation and the side chain cleavage pathways could enhance the AD (D) production, such as the deletion of 3-ketosteroid-9α-hydroxylase (Ksh) and 3-ketosteroid-1-dehydrogenase (KstD), or the overexpression of cholesterol oxidase and 3β-hydroxysteroid dehydrogenases (3β-HSDs) [3]. But little known the cofactor engineering could be used in *Mycobacterium* biocatalysis.

It has been known that cofactors (NAD (H), NADP (H), FAD (H_2), ATP, coenzyme A and its derivatives) are very important in the production of fermentation products. In the PS biotransformation process, it was speculated that side chain degradation of sitosterol resulted in the formation of propionyl-SCoA, $FADH_2$, NADH, and acetic acid. The propionates and acetates are utilized through the tricarboxylic acid cycle (TCA). The complete breakdown of sitosterol to AD (D) gave 21 molecules of NADH and 10 molecules of $FADH_2$ [4, 5]. It is crucially important for PS biotransformation that NADH can be oxidized to NAD^+.

Nicotinic acid phosphoribosyltransferase (NAPRTase), encoded by the *pncB* gene, is rate limiting in the salvage synthesis of NAD (H) from nicotinic acid (NA) [6]. Helena IMB [7] proved that the salvage pathway to synthesise NAD^+ was included in

L. Su · Y. Shen · T. Gao · L. Cui · J. Luo · M. Wang (✉)
Key Laboratory of Industrial Fermentation Microbiology, Ministry of Education, Tianjin Key Laboratory of Industrial Microbiology, College of Biotechnology, Tianjin University of Science and Technology, TEDA, No. 29, 13th Avenue, Tianjin 300457, People's Republic of China
e-mail: minw@tust.edu.cn

H. Liu et al. (eds.), *Advances in Applied Biotechnology*, Lecture Notes in Electrical Engineering 444, https://doi.org/10.1007/978-981-10-4801-2_36

Mycobacterium tuberculosis, and NAPRTase played an important role in cofactor salvage pathway. NAPRTase uses NA and α-D-5-phosphoribosyl-1-pyrophosphate (PRPP) with ATP hydrolysis to form nicotinate mononucleotide (NAMN), which is a precursor of NAD^+. Liang LY [6, 8] had reported that overexpression of the *pncB* gene in *Escherichia coli* (*E. coli*) could increase the NAD (H) levels and the NAD^+/NADH ratio. By overexpression of NAPRTase in *E. coli* NZN111, the NAD^+ concentration increased 6.2-fold [6]. After overexpression of the *pncB* and *nadE* (encoding NAD synthetase) in *E. coli*, 7-fold and 2-fold increased intracellular concentrations of NAD (H) and NADP (H), respectively [9].

In this work, to increase the availability of intracellular NAD^+ for AD (D) production, we constructed a recombinant strain, *MNR* M3-pMV306-*pncB*, and investigated the influence of modification of NAD (H) synthesis pathway on PS degradation process. The data obtained will improve the current understanding on the mechanism of the connection between intracellular cofactor levels with AD (D) biosynthesis, and will provide an effective approach for biocatalysis of other compounds in steroids biotransformation.

2 Materials and Methods

2.1 Strains and Plasmids Construction

Mycobacterium neoaurum TCCC 11028 M3 (*MNR* M3) was obtained from Tianjin University of Science and Technology Culture Collection Center (TCCC), Tianjin, China. *MNR* M3 was used as the template for cloning the *pncB* gene. The plasmids pMV261 and pMV306 were used to express the *pncB* gene. The sequences for the primers were as follows, which were designed based on the published sequence of the *pncB* gene (Gene ID: 17916325):

Sense: 5-CGCGGATCCGAGCACCGCGCTGCTGA-3.

Antisense: 5-CCCAAGCTTTCACCTGGCGGAAACC-3.

The underlined characters indicate the sites of *Bam*H I and *Hin*d III, respectively. Mycobacterial replicating vector pMV261 and integration vector pMV306 carrying the *Mycobacterium bovis* BCG Phsp60 promoter and kanamycin (*kan*) resistance were used. The plasmids pMV261 and pMV306 were the common plasmids used in Mycobacteria for gene expression [10]. The purified DNA was digested with *Bam*H I and *Hin*d III and ligated into the *Bam*H I–*Hin*d III double-digested pMV261 to construct pMV261-*pncB*. Then digested the vector pMV261-*pncB* by *Xba* I and *Hin*d III (with the Phsp60 promoter included) and inserted into the vector pMV306, with the vector pMV306-*pncB* generated also. The pMV306-*pncB* vector was imported into *MNR* M3 by electroporation. The transformants were screened by LB medium agar plate containing kan. The selected mutants were designated as *MNR* M3-pMV306-*pncB*, for further characterization.

2.2 *Media and Culture Conditions*

Substrate phytosterol (PS; 98.4% purity) was obtained from COFCO Tech Bioengineering Co. Ltd. (Tianjin). The standard of AD and ADD were purchased from Sigma Aldrich Co. (USA). All chemical solvents and salts used were of analytical grade or higher. The cultivation and bioconversion of microorganisms, as well as the preparation and analysis of transformation products, were performed following the procedures described by Shen et al. [1]. The PS and nicotinic acid (NA) concentration added in the fermentation system were 3 g/L and 10 mg/L, respectively.

2.3 *Analytical Methods and Enzyme Assays*

In the PS free medium, the growth of cell measured by optical density (OD), but it's difficult to measure the cell growth in the culture broth with the PS by this method. In this study, cell growth measurement in medium with PS was done as described by Paul RM [11].

To determine enzyme activities, cells were harvested by centrifugation at 10,000×*g* for 10 min at 4 °C, and washed with the buffer used in the subsequent enzyme assays. The suspended cells were disrupted by sonification on ice. Cell debris was removed by centrifugation at 10,000×*g* for 15 min at 4 °C. The crude cell extracts were used for the determination of enzyme activity. NAPRTase activity was measured by following the formation of ADP. The reaction mixture contained 4.5 mM ATP, 1.5 mM PRPP, 45 mM potassium phosphate, 45 mM Tris–HCl, 15 mM $MgCl_2$, 0.15 mM nicotinate, 4 mM PEP, 0.2 mM NADH, 13 U LDH, and 13 U pyruvate kinase (PYK) at pH 7.5. The reaction was started by the addition of nicotinate. The concomitant consumption of NADH (ε340 nm = 6.22 mM^{-1} cm^{-1}) was monitored spectrophotometrically. A unit of NAPRTase activity was defined as the amount of enzyme that catalyzed the oxidation of 1 μmol NADH to NAD^+ per minute. Protein concentrations were determined using the Bradford method. The intracellular concentrations of NADH and NAD^+ were assayed using a cycling method [12].

3 Results and Discussion

3.1 *Cloning and Overexpression of the Gene Encoding NAPRTase*

The complete sequence of the *pncB* gene from *MNR* M3 was obtained by PCR. The *pncB* gene sequence from *M. neoaurum* strain VKM Ac-1815D was used as a basis

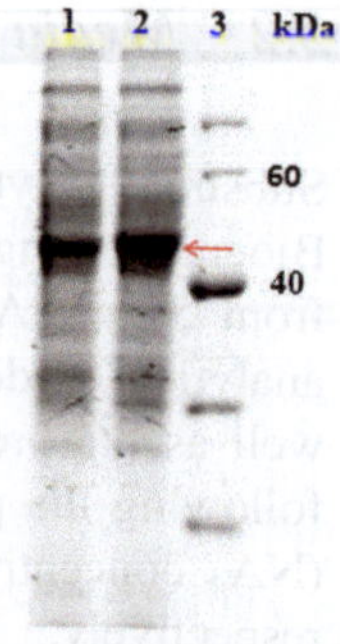

Fig. 1 SDS-PAGE analysis of cell extract of different *MNR* M3 strains. lane 1: *MNR* M3 with 10 mg/L NA; lane 2: *MNR* M3 pMV306-*pncB* with 10 mg/L NA; lane 3: molecular weight standards

to design the primers. The resulting fragment containing 1320 bp nucleotides is a complete open reading frame. The obtained sequences were BLAST searched in the NCBI. Results showed 100% query cover and 100% similarity to the *pncB* gene from *M. neoaurum* VKM Ac-1815D.

The recombinant plasmid pMV306-*pncB* was identified by single endonuclease digestion and double endonuclease digestion. The sequence of the cloned *pncB* gene was analyzed and the results indicated that the sequence was consistent with the reported data. The pMV306-*pncB* vector was imported into *MNR* M3 by electroporation. SDS-PAGE revealed the existence in *MNR* M3 pMV306-*pncB* of a most abundant protein with a molecular weight of 46 kDa that was not present in large amounts in *MNR* M3 (Fig. 1).

3.2 Effects of NAPRTase Overexpression on NAPRTase Activity and Cell Growth

The specific activity of NAPRTase was measured when cells entered end stage of exponential phase. As shown in Fig. 2, the NAPRTase activity in *MNR* M3 pMV306-*pncB* (0.024 U/mg) was 2-fold higher than in the parent strain (0.008 U/mg). The results demonstrated that NAPRTase was successfully overexpressed in *MNR* M3 pMV306-*pncB*.

To investigate the effect of NAPRTase overexpression on cell growth, the recombinant strain and the parent strain were cultured at the same conditions. As shown in Fig. 3, the recombinant strain grew to a higher OD_{600} as compared to the parent strain, indicating that the cell growth of *MNR* M3 was promoted due to the overexpression of NAPRTase. At the first 48 h, the recombinant strain entered the exponential phase while the control strain entered it later. The growth rates for both strains were then increased. At the end time of exponential phase of the control strain was still later compared to the recombinant strain and this trend lasted to the end.

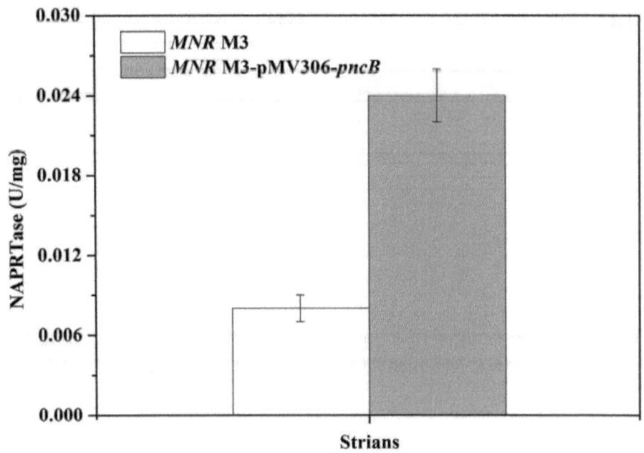

Fig. 2 The specific activity of NAPRTase in *MNR* M3 and *MNR* M3 pMV306-*pncB*

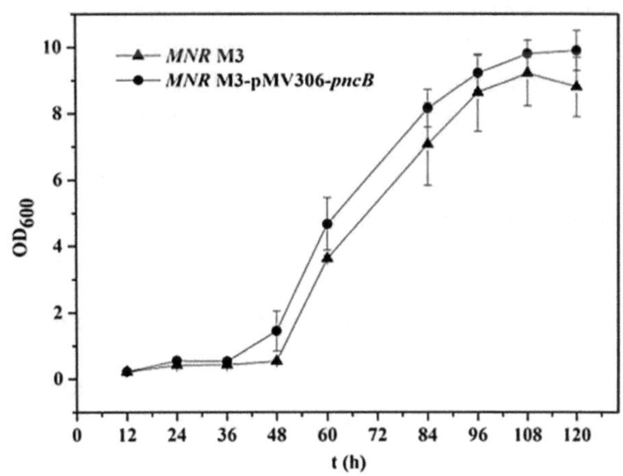

Fig. 3 The growth *curves* of *MNR* M3 and *MNR* M3 pMV306-*pncB*. *MNR* M3 (*triangle*), *MNR* M3-pMV306-*pncB* (*circle*)

3.3 *Effects of NAPRTase Overexpression on AD (D) Production in Fermentation Process*

3.3.1 Effect of the NAPRTase Overexpression on Intracellular Concentrations of NADH and NAD^+

To investigate the effect of NAPRTase overexpression on the total intracellular NAD (H) pool, time-courses of concentrations of intracellular NAD (H) were assayed. As shown in Fig. 4a, b, for both the parent and recombinant strains, the intracellular contents of NADH and NAD^+ were constantly changing with the fermentation time. The cofactor concentrations changing trend in both strains were the same. This was similar to the report of Ji [13]. However, overexpression of NAPRTase in *MNR* M3 led to higher levels of NAD^+ and NADH pool when compared to the parent strain. The highest NAD (H) concentration (27.31 μmol g^{-1}) was got in recombinant

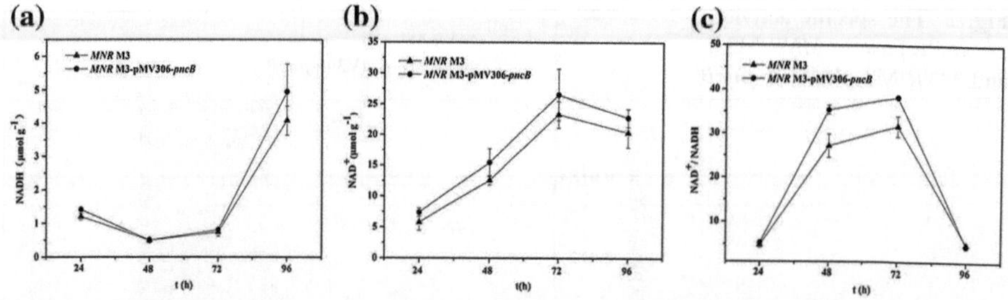

Fig. 4 Time-course of concentrations of intracellular NADH (**a**), NAD^+ (**b**), and NAD^+/NADH ratio (**c**) by *MNR* M3 and *MNR* M3-pMV306-*pncB*. *MNR* M3 (*triangle*), *MNR* M3-pMV306-*pncB* (*circle*)

strain, 13.8% higher than that in the control strain after 72 h. The difference of the NAD^+/NADH ratio between the parent and the recombinant strain was also observed (Fig. 4c), in the whole stages of fermentation, the higher NAD^+/NADH ratio was got in recombinant strain. The highest ratio in *MNR* M3-pMV306-*pncB* was 22.2% higher than that in *MNR M3.* This indicated that the redox status in *MNR* M3 was disturbed. The cofactor changes showed in *MNR* M3 was similar to that in engineered *E. coli* [6], in which the NAD (H) pool size and the NAD^+/NADH ratio were greatly increased by the overexpression of NAPRTase.

3.3.2 Effects of NAPRTase Overexpression on AD (D) Production

The AD (D) production of the parent and the recombinant strains were monitored (Fig. 5a). It could be observed that the AD (D) production was strongly influenced by the overexpressed NAPRTase. After 72 h of fermentation, a final AD (D) production of 94.9% was achieved, 9.6% higher than that without overexpression of NAPRTase which was 86.5%. What's more, the producing rate [mg/(L h)] of AD (D) in *MNR* M3-pMV306-*pncB* was also higher in the whole stages in PS conversion process than the parent strain. It was analysed that the process from β-sitosterol to AD uses NAD^+ as main cofactor with NADH are generated [4]. The results showed in this study, after overexpression of NAPRTase in *MNR* M3, more NAD^+ generated, which was helpful for the NAD^+-dependent PS degradation pathway. Therefore, the AD (D) yield was improved.

From the studies of Liang [6, 8], Ji [13] and Sun [14], manipulating the levels of intracellular cofactors had strongly effects on the redirection of glucose metabolism and the cell growth, with a high efficiency of producing the NAD^+-dependent microbial metabolite achieved. In the present study, the glucose consumption and cell growth were analysed as well.

As shown in Fig. 5b, c, in the fermentation process with PS, the overexpression of NAPRTase could accelerate the glucose consumption and cell growth to a certain extent, which was similar to the growth in PS free system (Fig. 3). It was known that the glucose degraded through glycolysis and TCA pathway, which were NAD^+-

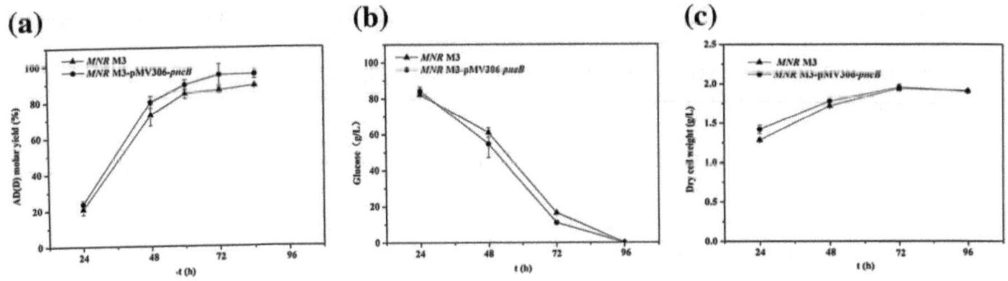

Fig. 5 Time-course of AD (D) molar yield (**a**), glucose (**b**), DCW (**c**) by *MNR* M3 and *MNR* M3-pMV306-*pncB*. *MNR* M3 (*triangle*), *MNR* M3-pMV306-*pncB* (*circle*)

dependent pathways. So the processes of glucose consumption could be enhanced by the overexpression of NAPRTase, with more NADH produced. The excess NADH would be channeled to the respiratory chain to form ATP, which was essential for cell growth (Fig. 5c). As the microorganism growth and PS transformation were completed simultaneously during transformation culture [15], the enhancement of cell growth had positive effect on PS degradation and AD (D) yield.

4 Conclusions

In conclusion, the overexpression of NAPRTase in *MNR* M3 is a powerful engineering tool to enhance the metabolic flux to desired metabolites. The present study showed that with no modification of AD synthesis pathway and only modification of NAD (H) synthesis pathway, overexpression of NAPRTase in *MNR* M3 led to the increase of the total NAD (H) levels and the NAD^+/NADH ratio. As expected, we observed that the yield of AD (D) was enhanced in *MNR* M3-pMV306-*pncB* accordingly. The idea developed in this study could be applied to the other similar industrial biotechnological process to achieve high product concentration.

Acknowledgments This work was supported by the National Natural Science Foundation of China (21276196, 21406167 and 21306138), Key Project of Chinese Ministry of Education (213004A), and Tianjin Programs for Science and Technology Development (15ZCZDSY00510).

References

1. Shen YB, Wang M, Li HN et al (2012) Influence of hydroxypropyl-b-cyclodextrin on phytosterol biotransformation by different strains of *Mycobacterium neoaurum*. J Ind Microbiol Biotechnol 39:1253–1259
2. Shao ML, Zhang X, Rao ZM et al (2015) Enhanced production of androst-1,4-diene-3,17-dione by *Mycobacterium neoaurum* JC-12 using three-stage fermentation strategy. PLoS ONE 10(9):1–13

3. Alok M, James G (2008) Androstenedione production by biotransformation of phytosterols. Biores Technol 99:6725–6737
4. Szentirmai A (1990) Microbial physiology of side chain degradation of sterols. J Ind Microbiol Biotechnol 6(2):101–115
5. Su LQ, Shen YB, Gao T et al (2017) Improvement of AD biosynthesis response to enhanced oxygen transfer by oxygen vectors in *Mycobacterium neoaurum* TCCC 11979. Appl Biochem Biotechnol. doi:10.1007/s12010-017-2418-3
6. Liang LY, Liu RM, Wang GM et al (2012) Regulation of NAD(H) pool and NADH/NAD$^+$ ratio by overexpression of nicotinic acid phosphoribosyltransferase for succinic acid production in *Escherichia coli* NZN111. Enzyme Microbiol Technol 51:286–293
7. Helena IMB, Xia X, Kapil T et al (2008) Biosynthesis and recycling of nicotinamide cofactors in *Mycobacterium tuberculosis*. J Biol Chem 283(28):19329–19341
8. Liang LY, Liu RM, Chen X et al (2013) Effects of overexpression of NAPRTase, NAMNAT, and NAD synthetase in the NAD(H) biosynthetic pathways on the NAD(H) pool, NADH/NAD$^+$ ratio, and succinic acid production with different carbon sources by metabolically engineered *Escherichia coli*. Biochem Eng J 81:90–96
9. Heuser F, Schroer K, Lütz S et al (2007) Enhancement of the NAD (P) (H) pool in *Escherichia coli* for biotransformation. Eng Life Sci 7:343–353
10. Wei W, Fan SY, Wang FQ et al (2014) Accumulation of androstadiene-dione by overexpression of heterologous 3-ketosteroid D1-dehydrogenase in *Mycobacterium neoaurum* NwIB-01. World J Microbiol Biotechnol 30:1947–1954
11. Paul RM, William RB, Lafras MS et al (1998) Novel method for rapid measurement of growth of *Mycobacteria* in detergent-free media. J Clin Microbiol 36(9):2752–2754
12. Leonardo MR, Dailly Y, Clark DP (1996) Role of NAD in regulating the *adhE* gene of *Escherichia coli*. J Bacteriol 178:6013–6018
13. Ji XJ, Xia ZF, Fu NH et al (2013) Cofactor engineering through heterologous expression of an NADH oxidase and its impact on metabolic flux redistribution in *Klebsiella pneumonia*. Biotechnol Biofuels 6:7
14. Sun JA, Zhang LY, Rao B et al (2012) Enhanced acetoin production by *Serratia marcescens* H32 with expression of a water-forming NADH oxidase. Biores Technol 119:94–98
15. Wang Z, Zhao F, Chen D et al (2006) Biotransformation of phytosterol to produce androsta-diene-dione by resting cells of *Mycobacterium* in cloud point system. Process Biochem 41:557–561

Study on the Different Replacing Groups of Trans-Paroxol for Enzymatic Resolution Using Molecular Simulations

Chuan Zhang, Yigang Jia, Chao Xu, He Huang and Yi Hu

1 Introduction

Paroxetine hydrochloride is a selective serotonin (5-HT) reuptake inhibitor, which is widely used to treat depression, obsession and anxiety symptom [1, 2]. Several synthetic strategies have been developed for this compound and its key intermediates, including the selective recrystallization of diastereomeric salts [3], chiral auxiliary assisted synthesis [4], biocatalytic resolutions [1, 5, 6], and the desymmetrization of a prochiral diester intermediate [7, 8]. Most synthetic routes of the key intermediates are based on (-)-trans-4-(4-fluorophenyl)-1-methyl-3-Hydroxymethylpiperidines ((-)-trans-paroxol). Recently, we attempted to synthesize this intermediate through direct emzymatic resolution of racemic (±)-trans-paroxol, the *candida antarctica* lipase B (CALB) showed the best result, while under the optimized reaction conditions, the E value (enantiomer selectivity) was only 6.20 [8]. (Scheme 1) In the previous report, Gonzalo de Gonzalo et al. showed the replacing groups on the piperidine ring of (±)-trans-paroxol had obvious influence on the enzymatic resolution, under the similar conditions as our procedure, the E value were 16, 45 and 46 corresponding to Alloc, Boc and Poc substituents respectively [1]. Up to now the E value for the enzymatic resolution of (±)-trans-paroxol is far from the practical application, and the molecular mechanism for different substituents on the emzymatic resolution of (±)-trans-paroxol has

This research was supported by National Natural Science Foundation of China (21676143), Qing Lan Project of Jiangsu Province, program for Innovative Research Team in University of Jiangsu Province and the Hi-Tech Research and Development Program of China (863 Program, 2011AA02A209).

C. Zhang · Y. Jia · C. Xu · H. Huang · Y. Hu (✉)
State Key Laboratory of Material-Oriented Chemical Engineering, College of Pharmacy, Nanjin Technical University, Nanjin 211816, China
e-mail: huyi@njtech.edu.cn

H. Liu et al. (eds.), *Advances in Applied Biotechnology*, Lecture Notes in Electrical Engineering 444, https://doi.org/10.1007/978-981-10-4801-2_37

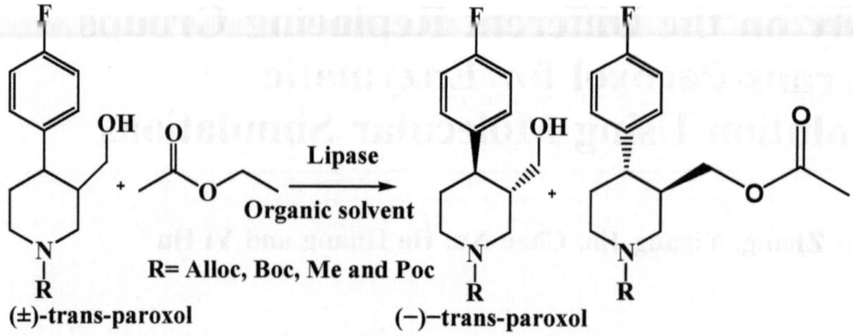

Scheme 1 Lipase-catalyzed acylation of (±)-trans-paroxol in organic solvent

not been revealed. Molecular dynamics (MD) simulations has been an important assistant tool in latest decade for revealing the interactions of enzyme with substrate and analyzing the molecular mechanism of enhancing the property of enzyme [9, 10]. Such as, Chen et al. [11] analysed the structural conformation of *Pseudomonas alcaligenes* lipase (PAL) using MD simulations, which indicated that stronger steric exclusion and structural rigidity facilitated diastereopreference. As well, Park et al. [12] found that the increasing of overall flexibility and the forming of hydrogen bonds were beneficial to understand the effect of tert-butanol on CALB catalyzed transesterification resolution by MD simulations.

In this study, we used molecular docking and MD simulations to analyze the influence of different replacing groups on the piperidine ring (Met, Alloc, Boc and Poc) of (±)-trans-paroxol for enzymatic resolution (Table 1). Generally, stabilized complex is always rigid with lower Root Mean Square Deviation (RMSD) and radius of gyration (Rg), and the stabilized state has more chance to stabilize the intramolecular interactions, including hydrogen bonds, electrostatic interactions and hydrophobic interactions [13, 14]. Herein, four substrates with different substituents were successful docked to the active site of CALB for MD simulations and analysis. We hope this research would be useful for understanding the molecular mechanism of enzymatic resolution of (±)-trans-paroxol and guiding to design the more appropriate substrate with higher E value.

2 Materials and Methods

2.1 Molecular Docking

The initial conformation of CALB solved 1.55 Å resolution was taken from X-ray diffraction (PDB entry code: 1TCA) [15]. CALB is composed of 317 amino acid residues, the ligands and waters in three-dimensional model were removed prior to the docking. Besides, the substrates were prepared with Autodock4.2 [16]. The coordinate of atom OG1 from residue Thr40 was set to the center of grid box.

Table 1 The structure of four different substrates for molecular docking and MD simulations and the E value of enzymatic resolution

Substrate	R-group	R-structure	E value
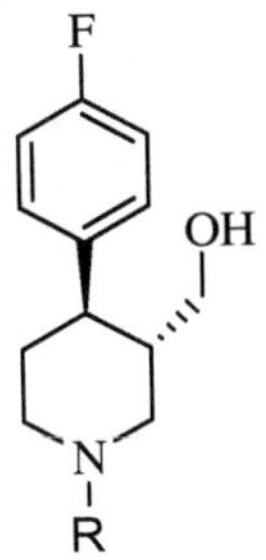	Alloc	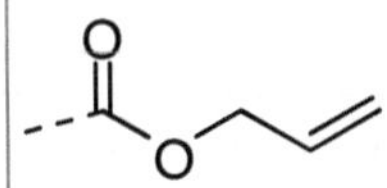	16
	Boc	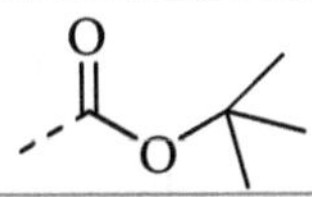	45
	Met	- -CH_3	6.20
	Poc	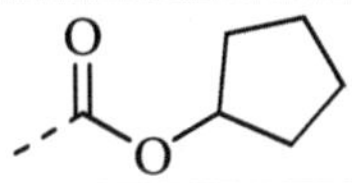	46

Spacing 0.375 Å, number of GA runs with 100. The binding energy was calculated with lamarckian GA. The figures of protein structure were carried out by PyMOL0.98 program [17].

2.2 *Molecular Dynamics Simulation*

All MD simulations were performed using GROMACS 4.5.4 package [18, 19] at Hp Z800 workstation, employing the GROMOS96 53a6 united atom protein force field [20]. Nonbonded interactions were calculated using a twin-range method [21] with short and long range cutoffs of 8 and 14 Å, respectively. The Particle Mesh Ewald (PME) method [22] was used to evaluate the long electrostatic interactions. Bond lengths of the solute were constrained with LINCS [23] and the ones of water with SETTLE [24]. The simple point change water model [25] was used in simulations. Considering a dielectric of 54 for simple point change water [26]. The snapshots of each complex system were saved every 1 ps for data analysis. Each initial system was subjected to energy minimization (1000 steepest descents and 1000 conjugate gradient) and 500 ps equilibration with position restrained. The protein, non-protein were coupled to an external heat bath [27] with temperature coupling constants of 0.1 ps. Then the MD simulations were carried out for 50 ns with a timestep of 2 fs at constant temperature (300 K) and pressure (1 atm). The MD trajectories were analyzed and calculated by GROMACS 4.5.4 program including the backbone-RMSD, Root Mean Square Fluctuation (RMS-Fluctuation) and the radius of gyration.

3 Results and Discussions

3.1 MD Simulations

The stabilized structure of protein is essential for catalysis. Generally, the stability of enzyme is determined by RMSD and radius of gyration for simulations [9]. For this, the MD simulations of the four complexes were addressed in 50 ns MD simulations at 300 K. The simulations were started from the optimized structure which obtained from cluster analysis using molecular docking. The backbone-RMSD value was given in Fig. 1a and the results showed that Alloc-cal, Boc-cal and Poc-cal were more stable than Met-cal. At the last 25 ns of simulations, the structure of Alloc-cal, Boc-cal and Poc-cal reached balance and the RMSD values stabilized around 0.25 nm. On the contrary, the RMSD value of Met-cal complex always kept upward trend and in balance about 0.4 nm. Unlike the RMSD's values changes of complex. We found that the Rg values changed slightly as shown in Fig. 1b. The Rg of Met-cal complex was 1.96 nm, and others were 1.90 nm. The compactness of Met-cal complex was worse than others obviously according to the result of simulations. Therefore, the result indicated that the stability and compactness of Alloc-cal, Boc-cal and Poc-cal complexes were better than Met-cal complex.

Meanwhile, through the analysis of RMS-Fluctuation (Fig. 2), we could find the fluctuation of residues for local area are very large. Combining the catalytic triad Ser105, Asp187 and His224 with the RMS-Fluctuation, the fluctuation of Met-cal is smaller than others. The reduction of flexibility made ligand not be easy to couple with the residues of active pocket, form strong interactions and steric exclusion. Therefore, the binding force and stabilized property of Met-cal complex were worse than other complexes, and reduced the efficiency of enzymatic resolution. And that also indicated the results of H-bond interaction at the following part.

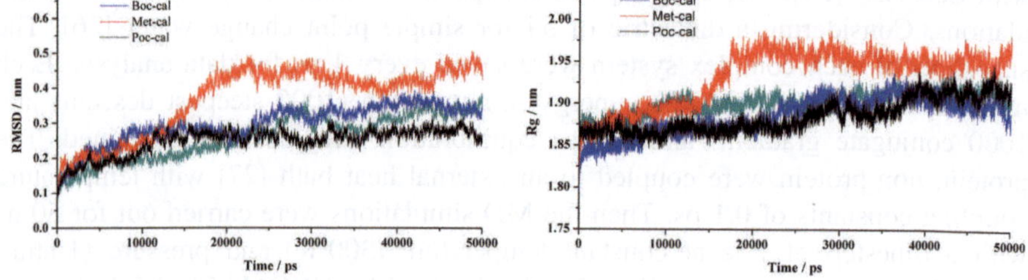

Fig. 1 The RMSD of backbone for different complex in 300 K (**a**), the radius of gyration of different complex in the balance of 15 ns simulations in 300 K of MD simulations (**b**)

3.2 The Distribution of Substrate in Active Pocket

Markus Doerr et al. [28] applied an enzyme–substrate docking protocol to model the MCCs of the deacylation reaction, which showed that both R- and S-propranolol accommodate within the binding pocket of CALB in two binding pockets. In this work, four different replacing substrates were docked to the active site of CALB using Autodock4.2 package. Viewed with the catalytic triad Asp224-His187-Ser105 oriented from left to right, the binding pocket is constituted by a large hydrophobic pocket (S1) above the catalytic triad and a medium size pocket (S2) below it (Fig. 3). The replacing groups (Alloc, Boc and Poc) of (-)-Trans-paroxol totally distributed in the S2 pocket. On the contrary, the group of Met was in the site of S1 pocket. The conformational inversion of space made the group of Met keep away from hydrophobic residues, and reduced the chance of forming hydrophobic interactions. As well, fluorophenyl can't form π–π bonds with the residues His224 and Trp104 [29]. Electrostatic interactions also play an important role in enzyme reactions [30]. The residues of active region indicated that S1 pocket is nonpolar and S2 pocket is polar. And the replacing group of Alloc, Boc and Poc could form electrostatic interactions with surrounding residues to enhance the interactions, one such set of interactions is likely to be the formation of an oxyanion hole. From the enzymology of serine proteases, the interactions with enzyme's oxyanion hole involving Thr40 which can stabilize the transition state of substrate [31]. Meanwhile, compared with the group of Alloc, the groups of Boc and Poc had a deflexion of distribution. This deflexion made them form hydrophobic interactions with the residues Val154 and Gln157, strengthened the steric exclusion and enhanced the property of enzymatic resolution. Therefore, compared with other complexes, the deflexion of distribution made the complex of Met-cal and Alloc-cal have a weak binding force and enzymatic resolution.

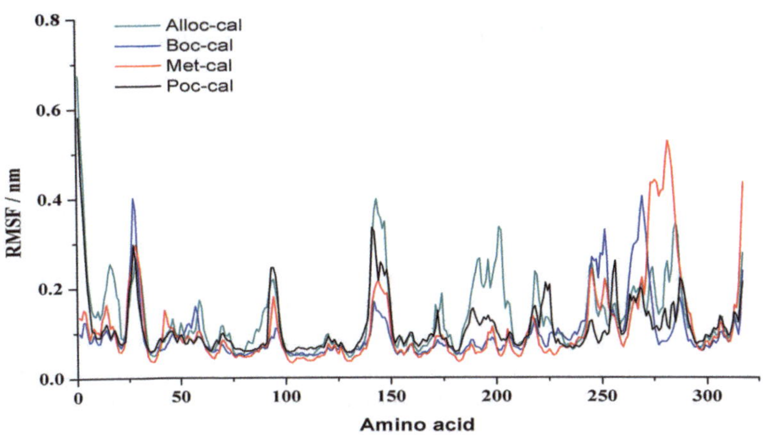

Fig. 2 The RMS-fluctuation of different complexes

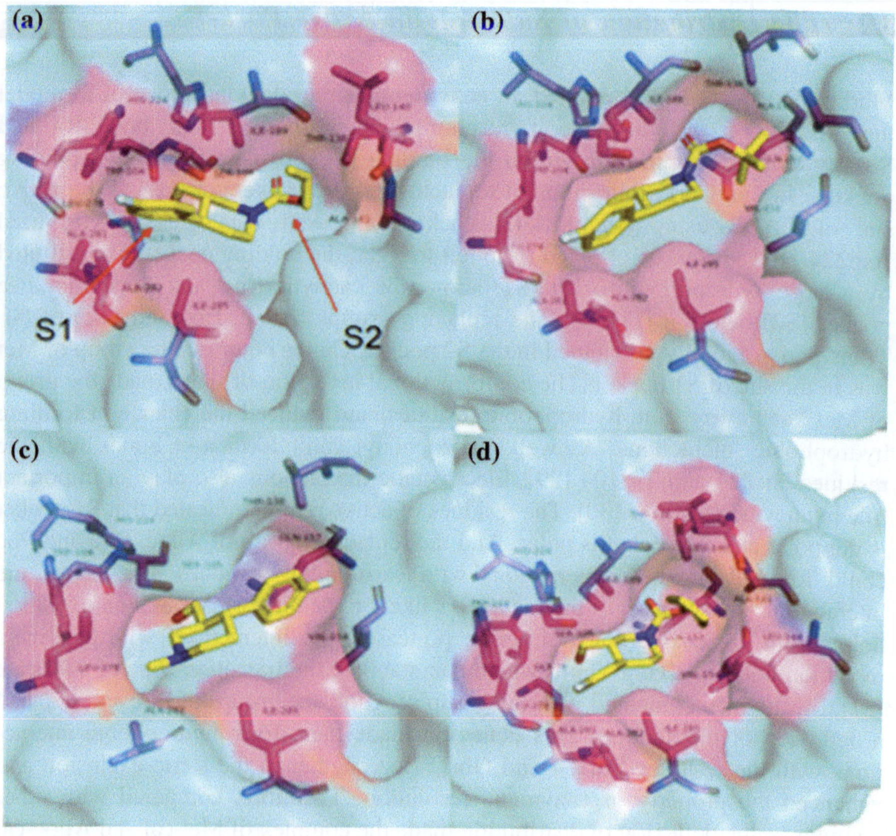

Fig. 3 The hydrogen bonds of different substrates for residues in the active pocket (the *imaginary lines* of *yellow* are hydrogen bonds)

3.3 Hydrogen Bond and Hydrophobic Interactions

Enzymatic reactions conducted in organic solvents have many advantages [32]. However, the organic solvent molecules may strip water molecules on the protein surface, which could lead to the denaturation of enzyme or change its reaction kinetics and substrate specificity [33–35]. Thus, it's important to maintain the stability of enzymes in organic solvents. Meanwhile, hydrogen bonds play an important role in maintaining the stability of complex structure and enzymatic reactions [36]. In addition, to check the reliability of the complexes identified by docking, they were subjected to 50 ns MD simulations in water. From the enzymology of serine proteases, it is known that the negative charge of the tetrahedral substrate intermediate must be stabilized by hydrogen bonds from the enzyme [37]. Through analyzing the hydrogen bonds of enzyme with substrates using molecular simulations, we found that all the replacing group of trans-paroxol formed three

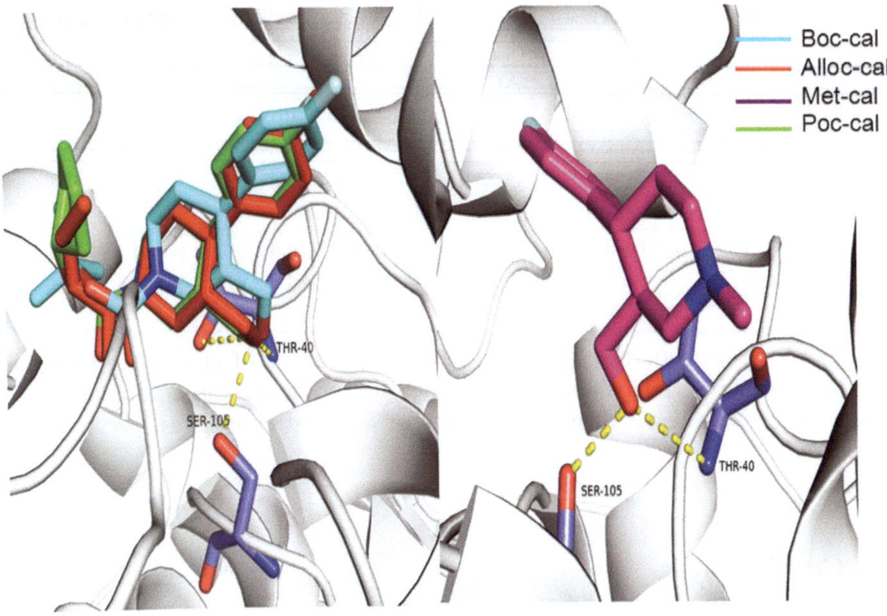

Fig. 4 The distribution of four substrates in the active pocket which were docked using Autodock4.2 package and the residues

hydrogen bonds except Met. In contrast, the ligand in Met-cal complex only formed hydrogen bonds with the atom OG of Ser105 and the atom N of Thr40 respectively (Fig. 4). This was because that the conformational inversion resulted in the phenomenon and made the atom O of Thr40 be far away from the atom O of Met-cal. Besides, we found that all the occupancy of hydrogen bonds were 100%, which means that the hydrogen bonds were quite stable and had a strong binding force. Although there is no rule for the distance of hydrogen bonds to expound the diversity of mechanism for enzymatic resolution (Table 2), the reduction of hydrogen bonds in number for Met-cal indicated that hydrogen bonds were an important element for the influence of enzymatic resolution's performance. Besides, hydrophobic interactions are also important for recognition effect and the interactions of substrate with enzyme. According to the analysis, we found that the complex had a large difference in interactions residues. Because of the conformational inversion and the small of steric exclusion for the Met ligand, it only had hydrophobic interactions with the residues Thr138, Val154, Gln157, Ile189, Leu278, Ala281 and Ile289 (Fig. 3). Based on the energy's analysis of hydrophobic interactions and hydrogen bonds for different complexes, the energy of four replacing groups(Alloc, Boc, Met and Poc) of trans-paroxol were −8.46, −7.62, −6.49 and −9.22 kcal/mol, Which were consistent with the number of residues of hydrophobic interactions. Through analyzing the residues, we considered that the residues of Thr40, Ser105, Thr138, Ile189, Leu278, Ala281 and Ile285 were

Table 2 The distance, donor H-X and acceptor of hydrogen bonds for four different complexes

Complex	Donor H-X	Acceptor	Distance (Å)	Complex	Donor H-X	Acceptor	Distance (Å)
Alloc-cal	Ser105.OG	Alloc.O1	2.99	Boc-cal	Ser105.OG	Boc.O1	3.27
	Thr40.N	Alloc.O1	2.94		Thr40.N	Boc.O1	3.05
	Thr40.O	Alloc.O1	3.10		Thr40.O	Boc.O1	3.01
Poc-cal	Ser105.OG	Poc.O1	2.97	Met-cal			
	Thr40.N	Poc.O1	2.89		Thr40.N	Met.O	2.64
	Thr40.O	Poc.O1	3.16		Ser105.OG	Met.O	2.98

essential for enzymatic resolution. Based on the analysis of above, the hydrogen bonds, hydrophobic interactions and the diversity for the stability of complex were major factors that resulted in the diversity of different complexes for enzymatic resolution.

4 Conclusions

Molecular dynamics simulations was performed to investigate the role of structure-function on enzyme-substrate complex. In this study, we analyzed the diversity of molecular mechanism for different complexes of enzymatic resolution. The stability of these hydrogen bonds throughout all the simulations are determining factors for the higher activity. An appropriate, stable structure state was very essential for catalysis. Hydrophobic interactions and electrostatic interactions also played an important role in enzymatic reaction. The illumination of mechanism will be useful for understanding the diversity of enzymatic resolution and guiding to design the appropriate substrates.

References

1. De GG, Brieva R, Sánchez VM, Bayod M, Gotor V (2001) Enzymatic resolution of trans-4-(4′-fluorophenyl)-3-hydroxymethylpiperidines, key intermediates in the synthesis of (-)-paroxetine. J Org Chem 6626:8947–8953
2. Su B, Bao Z, Xing H, Yang Y, Ren Q (2009) Enantioseparation of paroxetine intermediate on an amylose-derived chiral stationary phase by supercritical fluid chromatography. J Chromatogr A 121626:5140–5146
3. Christensen JA, Squires RF (1977) 4-Phenylpiperidine compounds. U.S. Patents 4007196, 8 Feb 1977
4. Amat M, Bosch J, Hidalgo J, Cantó M, Pérez M, Llor N et al (2000) Synthesis of enantiopure trans-3, 4-disubstituted piperidines. An enantiodivergent synthesis of (+)-and (-)-paroxetine. J Org Chem 6510:3074–3084
5. Gledhill L, Kell C (1998) Aminometyl oxooxazolidinyl benzene derivatives W.O. Patent 1999011642, 11 Mar 1998

6. Zepp M, Gas Y, Heefner D (1993) Method of preparing optically pure precursors of paroxetine, U.S. Patent 5258517, 2 Nov 1993
7. Marvin SY, Lantos I, Peng ZQ, Yu J, Cacchio T (2000) Asymmetric synthesis of (−)-paroxetine using PLE hydrolysis. Tetrahedron Lett 4130:5647–5651
8. Jia YG, Liu WM, Ni X, Huang H, Hu Y (2015) Kinetic resolution of transesterification reaction of trans-paroxol catalyzed by lipase in organic media. Chem Ind Eng Prog 34 (S1):116–120
9. Liu J, Yu H, Shen Z (2008) Insights into thermal stability of thermophilic nitrile hydratases by molecular dynamics simulation. J Mol Graph Model 274:529–535
10. Pikkemaat MG, Linssen AB, Berendsen HJ, Janssen DB (2002) Molecular dynamics simulations as a tool for improving protein stability. Protein Eng 153:185–192
11. Chen H, Wu J, Yang L, Xu G (2013) A combination of site-directed mutagenesis and chemical modification to improve diastereopreference of Pseudomonas alcaligenes lipase. BBA-Protein Proteom 183412:2494–2501
12. Park HJ, Park K, Yoo YJ (2013) Understanding the effect of tert-butanol on Candida antarctica lipase B using molecular dynamics simulations. Mol Simulat 398:653–659
13. Wang Y, Wang X (2015) Binding, stability, and antioxidant activity of quercetin with soy protein isolate particles. Food Chem 188:24–29
14. Li PY, Chen XL, Ji P, Li CY, Wang P, Zhang Y et al (2015) Interdomain hydrophobic interactions modulate the thermostability of microbial Esterases from the Hormone-Sensitive Lipase Family. J Biol Chem 29017:11188–11198
15. Jung JH, Yoon DH, Kang P, Lee WK, Eum H, Ha HJ (2013) CAL-B catalyzed desymmetrization of 3-alkylglutarate: "olefin effect" and asymmetric synthesis of pregabalin. Org Biomol Chem 1122:3635–3641
16. Morris GM, Huey R, Lindstrom W, Sanner MF, Belew RK, Goodsell DS et al (2009) AutoDock4 and AutoDockTools4: automated docking with selective receptor flexibility. J Comput Chem 3016:2785–2791
17. DeLano WL (2002) The PyMOL molecular graphics system, version 1.5.0.4. Schrödinger, LLC, Portland
18. Van Der Spoel D, Lindahl E, Hess B, Groenhof G, Mark AE, Berendsen HJ (2005) GROMACS: fast, flexible, and free. J Comput Chem 2616:1701–1718
19. Pronk S, Páll S, Schulz R, Larsson P, Bjelkmar P, Apostolov R et al (2013) GROMACS 4.5: a high-throughput and highly parallel open source molecular simulation toolkit. Bioinformatics: btt055
20. Oostenbrink C, Soares TA, Van der Vegt NF, Van Gunsteren WF (2005) Validation of the 53A6 GROMOS force field. Eur Biophys J 344:273–284
21. van Gunsteren WF, Berendsen HJ (1990) Computer simulation of molecular dynamics: methodology, applications, and perspectives in chemistry. Angew Chem Int Ed 299:992–1023
22. Darden T, York D, Pedersen L (1993) Particle mesh Ewald: An N log (N) method for Ewald sums in large systems. J Chem Phys 9812:10089–10092
23. Hess B, Bekker H, Berendsen HJ, Fraaije JG (1997) LINCS: a linear constraint solver for molecular simulations. J Comput Chem 1812:1463–1472
24. Miyamoto S, Kollman PA (1992) SETTLE: an analytical version of the SHAKE and RATTLE algorithm for rigid water models. J Comput Chem 138:952–962
25. Hermans J, Berendsen HJ, Van Gunsteren WF, Postma JP (1984) A consistent empirical potential for water–protein interactions. Biopolymers 238:1513–1518
26. Smith PE, van Gunsteren WF (1994) Consistent dielectric properties of the simple point charge and extended simple point charge water models at 277 and 300 K. J Chem Phys 1004:3169–3174
27. Berendsen HJ, Postma JV, van Gunsteren WF, DiNola A, Haak J (1984) Molecular dynamics with coupling to an external bath. J Chem Phys 818:3684–3690

28. Escorcia AM, Daza MC, Doerr M (2014) Computational study of the enantioselectivity of the O-acetylation of (R, S)-propranolol catalyzed by Candida antarctica lipase B. J Mol Catal B Enzym 108:21–31
29. Matsumura H, Yamamoto T, Leow TC, Mori T, Salleh AB, Basri M et al (2008) Novel cation-π interaction revealed by crystal structure of thermoalkalophilic lipase. Proteins Struct Funct Bioinf 702:592–598
30. Yao J, Xu Q, Guo H (2013) QM/MM and free-energy simulations of deacylation reaction catalysed by sedolisin, a serine-carboxyl peptidase. Mol Simulat 393:206–213
31. Hamberg A, Magnusson AO, Hu FJ, Hult K (2013) Selective Monoacylation of diols by substrate assisted catalysis in T40A Candida antarctica Lipase B. Chem Cat Chem 53:743–747
32. Klibanov AM (2001) Improving enzymes by using them in organic solvents. Nature 4096817:241–246
33. Zaks A, Klibanov AM (1988) The effect of water on enzyme action in organic media. J Biol Chem 26317:8017–8021
34. Valivety RH, Halling PJ, Macrae AR (1992) Reaction rate with suspended lipase catalyst shows similar dependence on water activity in different organic solvents. Biochim Biophys Acta Protein Struct Mol Enzymol 11183:218–222
35. Affleck R, Haynes CA, Clark DS (1992) Solvent dielectric effects on protein dynamics. Proc Nat Acad Sci 8911:5167–5170
36. Park HJ, Joo JC, Park K, Yoo YJ (2012) Stabilization of Candida antarctica lipase B in hydrophilic organic solvent by rational design of hydrogen bond. Biotechnol Bioproc Eng 174:722–728
37. Carter P, Wells JA (1990) Functional interaction among catalytic residues in subtilisin BPN′. Proteins: Struct. Funct Bioinf 74:335–342

Relationship Between Coenzyme Q_{10} Synthesis and Cytochrome Accumulation in *Rhodobacter sphaeroides* 2.4.1

Penghao Li, Dan Gao, Junqian Gao, Hao Liu and Zhengliang Qi

1 Introduction

Coenzyme Q_{10} (CoQ_{10}) is a vital electron transporter in redox respiration chain in most of the living things. It can obliterate free radicals to protect cell membrane from oxidation damage, which has been widely used in cosmetics industrial products and clinical treatment [1]. Manufacturing methods for CoQ_{10} include chemical synthesis, semi-chemical synthesis and microbial biosynthesis, and among them microbial fermentation is the primary way for producing currently [2, 3]. Because of its important physiology function, CoQ_{10} has been paid growing attention to improve its production which could satisfy the market requirement.

CoQ_{10} synthesis in bacteria involved three major branched metabolic pathways including MEP pathway, shikimate pathway and ubiquinone pathway. The former two pathways can generate decaprenyl-PP and hydroxybenzoic acid, which are the two precursors necessary for CoQ_{10} formation (Fig. 1). Many strategies have been conducted to increase the production of CoQ_{10}, a majority of which focus on the reconstruction of the metabolic pathway by numerous biological techniques. Currently, CoQ_{10} producing strains cover *R. sphaeroides*, *R. capsulatus* and *Agrobacterium tumefaciens* [4, 5]. Some researchers remoulded a natural CoQ_{10} producer *R. sphaeroides* by advancing the decaprenyl diphosphate supplement with the methods of overexpression *dxs*, *dxr*, *idi* and *ispD*, together with applying a self-regulation system combining a set of RBSs (ribosome binding sites) for adjusting the expression of the $LacI^q$ protein. Finally, they obtain a mutant strain yielding 138 ± 2.64 mg/L CoQ_{10} which was about 2-fold higher than that of the wild-type [6, 7]. Some people also attempted to improve the CoQ_{10} production in CoQ_{10} non-natural producer. For example, researchers deleted an endogenous gene

P. Li · D. Gao · J. Gao · H. Liu · Z. Qi (✉)
MOE Key Laboratory of Industrial Fermentation Microbiology, College of Biotechnology, Tianjin University of Science and Technology, Tianjin 300457, China
e-mail: qizhengliang@tust.edu.cn

H. Liu et al. (eds.), *Advances in Applied Biotechnology*, Lecture Notes in Electrical Engineering 444, https://doi.org/10.1007/978-981-10-4801-2_38

ispB in *Escherichia coli* accompany with overexpressing a *dps* gene derived from *A. tumefaciens*, resulting in a CoQ_{10} generating strain ultimately [8]. Moreover, mutagenesis has also been an efficient way to acquire CoQ_{10} high-yield strains. Researchers executed chemical mutagenesis to the *A.tumefaciens* AJ-24 strain using NTG with biotin K_3 as the selection marker. Finally, they obtain a mutant strain producing 20% higher CoQ_{10} yield than the original strain [9].

CoQ_{10} synthesis in *R. sphaeroides* is influenced by many aspects, such as the accumulation of cytochrome which competes in the common precursor (E, E)-farnesyl-PP with the formation of CoQ_{10}. The synthesis of cytochrome is regulated by many global transcription regulators. The amount of cytochrome can be decided by the level of photosynthetic (PS) apparatus including bacteriochlorophyll (Bchl), carotenoids (Crt), membrane proteins of the reaction center (RC) and two light-harvesting (LH) complexes as well as assembly factors [10, 11]. The PrrA/PrrB and AppA/PpsR two-component systems are two significant regulators which are directly related to cytochrome accumulation in *R. sphaeroides*. For the PrrA/PrrB two-component system, PrrB delivers the phosphate group (Pi) to the PrrA firstly before a self-phosphorylation, and then the PrrA-Pi complex facilitates the transcription of PS genes [12, 13]. The AppA/PpsR two-component system can

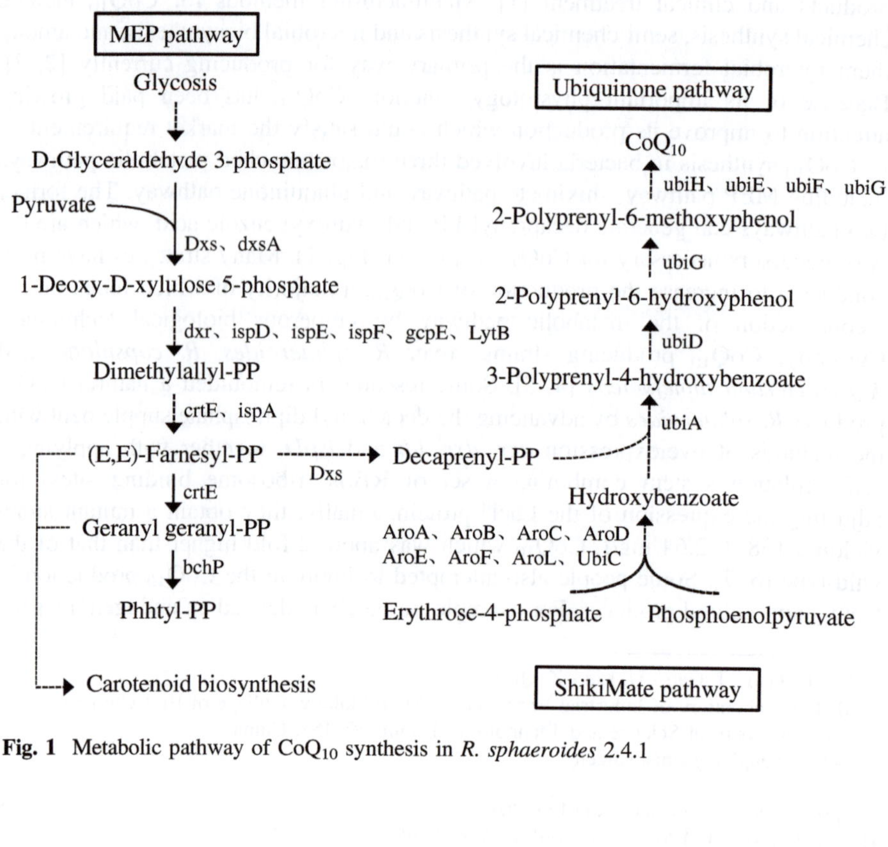

Fig. 1 Metabolic pathway of CoQ_{10} synthesis in *R. sphaeroides* 2.4.1

inhibit the transcription of *crt*, *bch* and *puc* genes directly with PpsR combining to their promoter regions and restraining *puf* indirectly by interaction with other regulatory proteins. Afterwards, the AppA can specifically bind with PpsR and liberate it from the operon, thus activating the transcription procession ultimately [14]. Other regulators such as FnrL, TrxA, Spb and IHF also make a effect on the synthesis of cytochrome [15].

Former studies mostly concentrated on the reconstruction of the metabolic pathway of CoQ_{10} synthesis in *R.sphaeroides*. In the present study, with attempting to address the relationship between cytochrome accumulation and synthesis of CoQ_{10}, we deleted the two key genes in the PrrA/PrrB and AppA/PpsR two-component systems, respectively, to alter the accumulated amount of intra-cellular pigments. Subsequently, shaking-flask fermentation was followed to evaluate the cell growth and CoQ_{10} synthesis [16, 17].

2 Materials and Methods

2.1 Strains, Plasmids, and Culture Media

All the strains and plasmids used in this study were summarized in Table 1. *E.coli* JM109 was used for plasmids construction and preparation. *E.coli* S17-1 was a kind gift of Prof. Dr. Hongwei Yu being used for di-parental conjugation. *R.sphaeroides* strains 2.4.1 (WT), Δ*prrA* and Δ*ppsR* were used for CoQ_{10} fermentation. Plasmid pLH146 was used for gene deleting in *R. sphaeroides* 2.4.1.

Table 1 Strains and plasmids used in this study

Strains			
Number	Name	Description	Source
S170	*R. sphaeroides* 2.4.1	WT	From Zhejiang University
S302	Δ*prrA*	*prrA* markerless deletion mutant	This study
S356	Δ*ppsR*	*ppsR* markerless deletion mutant	This study
Plasmides			
pLH146	pK18*mobsacB*	Markerless deletion plasmid	This study
pLH320	pK18*mobsacB*::*prrA*-L-R	*prrA* markerless deletion plasmid	This study
pLH358	pK18*mobsacB*::*ppsR*-L-R	*ppsR* markerless deletion plasmid	This study

E.coli strains were cultivated in LB (Luria-Bertani) medium (trypton 10 g/L, yeast extract 5 g/L, NaCl 10 g/L) at 37 °C supplemented with 100 μg/mL of kanamycin when necessary. For routine cultivation and genome preparation, *R. sphaeroides* strains were cultivated in medium A containing 3 g/L glucose, 2 g/L NaCl, 8 g/L yeast extract, 0.256 g/L $MgSO_4 \cdot 7H_2O$,1.3 g/L KH_2PO_4, 15 μg/L biotin, 1 mg/L thiamine hydrochloride, 1 mg/L nicotinic acid at 32 °C with 25 μg/mL kanamycin and 150 μg/mL K_2TeO_3 when necessary [6].

2.2 *Plasmid Construction*

For construction of the plasmid pLH320, we separately amplified an 898 and an 858 bp fragment from *R. Sphaeroides* genome as the left arm and right arm of *prrA* gene using the primer 751/752 and primer 753/754. Then, we jointed the two arms together by overlapping PCR resulting in a 1756 bp fragment, which was subsequently cloned into the plasmid pLH146 after digestion with *Eco*R I and *Pst* I. For construction of plasmid pLH358, its left arm and right arm were ligated into plasmid pLH146 stepwise. The primers and restriction enzymes used in this study were listed in Table 2.

2.3 *Di-parental Conjugation*

E.coli S17-1 and *R. Sphaeroides*, as donor and receptor strain respectively, were harvested and washed twice with fresh LB medium after cultivating to the early log phase. Then, mixing the suspension of *E.coli* S17-1 and *R. Sphaeroides* at a 1:10 ratio, which was subsequently added to a filter disc on the medium A agar plate and cultivated at 32 °C for 24 h. Ultimately, washing off the lawn on the plate with 400 μL sterile water and spreading it on the medium A agar plate supplemented with 25 μg/mL kanamycin and 150 μg/mL K_2TeO_3. The plate was cultivated at 32 °C until transformants presenting.

2.4 *Shaking-Flask Cultivation*

Shaking-flask experiments were performed in the 500 mL Erlenmeyer flasks with 100 mL fermentation medium containing 4 g/L corn steep liquor, 40 g/L glucose, 2.8 g/L NaCl, 3 g/L sodium glutamate, 3 g/L $(NH_4)_2SO_4$, 12.92 g/L $MgSO_4 \cdot 7H_2O$, 3 g/L KH_2PO_4, 15 μg/L biotin, 1 mg/L thiamine hydrochloride, 1 mg/L nicotinic acid. The cultures were cultivated in the dark at 32 °C for 96 h in a 200 rpm rotary

Table 2 Primers and restriction enzymes used in this study

Primers	Sequences (5′-3′)	Underlined sequence	Utilization
Primer751	CCGGAATTCGGAATATGTCATCGGAGACA	*Eco*R I	For plasmid pLH320 construction
Primer752	ATGGGCGACAATCATGGCTGACCAGTCGCACCGGCGGAAC		
Primer753	GTTCCGCCGGTGCGACTGGTCAGCCATGATTGTCGCCCAT		
Primer754	AAAACTGCAGGATCAGGGTGCGCAGTCGCA	*Pst* I	
Primer783	CCGGAATTCGATCGATGGAATCCACGT	*Eco*R I	For plasmid pLH358 construction
Primer784	CCGGAATTCGATCGATGGAATCCACGT	*Sal* I	
Primer785	ACGCGTCGACCTCAATGTCCTACCGATCGT	*Sal* I	
Primer786	ACGCGTCGACCTCAATGTCCTACCGATCGT	*Sph* I	

shaking incubator. The cell density was measured at OD_{600} each 12 h and the CoQ_{10} production was detected each 24 h during the entire course of fermentation.

2.5 *Metabolism Analytical Methods*

Biomass was determined by spectrophotometer (Mapada instruments UV-1800, China) and calculated using a calibration curve. The glucose consuming was followed by a SBA biosensor (Biology Institute of Shandong Academy of Sciences, China). In order to detect the cytochrome in *R. Sphaeroides*, 1 mL thallus were collected by centrifuging at 8000 rpm for 5 min, after which 2 mL acetone was added for ultrasonication (ultrasonic 3 min and spacing 2 min). Whereafter, the superstratum liquid was conducted with spectrum scanning at a 350–900 nm wavelength coverage. For extraction and quantification of CoQ_{10}, 1 mL cells were harvested with a 8000 rpm centrifuge for 5 min. The cell pellets were resuspended with 4 mL extraction solution (ethyl acetate/ethanol = 5/3) and incubated for 1 h at the room temperature in dark, during which the mixture should be vortexed vigorously each 15 min. Then the supernatant was disposed using a 0.22 μm filter after centrifuging as earlier and injected into a HPLC machine equipped with a Wondasil C18 column. The sample was detected at 40 °C temperature, a UV detector of 275 nm, a 25 μL sample size and methyl alcohol:isopropanol = 3:1 as mobile phase at a flow rate of 1 mL/min.

3 Results and Discussion

3.1 *Phenotype Analysis for WT,* ΔprrA *and* ΔppsR *Mutant Strains*

In *R. sphaeroides* 2.4.1 (WT), PrrA/PrrB and AppA/PpsR are the two two-component systems directly regulate cytochrome synthesis. Generally, PpsR (*ppsR*) acts as a repressor for cytochrome synthesis in *R. sphaeroides* under oxygenic status, while PrrA (*prrA*) can activate cytochrome synthesis under semi-aerobic condition (dissolved oxygen below 3 ppm). To change *R. sphaeroides* cytochrome synthesis level, we deleted *prrA* gene and *ppsR* gene in WT to construct two mutants (Δ*prrA* and Δ*ppsR*), respectively. As Fig. 2 shown, phenotype revealed no obvious variation except the colony color among the WT, Δ*prrA* and Δ*ppsR* after aerobic culture in the medium A agar plate for 96 h. For Δ*ppsR*, colony color became deep red than the WT which generally presented orange color under aerobic cultivation. In contrary, Δ*prrA* displayed white color under the same culture condition. Thus, we got three *R. sphaeroides* strains with different cytochrome synthesis levels for subsequent study.

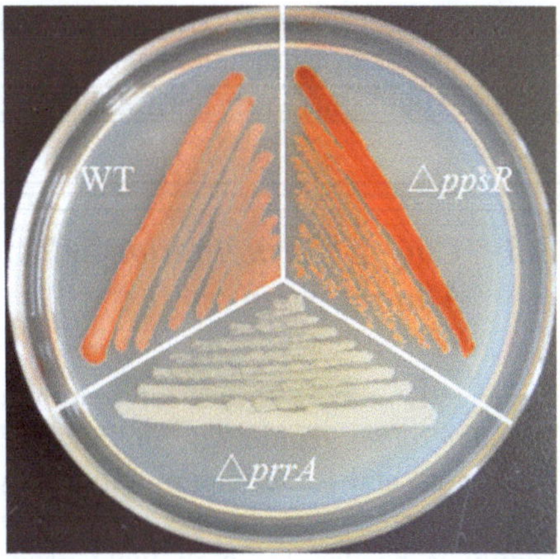

Fig. 2 Phenotype of WT, Δ*prrA* and Δ*ppsR* under aerobic cultivation

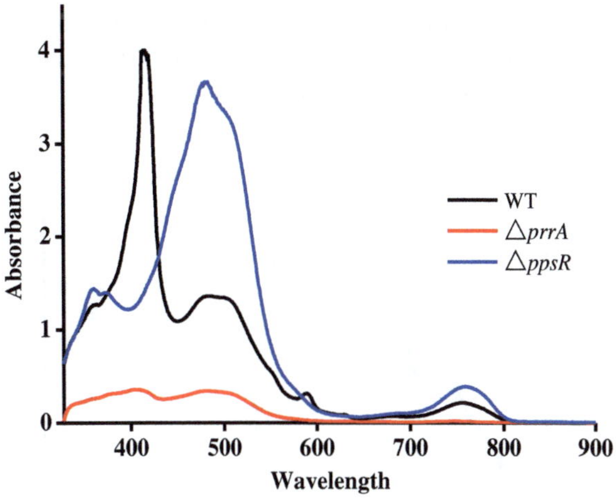

Fig. 3 Wavelength scanning for WT, Δ*prrA* and Δ*ppsR* with intact cells

3.2 *Spectrum Change for WT,* **ΔprrA** *and* **ΔppsR** *Mutant Strains*

Besides carotenoid and bacteriochlorophyll, there are some other intracellular compounds make cell with color such as heme. Although Δ*prrA* and Δ*ppsR* revealed obvious phenotype change in color, whether these phenomena caused by the change of carotenoid and bacteriochlorophyll was uncertain. Thus, wavelength scanning with intact cell was performed on the mutant strains Δ*prrA* and Δ*ppsR*, and WT was set as control. As Fig. 3 depicted, spectrum revealed apparent variation among these strains. The spectrum for Δ*ppsR* had a dramatic enhancement at 475 nm and a certain increase at 785 nm than that for WT, while Δ*prrA* displayed an adverse tendency as that of Δ*ppsR*. Generally, absorption peak at 475 and 785 nm separately stands for carotenoid and bacteriochlorophyll in photosynthetic bacteria. Hence, disruption of prrA or ppsR was directly related to synthesis of carotenoid and bacteriochlorophyll in *R. sphaeroides*. The result in this section could guarantee the validity and rationality of subsequent study.

3.3 CoQ_{10} Fermentation with WT, ΔprrA and ΔppsR Mutant Strains

In Fig. 1, the synthesis of carotenoid, bacteriochlorophyll and CoQ_{10} in *R. sphaeroides* requires a common precursor (E,E)-farnesyl-PP. Synthesis of the carotenoid and bacteriochlorophyll may competing with CoQ_{10} synthesis in such common precursor. In the view of metabolic engineering, adequate precursor supply is needed for high yield product fermentation. To achieve a high yield CoQ_{10} fermentation, we tried to disclose the relationship between synthesis of CoQ_{10} and cytochrome (carotenoid and bacteriochlorophyll). In this work, CoQ_{10} fermentation was carried out with WT, Δ*prrA* and Δ*ppsR* in shaking flasks under dark and result was shown in Fig. 4.

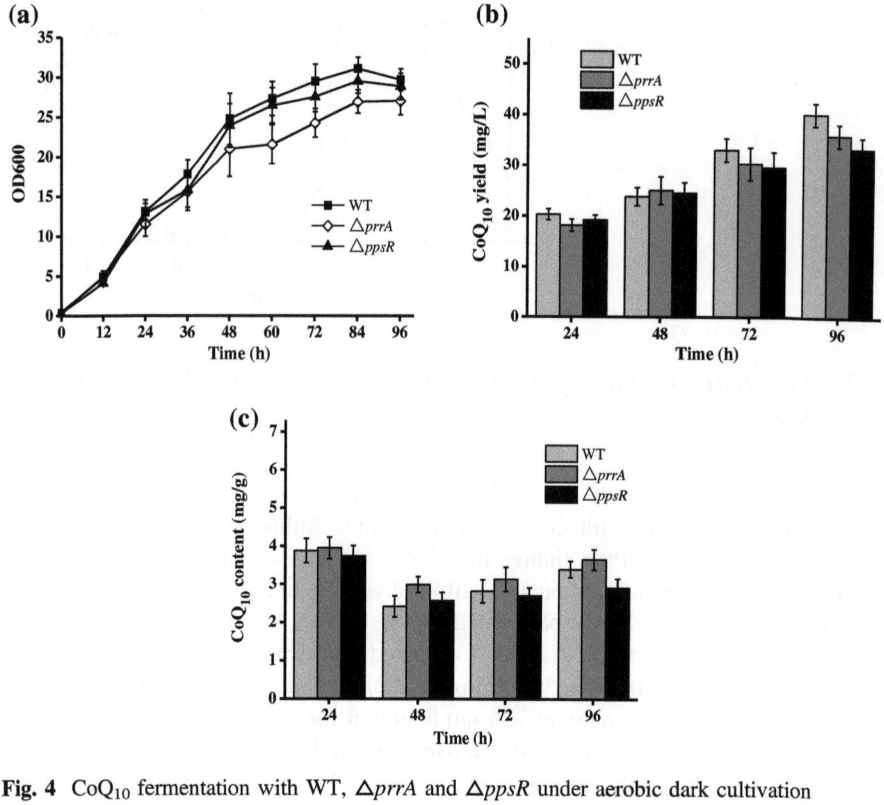

Fig. 4 CoQ_{10} fermentation with WT, Δ*prrA* and Δ*ppsR* under aerobic dark cultivation

Comparing to WT, Δ*prrA* growth revealed a decrease tendency after fermenting for 24 h and the final biomass accumulated was reduced by 15.6%. Furthermore, for Δ*prrA*, dynamic change of CoQ_{10} yield was consistent with the growth status and final CoQ_{10} yield was only 36 ± 0.58 mg/L which was 85.7% of the yield of WT. In the aspect of CoQ_{10} content, *prrA* mutation promoted CoQ_{10} content during fermentation and the final content increased than the WT. Deleting *prrA* might result in excess intermediate metabolite accumulation in the MEP pathway which was toxic to bacteria. In this sense, Δ*prrA* growth was repressed during fermentation. However, CoQ_{10} content for Δ*prrA* increased than WT which meant repressing synthesis of carotenoid and bacteriochlorophyll was beneficial to CoQ_{10} synthesis.

ppsR mutation showed a little influence on bacterial growth during CoQ_{10} fermentation. CoQ_{10} yield of Δ*ppsR* was reduced during fermentation and the final yield was decreased by 23.8% than that of the WT. Also, comparing to WT, CoQ_{10} content fell down during fermentation by Δ*ppsR*. Finally, the CoQ_{10} content for Δ*ppsR* was reduced by 20% than that for WT. *ppsR* defection significantly enhanced cellular carotenoid synthesis which might reduce the flux to CoQ_{10} synthesis pathway. As a result of the flux reduction, CoQ_{10} synthesis reduced which might weak oxidative respiration activity and repress bacterial growth. Carotenoid synthesis and bacteriochlorophyll synthesis may compete with CoQ_{10} synthesis in the common precursor (E,E)-farnesyl-PP (Fig. 1). In this sense, we proposed that the accumulation change of intracellular carotenoid and bacteriochlorophyll might directly related to CoQ_{10} synthesis. It has been confirmed that carotenoid synthesis, bacteriochlorophyll synthesis and CoQ_{10} synthesis in *R. sphaeroides* 2.4.1 required a common precursor (E,E)-farnesyl-PP (Fig. 1).

4 Conclusion

prrA and *ppsR* are directly related to the synthesis of carotenoid and bacteriochlorophyll in *R. sphaeroides* 2.4.1. *ppsR* mutation apparently activates carotenoid synthesis under aerobic cultivation, while *prrA* defection obviously inhibits synthesis of carotenoid and bacteriochlorophyll under the same condition. Comparing to WT, both growth and CoQ_{10} yield are repressed during aerobic CoQ_{10} fermentation by Δ*prrA* and Δ*ppsR*, but CoQ_{10} content reveal different phenomena that CoQ_{10} content for Δ*prrA* is increased and for *ppsR* is reduced. Therefore, the synthesis of cytochrome such as carotenoid and bacteriochlorophyll competes with the CoQ_{10} synthesis and weakening the related cytochrome synthesis pathways may be beneficial to CoQ_{10} accumulation.

References

1. Ndikubwimana JDD, Lee BH (2014) Enhanced production techniques, properties and uses of coenzyme Q_{10}. Biotech Lett 36(10):1917–1926
2. Negishi E, Liou SY, Xu C et al (2002) A novel, highly selective, and general methodology for the synthesis of 1,5-diene-containing oligoisoprenoids of all possible geometrical combinations exemplified by an iterative and convergent synthesis of coenzyme Q(10). Org Lett 4(2):261–264
3. Lipshutz BH, Lower A, Berl V et al (2005) An improved synthesis of the "miracle nutrient" coenzyme Q_{10}. Org Lett 7(19):4095–4097
4. Gin P, Hsu AY, Rothman SC et al (2003) The *Saccharomyces cerevisiae* CoQ_6 gene encodes a mitochondrial flavin-dependent monooxygenase required for coenzyme Q biosynthesis. J Biol Chem 278(28):25308–25316
5. Nguyen TPT, Theresa PT, Clarke, et al (2013) *S. cerevisiae* coq5 null mutants require over-expression of Coq8 kinase for rescue by *E. coli* CoQ_5 homolog *ubiE.* FASEB J 27(12):585–616
6. Lu W, Shi Y, He S et al (2013) Enhanced production of CoQ_{10} by constitutive overexpression of 3-demethyl ubiquinone-9 3-methyltransferase under tac promoter in *Rhodobacter sphaeroides*. Biochem Eng J 72:42–47
7. Lu W, Ye L, Xu H et al (2013) Enhanced production of coenzyme Q_{10} by self-regulating the engineered MEP pathway in *Rhodobacter sphaeroides*. Biotechnol Bioeng 111(4):761–769
8. Choi JH, Ryu YW, Park YC et al (2009) Synergistic effects of chromosomal *ispB* deletion and *dxs* overexpression on coenzyme Q_{10} production in recombinant *Escherichia coli* expressing *Agrobacterium tumefaciens dps* gene. J Biotechnol 144(1):64–69
9. Yoshida H, Kotani Y, Ochiai K et al (1998) Production of ubiquinone-10 using bacteria. J Gen Appl Microbiol 44(1):19–26
10. Matthews PD, Wurtzel ET (2000) Metabolic engineering of carotenoid accumulation in *Escherichia coli* by modulation of the isoprenoid precursor pool with expression of deoxyxylulose phosphate synthase. Appl Microbiol Biotechnol 53(4):396–400
11. Happ HN, Braatsch S, Broschek V et al (2005) Light-dependent regulation of photosynthesis genes in *Rhodobacter sphaeroides* 2.4.1 is coordinately controlled by photosynthetic electron transport via the PrrBA two-component system and the photoreceptor AppA. Mol Microbiol 58(3):903–914
12. Kim YJ, Ko IJ, Lee JM et al (2007) Dominant role of the cbb_3 oxidase in regulation of photosynthesis gene expression through the PrrBA system in *Rhodobacter sphaeroides* 2.4.1. J Bacteriol 189(15):5617–5625
13. Abada E, Balzer A, Jäger A et al (2002) Bacteriochlorophyll-dependent expression of genes for pigment-binding proteins in *Rhodobacter capsulatus* involves the RegB/RegA two-component system. Mol Genet Genomics Mgg 267(2):202–209
14. Jäger A, Braatsch S, Haberzettl K et al (2007) The AppA and PpsR proteins from *Rhodobacter sphaeroides* can establish a redox-dependent signal chain but fail to transmit blue-light signals in other bacteria. J Bacteriol 189(6):2274–2282
15. Li K, Pasternak C, Klug G (2004) Expression of the *trxA* gene for thioredoxin 1 in *Rhodobacter sphaeroides* during oxidative stress. Arch Microbiol 180(6):484–489
16. Tucker JD, Siebert CA, Escalante M et al (2010) Membrane invagination in *Rhodobacter sphaeroides* is initiated at curved regions of the cytoplasmic membrane, then forms both budded and fully detached spherical vesicles. Mol Microbiol 76(4):833–847
17. Niederman RA (2006) Structure, function and formation of bacterial intracytoplasmic membranes. Springer, Berlin, pp 193–227

Optimization of the One-Step and Two-Step Transformation Methods of Mannitol by *Lactobacillus buchneri*

Hongbin Wang, Yu Li, Yongshuai Wang, Yan Chen, Yong Zhang, Yongrui Feng, Haichao Han, Shuqi Gui and Fuping Lu

1 Introduction

Mannitol, an important functional sugar alcohol, is widely used in the pharmaceutical, food, painting and textile industry [1, 2]. The current study showed that lactic acid bacteria, yeasts, fungi and other microbial strains could synthesize mannitol [3]. However, current industrial microbial production of mannitol is not economically viable [4]. Thus, development of commercial microbial production of mannitol requires screening for more potential mannitol-producing strains.

Many studies on microbial production of mannitol have been reported, but so far no industrialization of microbial method was realized due to the problem of high cost [5–7]. Many microorganisms have been found to be able to synthesize mannitol, and the hetero-fermentative lactic acid bacterium received more attention [8–11]. Most studies focused on screening of more efficient production strains and based on the one-step batch transformation method that performing cell growth and cell catalysis at the same time [5, 6, 12]. But this method has many defects: first, it is difficult to separate mannitol from the impurities from culture medium, which increases the separation complexity and cost; second, the cells are used only one time for transformation, bringing higher cell cultivation cost [13, 14]. To solve the problem, a two-step transformation strategy was suggested. It separates the cell growth step with the biotransformation step for production of mannitol. Successful application of the two-step transformation method has been demonstrated, e.g., for the production of lactic acid [15] and propionic acid [16]. As far as we know,

H. Wang · Y. Li · Y. Wang · Y. Chen · Y. Zhang · Y. Feng · H. Han · S. Gui · F. Lu (✉)
Key Laboratory of Industrial Fermentation Microbiology, Ministry of Education;
the College of Biotechnology, Tianjin University of Science and Technology;
National and Local United Engineering Lab of Metabolic Control Fermentation Technology,
College of Biotechnology, Tianjin University of Science and Technology,
TEDA, No. 29, Ave 13, Tianjin 300457, China
e-mail: lfp@tust.edu.cn

H. Liu et al. (eds.), *Advances in Applied Biotechnology*, Lecture Notes
in Electrical Engineering 444, https://doi.org/10.1007/978-981-10-4801-2_39

no studies for mannitol production with the two-step transformation strategy were reported. In the theory, the strategy could reduce the complexity of product separation and increase the repeat times of cell use, which will help to reduce the producing cost of mannitol.

2 Materials and Methods

2.1 *Materials*

Peptone, beef extract and yeast extract were products of Aobox Biotechnology. Glucose and Fructose were purchased from Beijing solarbio. Ammonium citrate, Anhydrous sodium acetate, K_2HPO_4, $MgSO_4{\cdot}7H_2O$, $MnSO_4{\cdot}4H_2O$ and twain-80 were purchased by Tianjin north day medical chemical reagent factory, Corn steep liquor was supplied by Cargill. Mannitol standard product was purchased by Sigma.

2.2 *One-Step Transformation Method of Mannitol*

The activated *L. buchneri* CGMCC 7300 cells were added by 2% cell quantity to different MRS-improved fermentation medium with different total concentration of fructose (F) and glucose (G) (60, 90, 120, 150, 180 g/L) and different ratios of them (F:G = 1:2, 1:1, 3:2, 2:1) at 30 °C for 72 h.

2.3 *Two-Step Transformation Method of Mannitol*

The cells were added by 2% (v/v) quantity to MRS-improved medium to study the effects of different temperature (25, 30, 32, 35, 37, 40, 45 °C) and different initial pH (3.5, 4.0, 4.5, 5.0, 5.5, 6.0, 6.5, 7.0, 7.5, 8.0) on the growth condition of *L. buchneri* CGMCC 7300, and then conform the best growth condition.

The cells were cultured at the best growth condition, and then collected by centrifugation at 6000 r/min for 10 min and washed twice with sterile water. Then suspended with the mixture of fructose and glucose at different concentrations (90, 120, 150 g/L) and 0.5% yeast powder for biotransformation at different pH (3.0, 4.0, 5.0, 4.0, 5.0) and different temperature (25, 28, 30, 32, 35, 37, 40 °C) for 48 h. The concentration of mannitol and residual sugar were detected by high performance liquid chromatography (HPLC).

2.4 HPLC Analysis

Sugars and mannitol were analyzed by HPLC (Agilent, USA). An aminex HPX-87P column with a Carbo-P micro-guard column was used. The column was maintained at 85 °C, and the sugars and mannitol were eluted with deionized water (Milli-Q, Millipore Corp., Bedford, MA) at a flow rate of 0.6 mL/min. Fructose, glucose, and mannitol were identified by refractive index and quantified by comparison to retention time of authentic standards.

3 Results and Discussion

3.1 One-Step Transformation Method for Mannitol

3.1.1 The Effects of the Ratios of Fructose and Glucose on the Transformation of Mannitol

In one-step transformation method, the total concentration of sugars was 90 g/L. The production of mannitol was influenced by the ratio of fructose and glucose. It is evident from the results presented in Fig. 1. When carbon source was only glucose, the production of mannitol was very low, while using fructose alone, the output of mannitol reached 35.92 g/L and the conversion rate was 39.91%. When use the mixture of fructose and glucose as carbon source, the ratio of fructose and glucose 3:2 proved to be the best proportion, and mannitol production could reach 49.98 g/L and the conversion reached 55.53%.

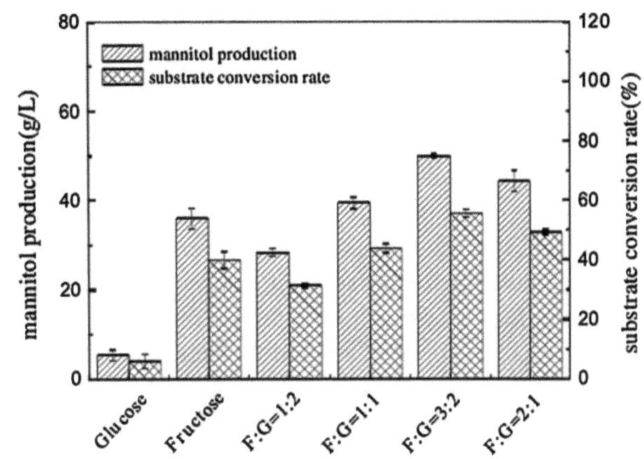

Fig. 1 Effect of different fructose/glucose ratios on the production of mannitol (n = 3)

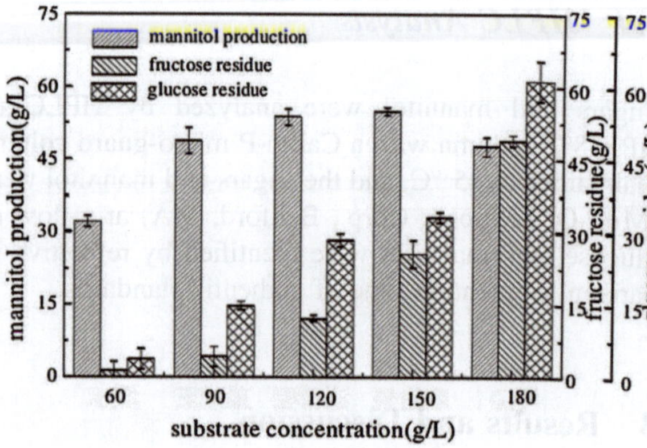

Fig. 2 Effects of different substrate concentration on the transformation of mannitol (n = 3)

3.1.2 The Effects of the Substrate Concentration on the Transformation of Mannitol

Under the condition of the ratio of fructose and glucose set to be 3:2, change the concentration of total sugar in the medium (60, 90, 120, 150, 180 g/L) and investigate their effects on the mannitol transformation. As can be seen in Fig. 2, high substrate concentration up to 180 g/L limited the transformation, while as the concentration of substrate increased from 60 to 150 g/L, the mannitol concentration also increased. When total sugar concentration was 150 g/L, the highest yield of mannitol 55.01 g/L was achieved, but the residual amount of fructose and glucose also increased to be 25.49 and 58.01 g/L respectively. Overall consideration, the substrate concentration of 90 g/L was the best, in which mannitol production reached 48.82 g/L, while fructose residual decreased to be 4.15 g/L, and conversion rate reached 50.43% and the proportion of mannitol in the total sugar reached 72.18%.

3.2 *Two-Step Transformation Method for Mannitol*

The two-step transformation method includes the first step of cell growth and the second step of cell catalysis. The essence of cell catalysis is the enzyme catalysis in biological system. Compared with the extraction of enzyme, whole-cell catalysis can use cellular cofactor and other enzymes for reducing cost and improving the efficiency of biological catalysis. As catalyst, the concentration of the cells determines the transformation speed of mannitol. Therefore, it is critical for transformation to optimize the growth conditions to maximize the cell concentration. In addition, it is also important to optimize the cell catalysis conditions for transformation.

3.2.1 Optimization of the Cell Growth Step

Except for the composition of medium, temperature and pH also affected the cell growth of *L. buchneri* CGMCC 7300. The effects of temperature on growth at 25, 30, 32, 35, 37, 40, and 45 °C were investigated. As can be seen in Fig. 3, cell concentration achieved maximum at 37 °C for OD600 0.4. The effects of different pH from 3.5 to 8.0 on the growth were also compared. Shown in Fig. 4, pH from 5.5 to 6.5 were suitable for the growth of L. buchneri CGMCC 7300, and the maximum cell concentration was reached at pH 6.5. The growth was obviously restrained at the pH lower than 4.0 or higher than 7.5.

3.2.2 The Optimization of the Cell Catalysis Step

The medium was removed by centrifuge followed by the catalysis step. The conditions in this step needs to satisfy the cell catalysis, but not the cell growth. So, the conditions optimized in the one-step transformation method or the first cell growth

Fig. 3 Effect of temperature on growth of *L. buchneri* CGMCC 7300 (n = 3)

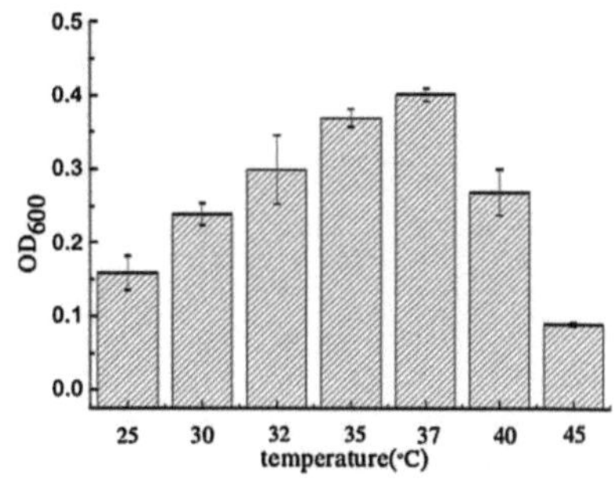

Fig. 4 Effect of different initial pH on growth of *L. buchneri* CGMCC 7300 (n = 3)

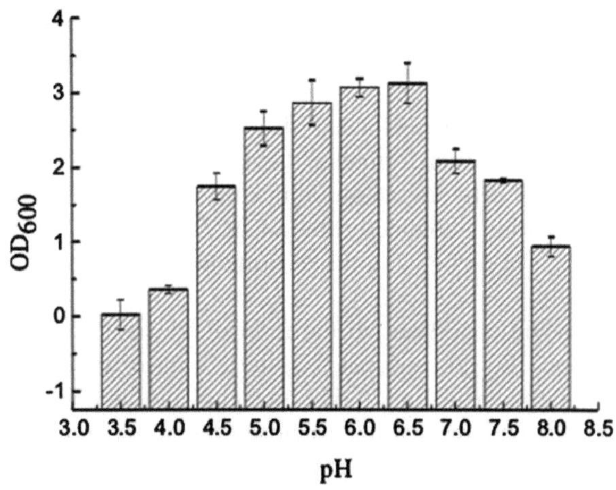

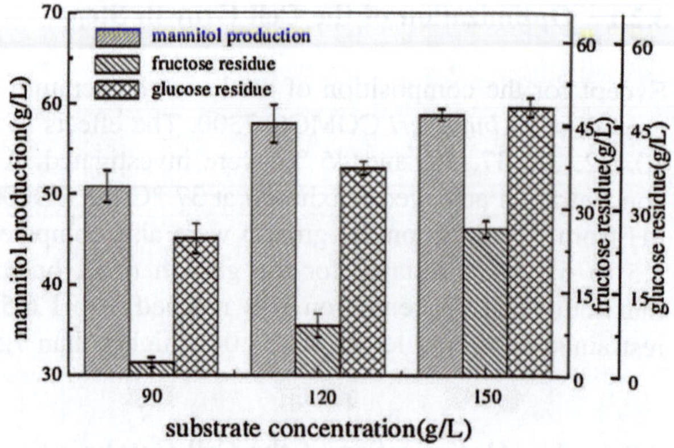

Fig. 5 Effect of substrate concentration on cell catalysis (n = 3)

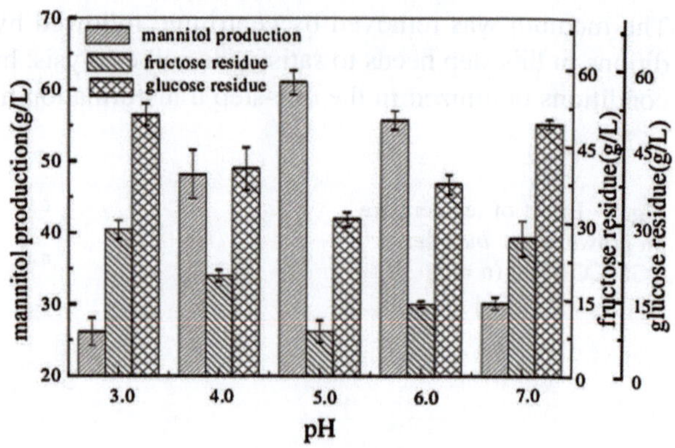

Fig. 6 Effect of pH on cell catalysis of mannitol (n = 3)

step may not suit the second cell catalysis step. The optimal substrate concentration, pH and temperature for cell catalysis were optimized in this study.

Taking both the yield and the conversion rate in account, the substrate concentration 120 g/L was considered the best, as was shown in Fig. 5. At 120 g/L, the mannitol concentration reached 57.81 g/L, and the residual fructose was only 8.93 g/L.

For the whole-cell catalysis of mannitol, the mannitol dehydrogenase plays a key role in the transformation [15]. Therefore, a suitable pH was required for the mannitol dehydrogenase to maximize the mannitol output. As shown in Fig. 6, the best pH for whole-cell catalysis by *L. buchneri* CGMCC 7300 was 5.0. Also, when pH $\geq$ 7.0 or $\leq$ 3.0, the enzyme activity of the mannitol dehydrogenase reduced greatly.

In the conditions of the ratio of fructose and glucose 3:2, total sugar concentration 120 g/L and pH 5.0, the effects of the temperature on the transformation were shown in Fig. 7. The mannitol production reached the highest 63.71 g/L at 32 °C, and meanwhile fructose was completely consumed. It indicated that the optimal temperature for transformation was different from that for cell growth.

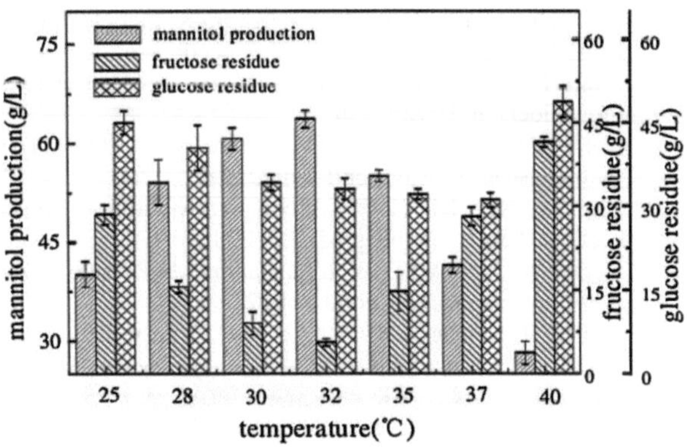

Fig. 7 Effect of different temperature on cell biocatalyst (n = 3)

4 Conclusions

In this study, both the one-step method and two-step method were used for the transformation of mannitol by *L. buchneri* CGMCC 7300 with the mixture of fructose and glucose as substrate, and their conditions were optimized respectively. The best conditions for one-step method were as follows: the ratio of fructose and glucose 3:2, substrate concentration 90 g/L. After fermentation at these conditions for 60 h, mannitol concentration reached 48.82 g/L and the conversion reached 50.43%. The best conditions of the two-step transformation method were as follows: in the first step, the initial pH was 6.5 and the temperature 37 °C; in the next cell catalysis step, substrate concentration 120 g/L, catalysis at 32 °C and pH 5.0 for 36 h. At these conditions, mannitol concentration reached 63.71 g/L with the residual fructose only 6.67 g/L and the conversion rate up to 50.24%. The yield of mannitol increased 30.5% more than that of the one-step transformation method, showing that the two-step transformation method had a great advantage in mannitol production. The results of this study will be significant for further industrial-scale production of mannitol.

Acknowledgements This study was supported by the National Natural Science Foundation (21306140).

References

1. Saha BC, Racine FM (2011) Biotechnological production of mannitol and its applications. J Appl Microbiol Biotechnol 89(4):879–891
2. Soetaert W, Buchholz K, Vandamme E (1995) Production of D-mannitol and D-lactic acid by fermentation by *Leuconostoc mesenteroides*. J Agro Food Ind Hi-Tech 6(1):41–44
3. Saha BC (2003) Production of mannitol by fermentation. In: Saha BC (ed) Fermentation biotechnology. American Chemical Society, Washington, DC, pp 67–85

4. Itoh Y, Tanaka A, Araya H, Ogasawara K, Inabi H, Sakamoto Y, Koga J (1992) *Lactobacillus* B001 for manufacture of mannitol, acetic acid, and lactic acid. European Patent 486–024
5. Erten H (1998) Metabolism of fructose as an electron acceptor by *Leuconostoc mesenteroides*. Proc Biochem 33:735–739
6. Korakli M, Schwarz E, Wolf G, Hammes WP (2000) Production of mannitol by *Lactobacillus* sanfranciscensis. Adv Food Sci 22:1–4
7. Yun JW, Kim DH (1998) A comparative study of mannitol production by two lactic acid bacteria. Ferment Bioeng 85:203–208
8. Korakli M, Schwarz E, Wolf G, Hammes WP (2000) Production of mannitol by *Lactobacillus sanfranciscensis*. Adv Food Sci 22:1–4
9. Richter H, Hamann I, Unden G (2003) Use of the mannitol pathway in fructose fermentation of Oenococcus oeni due to limiting redox regeneration capacity of the ethanol pathway. Arch Microbiol 179:227–233
10. Von Weymarn N, Hujanenmn M, Leisola M (2002) Production of D-mannitol by heterofermentative lactic acid bacteria. Process Biochem 37:1207–1213
11. Yun JW, Kim DH (1998) A comparative study of mannitol production by two lactic acid bacteria. J Ferment Bioeng 85:203–208
12. Saha BC, Nakamura LK (2003) Production of mannitol and lactic acid by fermentation with Lactobacillus intermedius NRRL B-3693. Biotechnol Bioeng 82:864–871
13. Major NC, Bull AT (1989) The physiology of lactate production by *Lactobacillus delbrueckii* in a chemostat with cell recycle. Biotechnol Bioeng 34:592–599
14. Vick Roy TB, Mandel DK, Dea DK, Blanch HW, Wilke CR (1983) The application of recycle to continuous fermentative lactic acid production. Biotechnol Lett 5:665–670
15. Mulchandani A, Rogers KR (1998) Enzyme and microbial biosensors. Humana Press, M USA
16. Major NC, Bull AT (1989) The physiology of lactate production by Lactobacillus delbrueckii in a chemostat with cell recycle. Biotechnol Bioeng 34:592–599

Optimization of Electroporation Conditions for *Arthrobacter simplex*

Jianmei Luo, Haijie Xue, Shengnan Zhao, Xiangsheng Liu, Xiangrong Chen, Jiajia Liu, Yanbing Shen and Min Wang

1 Introduction

Steroid hormones constitute a class of important drug intermediates, and the dehydrogenated steroids have better physiological activity than their precursors, and have fewer side effects. The use of microbes in dehydrogenation exhibits advantages such as few processing steps, simple procedure, and high product purity for steroids [7]. Among of these microorganisms capable of 1-en-dehydrogenation ability, *A. simplex*, a gram positive strain, is extensively used in steroid industry because of its superior bioconversion and high specificity [6]. It is clear that strain improvement is very necessary for the enhancement of biotransformation efficiency. However, genetic manipulation of microorganism has been obstructed because there was no versatile and efficient transformation protocol that suitables for *A. simplex* reported to date.

Electroporation, in which a high intensity pulse produces transient pores in the cell membrane and hence facilitates DNA uptake, is a rapid and widely used transformation method [8, 14, 16]. To date, there were only a few reports of electroporation of *Arthrobacter* strains. Morikawa firstly reported the instance of *Arthrobacter* sp.MIS38 electroporation and the transformation efficiency reached 3×10^5 transformants/μg plasmid DNA, which was used as a starting point for subsequent DNA recombination work and genetic engineering of *Arthrobacter* strain [9]. Zhang Haihong has developed the electroporation method for

H. Xue · S. Zhao · X. Liu · X. Chen · J. Liu · Y. Shen · M. Wang
Key Laboratory of Industrial Fermentation Microbiology (Tianjin University of Science and Technology), Ministry of Education, Tianjin Key Lab of Industrial Microbiology, College of Biotechnology, Tianjin University of Science and Technology, Tianjin 300457, People's Republic of China

J. Luo (✉)
TEDA, No. 29, 13th Avenue, Tianjin 300457, People's Republic of China
e-mail: luojianmei@tust.edu.cn

H. Liu et al. (eds.), *Advances in Applied Biotechnology*, Lecture Notes in Electrical Engineering 444, https://doi.org/10.1007/978-981-10-4801-2_40

Arthrobacter sp. A3 by optimizing the concentration of DNA, recovery time period, electrical parameters and some other factors, and an efficiency of 6.8×10^7 transformants/μg plasmid DNA was finally achieved [19]. Some exogenous genes have been transformed into a few *Arthrobacter* species, such as *Arthrobacter aureus* C70 [5], *Arthrobacter* sp. DS7 [13] and *Arthrobacter aurescens* DSM 3747 [18], but the detailed conditions of electroporation and efficiency values were not published.

In this paper, the entire electroporation process for *A. simplex* was firstly developed and optimized. Factors affecting electroporation efficiency were examined, including the cell growth phase, wall-weakening treatment conditions, DNA amounts and electroporation parameters. Furthermore, the transformation efficiency of our final protocol was also evaluated by the green fluorescent protein (GFP) in *A. simplex* and other type of *Arthrobacter* strains, such as *A. globisformis* ATCC 8010.

2 Materials and Methods

2.1 *Strains, Plasmid, and Media*

The *A. simplex* CPCC 140451 was obtained from China Pharmaceutical Culture Collection (CPCC). The wide type *A. globiformis* ATCC 8010 was obtained from American Type Culture Collection (ATCC). *Escherichia coli* strains DH5α was used for construction, copy, and storage plasmid DNA, which was purchased from Transgene Biotech (Beijing, China).

The plasmid pART2 and pART2-*gfp* [11] were respectively used to transform *A. simplex* CPCC 140451, *E. coli* and *A. globiformis* ATCC 8010, which were kindly provided by Cristinel Sandu (University of Rockefeller, USA). The plasmid pART2 has a size in 4634 bp, which includes the following features: pCG100 and ColE1 origins of replication; Kan^r, kanamycin resistance gene; P, promoter/operator of the hdnO gene (hdnOp); MCS; His8, eight-histidine tag coding sequence. The plasmid pART2-*gfp* was constructed by ligating the green fluorescent gene *(gfp)* in a size of 717-bp into the pART2 vector.

A. simplex and *A. globisformis* were routinely grown in LB broth at 32 °C. *E. coli* was grown in LB broth at 37 °C. When appropriate, kanamycin was added to medium at the final concentration 50 μg/ml for recombinant strains.

2.2 *Material*

The restriction enzymes, DNA-modifying enzymes, DNA markers and T4 Ligases were purchased from TaKaRa (Dalian, China). The kits for DNA purification, gel recovery and plasmid mini-prep were purchased from Omega Bio-Tek (Wisconsin, USA). Penicillin, glycine, threonine, and lysozyme were purchased from solarbio

(Beijing, China). Protein Marker II was purchased from Biomed (Beijing, China). Taq DNA polymerase, dNTPs was purchased from Sangon (Shanghai, China).

2.3 *Preparation of Electro-Competent Cells*

The optimized procedure for the preparation of electro-competent cells was as follows: A 1% inoculum from the cultures at the exponential growth phase of the first stage of *A. simplex* CPCC 140451 was inoculated into fresh LB broth. The cells were cultivated at 32 °C on a rotary shaker (160 rpm) until the optical density (OD) approached approximately 1.0 at 600 nm. Different kinds of wall-weakening agents, including glycine, threonine, lysozyme and penicillin G was respectively added at different final concentrations, and continued to cultivate for an additional 1 h. The cultures were quickly placed on ice-water for at least 10 min and were harvested by centrifugation (2000 g, 15 min, and 4 °C). Following three washes in ice-cold electroporation buffer (10% glycerol containing 0.5 M sorbitol), the cells were concentrated 100-fold. Finally, the suspension was distributed into 60 μl aliquots and stored at −80 °C.

2.4 *Electroporation Procedures to* A. Simplex *and* A. Globiformils

The optimized electroporation procedure was as follows: After thawing the electro-competent cells on ice, 60 μl of the cells were mixed with 200 ng of DNA (prepared from *E. coli* DH5/pART2-*gfp*) and this mixture was then transferred into a pre-chilled electroporation cuvette (0.1 cm electrode, Bio-Rad Laboratories, CA, USA). The mixture was subjected to a single pulse using a Bio-Rad X cell Gene Pulser (Bio-Rad Laboratories) with settings of 22 kV/cm, 25 μF, and 400 Ω. Immediately after the electric pulse, cell suspensions were transferred into an Eppendorf tube containing 800 μl of room temperature recovery medium (LB broth containing 0.5 M sorbitol) and were incubated with shaking at 120 rpm for 8 h at 32 °C. For selection of transformed cells, serial dilutions of bacteria were spread on LB agar plates containing kanamycin. With each set of experiment, negative controls were performed by omitting either the electric pulse or the addition of plasmid DNA; no transformants arose on these negative control plates. Transformation efficiency was defined as the number of transformants per microgram of DNA used. For each parameter tested, three independent transformation assays were conducted. Consensus results were acquired for each parameter.

2.5 Confirmation of Transformants

To confirm transformants, random transformants grown on the LB plates were selected and inoculated into LB broth (containing 50 μg/ml kanamycin) at 32 °C on a rotary shaker (160 rpm) for 30 h. The cultures were harvested and used for observing the green fluorescence by fluorescence microscope under blue light. The strain only harboring the plasmid pART2 was used as a negative control.

2.6 The Transformation of Green Fluorescent Protein Gene (gfp) in A. Simplex and A. Globisformis

The plasmid pART2-*gfp* was transformed into *A. simplex* CPCC 140451 using the above optimized electroporation protocol. The strains only containing the plasmid pART2 was used as the control strain.

3 Results and Discussion

3.1 Optimization of Cell Growth Phase

The cell growth curve of *A. simplex* was shown in Fig. 1. The strain was in the lag phase when cultured for 8 h with the OD_{600} below 1.0. *A. simplex* was in the logarithmic phase when it continued to cultivate until 18 h. It was previously reported that cells collected in the exponential growth phase produced better electroporation efficiency. In this study, cells at different phase were harvested and used to prepare the competent cells. As shown in Fig. 2, the resultant transformation efficiency varied slightly between an OD_{600} of 0.2–0.6, but a rapid increase in transformation efficiency was observed as OD_{600} value was further increased. The top transformation efficiency (4.83×10^3 transformants/μg DNA) was reached at OD_{600} of 1.0. With the continue increasing of OD_{600}, the transformation efficiency exhibited slight decrease tendency. Therefore, the best cell growth stage was selected at the OD_{600} of 1.0.

3.2 Optimization of Wall-Weakening Treatment Conditions

A. simplex, as a kind of Gram-positive bacteria with high GC content, its intact cell wall is always considered to be a strong barrier to the foreign DNA [2]. The cell

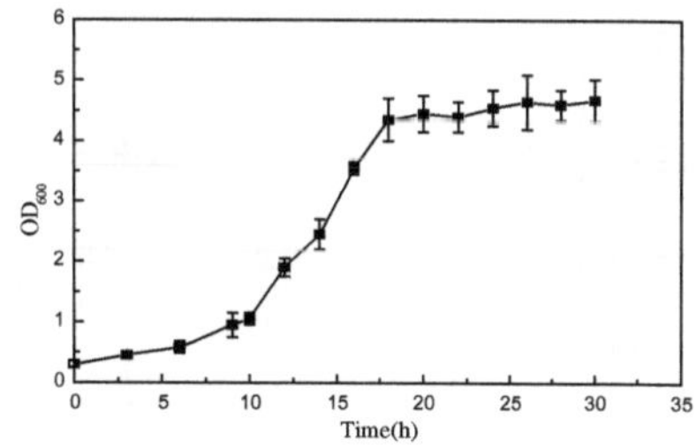

Fig. 1 Growth curve of *A. simplex* CPCC 140451. Data are means ± SD (n = 3) ($P < 0.05$)

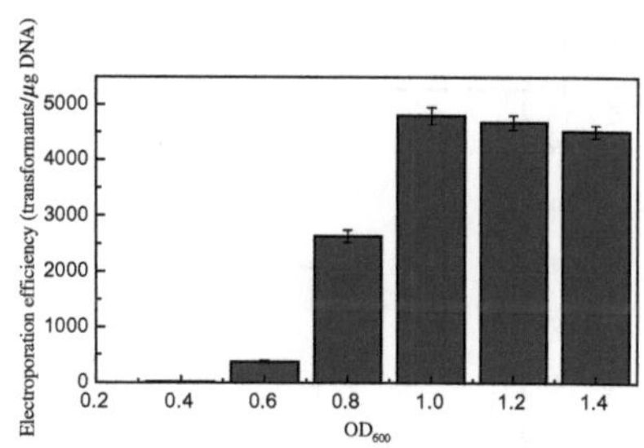

Fig. 2 Effect of cell growth stages on the electroporation efficiency of *A. simplex* CPCC 140451. Data are means ± SD (n = 3) ($P < 0.05$)

wall-weakening treatment is one of the most effective methods for the enhancement on the electroporation efficiency [15]. The effects of four wall-weakening agents, including glycine, threonine, lysozyme, and penicillin on the transformation efficiency of *A. simplex* were investigated and results were shown in Fig. 3.

When the OD_{600} of the culture reached 1.0, different kinds of wall-weakening agent were added to prepare the competent cells respectively, including 0–2.0% (w/v) threonine, 0–2.0% (w/v) glycine, 0–10 μg/ml lysozyme, or 10–110 μg/ml penicillin. An electrical transformation system, including 500 ng of plasmid DNA and 60 μl of competent cells was then constructed. As shown in Fig. 3, The transformation efficiency was about 20 transformants/(μg DNA) without any treatment. The transformation efficiency was greatly increased after treatment by different wall-weakening agents. For threonine, the highest transformation efficiency (207 transformants/(μg DNA)) was obtained at the concentration of 1.5%; For glycine, the highest transformation efficiency (314 transformants/(μg DNA)) was obtained at the concentration of 1.0%; For lysozyme, the highest transformation efficiency (352 transformants/(μg DNA)) was obtained at the concentration of 6 μg/ml; For penicillin, the highest transformation efficiency (5236 transformants/(μg DNA)) was obtained at the concentration of 70 μg/ml. Therefore, the best cell wall-weakening treatment condition was 70 μg/ml penicillin treated for 1 h.

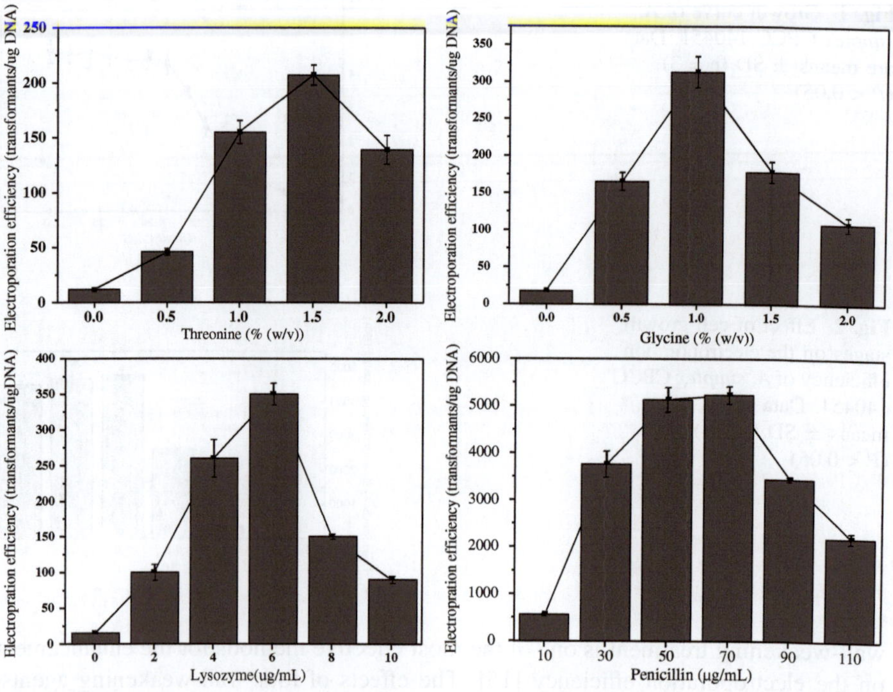

Fig. 3 Influence of cell wall-weakening treatments on the electroporation efficiency of *A. simplex* CPCC 140451. Data are means ± SD (n = 3). ($P < 0.05$)

3.3 *Selection of Optimal DNA Concentration*

The DNA concentration is one of the key factors for electroporation efficiency. Different qualities of DNA (1–2000 ng) were transformed into *A. simplex* CPCC 140451, respectively. Two particular trends were indicated in Fig. 4. When the quantity of DNA was no more than 500 ng, the number of transformants increased along with the augment of DNA quantity but no obvious increase in the transformants number was observed as DNA quantity was further increased. However the highest transformation efficiency (2531 transformants/(μg DNA)) occurred at 200 ng, further increase in DNA concentration resulted in significant decrease in transformation efficiency. Similar results were reported by previous researcher [1, 10]. In conclusion, 200 ng of DNA was selected for our follow-up experiments.

Fig. 4 Effect of DNA amount on the transformants number and electroporation efficiency of *A.simplex* CPCC 140451. Data are means ± SD (n = 3) ($P < 0.05$)

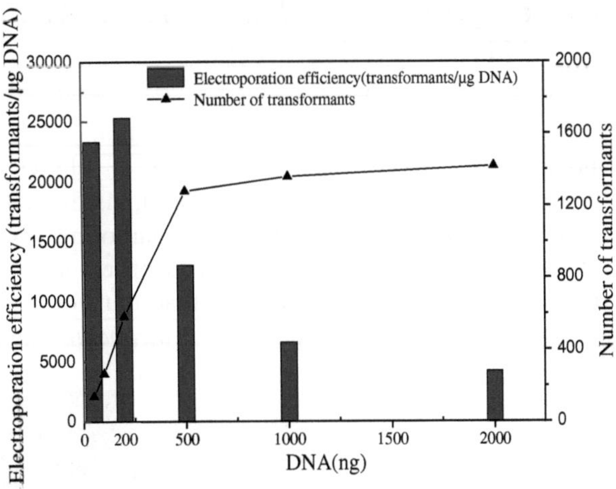

Fig. 5 Effect of voltages on the transformation efficiencies of *A.simplex* CPCC 140451. Data are means ± SD (n = 3) ($P < 0.05$)

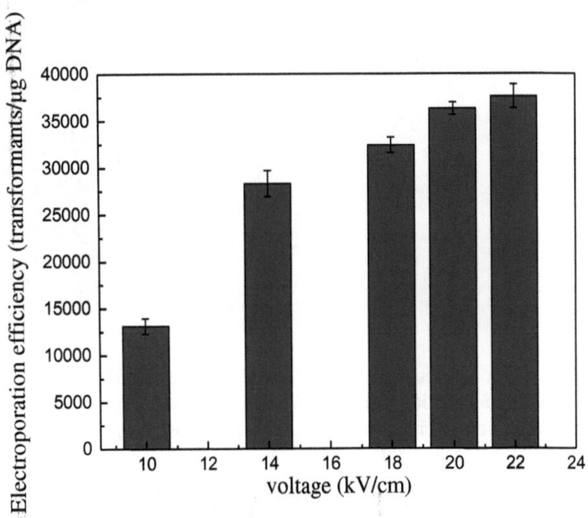

3.4 Optimization of Electroporation Condition

60 μl of competent cells and 200 ng DNA were added into 0.1 cm cuvettes to electroporation. The effects of different voltages on electroporation efficiency were examined. As shown in Fig. 5, the transformation efficiency increased with the raise of voltage ranging from 10 to 22 kV/cm, and the highest value reached 3.76×10^4 transformants/μg DNA at 22 kV/cm.

3.5 Confirmation of Transformants by Fluorescence Microscopy Observation

To validate the authenticity of transformation, random transformants were selected and tested by fluorescence observation. As shown in Fig. 6, the obvious green fluorescent was observed in each tested transformant but no green fluorescent was detected in control strain which only harboring pART2 plasmid by fluorescence microscopy. Similar results were also observed in *A. globisformis* ATCC 8010 (Fig. 7). These results confirmed that transformation proposed procedure was successful for *A. simplex* and other *Arthrobacter* strains, such as *A. globisformis.*

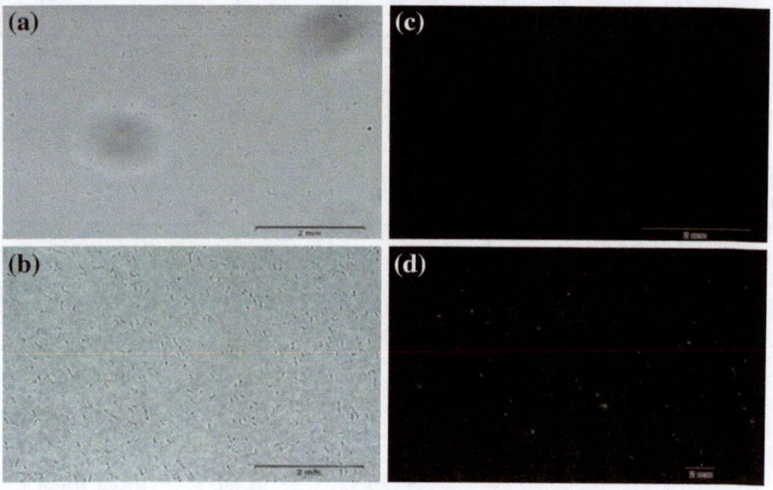

Fig. 6 Conformation of the strain *A. simplex*/pART2-*gfp* by fluorescence microscopy **a**. Control strain under white light; **b**. *A. simplex*/pART2-*gfp* under white light; **c**. Control strain under fluorescence microscopy; **d**. *A. simplex*/pART2-*gfp* under fluorescence microscopy

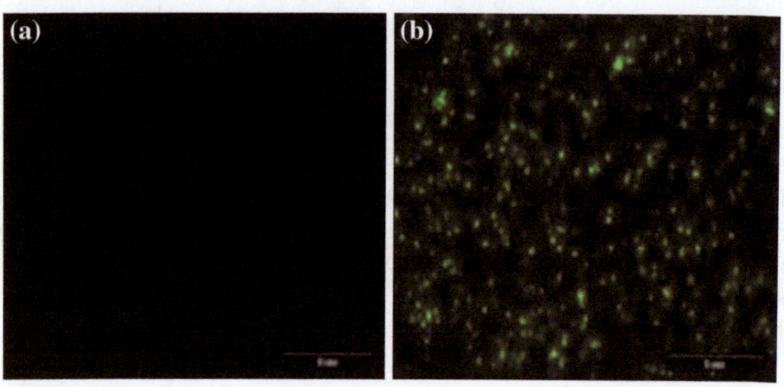

Fig. 7 Confirmation of *A. globisformis*/pART2-*gfp* by fluorescence microscopy **a**. Control strain under fluorescence microscopy; **b**. *A. globisformis*/pART2-*gfp* under fluorescence microscopy

4 Conclusions

In this study, we firstly describe the electroporation protocol for *A. simplex* CPCC 140451 with results of 3.76×10^4 transformants/μg DNA. To obtain high transformation efficiency, the parameters of two aspects were optimized. The first aspect is to optimize pre-treatment conditions of competent cells, including the cell growth phase and wall-weakening treatment conditions. Our results indicate that cells collected in the early stage of exponential phase give rise to better electroporation efficiencies than those collected in the middle or late period of log phase. The cell wall modification by using different kinds cell wall-weakening agents (glycine, threonine, lysozyme and penicillin) indeed promote electroporation efficiency to some extent and the best result is obtained at the penicillin treatment, which inhibitor the synthesis of peptidoglycan and has previously been shown to stimulate electro-competence in various bacteria [4, 12, 17], The second aspect is to optimize the factors of transformation process. Our results prove that both the quantity of DNA and high-voltage pulses are important parameters for electroporation efficiency. The optimized electroporation protocol can be well applied to other different strains (such as *A.globisformis* ATCC 8010), which provide feasible means for genetic manipulations in *A. simplex* CPCC 140451.

Acknowledgements This work was supported by the National Natural Science Foundation of China (No. 21646017 and 21306138) and the Lab Innovation Foundation of Tianjin University of Science and Technology (No. 1504A205X).

References

1. Dorella, FA, Estevam EM, Cardoso PG, Savassi BM, Oliveira SC, Azevedo V (2006) An improved protocol for electrotransformation of Corynebacterium pseudotuberculosis. Vet Microbiol 114(3–4):298–303
2. Dunny GM, Lee LN, LeBlanc DJ (1991) Improved electroporation and cloning vector system for Gram-positive bacteria. Appl Environ Microbiol 57(4):1194–1201
3. Fernandes P, Cruz A, Angelova B, Pinheiro HM, Cabral JMS (2003) Microbial conversion of steroid compounds: recent developments. Enzyme Microb Technol 32(6):688–705
4. Kurusu Y, Kainuma M, Inui M, Satoh Y, Yukawa H (1990) Electroporation–transformation system for coryneform bacteria by auxotrophic complementation. Agric Biol Chem 54 (2):443–447
5. Le Marrec C, Michotey V, Blanco C, Trautwetter A (1994) ΦAAU2, a temperate bacteriophage specific for "Arthrobacter aureus", whose integrative functions work in other corynebacteria. Microbiol 140(11):3071–3077
6. Luo JM, Ning J, Wang YX, Ning J, Wang YX, Cheng YX, Zheng Y, Shen YB, Wang M (2013) The effect of ethanol on cell properties and steroid 1-en-dehydrogenation biotransformation of Arthrobacter simplex. Biotechnol Appl Biochem 61(5):555–564
7. Marina VD, Olga VE (2012) Microbial steroid transformations: current state and prospects. Appl Microbiol Biotechnol 94(6):1423–1447
8. McDonald IR, Riley PW, Sharp RJ, McCarthy AJ (1995) Factors affecting the electroporation of Bacillus subtilis. J Appl Microbiol 79(2):213–218

9. Morikawa M, Daido H, Pongpobpibool S, Imanaka T (1994) Construction of a new host-vector system in Arthrobacter sp. and cloning of the lipase gene. Appl Microbiol Biotechnol 42(2):300–303
10. Powell IB, Achen MG, Hillier AJ, Davidson BE (1998) A simple and rapid method for genetic transformation of lactic streptococci by electroporation. Appl Environ Microbiol. 54(3):655–660
11. Sandu C, Chiribau CB, Sachelaru P, Brandsch R (2005) Plasmids for nicotinedependent and -independent gene expression in Arthrobacter nicotinovorans and other Arthrobacter species. Appl Environ Microbiol 71(12):8920–8924
12. Satoh Y, Hatakeyama K, Kohama K, Kobayashi M, Kurusu Y, Yukawa H (1990) Electrotransformation of intact cells of Brevibacterium flavum MJ-233. J Ind Microbiol 5(2):159–165
13. Serbolisca L, de Ferra F, Margarit I (1999) Manipulation of the DNA coding for the desulphurizing activity in a new isolate of Arthrobacter sp. Appl Microbiol Biotechnol 52(1):122–126
14. Shigekawa K, Dower WJ (1988) Electroporation of eukaryotes and prokaryotes: a general approach to the introduction of macromolecules into cells. Biotechniques 6(8):742–751
15. Sunairi M, Iwabuchi N, Murakami K, Watanabe F, Ogawa Y, Murooka H, Nakajima M (1996) Effect of penicillin G on the electroporation of Rhodococcus rhodochrous CF222. Lett Appl Microbiol 22(1):66–69
16. Weaver JC (1995) Electroporation theory. Concepts and mechanisms. Meth Mol Biol 48:3–28
17. Wei MQ, Rush CM, Norman JM, Hafner LM (1995) An improved method for the transformation of Lactobacillus strains using electroporation. J Microbiol Meth 22(1):66–69
18. Wiese A, Syldatk C, Mattes R, Altenbuchner J (2001) Organization of genes responsible for the stereospecific conversion of hydantoins to α-amino acids in Arthrobacter aurescens DSM 3747. Arch Microbiol 176(3):187–196
19. Zhang HH (2010) Optimization of electroporation conditions for Arthrobacter with plasmid PART2. J Microbiol Meth 84(1):114–120

Medium Optimization for γ-Aminobutyric Acid Production by Response Surface Methodology

Chuan-You Chang, Shen-Xi Ma, Jun Zhang and Qiang Gao

1 Introduction

γ-Aminobutyric acid (GABA) exists widely among animals, plants and microorganisms as a natural non-protein amino acid [1–3]. Also, GABA acts as major inhibitory neurotransmitter in the mammalian central system. Due to its involvement in many important physiological functions, GABA has broad potential for application as a bioactive additive in the food and pharmaceutical industries [4–7].

GABA can be produced by chemical and biological syntheses [8–10]. However, GABA produced by chemical methods cannot be used in food industry due to its unsafety. On the contrary, many kinds of microorganisms, such as lactic acid bacteria (LAB) and yeasts, are generally recognized as safe (GRAS). Moreover, some GABA-producing LABs have been reported in recent years, including *Lactobacillus brevis, Lactococcus lactis* and *Lactobacillus buchneri* [11–13]. Besides, LABs exhibit a great promising for using as a starter in fermented food, such as yoghurt, bread and cheese [14, 15].

Culture conditions, such as medium, exhibit a great effect on GABA production in various strains. Response surface methodology (RSM) is one of the best experimental strategies to seek optimal conditions for multivariable system and have been successfully employed for optimizing the medium composition [16–18]. To improve the GABA production of *Enterococcus raffinosus* TCCC11660, RSM protocol was used for optimizing conditions for fermentation medium composition in the present work.

C.-Y. Chang · S.-X. Ma · J. Zhang · Q. Gao (✉)
Key Laboratory of Industrial Fermentation Microbiology, Ministry of Education,
College of Biotechnology, Tianjin University of Science and Technology,
Tianjin 300457, People's Republic of China
e-mail: gaoqiang@tust.edu.cn

H. Liu et al. (eds.), *Advances in Applied Biotechnology*, Lecture Notes
in Electrical Engineering 444, https://doi.org/10.1007/978-981-10-4801-2_41

2 Materials and Methods

2.1 *Strain and Materials*

Enterococcus raffinosus strain TCCC11660 (CGMCC No. 5584) used in this work was isolated from Chinese traditional pickled vegetables in our previous work, and the initial medium and cultural condition were the same as previously described [19].

Pyridoxal 5′-phosphate (PLP) and GABA standard were bought from Sigma-Aldrich (St Louis, MO, USA). Monosodium glutamate (MSG, purity 99%) was bought from Neimenggu Fufeng Biotechnologies Co., Ltd. (Hohhot, Inner Mongolia, China). All other chemicals were of analytical or biochemical grades.

2.2 *Qualitative and Quantitative Analysis of GABA*

GABA concentrations in the fermentation broth were assayed using high performance liquid chromatography (HPLC) method as previously described [6, 20].

2.3 *Evaluation of the Key Medium Ingredients*

The Plackett-Burman statistical design is a particularly effective method for screening a large numbers of factors, which provides sufficient information for investigating the significant factors in a given response. In this work, in order to complete an evaluation of the medium ingredients, the 9 variables of citric acid (A), glycine (B), PLP (C), arginine (D), Na_2MoO_4 (E), asparagine (F), VB_2 (G), VB_3 (H) and VB_6 (I) were evaluated using this design (Table 1). The selected ingredients were reported as comprising a key medium supplement for GABA production by LABs. The factors investigated in the design was utilized two levels: the high level (+1) and the low level (−1). Meanwhile, the yield of GABA was used as response for statistical analysis of the Plackett-Burman design. According to the results of ANOVA analysis, variables on the response were considered to be significant while the P-value < 0.05.

2.4 *Response Surface Methodology Design*

The Box-Behnken design of the response surface methodology (RSM) was proposed in order to analyze the yield of GABA by *E. raffinosus* TCCC11660, as a function of the chosen factors on the response surfaces of the investigated region.

Table 1 Levels of independent variables in Plackett-Burman design

Variables	Level	
	Low (−1)	High (+1)
Citric acid (g/L)	8	10
Glycine (g/L)	0.5	1
PLP (mmol/L)	0.1	0.3
Arginine (g/L)	15	20
Na_2MoO_4 (mmol/L)	4	6
Aspartic acid (mg/L)	8	10
VB_2 (mg/L)	3	5
VB_3 (mg/L)	3	5
VB_6 (mg/L)	5	10

Table 2 The factor levels of Box-Behnken design

Factor	Coded values and corresponding concentrations		
	1	0	−1
Citric acid (g/L)	8.0	9.0	10
Glycine (g/L)	0.6	0.8	1.0
PLP (mmol/L)	0.1	0.2	0.3

The screened factors in the Plackett-Burman design were used for the Box-Behnken design, and each factor was examined at three different coded levels (Table 2). The Design-Expert (Version 8.0.6) software was used for RSM design and regression analysis. The *P*-value < 0.05 was considered to be significant.

3 Results and Discussion

3.1 Evaluation of Significant Medium Components Affecting GABA Production

Nine components were analyzed for their effects on GABA production using the Plackett-Burman design. The results were presented in Table 3 and the analysis of variance (ANOVA) for GABA production was performed in Table 4. Based on the ANOVA results, citric acid, PLP, Na_2MoO_4 and glycine showed significant effects (*P*-value < 0.05) on GABA production while the rest of the other ingredients had no statistically significant effects (*P*-value > 0.05) on it. Besides, the analysis of coefficients indicated that citric acid, glycine and PLP had positive effects on GABA production while Na_2MoO_4 exerted negative effect. The most significant variable was citric acid, which had the highest coefficient, followed by glycine and PLP. Then the significant variables were used for further analysis.

Table 3 Plackett-Burman experimental design and results

No.	A	B	C	D	E	F	G	H	I	GABA (g/L)
1	1	−1	1	−1	−1	−1	1	1	1	67.53
2	−1	1	1	−1	1	−1	−1	−1	1	69.80
3	1	1	1	1	−1	−1	−1	1	−1	71.49
4	1	1	−1	−1	1	1	−1	1	1	70.06
5	−1	−1	−1	−1	−1	−1	−1	−1	−1	71.23
6	−1	1	1	1	−1	1	1	−1	1	65.91
7	−1	−1	1	1	1	1	−1	1	−1	67.79
8	1	−1	−1	1	−1	1	−1	−1	1	66.72
9	1	−1	1	−1	1	1	1	−1	−1	64.27
10	−1	−1	−1	1	1	−1	1	1	1	67.24
11	1	1	−1	1	1	−1	1	−1	−1	66.29
12	−1	1	−1	−1	−1	1	1	1	−1	67.09

Table 4 ANOVA analysis of the main effects for Plackett-Burman design

Variables	Effect	Coefficient	Stand error	T-value	*P*-value
Constant	–	67.9579	0.1181	575.40	0.000
A	2.672	1.336	0.1181	11.31	0.008
B	2.0779	1.0389	0.1181	8.80	0.013
C	1.4309	0.7155	0.1181	6.06	0.026
D	1.5739	0.7869	0.1181	6.66	0.052
E	−0.0965	0.0483	0.1181	0.41	0.021
F	0.6315	0.3157	0.1181	2.67	0.722
G	1.4335	0.5314	0.1181	3.53	0.648
H	−0.9685	−0.4843	0.1181	−4.10	0.055
I	0.3070	0.1536	0.1181	1.30	0.323

3.2 Determination of Optimal Concentration of the Positive Significant Factors

3.2.1 The Effect of Citric Acid Concentration on GABA Production

In three independent experiments, different concentrations of citric acid were added to the medium and the other components in the medium were the same as the initial medium. As shown in Fig. 1, GABA concentration increased rapidly with the increase of citric acid concentration from 5 to 10 g/L and reached maximum at 10 g/L, but decreased when citric acid was above 10 g/L.

Fig. 1 The effect of citric acid concentration on GABA production

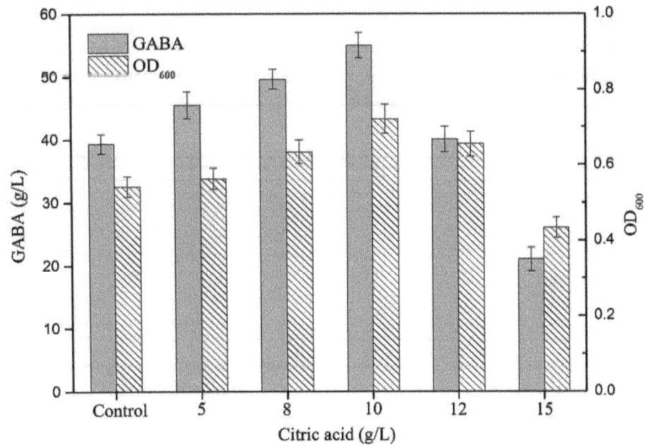

Fig. 2 The effect of glycine concentration on GABA production

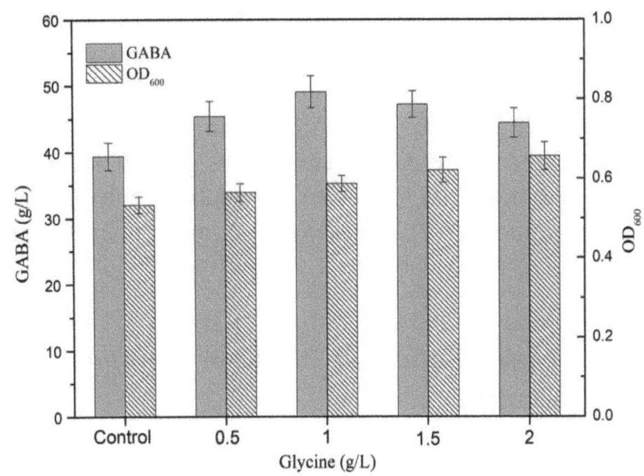

3.2.2 The Effect of Glycine Concentration on GABA Production

In this experiment, different concentrations of glycine were added to the fermentation medium. The other components in the medium were fixed as follows (g/L): sucrose 30, yeast extract 30, MSG 40, $MgSO_4 \cdot 7H_2O$ 0.04, citric acid 10, Tween-80 2 mL/L. The GABA concentration increased with the increase in glycine concentration from 0.5 to 2.0 g/L and reached the maximum at 1.0 g/L, but slightly declined when glycine exceeded 1.5 g/L (Fig. 2).

3.2.3 The Effect of PLP Concentration on GABA Production

Glutamate decarboxylase (GAD) is the sole enzyme that synthesizes GABA through the irreversible decarboxylation of L-glutamate. As a PLP-dependent enzyme, GAD activity can be significantly improved by PLP. The concentration of

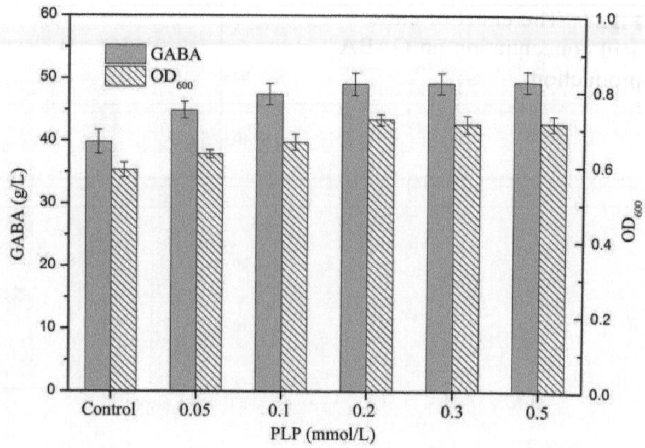

Fig. 3 The effect of PLP concentration on GABA production

PLP added in the medium from 0.05 to 0.5 mmol/L and the other components in the medium were fixed as follows (g/L): sucrose 30, yeast extract 30, L-MSG 40, $MgSO_4 \cdot 7H_2O$ 0.04, K_2HPO_4 2.0, $Al_2(SO_4)_3$ 0.03, citric acid 10, glycine 1.0, and Tween-80 2 mL/L. As indicated in Fig. 3, GABA production increased significantly when different concentrations of PLP were added. The results revealed that PLP was essential to GABA production, but the GABA titer leveled off when the concentration of PLP was equal or more than 0.2 mmol/L.

3.3 *Optimization of Medium Composition by Box-Behnken Design*

In order to approach the GABA yield, significantly positive (P-value < 0.05) variables (citric acid, glycine and PLP) were further studied. The suitable concentration ranges of these variables were also preliminarily determined at three levels each according to the Box-Behnken design (Table 2). As shown in Table 5, the real values of the coded variables by Box-Behnken design together with the results of the GABA yield. The results were analyzed by linear multiple regression using the Design-Expert and the model equation for GABA yield was obtained.

The regression equation was obtained as following:

$$Y_{\mathrm{GABA}} = 77.50 - 0.36 * A - 0.68 * B + 0.42 * C - 0.047 * AB + 0.58 * AC - 0.49 * BC - 1.37A^2 - 2.30 * B^2 - 2.77 * C^2$$

where A, B and C are the symbols of concentration of citric acid, glycine and PLP, respectively.

The results of the response surface model were analyzed using ANOVA to validate the regression coefficient (Table 6). The F-test showed the significance of the model (P-value = 0.0019). Besides, the lack of fit test measures the failure of the model to

Table 5 Designs and responses of Box-Behnken design

No.	A (citric acid)	B (PLP)	C (glycine)	GABA (g/L)
1	1	−1	0	74.24
2	0	0	0	77.47
3	0	0	0	77.71
4	0	0	0	77.32
5	−1	0	1	73.58
6	−1	1	0	73.51
7	−1	0	−1	74.39
8	1	0	1	73.49
9	−1	−1	0	74.33
10	0	1	1	71.68
11	1	1	0	73.23
12	0	−1	1	74.49
13	0	1	−1	71.36
14	0	−1	−1	72.20
15	1	0	−1	71.97

Table 6 Analysis for regression equation

Source	*df*	Sum of squares	Mean square	F-value	*P*-value
Model	9	56.79	6.31	20.82	0.0019
A	1	1.04	1.04	3.42	0.1236
B	1	3.75	3.75	12.39	0.0169
C	1	1.38	1.38	4.55	0.0862
A*A	1	6.97	6.97	22.99	0.0049
B*B	1	19.51	19.51	64.39	0.0005
C*C	1	28.31	28.31	93.41	0.0002
A*C	1	1.36	1.36	4.48	0.0879
A*B	1	9.025×10^{-3}	9.025×10^{-3}	0.030	0.8698
B*C	1	0.97	0.97	3.20	0.1336
Lack of fit	3	1.44	0.48	12.38	0.0756

represent data in the experimental domain at points that are not included in the regression. The *P*-value of lack of fit was 0.0756, implying that the lack of fit test is not significant. A regression model is considered as having a high correlation when the R^2 value of the model is higher than 0.9 [21], and the R^2 value of the present model is 0.99. Thus, it is reliable to use the regression model for further analysis.

The 3D response surface plots serve as graphical representations of the regression model equation for GABA production. The graphical representation provides a method to demonstrate the interaction effects of two parameters when the third one is kept constant at a center point. Figure 4 demonstrated the relationship between

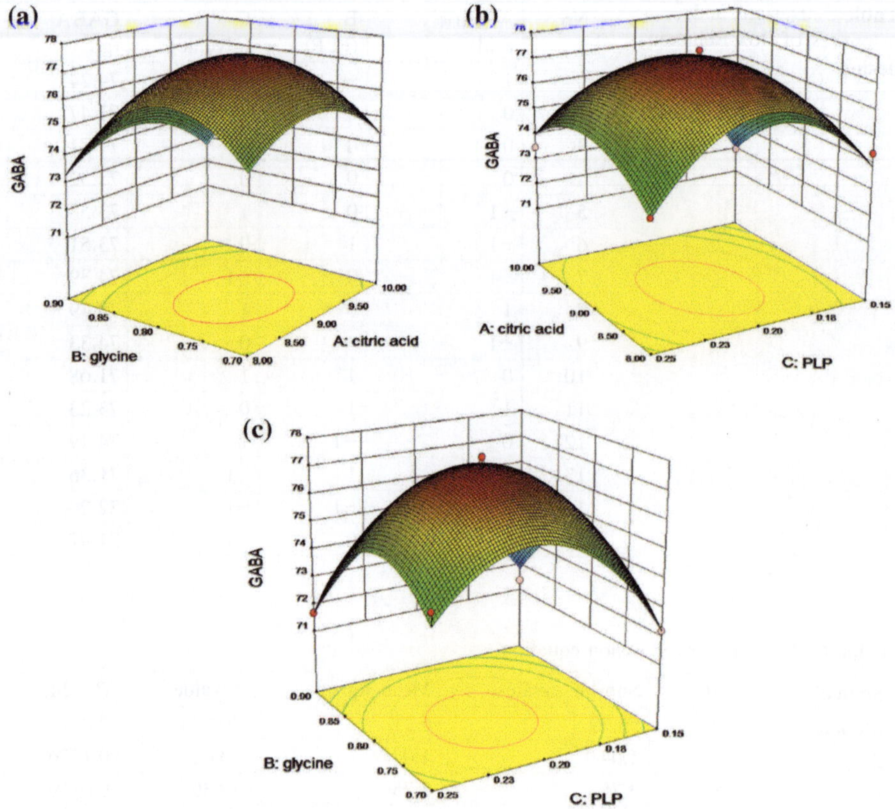

Fig. 4 Three-dimensional surface plots showing the effect of the concentrations of citric acid, glycine and PLP on GABA production

the response and design levels of each variable and the type of interactions between test variables to deduce the optimum conditions on GABA production. Therefore, the optimal medium composition was found as the follows (g/L): citric acid 8.9, glycine 0.82 and PLP 0.21 mmol/L. To verify the proposed medium for GABA production, fermentation experiments were implemented with the optimized nutrients levels. The maximum value of GABA production was 77.5 g/L and 103.9% higher than the initial medium. The results suggested that the tested strain has a great potential for industry applications.

4 Conclusion

This work presents a substantial increase in GABA production via an efficient strategy of optimization of the fermentation medium of *E. raffinosus* TCCC11660. By an evaluation of Plackett-Burman design, the significantly positive effects of sucrose, glycine and PLP were determined and used for Box-Behnken design which is time-saving and inexpensive for optimization of medium. The final key medium composition was found to be citric acid 8.9 g/L, glycine 0.82 g/L and PLP 0.21 mmol/L. After the optimization, the GABA yield achieved 77.5 g/L, which is 103.9% higher than the initial medium. In summary, this study revealed an inexpensive key medium composition for GABA production by *E. raffinosus* TCCC11660.

Acknowledgements This work was financially supported by the National Basic Research Program (973 Program) of China (2013CB734004), the National Natural Science Foundation of China (31370075, 31471725 & 61603273), and the Youth Innovation Fund of Tianjin University of Science & Technology of China (2014CXLG28).

References

1. Cho YR, Chang JY, Chang HC (2007) Production of gamma-aminobutyric acid (GABA) by *Lactobacillus buchneri* isolated from kimchi and its neuroprotective effect on neuronal cells. J Microbiol Biotechnol 17:104–109
2. Sandmeier E, Hale TI, Christen P (1994) Multiple evolutionary origin of pyridoxal-5′-phosphate-dependent amino acid decarboxylases. Eur J Biochem 221:997–1002
3. Trobacher CP, Zarei A, Liu J et al (2013) Calmodulin-dependent and calmodulin-independent glutamate decarboxylases in apple fruit. BMC Plant Biol 13:144
4. Ding JZ, Yang TW, Feng H et al (2016) Enhancing contents of gamma-aminobutyric Acid (GABA) and other micronutrients in dehulled rice during germination under normoxic and hypoxic conditions. J Agric Food Chem 64:1094–1102
5. Park KB, Oh SH (2007) Cloning, sequencing and expression of a novel glutamate decarboxylase gene from a newly isolated lactic acid bacterium, *Lactobacillus brevis* OPK-3. Bioresour Technol 98:312–319
6. Gao Q, Duan Q, Wang DP, Zhang YZ et al (2013) Separation and purification of γ-aminobutyric acid from fermentation broth by flocculation and chromatographic methodologies. J Agric Food Chem 61:1914–1919
7. Shi XF, Chang CY, Ma SX et al (2016) Efficient bioconversion of L-glutamate to γ-aminobutyric acid by *Lactobacillus brevis* resting cells. J Ind Microbiol Biotechnol. doi:10.1007/s10295-016-1777-z
8. Plokhov AY, Gusyatiner MM, Yampolskaya TA et al (2000) Preparation of γ-aminobutyric acid using *E. coli* cells with high activity of glutamate decarboxylase. Appl Biochem Biotechnol 88:257–265
9. Komatsuzaki N, Shima J, Kawamoto S et al (2005) Production of γ-aminobutyric acid (GABA) by *Lactobacillus paracasei* isolated from traditional fermented foods. Food Microbiol 22(6):497–504
10. Choi SI, Lee JW, Park SM et al (2006) Improvement of gamma-aminobutyric acid (GABA) production using cell entrapment of *Lactobacillus brevis* GABA 057. J Microbiol Biotechnol 16:562–568

11. Li H, Gao DD, Cao YS et al (2008) A high γ-aminobutyric acid-producing ability: *Lactobacillus brevis* isolated from Chinese traditional paocai. Ann Microbiol 58:649–653
12. Siragusa S, Angelis M, Cagno R et al (2007) Synthesis of γ-aminobutyric acid by lactic acid bacteria isolated from a variety of Italian cheeses. Appl Environ Microbiol 73:7283–7290
13. Park KB, Oh SH (2006) Isolation and characterization of *Lactobacillus buchneri* strains with high gamma-aminobutyric acid producing capacity from naturally aged cheese. Food Sci Biotechnol 15:86–90
14. Nomura M, Kimoto H, Someya Y et al (1998) Production of gamma-aminobutyric acid by cheese starters during cheese ripening. J Dairy Sci 81:1486–1491
15. Skeie S, Ardo Y (2000) Influence from raw milk flora on cheese ripening studied by different treatments of milk to model cheese. LWT Food Sci Technol 33:499–505
16. Li H, Qiu T, Gao D (2010) Medium optimization for production of gamma-aminobutyric acid by *Lactobacillus brevis* NCL912. Amino Acids 38:1439–1445
17. Izmirlioglua G, Demircia A (2016) Strain selection and medium optimization for glucoamylase production from industrial potato waste by *Aspergillus niger*. J Sci Food Agric 96:2788–2795
18. Triguerosa DEG, Fioresea ML, Kroumov AD et al (2016) Medium optimization and kinetics modeling for the fermentation of hydrolyzed cheese whey permeate as a substrate for *Saccharomyces cerevisiae* var. *boulardii*. Biochem Eng J 110:71–83
19. Chen HM, Gao Q, Su Z et al (2012) Screening, identification and flask fermentation optimization of a high-yield γ-aminobutyric acid *Enterococcus raffinosus* strain. Microbiol China 39:1642–1652 (in Chinese)
20. Shi XF, Zheng B, Chang CY et al (2015) Enzymatic bioconversion for γ-aminobutyric acid by *Lactobacillus brevis* CGMCC No. 3414 resting cells. Lect Notes Electr Eng 333:609–617
21. Rastogi NK, Rashmi KR (1999) Optimisation of enzymatic liquefaction of mango pulp by response surface methodology. Eur Food Res Technol 209:57–62

A Concise and Diversity-Oriented Strategy for the Synthesis of Substituted 2-Pyrrolidinones via an Ugi/Intramolecular Inverse Electron Demand Diels-Alder Sequence

Yan Liu, Tianyi Shang, Zhong Chen, Yaozhou Zhang, Wei Sun, Junjie Yang and Yan Pan

1 Introduction

Multicomponent reaction (MCR) is considered to be one of the most promising and important methods in diversity-oriented synthesis (DOS) and even undoubtedly can be utilized as a tool to discover drugs [1]. Isocyanide-based multicomponent reactions (IMCRs) have already been a research hotspot in the field of organic synthesis for decades due to its easy access to potential drug molecules with considerable structural diversity in a rapid and efficient manner [2]. Furthermore, MCRs are often associated with tandem reactions or domino processes to synthesize biologically active nature products and pharmaceuticals [1]. Among the named IMCRs, Ugi four-component reaction (Ugi-4CR) which can build up α-acylaminoamides from a mixture of aldehydes or ketones, amines, carboxylic acids and isocyanidesis is well known for its atom-economy and convenient synthetic operations [3, 4]. Clearly, a proper combination of Ugi with another reaction cascades in a sequence including Ugi-Smiles reaction [5], Ugi-Micheal reaction [6], Ugi-Diels-Alder reaction [7], etc. has been developed and will broaden its application in DOS.

Among pharmaceutically related heterocycles, 2-pyrrolidinones has attracted much attention as a core structure in many biologically active compounds such as Phosphodiesterase Inhibitor [8], antiarrhythmic agent [9], antihypertensive agent [9] and anticancer agents [10]. So our group is trying to develop synthetic methodologies to construct polysubstituted 2-pyrrolidinones via Ugi tandem reaction. Meanwhile we became interested in combining the Ugi reaction with an

Y. Liu (✉) · T. Shang · Z. Chen · Y. Zhang · W. Sun · J. Yang · Y. Pan
College of Biological and Pharmaceutical Engineering,
Xinyang College of Agriculture & Forestry, 24th Avenue,
Yangshan Development Area, Xinyang 464000, China
e-mail: liuyan3020@yeah.net

H. Liu et al. (eds.), *Advances in Applied Biotechnology*, Lecture Notes in Electrical Engineering 444, https://doi.org/10.1007/978-981-10-4801-2_42

inverse electron demand Diels-Alder reaction [11] to develop a novel multicomponent synthesis of 2-pyrrolidinones.

One possible Ugi/intramolecular inverse electron demand Diels-Alder reaction was illustrated by Irini Akritopoulou-Zanze et al to afford 2-pyrrolidinones (Scheme 1) [12] and another additional reaction had already shown that MCR/IMDA reaction cascades with electron-deficient oxadiazole as dienophile was possible leading to [4 + 2] cycloaddition product (Scheme 2) [13].

In order to further apply the illustrated strategy to diversity-oriented synthesis, we set out to examine the previously unexplored Ugi reaction subsequentially combined with intramolecular inverse electron demand Diels-Alder cycloaddition. We initially selected benzaldehyde (**1a**), 1,3,4-oxadiazol-2-amine (**2a**), (Z)-4-(benzyloxy)but-3-enoic acid (**3a**) and tert-butyl isocyanide (**4a**) as substrates for the

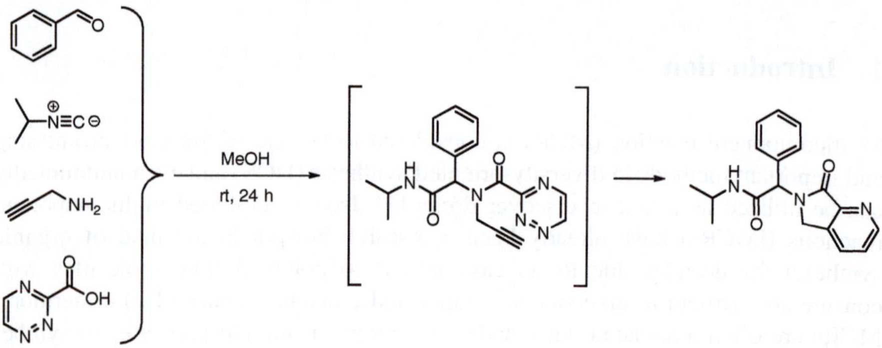

Scheme 1 Construction of heterocycle via sequential Ugi/inverse electron demand Diels-Alder reaction

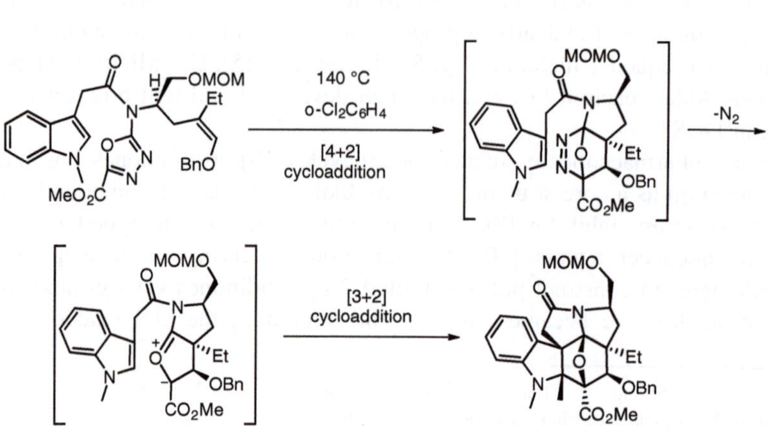

Scheme 2 Construction of 2-pyrrolidinones via MCR/inverse electron demand Diels-Alder sequence

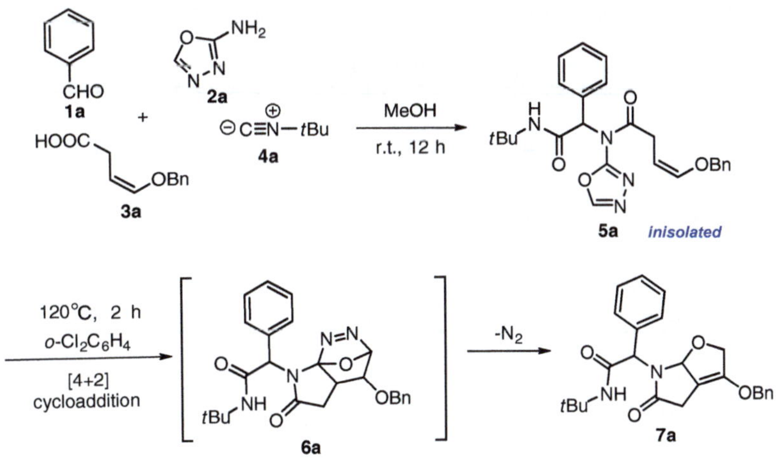

Scheme 3 Synthesis of 2-pyrrolidinones via an Ugi/Intramolecular inverse electron demand Diels-Alder sequence

designed Ugi/Diels-Alder cascade reaction. The reaction proceeded to **5a** in protic solvents such as MeOH. When heated to 60 °C, there was still no desired intramolecular Diels-Alder reaction product being detected. Once the formation of **5a** deemed completed by TLC analysis, the reaction was stoped. Then mixture was cooled to room temperature and concentrated under reduced pressure to give a residue. Put *o*-$C_6H_4Cl_2$ into the residue, then elevated temperature to 120 °C for 2 h. We had found that the Diels-Alder product **7a** was cleanly produced in 47% yield and its structure was verified by NMR (Scheme 3). No other identifiable products were observed or isolated.

From results of the experiment, a sequence of reactions involving Ugi and intramolecular Diels-Alder cycloaddition was proposed to explain the formation of adduct **7a**. In this process, a traditional oxo-bridged tricycle intermediate **6a** was formed via the sequence. Herein, we reported the realization of the designed Ugi/intramolecular inverse electron demand Diels-Alder sequence to yield diversified 2-pyrrolidinones.

2 Result and Discussion

Initially, the Diels-Alder cycloaddition was performed in *o*-$C_6H_4Cl_2$ with **5a** as the substrate, and the yield was 47% (Table 1, entry 1). A solvent screen found that *o*-$C_6H_4Cl_2$ was the best solvent (Table 1, entries 1–6). Lowering the temperature to 100 °C decreased the yield. Although heating the reaction could shorten the reaction time, it gave a lower yield (Table 1, entries 7 and 8). The reaction is also tolerant to a variety of reaction times from 1 to 4 h (Table 1, entries 9 and 10), but

Table 1 Reaction conditions varying the solvents, temperatures and reaction times[a]

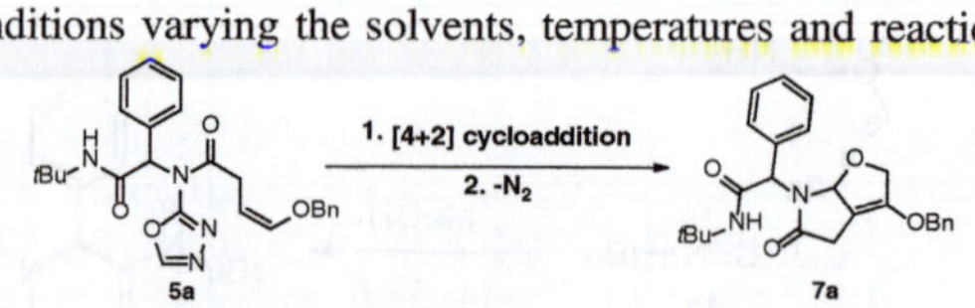

Enty	Sovelution	[5a] (M)	Temp. (°C)	Time (h)[b]	Yield (%)[c]
1	**o-$C_6H_4Cl_2$**	**1.0**	**120**	**2**	**47**
2	p-$C_6H_4Cl_2$	1.0	120	2	21
3	m-$C_6H_4Cl_2$	1.0	120	2	16
4	Toluene	1.0	Reflux	2	0
5	o-xylene	1.0	120	2	0
6	DMSO	1.0	120	2	0
7	o-$C_6H_4Cl_2$	1.0	100	2	30
8	o-$C_6H_4Cl_2$	1.0	140	2	36
9	o-$C_6H_4Cl_2$	1.0	120	1	29
10	o-$C_6H_4Cl_2$	1.0	120	4	45

[a]Reaction conditions: **5a** (1.0 mmol) heated in 2 mL solvent
[b]The reaction was monitored by TLC
[c]Isolated yield

long reaction time did not led to increased yield. Generally, the optimized condition of the Diels-Alder additional reaction to generate **7a** used 1.0 equiv of **5a**, in o-$C_6H_4Cl_2$ at 120 °C for 2 h (Table 1, entry 1).

With the conditions optimized, we further endeavoured to explore the substrate scope and generosity of this Ugi/Intramolecular inverse electron demand Diels-Alder sequence using commercially available substrates to build up a diversified and substituded 2-pyrrolidinones library. Table 2 summarizes the results of additional examples employing different substrate. As expected, the optional use of benzylamine **1b** (Table 1, entry 2), a double bond input **3c** (Table 1, entry 3) and isocyanide with little steric hindrance **4d** (Table 1, entry 4) resulted in the similar scaffold as the one in entry 1, Table 2 via air oxidation of the corresponding Diels-Alder intermediate. 1,3,4-oxadiazol-2-amine (**2a**) were required for complete conversion to the final products (Table 1, entry 1–4).

In conclusion, we have developed an Ugi/intramolecular inverse electron demand Diels-Alder reaction sequence, which provided access to a concise and diversity-oriented strategy for the synthesis of substituded 2-pyrrolidinones in a two-step procedure. The procedure is general and efficient and the substrates are easily available. Considering the importance of 2-pyrrolidinone derivatives, this method may find wide application in the synthesis of more complex polycyclic heterocycles.

Table 2 Substrate scope of the Ugi/Diels-Alder cascade reaction[a]

Enty	Aldehyde	Amine	Acid	Isocyanide	Isolated product	Yield (%)[b]
1	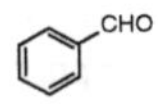	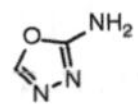	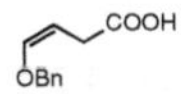		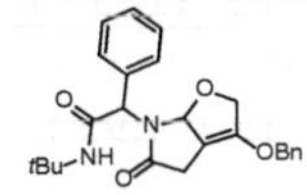	**7a**, 47
2	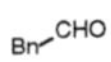	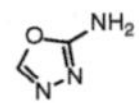	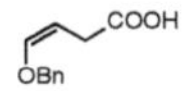		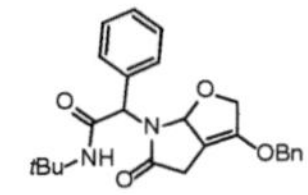	**7b**, 42
3	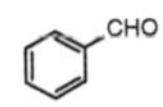	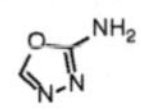	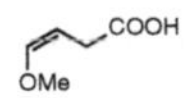	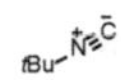	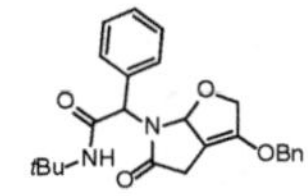	**7c**, 43
4	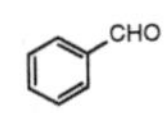	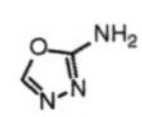	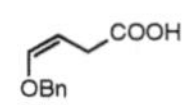	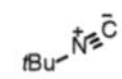	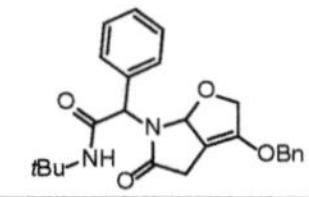	**7d**, 39

[a]Reaction conditions: amine (1.0 mmol), acid (1.0 mmol), aldehyde (1.0 mmol) and isocyanide (1.0 mmol) in MeOH (2 mL) at r.t. for 12 h; then in *o*-$C_6H_4Cl_2$ (5 mL) at 120 °C for 2 h

[b]Isolated yield

3 Experimental

3.1 *General Procedure for the Synthesis of 2-Pyrrolidinones 7a, 7b, 7c and 7d*

A mixture of benzaldehyde (**1a**) (106 mg, 1.0 mmol), 1,3,4-oxadiazol-2-amine (**2a**) (85 mg, 1.0 mmol), (Z)-4-(benzyloxy)but-3-enoic acid (**3a**) (192 mg, 1.0 mmol) and tert-butyl isocyanide (**4a**) (83 mg, 1.0 mmol) was added to a round bottom flask equipped with 2 mL methanol, which was stirred at r.t. for 12 h. The reaction was monitored by TLC. When it completely finished, stopped the reaction. Then the mixture was cooled to r.t. and solvent was distilled under reduced pressure to give a residue. Without any isolation, the residue was add to 5 mL *o*-$C_6H_4Cl_2$, then elevated temperature to 120 °C for 2 h. Once finished, *o*-$C_6H_4Cl_2$ was removed to give another residue which was purified by silica gel column chromatography with Petroleum ether/ethyl acetate/DCM as the mobile phase.

3.2 *Characterization of Compound 7*

Compound 7a (197 mg, 47%) was obtained as white solid. ^{1}H NMR (400 MHz, d_6-DMSO): δ 8.18 (s, 1H), 7.27–7.33 (m, 10H), 6.53 (s, 1H), 5.82 (s, 1H), 4.93 (s, 2H), 4.53 (s, 2H), 2.85 (s, 2H), 1.15 (s, 9H). ^{13}C NMR (100 MHz, d_6-DMSO):

δ 27.4, 29.2, 60.0, 64.6, 71.7, 83.1, 99.8, 127.1, 127.6, 128.9, 129.6, 136.1, 151.3, 168.6, 169.7. LRMS (ESI) m/z: 421.2 $[M + H]^+$.

Compound 7b (183 mg, 42%) was obtained as white solid. ^1H NMR (400 MHz, d_6-DMSO): δ 8.18 (s, 1H), 7.32–7.33 (m, 5H), 7.14–7.19 (m, 5H), 6.53 (s, 1H), 5.82 (s, 1H), 4.93 (s, 2H), 4.92 (t, 2H), 4.53 (s, 2H), 3.33 (d, 2H), 2.85 (s, 2H), 1.15 (s, 9H). ^{13}C NMR (100 MHz, d_6-DMSO): 29.2, 35.6, 57.2, 60.0, 71.7, 127.1, 127.6, 127.7, 128.6, 128.9, 136.1, 136.6, 171.1. LRMS (ESI) m/z: 435.2 $[M + H]^+$.

Compound 7c (148 mg, 43%) was obtained as white solid. ^1H NMR (400 MHz, d_6-DMSO): δ 8.18 (s, 1H), 7.27–7.31 (m, 5H), 6.53 (s, 1H), 5.82 (s, 1H), 4.53 (s, 2H), 3.51 (s, 3H), 2.85 (s, 2H), 1.37 (s, 9H). ^{13}C NMR (100 MHz, d_6-DMSO): δ 27.4, 29.2, 56.1, 60.0, 71.4, 83.1, 99.8, 127.6, 128.9, 129.2, 129.6, 138.8, 151.3, 168.6, 169.7. LRMS (ESI) m/z: 345.1 $[M + H]^+$.

Compound 7d (164 mg, 39%) was obtained as white solid. ^1H NMR (400 MHz, d_6-DMSO): δ 8.01 (s, 1H), 7.27–7.33 (m, 10H), 6.53 (s, 1H), 5.82 (s, 1H), 4.93 (s, 2H), 4.53 (s, 2H), 3.18 (t, 2H), 2.85 (s, 2H), 1.49 (m, 2H), 1.29 (m, 2H), 0.89 (t, 3H). ^{13}C NMR (100 MHz, d_6-DMSO): δ 13.8, 19.8, 32.2, 38.9, 64.0, 71.7, 83.1, 99.8, 127.1, 127.6, 128.9, 129.2, 129.6, 136.1, 136.8, 151.3, 169.2, 169.7. LRMS (ESI) m/z: 421.2 $[M + H]^+$.

References

1. Zhu J, Bienaymé H (2006) Multicomponent reactions. Wiley, Weinheim
2. Dömling A (2006) Recent developments in isocyanide based multicomponent reactions in applied chemistry. Chem Rev 106(1):17–89
3. Dömling A (2002) Recent advances in isocyanide-based multicomponent chemistry. Curr Opin Chem Biol 6(3):306–313
4. Zhu J (2003) Recent developments in the isonitrile-based multicomponent synthesis of heterocycles. Eur J Org Chem 2003(7):1133–1144
5. El Kaim L, Grimaud L (2014) The Ugi-smiles and Passerini-smiles couplings: a story about phenols in Isocyanide-based multicomponent reactions. Eur J Org Chem 2014(35): 7749–7762
6. Santra S, Andreana PR (2011) A bioinspired Ugi/Michael/Aza-Michael cascade reaction in aqueous media: natural-product-like molecular diversity. Angew Chem Int Ed 50(40): 9418–9422
7. Huang X, Xu J (2009) One-pot facile synthesis of substituted isoindolinones via an Ugi four-component condensation/Diels-Alder cycloaddition/deselenization-aromatization sequence. J Org Chem 74(22):8859–8861
8. Baures PW, Eggleston DS, Erhard KF et al (1993) Crystal structure, absolute configuration, and phosphodiesterase inhibitory activity of (+)-1-(4-bromobenzyl)-4-[(3-cyclopentyloxy)-4-methoxyphenyl]-2-pyrrolidinone. J Med Chem 36(22):3274–3277
9. Malawska B, Kulig K, Filipek B et al (2002) Synthesis, antiarrhythmic, and antihypertensive effects of novel 1-substituted pyrrolidin-2-one and pyrrolidine derivatives with adrenolytic activity. Eur J Org Chem 37(3):183–195

10. Ramachandran G, Karthikeyan NS, Giridharan P et al (2012) Efficient iodine catalyzed three components domino reaction for the synthesis of 1-((phenylthio)(phenyl) methyl) pyrrolidin-2-one derivatives possessing anticancer activities. Org Bio Chem 10(28): 5343–5346
11. Boger DL (1986) Diels-Alder reactions of heterocyclic aza dienes. Scope and applications. Chem Rev 86(5):781–793
12. Akritopoulou-Zanze I, Wang Y, Zhao H et al (2009) Synthesis of substituted fused pyridines, pyrazines and pyrimidines by sequential Ugi/inverse electron demand Diels-Alder transformations. Tetrahedron Lett 50(42):5773–5776
13. Pellissier H (2012) Stereocontrolled domino reactions. Chem Rev 113(1):442–524

Optimization of Medium Components for the Production of Flavonoids and Soluble Protein with *Phellinus igniarius* in Liquid Culture

Yingchao Wang, Zhouli Yuan, Xinrong Tan, Zaohong Ran and Hong Jin

1 Introduction

The submerged fermentation of edible fungi not only shows the higher growth speed of mycelium and higher utilization rate of the nutrition, the mycelia obtained have a variety of nutrients and bioactive substances, but also the submerged ferments contain a wealth of primary metabolic products, such as proteins, polysaccharides and secondary metabolism products [1–4], such as flavonoids, antibiotics, etc. In our previous studies also confirmed that edible fungi of *Phellinus igniarius* and *Grifola frondosa* produced a lot of secondary metabolites such as polysaccharides, flavonoids and soluble protein in liquid fermentation broth [5, 6]. These substances can enhance the ability of human and animal to resist, increase nutrition and improve the taste. The study will provide important new resources for the development of new feed and functional drinks rich in mycelium body, proteins and flavonoids [7].

Phellinus igniarius was a kind of edible fungus with an important medicinal value, and its submerged ferments contained a variety of bioactive substances, such as soluble protein, vitamin C and flavonoids etc. [8]. Corn silk had the effects of diuresis, detumescence and lowering blood sugar, cholesterol and blood pressure in traditional Chinese medicine, and be treated as health care tea because of its rich nutrients [9, 10]. However, only a little corn silk was used in Chinese medicine in our country, most of it is discarded as an agricultural waste after harvest. If it is

Y. Wang
College of Basic Science,
Tianjin Agriculture University, Tianjin 300384, China
e-mail: shibiology@163.com

Z. Yuan · X. Tan · Z. Ran · H. Jin (✉)
College of Agronomy and Resource Environment,
Tianjin Agriculture University, Tianjin 300384, China
e-mail: jinhongtj@126.com

H. Liu et al. (eds.), *Advances in Applied Biotechnology*, Lecture Notes in Electrical Engineering 444, https://doi.org/10.1007/978-981-10-4801-2_43

collected, utilized and added in the fermentation culture medium of *Phellinus igniarius*, will have the important significance to preparing new functional feed and drink by submerged fermentation of producing flavonoids and protein. Similar researches were rarely reported [11]. In our study, the corn silk and wheat bran as materials of *Phellinus igniarius* submerged fermentation, and contents of flavonoids and soluble protein in ferments were determined. The optimum medium formula was obtained by through orthogonal test and overall balance analysis in order to provide research basis for corn silk further development and application.

2 Materials and Methods

2.1 *Materials*

Corn silk, collected by ourselves, the corn silk was screened by the same maturity degree, dried at 60 °C, crushed and reserved over 60 mesh sieve. *Phellinus igniarius* strain was obtained from the biotechnology laboratory of Tianjin Agricultural University. It was maintained on potato agar and incubated at 28 °C for 7 days. Formula of liquid seed medium: 3% glucose, 0.3% peptone, 0.05% $MgSO_4$, 0.1% KH_2PO_4, pH unadjusted. Formula of basal liquid submerged fermentation medium (100 mL): 6 g wheat bran, 0.2 g KH_2PO_4, 0.4 g yeast extract, 0.1 g $MgSO_4$, pH unadjusted [12–14].

2.2 *Experimental Methods*

2.2.1 Inoculation and Culture

The seed culture was cultivated for 5 days in 100 mL of liquid seed medium that was inoculated with 3–4 strain blocks of the activated *Phellinus igniarius* with the size of 0.5 cm^2 at 25 °C and 150 rpm in a shaking incubator. 10 mL of the seed culture was transferred into 100 mL of fresh liquid submerged fermentation medium in a 250 mL flask. They were then cultivated for another 5 days at 25 °C and 150 rpm in a shaking incubator. After 5 days' cultivation, the culture broth was used as the culture strains of further fermentation.

2.2.2 Determination of the Contents of Flavonoid and Soluble Protein in the Fermentation Liquid [15, 16]

The fermentation broth was centrifuged at 3000 rpm for 10 min, the supernatant was used to determine the contents of flavonoid and soluble protein by the method from Liu [8] the absorbance of flavonoids was determined at 285 nm wavelength.

With thc concentration of quercetin (mg/mL) as the horizontal coordinate, the absorbance as the vertical coordinate, the standard curve was drawn, the equation was $y = 36.800x + 0.0074$, $R^2 = 0.9967$ with a good linear relationship. With the protein concentration (mg/mL) as the horizontal coordinate, the absorbance is the vertical coordinate, and the standard curve is drawn at the 595 nm wavelength. The linear regression equation was obtained as follows: $R^2 = 0.9971$; $y = 5.8514x + 0.0021$, with a good linear relationship. The content of flavonoids (mg/mL) in the fermentation liquid = concentration of extract (mg/mL) × dilution ratio. Soluble protein content (mg/mL) in fermentation broth = concentration of extract (mg/mL) × dilution ratio.

2.2.3 Determination of Culture Time

10 mL of the seed culture was transferred into 100 mL of fresh liquid submerged fermentation medium in a 250 mL flask. They were then cultivated for 7 days at 25 °C and 150 rpm in a shaking incubator. Fermentation broth was sampled for once every 24 h and the contents of flavonoid and soluble protein from the broth were determined. The strain growth curves were drawn to find out when the contents of flavonoid and soluble protein were the highest with the cultivation.

2.2.4 The Effect of Different Ratios of Corn Silk and Wheat Bran

Different ratio of corn silk and wheat bran was added into the basal liquid submerged fermentation medium. The ratios of corn silk and wheat bran were respectively: corn silk to wheat bran (0–6%), (3–6%), (6–6%), (9–6%), (12–6%), (15–6%) and (18–6%). 10 mL of the seed culture was transferred into 100 mL of fresh such liquid submerged fermentation medium in a 250 mL flask. They were then cultivated at 25 °C and 150 rpm in a shaking incubator and ended at the day which had been found to be the optimal cultivation. Contents of flavonoids and soluble protein from the broth were determined to find the optimal ratio at which contents of flavonoids and soluble protein were the highest.

2.2.5 Effects of Different Inoculation Amount

On the basis of optimal ratio, three different inoculation amounts (1.5, 3, 4.5%) were studied to find the optimal one.

2.2.6 Orthogonal Test

In order to obtain the optimal formula of culture medium whose ferment broth showed the highest levels of flavonoids and soluble protein, based on the above

Table 1 The orthogonal test design of four factors and four levels

Levels	Factors			
	Ratio of corn silk to bran	KH_2PO_4 (%)	Yeast extract (%)	$MgSO_4$ (%)
1	6% + 3%	0.10	0.2	0.05
2	5% + 6%	0.15	0.3	0.10
3	4% + 9%	0.20	0.4	0.15
4	3% + 12%	0.25	0.5	0.20

single factor experiment, ratio of corn silk to bran, yeast extract, KH_2PO_4, and $MgSO_4$ were selected as factors to design the orthogonal test of four factors and four levels, the factors and levels were shown in Table 1. The optimal culture medium formula was obtained through comprehensive equilibrium analysis.

2.2.7 Statistical Analysis Method

All the data were analyzed by the software of SPSS 17.1.

3 Results and Analysis

3.1 *The Change Curve of the Contents of Flavonoids and Soluble Protein*

The change curve of the content of flavonoids with culture time was shown in Fig. 1. During the process of submerged fermentation, there was a linear relationship between the content of metabolite flavonoids and the culture time, and content of flavonoids was increased with the time increasing, until the 4th day it reached the maximum and thereafter was decreased with the increase of time. Flavonoids are secondary metabolites, which were mainly produced in the stable phase. At early stage of infection, mycelium was just added into the new medium, and grew slowly for adapting to the new culture medium, so, the content of metabolites was lower. With the increase of days of culture, *Phellinus igniarius* was growing to the exponential phase, content of metabolites was accumulated, then to a period of decline, the death of the fungus, reduced metabolites and accumulation would appear and continue to decline. Predictably, mycelium in the growth process, continued to use the nutrients for its breeding and production of products until the nutrients could not maintain biological activity and metabolism, so, the incubation time was the 4th day after incubation when flavonoid content was reached the highest.

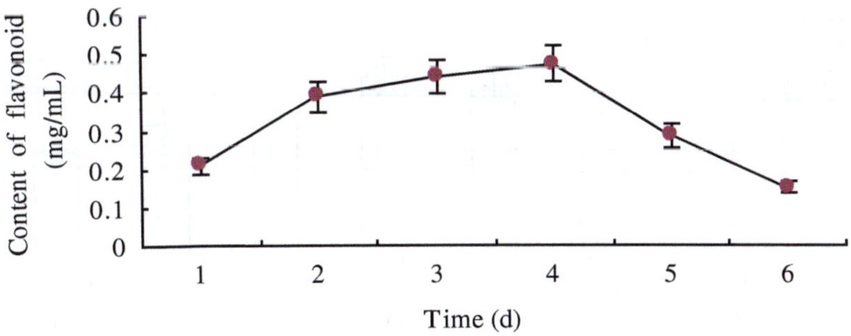

Fig. 1 The change curve of the content of flavonoids during the process of *Phellinus igniarius* submerged fermentation

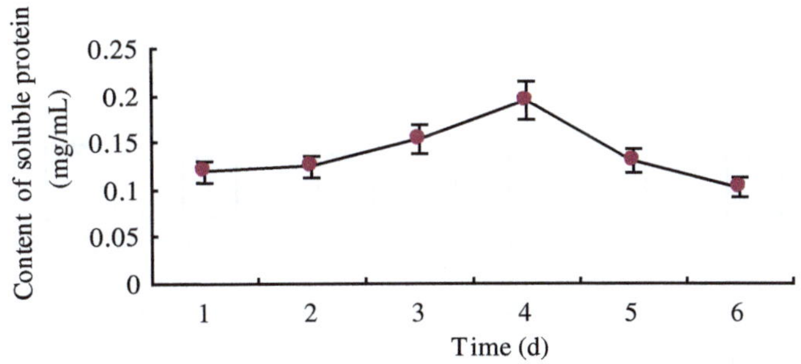

Fig. 2 The change curve of content of soluble protein during the process of *Phellinus igniarius* submerged fermentation

The change curve of soluble protein content was shown in Fig. 2. The circumstance as for change of soluble protein content with culture time was similar to that of the flavonoids. at the 4th day of incubation, content of soluble protein reached maximum value.

As a whole, at the 4th day of incubation, contents of metabolites of flavonoids and soluble protein were determined to reach the highest level. In the following test, the stop incubation time was set at the 4th of incubation.

3.2 *Effect of Different Concentration Ratio of Corn Silk and Wheat Bran*

The effect curves of corn silk and wheat bran with different ratio on the contents of flavonoids and soluble protein in the fermentation liquid were shown in Figs. 3 and 4. Ratios 1–7 were established on the basic 6% wheat bran and added different

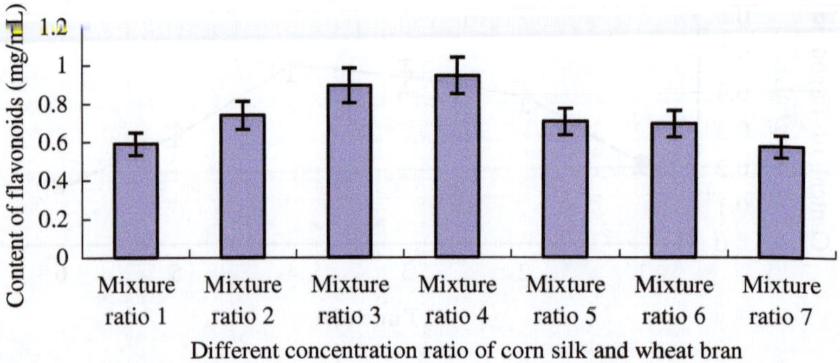

Fig. 3 Effect of different concentration ratio of corn silk and wheat bran on content change of flavonoids

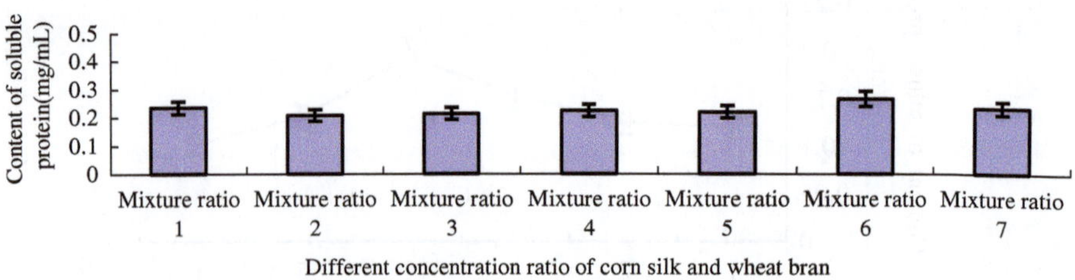

Fig. 4 Effect of different concentration ratio of corn silk and wheat bran on content change of soluble protein

concentration of corn silk (0, 3, 6, 9, 12, 15 and 18%), which was shown in Fig. 3. In the process of submerged fermentation of *Phellinus igniarius*, content of flavonoids varied greatly with different conc. of corn silk, the ratio 4 (9% corn silk, 6% wheat bran) showed the highest level of flavonoid content.

That meant that corn silk might play an active supplement role on the synthesis of flavonoids in the process of submerged fermentation of *Phellinus igniarius*. Over a certain concentration, metabolite synthesis showed a downward trend. According to Han ping' report [17], the main chemical constituents in corn silk were volatile fatty oils, volatile oils, saponins, polysaccharides, vitamins, various organic acids, and so on, thus, corn silk in culture medium was mainly used as carbon source, and wheat bran also as carbon source, whose main component was cellulose and hemicellulose. From the point of nutritional components, the nutrition of corn silk was more comprehensive than that of wheat bran, to some extent, it was a good supplement to the wheat bran, but the nutritional carbon source factor of corn silk was not enough, reasonable collocation of both would be more beneficial to the fermentation of *Phellinus igniarius*. How to use the waste of corn silk instead of wheat bran, reduce the cost of fermentation was the purpose of our study. Therefore, the optimal ratio of corn silk and wheat bran for flavonoids was ratio 4, namely 9% corn silk, 6% wheat bran.

It was shown in Fig. 4, based on the fixed concentration of wheat bran, different ratios of corn silk and wheat bran had slight effect on the content of soluble protein in fermentation liquid. To comprehensive consider as for the contents of flavonoids and soluble protein, the optimal ratio was ratio 4, it was 9% corn silk, 6% wheat bran.

3.3 The Determination of Inoculum Amount

In the process of *Phellinus igniarius* submerged fermentation, different inoculation amount affected the yields of metabolites in the fermentation broth. yields of flavonoids and soluble protein increased with the increase of inoculum amount, when the inoculum amount reached 3%, yields of flavonoids and soluble protein reached the maximum value and then began to decline. But the specific trends of the changes in the yield of both were different. The yield of flavonoids increased very fast under lower inoculation volume, and decreased little slower under high inoculation amount, as for soluble protein, the circumstance was vice versa.

The secondary metabolites are the metabolites that were synthesized after the growth of edible fungi to the stable stage. Inoculum size was smaller, making fewer mycelia in the fermentation broth, slowed the growth of mycelium, extended the growing season, delayed the stabilization period, the formation time of secondary metabolites was delayed, so total volume of products was reduced; inoculum size was larger, there were more hyphal fragments in medium, fungi thrived, at the of steady stage there were a large number of secondary metabolites of the cell, thereby increasing the total amount of metabolites in the fermentation broth. If Inoculum size was too big, mycelium growth was too fast and uneven, produced a lot of small pellets, broth viscosity was increased, causing the lack of oxygen, instead, the metabolites synthesis decreased. In conclusion of Figs. 5 and 6, the optimal inoculation amount was determined to be 3%.

Fig. 5 The effect of different inoculum amount on the content of flavonoids in the fermentation liquid

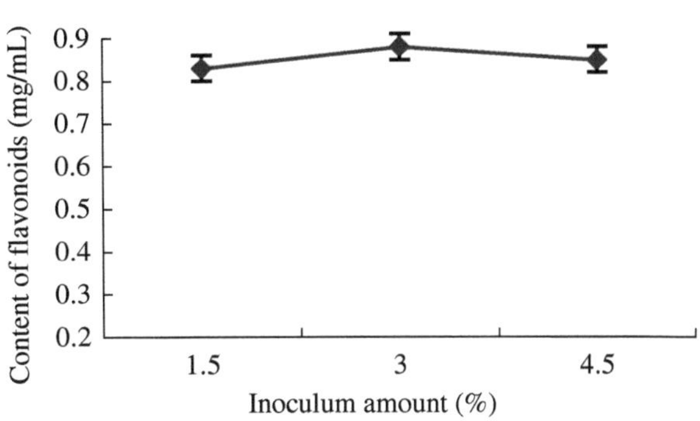

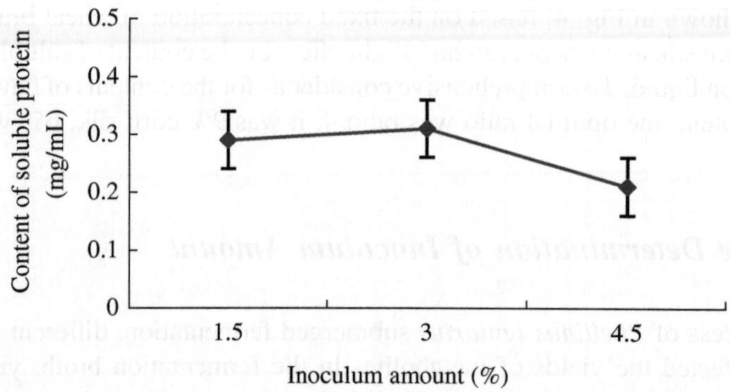

Fig. 6 The effect of different inoculum amount on the content of soluble protein in the fermentation liquid

3.4 Orthogonal Test

In order to obtain the optimal formula of culture medium whose ferment broth showed the higher levels of both flavonoid and soluble protein, based on the above single factor experiment, ratio of corn silk to bran, yeast extract, KH_2PO_4, and $MgSO_4$ were selected as factors to design the orthogonal test of four factors and four levels according to Table 1. Results were shown in Table 2. From Table 2, the best culture medium formula based on the content of flavonoids was $A_4B_2C_3D_1$. According to the analysis of range, size order of each factor affecting the content of flavonoids in fermentation liquid was: ratio of corn silk to wheat bran > KH_2PO_4 > yeast extract > $MgSO_4$. The best medium formula based on the content of soluble protein was $A_4B_1C_4D_1$, the size order was ratio of corn silk to wheat bran > yeast extract > $MgSO_4$ > KH_2PO_4. The two schemes were not the same, so, the comprehensive balance method [18] was used to carry out the analysis of multi-indexes to obtain the final optimal formula of medium with both higher levels of flavonoids and soluble protein.

From Table 2, factor A (the concentration ratio of corn silk to wheat bran) showed significant effects on contents of flavonoid and soluble protein with bigger ranges, indicating that factor A was the biggest effective factor. In terms of the content of flavonoids, A_4 level was the best, as for the content of soluble protein, A_4 level was also the best. Overall, A_4 level was selected. In terms of the content of flavonoids, B_2 level (KH_2PO_4) was the best. In terms of soluble protein content, B_1 level was the best, but B_2 level was not bad. Aiming to both of indexes, B_2 level was selected. Factor C (yeast extract) showed smaller range to the content of flavonoids, but bigger range of soluble protein content. That meant yeast extract was the major influence factor on the soluble protein content. In terms of the content of flavonoids, C_3 level was the best, C_4 level is not too bad; for soluble protein content, C_4 level was the best. In summary, the level of the best yeast

Table 2 Results of orthogonal test

Test no.	A	B	C	D	Content of flavonoid (mg/mL)	Content of soluble protein (mg/mL)
1	1	3	2	3	0.529	0.437
2	2	1	4	2	0.955	0.607
3	3	3	4	4	1.162	0.687
4	4	1	2	1	1.496	0.690
5	1	1	3	4	0.654	0.430
6	2	3	1	1	0.874	0.542
7	3	1	1	3	1.222	0.603
8	4	3	3	2	1.453	0.660
9	1	4	1	2	0.491	0.337
10	2	2	3	3	0.983	0.576
11	3	4	3	1	1.330	0.670
12	4	2	1	4	1.488	0.656
13	1	2	4	1	0.722	0.453
14	2	4	2	4	0.811	0.535
15	3	2	2	2	1.224	0.625
16	4	4	4	3	1.434	0.723
K_{H1}	0.599	1.082	1.019	1.106		
K_{H2}	0.906	1.104	1.015	1.031		
K_{H3}	1.235	1.005	1.105	1.042		
K_{H4}	1.468	1.017	1.068	1.029		
R_H	0.869	0.099	0.090	0.077		
K_{P1}	0.414	0.583	0.535	0.589		
K_{P2}	0.565	0.578	0.572	0.557		
K_{P3}	0.646	0.582	0.584	0.585		
K_{P4}	0.682	0.566	0.618	0.577		
R_P	0.268	0.017	0.083	0.032		

extract was C_4. The impact of Factors D ($MgSO_4$) on both of indicators: Factor D showed smaller ranges to the contents of both flavonoids and soluble protein. That meant $MgSO_4$ was the minor effect factor on the soluble protein content. In terms of the content of flavonoids and soluble protein, D_1 level was the best, so, D_1 level was selected.

In conclusion, after the overall balance, the optimal scheme of medium was gotten to be $A_4B_2C_4D_1$, namely corn silk 12%, wheat bran 3%, KH_2PO_4 0.15%, yeast extract 0.5%, $MgSO_4$ 0.05%. Three sets of parallel experiments were confirmed by using the optimal medium formula. The content of flavonoids in fermentation broth was 1.551 mg/mL, and the soluble protein content was 0.691 mg/mL.

4 Conclusion

Through orthogonal test and comprehensive balance analysis method, the optimal medium composition was determined: 3% wheat bran, 12% corn silk, 0.5% yeast extract, 0.15% KH_2PO_4, 0.05% $MgSO_4$. The best inoculation amount was 3%. With the optimal formula. The confirmation experiment was that the content of flavonoids in fermentation broth was 1.551 mg/mL, and the soluble protein content was 0.691 mg/mL.

Acknowledgements Thank professor Litong BAN from dept. of biotechnology in Tianjin Agricultural University for the donation of the strain of *Phellinus igniarius.*

References

1. Wang W, Zhu J (1998) Research progress and prospect of edible fungus liquid deep fermentation technology. Edible Fungi China 17(2):11–13
2. Guo S, Wang H (2013) Submerged fermentation technology of edible fungus and its application. J Shanxi Agric Sci 41(8):885–888
3. Qin J, Chen M, Chen H et al (2004) Prospect and current studies on edible and pharmaceutical fungi polysaccharides. Edible Fungi China 23(2):6–9
4. Chen X, Liu Q, Chen B (2008) Preliminary study of antimicrobial activities of *Russula vinosa* lindbl mycelium and its submerged culture fermentation broth. Food Sci 29(7):260–262
5. Jin H, Yang X, Ren C et al (2015) The comparison of the contents of bioactive substances in edible fungus by submerged ferments. Food Res Dev 36(3):93–96
6. Yang X, Liu M, Wu X et al (2015) Study on optimizing and improving polysaccharide production of *Grifola frondosa* in submerged ferments. Food Res Deve 36(24):51–55
7. Su Y, Lu P, Guo Q et al (2016) Progress in edible fungus by submerged fermentation. J Food Saf Qual 7(2):645–650
8. Liu W (2012) Study on the metabolic regulation of flavones produced by a medicinal fungus *Phellinus igniarius.* China University of Petroleum, Beijing, pp 1–23
9. She M (2014) Overview of clinical application research of Maize. China Health Care Nutr (4):2383–2384
10. Chen Y, Ma Q (2014) Study on blood sugar and blood pressure, blood lipid, liver function of corn silk. Tradit Chin Med Res 27(3):78–80
11. Cao S, Zheng H, Chen L et al (2014) Optimized conditions of botway solid FERMENTATION for transforming inonotus obliquus with stigma maydis. Inform Tradit Chin Med 31(4):61–64
12. Xu Q (2015) Optimization of liquid culture medium for fermenting mycelial flavone in *Phellinus igniarius*. Food Mach 31(1):190–193
13. Lima LFO, Habu S, Gern JC et al (2008) Production and characterization of the exopolysaccharides produced by *Agaricus brasiliensis* in submerged fermentation. Appl Biochem Biotechnol 151(2):283–294
14. Kwon JS, Lee JS, Lee KE et al (2009) Optimization of culture conditions and medium components for the production of mycelial biomass and exo-polysaccharides with *Cordyceps militarisin* liquid culture. Biotechnol Bioprocess Eng 14(6):756–762

15. Deng L, Pan X, Sheng J et al (2012) Optimization of experimental conditions for the determination of water soluble protein in apple Pulp using coomassie brilliant blue method. Food Sci 33(24):185–189
16. Liu Y, Yang Y, Jia W et al (2006) Study on the determination method of total flavonoids from *Phellinus igniarius*. Acta Edulis Fungi 13(2):45–48
17. Han P, Li H, Hou C (2009) The chemical component of maize silk and its application research progress. Mod Agric Sci Technol (18):19–21
18. Zhang J, Ma Y (2016) Optimization of processing technology of nine steaming nine drying of *Rehmannia glutinosa* by multiple indexes comprehensive balance method-orthogonal test. China Pharm 27(7):962–965

Application of Sun Light Conversion Film on the Outdoor Culture of *Nostoc flagelliforme*

Shigang Shen, Peipei Han, Shunyu Yao, Rongjun Guo and Shiru Jia

1 Introduction

Nostoc flagelliforme is an edible filamentous terrestrial cyanobacterium with great food and herbal values, which is distributed in arid and semi-arid areas. The extracellular polysaccharide (EPS) of *N. flagelliforme* has been proved to possess a variety of properties, such as antivirus, antioxidant, and anti-tumor [1–3]. Besides, it has been reported with high intrinsic viscosity, good emulsification activity, and excellent flocculation capability, indicating it was a very promising candidate for numerous industrial applications [4]. However, the poor growth rate of *N. flagelliforme* plus increased market demands causes the endangered status of this specie [5]. Therefore, the Chinese government designated *N. flagelliforme* a Category II Protected Plant in 2000. Under the protection regime, picking and trading of wild *N. flagelliforme* are strictly prohibited. To conserve the endangered resource, the cultivation of *N. flagelliforme* cells isolated from colonial filaments in liquid suspension culture has been successfully established. And the cultivation of *N. flagelliforme* cells isolated from colonial filaments in liquid suspension culture has been successfully established. Our previous work has demonstrated that exopolysaccharide of wild *N. flagelliforme* and liquid-cultured *N. flagelliforme* showed great similarity, which provided the basis for industrial use of *N. flagelliforme* [6].

Currently, the open pond raceway and closed tubular photobioreactors are the main ways for the large-scale cultivation of microalgal. Photoautotrophic mode is commonly used, and light is the only source for the generation of energy and reduced carbon under photoautotrophic conditions. Changes in light quality and light intensity have been found with significant influence on the biomass and EPS

S. Shen · P. Han · S. Yao · R. Guo · S. Jia (✉)
Key Laboratory of Industrial Microbiology, Ministry of Education,
College of Biotechnology, Tianjin University of Science and Technology,
Tianjin 300457, People's Republic of China
e-mail: jiashiru@tust.edu.cn

H. Liu et al. (eds.), *Advances in Applied Biotechnology*, Lecture Notes
in Electrical Engineering 444, https://doi.org/10.1007/978-981-10-4801-2_44

production of cyanobacteria [7–9]. In our previous work investigating the effects of different wavelengths of light on *N. flagelliforme* culture, it was found that blue and red lights could promote biomass accumulation and EPS production [10]. Therefore, efficient use of the light source is essential to cultivate the *N. flagelliforme* in the outdoors, especially to improve the biomass and EPS production.

Natural sunlight is the only energy in the large scale culture of microalgal, but the sunlight and ultraviolet radiation is serious at noon, which has a significant inhibitory effect on the growth of algal cells. Sunlight conversion film (SCF) could make light quality conversion (convert violet and green light into red light) and thus improve light utilization rate. The results of the fluorescence test showed that SCF could strengthen the intensity mainly in red light range (630–680 nm) and a little in blue light range, which is in favor of the algal growth. However, SCF was usually used in agriculture, and there are few studies about the application of SCF on microalgae culture. Considering the SCF function and adaptability of algae to light quality (red and blue light), application of SCF on microalgae culture can improve the photosynthetic efficiency of microalgae, thereby increasing the productivity of biomass and EPS. Therefore, in this study, the effects of different types of SCF on *N. flagelliforme* culture were examined.

2 Materials and Methods

2.1 Strain and Culture Conditions

The *N. flagelliforme* cells (TCCC11757) cultivated in BG-11 medium [11] were obtained from the Tianjin Key Lab of Industrial Microbiology (Tianjin, China). Inoculums were prepared by transferring the cells from stock culture, and incubated aseptically in 2 L Erlenmeyer flasks containing 1 L BG-11 medium. Cultures were bubbled with sterilized-filtered air, and the air flow was controlled by a rotameter (FMA3306, Omega Engineering Limited, Manchester, UK). Photoautotrophic culture of *N. flagelliforme* was performed in closed Erlenmeyer flasks under outdoor conditions, and the light intensity and daytime temperature in the experiment process were natural. The cells grown under common polyethylene film were treated as control. The surface of Erlenmeyer flasks was covered with seven different types of SCF, which was prepared by adding 3‰ sunlight conversion agent (SCA) on the basis of common polyethylene film. SCA can convert violet and green light into red light. The model numbers of seven kinds of SCF were 34#, 36#, 39#, 40#, 41#, 43#, 47#, and the brands of added SCA were RBM-4565, VTR-0617, HNND-14-1, Zhuhai, Yuanfeng, GTR-0645, SDAU, respectively. Among them, both the SCF 41# and 43# added an extra 5‰ red color masterbatch during the process of manufacturing.

2.2 The Measurement of Cell Growth, EPS Production, and Protein Content

The growth increment of the *N. flagelliforme* cells was measured with dry mass method. The EPS and CPS production was performed via the modified phenol–sulphuric acid method using glucose as standard [12]. Crude protein content was determined by Automatic Kjeldahl Apparatus (JK9830) in accordance with the method of the determination of protein in foods (GB 5009.5-2010).

2.3 The Measurement of Photosynthetic Pigment

To determine pigment content, 5 mL of sample was taken and centrifuged at 4000×*g* for 10 min. Subsequently, the supernatant was discarded and then the cells were re-suspended in 5 mL of methanol in a closed centrifuge tube and stored at 4 °C for 12 h under dark conditions. After that, the cells were centrifuged at 10,000×*g* for 10 min, supernatant was collected and absorbance was measured at 665 nm. The content of chlorophyll *a* and carotenoid were calculated using the following equations, respectively [13]:

$$\text{Chlorophyll}\,a\,(\text{mg/L}) = 13.43 \times A_{665} \times V_{\text{methanol}}/V_{\text{cell}} \tag{2.1}$$

$$\text{Carotenoid}\,(\text{mg/L}) = (1000 \times A_{470} - 44.76 \times A_{665})/221 \times V_{\text{methanol}}/V_{\text{cell}} \tag{2.2}$$

2.4 Statistical Analysis

To ensure reproducibility of results, each experiment was performed with three replications, and the values were expressed as the mean ± standard deviation. The data were analyzed by analysis of variance (ANOVA) or Independent-samples t Test using the SPSS statistical software (version 20.0). The significance level was set at $P < 0.05$.

3 Results and Discussion

3.1 Effects of SCF on the Biomass Accumulation

Effects of seven different types of SCF on the biomass production were investigated, and the results were shown in Fig. 1. Compared to the control group (under

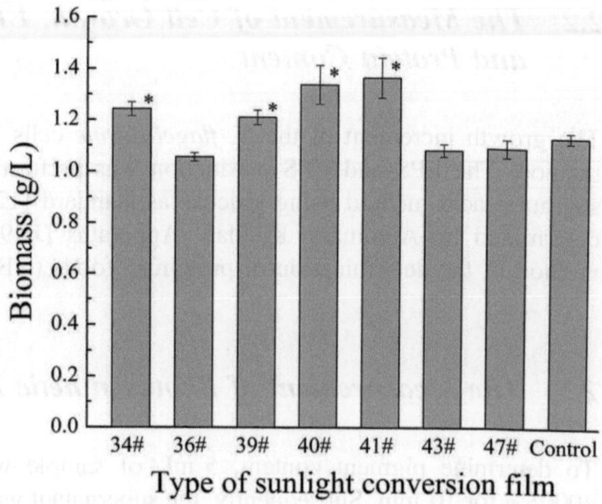

Fig. 1 Effects of SCF on the biomass production of *N. flagelliforme*. *Asterisk* indicate significant differences ($P < 0.05$ versus control). Control (the cells grown under common polyethylene film)

the use of common polyethylene film), the biomass was greatly increased under the use of SCF 34#, 39#, 40# and 41#, but there were no obvious difference under the use of SCF 36# and 47#. After 14 days of culture, under the use of SCF 41#, the highest biomass was achieved with the value of 1.36 g/L, which was increased by 20.35% compared with control group. The results showed that SCF 40# and 41# can significantly promote the biomass accumulation ($P < 0.05$), which indicted that SCF 40# and 41# were more suitable for cell growth.

3.2 *Effects of SCF on EPS Production*

As we can see from Fig. 2, under the use of different types of SCF, the significant difference was observed in the EPS production. After 14 days of culture, EPS production was higher in the order of 41#, 36#, 39#, 40#, 43#, 34#, 47# and control, which reached 46.53, 36.64, 36.61, 35.35, 34.03, 30.71, 30.32 and 28.52 mg/L, respectively. Under cultivation of SCA 41#, the EPS production was 1.63 times to that of control group. The results showed that SCF 41# significantly stimulated the EPS production, which may be related to the properties of adding red color masterbatch ($P < 0.01$). Red color masterbatch is originally used for agricultural production, and it has the function of promoting fruit ripening.

From Figs. 1 and 2, it was found that both the biomass and EPS production reached the highest under cultivation of SCF 41#, which increased by 20.35 and 63.15% compared with control group, respectively. Results of the experiment indicated SCF could convert useless wavelengths to be useful for *N. flagelliforme* growth, thereby increasing the productivity of biomass and EPS, which could bring up remarkable economy benefits. Taken biomass and EPS into consideration, SCF 41# was more suitable for culture of *N. flagelliforme*.

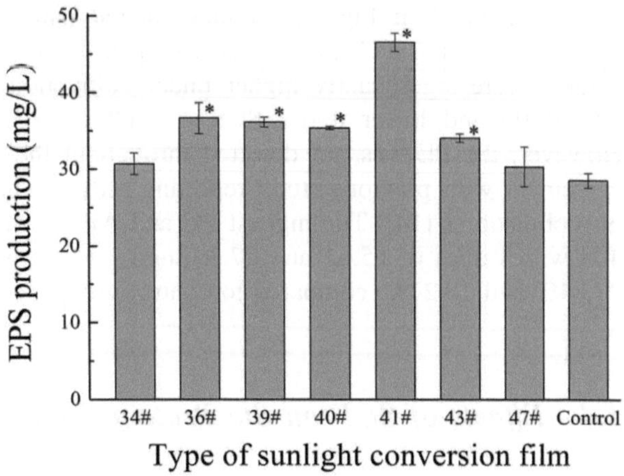

Fig. 2 Effects of SCF on the EPS production of *N. flagelliform. Asterisk* indicate significant differences ($P < 0.05$ versus control). Control (the cells grown under common polyethylene film)

3.3 *Effects of SCF on the Photosynthetic Pigment Content*

In order to explore the effect of SCF on the photosynthetic pigments of *N. flagelliforme*, the photosynthetic pigment contents were measured after 14 days of culture. As shown in Fig. 3a, compared with control group, the carotenoid content was greatly increased under cultivation of SCF 40#, 34#, 39# ($P < 0.05$), but there were no obvious difference among the treatments of SCF 36#, 47#, 41#, 43#. While for chlorophyll *a*, it was higher under cultivation of SCF 40#, 34#, 39# and lower under 36#, 47#, 41#, 43#, than that of control group. Under cultivation of SCF 40#, the highest content of carotenoid and chlorophyll *a* was obtained with values 3.87 and 8.69 mg/L, which was an increase of 19.44 and 10.42% compared with control group, respectively.

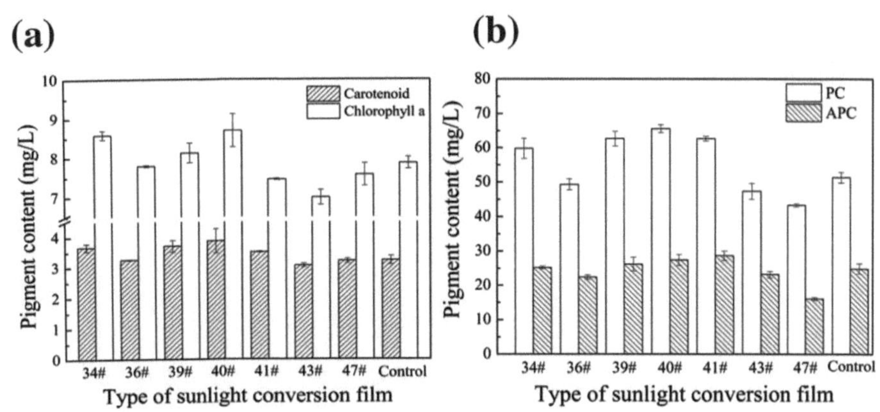

Fig. 3 Effects of SCF on the photosynthetic pigment content of *N. flagelliforme*. Control (the cells grown under common polyethylene film)

As presented in Fig. 3b, similar change patterns of phycocyanin (PC) and allophycocyanin (APC) were observed under the use of SCF. Both the PC and APC content were significantly higher under cultivation of SCF 40#, 39#, 41#, 34# ($P < 0.01$) and lower under SCF 47# ($P < 0.01$), than under SCF 36#, 43#. However, the PE was not detected throughout the culture period, which was in agreement with previous study reporting only PC and APC in the isolated intact phycobilisomes [14]. The highest PC and APC content were obtained under SCF 40# with values of 65.61 and 27.39 mg/L, which were respectively increased by 27.84% and 10.27%, compared to control group.

3.4 *Effects of SCF on the Protein Content*

After 14 days of cultivation, the protein content of *N. flagelliforme* under cultivation of different SCFs was measured, and the results were shown in Fig. 4. Notably, under cultivation of SCF 34# and 36#, although the crude protein content was slightly higher than that of the control group, it was much higher than that of the other SCFs treatments ($P < 0.05$). Particularly, compared with control group, under cultivation of 39#, the protein content was significantly decreased ($P < 0.05$). The results showed that SCF 34# and 36# were beneficial for protein synthesis.

4 Conclusions

In this study, different types of SCF were applied to culture the *N. flagelliforme* in order to improve the production of biomass and EPS. The results demonstrated clearly that SCF could significantly increase the biomass and EPS production.

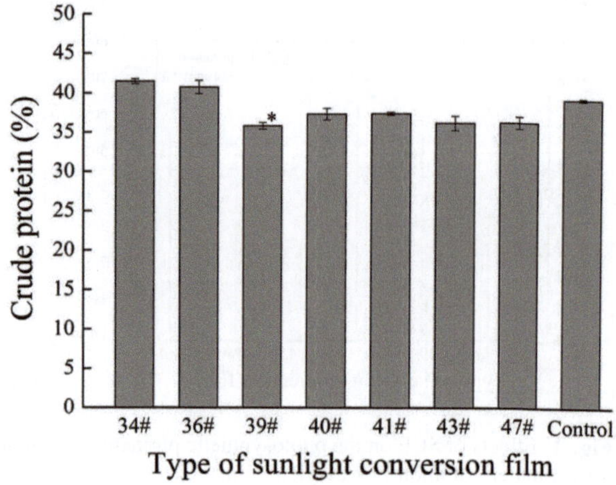

Fig. 4 Effects of SCF on the crude protein of *N. flagelliforme*. *Asterisk* indicate significant differences ($P < 0.05$ versus control). Control (the cells grown under common polyethylene film)

Particularly, under the use of SCF 41#, both the biomass and EPS production reached the highest, which were increased by 20.35 and 63.15% compared with control group, respectively. Based on this study, application of SCF on the microalgal culture was feasible and these results will shed light on novel culture strategies for microalgal outdoor cultivation.

Acknowledgements This work was supported by the National Natural Science Foundation of China (Grant No. 31671842 and 31201405) and Changjiang Scholars and Innovative Research Team in University (Grant No. IRT1166), which are all acknowledged.

References

1. Jia SR, Yao J, Tan ZL et al (2008) The preparation method for *N. flagelliforme* cells and *N. flagelliforme* extract with anti-oxidant activity. Chinese granted patent ZL200810152297.8
2. Kanekiyo K, Lee JB, Hayashi K et al (2005) Isolation of an antiviral polysaccharide nostoflan from a terrestrial cyanobacterium *Nostoc flagelliforme*. J Nat Prod 68:1037–1041
3. Jia SR, Yu HF, Lin YX et al (2007) Characterization of Exocellular Polysaccharide of *Nostoc flagelliforme* in Liquid Suspension Culture. Biotechnol Bioproc E 12(3):271–275
4. Han PP, Sun Y, Wu XY et al (2014) Emulsifying, flocculating, and physicochemical properties of exopolysaccharide produced by cyanobacterium *Nostoc flagelliforme*. Appl Biochem Biotechnol 172(1):36–49
5. Gao KS (1998) Chinese studies on edible blue-green alga, *Nostoc flagelliforme*: a review. J Appl Phycol 10:37–49
6. Yu HF, Jia SR, Dai YJ (2010) Accumulation of Exopolysaccharides in Liquid Suspension Culture of *Nostoc flagelliforme* Cells. App Biochem Biotechnol 160:552–560
7. Mishra SK, Shrivastav A, Maurya RR et al (2012) Effect of light quality on the C-phycoerythrin production in marine cyanobacteria *Pseudanabaena* sp. isolated from Gujarat coast, India. Protein Express Purif 81:5–10
8. Otero A, Vincenzini M (2003) Extracellular polysaccharide synthesis by *Nostoc* strains as affected by N source and light intensity. J Biotechnol 102:143–152
9. Yang Z, Geng LL, Wang W et al (2012) Combined effects of temperature, light intensity, and nitrogen concentration on the growth and polysaccharide content of *Microcystis aeruginosa* in batch culture. Biochem Syst Ecol 41:130–135
10. Han PP, Sun Y, Jia SR et al (2014) Effects of light wavelengths on extracellular and capsular polysaccharide production by *Nostoc flagelliforme*. Carbohyd Polym 105:145–151
11. Liu XJ, Jiang Y, Chen F (2005) Fatty acid profile of the edible filamentous cyanobacterium *Nostoc flagelliforme* at different temperatures and developmental stages in liquid suspension culture. Process Biochem 40:371–377
12. Dubois M, Gilles KA, Hamilton JK et al (1956) Colorimetric method for determination of sugars and related substances. Anal Chem 28:350–356
13. Wellburn AR (1994) The spectral determination of chlorophylls *a* and *b*, as well as total carotenoids, using various solvents with spectrophotometers of different resolution. Plant Physiol 144(3):307–313
14. Yi ZW, Huang H, Kuang TY et al (2005) Three-dimensional architecture of phycobilisomes from Nostoc flagelliforme revealed by single particle electron microscopy. FEBS Lett 5793:569–3573

Determination of *Ginkgolic* Acids in *Ginkgo* Seeds Using HPLC-MS in the Presence of Lipids

Yanying Hu, Guojuan Sun, Huitu Zhang, Liming Zhang, Tongcun Zhang and Yujie Dai

1 Introduction

Ginkgolic acids are a class of 2-hydroxy-6-alkyl benzoate compounds existing in *ginkgo* leaves, seeds, roots and barks. Commonly there are 5 kinds of *ginkgolic* acids, C13:0, C15:0, C15:1, C17:1, and C17:2, detected in previous reports. They have the potential to sensitization and mutagenicity, and strong cytotoxicity [1–3]. They can cause allergic reaction, gene mutation, nerve damage, which leads to nausea, heartburn, allergic shock, allergic purpura, exfoliative dermatitis, gastrointestinal mucosal allergy, convulsion, paralysis and other adverse reactions. Their content is strictly limited in a variety of *ginkgo* related phytopharmaceuticals [2]. In China, because there are no strict limits of *ginkgolic* acids for *ginkgo* seeds used as food, some adverse reactions are usually taken place when they are overdosed. It is necessary to determine the content of *ginkgolic* acids with a sensitive and selective analytical method.

There are many methods reported for the determination of *ginkgolic* acids in *ginkgo* biloba leaves, such as TLC, GC, GC-MS, HPLC etc. [4–6]. TLC method is

---

Yanying Hu, Guojuan Sun—These two authors contributed to the work equally and are treated as co-first authors.

Y. Hu · G. Sun · H. Zhang · L. Zhang · T. Zhang · Y. Dai (✉)
Key Laboratory of Industrial Fermentation Microbiology, Ministry of Education, College of Bioengineering, Tianjin University of Science and Technology, Tianjin 300457, People's Republic of China
e-mail: yjdai@126.com

Y. Hu
Department of Life Science and Engineering, Jining University, Qufu 273155, Shandong Province, People's Republic of China

H. Liu et al. (eds.), *Advances in Applied Biotechnology*, Lecture Notes in Electrical Engineering 444, https://doi.org/10.1007/978-981-10-4801-2_45

fast, cheap and easy to operate, but only suitable for qualitative analysis because of its large errors and limited precision. GC and GC-MS need a derivatization process for the determination of *ginkgolic* acids, which is relatively cumbersome [7], further, it is not easy to judge the results because the *ginkgo* acid derivatives, not *ginkgolic* acids themselves are directly detected for this process. Duo to the high resolution and no need of derivation process, HPLC was reported recently to be used for the quantitative analysis of *ginkgolic* acids in *ginkgo* biloba leaves and their extracts. HPLC-MS has the advantages of high efficiency, fast speed and high sensitivity, which is suitable for the determination of components with trace content, difficulty for separation, or variability. It also gives rich information of the samples such as retention time, online UV spectra, molecular weight and characteristics of structural fragments of the components synchronously. When these methods were used for the determination of *ginkgolic* acid contents in *ginkgo* seeds [8], an extraction process using a Soxhlet extractor, which is complicated and time-consuming, was usually conducted. Compared with the extract of *ginkgo* leaves, the extracts of *ginkgo* seeds or *ginkgo* sarcotesta contain higher content of lipid substances [9]. On the other hand, strong polar solvents such as water, methanol or acetic acid were usually used as the mobile phase in HPLC analysis [4–6]. Because the containing lipids are immiscible in these mobile phases and are similar in properties to *ginkgolic* acids, which may give negative influence on the determination of *ginkgolic* acids, the pretreatment of silica gel column chromatography was usually conducted in order to get rid of them [10]. It is a time-consuming process, which may decrease the accuracy of the qualitative determination of *ginkgolic* acids with HPLC method due to the inevitable sample loss.

In order to simplify the pretreatment processes and improve the accuracy of *ginkgolic* acids' analysis, isopropanol, which has a better solubility to *ginkgo* acids, was used in this study as extraction solvent and a component of mobile phase and HPLC-MS, which has the high sensitivity and selectivity, was employed to determine the content of *ginkgolic* acids in *ginkgo* seeds.

2 Materials and Methods

2.1 Samples and Reagents

The fresh *ginkgo* seeds were purchased from Taixing (Jiangsu Province, China). Standard samples (>99.0%) of *ginkgo* acids C15:1, and *ginkgolic* acid mixture (C13:0, C15:0, C15:1, C17:1 and C17:2) were purchased from Shanghai Tongtian Biotechnology Incorporated Company (China), batch number were 12101935, 12102306 and 12102406 respectively.

2.2 Methods and Procedure

2.2.1 Preparation of the Sample Solution

Before extraction, the hulled *ginkgo* seeds were dried to constant weight, grinded to powder, and then screened with a 200 mesh sieve. 2.0 g of the powder was weighted accurately into a 50 mL conical flask with plug and placed on a thermostatic shaker (Jintan Science Analysis Instrument Co. Ltd, THZ-82, China). 25 mL isopropanol was added to extract the *ginkgolic* acids and the shaker was shaking for a period (the shaking speed is 124 r min^{-1} at 28 °C). Then the liquid was filtered with 0.22 μm microporous membrane (Tianjin Keyilong Experiment Equipment Co., Ltd., China) and condensed with a rotary evaporator (Shanghai Yarong Biochemistry Instrument Factory, RE-3000, China), finally the volume of the extraction was adjusted to 2.5 mL to obtain the sample solution for the determination of *ginkgolic* acids.

2.2.2 Preparation of the Standard Solution

1.0 mg of the standard, *ginkgolic* acid C15:1, was weighted accurately into a 10 mL volumetric flask and isopropanol was added to adjust the volume of *ginkgolic* acid solution to 10 mL as the stock solution. Before the determination, 1.0 mL of the stock solution was then diluted for four times and filtered with 0.22 μm microporous membrane.

2.2.3 Analyses of *Ginkgolic* Acids by HPLC-MS

Quadrupole LC/MS system (1260 Agilent Technologies, USA) with an Agilent ZORBAX SB-C18 column (4.6 mm × 75 mm, 3.5 μm, Agilent Technologies, USA) was employed for the determination of *ginkgolic* acids in *ginkgo* seeds. Acetonitrile, isopropanol and methanol were chromatographic grade and the high purity water was bought from Wahaha Tianjin Bottled Water Co. Ltd. (China). Column temperature was 30 °C. The mobile phase was isopropanol/acetonitrile/acetic acid (42:57:1, v/v). The flow rate was 0.25 mL min^{-1}. The column was equilibrated for 12 min before the sample injections. UV spectra were recorded over a range of 190–400 nm and the chromatographic elution curve was acquired at 312 nm. The LC elution was introduced directly into the ESI interface without splitting. Object compounds were analyzed in negative ion mode. The nebulizer pressure was 60 psi. The capillary voltage was 3 kV. The dry gas flow was 6.6 L min^{-1} and the temperature of the dry gas was 350 °C. The analysis was carried out using selected ion monitoring (SIM) at 347, 345, 373 and 371 for *ginkgolic* acid C15:0, C15:1, C17:1 and C17:2 respectively. The peak width is 0.1 min and the injection volume was 5.0 μL.

2.2.4 Preparation of Working Curve (Linearity)

Eight levels of concentration within the range 0.0008–0.1 mg mL^{-1} were prepared and injected into the HPLC-MS system respectively, and the peak areas were obtained.

2.2.5 Validation of the Determination Method

The limit of detection (LOD) was determined by the ratio of signal to noise, the ratio was detected using the lowest concentration of the standard curve through Agilent ChemStation. And the levels in which the response was 3× higher than the ratio of signal to noise was taken as LOD. The limit of quantitation (LOQ) for *ginkgolic* acids C15:1 was identified from the LOD as follows:

$$LOQ = (10 \times LOD)/3 \quad (1)$$

Each sample was injected into the chromatography system for six parallel times and chromatograms were recorded. The injection volume was 5.0 μL. The peak area values were recorded and the content of the *ginkgolic* acid was calculated, the average value and RSD (Relative Standard Deviation) were all calculated.

The recovery of the determination method was examined by adding standard solutions at three different concentrations: 9.4, 12.5, and 15.6 μg mL^{-1}, respectively to the container of 2.00 g *ginkgo* seed powder, followed by subsequent extraction using the same process employed for the samples, which consisted of filtration and direct analysis of the extraction solution by HPLC-MS.

3 Results and Discussion

3.1 Selection of the Extraction Solvent and Mobile Phase

Ginkgo seeds contain large amounts of lipids. When the strong polar solvents such as methanol or ethanol were used for the extraction of *ginkgolic* acids, these lipids may form a separate phase from the solvent, which is not conducive to constant volume process before the determination. However, when the nonpolar solvents were employed as extraction solvent, *ginkgo* seeds must be fully dried before the extraction, otherwise, these solvents were difficult to penetrate into the wet *ginkgo* seeds, resulting in the low extraction ratio. In addition, the nonpolar solvents such as cyclohexane and petroleum ether, cannot be completely miscible with the common HPLC mobile phase with strong polarity (for example, the acetonitrile). Through our experiments, it was found that isopropanol not only has good miscibility with lipids and *ginkgolic* acids, but it also miscible with methanol and acetonitrile, which are often used as a mobile phase in HPLC. Isopropanol itself can also usually used as a HPLC mobile phase component. Thus, in our study isopropanol was selected as the extraction solvent and the main component of the HPLC mobile phase, which facilitates the extraction and HPLC determination of *ginkgolic* acids. In order to improve the separation ability of the HPLC technique, acetonitrile was used to adjust the polarity of the mobile phase and a small amount of acetic acid was introduced to reduce the trailing behavior. Thus, in our experiments, isopropanol/acetic acid/acetonitrile (42:1:57, v/v) was used as the mobile phase.

3.2 Determination of the Extraction Time

2.0 g of the dried *ginkgo* seed powder was taken for the extraction into a 50 mL conical flask with plug and placed on the thermostatic shaker with the shaking speed of 124 r min^{-1} at 28 °C. 25 mL of isopropanol was used as the extraction solvent for the extraction of *ginkgolic* acids from *ginkgo* seeds. In order to ensure the *ginkgolic* acids in *ginkgo* seeds can be adequately extracted, the extraction time was optimized with single factor experiment. The results are shown in Fig. 1. It can be seen that the content of *ginkgolic* acids C15:1 in the extract solvent nearly doesn't change after the extraction time is longer than 120 min. Thus, all the extraction for subsequent experiment was kept for 120 min.

Fig. 1 Changes of the content of C15:1 with extraction time

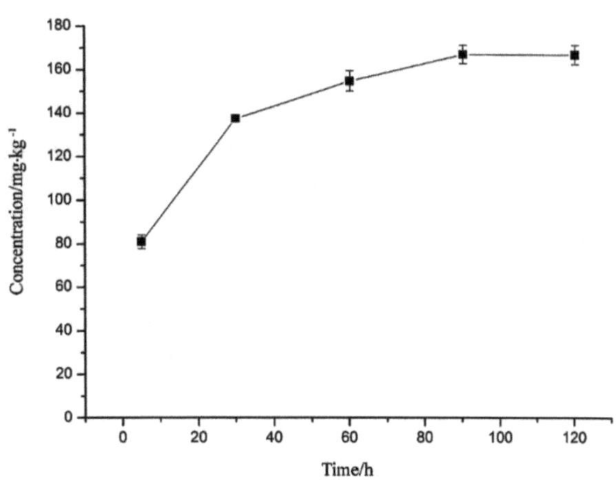

3.3 HPLC and HPLS-MS Flow Curves of the Sample Solution

HPLC is widely applied to the separation and determination of natural products. However, because of the complexity of the natural products, directly injecting the samples of the solvent extract into the HPLC apparatus can't give an ideal result. It is difficult to make all the components in the extract to be separated completely in the flow curve. Generally, the crude extracts were further separated and purified, which enhances the complexity of the determination of the components in it. For the extract of *ginkgo* seeds, it has plenty of lipids, which are similar to *ginkgolic* acids in many properties. These lipids may interrupt the determination of *ginkgolic* acids. Previous investigations used the silica gel chromatography as a pretreatment to remove these lipids [10]. It is inevitable to introduce extra errors for the quantitative determination *ginkgolic* acids. Recently, HPLC-MS has been extensively applied to the qualitative and quantitative analysis of complex mixtures. Its unique advantage is that it completely omitted the interference of the substances with different molecular weights. It can give a quite satisfactory flow curve for qualitative and quantitative determination of object molecules. There are two modes for the HPLC-MS analysis, Scan mode and SIM mode. The Scan mode is for scanning the sample extracts for expected and unexpected substances as a qualitative analysis while the SIM mode yields good sensitivity for quantitative analysis of the target compounds. In this research, in order to determine the retention times and characteristic mass fragments of *ginkgolic* acids, the primary electron ionization mass spectra of the analysts were recorded in full-scan mode (*m/z* 190–400, scan frequency 5 Hz). Simultaneously, for quantitative analysis SIM mode was applied with target masses of *m/z* 347.0, 345.0, 373.0 and 371.0 for *ginkgolic* acids C15:0, C15:1, C17:1, and C17:2, respectively.

Figure 2 is the HPLC and Selected ion chromatograms (*m/z* 345 for detecting C15:1) of the extract of *ginkgo* seeds. It can be seen that HPLC can't give independent peaks for all the components with direct injection of isopropanol extract sample, indicating that it can't give isolated peaks for the extract composition by using direct injection of the sample solution without any pretreatment. However, HPLC-MS shows a distinct regular peak at 5.181 min, representing the component with nucleocytoplasmic ratio of 345, which accords with the peak of C15:1. Other *ginkgolic* acid components can also give distinct regular peaks by adjusting nucleocytoplasmic ratio to corresponding values of their molecular weights.

3.4 Validation of the Determination Method

An analytical procedure should be demonstrated that it is suitable for its intended purpose, and that an analytical method should meet the requirements of analytical applications, so that the reliability and credibility of the results should be ensured.

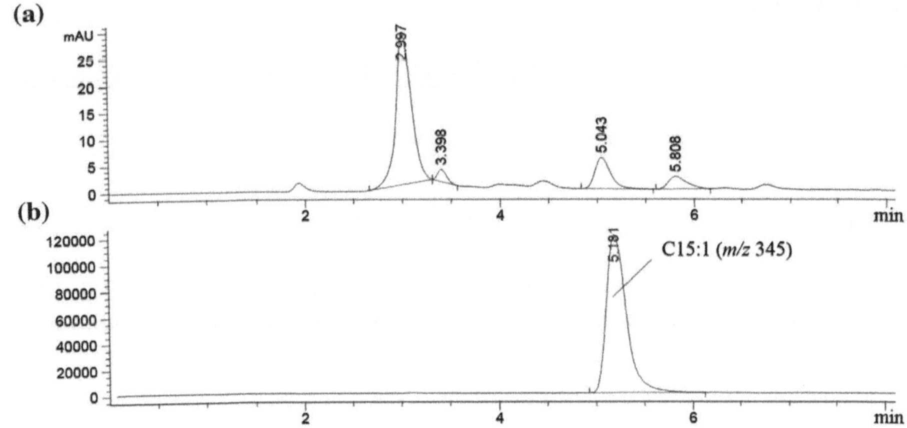

Fig. 2 HPLC and HPLC-MS analysis of the extract of *ginkgo* seeds. **a** HPLC determination of *ginkgolic* acids in the isopropanol extract injected directly. **b** HPLC-MS analysis (SIM mode *m/z* 345) of *ginkgolic* acid C15:1 in the isopropanol extract of *ginkgo* seeds

The following parameters were evaluated for the method validating herein: linearity, limits of detection and quantitation, precision, and recovery. To this end, a standard of C15:1 and *ginkgolic* acid mixture (C15:1, C17:1, C17:2, C15:0) were employed. The results are presented below.

3.4.1 Linearity/Range

The linearity of an analytical methodology indicated that the obtained results of peak area are directly proportional to the concentration of the analyte in the standard sample, within a specified range. In this range, the concentrations that are expected to be detected during analysis of the samples. In the present case, concentrations in the 0.0008–0.1 mg mL^{-1} range were taken into consideration for the standard C15:1. The regression equations and the squared correlation coefficients were:

$$Y = 3.5 \times 10^{-8} X - 0.0011 (r^2 = 0.9997) \quad (2)$$

While X is the sample weight (μg) and Y is the peak area. It correlates well in the range of 0.008–1 μg of *ginkgolic* acid [11].

3.4.2 Limits of Detection and Quantitation

The ratio that was detected using the lowest concentration of the standard curve through Agilent ChemStation was 65.7, and the LOD and LOQ are 0.354×10^{-3} μg and 1.179×10^{-3} μg respectively.

3.4.3 Precision Test

Precision values are expressed as the coefficient of variation (CV), according to the formula:

$$CV(\%) = SD/DAC \times 100 \quad (3)$$

where SD is the standard deviation and DAC is the determined average concentration. As for repeatability, the sample solution was determined for six times. Correlation between the results was determined within a short period and the data measured by the same analyst by means of the same instrumentation. The average concentration was 170.95 mg kg^{-1} and CV was 3.3%.

3.4.4 Recovery Test

Recovery, a measure of the extraction efficiency of the analytical method, was analyzed by adding known standards to the samples. In our experiments, three different amounts of *ginkgolic* acid C15:1 was added into 3 samples each containing 2.0 g of dry *ginkgo* seed powder. It was extracted and determined with HPLC-MS as described previously. Each sample was determined for three times and the results were listed in Table 1. The average recovery is 102.29%, RSD = 5.5%. All the values were satisfactory for the purposes of this work.

3.5 *Determination of Content of Ginkgolic Acids in Samples*

In this study, UV detector was set at 312 nm. The results are listed in Table 2. The results show that there are four kinds of *ginkgo* acids in sample and the *ginkgolic* acid C15:1 has the most of 170.95 ± 0.033 mg kg^{-1}. The *ginkgolic* acid of C15:0 may have

Table 1 The recovery of HPLC-MS method was calculated for ginkgo seed powder spiking with different amounts of ginkgolic acid C15:1

	Sample (mg)	Standard (mg)	Spiked sample (mg)	Recovery (%)
A1	2.08×10^{-5}	3.34×10^{-5}	5.57×10^{-5}	104.50
A2	2.10×10^{-5}	3.34×10^{-5}	5.60×10^{-5}	104.80
A3	2.16×10^{-5}	3.34×10^{-5}	5.79×10^{-5}	108.68
B1	2.11×10^{-5}	2.62×10^{-5}	4.83×10^{-5}	103.82
B2	2.20×10^{-5}	2.64×10^{-5}	4.93×10^{-5}	103.41
B3	2.14×10^{-5}	2.77×10^{-5}	4.81×10^{-5}	96.39
C1	2.07×10^{-5}	2.24×10^{-5}	4.34×10^{-5}	101.34
C2	2.15×10^{-5}	2.32×10^{-5}	4.25×10^{-5}	90.52
C3	2.09×10^{-5}	2.23×10^{-5}	4.48×10^{-5}	107.17

Table 2 The content of main components of *ginkgolic* acids

Components	Content/mg kg^{-1}
C15:0	12.16 ± 0.033
C15:1	170.95 ± 0.033
C17:2	9.63 ± 0.033
C17:1	26.34 ± 0.033

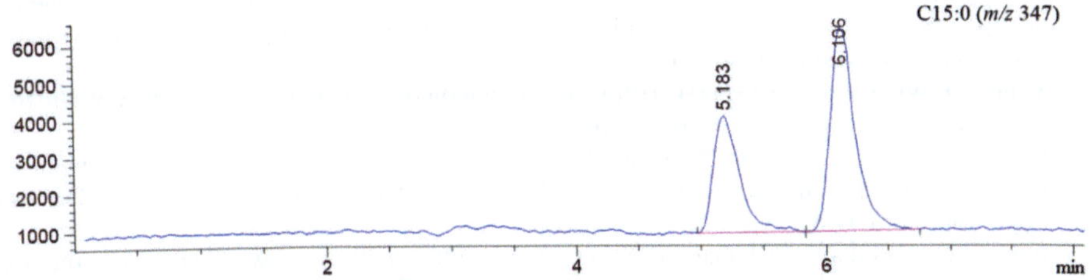

Fig. 3 The HPLC-MS (SIM mode, *m/z* 347) analysis of ginkgolic acid C15:0

isomers because there are two adjacent peaks with the same *m/z* appeared (Fig. 3). The peak time of sample and the *ginkgolic* acid mixture standard are the same.

4 Conclusions

In this study, a determination method using HPLC-MS for four kinds of *ginkgolic* acids in *ginkgo* seeds C15:0, C15:1, C17:1, C17:2 was established with isopropanol as the extraction solvent and a component of the mobile phase. The concentrations of *ginkgolic* acids in the extract can be determined using HPLC-MS by the direct injection of the extract solution without any lipid removal process. The coefficient of variation and the recovery of the determination method are 3.3% and 102.29% respectively. It has good linearity within the range of 0.0008–0.1 mg mL^{-1}. This paper provides a simple and convenient method for the determination of *ginkgolic* acids in *ginkgo* seeds with the presence of a large amount of lipids.

References

1. Koch E, Jaggy H, Chatterjee SS (2000) Evidence for immunotoxic effects of crude *Ginkgo biloba* L. leaf extracts using the popliteal lymph node assay in the mouse. Int J Immunopharmacol 22(3):229–236
2. Baron-Ruppert G, Luepke NP (2001) Evidence for toxic effects of alkylphenols from Ginkgo biloba in the hen's egg test (HET). Phytomed Int J Phytotherapy Phytopharmacol 8(2):133–138. doi:10.1078/0944-7113-00022

3. Hecker H, Johannisson R, Koch E, Siegers CP (2002) In vitro evaluation of the cytotoxic potential of alkylphenols from *Ginkgo biloba* L. Toxicology 177(2–3):167–177
4. Kong YJ, Li CK, Wang PP, Song JY (2010) Study on separation of ginkgolic acids by resh adsorption. J Liaoning Univ Nat Sci Ed 37(01):57–60
5. Ndjoko K, Wolfender JL, Hostettmann K (2000) Determination of trace amounts of ginkgolic acids in *Ginkgo biloba* L. leaf extracts and phytopharmaceuticals by liquid chromatography-electrospray mass spectrometry. J Chromatogr B 744(2):249–255
6. Yin XL (2003) Study on separation technology of ginkgolic acids from the sarcotesta of *Ginkgo biloba* L. D, Jiangsu University
7. Gao ZS, Xian QM, Han XD, Pan XP, Feng JF, Yu HX (2008) Determination of nonylphenol in cells by pre-column derivatization and gas chromatography coupled with mass spectrometry. Environ Chem 27(6):819–822
8. Yang LQ, Wu XY, Chen J (2004) HPLC determination of ginkgolic acids in fruits of *Ginkgo biloba*. Chin J Pharm Anal 24(6):636–639
9. Chen WY, Wang CZ, Gao CX, Zhang SF, Zheng GY (2006) The study on the content of flavonoids and the chemical composition of lipids in seeds of *Ginkgo biloba* L. Biomass Chem Eng 40(6):6–8
10. Itokawa H, Totsuka N, Nakahara K, Takeya K, Lepoittevin JP, Asakawa Y (1987) Antitumor principles from *Ginkgo biloba* L. Chem Pharm Bull 35(7):3016–3020
11. Gouvea DR, Meloni F, Ribeiro Ade B, Lopes JL, Lopes NP (2012) A new HPLC-DAD-MS/MS method for the simultaneous determination of major compounds in the crude extract of *Lychnophora salicifolia* Mart. (Brazilian arnicao) leaves: application to chemical variability evaluation. Anal Chim Acta 748:28–36. doi:10.1016/j.aca.2012.08.027

High Concentration Vinegar Production by Acetic Acid Bacteria Using Edible Alcohol as Feedstock

Xiao-Yan Yin, Rong Zhai, Wu-Kun Zhong, Jiao Huo and Zhong-Hua Yang

1 Introduction

Vinegar is a traditional oriental grain condiment with a sour taste and distinct fragrance. Now, it has been used as a condiment sauce worldwide [3]. The global vinegar demand is increasing significantly in the last few years [13]. Its main industrial production technology was semi-continuous fermentation with acetic acid bacteria (AAB), in which the acidity usually does not exceed 60 g/L. High acidity vinegar(acidity $\geq$ 90 g/L acetic acid) can save storage and transport cost [6]. So, production of high acidity vinegar is valuable [12].

However, there is some hardness in production of high concentration vinegar with AAB fermentation. The inferior quality white vinegar was mainly produced by blending the glacial acetic acid. This means that it is not healthy and poor taste. The traditional high quality vinegar was produced from ordinary fermentation vinegar by repeatedly freezing, however its productivity is very low [4]. Here, we try to produce the vinegar from edible alcohol by biocatalysis with acetic acid bacteria (AAB). This is a safe and feasible route.

AAB has powerful ability to oxidise ethanol. In the two-step reaction [14], the AAB acts as a cell factory, in which the ethanol is oxidized to acetic acid catalyzed

X.-Y. Yin · W.-K. Zhong · J. Huo · Z.-H. Yang (✉)
School of Chemistry and Chemical Engineering, Wuhan, University of Science and Technology, Wuhan 430081, China
e-mail: yangzh@wust.edu.cn

R. Zhai (✉)
College of Life Science and Chemistry, Wuhan Donghu University, Wuhan 430081, China

H. Liu et al. (eds.), *Advances in Applied Biotechnology*, Lecture Notes in Electrical Engineering 444, https://doi.org/10.1007/978-981-10-4801-2_46

by the two enzymes, alcohol dehydrogenase (ADH) and acetaldehyde dehydrogenase (ALDH). Simultaneously, some other flavor components are also synthesized [10].

Carbon source, nitrogen source, inorganic ions and vitamins are the key factors for ABB [11]. Berraud studied the composition of nutrient salt to the production of the high concentrated alcohol vinegar [1]. Yang also optimized the medium for alcohol vinegar fermentation [15]. But in their research, the AAB biomass was still in a low level, which limited the productivity further improvement. To improve the AAB biomass in the vinegar fermentation, the carbon source, nitrogen source and ethanol concentrations were optimized in this work.

2 Materials and Methods

2.1 *Microorganisms*

The bacteria *AAB.WUST-01* was applied as the production strain, which was preserved in our lab.

2.2 *The Original Medium and Culture Conditions*

The original medium composition for AAB culture were glucose (10 g/L), yeast extract (10 g/L), KH_2PO_4 (0.5 g/L), $MgSO_4$ (0.5 g/L) and NaH_2PO_4 (0.5 g/L), and ethanol (20 mL/L).

Carbon source optimization: The concentration of glucose was as follows (w/v) 0.2, 0.3, 0.4, 0.5, 0.6%. Others compositions are the same as the original medium.

Nitrogen source optimization: The concentration of yeast extract was as follows (w/v) 2.3, 2.4, 2.5, 2.6, 2.7%. Others compositions are the same as the original medium.

Ethanol concentrations optimization: In starting-up stage, the concentration of ethanol was as follows (v/v) 0.0, 1.0, 2.0, 3.0, 4.0%. The concentration of glucose and yeast extract were the optimized amount. Others compositions are the same as the original medium. In production acid stage, the concentration of ethanol was as 3.0, 4.0, 5.0, 6.0, 7.0% (v/v).

In the process, the AAB culture broth, which was at exponential growth phase, was inoculated into the fermenter with 10% of inoculum size. The fermenter was opened in 30 °C and 220 r/min agitation rate. The first 24 h culture is the starting-up stage for AAB biomass growth. The next stage is production stage, which is mainly production of vinegar.

2.3 Analysis Methods

The acidity of acetic acid was quantitatively determined by titration with 0.1 M NaOH solution using phenolphthalein as indicator [9].

The total biomass was determined by turbidity method based on the optical density measurements at 600 nm using a UV/visible spectrophotometer [14]. Samples were diluted to the appropriate concentration to keep the OD_{600nm} value between 0.2 and 0.8. The relationship between dry cell weight and OD_{600nm} was constructed by standard line [8].

To model the growth curve, the biomass data was analyzed with Logistic model (1), Gaussian Eq. (2) and Lorentz Eq. (3).

$$c = \frac{c_m(c_0 - c_m)}{c_0 \cdot e^{\mu_m \cdot t} - c_0 + c_m} + c_m \tag{1}$$

$$c = c_0 + \frac{A \cdot e^{\frac{-4\ln 2 \cdot (c - c_m)^2}{w^2}}}{w\sqrt{\frac{\pi}{4\ln 2}}} \tag{2}$$

$$c = c_0 + \frac{2A}{\pi} \cdot \frac{w}{4(c - c_m)^2 + w^2} \tag{3}$$

C_0: the biomass at the start in theory, C_m: the maximal biomass in theory, w: half-peak width, μ_m: the specific growth rate in theory [2, 5, 7].

3 Results and Discussion

3.1 Carbon SourceOptimization

Glucose was used as the carbon source. To optimize the glucose concentration, various glucose concentration was applied and the corresponding growth curve was determined (Fig. 1).

To find the relationship between C_m and glucose concentration, we fitting the curves with least square method. The result were given in Table 1.

The data were further fitted with Gaussian equation and Lognormal equation. The results were given in Fig. 2. It shows that Logistpk equation, Gaussian equation and Lognormal equation all have a good imitative effect in the relationship between C_m and glucose concentration. The fitting degree (R^2) are respectively 0.9860, 0.9871 and 0.9870. Based on the R^2, Gaussian equation is the most suitable. When C have the highest, the corresponding parameter values were $C_0 = 0.7620$,$C_m = 0.5066$,A = 0.0108,w = 0.1144. According it, the optimum value for glucose concentration in the medium is about 0.51% (w/v).

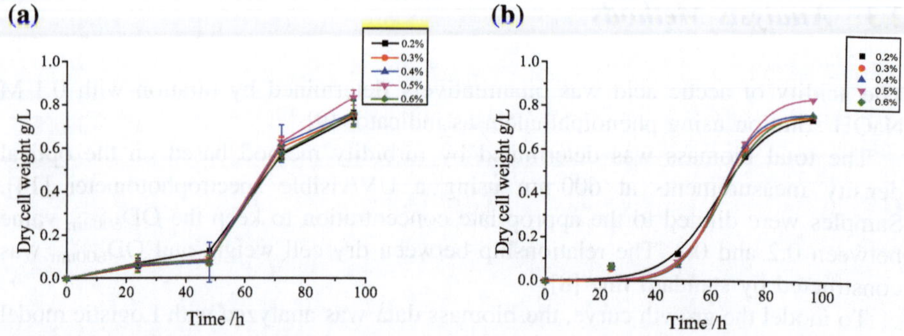

Fig. 1 Growth curves in different concentration of glucose **a** experiment data, **b** fitting curves of biomass with Logistic equation

Table 1 Parameter values of the logistic mode (glucose)

Glucose concentration g/100 mL	C_0 g/L	C_m g/L	μ_m/h	R^2
0.2	0.001	0.765	0.106	0.9849
0.3	0.001	0.759	0.145	0.9886
0.4	0.001	0.770	0.150	0.9871
0.5	0.001	0.850	0.129	0.9881
0.6	0.001	0.776	0.128	0.9831

Fig. 2 Fitting curves of C_m with Logistpk equation, Gaussian equation and Lognormal equation in different concentration of glucose

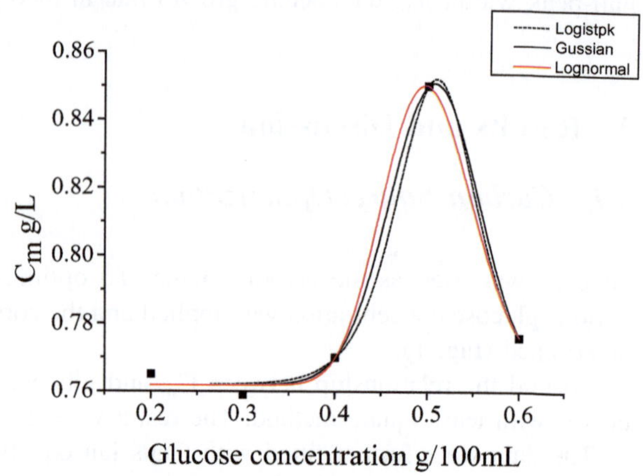

3.2 Nitrogen Source Optimization

Yeast extract was applied as the nitrogen source. Various yeast extract amount was applied and the corresponding growth curve was determined (Fig. 3).

To find the relationship between C_m and glucose concentration, we fitting the curves with least square method. The result were given in Table 2.

The data were further fitted with Gaussian equation and Lognormal equation. The results were given in Fig. 4. It shows that Lorentz equation, Gaussian equation and Logistic equation all have a good imitative effect in the relationship between C_m and yeast extract concentration. The fitting degree (R^2) are 0.9993, 0.9914 and 0.9945 for Lorentz equation, Gaussian equation and Logistic equation. According the fitting degree (R^2), the Lorentz equation is the best $y = y_0 + \frac{2A}{\pi} \cdot \frac{w}{4(x-x_c)^2 + w^2}$, The parameter values were C_0 = 0.8937, C_m = 2.6207, w = 0.1188, A = 0.0158. when C = C_m, C have the highest. So, the optimum value for yeast extract amount is about 2.62% (w/v).

3.3 Ethanol Concentrations Optimization

Ethanol is the feedstock to the vinegar product. Its concentration is the key factor to the production process. If the ethanol concentration is not enough, the process can't reach a high productivity. Simultaneously, the vinegar acidity will be very low. On the other extreme, the excessively high ethanol concentration will inhibit the AAB activity even cause the AAB death. The effect of ethanol concentration to the AAB

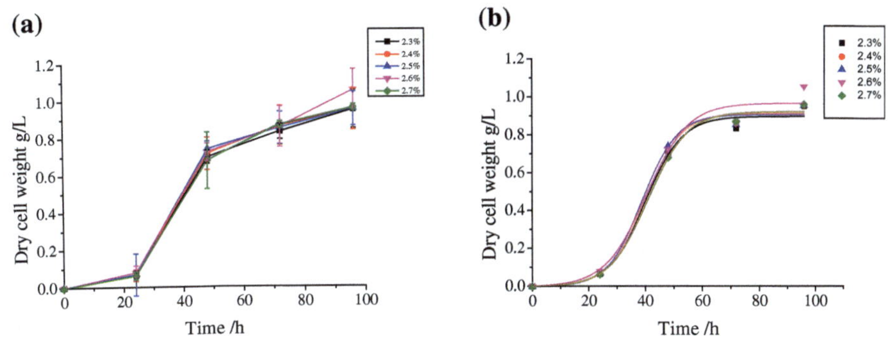

Fig. 3 Growth curves in various amount of yeast extract **a** experiment data; **b** fitting curves of biomass with Logistic equation

Table 2 Parameter values of the Logistic model (yeast extract)

Yeast extract g/100 mL	C_0 g/L	C_m g/L	μ_m/h	R^2
2.3	0.002	0.896	0.154	0.9841
2.4	0.001	0.900	0.158	0.9921
2.5	0.001	0.910	0.164	0.9883
2.6	0.003	0.969	0.136	0.9626
2.7	0.002	0.924	0.147	0.9918

Fig. 4 Fitting curves of C_m with Lorentz equation, Gaussian equation and Logistpk equation in different amount of yeast extract

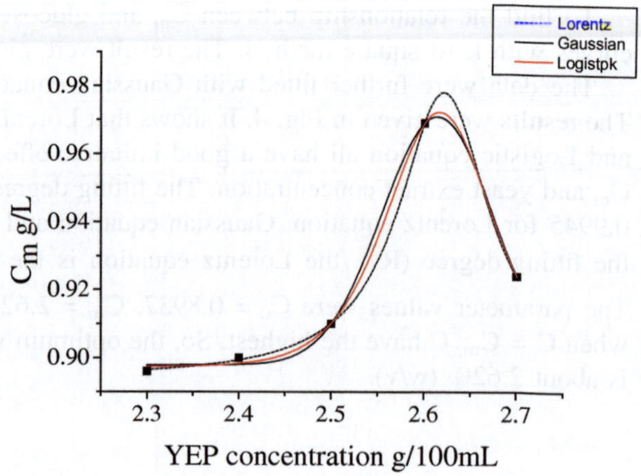

Fig. 5 Growth curves of AAB in various concentration of ethanol

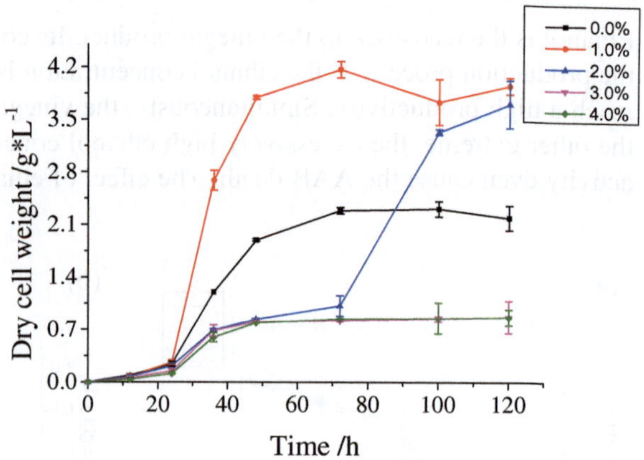

growth and the vinegar acidity were investigated. The corresponding results was shown in Figs. 5 and 6.

Figure 5 shows that with 1.0% (v/v) ethanol concentration dry cell weight could reach 4.16 g/L at 72 h. While in other conditions, the biomass are obviously lower. So in starting-up stage, 1.0% (v/v) ethanol concentration was favorable to improve the biomass.

Figure 6 shows that the total acid was increased with ethanol concentration increasing from 3 to 5% in 72 h. The high acidity 45.71 g/L was obtained with 5% ethanol and the yield could reach up to 88.76% in 72 h. Although the acid in 6 and 7% can also reached around 45 g/L in 96 h, but that means the productivity is lower.

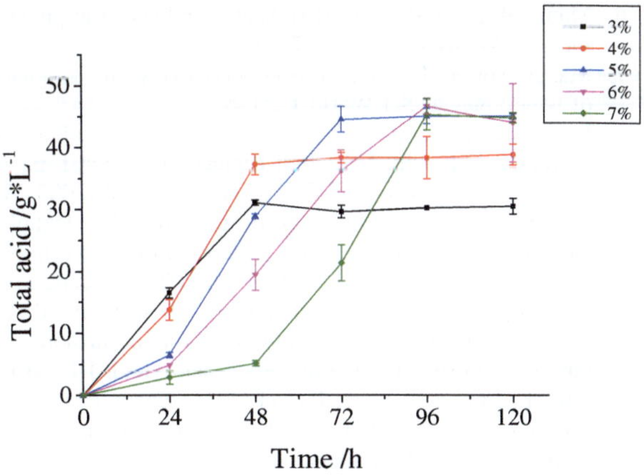

Fig. 6 Course curves of total acid in various ethanol concentration

4 Conclusions

The carbon source and nitrogen source were optimized through fitting model equations for the alcohol vinegar production. At the same time, the ethanol concentrations in starting-up stage and production of acid stage were also optimized. With the optimized condition, the biomass can increased 4.34 fold, and the yield of ethanol can reaches up to 88.76% in 5% ethanol concentration. This can provide basic information for alcohol vinegar production.

Acknowledgements The present work was financed by the National Natural Science Foundation of China (Grant no. 21376184), the Scientific Research Foundation for the Returned Overseas Chinese Scholars (State Education Ministry), Foundation from Educational Commission of Hubei Province of China (Grant no. D20121108) and the Innovative Team of Bioaugmentation and Advanced Treatment on Metallurgical Industry Wastewater.

References

1. Berraud C (2000) Production of highly concentrated vinegar in fed-batch culture. Biotechnol Lett 22:451–454
2. Dorini FA, Cecconello MS, Dorini LB (2016) On the logistic equation subject to uncertainties in the environmental carrying capacity and initial population density. Commun Nonlinear Sci Numer Simulat 33:160–173
3. Janja T, Francios B (2015) Updates on quick identification of acetic acid bacteria with a focus on the 16S–23S rRNA gene internal transcribed spacer and the analysis of cell proteins by MALDI-TOF mass spectrometry. Int J Food Microbiol 196:137–144
4. Lee CY, Nakano A, Naofumi S (2003) Effects of substrate feed rates on heterologous protein expression by *Pichia pastoris* in DO-stat fed-batch fermentation. Enzyme Microb Technol 33:358–365
5. Liu JZ, Weng LP, Zhang QL et al (2003) A mathematical model for gluconic acid fermentation by *Aspergillus niger*. Biochem Eng J 14:137–141

6. Mamlouk D, Gullo M (2013) Acetic acid bacteria: physiology and carbon sources oxidation. Indian J Microbiol 53(4):377–384
7. Mas A, Torija M, Troncoso A et al (2014) Acetic acid bacteria and the production and quality of wine vinegar. Sci World J 2:1–6
8. Mitsukura K, Uno T, Yoshida T (2007) Microbial asymmetric oxidation of 2-butyl-1, 3-propanediol. Appl Microbiol Biotechnol 76(1):61–65
9. Matsushita K, Kobayashi Y, Mizuguchi M et al (2008) A tightly bound quinone functions in the ubiquinone reaction sites of quinoprotein alcohol dehydrogenase of an acetic acid bacterium *Gluconobacter suboxydans*. Biosci Biotechnol Biochem 72(10):2723–2731
10. Natsaran S, Kazunobu M, Osao A et al (2015) Acetic acid bacteria: A group of bacteria with versatile biotechnological applications. Biotechnol Adv 33:1260–1271
11. Qi Z, Yang H, Yu X (2013) A protocol for optimization vinegar fermentation according to the ratio of oxygen consumption versus acid yield. J Food Eng 116:304–309
12. Sofia L, Efimia H, Maria P et al (2015) Beyond traditional balsamic vinegar: compositional and sensorial characteristics of industrial balsamic vinegars and regulatory requirements. J Food Compos Anal 43:175–184
13. Torija M, Mateo E, Vegas C (2009) Effect of wood type and thickness on acetification kinetics in traditional vinegar production. Int J Wine Res 1(1):155–160
14. Toshiharu Y (2010) Alcohol dehydrogenase of acetic acid bacteria: structure, mode of action, and applications in biotechnology. Appl Microbiol Biotechnol 86:1257–1265
15. Yang HL, Qi ZL, Xia XL et al (2011) An optimum medium designed and verified for alcohol vinegar fermentation. Afr J Biotechnol 10(42):8421–8427

Optimization of Process for Optical Resolution of DL-Pantolactone Using Immobilized D-Lactonohydrolase

Mei-juan Xuan, Jian-zhong Zhang, Hao-yu Yuan, Run Wei, Fu-ping Lu and Ming Li

1 Introduction

D-pantolactone, a γ-lactone of D-pantoic acid, is an important intermediate for the production of pantothenic acid, belonging to the group of B vitamins, panthenol and calcium D-pantothenate, an additive for animal feeds [1]. The conventional production of D-pantolactone was mainly through resolution of racemic pantolactone using expensive and toxic chemical agents such as quinine of brucine. However, the recovery of D-pantolactone was very difficult in the chemical resolution process. Since Kataoka et al. [2] found that D-pantolactone hydrolase from several filamentous fungi belonging to the genera Gibberella, Fusarium and Cylindrocarpon could selectively hydrolyze the D-pantolactone in DL-pantolactone to D-pantoic acid, which can easily be converted into D-pantolactone, instead of the chemical resolution process, several enzymatic processes for microbial resolution of DL-pantolactone using the microbes producing-D-lactonohydrolase have been reported. For example, Fusarium oxysporum [3, 4] and Fusarium moniliforme SW-902 [5–7] asymmetrically hydrolyzed D-pantolactone in DL-pantolactone into D-pantoic aicd which was separated and then was converted into D-pantolactone by heating. However, the efficiency of this process was low because of low biomass and low enzyme activity.

To improve the enzyme activity and the efficiency of enzymatic resolution process for DL-pantolactone, screening and optimizing of fermentation of high D-lactonohydrolase activity strain, enzymatic resolution process for DL-pantolactone using free and immobilized cells producing-D-lactonohydrolase have been studied

M. Xuan · J. Zhang · H. Yuan · R. Wei · F. Lu · M. Li (✉)
Key Lab of Industrial Fermentation Microbiology, Ministry of Education (TUST), National and Local United Engineering Lab of Metabolic Control Fermentation Technology (TUST), College of Biotechnology, Tianjin University of Science and Technology, Tianjin 300457, China
e-mail: liming09@tust.edu.cn

H. Liu et al. (eds.), *Advances in Applied Biotechnology*, Lecture Notes in Electrical Engineering 444, https://doi.org/10.1007/978-981-10-4801-2_47

widely [1, 8, 9]. In previous paper, we reported the immobilization of D-lactonohydrolase by weakly alkaline ion exchange resin D-380 to improve half-life and reusability of D-lactonohydrolase [10]. In this study, the process for optical resolution of DL-pantolactone using immobilized D-lactonohydrolase was investigated by optimization of reaction condition of immobilized D-lactonohydrolase and the operational stability of immobilized D-lactonohydrolase was tested.

2 Materials and Methods

2.1 *Materials*

Fusarium solani var. AS3.4489 was preserved in our lab. The weakly alkaline ion exchange resin D-380 donated by Nankai University (Tianjin, China) was previously washed with distilled water and then activated with 1 mol/L hydrochloric acid solution, with 1 mol/L sodium hydroxide solution and then washed again with distilled water and 0.2 mol/L Tris-HCl buffer (pH 7.5). Standard of DL-pantolactone, D-lactonohydrolase and D-pantoic acid was purchased from Sigma Company. The other reagents used were made in China and analytical grade.

2.2 *Preparation of the Rude D-Lactonohydrolase Solution*

Fusarium solani var. AS3.4489 was cultured in 500 mL flask containing 100 mL medium (1% corn steep liquor, 2% glycerin, 1% Peptone and 1% Yeast extract, pH 6.7) at 30 °C for 64 h with 180 r/min. The mycelium was harvested by filtration. 100 g mycelium was washed two times using Tris-HCl buffer (0.2 mol/L, pH 7.5) and was broken with high-pressure homogenizer, and then the supernatant was collected by centrifugation. The rude D-lactonohydrolase solution was obtained by ultra-filtration of the supernatant with ultrafiltration membrane of MWCO ≦ 2000 and MWCO ≧ 100000.

2.3 *Immobilization of D-Lactonohydrolase*

10 g D-380 resin was suspended in enzyme solution (30 ml) at 30 °C for 12 h under stirring, subsequently cross-linked with glutaraldehyde (0.05% v/v) at 4 °C for 2 h. The immobilized D-lactonohydrolase was separated from the supernatant by filter and washed with Tris-HCl buffer (0.2 mol/L, pH 7.5) to no activity in Tris-HCl buffer. The immobilized D-lactonohydrolase was assayed and stored in 0.2 mol/L Tris-HCl buffer (pH 7.5).

2.4 Optimization of Resolution Process of DL-Pantolactone by Immobilized Enzyme

The hydrolysis rate of DL-pantolactone was determined by changing temperature, pH, substrate concentration and reaction time to optimize the process for optical resolution of DL-pantolactone by immobilized D-lactonohydrolase. The reaction mixture of 20 mL in 0.2 mol/L Tris-HCl buffer comprising 200 mg (as weight of resin) immobilized enzyme and DL-pantolactone was placed in a 100 mL flask. It was then incubated at designed temperature for suitable time with shaking (120 r/min). After removal of the immobilized enzyme by centrifugation, the supernatant was analyzed for hydrolysis rate of DL-pantolactone by HPLC.

2.5 Reuse of Immobilized D-Lactonohydrolase for Resolution of DL-Pantolactone

To test the reusability of the immobilized D-lactonohydrolase, batch experiments were conducted in 1 L fermentor at 40 °C, pH 7.0 (adjusting with 11.8 mol/L ammonia solution), 150 r/min. The total volume of reaction system was 500 mL, substrate concentration was 20% and reaction time was 8 h. After each batch, the immobilized D-lactonohydrolase was recovered by centrifugation and was reused for the next batch reaction under the same conditions.

2.6 Analysis of DL-Pantolactone and D-Pantoic Acid

DL-pantolactone and D-pantoic acid were determined by HPLC on a reverse phase C18-chromatographic column (5 μm, 4.6 mm × 250 mm) monitored at 215 nm with an eluent of 10% (v/v) acetonitrile containing 0.02 mol/L KH2PO4 at a flow rate of 1 mL/min. Under these conditions, DL-pantolactone and D-pantoic acid were eluted at 7.2 min and 4.0 min, respectively.

2.7 Enzyme Assay

The reaction mixture of 2 mL in 0.2 mol/L Tris-HCl buffer (pH 7.5) comprising 20 mg (as weight of resin) immobilized enzyme and 500 mg DL-pantolactone was incubated at 40 °C for 1 h with gentle shaking. After reaction, the D-pantoic acid was quantified by HPLC. One unit of D-lactonohydrolade activity was defined as the amount of enzyme to form 1.0 μmol pantoic acid under the above described condition.

3 Results and Discussion

3.1 *Immobilization of* D*-Lactonohydrolase*

In order to confirm that D-lactonohydrolase was immobilized on D-380, the FT-IR spectra of D-380, D-lactonohydrolase and immobilized enzyme was studied. As shown in Fig. 1, the presence of an intense absorption band at 3720 cm^{-1} indicated that D-380 had an abundance of free C–OH groups [11]. After adsorption of D-lactonohydrolase, the intensity of this band decreased dramatically, suggesting an hydrogen bonding interaction between free C–OH groups and the N-H groups of D-lactonohydrolase. Furthermore, absorption bands associated with C–H stretching (between 2900 and 3000 cm^{-1}), C=O stretching (1651 cm^{-1}) and C–H deformations (around 1453 cm^{-1}) were observed in the spectra of D-lactonohydrolase and immobilized D-lactonohydrolase, but not in the spectra of D-380. These results confirmed that the D-lactonohydrolase had been successfully immobilized on the D-380.

3.2 *Optimization of Process for Optical Resolution of* DL*-Pantolactone Using Immobilized* D*-Lactonohydrolase*

3.2.1 Effect of Temperature on Hydrolysis Reaction

The effect of temperature on hydrolysis of DL-pantolactone was examined with the immobilized D-lactonohydrolase at the range of 20–70 °C. The results were shown in Fig. 2. The hydrolysis rate of DL-pantolactone increased with the temperature increase from 20 to 40 °C, and maximum hydrolysis rate was observed at 40 °C. On the contrary, when the temperature increased above 40 °C, hydrolysis rate of

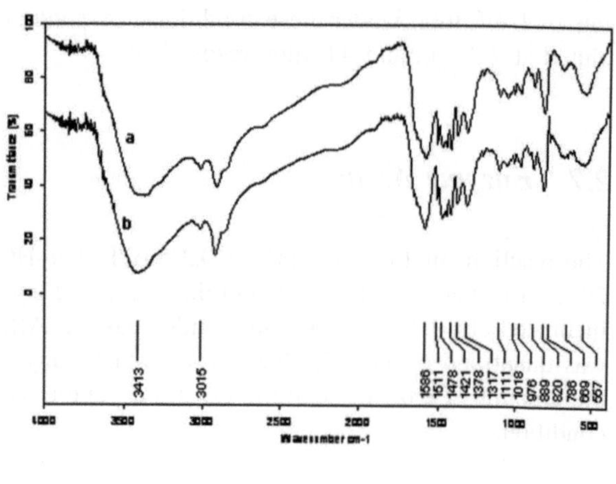

Fig. 1 FT-IR spectrums of the resin D-380 and immobilized D-lactonohydrolase

DL-pantolactone decreased with the temperature increase. It was also observed that the change of immobilized D-lactonohydrolase activity with tempreture was consistent with the change of hydrolysis rate of DL-pantolactone, and the change between them was proportional, showing that tempreture influenced the hydrolysis rate of DL-pantolactone by affecting the activity of the immobilized D-lactonohydrolase. Therefore, 40 °C was selected as the optimal temperature for hydrolysis reaction of DL-pantolactone.

3.2.2 Effect of pH on Hydrolysis Reaction

Because the pH of reaction system affected both the hydrolysis rate of DL-pantolactone and optical purity of product, we investigated the effect of pH on hydrolysis of DL-pantolactone with the immobilized D-lactonohydrolase. The result was shown in Fig. 3. At pH values between 5.0 and 7.5, the hydrolysis rate of DL- pantolactone increased in proportion to the pH of the reaction system, and the maximum DL-pantolactone hydrolysis activity was observed at pH 7.5. However, when pH was higher than 7.0, optical purity of product began to decrease, the same as the the report by Kataoka et al., which was not expected. According to the report by Kataoka et al., it was the chemical hydrolysis of L-isomer that resulted in the decrease of optical purity of product. So, the suitable pH for hydrolysis reaction was at pH 7.0.

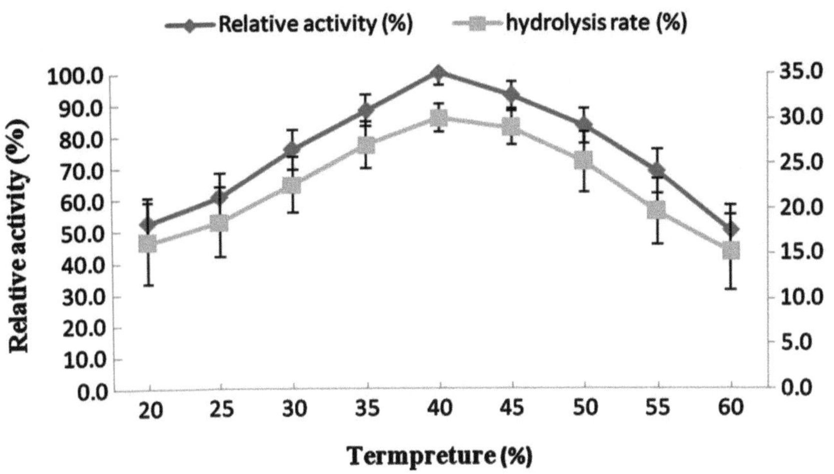

Fig. 2 The effect of temperature on hydrolysis reaction

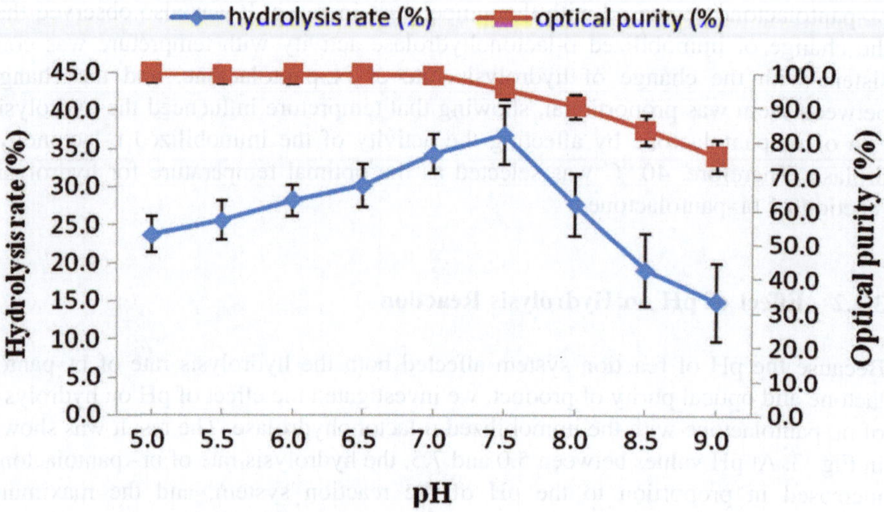

Fig. 3 The effect of pH on hydrolysis reaction

3.2.3 Effect of Substrate Concentration on Hydrolysis Reaction

The effect of substrate concentration on the hydrolysis rate of DL-pantolactone was investigated within the concentration range of 10–70% (w/v). The result was shown in Fig. 4. At substrate concentrates between 10 and 40%, D-pantoic acid increased with the increase of the substrate concentration, while when substrate concentration was more than 40%, the D-pantoic acid level didn't improve any more. On the contrary, hydrolysis rate of DL-pantolactone decreased with the increase of the substrate concentration at the same hydrolysis time. Considering about hydrolysis time and hydrolysis rate, the suitable substrate concentration should be 20%.

3.2.4 Time Course of Hydrolysis of DL-Pantolactone Using Immobilized D-Lactonohydrolase

The time course of hydrolysis of DL-pantolactone by immobilized D-lactonohydrolase was shown in Fig. 5 at 40 °C, pH 7.0 and substrate concentration 20%. The hydrolysis rate increased quickly during the initial 8 h, and hydrolysis reaction became slower after 8 h. Therefore, 8 h was the suitable reaction time with maximum hydrolysis rate 36.8%.

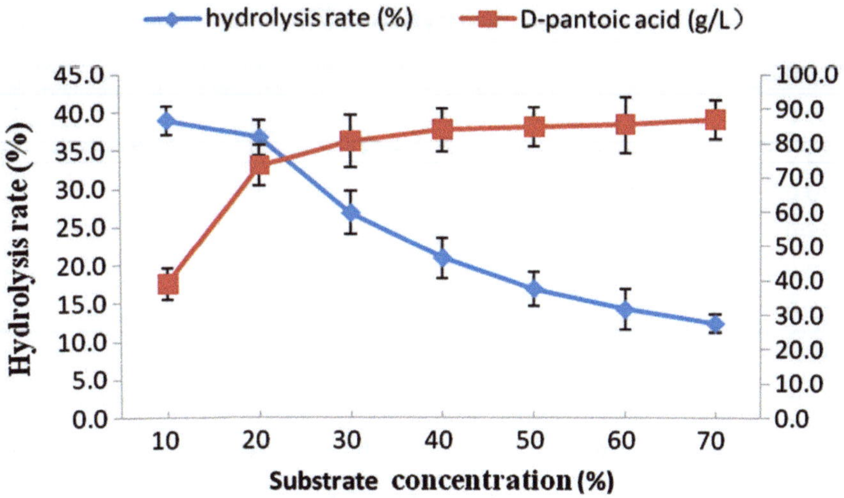

Fig. 4 The effect of substrate concentration on hydrolysis reaction

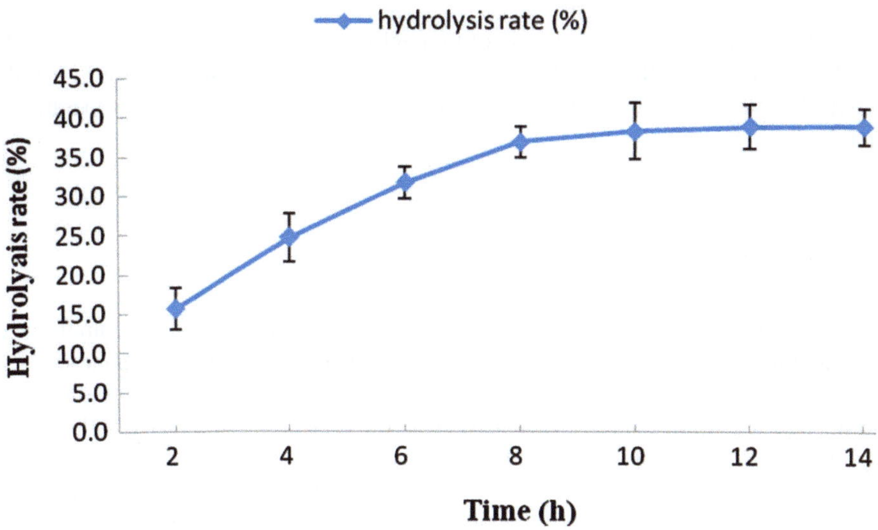

Fig. 5 Time course of hydrolysis of DL-pantolactone

3.3 Repeated Hydrolysis Reaction by Reusing Immobilized D-Lactonohydrolase

Figure 6 showed the hydrolysis process by repeated-batch operations to investigate the reusability and stability of the immobilized D-lactonohydrolase for hydrolysis of

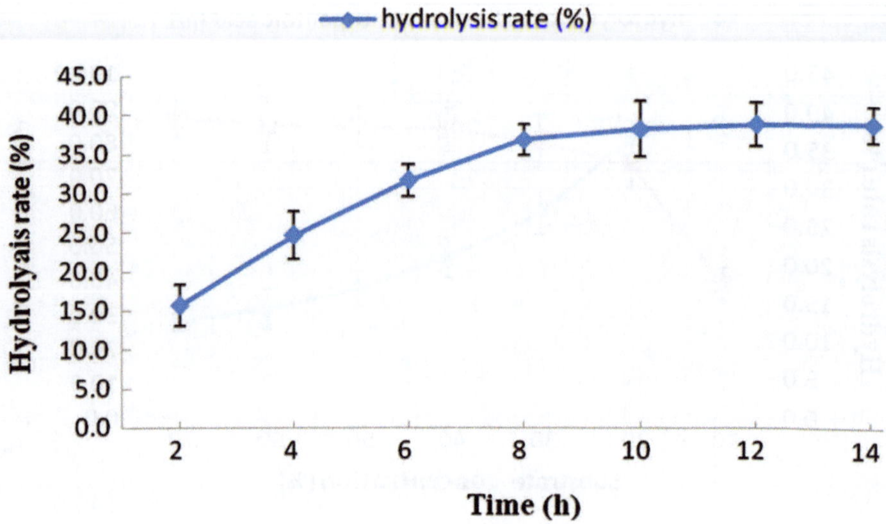

Fig. 6 Batch reactions of immobilized enzyme

DL-pantolactone. The hydrolysis rate of DL-pantolactone still reached 31.7% after 20 batches of reuse which remained at 85.6% of the original hydrolysis rate. It indicated that the immobilized D-lactonohydrolysis exhibited a high operational stability.

4 Conclusions

In order to improve the hydrolysis rate of DL-pantolactone using immobilized D-lactonohydrolase, the optimum process for optical resolution of DL-pantolactone using immobilized D-lactonohydrolase was established by optimizing the temperature, pH, substrate concentration and reaction time of hydrolysis reaction. The optimum process for optical resolution of DL-pantolactone was: temperature 40 °C, pH 7.0, substrate concentration 20%, hydrolysis time 8 h. Under the optimum condition, the hydrolysis rate of DL-pantolactone reached 36.8%, and the hydrolysis rate still remained at 85.6% of the original hydrolysis rate after 20 batches of reuse, indicating that immobilized D-lactonohydrolase had excellent operational stability. This work will be helpful for the production of D-pantolactone by D-lactonohydrolase.

Acknowledgements This work was supported by the National Natural Science Foundation of China (No. 21176190). We thank NanKai University for providing the resin D-380 kindly.

References

1. Yu MR, Tan TW (2005) Optical resolution of racemic pantolactone by *Fusarium sp.* BU-011 with high lactonohydrolase activity. Process Biochem 40(8):2609–2614
2. Kataoka M, Shimizu K, Sakamoto K et al (1995) Optical resolution of racemic pantolactone with a novel fungal enzyme, lactonohydrolase. Appl Microbiol Biotechnol 43:974–977
3. Sakamoto K, Yanada KH (1994) Process for the preparation of D-pantolactone. US 5275949
4. Kataoka M, Shimizu K, Sakamoto K (1995) Lactonohydrolase-catalyzed optical resolution of pantoyl lactone: selection of a potent enzyme producer and optimization of culture and reaction conditions for practical resolution. J Appl Microbiol Biotechnol 44:333–338
5. Tang YX, Sun ZH, Hua L (2002) Production of D-pantolactone hydrolase by *Fusarium moniliforme* SW902. J Acta Microbiol Sin 42(1):81–87
6. Tang YX, Sun ZH, Hua L (2001) Optical resolution of racemic DL-pantolactone by a fungal enzyme, D-lactonohydrolase. J Chin Ind Microbiol 31(3):1–5
7. Tang YX, Sun ZH, Hua L (2002) Kinetic resolution of DL-pantolactone by immobilized *Fusarium moniliforme* SW-902. Process Biochem 38:545–549
8. Hua L, Sun ZH, Zheng P (2004) Biocatalytic resolution of dl -pantolactone by glutaraldehyde cross-linked cells of *Fusarium moniliforme* CGMCC 0536. Enzym Microb Technol 35 (2):161–166
9. Hua L, Sun ZH, Leng Y (2005) Continuous biocatalytic resolution of DL-pantolactone by cross-linked cells in a membrane bioreactor. Process Biochem 40:1137–1142
10. Li M, Zhang JZ, Song XY et al (2011) Effect of pH and temperature on activity of immobilized lactonase. In: The 5th international conference on bioinformatics and biomedical engineering (iCBBE 2011), Wuhan, China, 13–15 May 2011
11. Bucke C (1987) Cell immobilization in calcium alginate. Meth Enzymol 135:175–189

Enzymatic Synthesis of L-Cysteine by *Escherichia coli* Whole-Cell Biocatalyst

Mingli Ma, Tao Liu, Heyun Wu, Fangqing Yan, Ning Chen and Xixian Xie

1 Introduction

L-cysteine is a very important sulfur-containing amino acid in respect of its applications in the medicine, food, fodder and cosmetic industries, and it is required for the biosynthesis of sulfur-containing compounds such as L-methionine, thiamine, biotin, coenzyme A and others [1, 2]. L-cysteine can be produced by three methods: chemical hydrolysis, microbial fermentation and enzymatic synthesis. Currently, L-cysteine is produced mainly via acid hydrolysis of human hair and animal feathers. However, the hydrolysis processes are associated with many disadvantages, for instance, the relatively low yield, the high cost as well as causing problems in waste-water treatment [3]. Microbial fermentation has been successfully used for commercial production of many amino acids [4–6]. Recently, researchers show an increasing interest in construction of L-cysteine-producing strains. However, high level production of L-cysteine with microorganisms hasn't yet been achieved to date due to the strong feedback regulation [7, 8], and toxic effects of high concentration of L-cysteine on cells [9, 10]. On the other hand, the supply of sulfur seems to be another major bottleneck for the fermentative production of L-cysteine.

The development of an efficient and environment-friendly method for L-cysteine production is challenging. Engineering of microbial catalysts for L-cysteine

M. Ma · T. Liu · H. Wu · F. Yan · N. Chen · X. Xie (✉)
College of Biotechnology, Tianjin University of Science and Technology,
Tianjin 300457, China
e-mail: xixianxie@tust.edu.cn

M. Ma · T. Liu · H. Wu · F. Yan · N. Chen · X. Xie
National and Local United Engineering Lab of Metabolic Control Fermentation
Technology, Tianjin 300457, China

H. Liu et al. (eds.), *Advances in Applied Biotechnology*, Lecture Notes
in Electrical Engineering 444, https://doi.org/10.1007/978-981-10-4801-2_48

production could be a promising alternative to current hydrolysis method. This method offers the advantages of mild reaction conditions, strong specificity and environmental compatibility. An enzymatic method for L-cysteine production has been first investigated utilizing chemically synthesized substrate DL-ATC, an intermediate compound in DL-cysteine chemosynthesis in the 1980s [11]. *Pseudomonas* sp. was reported to have the ability to hydrolyze DL-ATC to synthesize L-cysteine, and the pathways and their corresponding genes involved in DL-ATC biotransformation were identified subsequently [12, 13]. Tamura et al. reported that N-carbamoyl-L-cysteine is an intermediate in the conversion of DL-ATC to L-cysteine in *Pseudomonas* sp. ON4a and *Pseudomonas putida* AJ3865 [14]. Huai et al. reported that [15] *Pseudomonas* sp. TS1138 has the ability to convert DL-ATC to L-cysteine. Although enzymatic synthesis has the potential for L-cysteine production, there are still some problems to be solved in large-scale industrial application. In general, the activities of L-ATC hydrolases in wild *Pseudomonas* sp. are relatively low, and complicated medium compositions may not maintain relatively abundant enzyme sources. Another major obstacle to the L-cysteine biosynthesis is the low solubility of the substrate DL-ATC, which leads to a low conversion rate of DL-ATC [16].

As mentioned above, low substrate solubility and enzymatic activity affect the efficiency of catalytic reaction and the yield of L-cysteine. The solubility of DL-ATC can be enhanced in strong acid or base conditions, or with some solvents, however, these conditions often have a toxic effect on the metabolism of cells and easily inactivate the catalytic enzymes. In this study, a new method of enhancing the solubility of DL-ATC was proposed. In addition, a recombinant *E. coli* strain simultaneously overexpressing L-ATC hydrolase and L-NCC amidohydrolase from *Pseudomonas* sp. was constructed to increase the enzymatic activities of biocatalyst. To reduce the degradation of L-cysteine, the gene of major L-cysteine-degrading enzyme (tryptophanase/L-cysteine desulfhydrase) was deleted from *E. coli* BL21. Subsequently, the optimal conditions of the enzymatic conversion of DL-ATC to L-cysteine via *E. coli* whole-cell biocatalyst were investigated.

2 Materials and Methods

2.1 Strains and Plasmids

Escherichia coli BL21 and plasmid pET-his (Shenzhen Gene Power) were used as host strain and expression vector. The plasmids pKD46, pKD3 and pCP20 used in the λ Red recombination [17, 18] system are stored in this laboratory. All other chemicals were commercially available reagents of analytical grade.

2.2 Construction of the Strains

For heterologous expression of the *atcBC* gene cluster of *Pseudomonas* sp. in *E. coli*, the 1800-bp gene sequence was synthesized by GENEWIZ Bio Inc. (Suzhou, China), whereby the codon usage was adjusted to that preferred by *E. coli* using the proprietary Gene Optimizer software. The synthesized fragment was digested with *BamH* I and *EcoR* I (Takara, Japan) and cloned into pET-his (pET) to generate pET-his/*atcBC* (pET*atcBC*). The recombinant plasmid pET*atcBC* was transformed into *E. coli* BL21.

Deletion of the gene *tnaA* of *E. coli* BL21 was performed by homologous recombination using the λ Red recombination system. Both the upstream and downstream regions of the gene *tnaA*, were obtained by PCR using primer pairs: TnaA-1 (5′-CATTGTTACCACTCCTGTTATTCCT-3′) and TnaA-2 (5′-AT CGCTCAAGACGTGTAATGGATCCCTGTGAATATTACATCTGCT-3′), Tna A-3 (5′-TCTCAGTGCGTTACATCCCTGCCACTCTCTTACCCTACATCCT-3′) and TnaA-4 (5′-GTGCAGATCCCGACAATACG-3′), respectively. The chloramphenicol resistance (Cm^r) gene was obtained using primers Cm^{-1} (5′-ATGT AATATTCACAGGGATCCATTACACGTCTTGAGCGATTG-3′) and Cm^{-2} (5′-ATGTAGGGTAAGAGAGTGGCAGGGATGTAACGCACTGAGAAG-3′) with plasmid pKD3 as template. Overlapping PCR was used to fuse the fragments up- and down-stream of *tnaA* and Cm^r gene with the primers TnaA-1 and TnaA-4. The PCR products were transformed into the BL21 harboring pKD46. Clones were selected on LB agar plates containing chloramphenicol (30 μg/ml). Successful gene replacement with the antibiotic marker was confirmed by direct PCR. In order to remove the Cm^r gene from the integrated locus, the cells were transformed with temperature-sensitive plasmid pCP20 carrying the FLP recombinant gene.

2.3 Media and Culture Conditions

Recombinant strains were grown in 100 ml of LB medium containing 100 μg/ml of ampicillin at 37 °C for 12 h, and then used as seed cultures. Batch cultures were carried out in a 5 L bioreactor (BXBIO/MCGS BIOTECH-5BG-7000, BXBIO, China) containing 3 L medium. The medium contained (per liter) 3 g glucose, 5 g yeast extract, 10 g dried corn steep liquor powder, 5 g NaCl, 1 g KH_2PO_4, 0.5 g $MgSO_4$ and 1 mg/L V_H. The seed cultures were inoculated into the medium with a 10% inoculum size. The pH was kept constant at 7.0 by automated addition of NH_4OH (25%, v/v). Dissolved oxygen was maintained at about 30% by variation of the stirrer speed and the aeration rate. The temperature was adjusted to 28 °C when the OD_{600} was 20–25. At the same time, 0.1 mM IPTG was added into the medium as the inducer. When glucose was exhausted, 80% glycerol solution was added at an appropriate rate to maintain normal fermentation.

2.4 Preparation of Whole-Cell Solution

Recombinant strains were cultured and induced with IPTG for 12 h at 28 °C. Expression of recombinant proteins was evaluated by 12% SDS-PAGE. The cultivated cells were harvested by centrifugation at 6000 rpm and 4 °C for 5 min, and washed two times with phosphate buffered saline (PBS). Then 1 g cells (wet cells) were resuspended in 10 mL PBS buffer. The whole-cell solution was treated at −80 °C for 12 h, and then thawed at 4 °C.

2.5 Enzymatic Transformation of DL-ATC

The reaction was performed in 100 ml PBS containing 30 g/L cells (wet cells) and 3.0% (w/v) DL-ATC in a 250 ml flask with shaking at 200 rpm. To determine the whole-cell activity, the specific cell activity for L-cysteine synthesis was defined as the amount (μmol) of L-cysteine conversion for 1 min at 37 °C per unit cell mass (mg^{-1} wet cells) in the reaction mixture. For temperature optimization, the reaction temperature was varied from 35 to 41 °C. For pH optimization, pH of the reaction mixture was varied from 6.0 to 8.0. All measures were performed in triplicate and error bars denote standard deviation of the mean.

2.6 Analysis of L-Cysteine Concentration

The concentration of L-cysteine was determined by high performance liquid chromatography (HPLC, C18 150 mm*4.6 mm/5 μm, UV-VIS, Shimadzu), the mobile phase was 50% acetonitrile (v/v) and 0.42% (g/L) sodium acetate anhydrous solution, the detection wavelength was 360 nm, column temperature was 33 °C, the flow rate was 1 ml/min.

3 Results and Discussion

3.1 Construction of the Strains for L-Cysteine Production

For construction of an efficient biocatalyst, the codon-optimized genes *atcB* and *atcC* of *Pseudomonas* sp. were synthesized and cloned into *E. coli* BL21 for simultaneous expression of L-ATC hydrolase and L-NCC amidohydrolase. The theoretical molecular mass of L-ATC hydrolase and L-NCC amidohydrolase was estimated to be 19.7 and 45.0 kDa based on the amino acid sequences. As shown in

Fig. 1, the recombinant L-ATC hydrolase and L-NCC amidohydrolase were correctly expressed as soluble proteins after IPTG induction. Assay of whole-cell biocatalyst reaction was performed using DL-ATC as substrate. 13.1 mM of L-cysteine was accumulated with BL21/pET*atcBC* as the whole-cell biocatalyst, whereas only a small amount of L-cysteine (0.99 mM) was detected when using BL21/pET as the whole-cell biocatalyst (Fig. 2). These results showed that recombinant L-ATC hydrolase and L-NCC amidohydrolase could convert DL-ATC to L-cysteine efficiently.

Although the cells doesn't grow during the process of biocatalysis, L-cysteine desulfhydrase (CD) existing in cells catalyzes the degradation of L-cysteine to pyruvate, ammonia and sulfide [19, 20]. In *E. coli*, five proteins, encoded by *metC*, *malY*, *cysK*, *cysM* and *tnaA* respectively, have the ability of degrading L-cysteine, of which the tryptophanase/L-cysteine desulfhydrase encoded by *tnaA* was reported to play a major role in L-cysteine degradation [21]. In order to reduce L-cysteine degradation in the catalytic process, we knocked out the gene *tnaA* of *E. coli* BL21. The cultivation results demonstrated that the deficiency in *tnaA* had no effect on cell growth (data not shown). Compared with that of BL21/pET*atcBC*, the concentration of L-cysteine (18.8 mM) was increased by 54.4% by using BL21*ΔtnaA*/pET*atcBC* as the whole-cell biocatalyst (Fig. 2). Since the activity of L-cysteine desulfhydrase of BL21*ΔtnaA* had been significantly reduced, we didn't attempt to delete other four genes encoding L-cysteine desulfhydrase and the strain BL21Δ*tnaA*/pET*atcBC* was thus used for subsequent optimization efforts to boost L-cysteine production. Recombinant strain, BL21*ΔtnaA*/pET*atcBC*, was cultivated

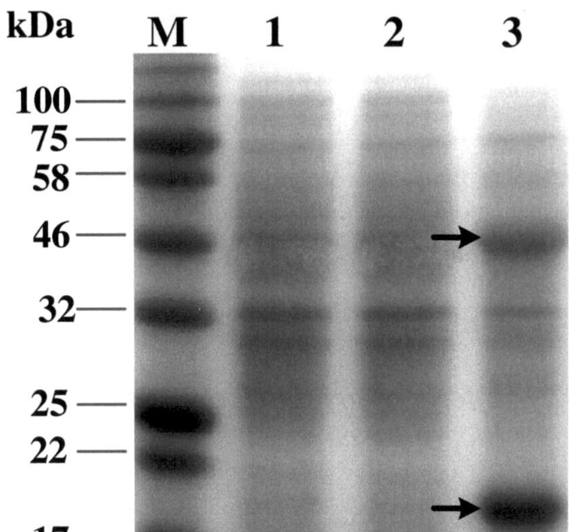

Fig. 1 Analysis of the recombinant proteins encoded by *atcB* and *atcC* with 12% SDS-PAGE. *Lane M* marker; *Lane 1 E. coli* BL21/pET; *Lane 2 E. coli* BL21/pET*atcBC* was not induced by IPTG; *Lane 3 E. coli* BL21/pET*atcBC* was induced by IPTG

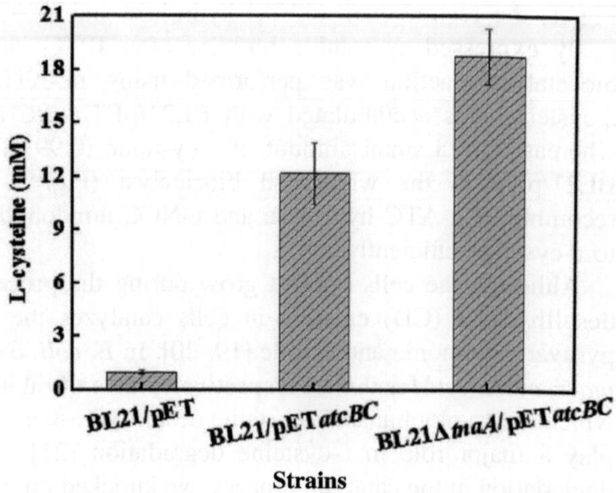

Fig. 2 Accumulation of L-cysteine by different strains

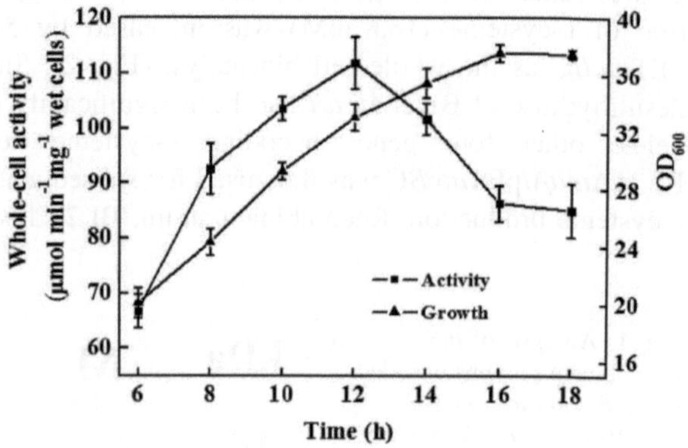

Fig. 3 Time courses of the growth (*triangle*) and whole cell activity (*square*) by the recombinant strain BL21Δ*tnaA*/pET*atcBC* in batch fermentations

in a 5 L bioreactor. As shown in Fig. 3, the specific whole-cell activity in batch fermentations reached the maximum (112 μmol min^{-1} mg^{-1} wet cells) at 12 h. The cells were harvested after 12 h of cultivation for subsequent optimization of catalytic conditions.

3.2 *Effects of pH and Temperature on* L*-Cysteine Biosynthesis*

Both pH and temperature could affect the efficiency of the whole-cell biocatalyst, including reaction rate and cellular maintenance. In whole-cell biocatalyst system for L-cysteine production, pH not only has an impact on enzymatic activities of L-ATC hydrolase and L-NCC amidohydrolase, but also the extent of L-cysteine

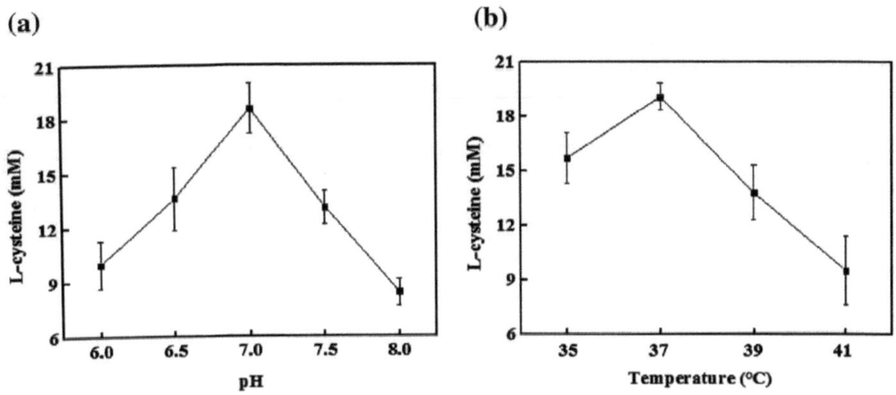

Fig. 4 Effects of pH (**a**) and temperature (**b**) on L-cysteine biosynthesis

oxidization. In general, L-cysteine is oxidized easily to L-cystine in alkaline environment instead of under acidic condition (Read et al., [22]. Our previous experiments showed that PBS was an ideal reaction buffer for L-cysteine production. In this study, the bioconversion of DL-ATC to L-cysteine at various pH values was measured. The results showed that the optimal pH for whole-cell biocatalyst was 7.0 (Fig. 4a).

The temperature affects not only the catalytic activity and catalytic reaction rate, but also the stability of the enzymes. The optimal temperature for whole-cell biocatalyst was determined by conducting the DL-ATC conversion in PBS (pH 7.0) ranging from 35 to 41 °C. As shown in Fig. 4b, the optimal temperature was around 37 °C.

3.3 *Effects of Glycerol on DL-ATC Solubility and L-Cysteine Biosynthesis*

DL-ATC, a chemical material, is a direct precursor of L-cysteine biosynthesis (Sano and Mitsugi, [23]. DL-ATC is extremely difficult to dissolve in water solution at room temperature. Low solubility of DL-ATC apparently affects the efficiency of catalytic reaction [24]. In this study, we found that the glycerol could help to increase the dissolution of DL-ATC substrate in the water solution, and the maximum DL-ATC (about 20 g/L) could dissolve in PBS containing 10% (v/v) glycerol. Furthermore, the addition of 10% glycerol led to a 9.1% increase in L-cysteine accumulation (Fig. 5). Although the dissolution of DL-ATC was enhanced with the increase of glycerol concentration, the fluidity of the mixture became worse and accumulation of L-cysteine also decreased significantly. We therefore hypothesized that high concentration of glycerol might have a negative effect on biocatalyst [25].

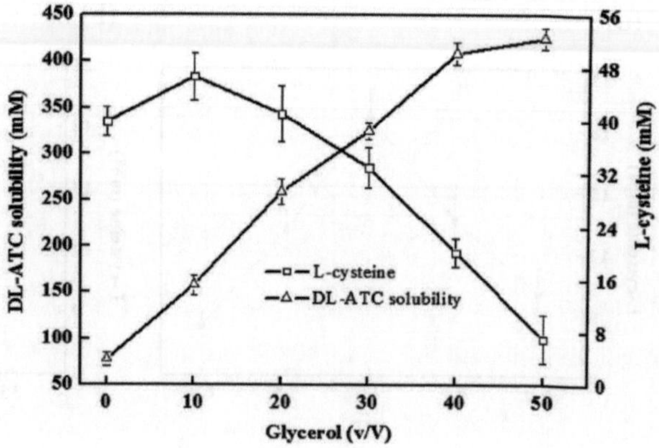

Fig. 5 Effect of glycerol on the solubility of DL-ATC and the biosynthesis of L-cysteine

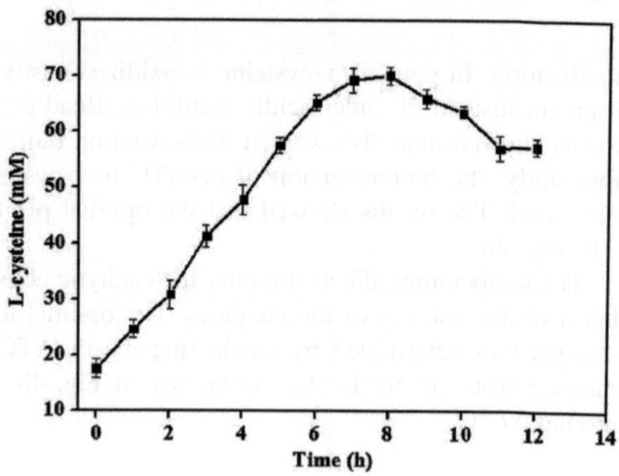

Fig. 6 Time course of L-cysteine production under optimized conditions with whole-cell catalysis

3.4 *Time Course Production of L-Cysteine with Whole-Cell Catalysis*

The time course production of L-cysteine from DL-ATC under the optimized conditions determined in the previous sections was carried out by *E.coli*BL21Δ*tnaA*/pET*atcBC*. After optimization, the production of L-cysteine reached the maximum concentration (70.2 mM) at 8 h (Fig. 6), which was a 5.36-fold increase in the yield. Chemosynthetic DL-ATC is an equimolar mixture of D-ATC and L-ATC, and D-ATC cannot be catalyzed by L-ATC hydrolase. The molar conversion rate from L-ATC to L-cysteine was 68.2%. AtcA encoded by *atcA* was reported to be an L-ATC racemase that can catalyze D-ATC to L-ATC [26, 27]. Thus, L-ATC racemase could be introduced in the whole-cell biocatalyst to enhance the production of L-cysteine and the substrate conversion rate in the further study.

4 Conclusions

In this study, the bioconversion process from DL-ATC to L-cysteine was developed by *E. coli* whole-cell biocatalyst, and the catalytic conditions were explored and optimized. Heterologous expression of the L-ATC hydrolase and L-NCC amidohydrolase in *E. coli* enabled us to construct an efficient whole-cell biocatalyst, and reduction of the activity of L-cysteine desulfhydrase in *E. coli* further increased the accumulation of L-cysteine. Effects of various reaction conditions, including pH, temperature and glycerol concentration on catalytic activities and L-cysteine production were examined, and the optimal conditions were determined. Under the optimized conditions, the production of L-cysteine reached up to 70.2 mM after 8 h whole-cell catalysis. This study provides a new economical and environment-friendly route for large-scale production of L-cysteine.

Acknowledgements This work was supported by Tianjin Science and Technology Support Program (NO. 16YFZCSY00770).

References

1. Kuśmierek K, Bald E (2008) Reduced and total glutathione and cysteine profiles of citrus fruit juices using liquid chromatography. Food Chem 106(1):340–344
2. Ismail NI, Hashim YZ, Jamal P et al (2014) Production of cysteine: approaches, challenges and potential Solution. Inter J Biotechfor Wellness Ind 3:95–101
3. Duan J, Zhang Q, Zhao H et al (2012) Cloning, expression, characterization and application of *atcA*, *atcB* and *atcC* from *Pseudomonas* sp. for the production of L-cysteine. Biotech Lett 34(6):1101–1106
4. Leuchtenberger W, Huthmacher K, Drauz K (2005) Biotechnological production of amino acids and derivatives: current status and prospects. Appl Microbiol Biot 69(1):1–8
5. Chen N, Huang J, Feng ZB et al (2009) Optimization of fermentation conditions for the biosynthesis of L-threonine by *Escherichia coli*. Appl Biochem Biotechnol 158(3):595–604
6. Park JH, Jang YS, Lee JW et al (2011) *Escherichia coli* W as a new platform strain for the enhanced production of L-valine by systems metabolic engineering. Biotechnol Bioeng 108(5):1140–1147
7. Grant GA, Hu Z, Xu XL (2001) Specific interactions at the regulatory domain-substrate binding domain interface influence the cooperativity of inhibition and effector binding in *Escherichia coli* D-3-phosphoglycerate dehydrogenase. J Biol Chem 276(2):1078–1083
8. Nakatani T, Ohtsu I, Nonaka G et al (2012) Enhancement of thioredoxin/glutaredoxin-mediated L-cysteine synthesis from S-sulfocysteine increases L-cysteine production in *Escherichia coli*. Microb Cell Fact 11(1):62
9. Park S, Imlay JA (2003) High levels of intracellular cysteine promote oxidative DNA damage by driving the fenton reaction. J Bacteriol 185(6):1942–1950
10. Sørensen MA, Pedersen S (1991) Cysteine, even in low concentrations, induces transient amino acid starvation in *Escherichia coli*. J Bacteriol 173(16):5244–5246
11. Sand K, Eguchi C, Yasuda N et al (1979) Metabolic pathway of L-cysteine formation from DL-2-amino-Δ^2-thiazoline-4-carboxylic acid by *Pseudomonas*. J Gen Appl Microbiol 43(11): 2373–2374

12. Yu Y, Liu Z, Liu C et al (2006) Cloning, expression, and identification of genes involved in the conversion of DL-2-Amino-Δ2-thiazoline-4-carboxylic acid to L-cysteine via S-Carbamyl-L-cysteine pathway in *Pseudomonas sp.* TS1138. Biosci Biotechnol Biochem 70(9):2262–2267
13. Tamura Y, Ohmachi T, Asada Y (2001) Induction of 2-amino-Δ^2-thiazoline-4-carboxylic acid hydrolase and N-carbamoyl-L-cysteine amidohydrolase by S-compounds in *Pseudomonas* putida AJ3865. J Gen Appl Microbiol 47(4):193–200
14. Tamura Y, Nishino M, Ohmachi T et al (1998) N-carbamoyl-L-cysteine as an intermediate in the bioconversion from DL-2-Amino-Δ^2-thiazoline-4-carboxylic acid to L-cysteine by *Pseudomonas* sp. ON-4a. Biosci Biotechnol Biochem 62(11):2226–2229
15. Huai L, Chen N, Yang W et al (2009) Metabolic control analysis of L-cysteine producing strain TS1138 of *Pseudomonas* sp. Biochemistry (Mosc) 74(3):288–292
16. Youn SH, Park HW. Shin CS (2012) Enhanced dissolution of the substrate DL-2-amino-Δ^2-thiazoline-4-carboxylic acid and enzymatic production of L-cysteine at high concentrations. Eng Life Sci. 12:514–517
17. Gu X, Li C, Cai Y, Dong H et al (2013) Construction of lactococcus lactis thyA-null using the Red recombination system. Ann Microbiol 63(3):951–956
18. Datsenko KA, Wanner BL (2000) One-step inactivation of chromosomal genes in *Escherichia coli* K-12 using PCR products. Proc Natl Acad Sci U S A 97(12):6640–6645
19. Awano N, Wada M, Kohdoh T et al (2003) Effect of cysteine desulfhydrase gene disruption on L-cysteine overproduction in *Escherichia coli*. Appl Microbiol Biotechnol 62(2):239–243
20. Dwivedi CM, Ragin RC, Uren JR (1982) Cloning, purification, and characterization of beta-cystathionase from *Escherichia coli*. Biochemistry 21(13):3064–3069
21. Awano N, Wada M, Mori H et al (2005) Identification and functional analysis of *Escherichia coli* cysteine desulfhydrases. Appl Environ Microbiol 71(7):4149–4152
22. Read JF, Bewick S, Graves CR et al (2000) The kinetics and mechanism of the oxidation of s-methyl-L-cysteine, L-cystine and L-cysteine by potassium ferrate. Inorg Chim Acta 303(2): 244–255
23. Sano K, Mitsugi K (1978) Enzymatic production of L-cysteine from DL-2-amino-Δ^2-thiazoline-4-carboxylic acid by *Pseudomonas thiazolinophilum*: optimal conditions for the enzyme formation and enzymatic reaction. Agric Biol Chem 42(12):2315–2321
24. Youn SH, Park HW, Choe D et al (2014) Preparation of eutectic substrate mixtures for enzymatic conversion of ATC to L-cysteine at high concentration levels. Bioprocess Biosyst Eng 37(6):1193–1200
25. Kumar A, Pawar SS (2012) High viscosity of ionic liquids causes rate retardation of Diels-Alder reactions. Sci China Chem 55(8):1633–1637
26. Wang P, He JY, Yin JF (2015) Enhanced biocatalytic production of L-cysteine by *Pseudomonas s*p. B-3 with in situ product removal using ion-exchange resin. Bioprocess Biosyst Eng 38(3):421–428
27. Wada M, Takagi H (2006) Metabolic pathways and biotechnological production of L-cysteine. Appl Microbiol Biotechnol 73(1):48–54

Analysis on Acid, Bile, and Heat Tolerance of Probiotics Strains in Maca-Probiotics Granule

Ya-Xin Jiang, Qing-Qing Dong, Ji-Ping Qu, Tong-Cun Zhang, Ya-Jian Song, Zhong-Yuan Li and Xue-Gang Luo

1 Introduction

Probiotics have been defined as "a live microbial food supplement which affects the host beneficially by improving the intestinal microbial balance" [1, 2]. As summarized in the World Health Organization (WHO) and Joint Food and Agriculture Organization (FAO), probiotics were defined as live microorganisms, providing a health benefit to the host, when consumed in a sufficient quantity as a part of food [3]. As "living micro-organisms", probiotics exert health affects beyond inherent basic nutrition after ingesting in certain numbers [4]. To provide health benefit, the suggested concentration of probiotics strains is 10^6 CFU/g [5]. A growing number of evidences have been shown that probiotics were associated with numerous health benefits, such as reduction of lactose intolerance [6], inhibition of pathogenic organisms [7, 8], decrease of blood glucose levels [9], anti-inflammation [10], increasing immune response [11], inhibition of cancer [12, 13], etc. To provide health benefits, probiotics must overcome chemical barriers, especially acidity and bile salt, in the gastrointestinal tract. In addition, probiotics in the commercial products should also have enough heat tolerance ability to ensure the survival during the long-time storage.

Lepidium meyenii walp, known as maca, grows over 4000 m altitude in the central Peruvian Andes. Proved by scientific research, maca has antioxidant properties in vitro and in vivo [14, 15], and can improve sperm production [16],

Y.-X. Jiang · Q.-Q. Dong · T.-C. Zhang · Y.-J. Song · Z.-Y. Li · X.-G. Luo (✉)
Key Laboratory of Industrial Fermentation Microbiology, Ministry of Education and Tianjin City, College of Biotechnology, Tianjin University of Science and Technology, Tianjin 300457, China
e-mail: luoxuegang@tust.edu.cn

J.-P. Qu
Aladdin Biotechnology Limited Company, Suzhou, China

H. Liu et al. (eds.), *Advances in Applied Biotechnology*, Lecture Notes in Electrical Engineering 444, https://doi.org/10.1007/978-981-10-4801-2_49

sexual performance parameters [17, 18], memory ability [19] and fatigue resistance [20]. Maca products have become one of the best-selling health food in the world.

In order to combine the functions of maca and probiotics, a compound granule consisted of maca and five probiotics, *Bifidobacterium, Lactobacillus plantarum, Lactobacillus rhamnosus, Lactobacillus acidophilus* and *Bifidobacterium longum*, was prepared in our previous study. However, whether the probiotics in the composite particles could survive in the long-term storage or after oral administration has not been researched. To address this problem, we analyzed the acid resistance, bile salt tolerance and heat tolerance of five probiotics strains in the compound granule, and provided the theoretical guidance for the storage and reasonable application of this healthy product.

2 Materials and Methods

2.1 Materials

The maca-probiotics compound granule in this study was offered by Aladdin Biotechnology Limited Company in Suzhou of China. To analyze the probiotics strains in the granule, the compound product (2 g) was dissolved with de Man, Rogosa, and Sharpe (MRS) broth at pH 7.4, and cultured in an anaerobic incubator at 32 °C for 48 h by gradient dilution and pour plate counts [21].

2.2 Analysis of Acid Resistance of Probiotics

The total lactobacillus in the maca-probiotics compound granule were firstly activated in MRS broth 5 mL (1%) and then the medium were adjusted to pH 4.0, 3.5, 3.0, 2.5 and 2.0 with 5 N HCl and incubated at 32 °C for 4 h. The initial bacterial concentration was 2.2×10^6 CFU. Strains were calculated by pour plate counts using 10-fold serial dilutions with MRS, and incubated at 32 °C for 48 h. Acid tolerance was determined by comparing the final plate counts after 4 h with the initial plate counts at 0 h [22]. The experiment and analysis were carried out in triplicate.

2.2.1 Acid Tolerance in Different Time at pH 3.0

The lactobacillus was grown in MRS broth 5 mL (1%) which was adjusted to pH 3.0. The viable number of strains were calculated by pour plate counts using 100-fold serial dilutions with MRS (pH 3.0) per hour for 1, 2, 3, 4 h and incubated at 32 °C for 48 h. Acid tolerance was determined by comparing the final plate counts after different hours (1, 2, 3, 4 h) with the initial plate counts at 0 h.

2.3 Analysis of Bile Salt Tolerance

2.3.1 Tolerance at Different Concentrations of Bile Salts

The total lactobacillus in the maca-probiotics compound granule was firstly activated in MRS broth 5 mL (1%) and then the cultures were treated by pig bile salt of 0.05, 0.1, 0.2, 0.3% (w/v) and cultivated at 32 °C for 2 h. Probiotics were enumerated by pour plate counts using 20-fold serial dilutions with MRS (bile salt of 0.05, 0.1, 0.2, 0.3%) incubated at 32 °C for 48 h.

2.3.2 Tolerance in Different Time at Bile Salts 0.3%

The lactobacillus was grown in MRS broth 5 mL (1%) which containing bile salt of 0.3%. Samples were taken at 0.5, 1, 1.5, 2 h and the viable number of strains were calculated by pour plate counts using 100-fold serial dilutions with MRS (bile salt of 0.3%) and incubated at 32 °C for 48 h. Bile salts tolerance was determined by comparing the final plate counts after different hours (0.5, 1, 1.5, 2 h) with the initial plate counts at 0 h.

2.4 Analysis of Heat Tolerance

Total lactobacillus was cultured at 37, 42, 47, 52 °C for 10 min. Strains were enumerated by pour plate counts using 20-fold serial dilutions with MRS (pH 7.4) incubated at 32 °C for 48 h.

3 Results

3.1 Analysis of Acid Tolerance of Probiotics in the Maca-Probiotics Compound Granule

Stomach has the highest acidity (pH 1–4) in the gastrointestinal tract [23]. In order to exhibit the beneficial functions, the probiotics must be able to survive under such rigorous conditions.

As shown in Fig. 1, the survival rate of probiotics strains in the maca-probiotics granule was 100% at pH 4.0 for 4 h, whereas the survivors decreased when the pH was below 3.5, and there was almost no lactic acid bacteria survived at pH $\leq$ 2.5. These results suggested that probiotics strains in the maca-probiotics compound granule had certain acid-resistance ability. However, in order to obtain viable probiotics as much as possible, this product should be taken after meals to reduce the damage of gastric acid on probiotics.

Figure 2 showed the viability of probiotics strains in the maca-probiotics granule in different time at pH 3.0. The viable counts of probiotics began to decrease after 2 h, and the survival rate was 50% when cultured for 4 h. These results suggested that a reasonable diet to increase gastric emptying rate and reduce the residence time of probiotics in the stomach would help to increase the viable number of probiotics in the maca-probiotics compound granule after oral administration.

3.2 *Analysis of Bile Salt Tolerance of Probiotic Strains in the Maca-Probiotics Compound Granule*

Bile salt in the intestinal tract is an important factor to kill probiotics and other microorganisms. The physiological concentration of human bile is different in

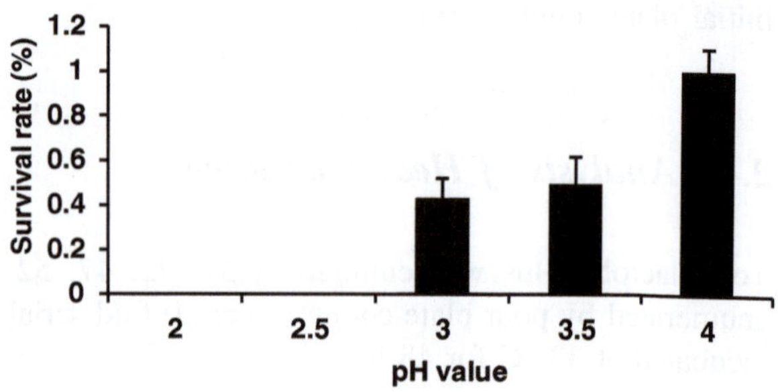

Fig. 1 Survival rate of probitics strains in the maca-probiotics granule under different acidity conditions

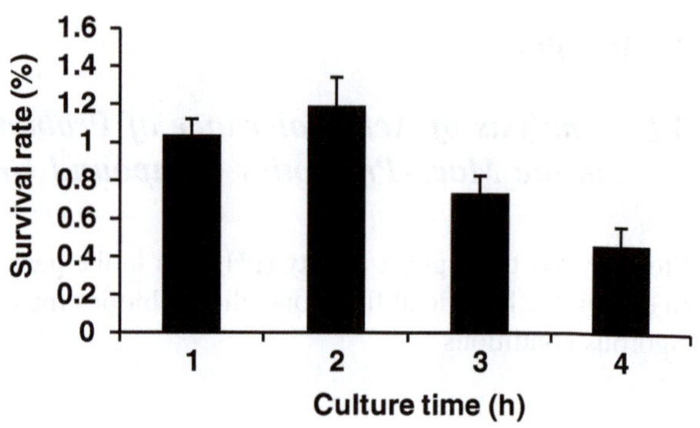

Fig. 2 Survival rate of probitics strains of maca-probiotics granule in different time at pH 3.0

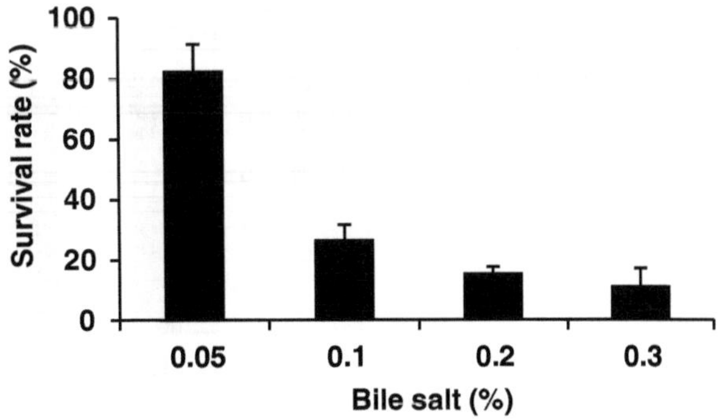

Fig. 3 Survival rate of probitics strains of maca-probiotics granule in bile salt at different concentration

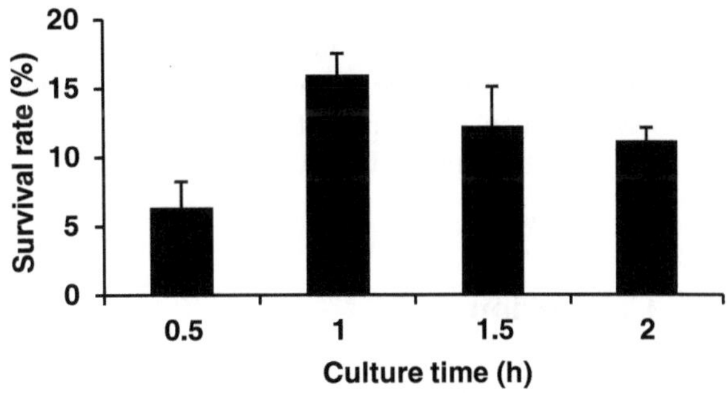

Fig. 4 Survival rate of probitics strains of maca-probiotics granule at different time in 0.3% bile salt

different regions of the human body and usually below 0.3%. Therefore, resistance to bile acid of such concentration is an important characteristic that enables probiotics to survive in the small intestine.

The result described in Fig. 3 demonstrated that the probiotics strains in the maca-probiotics granule could be viable in 0.05, 0.1, 0.2, 0.3% bile salts for 2 h, but the survivors would be less than 20% when the concentration of bile salts was over 0.1%.

As shown in the Fig. 4, it is apparently that the probiotics strains in the maca-probiotics granule could survive at 0.3% bile salt during 3 h. They decreased sharply in the first 0.5 h, but then recovered lately, indicating that these probiotics strains had ability to adapt to the damage of bile salts.

3.3 *Analysis of Temperature Tolerance of Probiotic Strains in the Maca-Probiotics Compound Granule*

The heat tolerance of probiotics incubated at different temperature such as room temperature (RT, 25 °C), 37, 42, 47 and 52 °C for 10 min was shown in Fig. 5. The results suggested that the number of survivors above 85% at the temperature

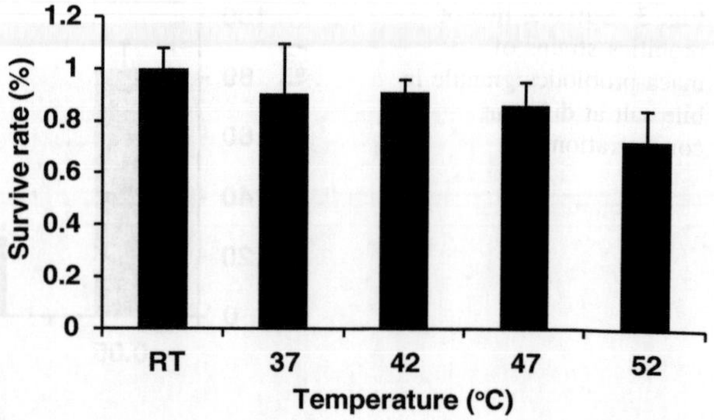

Fig. 5 Heat tolerance of probitics strains in the maca-probiotics granule

below 47 °C, whereas the number of probiotics decreased with the increase of temperature. The decline in the number of viable bacteria will be significantly accelerated when the temperature exceed 47 °C. These results implied that this product should be taken with water below 47 °C.

4 Conclusion

To combine the efficacy of probiotics and Maca together, the compound granule was prepared with maca and five types of probiotics such as *Bifidobacterium*, *Lactobacillus plantarum*, *Lactobacillus rhamnosus*, *Lactobacillus acidophilus* and *Bifidobacterium longum*. The present study further confirmed that the probiotics strains in the maca-probiotics compound granule still have acceptable ability of acid resistance, bile salt tolerance and heat tolerance. Therefore, this healthy product could exhibite expected functions after oral administration. In addition, in order to obtain the best application effect, the product should be storage at low temperature and be taken after meals with water below 47 °C.

Acknowledgements This work was supported by the 863 (Hi-tech research and development program of China) program under contract NO. 2012AA022108, Tianjin Research Program of Application Foundation and Advanced Technology (14JCZDJC33200) and the Laboratory Open Foundation of Tianjin University of Science and Technology (1504A303).

References

1. Chou LS, Weimer B (1999) Isolation and characterization of acid- and bile-tolerant isolates from strains of *Lactobacillus acidopHilus*. J Dairy Sci 82:23–31
2. Fuller R (1989) A review: probiotics in man and animals. J Appl Bacteriol 66:365–378
3. Servin AL, Coconnier MH (2003) Adhesion of probiotic strains to the intestinal mucosa and interaction with pathogens. Best Pract Res Clin Gastroenterol 17:741–754

4. Guarner F, Schaafsma GJ (1998) Probiotics. Int J Food Microbiol 39:237–238
5. Lankaputhra WEV, Shah NP (1995) Survival of *Lactobacillus acidopHilus* and *Bifidobacterium* spp. in the presence of acid and bile salts. Cultured Dairy Prod J 30:2–7
6. Gilliland SE, Kim HS (1984) Effect of viable starter culture bacteria in yogurt on lactose utilization in humans. J Dairy Sci 67:1–6
7. Hammilton-Miller JMT (2003) The role of probiotics in the treatment and prevention of *H. pylori* infection. Int J Antimicrob Agents 22:360–366
8. Sgouras D, Maragkoudakis P et al (2004) In vitro and in vivo inhibition of Helicobacter pylori by *Lactobacillus casei* strain Shirota. Appl Environ Microbiol 70:518–526
9. Huang CY, Yang AL et al (2012) Anti-apoptotic and pro-survival effects of exercise training on hypertensive hearts. J Appl Physiol 112:883–891
10. Fedorak RN, Feagan BG et al (2015) The probiotic VSL#3 has anti-inflammatory effects and could reduce endoscopic recurrence after surgery for Crohn's Disease. Clinical gastroenterology and hepatology: the official clinical practice. Clin Gastroenterol Hepatol 13:928–935
11. Matsuzaki T, Yamazaki R (1998) The effect of oral feeding of *Lactobacillus casei* strain Shirota on immunoglobulin E production in mice. J Dairy Sci 81:48–53
12. Hirayama K, Rafter J (2000) The role of probiotic bacteria in cancer prevention. Microbes Infect 2:681–686
13. Ma EL, Choi YJ et al (2010) The anticancer effect of probiotic Bacillus polyfermenticus on human colon cancer cells is mediated through ErbB2 and ErbB3 inhibition. Int J Cancer 127:780–790
14. Sandoval M, Okuhama NN et al (2002) Antioxidant activity of the cruciferous vegetable Maca *Lepidium meyenii*. Food Chem 61:207–213
15. Lee KJ, Dabrowski K et al (2005) Activity-guided fractionation of phytochemicals of Maca meal, their antioxidant activities and effects on growth, feed utilization and survival in rainbow trout (*Oncorhynchus mykiss*) juveniles. Aquaculture 244:293–301
16. Gonzales GF, Cordova A et al (2001) *Lepidium meyenii* (Maca) improved semen parameters in adult men. Asian J Androl 3:301–303
17. Zheng BL, He K et al (2001) Effect of a lipidic extract from *Lepidium meyenii* on sexual behavior in mice and rats. Urology 55:598–602
18. Rubio J, Caldas M et al (2006) Effect of three different cultivars of *Lepidium meyenii* (Maca) on learning and depression in ovariectomized mice. BMC Complement Altern Med 23:6–23
19. Yang Q, Jin W et al (2016) Effects of macamides on endurance capacity and anti-fatigue property in prolonged swimming mice. Pharm Biol 54:827–834
20. Minyu S, Bohyun Y (2015) Characterization of selected Lactobacillus strains for use as probiotics. Kor J Food Sci Anim Resour 35:551–556
21. Ding WK, Shah NP (2009) An improved method of microencapsulation of probiotic bacteria for their stability in acidic and bile conditions during storage. J Food Sci 74:53–61
22. Dunne C, O'Mahony L (2001) In vitro selection criteria for probiotic bacteria of human origin: correlation with in vivo findings. Am J Clin Nutr 73:386–392
23. Hyronimus B, Le Marrec C (2000) Acid and bile tolerance of spore-forming lactic acid bacteria. Int J Food Microbiol 61:193–197

Near-Infrared Spectroscopy for the Monitoring of Leucine Fermentation in *Corynebacterium glutamicum*

Hongbo Wei, Tuo Shi, Lihong Du, Qixin Chen, Yuechao Ma, Quanwei Zhang, Qian Ma and Ning Chen

1 Introduction

Near-infrared spectroscopy (NIRS) technique could be used to monitor the absorbance at wavelengths between 700 and 2500 nm by detecting the vibrations of the –CH, –OH, –NH, and –SH bonds [1]. It has the advantages of fast detection, low cost and non-destruction, and is suitable for analyzing different forms of samples including solid, liquid, gel etc. NIRS has been successfully applied in different fields including food, petroleum chemical industry, drug composition analysis, biological industry, etc. [2–7]. Xiao et al. [8] used the technology and partial least-squares regression to determine the main categories of components in corn steep liquor. Arnold et al. [9] used NIRS to monitor the biomass of *Escherichia coli* in an industrial fed-batch process. Fazenda et al. [10] investigated the potential of NIRS as an input generating for metabolic flux analysis models, contributing to the optimization of cell factory metabolism in real-time. All these studies suggested NIRS is powerful in the monitoring and controlling of industrial processes.

NIRS technique is mainly consisted of the establishment of a calibration model and the subsequent analysis of unkown samples. Various pretreatment methods, such as the first derivative, second derivative, standard normal variate, and fourier transform etc., could be used to reduce the noise of raw NIR spectra. The calibration model could be established by multivariate analysis based on the pretreated NIR spectra information and the true experimental concentrations of samples. The obtained model could be used in the monitoring and controlling of different processes.

H. Wei · T. Shi · L. Du · Q. Chen · Y. Ma · Q. Zhang · Q. Ma · N. Chen (✉)
Key Laboratory of Industrial Fermentation Microbiology, Ministry of Education,
Tianjin Key Lab of Industrial Microbiology, Tianjin University of Science and Technology,
Tianjin 300457, People's Republic of China
e-mail: ningch@tust.edu.cn

H. Liu et al. (eds.), *Advances in Applied Biotechnology*, Lecture Notes
in Electrical Engineering 444, https://doi.org/10.1007/978-981-10-4801-2_50

Corynebacterium glutamicum has been playing a critical role in the industrial production of amino acids, organic acids, nucleosides and related derivatives [11–14]. As an essential branched-chain amino acid, leucine is widely used in food and medicine industries [15]. In this study, NIRS technique was used to monitor the leucine fermentation in *C. glutamicum* CP. A model for the prediction of leucine concentration was established, and the results showed that the method could accurately predict and monitor the fermentation process.

2 Materials and Methods

2.1 Microorganism and Medium

The leucine-producing strain used throughout this study was *C. glutamicum* CP, which was stored at the Culture Collection of Tianjin University of Science and Technology. The slope medium contained: yeast extract 5 g/L, tryptone 10 g/L, beef extract 10 g/L, NaCl 2.5 g/L, MgSO4 0.5 g/L, KH_2PO_4 1 g/L, agar 20 g/L. The seed medium contained: glucose 40 g/L, corn steep liquor 20 mL/L, KH_2PO_4 2 g/L, $MgSO_4{\cdot}7H_2O$ 2 g/L, $MnSO_4$ 2.5 mg/L, $FeSO_4$ 5 mg/L, vitamin B_1 0.2 mg/L, vitamin H 0.5 mg/L. The fermentation medium contained: glucose 50 g/L, corn steep liquor 20 mL/L, soybean protein hydrolysate 20 mL/L, $MgSO_4$ 2 g/L, $MnSO_4$ 30 mg/L, K_2HPO_4 3 g/L, $FeSO_4$ 30 mg/L, vitamin H 0.3 mg/L, vitamin B_1 0.3 mg/L. The pH of each medium was adjusted to 7.0–7.2 with 4 mol/L NaOH.

2.2 Fermentation Process

C. glutamicum CP was first cultured on the slope medium, and then cultured in 100 mL seed medium in a 1000 mL baffled flask at 32 °C and 200 rpm for 14 h.

An inoculum of 300 mL seed culture was transferred to a 5 L fermentor (New Brunswick controller, BioFlo®/CellGen®115, Germany) containing 2.7 L fermentation medium, and was cultivated at 32 °C for 32 h. The dissolved oxygen (DO) level was controlled by adjusting the aeration rates, tank pressure and stirring speed. The DO was maintained at 20% during the growing stage, and 10% during the acid producing stage. The pH was adjusted to 7.0 with 25% (v/v) ammonia during the overall cultivation process.

2.3 Sampling and Analytical Methods

During the fermentation process, samples were collected every half hour from the beginning of the fermentation. 189 samples obtained from three batches were centrifugated at 13,000 rpm for 2 min. Collect the supernatants to determine the concentrations of leucine using high performance liquid chromatography analysis system.

2.4 NIR Spectra Acquisition and Model Building

The near infrared spectra of the fermentation samples were obtained by a NIR analyzer Tensor 37 spectrophotometer (Bruker Optics, Ettlingen, Germany) with a 2 mm cuvette, which was filled with 1 mL fermentation broth. The scanning was from 833 to 2500 nm at 25 °C. Air was used as the reference, and one measurement was the accumulation of 32 scans. The near infrared spectra obtained were first pretreated by the first derivative, second derivative, standard normal variate, or fourier transform etc., and then further processed to establish a near infrared calibration model by the Partial Least Squares (PLS) regression method.

3 Results and Discussion

3.1 Raw NIR Spectra Analysis of the Samples

The NIR spectra of 189 samples from 3 batches of fermentation were scanned by Tensor 37, and the raw NIR spectra were shown in Fig. 1. NIR peaks appeared at 5500–6200, 7000–9000 and 9500–11,000 cm^{-1}. From the figure, it could be seen that the NIR peaks at 5500–6200 cm^{-1} were unstable, and the absorbency at 9500–11,000 cm^{-1} was quite low, while, the NIR peaks at 7000– 9000 cm^{-1} were significant and stable. Thus, we chose this NIR spectra range for further modelling.

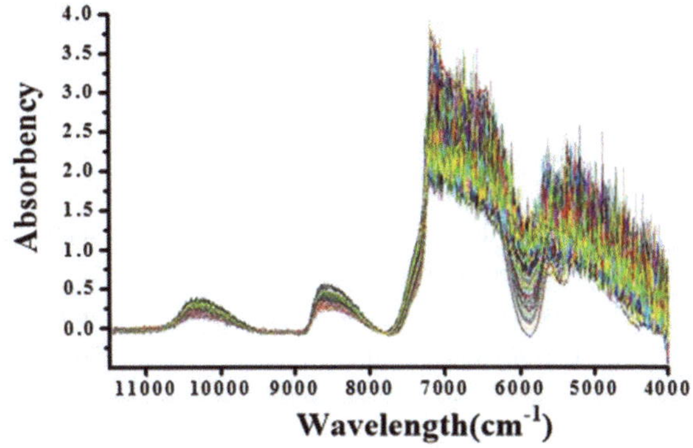

Fig. 1 The raw NIR spectra of samples from the leucine fermentation process

3.2 PLS Model for NIRS Calibration for Leucine Concentration Determination

The 189 samples were divided into two groups, the calibration set and the validation set (Table 1). The concentration range of leucine in the calibration set was 0.38–31.83 g/L, and that for the validation set was 0.47–30.79 g/L. The concentration range of leucine in the validation set was in the range of the calibration set, and the standard deviations of the two sets were close, thus, these two sets were suitable for further NIRS modelling.

Different spectral pretreatment methods were tested for the PLS calibration modelling of leucine concentration using the software OPUS 7.0 (Bruker Optics). The results of these pretreatments were shown in Table 2. The root-mean square error of calibration and validation (RMSECV) value was used for the performance evaluation of the PLS model. As the R^2 with all these pretreatment methods were pretty low (<0.8), the three pretreatment methods with lower RMSECV, namely model 1, 2, and 4, were selected for further optimization.

The three pretreatment methods: no spectral data preprocessing, straight line substraction and min-max normalization were further optimized using OPUS 7.0.

Table 1 Parameters of leucine in calibration set and validation set

Sample classification	Sample quantity	Range (g/L)	Mean (g/L)	SD (g/L)
Calibration set	169	0.38–31.83	13.04	7.14
Validation set	20	0.47–30.79	12.86	7.05

SD standard deviation

Table 2 Optimal model of leucine concentration under different spectral pretreatment methods

Model number	Pretreatment methods	Range/cm^{-1}	RMSECV (g/L)	RPD	R^2
1	No spectral data preprocessing	7000–10,000	6.79	1.53	0.7559
2	Straight line subtraction	7000–10,000	6.94	1.49	0.7435
3	Vector normalization(SNV)	7000–10,000	7.94	1.3	0.6378
4	Min-Max normalization	7000–10,000	6.85	1.5	0.7479
5	First derivative	7000–10,000	7.55	1.36	0.6985
6	Second derivative	7000–10,000	8.51	1.21	0.5677
7	First derivative + Straight line subtraction	7000–10,000	7.05	1.46	0.7402
8	First derivative + SNV	7000–10,000	8.47	1.22	0.5724
9	First derivative + MSC	7000–10,000	8.43	1.22	0.577

Table 3 Optimal model of leucine concentration with further treatment

Model number	Pretreatment methods	Range/cm^{-1}	RMSECV (g/L)	RPD	R^2
10	No spectral data preprocessing	9403.4–7489.1	3.32	3.1	0.9507
11	Straight line subtraction	9403.4–7489.1	1.34	7.22	0.9905
12	Min-Max normalization	9403.4–7489.1	3.6	2.89	0.9396

Table 4 Parameters of calibration model for the correction of leucine concentration

	Calibration set	Validation set
Dimension	8	
Range	9403.4–7489.1 cm^{-1}	
Pretreatment conditions	Straight line subtraction	
RMSECV (g/L)	1.34	–
RMSEP (g/L)	–	2.31
R^2	0.9905	0.9782
Offset	0.429	0.610
RPD	7.22	4.82

The optimal models were shown in Table 3. The straight line substraction method had the lowest RMSECV (1.34 g/L) and the highest R^2 (0.9905), and was thus chosen as the optimal pretreatment method. After pretreatment, the parameters of the calibration model for the correction of leucine concentration were illustrated in Table 4. The NIR peak range was 9403.4 ~ 7489.1 cm^{-1}, root-mean square error of prediction (RMSEP) was 2.31, with a 0.9782 R^2. The NIR spectra before and after pretreatment were compared in Fig. 2, which suggested that the pretreatment could reduce quite a number of backgrounds and noises. The relationships between the experimental values and the predictive values of leucine concentration in calibration set and validation set during the fermentation of *C. glutamicum* were shown in Fig. 3. This figure indicated that there was a good linear relationship between the experimental value and the predictive value, in other words, the model had good predictive capacity.

3.3 *Verification of the NIRS Model for Leucine Concentration Determination*

A new batch of leucine fermentation of *C. glutamicum* was used as an external validation to further evaluate the stability of the NIRS model. The experimental value and the predictive value of leucine concentration during the fermentation process were depicted in Fig. 4, from which we could see that the prediction value well reflected the true value. The results all suggested that the NIRS model had a

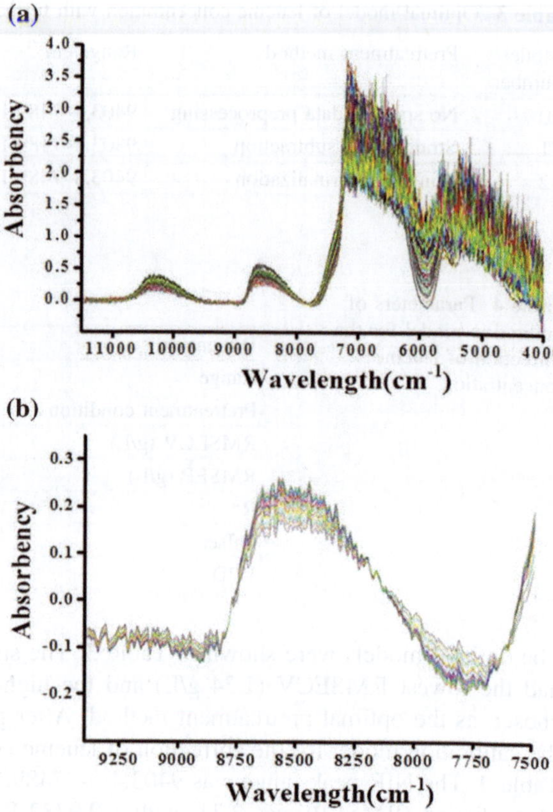

Fig. 2 Original NIR spectra **(a)** and pretreatmented NIR spectra **(b)**

good predictive ability, and could be used for the leucine concentration determination during fermentation.

4 Conclusion

In this study, near-infrared spectroscopy was used to monitor the leucine concentration in the fermentation of *C. glutamicum* CP. A PLS model for the prediction of leucine concentration was established, and the results showed that the method could predict and monitor the leucine concentration during the fermentation process accurately. The on-line monitoring of substrate and product concentrations could be conveniently conducted by NIRS, based on which prompt control strategies could be implemented. In summary, NIRS is quite a helpful tool for the monitoring and controlling of in-process parameters in industry.

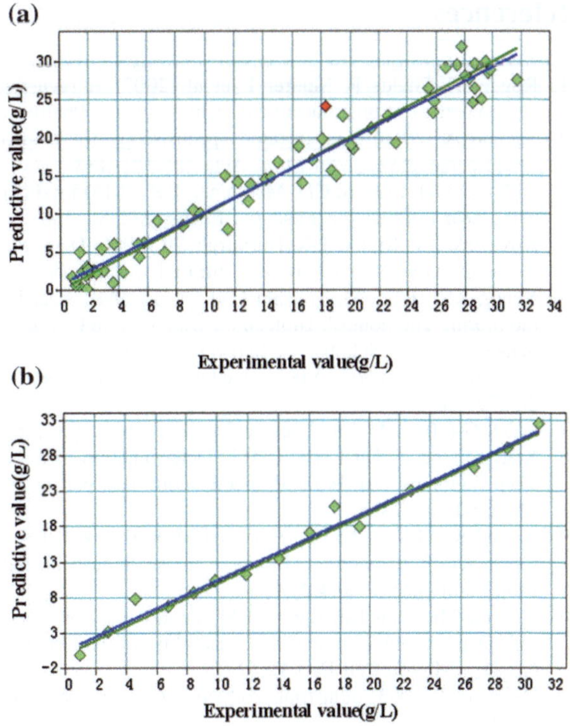

Fig. 3 The relationship between the experimental value and the predictive value of the leucine concentration in Calibration set (**a**) and Validation set (**b**)

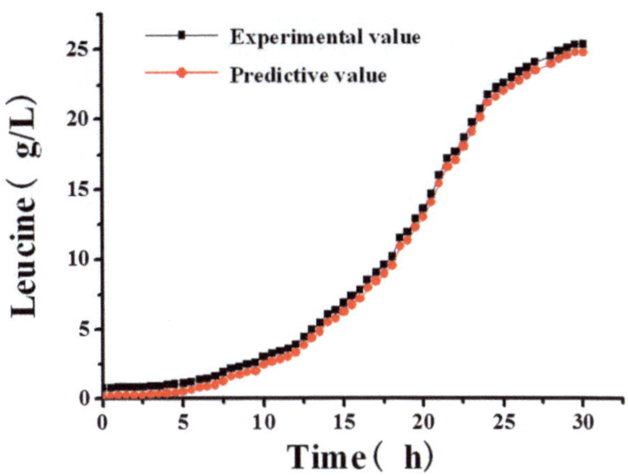

Fig. 4 Predictive and experimental values of leucine concentrations during the fermentation process

Acknowledgements The authors are very grateful for the financial support from the Foundation (No.2016IM104) of Key Laboratory of Industrial Fermentation Microbiology of Ministry of Education and Tianjin Key Lab of Industrial Microbiology (Tianjin University of Science & Technology), and the Foundation for Young Teachers of Tianjin University of Science and Technology (Grant No.2016LG11).

References

1. Roggo Y, Chalus P, Maurer L et al (2007) A review of near infrared spectroscopy and chemometrics in pharmaceutical technologies. J Pharmaceut Biomed 44:683–700
2. Sakudo A (2016) Near-infrared spectroscopy for medical applications: current status and future perspectives. Clin Chim Acta 455:181–188
3. Gillespie GD, Everard CD, Mcdonnell KP (2015) Prediction of biomass pellet quality indices using near infrared spectroscopy. Energy 80:582–588
4. Kovalenko IV, Rippke GR, Hurburgh CR (2006) Determination of amino acid composition of soybeans (Glycine max) by near-infrared spectroscopy. J Agr Food Chem 54:3485–3491
5. Yeung KS, Hoare M, Thornhill NF et al (1999) Near-infrared spectroscopy for bioprocess monitoring and control. Biotechnol Bioeng 63:684–693
6. Scheeren TWL, Schober P, Schwarte LA (2012) Monitoring tissue oxygenation by near infrared spectroscopy (NIRS): background and current applications. J Clin Monit Comput 26:279–287
7. Reich G (2016) Mid and near infrared spectroscopy. Anal Tech Pharm Sci 61–138
8. Xiao X, Hou Y, Du J et al (2012) Determination of main categories of components in corn steep liquor by near-infrared spectroscopy and partial least-squares regression. J Agr Food Chem 60:830–7835
9. Arnold SA, Gaensakoo R, Harvey LM et al (2002) Use of at-line and in-situ near-infrared spectroscopy to monitor biomass in an industrial fed-batch *Escherichia coli* process. Biotechnol Bioeng 80:405–413
10. Fazenda ML, Dias JM, Harvey LM et al (2013) Towards better understanding of an industrial cell factory: investigating the feasibility of real-time metabolic flux analysis in *Pichia pastoris*. Microb Cell Fact 12:51
11. Hermann T (2003) Industrial production of amino acids by coryneform bacteria. J Biotechnol 104:155–172
12. Jojima T, Noburyu R, Sasaki M et al (2015) Metabolic engineering for improved production of ethanol by *Corynebacterium glutamicum*. Appl Microbiol Biot 99:1165–1172
13. Eggeling L, Bott M (2015) A giant market and a powerful metabolism: l-lysine provided by *Corynebacterium glutamicum*. Appl Microbiol Biot 99:3387–3394
14. Otten A, Brocker M, Bott M (2015) Metabolic engineering of *Corynebacterium glutamicum* for the production of itaconate. Metab Eng 30:156–165
15. Gui Y, Ma Y, Xu Q et al (2016) Complete genome sequence of *Corynebacterium glutamicum* CP, a Chinese l-leucine producing strain. J Biotechnol 220:64–65

Modification of *Corynebacterium glutamicum* YILW for Isoleucine Production Improvement

Ning Xue, Zhixiang Li, Junjie Zhan, Jie Ma, Qingyang Xu, Chenglin Zhang and Ning Chen

1 Introduction

As an essential amino acid, L-isoleucine is widely used in food, pharmaceutical and animal feed industries. Current methods for L-isoleucine production is fermentation by *Corynebacterium glutamicum* [1, 2]. As showed in Fig. 1, *C. glutamicum* synthesises L-isoleucine in a split pathway from oxaloacetate [3, 4]. In the pathway, pyruvate carboxylase, aspartate kinase, threonine dehydrogenase (TD, encoded by *ilvA*) and acetohydroxy acid synthase(AHAS, encoded by *ilvB* and *ilvN*) are the key enzymes (Fig. 1) [5, 6].

Present isoleucine producing strains used in industry were mainly obtained through random mutagenesis [7]. However, such methods usually result in colonies that inevitably accumulate numerous unidentifiable and unwanted mutations, which may occur alongside the improvements [8]. Recently, rational metabolic engineering has been considered as a method to overcome the disadvantages of the classical method of strain development [9].

In this study, *ilvA*, *ilvBN* and isoleucine exporter *brnEF* were rationally over-expressed, leading to 41.1 and 40.0% higher isoleucine production and yield by

N. Xue · Z. Li · J. Zhan · J. Ma · Q. Xu · C. Zhang (✉) · N. Chen (✉)
College of Biotechnology, Tianjin University of Science and Technology, Tianjin 300457, China
e-mail: zcl@tust.edu.cn

N. Chen
e-mail: ningch@tust.edu.cn

Q. Xu · C. Zhang · N. Chen
National and Local United Engineering Lab of Metabolic Control Fermentation Technology, Tianjin 300457, China

H. Liu et al. (eds.), *Advances in Applied Biotechnology*, Lecture Notes in Electrical Engineering 444, https://doi.org/10.1007/978-981-10-4801-2_51

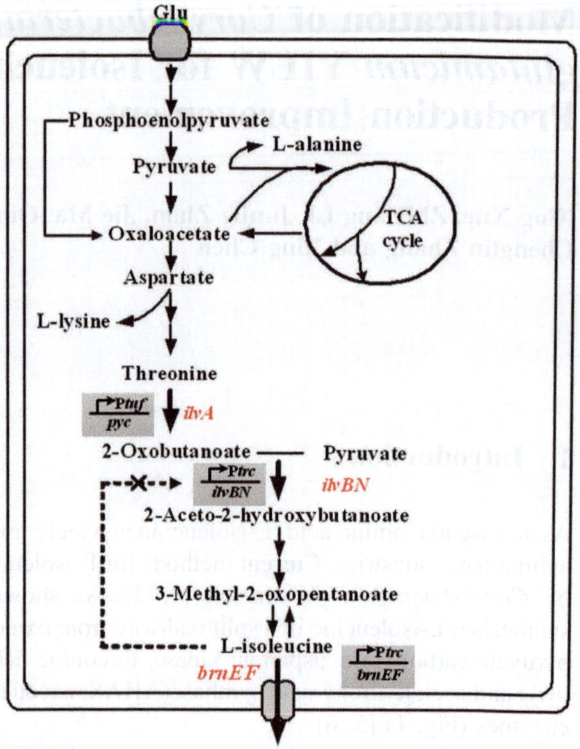

Fig. 1 Biosynthesis pathway of isoleucine in *C. glutamicum,* The *solid lines* represent metabolic conversions, the *dotted line* depicts feedback control, and the *thick arrows* indicate the increased flux by directly overexpressing the corresponding genes. The × indicates the removal of the feedback control

YILW-C. The design principles described in this study would be useful to construct strains for producing other similar biological products.

2 Materials and Methods

2.1 Strains and Plasmids

Strains, plasmids and primers used in this work were listed in Tables 1 and 2.

2.2 Media and Culture Condition

E. coli was grown at 37 °C in Luria-Bertani medium [LB, 1% (w/v) tryptone, 0.5% (w/v) yeast extract and 1% (w/v) NaCl]. Kanamycin (50 μg/mL) was added for strain selection.

Table 1 Strains and plasmids

Strain/plasmid	Relevant characteristics	Source
Strains		
Corynebacterium glutamicum YILW	The L-isoleucine producing strain (Leu^L +AHV^r + SG^r + Leu-ME^r)	Laboratory stock
YILW-A	YILW with replacement of the native promoter of *ilvA* gene with the promoter of the *tuf* gene	This work
YILW-B	YILW-A with replacement of native *ilvBN* promoter with the promoter of the *tuf* gene	This work
YILW-C	YILW-C with replacement of native *brnEF* promoter with Ptac	This work
Escherichia coli DH5αMCR	F^- φ80d*lacZ*ΔM15 Δ(lac*ZYA-argF*)U169 *recA1 endA1 hsdR17* (r_K^-,m_K^+) *deoR supE44* λ^- *thi-1 gyrA*96 *relA1*	Laboratory stock
Plasmids		
pXMJ19	*C. glutamicum*/*E. coli* shuttle vector, 6.6 kb, ptac, lacIq, Cm	Laboratory stock
pK18*mobsacB*	Integration vector, Km^r, *oriV*$_{Ec}$*oriT*, *sacB*	Laboratory stock
pK18*mobsacB*Ptuf-A	Integration vector for institution of *ilvA* gene promoter with the *tuf* promoter	This work
pK18*mobsacB*Ptuf-BN	Integration vector for institution of *ilvBN* gene promoter with the *tuf* promoter	This work
pK18*mobsacB*Ptac	Integration vector for institution of native *brnEF* promoter with Ptac	This work

C. glutamicum cells was inoculated from seed culture (glucose 30 g/L, yeast extract 5 g/L, $(NH_4)_2SO_4$ 3 g/L, $KH_2PO_4{\cdot}3H_2O$ 1.5 g/L, $MgSO_4{\cdot}7H_2O$ 0.6 g/L, $FeSO_4{\cdot}7H_2O$ 0.01 g/L, $MnSO_4{\cdot}H_2O$ 0.01 g/L, corn steep liquor 30 mL/L, soybean hydrolysate 30 mL/L.) and cultured to exponential growth period (for quantitative RT-PCR and intracellular metabolites detection) or to 48 h (for isoleucine fermentation) in 27 mL fermentation medium (glucose 80 g/L, $(NH_4)_2SO_4$ 4 g/L, $FeSO_4{\cdot}7H_2O$ 0.015 g/L, $MgSO_4{\cdot}7H_2O$ 0.5 g/L, $MnSO_4{\cdot}H_2O$ 0.015 g/L, $KH_2PO_4{\cdot}3H_2O$ 1.5 g/L, $K_2HPO_4{\cdot}3H_2O$ 3 g/L, biotin 100 μg/L, vitamin B_1 5 mg/L, soybean hydrolysate 20 ml/L, and corn syrup 15 mL/L) in 500-mL shake flasks with 10% (v/v) inoculum size at 35 °C with 200 rpm [9].

2.3 Quantitative RT-PCR

Total RNA was isolated from *C. glutamicum* cells and transcription level of selected genes was measured by quantitative RT-PCR according to the manufacturer instructions of One Step SYBR® PrimeScript™ RT-PCR Kit(Takara, Japan).

Table 2 Primers

Primers[a]	Sequence (5′-3′)	Description
IlvA-1	TCTAGACTGCAGTACCGCAAAAGGAAT (*Xba* I)	Amplifying upstream fragment of *ilvA* promoter
IlvA -2	TTTCCTTCGGATCTAAACGATCTCAATACTGTGGTTGTGGCACTACA	
IlvA -3	ACCACGAAGTCCAGGAGGACATACAATGAGTGAAACATACGTGTCTGAGAA	Amplifying downstream fragment of *ilvA* promoter
IlvA -4	GGATCCGGCTCGATCAGCGTTGC (*Bam*H I)	
IlvA -5	AGATCGTTTAGATCCGAAGGAAA	Amplifying promoter of *tuf* gene
IlvA -6	TGTATGTCCTCCTGGACTTCGTGGT	
IlvB-1	AAGCTTCCGGCGAACGCGAGCTG (*Hind* III)	Amplifying upstream fragment of *ilvBNC* promoter
IlvB-2	TTTCCTTCGGATCTAAACGATCTTCTGTTGCAAAGTTGGCTACTTG	
IlvB-3	ACCACGAAGTCCAGGAGGACATACAGTAAAGGAGCCAGAAAGTCGTGAATGTG	Amplifying downstream fragment of *ilvN* promoter
IlvB-4	GGATCCTTAGATCTTGGCCGGAGCCATG (*Bam*H I)	
Brn-1	AAGCTTTTTCATAGCAAAACGTGATGAGATC (*Hind* III)	Amplifying upstream fragment of *brnEF* promoter
Brn-2	CACGAATTATGCAGTGATTTACGAACTACAATCATCACACAATTGCCG	
Brn-3	TTTCACACAGGAAACAGAATTAATTCGTGTGTGCAAAAAACGCAAGAGATTC	Amplifying downstream fragment of *brnEF* promoter
Brn-4	GGATCCCCGCCGAATACCCAGTAGG (*Bam*H I)	
Brn-5	TCGTAAATCACTGCATAATTCGTG	Amplifying tac promoter
Brn-6	ACACGAATTAATTCTGTTTCCTGTGTGAAA	
RT-7F	TCAGGGACGCATCTATGTTCCTGTGCAGAC	RT-PCR for *ilvA*
RT-7R	GCTTCGTCGAAGTTATTGCCAGTGACCACC	
RT-8F	CCCCATTAAGATCGCACTAATCAACAACGGAAA	RT-PCR for *ilvB*
RT-8R	CCTCAGAAAGGGTAACAAAGTCGGGCATGTACT	
16S-F	GTGGTTTGTCGCGTCGTCTGT	Internal reference for RT-qPCR
16S-R	TGCCTTCGCCATTGGTGTTC	

[a]Recognition sites are underlined and restriction enzymes are shown in parentheses

Data were analyzed using the $2^{-\Delta\Delta CT}$ method(16S rDNA was used as internal control) [10].

2.4 Construction of YILW-A, YILW-B and YILW-C Strains

Fragment, in which the promoter of *tuf* gene [11] was flanked by regions upstream and downstream the native promoter of *ilvA* was generated in two rounds of overlap PCR using primers indicated in Table 2 (Fig. 2). Fragment containing regions upstream and downstream of the native promoter of *ilvBNC* or *brnEF* operon as well as *tuf* or *tac* promoter [12] was obtained by the same way. And then the fragments were inserted into pK18mobsacB after digesting with respective enzyme, using *E. coli* DH5αMCR as host. The plasmids were designated pK18*mobsacB*Ptuf-A, pK18*mobsacB*Ptuf-BN and pK18*mobsacB*Ptac, respectively (Fig. 2).

The plasmid pK18*mobsacB*Ptuf-A was electroporated into *C. glutamicum* YILW where Ptuf replaced the native promoter of *ilvA* was constructed by two-step double homologous as reported (Fig. 2) [13, 14]. Strain YILW-B where *tuf* promoter replaced the native promoter in YILW-A was constructed with the same method using pK18*mobsacB*Ptuf-BN. Finally, pK18*mobsacB*Ptac was introduced into YILW-B in the same way, resulting YILW-C.

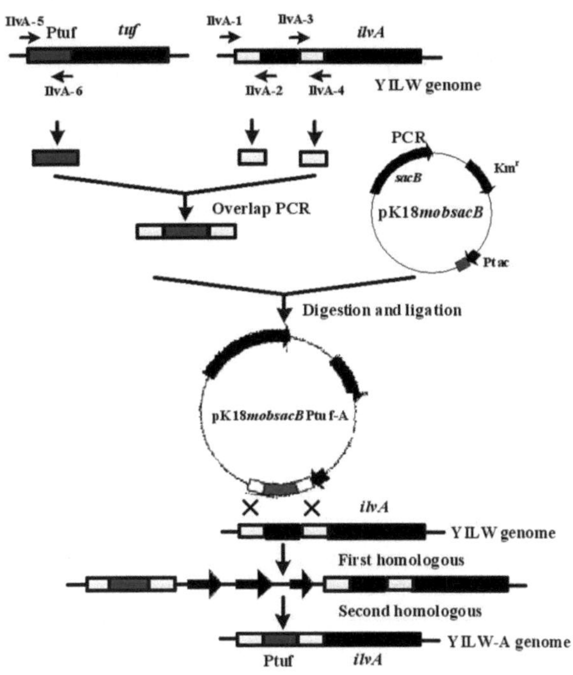

Fig. 2 Construction of YILW-A

2.5 Extraction of Intracellular Metabolite

5 mL of samples were injected into 15 mL of quenching solutions (70 mmol/L HEPES in 60% aqueous methanol (v/v), −50 °C) and centrifuged at 6000 × g at −20 °C. Cell pellets were collected and resuspended in 35% (v/v) perchloric acid. After one freeze (−80 °C)-thaw cycle, the sample was neutralized by 5 mmol/L K_2CO_3 and the precipitate was removed by another centrifugation. The resulting supernatants were stored at −80 °C until analysis [13].

2.6 Analytical Procedure

Cell growth was detected by measuring OD_{600} and was converted to the corresponding cell dry weight (DCW (g/L) = 0.24 × OD_{600} − 0.01). Glucose concentration was determined by an SBA biosensor analyzer (Institute of Biology of Shandong Province Academy of Sciences, Shandong, China). Amino acids and intracellular metabolite were analyzed by high-performance liquid chromatography (HPLC).

3 Results and Discussions

3.1 Effect of Pyc Overexpression on Isoleucine Production

Threonine dehydrogenase(TD) encoded by *ilvA* catalyzes 2-ketobutyric acid synthesis from threonine, which is one of key enzyme for isoleucine synthesis. Considering the potential metabolic burden caused by plasmid and further engineering modification, *ilvA* was overexpressed via replacement of its native promoter by promoter of *tuf* gene (resulting in YILW-A) which has been proved to be a strong promoter in *C. glutamicum* [11]. Transcription level of *ilvA* was detected and found to be significantly increased by 18.9 times in YILW-A. To assess the effects of *ilvA* overexpression on isoleucine production, batch cultivations were carried out. Unfortunately, isoleucine production and yield by YILW-A was not significantly increased.

Intracellular threonine and 2-ketobutyrate concentration as main intermediate metabolites was detected. Compared with YILW, threonine(1.67 μmol/g (DCW)) in YILW-A was 27.5% lower and 2-ketobutyrate(2.52 μmol/g (DCW)) was 92.4% higher. So it could be deduced that overexpression of *ilvA* resulted in increased 2-ketobutyrate pool.

Concentration of by-products during isoleucine fermentation was detected. Alanine, lysine and valine decreased from 0.76, 0.89 and 0.95 to 0.66, 0.61 and 0.72 g/L in YILW-1 (Table 4). Pyruvate is the precursor of the both by-products

[15], so overexpression of *ilvA* might draw more flux to 2-ketobutyrate synthesis and consequently result in decreased alanine, lysine and valine accumulation.

3.2 *Effect of IlvBN Overexpression on Isoleucine Production by YILW-B*

Higher 2-ketobutyrate accumulation in YILW-A indicated the necessity for improvement of flux from 2-ketobutyrate to isoleucine (Table 4). In *C. glutamicum*, AHAS catalyzing the reaction from 2-ketobutyrate to 2-aceto-2-hydroxybutyrate is encoded by *ilvBN*, which form an operon together with *ilvC* gene. Transcription of the *ilvBNC* operon is controlled by transcriptional attenuation mediated by isoleucine [16–18]. So defeating transcriptional attenuation of *ilvBNC* operon by isoleucine is urgent. The native promoter of *ilvBN* in YILW-A was replaced by Ptuf, which resulted in 21.31 times higher of *ilvB* transcription in YILW-B.

Batch cultivations were performed and it was found that the isoleucine production and yield by YILW-B was 13.7 and 30.0% higher than that by YILW-A (elevated from 5.11 g/L and 0.10 g/g glucose to 6.04 g/L and 0.13 g/g glucose, Table 3).

Moreover, intracellular 2-ketobutyrate concentration was found to be significant decreased to 1.49 μmol/g (DCW) in YILW-B (Table 4). Alanine, lysine and valine accumulation by YILW-B was decreased to 0.44, 0.53 and 0.59 g/L, which was 33.3, 13.1 and 18.1% lower than those by YILW-A. It should be notable that *ilvC* in the operon was together overexpressed and its translation product, acetohydroxy acid isomeroreductase, needs NADPH, so followed work will focus on improvement of NADPH supply within cell.

Table 3 Metabolic characterization of *C. glutamicum* strains

Parameters	*C. glutamicum* strains			
	YILW	YILW-A	YILW-B	YILW-C
Isoleucine production (g/L)	4.97 ± 0.03	5.31 ± 0.01	6.04 ± 0.01$^{*\Delta}$	7.48 ± 0.02$^{*\Delta}$
Biomass [g (DCW)/L]	8.21 ± 0.05	8.19 ± 0.11	8.11 ± 0.08	8.0. ± 0.13
Yield (g/g glucose)	0.10 ± 0.01	0.10 ± 0.02	0.13 ± 0.01$^{*\Delta}$	0.14 ± 0.01$^{\Delta}$
Alanine (g/L)	0.76 ± 0.01	0.66 ± 0.01$^{*\Delta}$	0.44 ± 0.01$^{*\Delta}$	0.21 ± 0.02$^{*\Delta}$
Lysine (g/L)	0. 89 ± 0.01	0. 61 ± 0.01$^{*\Delta}$	0.53 ± 0.01$^{*\Delta}$	0.41 ± 0.02$^{*\Delta}$
Valine (g/L)	0.95 ± 0.01	0.72 ± 0.01$^{*\Delta}$	0.59 ± 0.02$^{*\Delta}$	0.44 ± 0.01$^{*\Delta}$

*Significantly different compared with respective starting strain ($P < 0.05$)
$^{\Delta}$Significantly different compared with YILW ($P < 0.05$)

Table 4 Intracellular metabolite pool of *C. glutamicum* strains

Concentration [μmol/g (DCW)]	*C. glutamicum* strains			
	YILW	YILW-A	YILW-B	YILW-C
Threonine	1.67 ± 0.01	1.21 ± 0.02	–	–
2-Ketobutyrate	1.31 ± 0.02	2.52 ± 0.03*	1.49 ± 0.02*	–
Isoleucine	11.21 ± 0.11	11.35 ± 0.10*	14.26 ± 0.11*	10.09 ± 0.06*

* Significantly different compared with respective starting strain($P < 0.05$)

3.3 *Effect of Enhancing Isoleucine Efflux on Isoleucine Production by YILW-C*

Efflux is often overlooked as a bottleneck in metabolic pathways [19]. The last reaction for isoleucine synthesis by branched-chain amino acid aminotransferase is almost freely reversible [18]. Efflux of isoleucine out of cell could alleviate feed-back inhibition. Intracellular isoleucine concentration in YILW-B (14.26 μmol/g (DCW)) was found to be higher in YILW-A (11.35 μmol/g (DCW)) and YILW (11.21 μmol/g (DCW)) than that in *C. glutamicum* YILW. So enhancement of isoleucine efflux could increase isoleucine production.

Isoleucine exporter in *C. glutamicum* is a two-component permease, encoded by *brnE* and *brnF* [20]. Early studies of our group reported that overexpression of *brnEF* resulted in a significantly increased export rate and production of isoleucine [21]. To enhance isoleucine export, *brnEF* was overexpressed driven by promoter of *tuf* gene (resulting in YILW-C).

Batch cultivations were performed to assess the effects of *brnEF* overexpression on isoleucine production. As shown in Table 3, isoleucine production and yield by YILW-C was 23.3 and 7.7% higher (elevated from 6.04 g/L and 0.12 g/g glucose to 7.48 g/L and 0.14 g/g glucose) than that by YILW-B. Furthermore, alanine, lysine and valine accumulation by YILW-C was decreased to 0.21, 0.41 and 0.44 g/L, which was 52.3, 22.6 and 25.1% lower than those by YILW-B.

4 Conclusion

In this study, *ilvA* overexpressed in isoleucine producer *C. glutamicum* YILW made no contribution to isoleucine improvement but resulted in increase of intracellular 2-ketobutyric acid. Overexpression of *ilvBNC* operon effectively enhance the metabolic flux from 2-ketobutyric acid to isoleucine. Moreover, enhancing isoleucine export resulted in its production. The strategy lead to 50.5% higher isoleucine production (from 4.97 to 7.48 g/L) and 40.0% higher yield (from 0.10 g/g glucose to 0.14 g/g glucose) by YILW-C than that by *C. glutamicum* YILW. Strategy used in this study had potential applications for rational modification of industrial microorganisms.

Acknowledgements This work was supported by National High Technology Research and Development Program 2013AA102106), by the National Natural Science Foundation of China (31300069); Tianjin Municipal Science and Technology Commission (grant No. 15JCTPJC62800); Tianjin Undergraduate Training Program for Innovation and Entrepreneurship (201510057063). Innovation training program for college students in Tianjin (201710057039).

References

1. FIkeda S, Fyhuta I, Yoshinage F (1976) Screening of L-isoleucine producers among ethionine resistant mutants of L-threonine producing Bacteria. Agric Biol Chem 40(3):511–516
2. Park JH, Lee SY (2010) Fermentative production of branched chain amino acids: a focus on metabolic engineering. Appl Microbiol Biotechnol 85(3):491–506
3. Shiio I, Miyajima R (1969) Concerted inhibition and its reversal by end products of aspartate kinase in *Brevibacterium flavum*. J Biochem 65(6):849–859
4. Yin L, Hu X, Xu D et al (2012) Co-expression of feedback-resistant threonine dehydratase and acetohydroxy acid synthase increase L-isoleucine production in *Corynebacterium glutamicum*. Metab Eng 14(5):542–550
5. Miyajima R, Shiio I (1970) Regulation of aspartate family amino acid biosynthesis in *Brevibacterium flavum* III. Properties of homoserine dehydrogenase. J Biochem 68(3): 311–319
6. Eikmanns BJ, Metzger M, Reinscheid D et al (1991) Amplification of three threonine biosynthesis genes in *Corynebacterium glutamicum* and its influence on carbon flux in different strains. Appl Microbiol Biotechnol 34(5):617–622
7. Eggeling L, Bott M (2005) Handbook of *Corynebacterium glutamicum*. CRC Press, Boca Raton, pp 520–521
8. Bailey JE (1991) Toward a science of metabolic engineering. Science 252(5013):668–1675
9. Zhang C, Du S, Liu Y et al (2015) Strategy for enhancing adenosine production under the guidance of transcriptional and metabolite pool analysis. Biotech Lett 37(7):1361–1369
10. Livak KJ, Schmittgen TD (2001) Analysis of relative gene expression data using real-time quantitative PCR and the 2(-Delta Delta C(T)) method(2001). Methods 25(4):402–408
11. Becker J, Klopprogge C, Zelder O et al (2015) Amplified expression of fructose 1, 6-bisphosphatase in *Corynebacterium glutamicum* increases in vivo fluxthrough the pentose phosphate pathway and lysine production on different carbon sources. Appl Environ Microbiol 71(12):8587–8596
12. Cheng Y, Zhou Y, Yang L et al (2013) Modification of histidine biosynthesis pathway genes and the impact on production of L-histidine in *Corynebacterium glutamicum*. Biotech Lett 35 (5):735–741
13. Becher J, Zelder O, Hafner S et al (2011) From zero to hero-design-based systems metabolic engineering of *Corynebacterium glutamicum* for L-lysine production. Metab Eng 13(2): 159–168
14. Buchholz A, Hurlebaus J, Wandery C et al (2012) Metabolomics: quantification of intracellular metabolite dynamics. Biomol Eng 19(1):5–15
15. Wang J, Wen B, Wang J et al (2013) Enhancing L-isoleucine production by *thrABC* overexpression combined with *alaT* deletion in *Corynebacterium glutamicum*. Appl Biochem Biotechnol 171(1):20–30
16. Morbach S, Junger C, Sahm H et al (2000) Attenuation control of *ilvBNC* in *Corynebacterium glutamicum*: evidence of leader peptide formation without the presence of a ribosome binding site. J Biosci Bioeng 90(5):501–507
17. Elisakova V, Patek M, Holatko J et al (2005) Feedback-resistant Acetohydroxy acid synthase increases valine production in *Corynebacterium glutamicum*. Appl Environ Microbiol 71 (1):207–213

18. Wilhelm C, Eggeling I, Nassenstein A et al (1989) Limitations during hydroxybutyrate converison to isoleucine with *Corynebacterium glutamicum*, as analysed the formation of byproducts. Appl Microbiol Biotechnol 31(5–6):458–462
19. Jones CM, Hernández Lozada NJ, Pfleger BF (2015) Efflux systems in bacteria and their metabolic engineering applications. Appl Microbiol Biotechnol 99(22):9381–9393
20. Kennerknecht N, Sahm H, Yen MR et al (2002) Export of L-isoleucine from *Corynebacterium glutamicum*: a two-gene-encoded member of a new translocator family. J Bacteriol 184(14):3947–3956
21. Xie X, Xu L, Shi J et al (2012) Effect of transport proteins on L-isoleucine production with the L-isoleucine-producing strain *Corynebacterium glutamicum* YILW. J Ind Microbiol Biotechnol 39(10):1549–1556

Liquid-Liquid Extraction of Hydrocortisone from the Fermentation Liquor Contain Hydroxropyl-β-Cyclodextrin

Dongchao Yuan, Yanling Dong, Yanbing Shen, Qing Zhao and Min Wang

1 Introduction

Hydrocortisone is an important hormone drug usually used as antianaphylaxis or anti-inflammatory and principal in the therapy for chronic adrenocortical insufficiency disease, which can be obtained via the hydroxylation at C11 of 17α-hydroxy-pregnant-4-ene-3, 20-diketone-21-acetate (RSA) by *Absidia orchidis* [2, 5, 9]. Low solubility of steroid hormone is a big problem limiting biotransformation. Cyclodextrins (CDs) are water-soluble cyclic carbohydrate compounds with a hydrophobic core which is a unique property can improve the solubility of steroids [1, 3, 4]. The efficiency of RSA biotransformation improved in presence of CDs. The CDs improved the solubility of both substrate RSA and product hydrocortisone, which caused difficulties to the later product extraction process by organic solvent, because that final product existed in the fermentation liquor in various forms, such as free of CDs and hydrocortisone, as well as inclusion complex of hydrocortisone with CDs. Thus, it is necessary to research the product extraction technology to establish product separation of extracting process in CDs fermentation media, set a basis for steroidal drugs and its intermediates high concentrations of industrialization technology with CDs media.

Liquid–liquid extractions consist of an interesting purification alternative since several features of the early processing steps can be combined into a single operation [7]. Because the crosscurrent extraction process has many advantages, this paper adopts a continuous countercurrent extraction process method and uses the

D. Yuan · Y. Dong
Jinjin Pharmaceutical Co., Ltd, Tianjin, China

Y. Shen · Q. Zhao · M. Wang (✉)
Key Laboratory of Industrial Fermentation Microbiology, Ministry of Education,
College of Biotechnology, Tianjin University of Science and Technology,
Tianjin 300457, People's Republic of China
e-mail: minw@tust.edu.cn

H. Liu et al. (eds.), *Advances in Applied Biotechnology*, Lecture Notes
in Electrical Engineering 444, https://doi.org/10.1007/978-981-10-4801-2_52

principle of split the subject and object by using organic solvent competitive replace for the recovery of hydrocortisone from the fermentation liquor to avoid the above-mentioned drawbacks [6].

2 Materials and Methods

2.1 *Reagents*

The fermentation liquor of hydrocortisone was obtained from Tianjin Key Lab of Industrial Microbiology, Tianjin University of Science and Technology. The major composition of the fermentation liquor is shown in Table 1. HP-β-CD (Mw: 1522, average degree of substitution: 6.5) were obtained from Wacker Biochem. Corp. (Munich, Germany). All chemical solvents and salts used were of analytical grade or higher.

2.2 *Equilibrium Experiments*

The contact between aqueous and organic phases was carried out in closed tubes with a maximum capacity of 50 mL, at 25 °C. The experimental procedure involved measuring and adding to the tubes the proper volume of each phase according to the selected volume ratios A/O = 1:1. The tubes were shaken in an orbital stirrer at 170 r min^{-1} for 5 min. At the end of the contact time, the phases were separated by centrifugation at 5000 r min^{-1} for 10 min. Final volumes were measured and the aqueous phase or the organic phase was sampled for analysis. All experiments were conducted on a replicate basis.

For analysis of the influence of the manipulate condition on the extraction equilibrium, the extraction of hydrocortisone in a butyl acetate/water system. A batch stagewise test was used to simulate the continuous countercurrent process. Before the test, a simulation calculation of the separation process was conducted to determine the optimum flow ratio and equilibrium stage number based on the data obtained in the distribution equilibrium test for the hydrocortisone in fermentation liquor/water system. The optimum results were then evaluated by the batch stagewise test. Each test tube simulated a shakeout unit in the stagewise phase equilibrium.

Table 1 The major composition of fermentation liquor

Component	Concentration (g/L)
Hydrocortisone	1.10–1.50
Epi-hydrocortisone	0.61–0.65
HP-β-CD	18.26

2.3 *Analytical Methods*

The hydrocortisone contents of the aqueous phase and the organic phase in each phase were determined using HPLC with an ODS Hypersil-C18 reversed phase column using the following conditions: mobile phase water/methanol (20:80 v/v), flow rate 0.8 mL min^{-1}, a UV detector set at 254 nm. The HP-β-CD was determined using HPLC analysis was performed using refractive index detection and a Mark-NH_2 column (250 × 4.6 mm) with mobile phase water/acetonitrile (40:60 v/v) as the mobile phase.

3 Results and Discussion

3.1 *Distribution of Hydrocortisone in the Butyl Acetate/Water System*

The Effect of the HP-β-CD concentration on the distribution of hydrocortisone in the butyl acetate/water system at 25 °C can be described by Fig. 1. Experimental results showed that the distribution ratios of hydrocortisone reduced with increasing the HP-β-CD concentration in the aqueous phase, until the HP-β-CD concentration was 15 mM were essentially unchanged. The extraction efficiency decreased with increasing the HP-β-CD concentration in the aqueous phase, the results were given in Fig. 2. Because the addition of HP-β-CD into the water can increase the solubility of hydrocortisone in the aqueous phase, the extraction efficiency and distribution ratios of hydrocortisone in the Butyl Acetate/Water System both reduced.

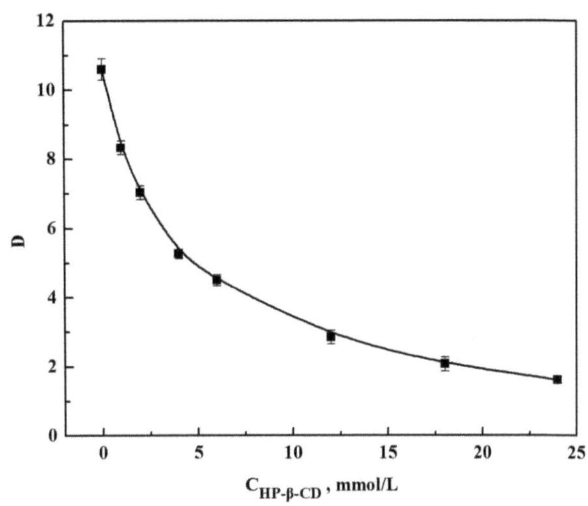

Fig. 1 Effect of HP-β-CD concentration on the distribution in a butyl acetate/water system at 25 °C

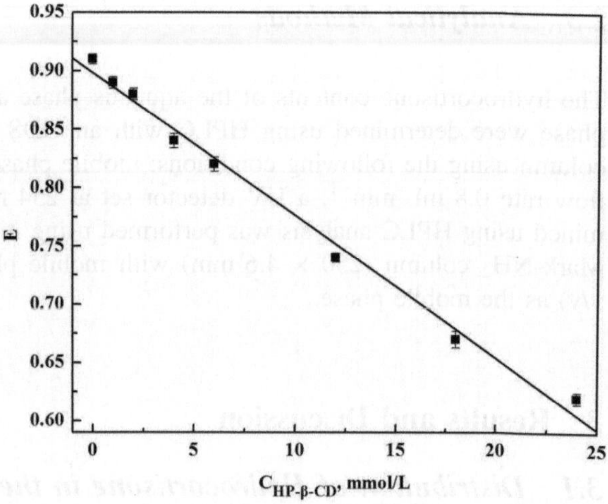

Fig. 2 Effect of HP-β-CD concentration on the extraction in a butyl acetate/water system at 25 °C

The temperature and pH are two important parameters for extraction process. The results showed that the distribution ratio of hydrocortisone increased with increasing temperature and pH, but the separation factor and the extraction efficiency were almost unchanged (data not shown). The multi-level of the extraction process implied that volumetric ratios of organic phase and water phase are the same important as temperature's effect on the phase equilibrium. And the flow ratio of extraction process is according to the technics requirement and selected operation condition. Obviously, the greater flow rate and the amount of organic phase, the less theoretical stages of extraction theory series was performed under the other operation condition.

The addition of some ammonium sulfate into the aqueous phase significantly increased the distribution ratio, seen in Table 2. In general, the salting–out effect resulted from the reduced water activity due to the salt addition. In addition, the strong electrolyte reduced the extractant solubility in the aqueous phase, which decreased solvent losses. But the salt is difficult to recover and reuse, thus commercial application of the salting-out effect would be limited.

Table 2 The effect of ammonium sulfate on the distribution and extraction in a butyl acetate/water system with 12 mM HP-β-CD. (pH = 7, T = 25 °C)

Ammonium sulfate content in water (wt%)	Distribution ratio of hydrocortisone	Separation factor	Extraction ratio of hydrocortisone
0	2.620 ± 0.018	2.361 ± 0.024	0.721 ± 0.010
5	2.693 ± 0.026	2.383 ± 0.019	0.732 ± 0.006
8	2.738 ± 0.035	2.335 ± 0.022	0.733 ± 0.008
10	3.172 ± 0.040	2.430 ± 0.031	0.760 ± 0.011

3.2 *Mass Transfer Rate of Hydrocortisone in the Butyl Acetate/Fermentation Liquor System*

Effect of mixing time on the hydrocortisone concentration in the organic phase and the aqueous phase were presented in Table 3. It was shown that when the mixing time was more than 3 min, the hydrocortisone concentration in the organic and aqueous phase does not change with the mixing time, and the extraction process reached an equilibrium. Distribution of hydrocortisone in butyl acetate/fermentation liquor systems at 25 °C was showed in Fig. 3. The apparent distribution equilibrium can be described by Eq. (1).

$$y = 0.038 + 4.256x - 17.475x^2 + 113.456x^3 - 184.129x^4 + 89.109x^5 \quad (1)$$

The Janecke diagram mode shows that the extraction of hydrocortisone from the fermentation liquor was a continuous multistage countercurrent extraction process with 6 stages. A over 18-row transfer process was needed for a flow ratio of 1 to extraction efficiency over 98%.

Table 3 Variation of the hydrocortisone concentration in the organic phase and the aqueous phase with the mixing time (25 °C)

Mix time (min)	1	3	5	7	9
Concentration in the organic phase (g/L)	0.725 ± 0.006	0.738 ± 0.012	0.736 ± 0.020	0.740 ± 0.009	0.739 ± 0.018
Concentration in the aqueous phase (g/L)	0.182 ± 0.004	0.176 ± 0.002	0.174 ± 0.005	0.176 ± 0.005	0.175 ± 0.002

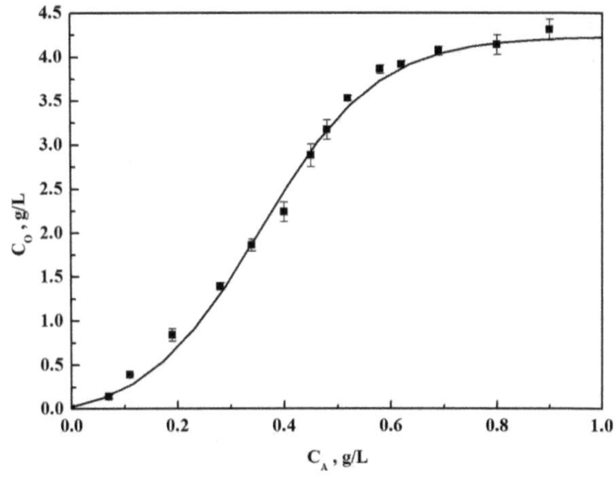

Fig. 3 Distribution of hydrocortisone in butyl acetate/fermentation liquor systems at 25 °C

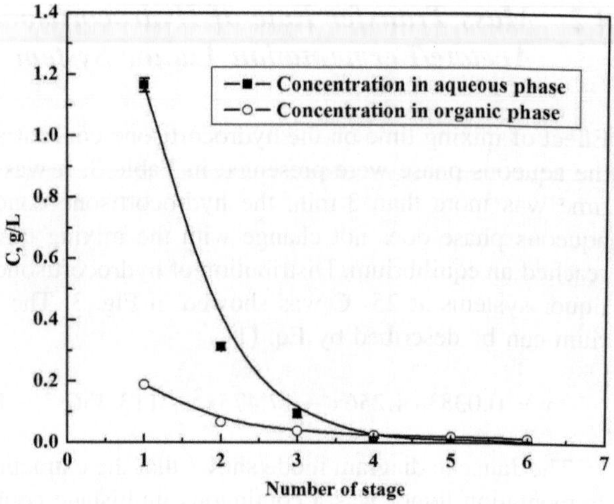

Fig. 4 Concentration profiles of hydrocortisone for the crosscurrent extraction process

Table 4 The result of analogy countercurrent extract hydrocortisone method

Extraction series	The extraction phase concentration of hydrocortisone	The raffinate phase concentration of hydrocortisone
1	1.165	0.189
2	0.313	0.067
3	0.094	0.038
4	0.017	0.024
5	0.010	0.019
6	0.005	0.015

The concentration profiles in the extraction process are plotted in Fig. 4. After the 24-row transfer for 6 stages and O/A 1, with an initial concentration of 1.179 g/L, the determined concentration profiles were in good agreement with the data predicted by Eq. (1). Xien Hu et al. had studied the extraction of steroid by-products in the fermentation liquor. The recovery of hydrocortisone and epi-hydrocortisone through the extraction process was 95% with a flow ratio of 1.1 and 6 stages [8]. By contrast, the extraction efficiency of hydrocortisone through the continuous multistage countercurrent extraction process was 98.80% with a flow ratio of 1 and 6 stages in our research (Table 4).

4 Conclusions

A continuous counter current extraction process for there extraction of hydrocortisone from the fermentation liquor has been developed with 12 mM/L HP-β-CD. The distribution ratios of hydrocortisone increased with increasing temperature, the distribution was also affected by the ammonium sulfide content, the greater flow rate and the amount of organic phase, the less theoretical stages of extraction theory series was performed under the other operation condition. When the hydrocortisone was extracted from the fermentation liquor with the butylacetate, the distribution ratio reduced with the increasing of high concentration of the hydrocortisone. The recovery of hydrocortisone and epi-hydrocortisone through the extraction process was 98.8% with a flow ratio of 1 and 6 stages.

Acknowledgements The current work was supported by the National Natural Science Foundation of China (21276196, 21406167) and Tianjin Programs for Science and Technology Development (15ZCZDSY00510).

References

1. Arun RA, Kumar ACK, Sravanthi VVNSS (2008) Cyclodextrins as drug carrier molecule: a review. Sci Pharm 76:567–598
2. Ma B, Shen Y, Fan Z et al (2011) Characterization of the inclusion complex of 16,17a-epoxyprogesterone with randomly methylated b-cyclodextrin in aqueous solution and in the solidstate. J Incl Phenom Macrocycl Chem 69:273–280
3. Donova MV, Nikolayeva VM, Dovbnya DV et al (2007) Methyl-β-cyclodextrin alters growth, activity and cell envelope features of sterol-transforming mycobacteria. Microbiology 153:1981–1992
4. Hapiot F, Ponchel A, Tilloy S et al (2011) Cyclodextrins and their applications in aqueous-phase metal-catalyzed reactions. ChemInform 14:149–166
5. Zhou J, Duan W, Zhou X et al (2010) Extraction of hydrocortisone from the fermentation liquor with annular centrifugal contactors. Sep Sci Technol 41(3):573–581
6. Loftsson T, Hreinsdóttir D (2007) The complexation efficiency. J Incl Phenom Macrocycl Chem 57(1):545–552
7. Mazzola PG, Lopes AM, Hasmann FA et al (2008) Liquid-liquid extraction of biomolecules: an overview and update of the main techniques. J Chem Technol Biotechnol 83:143–157
8. Hu X, Tang Z, Zhu Y (2001) Separation of hydrocortisone and its steroisomer from fermentation liquor by chloroform extraction process. Sep Sci Technol 36(7):1421–1435
9. Shen Y, Ma B, Zheng Y et al (2010) The mechanism of β-cyclodextrin on the 11β-hydroxylation biotransformation of steroid. In: Proceedings of the 2010 3rd international conference on biomedical engineering and informatics (BMEI 2010), vol 5, pp 1964–1967

Conversion of Food Waste and Feldspar into Biofertilizer Using a Stress-Tolerant Keldspar-Solubilizing *Bacillus Subtilis* Xue-113168

Shengping Xue, Liangtian Miao, Jinjun Xue, Hanbo Yan and Guiqin Li

Abbreviations

CMF	Compound Microbial Fertilizer
PGPR	Plant growth-promoting rhizosphere
SSF	Solid-state fermentation
SMF	Submerged fermentation
SMS	spent mushroom substrate
KSM	Potassium dissolving microorganism
LiCl	Lithium chloride
KCl	Kalium chloratum

1 Introduction

With the exhaustion of non-renewable resources like coal and peat, the advancing of technology also prompt the development of new sustainable sources of humic products (e.g. organic wastes) [1]. Biofertilizer without chemical fertilizer has a slow effect on crops, Compound Microbial Fertilizer (CMF) with stress-tolerant microorganisms combine the advantages of both biofertilizer and chemical fertilizer.

Microbial fertilizers promote biological fertility of soil, maintain and slow the release of chemical fertilizer, reduce the amount of nitrate nitrogen, heavy metals,

S. Xue (✉) · H. Yan · G. Li
College of Bioscience and Bioengineering, Hebei University of Economic and Business, Shijiazhuang, China
e-mail: xsp6210@163.com

L. Miao
Institute of Microbiology, The Academy of Sciences of China, Beijing, China

J. Xue
College of Horticulture, Guangxi University Nanning, Nanning, China

H. Liu et al. (eds.), *Advances in Applied Biotechnology*, Lecture Notes in Electrical Engineering 444, https://doi.org/10.1007/978-981-10-4801-2_53

and pesticides in crops, and reduce the occurrence of crop diseases [2]. Biofertilizers could mitigate crisis, e.g., the energy crisis, scarcity of resources, and environmental pollution [3]. Plant growth-promoting rhizosphere (PGPR) raises the resistance to pathogen by producing antibiotics and System resistance, etc. The CMF can palliate Jujube and Selenium-enriched Jujube from rust and anthracnose. Solid state fermentation using waste is a popular way of biofertilizer manufacturing. Low water activity in solid state fermentation (SSF) medium facilitates spore forming on the solid substrate. Unlike in submerged fermentation (SMF) processes, humidity is an important parameter of SSF. Microbial releases of potassium from K-bearing minerals have two steps, fermentation and bioleach [4]. The solubilization of potassium-bearing rock powder by Aspergillus niger in submerged fermentation lasted for 35-day period [5]. This study achieves it in one process for 7 days, saving production cycle and enhancing robustness. The imbalance of soil nutrients in China is the consequence of available potassium deficiency in the soil and shortages of potassium fertilizer, although K-bearing minerals in soil and mine, such as K-feldspar is rich, the amount of potassium dissolved by bacteria in soil is limited.

Keldspar can be dissolved by silicate bacteria such as *B. mucilaginosus, B. edaphicus, B. circulans*, *B. subtilis* which are part of the PGPR [6, 7]. Humus and Humic acids, which exist in spent mushroom substrates and biofertilizer, can complex with the potassium released from K-feldspar to improve its dissolving efficiency. Conversion of food waste and feldspar into biofertilizer using a stress-tolerant keldspar-solubilizing *Bacillus subtilis* Xue-113168 make the best use of agricultural and mineral resources, enhance crop yields and play an important role in the sustainable development.

2 Materials and Methods

2.1 Food Waste and Feldspar for SSF

In an SSF process, the solid substrate not only supplies anchorage for the cells, but also nutrients and air channel. A number of cheaply available raw substrates have been screened for the SSF. Shrimp shell, spent mushroom substrate (SMS) and corn bran of 0.1 mm, are all support material for ventilation in SSF. Chitin in shrimp shell and SMS can induce chitinase. K-feldspar containing 10.0% total potassium and insoluble potassium, was purchased from Lingshou County, Hebei, P.R. China. The shrimp shell was from Whiteleg. SMS of Pleurotus ostreatus, constitute of 4.45% organic matter, 1.58% total N, C/N ratio 28/1, water absorption rate 78–80%, humic acid content 20–30%, obtained from a local market. Corn bran constitute of 11.8% protein, 32% glucose, 1% cellulose, purchased from the North China Pharmaceutic Corporation, Hebei, P.R. China.

2.2 *Potassium Dissolving Microorganism (KSM)*

KSM are isolated from medium containing feldspar as the only potassium source (Hutchens et al. 2003). Medium's ingredients are as follows (g/L): starch, 5.0; yeast extract, 1.0; $MgSO_4 \cdot 7H_2O$, 0.5; $CaCO_3$, 0.1; $FeCl_36H_2O$, 5 mg, feldspar 10, pH 7.5. After 1-day incubation at 30 °C, the transparent colonies, which resemble half round glass bead, were picked and purified by growth of K-feldspar medium.

Stress-tolerance induction was carried out by semi-continuous culture in 50-ml shaking flasks at 30 °C on a reciprocal shaker (150 rpm). The cells were transferred to the induction solution which contains KCl, monosodium glutamate 4.27 mg%. Concentration of k is increasing at a rate of about 2.5% per circulation, from 2.5 to 15%.

This isolates were originally named *Bacillus circulans*, later identified as *Bacillus subtilis* by culturing characteristics and morphological observation, physiological and biochemical characteristics and identification of genetic characterization by Institute of Microbiology, Chinese Academy of Sciences in December 2016. The potassium dissolving rate was determined as = (St–Sc)/It*100%, where "St" and "Sc" are the water-soluble potassium in the treatment and matrix, respectively, and "It" = the total potassium in the treatment.

For UV's convenience and efficacy, breeding a high potassium dissolving rate *B. subtilis* strain had been performed using UV mutagenesis, screen in a fermentation broth using the potassium tetraphenylborate spectrophotometric method. After a series of mutation steps (Fig. 1), *B. subtilis* Xue-113168 has been obtained, which have chitinase and dissolving potassium function, preserved as patent strain in the Chinese General Microbiological Culture Collection Center (CGMCC) with the accession number 5155 [8].

2.3 *CMF Preparation by Solid-State Fermentation (SSF)*

The biofertilizer was made from SMS, shrimp shell, corn bran and feldspars using a simultaneously dissolving K-feldspar and SSF. Besides solid material, flour and feldspar were used as substrates at 10 and 2% (w/v), respectively, medium are sterilized in 121 °C, 0.1 MPa, for 30 min. An initial moisture content is 60–65%, and culture temperature is 30 °C. The SSF was stirred every 48 h for 7 days. The pH was nature during the fermentation of the SSF with *B. subtilis* Xue-113168. After 168 h of fermentation, the spore number and dissolving K rate are tested.

Fig. 1 Breeding pedigree of *B. subtilis* Xue-113168

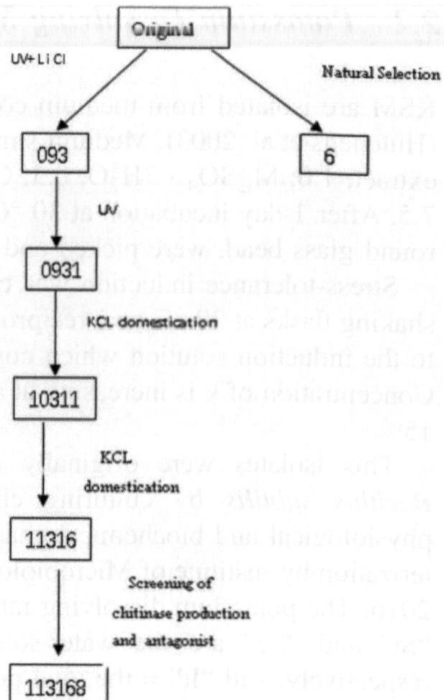

2.4 Testing the Quality Index of the CMF

The amount of viable *B. subtilis* Xue-113168 in the CMF was calculated by counting colonies of the medium: (g/L) sucrose (5.0), $MgSO_4$ (1.0), $FeCl_3$ (0.2), yeast extract (0.2), $(NH_4)_2SO_4$ (0.5), KH_2PO_4 (1), pH 7.0. The moisture content was measured using the vacuum oven method. PH of Solid-water suspension (1:1) was determined using a pH meter. Reagent of digested CMF is H_2SO_4–H_2O_2. The total N K_2O P_2O_5 content was determined by the Kjeldahl method, potassium tetraphenylborate gravimetric method, molybdenum antimony anti-colorimetry respectively. The organic matter was digested and tested with H_2SO_4–$K_2Cr_2O_7$ method.

2.5 Statistical Analysis

Variance (ANOVA) of experimental data was analysed by a significance value of $p < 0.05$ and to Duncan's multiple range test (Duncan's) using SPSS 18.

3 Results and Discussion

3.1 *Isolation and Breeding of Bacillus Subtilis Xue-113168*

Bacillus subtilis Xue-113168 was isolated from a corn rhizosphere and breed by UV, UV+LiCl, natural selection, breed genealogy is showed in Fig. 1. Sieved by feldspar as the only source of potassium and colloid chitin for carbon. We use monosodium glutamate as osmotic protection, culture the stress-tolerance bacteria with a stepwise increase of KCl concentration, which are different from sorbitol [9], NaCl, glycerol and glucose as osmotic protection [10]. The biofertilizer are manufactured by *B. subtilis* Xue-113168 for dissolve potassium mineral substance in soil. Potassium dissolving microorganisms (KSM) is able to solubilize 'unavailable' forms of K-bearing minerals, such as micas, illite and keldspar, by excreting organic acids that either directly dissolves rock K or chelate silicon ions to bring the K into solution [11].

B. subtilis Xue-113168 has a high potassium dissolving rate (41%), also produces chitinase. Due to superiority alive in 8% K_2O of *Bacillus subtilis* Xue-113168, CMF had the combining advantage of both chemical fertilizer and biological fertilizer.

B. subtilis Xue-113168 colon morphology changes into opaque, *B. subtilis* Xue-10311 still transparent and *B. subtilis* Xue-103116 stay thickness and viscous (Fig. 2). The changes of colon morphology indicate the potential increase of stress-tolerant capacity.

Morphology of *B. subtilis* Xue-113168 is Gram-positive, rod shape showed in Fig. 3.

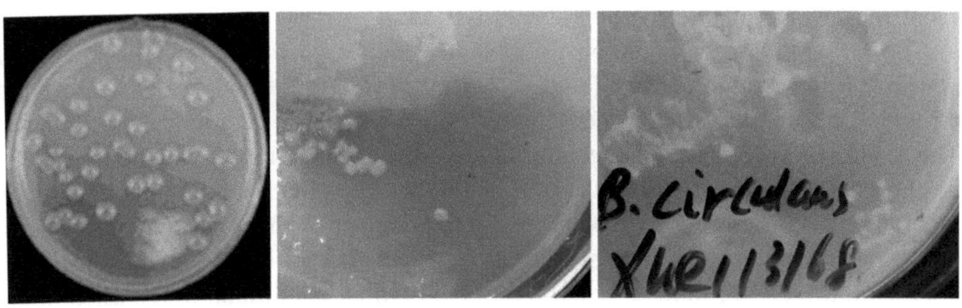

Fig. 2 Colon morphology of *B. subtilis* Xue-10311 (*left*), *B. subtilis* Xue-11316 (*middle*), *B. subtilis* Xue-113168 (primitive name *B. circulans*) (*right*)

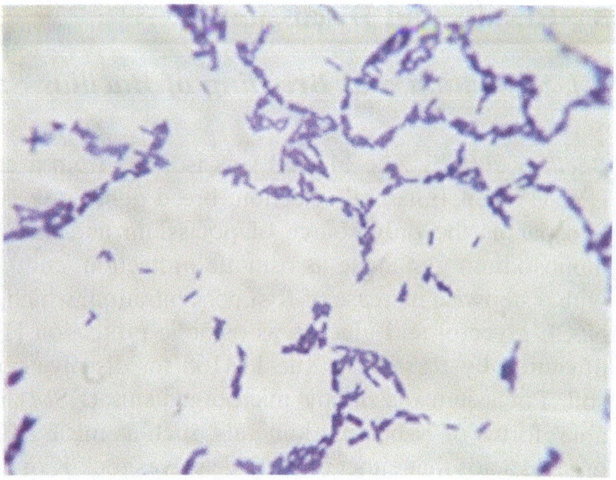

Fig. 3 Morphology of *B. subtilis* Xue-113168(*1600) by Gram Stain

3.2 *Formulation of the Biofertilizer Produced by the SSF*

The component of formulation is a key factor of the SSF and influences the overall cost significantly.

The formulation of SSF could be similar to that of a submerged fermentation (SMF) [12]. But sucrose was replaced with wheat flour for the SSF in this study and the spore formation rate of SSF is higher than SMF because these two processes have state and component of substance. The combination of corn bran, shrimp shell, and spent mushroom substrate was the best medium for the SSF. The K-dissolving rate (43.00%) and spore formation rate (83.18%) achieved were the best among all the groups, and the live bacteria amount of 9.16 lg (CFU) was second only to the spent mushroom substrate group (10.08 lg(CFU)).

The fermentation formulation for obtain a higher concentration of spores and potassium dissolving capacity for *B. subtilis* Xue-113168 was initially determined through a single factor experiment, then further through an orthogonal experiment. The range of $(NH_4)_2SO_4$ concentrations was the most important variable, suggesting that in the *B. subtilis* Xue-113168 culture, the effects of the three factors of the potassium dissolving ratio were $(NH_4)_2SO_4$ content > $MgSO_4$ content > wheat flour content by range analysis. The above three factors have evident effects on potassium dissolving ratio ($P < 0.05$) and the optimal conditions were: wheat flour content, 0.75%; $(NH_4)_2SO_4$ content, 0.1%; and $MgSO_4$ content, 0.2%.

As for the spore concentration, it can be concluded that the ranges of $(NH_4)_2SO_4$ and wheat flour concentrations were larger compared to the $MgSO_4$ concentration. Thus, the effects of the three factors of the spore concentration were $(NH_4)_2SO_4$ content > wheat flour content > $MgSO_4$ content, $MgSO_4$ is not significant factor for spore concentration ($p > 0.05$). Optimal conditions were: wheat flour content, 0.75; $(NH_4)_2SO_4$ content, 0.1%; and $MgSO_4$ content, 0.2%. The effect of ingredient composition and concentration in medium for spore formation is the same with potassium dissolving condition. The K-dissolving rates are different between the SMS and Corn bran + Shrimp shell + SMS in the SSF. The humic acids in the SMS can also dissolve the K-feldspar, which is complementary to the effect of *B. subtilis* Xue-113168.

Organic matter is 250 g kg^{-1} for the *B. subtilis* Xue-113168 high decompose performance to food waste. Organic matter reach 532 g kg^{-1} by adding humic acid.

3.3 Process Index of SSF

We observed some changes in pH in the reactor with different medium during the experiment. The pH decreased from 7.0 to 5.0 during the first day, recovered to 5.0–6.0 in a few days and then increased and maintained around 7.6 to the end of the experiment.

The Spore formation was affected by varietal spectrum of humidity and temperature. Controlling the moisture content of the SSF process was performed as follows: Initially the moisture content was adjusted to approximately 60–65%, at the end of the fermentation, the moisture content was 40–50%. The moisture content was controlled by environmental moisture and also by controlling the temperature in three phrases. During the prophase, the temperature was 33–35 °C, whereas, at the metaphase, the temperature dropped to 30–33 °C. Finally, it increased to 33–35 °C during the anaphase.

The fermentation time was also an important parameter. By the time that the rate of spores is 80%, the processes had typically ended. An appropriate temperature and humidity ensured the formation of the spores. When the fermentation last for 4 days, the potassium dissolving rate was 24.09%. After the fermentation time was extended to 7 days, the potassium dissolving rate increased to 41.53%.

The process of potassium dissolving from feldspar is a synthesized multifaceted result. There is no single gene that encodes for a potassium dissolving ability in bacteria, which is similar to the process of dissolving phosphate. It is well known that there are many differences between SSF and SMF in physiology, such as Aw (water activity) and metabolite. The pH range for *B. subtilis* Xue-113168 is between 5 and 9, and the temperature range is 28–37 °C. Moreover, the indispensable moisture required for SSF was low. All the characteristics described above make *B. subtilis* Xue-113168 suitable for SSF. Sheng demonstrated that the potassium dissolving rate of Bacillus edaphicus is low, which also has been proven to have a low potential for commercial use [13]. Tan reported that 9–28% of the potassium that they measured was released by humic acids [14]. It has been hinted that the potassium dissolving rate (41.53%) of *B. subtilis* Xue-113168 may include dissolved potassium from K-feldspar through the action of humic acids. Consistent and sustained product of SSF can be achieved in different growing media as long as the following variables are being controlled: consistency of the parent organic material, moisture content, carbon to nitrogen ratio, and process parameters.

During SSF with feldspar, spore concentrations can reach 2×10^9 cfu/g and soluble potassium content is 1%, *B. subtilis* Xue-113168 ferments and dissolves K simultaneously. *B. subtilis* Xue-113168 displayed the highest K-solubilizing activity, as high as 41.53% after 7d in culture with K-feldspar powder. This cycle is shorter than Lian and Ciceri [4, 15], who performed SSF and SMF for 15 and 35 d respectively. Bioleaching combined with SSF is more advantageous, more economical and more timesaving. Stephanie. K gets mutant bacteria for a high efficient bioleaching also [16].

3.4 Quality Index of the CMF

The quality index plays a vital role in measuring the efficiency of a fertilizer. The indexes are consistent with the Compound Microbial Fertilizer Standard (NY/T 798-2015, China): $N + P_2O_5 + K_2O \geq 80\text{–}250\ g\ kg^{-1}$, number of viable cells is 0.2 (one hundred million/g) [17]. Compound Microbial Fertilizer (CMF), formulated by humic acid and K_2HPO_4 with biofertilizer, have a quick and long-lasting effect. The stress-tolerance towards K_2HPO_4 of *B. subtilis* Xue-113168 is a key factor, the humic acid's chelating functions also lessens the damage to bacteria done by K_2HPO_4. Product Quality index of the CMF were examined by Quality supervision, inspection and Testing Center for microbial products of Ministry of Agriculture in WuHan in December 2016. Provisional registration certificate of CMF is No. 3878 by Ministry of Agriculture of the People's Republic of China in March 15, 2017. Product Quality index of the CMF is different from the that of pilot sample showed in Table 1, such as content of P_2O_5, inferring phosphorus as massive element in cell is less harmful to microorganism. Function of PGPR dependent on the clone in zone of root [18]. Organic matter in CMF offer nutrient, shelter to adverse situation for PGPR in field.

Table 1 Results of the quality index determination experiment

Item	Effective viable cells a hundred million (g)	Total primary nutrient ($g\ kg^{-1}$)	Organic matter ($g\ kg^{-1}$)	N Nitrogen ($g\ kg^{-1}$)	P_2O_5 (%)	Available potassium ($g\ kg^{-1}$)
NY/T798-2015	0.2	80–250	200			
CMF (pilot)	2	130	250	12	38	80
CMF (product)	2.8	174	532	35	91	48

4 Conclusions

This CMF is formulated by humic acid, K_2HPO_4, and biofertilizer which is SSF by chemical fertilizer-tolerant feldspar-solubilizing *B. subtilis* Xue-113168. Provisional registration certificate of CMF is No. 3878 by Ministry of Agriculture of the People's Republic of China in March 15, 2017.

Bacillus subtilis Xue-113168 converts K-feldspar, corn bran, shrimp shell, and spent mushroom substrate to a biofertilizer that have a 41% potassium dissolving rate. The CMF used in this study can enhance the efficiency of fertilizers and utilize recycled waste resource. It also promotes a sustainable agriculture industry, complex formulations and process of the SFF are suitable for factory, especially farmer uses in the fields.

Acknowledgements This work was supported by the Ministry of Science, People's Republic of China under Grant [number 2008GA620020]; Hebei Province natural fund under Grant [number C2015207019]. Hebei University of Economics and Business fund under Grant [number 2016kyz01]. Hebei Provincial department of education under grant [number ZD2017229].

Thanks to professor Yu Li, Wenhang Wang, Guorong He, Yingjie Pan for review the paper.

References

1. Placido J, Capareda S, Karthikeyan R (2016) Production of humic substances from cotton stalks biochar by fungal treatment with ceriporiopsis subvermispora. Sustain Energy Technol Assess 13:31–37
2. Dilfuza E, Elsayed FAA, Jaime A. Teixeira da S (2016) Microbially Assisted Phytoremediation of Heavy Metal–Contaminated Soils. In: Ahmad P (ed) Plant metal interaction. 1st. Elsevier Inc.
3. Reddy KG, Madhavi GB, Reddy ASR, Yellareddygari SK, Reddy MS, Reddy MS et al (2011) Current status of biofertilizers development, farmers acceptance, utilization and future perspective in Andhra Pradesh, india. Appl Acoust 63(12):1301–1313
4. Lian B, Wang B, Pan M et al (2008) Microbial release of potassium from K-bearing minerals by thermophilic fungus Aspergillus fumigates. Geochim Cosmochim Ac 1:87–98
5. Lopes-Assad ML, Avansini SH, Rosa MM, de Carvalho JR, Ceccato-Antonini SR (2010) The solubilization of potassium-bearing rock powder by aspergillus niger in small-scale batch fermentations. Can J Microbiol 56(56):598–605
6. Esitken A, Yildiz HE, Ercisli S, Donmez MF, Turan M (2010) Effects of plant growth promoting bacteria (PGPB) on yield, growth and nutrient contents of organically grown strawberry. Sci Hortic 124:62–66
7. Adesemoye AO, Kloepper JW (2009) Plant-microbe interaction in improving fertilizer use efficiency. Appl Microbiol Biotechnol 1:1–12
8. Xue S (2011) Biological compound potash fertilizer and preparation method. China. Patent 201,110,256,623.1
9. Ikeuchi T, Ishida A, Tajifi M, Nagata S (2003) Induction of salt tolerance in bacillus subtilis ifo 3025. J Biosci Bioeng 96(2):184–186
10. Palazzini JM, Ramirez ML, Alberione EJ, Torres AM, Chulze SN (2009) Osmotic stress adaptation, compatible solutes accumulation and biocontrol efficacy of two potential biocontrol agents on fusarium head blight in wheat. Biol Control 51(3):370–376

11. Sheng X (2005) Growth promotion and increased potassium uptake of cotton and rape by a potassium releasing strain of *Bacillus edaphicus*. Soil Biol Biochem 37:1918–1922
12. Tan KH (1978) Effects of humic and fulvic acids on release of fixed potassium. Geoderma 21 (1):67–74
13. Barrios-González J (2012) Solid-state fermentation: physiology of solid medium, its molecular basis and applications. Process Biochem 47:175–185
14. Vaneeckhaute C, Janda J, Vanrolleghem PA, Tack FMG, Meers E (2016) Phosphorus use efficiency of bio-based fertilizers: bioavailability and fractionation. Pedosphere 26(3): 310–325
15. Ciceri D, Allanore A (2015) Microfluidic leaching of soil minerals: release of k^+ from k feldspar. PLoS One 10(10):1–10
16. Kraft S, Obst U, Schwartz T (2011) Immunological detection of uv induced cyclobutane pyrimidine dimers and (6–4) photoproducts in dna from reference bacteria and natural aquatic populations. J Microbiol Methods 84(3):435–441
17. Ministry of Agriculture of China (2015) Compound microbial fertilizer standard NY/T 798-2015. Standard Press
18. Dutta S, Podile AR (2010) Plant growth promoting rhizobacteria (PGPR): the bugs to debug the root zone. Crit Rev Microbiol 36(3):232–244

Part III
Biological Separation and Biological Purification

Biological Degradation of Aflatoxin B1 by *Emericellopsis* sp. 1912 and *Sarocladium* sp. 10A

Hui Wang, Zhongyuan Li, Haiyan Qiu, Shuang Li and Tongcun Zhang

1 Introduction

Aflatoxins are highly toxic secondary metabolites predominantly produced by some fungi of *Aspergillus flavus*, *Aspergillus nomius*, and *Aspergillus parasiticus* [1–3]. Different types of aflatoxins were found, such as B1, B2, G1, G2, M1, M2 [4]. Among them, aflatoxin B1 (AFB1) is extremely toxic, teratogenic, carcinogenic, mutagenic to the health of human and animals [5], especially it plays the key role on high incidence of primary liver cancer [6, 7].

Although different physical and chemical methods have been used to reduce and detoxicate the aflatoxins, these methods have many limitations, and most of these methods had some negative influences, such as part of the product nutrient loss, organoleptic qualities, undesirable health effects, and high cost of equipments [8, 9]. Compared with these methods, biological degradation has been proved to be a feasible and efficient method because of its specificity, efficiency and environmental friendliness [10].

Although some microbes have been found to degrade aflatoxin, most microorganisms are achieved by adsorption, can not really remove aflatoxin. It's urgent to explore the rapidly and effectively biodegradation of AFB1 of microorganisms.

H. Wang · Z. Li · H. Qiu · S. Li · T. Zhang (✉)
College of Biotechnology, Tianjin University of Science and Technology, Tianjin 300457, People's Republic of China
e-mail: lizhongyuan@tust.edu.cn

H. Liu et al. (eds.), *Advances in Applied Biotechnology*, Lecture Notes in Electrical Engineering 444, https://doi.org/10.1007/978-981-10-4801-2_54

The basic structure of AFB1 is consist of a furan ring and an oxanaphthalene O-keto (coumarin), and previous studies have successfully screen AFB1 degradation microorganisms with coumarin as the only carbon source and energy source. Therefore, in this study, we used coumarin instead of AFB1 to isolate AFB1 degradation microorganisms from moldy soil of Tianjin.

2 Materials and Methods

2.1 Sample, Medium and Chemicals

In this study, the moldy soils from Tianjin of China were used to screen the AFB1 degradation microorganisms. AFB1 was purchased from Sigma (St. Louis, MO) and it was dissolved with methanolto a stock solution of 100 μg ml^{-1}, and then the storage solution was diluted with double distilled water to the working concentration of 1 ppm. Nutrient borth (NB) medium was used for strains cultivation and it contained the following ingredients (per liter): 3.0 g yeast extract, 5.0 g peptone, 6.0 g glucose, 10 g NaCl. Coumarin medium (CM) was used for screening the AFB1 degradation strains, and it contained the following ingredients (per liter): 1.0 g coumarin (Aladdin, Shanghai, China), 2.0 g $(NH_4)_2SO_4$, 0.2 g $MgSO_4.7H_2O$, 1.5 g $Na_2HPO_4.12H_2O$, 0.001 g $FeSO_4.7H_2O$, 0.01 g $CaCl_2.2H_2O$ and 1.5 g KH_2PO_4.

2.2 Isolation and Screening the Strain for Degrading AFB1

Coumarin medium (CM) was used to isolate the AFB1 degradation strains from the moldy peanut and moldy corn according to Guan et al. [11] with modification. About 2 g of each sample was grinded and dissolved in 10 ml 0.9% (w/v) NaCl, then these samples were transferred into 50 ml NB medium and incubated at 30 °C (n = 220 rpm) for 4 h. Then the cells was collected by centrifuged at 5000 rpm for 10 min and transferred the cells into 50 ml coumarin medium cultured at 30 °C for 7 days. Then, the strains which could grow in CM were transferred into the fresh CM for the further screening. This process was repeated three times. In order to obtain the purified strains, the liquid culture (0.3 ml) was spread on CM plates and incubated at 30 °C until visible colonies appeared. In order to avoid the effect of agar to the screening results, each single colony was transferred into 50 ml fresh CM and cultivated at 30 °C for further screening.

2.3 Analysis of AFB1 Degradation Ability of the Isolated Microorganisms by High-Performance Liquid Chromatography

The isolated strains were incubated in 100 ml NB medium at 30 °C and 220 rpm for 48 h, and the cells of microbes were removed by centrifugation at 12,000 rpm for 10 min, then 4.5 ml culture supernatant was taken out and incubated with 0.5 ml AFB1 (1000 ppb) solution on a rotary shaker (n = 220 rpm) at 30 °C in dark for 48 h. AFB1 was extracted from reaction mixtures by immunoaffinity chromatography (Huaan Maghech B-Tech Co, China), then evaporated by dryer at 60 °C and derived by incubating with N-hexane (200 μl) and trifluoroacetic acid (50 ul) for 5 min at 40 °C. The derivatized was dissolved in 1.95 ml of 10% (v/v) acetonitrile by mixing vigorously for 30 s and standing for 10 min, and then filtered with filter membrane (Millex-GV, Durapore, 0.22 μm), analyzed using high-performance liquid chromatography.

High-performance liquid chromatography was performed using a C18 column (250 mm × 4.6 mm, 5 um, Agilent) equipped with a fluorescence detector (excitation wavelength: 350 nm; detection wavelength: 450 nm). The injection volume of each sample was 20 μl and the mobile phase was water: methanol: acetonitrile (70:17:17, v/v) at a flow rate of 0.6 ml/min. The reaction with NB medium instead of microbial supernatant was used as the control. According to Guan et al. [11], the percentage of AFB1 degradation was calculated using the following formula: (1-AFB1 peak area in treatment/AFB1 peak area in control) × 100%.

2.4 Identification of the AFB1 Degradation Strains 1912 and 10A

The isolate strains 1912 and 10A were cultivated in NB medium at 30 °C for 48 h, the cells were collected by centrifugation at 12,000 rpm for 10 min, and the genomic DNA was extracted by CTAB-SDS method [12]. The genes encoding the ribosomal rRNAs in the eukaryotic genome include 28S rDNA, 5S rDNA, 18S rDNA, and 5.8S rDNA. The internal transcribed spacer Its1 and Its4 were used for strains classification and identification because of their specificity of interspecific and intraspecific conservative [13, 14]. In our study, the ITS rDNA sequences of strain 1912 and 10A were amplified by PCR using primers Its1 and Its4 (5′-TCCGTAGGTGAACCTGCGC-3′; 5′-TCCTCCGCTTATTGATATGC-3′), and the genomic DNA of strain1912 and strain 10A was used as templates for amplification, respectively. Then, the PCR products were analyzed by agarose gel electrophoresis and ligated into the vector pMD-19T for sequencing by Genewiz (Beijing, China).

2.5 Phylogenetic Analysis

The ITS rDNA sequences of strains 1912 and 10A were analyzed by BLAST search in the GenBank database (http://blast.ncbi.nlm.nih.gov/Blast.cgi), and the high the homology nucleic acid sequences of each strain were downloaded as the reference sequences. A phylogenetic tree was constructed using Molecular Evolutionary Genetics Analysis, version 7, (MEGA 7), software (www.megasoftware.net) with the neighborjoining method. The tree was evaluated by bootstrap analysis based on 1000 resamplings by MEGA 7.

3 Results and Discussion

3.1 Isolation of AFB1 Degradation Strains

The single colonies picked from the soild coumarin medium (CM) were cultured at 30 °C for 7 days, strains 1912 and 10A could grow on liquid culture medium (CM) as shown in Fig. 1, and these two strains were used as the candidates for the further study. The mycelial morphology was shown in Fig. 1.

3.2 High-Performance Liquid Chromatography Analysis of AFB1 Degradation Ability by the Isolation Strains 1912 and 10A

The cell-free supernantant of each strains was collected by centrifugation at 4 °C, then 4.5 ml culture supernatant was taken out and incubated with 0.5 ml AFB1 (1000 ppb) solution on a rotary shaker (n = 220 rpm) at 30 °C in dark for 48 h together. AFB1 were extracted from reaction mixtures and derived by incubating with N-hexane and trifluoroacetic acid for high-performance liquid chromatography analysis. The ratio of degradation of AFB1 were 60.49 and 45.51% by strains 1912 and 10A, respectively (Fig. 2).

3.3 Identification and Phylogenetic Analysis of Strains 1912 and 10A

The ITS rDNA sequences of strains 1912 and 10A were successfully obtain by PCR, and the PCR products were detected by electrophoresis (Fig. 3). Based on the

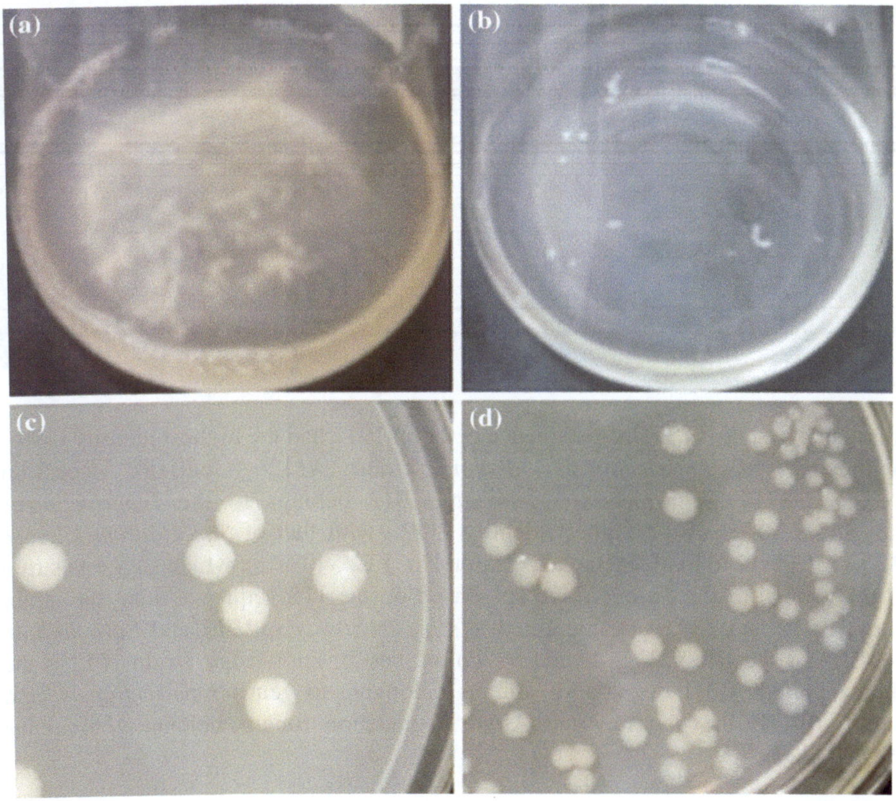

Fig. 1 Morphological characteristics of strains 1912 and 10A. **a** strain 1912 grew in liquid CM. **b** strain 10A grew in liquid CM. **c** strain 1912 grew on the solid NB plate. **d** strain 10A grew on the solid NB plate

Fig. 2 The degradation ratio of AFB1. The AFB1 degradation ratios were 60.49 and 45.51% by strains1912 and 10A, respectively

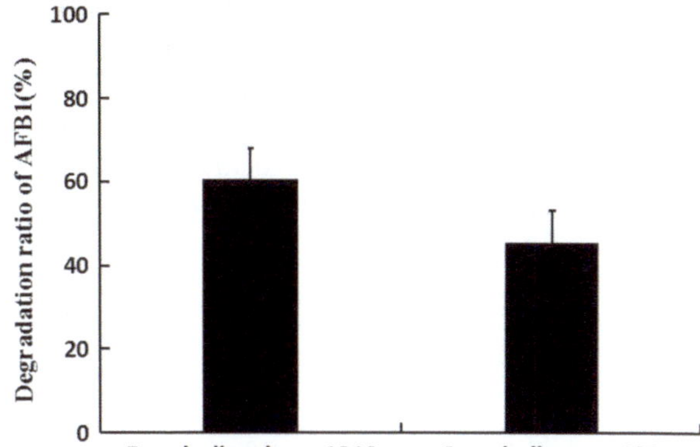

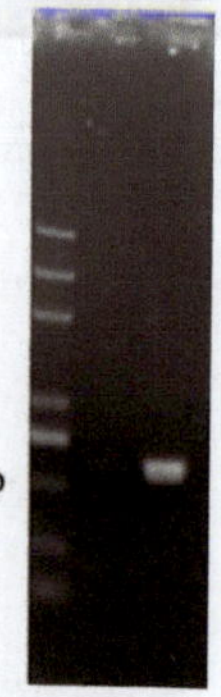

Fig. 3 ITS rDNA products of strains 1912 and 10A. *M* Marker. *1* PCR product of ITS rDNA sequence of strain 10A. *2* PCR product of ITS rDNA sequence of strain 1912

blast analysis, the ITS rDNA sequence of strain 1912 had the highest identity (99%) with that of *Emericellopsis alkalina* strain A112 (GenBank accession No. KC987149.1), which suggests strain1912 belongs to the *Emericellopsis* sp. Strain 10A had the highest identity (99%) with that of *Sarocladium kiliense* UOA/HCPF 12768A (GenBank accession No. KC254087.1), which suggests strain 10A belongs to the *Sarocladium* sp. Based on the ITS sequence analysis, strain strain 1912 and 10A were identified as *Emericellopsis* sp.1912 and *Sarocladium* sp. 10A, respectively (Fig. 4a and b). To our best of knowledge, strain 1912 is the first reported AFB1 degradation fungus belonging to genus *Emericellopsis*, and strain 10A is the first reported AFB1 degradation fungus belonging to genus *Sarocladium*.

3.4 Conclusion

In this research, we had screened the strains 1912 and 10A. Based on BLAST analysis, strain 1912 belongs to the *Emericellopsi* sp. and strain 10A belongs to the *Sarocladium* sp. The AFB1 degradation ratios were 60.49 and 45.51% in *Emericellopsi* sp. 1912 and *Sarocladium* sp. 10A, respectively. Moreover, *Emericellopsi* sp. 1912 was the first to be found in *Emericellopsi* sp. that has the ability to degrade AFB1 and *Sarocladium* sp. 10A was the first strain that has the ability to degrade AFB1 of *Sarocladium* sp.

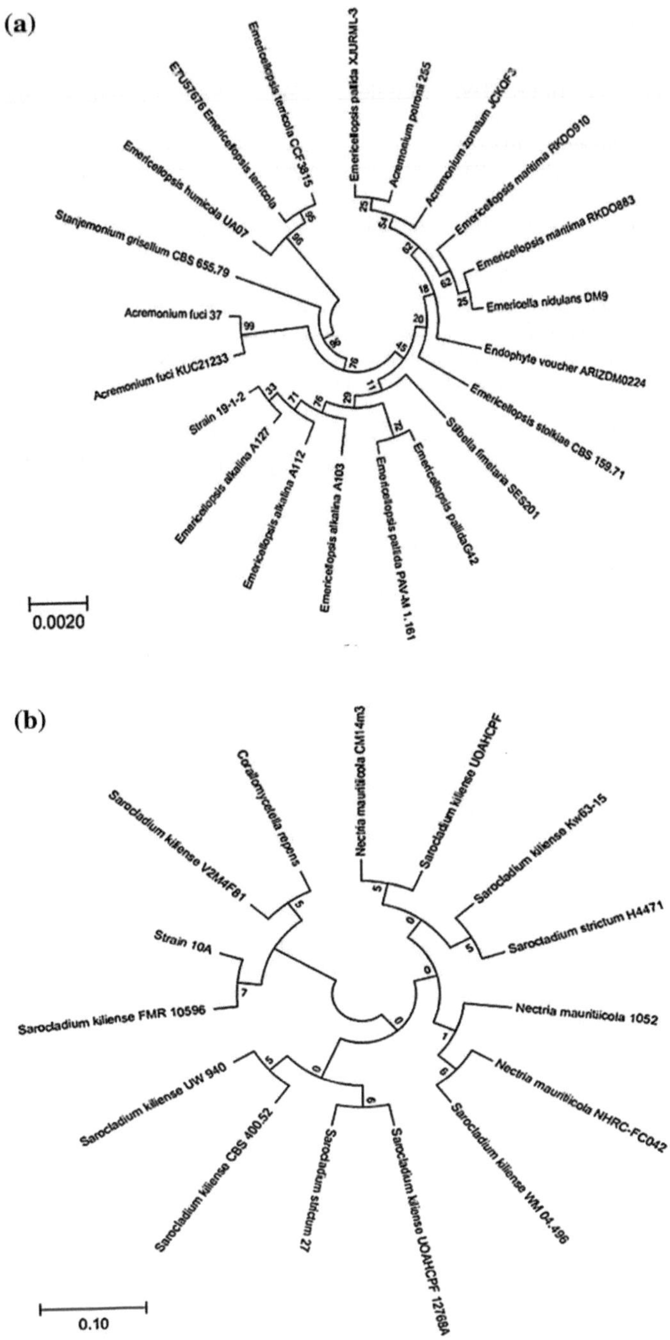

Fig. 4 Phylogenetic analysis of the isolation strains 1912 (**a**) and 10A (**b**) by MEGA 7 software

References

1. Kurtzman CP, Horn BW, Hesseltine CW (1987) Aspergillus nomius, a new aflatoxin-producing species related to Aspergillus flavus and Aspergillus tamarii. Antonie Van Leeuwenhoek 53:147–158
2. Hong HA, Le HD, Cutting SM (2005) The use of bacterial spore formers as probiotics. FEMS Microbiol Rev 29:813–835
3. Cotty PJ, Bhatnagar D (1994) Variability among atoxigenic Aspergillus flavus strains in ability to prevent aflatoxin contamination and production of aflatoxin biosynthetic pathway enzymes. Appl Environ Microb 60:2248–2251
4. Eaton DL, Gallagher EP (1994) Mechanisms of aflatoxin carcinogenesis. Annu Rev Pharmacol 34:135–172
5. EI-Nezami H, Kankaanpaa P, Salminen S, Ahokas J (1998) Ability of dairy strains of lactic acid bacteria to bind a common food carcinogen, Afatoxin B1. Food Chem Toxicol 36:321–326
6. Alberts JF, Engelbrecht Y, Steyn PS, Holzapfel WH, van Zyl WH (2006) Biological degradation of aflatoxin B1 by Rhodococcus erythropolis cultures. Int J Food Microbiol 109:121–126
7. Pitt JI (2000) Toxigenic fungi and mycotoxins. Brit Med Bull 56:184–192
8. Teniola OD, Addo PA, Brost IM, Fa ¨rber P, Jany KD, Alberts JF, Van Zyl WH, Steyn PS, Holzapfel WH (2005) Degradation of aflatoxin B1 by cell-free extracts of Rhodococcus erythropolis and Mycobacterium fluoranthenivorans sp. nov. DSM44556T. Int J Food Microbiol 105:111–117
9. Basappa SC, Shantha T (1996) Methods for detoxification of aflatoxins in foods and feeds—a critical appraisal. J Food Sci Technol 33:95–107
10. Jard G, Liboz T, Mathieu F, Guyonvarch A, Lebrihi A (2011) Review of mycotoxin reduction in food and feed: from prevention in the field to detoxification by adsorption or transformation. Food Addit Contam Part A Chem Anal Control Expo Risk Assess 28:1590–1609
11. Guan S, Ji C, Zhou T, Li J, Ma Q, Niu T (2008) Aflatoxin B1 degradation by Stenotrophomonas maltophilia and other microbes selected using coumarin medium. Int J Mol Sci 9:1489–1503
12. Brady SF (2007) Construction of soil environmental DNA cosmid libraries and screening for clones that produce biologically active small molecules. Nat Protoc 2:1297–1305
13. Pryce TM, Palladino S, Price DM et al (2006) Rapid identification of fungal pathogens in Bac T/ALERT, BACTEC, and BBL MGIT media using polymerasechain reaction and DNA sequencing of the internal transcribed spacer regions. Diagn Microbiol Infect Dis 54(4):289–297
14. Campa D, Tavanti A, Gemignani F et al (2008) DNA microarray based on arrayed primer extension technique for identification of pathogenic fungi responsible for invasive and superficial mycoses. J Clin Microbiol 46(3):909–915

Study on Ultrasonic-Assisted Extraction of Star Anise Oleoresin from the Fruits of *Illicium verum* and Preliminary Investigation of Its Antimicrobial Activity

Ping Li, Zhan Shu, Lan Zhang, Tao Li and Lin Tian

1 Introduction

In recent years, public concerns regarding safety of food have brought scientists to focus on the application of many spices and herbs extracts to preserve the quality of food [1, 2]. Application of plant-origin essential oils as food preservatives has received more and more attention since many works have demonstrated their antimicrobial and antioxidant activities [3–5]. Vatavali et al. [6] reported that the combination of chitosan and oregano essential oil was very effective on extending the shelf life of red porgy stored in ice. Kapoor et al. [7] demonstrated that black pepper essential oil had better preservative effect for orange juice. Cristina et al. [8] revealed that cinnamon bark oil and citrus extracts, even at low concentrations, exhibited high anti-yeast effects and could increase shelf life of ready-to-eat table grape compared to the control. In addition, cinnamon oil combined sand-based storage at 12 °C could prolong storage time for baby ginger [9].

Star anise (*Illicium verum*) is an aromatic evergreen tree and it is widely cultivated in China, especially in Guangxi and Guangdong provinces. Its fruit is a traditional medicine as well as a spice widely used in foods and beverages [10]. Many kinds of active compounds have been identified in *Illicium verum* such as essential oils, oleoresin, and shikimic acid [11]. Among these active compounds, star anise oleoresin, the major flavoring ingredient in star anise, is of particular importance. Star anise oleoresin contains volatile essential oil, less volatile components, pigments, fatty acids and resins. The taste and aroma of star anise oleoresin is more balanced and complete than the essential oil in terms of food seasoning and it has been allowed using as an additive in food processing industry in many

P. Li (✉) · Z. Shu · L. Zhang · T. Li · L. Tian
College of Basic Science, Tianjin Agricultural University,
300384 Tianjin, People's Republic of China
e-mail: liping790520@126.com

H. Liu et al. (eds.), *Advances in Applied Biotechnology*, Lecture Notes
in Electrical Engineering 444, https://doi.org/10.1007/978-981-10-4801-2_55

countries [12]. Traditional methods for obtaining extracts from *Illicium verum* fruits are hydro-distillation, steam distillation and solvent extraction, which have some disadvantages. For hydro-distillation and steam distillation, some thermal-sensitive constituents may decompose during the prolonged extraction time [13]. Solvent extraction, such as heat reflux and soxhlet extraction takes a long time and high energy cost. These shortcomings have led to consideration of the use of innovative and environment-friendly techniques in active compounds extraction, such as ultrasound-assisted extraction [14], supercritical fluid extraction [15, 16] and microwave extraction [17]. Recently, ultrasonic-assisted method has been successfully applied in the extraction of many active extracts and this method can offer many important advantages over traditional alternatives, viz. shorter extraction time, simple sampling, better yields, low energy cost, cleaner features and provide a more valuable extracts [11, 18]. Many reports were published for the identification and application of essential oil of *Illicium verum* and its antimicrobial activity [3, 5, 10]. However, rarely studies have been conducted on the extraction of star anise oleoresin, especially ultrasonic-assisted extraction, and its antimicrobial activity. The aim of this study was to optimize the conditions for ultrasound-assisted extraction of star anise oleoresin from *Illicium verum* fruits using orthogonal experiment design, as well as investigate its antimicrobial activity in order to provide a reference for the extraction and application of star anise oleoresin.

2 Materials and Methods

2.1 *Materials and Stains*

Illicium verum fruits were purchased from a vanguard supermarket (Nankai District of Tianjin, China). The fruits were dried at 50 °C for 24 h and pulverized by a grinder and then passed through different mesh sieves. The bacterial stains used in this study were *Escherichia coli* and *Bacillus subtilis*, which were provided by College of Agronomy & Resources and Environment, Tianjin Agricultural University. After activated, the stains were inoculated into liquid standard nutrient broth medium, and cultured at 37 °C with regularity shaking at 180 rpm for 18 h. Series dilution were carried out to obtain the final concentration of colonies was 10^6 CFU/mL by using sterile distilled water. All the other chemicals used in this study were analytical reagent grade.

2.2 *Ultrasonic-Assisted Extraction Procedure*

Star anise oleoresin was prepared by ultrasonic-assisted extraction using ultrasonic cleaning device (P180H model, Elma, Germany). Briefly, star anise powders were put into a beaker with a specific volume of solvent and subjected to ultrasonic-assisted extraction under settled conditions. After extraction, the solution was filtered, and the filtrate was concentrated by a vacuum rotary evaporator, then the oleoresin was obtained and the yield was calculated using the following formula:

$$\text{Yield}(\%) = \frac{\text{oleoresin weight (g)}}{\text{star anise powder weight(g)}} \times 100 \quad (1)$$

2.3 *Experimental Design*

The influences of many parameters on the yield of the oleoresin were investigated by single factor experiment. On the basis of the results of single factor experiment, optimization experiments were carried out according to the L_{16} (4^5) orthogonal experiment design (Table 1).

2.4 *Purification of Star Anise Oleoresin*

Since star anise oleoresin obtained by ultrasonic-assisted extraction was a dark green oily liquid and contained wax, colloid and other impurities, purification of the obtained oleoresin was also carried out by column separation (silica gel, 300–400 mesh, petroleum ether and ethyl acetate (v/v, 1:15) as the eluent). The received solution after elution was concentrated by a vacuum rotary evaporator and the purified star anise oleoresin obtained was a pale yellow oily liquid.

2.5 *GC-MS Analysis*

For GC-MS analysis, the samples were injected into a HP-5MS capillary column (30 m × 0.25 mm i.d with 0.25 μm film thickness, Agilent United States,

Table 1 Factors and levels of orthogonal experiment design

Levels	Factors			
	Ratio of material to solvent (g/mL)	Ultrasonic time (min)	Powder granularity (mesh)	Ultrasonic power (W)
1	6:50	20	20–40	231
2	6:60	30	40–60	264
3	6:70	40	60–80	297
4	6:80	50	80–100	330

7890A/5975C model). Chromatographic conditions were as follows: initial oven temperature was 60 °C and hold 1 min, then increased to 160 °C at a rate of 5 °C/min, maintaining 4 min, and raised to 210 °C at a rate of 5 °C/min, holding 5 min. Injector temperature was set at 220 °C and injection volume was 0.5 μL with split ratio of 20:1; high-purity helium was used as a carrier gas at a flow rate of 1.0 mL/min. MS conditions were as follows: ionization voltage was 70 eV. Temperatures of transfer line and ion source were set at 280 and 230 °C, respectively. In addition, four pole was 150 °C and mass range was 30.0–550.0 mass units. The compounds were identified by comparison of their mass spectra with the NIST2008 mass spectral database and literatures.

2.6 *Antimicrobial Activity Assay*

Initial concentration of the oleoresin was 1.00 mg/mL and it was diluted twofold with N, N-dimethylformamide (DMF) as a solvent to obtain a series of different concentrations of samples.

Antimicrobial activity was determined by disc diffusion method described by Wang et al. [19]. DMF was used as a negative control. Three replicates were performed for each concentration. After incubation, diameters of the inhibition zone were measured and inhibition (%) was calculated according to the following formula:

$$\text{Inhibition}(\%) = \frac{M_1 - M_2}{M_1} \times 100 \tag{2}$$

where M_1 is the diameter of inhibition zone treated with oleoresin; M_2 is the diameter of the control.

3 Results and Discussion

3.1 *Single Factor Experiment for the Extraction of Star Anise Oleoresin*

3.1.1 Extraction Solvent Selection

Effects of different extraction solvents on the yield of star anise oleoresin were shown in Fig. 1. It could be seen that ethyl acetate was the best extraction solvent with the highest extraction yield (19.93%), while the other conditions were kept the same and the obtained oleoresin was a dark green oily liquid with significant aromatic odor.

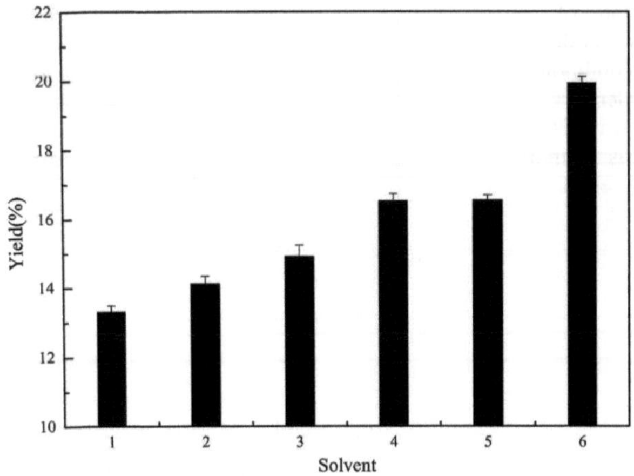

Fig. 1 Effect of different solvents on the yield; *1* petroleum ether (b.p. 30–60 °C); *2* petroleum ether (b.p. 60–90 °C); *3* acetone; *4* anhydrous ethanol; *5* cyclohexane; *6* ethyl acetate; extraction conditions: 40–60 mesh powder; ratio of material to solvent was 6:60 (g/mL); ultrasonic power was 330 W; ultrasonic time was 30 min; room temperature

3.1.2 Effect of Powder Granularity on the Yield of Star Anise Oleoresin

Powder granularity is one of the most important factors that can affect the oleoresin yield. It could be seen from Fig. 2, as the powder granularity increasing, the yield increased rapidly, and the maximum yield (23.31%) could be obtained at 60–80 mesh, and then declined. One possible explanation was that fine materials was benefit for the extraction due to the increased surface area, yet too small powders might cause adhesion between materials, then impeded the dissolution of the oleoresin, and our result was in agreement with that of the previous observation [20].

Fig. 2 Effect of powder granularity on the yield; extraction conditions: ratio of material to solvent was 6:60 (g/mL); ultrasonic power was 330 W; ultrasonic time was 30 min; room temperature; ethyl acetate as a solvent

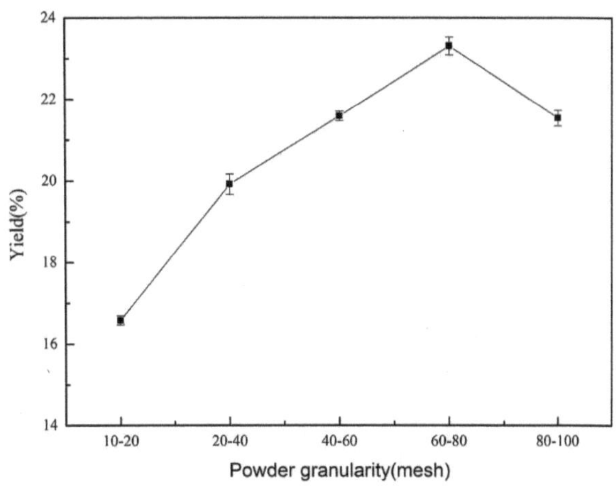

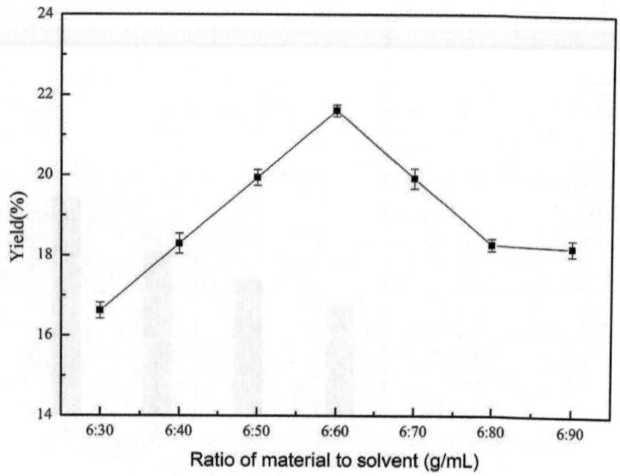

Fig. 3 Effect of ratio of material to solvent on the yield; extraction conditions: 60–80 mesh powder; ultrasonic power was 330 W; ultrasonic time was 30 min; room temperature; ethyl acetate as a solvent

3.1.3 Effect of Ratio of Material to Solvent on the Yield of Star Anise Oleoresin

Ratio of material to solvent is another crucial factor that can affect the yield of oleoresin. As shown in Fig. 3, the result indicated that the increase of ratio elevated oleoresin yield during 6:30–6:60 (g/mL), and the yield reached maximum (21.61%) at 6:60 (g/mL), and then declined. The reason might be that with the increasing of ratio, ingredients could be fully dissolved in solvent, however, excessive solvent might absorb a large amount of ultrasonic energy, which resulting in the decrease of absorption of ultrasound intensity for raw materials in extracted system and yield reduction.

3.1.4 Effect of Ultrasonic Temperature on the Yield of Star Anise Oleoresin

Effects of different ultrasonic temperatures on the oleoresin yield were shown in Fig. 4. It could be seen that the highest yield (21.60%) was achieved at 30 °C and then reduced. It was mainly attributed to the serious evaporation and degradation of some thermal-sensitive constituents under high temperature. In addition, we also observed that the influence of temperature on the extraction yield was relatively small, so temperature was not considered in the next orthogonal experiment design.

3.1.5 Effect of Ultrasonic Power on the Yield of Star Anise Oleoresin

Besides, ultrasonic power is also a primary factor that can significantly influence the oleoresin yield. Figure 5 showed that with the increasing of ultrasonic power, the yield increased and reached maximum (19.96%) at 264 W, and then turned to

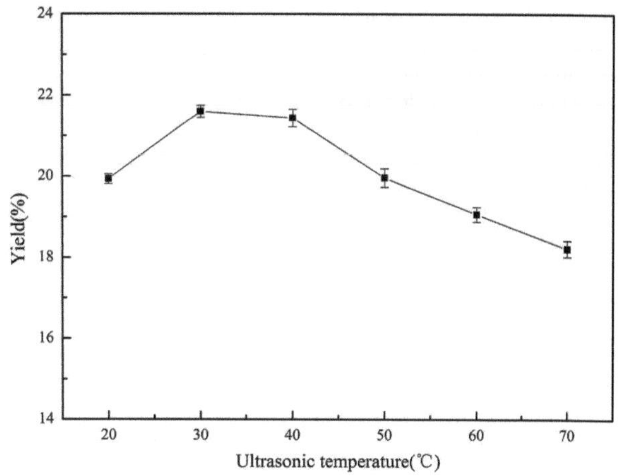

Fig. 4 Effect of ultrasonic temperature on the yield; extraction conditions: 60–80 mesh powder; ratio of material to solvent was 6:60 (g/mL); ultrasonic power was 330 W; ultrasonic time was 30 min; ethyl acetate as a solvent

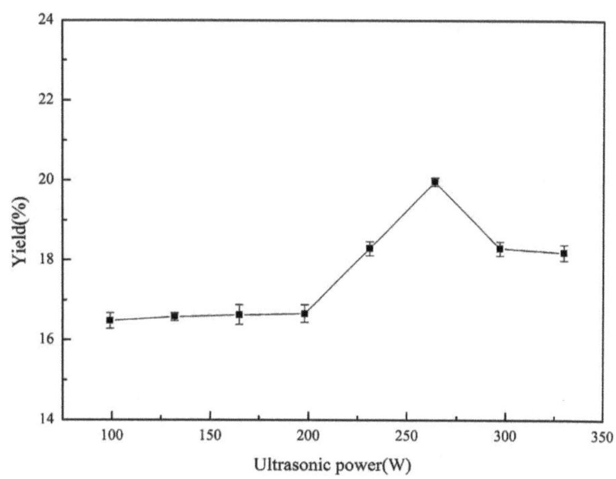

Fig. 5 Effect of ultrasonic power on the yield; extraction conditions: 60–80 mesh powder; ratio of material to solvent was 6:60 (g/mL); ultrasonic time was 30 min; ultrasonic temperature was 30 °C; ethyl acetate as a solvent

decline, which indicated that any excess power was useless. One reason was that with the power increasing, temperature in the extraction system enhanced, which resulting in some losses of volatile components in oleoresin and yield reduction.

3.1.6 Effect of Ultrasonic Time on the Yield of Star Anise Oleoresin

As shown in Fig. 6, oleoresin yield increased with ultrasonic time from 10 to 30 min, reaching maximum (21.63%) at 30 min, and then decreased. One possible explanation was that ultrasonic extraction was a diffusion and mass transfer process, and before the process reaching equilibrium, extraction yield could increase. However, long extraction time could result in the losses of some easily volatile substances to the atmosphere, which caused the decline of yield.

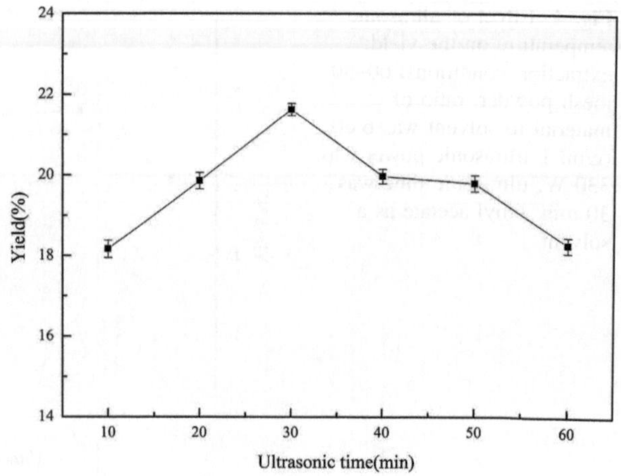

Fig. 6 Effect of ultrasonic time on the yield; extraction conditions: 60–80 mesh powder; ratio of material to solvent was 6:60 (g/mL); ultrasonic power was 264 W; ultrasonic temperature was 30 °C; ethyl acetate as a solvent

3.2 Orthogonal Experiments for the Extraction of Star Anise Oleoresin

L_{16} (4^5) orthogonal experimental design was employed to optimize ultrasonic-assisted extractions of star anise oleoresin from *Illicium verum* fruits (Table 2). Results showed that the optimum extraction condition was $A_2B_2C_3D_2$, that was, ratio of material to solvent was 6:60 (g/mL), ultrasonic time was 30 min, 60–80 mesh powder granularity, and ultrasonic power was 264 W. Range analysis showed that powder granularity had the largest effect on the oleoresin yield, followed by ratio of material to solvent, ultrasonic time and ultrasonic power.

Three verification experiments were carried out at the optimum extraction conditions and the yields were 23.90, 23.90 and 23.88%, respectively. The mean and relative mean deviation were 23.89 and 0.04%, respectively, which demonstrated that the optimum extraction conditions had a good stability and ultrasonic-assisted extraction of oleoresin from *Illicium verum* fruits was feasible.

3.3 Antimicrobial Activity of Star Anise Oleoresin

The results of antimicrobial activities of unpurified and purified star anise oleoresin were shown in Fig. 7. It could be seen that all the obtained oleoresins, either purified or not, exhibited good inhibitory effects on the two tested stains and antimicrobial activities positively correlated with the concentrations. The antimicrobial activity of the oleoresin could be attributed to its high content of *trans*-anethole documented by Huang et al. [21]. It could be seen that the inhibitory effect of unpurified oleoresin was higher than purified oleoresin at all the tested concentrations, which could be well interpreted by the GC-MS results that the content

Table 2 Results of orthogonal experiment design for ultrasonic-assisted extraction of star anise oleoresin

Number	Factors				Yield (%)
	A	B	C	D	
1	1	1	1	1	18.24
2	1	2	2	2	21.59
3	1	3	3	3	21.56
4	1	4	4	4	14.98
5	2	1	2	3	19.93
6	2	2	1	4	18.27
7	2	3	4	1	19.95
8	2	4	3	2	23.28
9	3	1	3	4	21.66
10	3	2	4	3	19.91
11	3	3	1	2	14.95
12	3	4	2	1	19.99
13	4	1	4	2	19.96
14	4	2	3	1	21.58
15	4	3	2	4	19.91
16	4	4	1	3	18.28
k_1	19.09	19.95	17.44	19.93	
k_2	20.36	20.34	20.35	19.94	
k_3	19.13	19.09	22.02	19.92	
k_4	19.93	19.13	18.70	18.70	
R	1.27	1.25	4.58	1.24	

k_1, k_2, k_3, k_4 are the mean values of the sum of deviation under the four same level, respectively; R is the range value

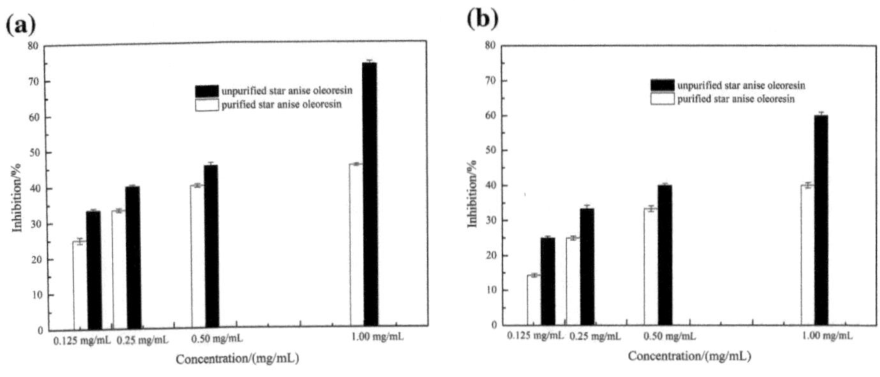

Fig. 7 Antimicrobial activities of star anise oleoresin and purified star anise oleoresin against *Escherichia coli* (**a**) and *Bacillus subtilis* (**b**)

of *trans*-anethole in unpurified oleoresin (90.91%) was higher than that of in purified oleoresin (83.72%)(other data not shown). Therefore, it was not recommended for star anise oleoresin with purification treatment in terms of antimicrobial activity. Moreover, star anise oleoresins were obviously more effective against *Escherichia coli* than *Bacillus subtilis*. Besides, Oleoresin is a very complex mixture, it is not sure that *trans*-anethole is the only component reflecting antibacterial activity in star anise oleoresin. Many reports have also proved that phenols and aldehydes are also constituents in the oleoresin with high antibacterial activities [22, 23]. In addition, many researchers have been extensively studied the inhibitory mechanism which indicating that the hydrophobicity of essential oils and oleoresins may enable them to partition in the lipids of bacterial cell membrane and render them more permeable, leading to leakage of ions and cell contents, and extensive loss of critical molecules and ions which causing the death of cells [24, 25]. More in-depth investigations of the mechanism of inhibition are further required in future research works.

4 Conclusions

Ultrasonic-assisted technology could be successfully applied in the extraction of oleoresin from the fruits of *Illicium verum* in our study, which provided shorter extraction time and saved energy compared to traditional methods. The optimum extraction conditions were as follows: ratio of material to solvent was 6:60 (g/mL, ethyl acetate as a solvent), ultrasonic time was 30 min, 60–80 mesh powder granularity, ultrasonic power was 264 W and ultrasonic temperature was 30 °C. Under these conditions, the yield could reach 23.89% and the content of *trans*-anethole in the oleoresin was 90.91%. Star anise oleoresin exhibited a good inhibitory effect against *Escherichia coli* and *Bacillus subtilis* and antimicrobial activity positively correlated with the concentrations. Besides, the antimicrobial activity of unpurified oleoresin was stronger than purified oleoresin and it was not recommended for star anise oleoresin with purification treatment in terms of antimicrobial activity. In conclusion, star anise oleoresin extracted by ultrasonic-assisted method was feasible and the oleoresin can be developed as a natural source of antimicrobial agent in many areas.

Acknowledgements This research was financially supported by Tianjin University Students Innovation and Entrepreneurship Training Program (No. 201610061067).

References

1. Antunes MDC, Cavaco AM (2010) The use of essential oils for postharvest decay control. A review. Flavour Frag J 25:351–366
2. Gracia CM, Bermudez CAG, Valcarcel AMC et al (2015) Use of herbs and spices for food preservation: advantages and limitations. Curr Opin Food Sci 6:38–43

3. Vijayakumar A, Duraipandiyan V, Jeyaraj B et al (2012) Phytochemical analysis and in vitro antimicrobial activity of *Illicium griffithii* Hook. f. & Thoms extracts. Asian Pac J Trop Dis 2:190–199
4. Singh G, Maurya S, delampasona MP et al (2006) Chemical constituents, antimicrobial investigations and antioxidative potential of volatile oil and acetone extract of star anise fruits. J Sci Food Agr 86:111–121
5. Yang CH, Chang FR, Chang HW et al (2012) Investigation of the antioxidant activity of *Illicium verum* extracts. J Med Plants Res 6:314–324
6. Vatavali K, Karakosta L, Nathanailides C et al (2013) Combined effect of chitosan and oregano essential oil dip on the microbiological, chemical, and sensory attributes of red porgy (*Pagrus pagrus*) stored in ice. Food Bioprocess Technol 6:3510–3521
7. Kapoor IPS, Singh B, Singh S et al (2014) Essential oil and oleoresins of black pepper as natural food preservatives for orange Juice. J Food Process Preserv 38:146–152
8. Cristina C, Annalisa L, Amalia C et al (2013) In vitro and in vivo application of active compounds with anti-yeast activity to improve the shelf life of ready-to-eat table grape. World J Microbiol Biotechnol 29:1075–1084
9. Liu J, Sui GL, He YZ et al (2014) Prolonging storage time of baby ginger by using a sand-based storage medium and essential oil treatment. J Food Sci 99:M593–M599
10. Aly SE, Sabry BA, Shaheen MS et al (2016) Assessment of antimycotoxigenic and antioxidant activity of star anise (*Illicium verum*) in vitro. J Sau Soc Agr Sci 15:20–27
11. Cai M, Luo YL, Chen J et al (2014) Optimization and comparison of ultrasound-assisted extraction and microwave-assisted extraction of shikimic acid from Chinese star anise. Sep Purif Technol 133:375–379
12. Wang Q, Lei HW, Jiang L et al (2014) Optimization and evaluation of microencapsulation of star anise oleoresin. J Food Process Preserv 38:2129–2136
13. Ferhat MA, Meklati BY, Chemat F (2007) Comparison of different isolation methods of essential oil from *Citrus* fruits: cold pressing, hydrodistillation and microwave 'dry' distillation. Flavour Frag J 22:494–504
14. Lee AL, Kim HS, Choi G (2014) Optimization of ultrasonic-assisted extraction of active compounds from the fruit of star anise by using response surface methodology. Food Anal Methods 7:1661–1670
15. Rodrigues VM, Rosa PTV, Marques MOM et al (2003) Supercritical extraction of essential oil from aniseed (*Pimpinella anisum* L) using CO_2: solubility, kinetics, and composition data. J Agric Food Chem 51:1518–1523
16. Guan WQ, Li SF, Yan RX et al (2007) Comparison of essential oils of clove buds extracted with supercritical carbon dioxide and other three traditional extraction methods. Food Chem 101:1558–1564
17. Cai M, Guo XY, Liang HHJ et al (2013) Microwave-assisted extraction and antioxidant activity of star anise oil from *Illicium verum Hook.f.* Int J Food Sci Technol 48:2324–2330
18. Anal AK, Jaisanti S, Noomhorm A (2014) Enhanced yield of phenolic extracts from banana peels (*Musa acuminata Colla AAA*) and cinnamon barks (*Cinnamomum varum*) and their antioxidative potentials in fish oil. J Food Sci Technol 51:2632–2639
19. Wang Q, Ou ZB, Lei HWJ et al (2012) Antimicrobial activities of a new formula of spice water extracts against foodborne bacteria. J Food Process Preserv 36:374–381
20. Kuang X, Bin L, Rui K et al (2011) Granularity and antibacterial activities of ultra-fine cinnamon and clove powders. J Food Saf 31:291–296
21. Huang YF, Zhao JL, Zhou L et al (2010) Antifungal activity of the essential oil *Illicium verum* fruit and its main component *trans*-anethole. Molecules 15:7558–7569
22. Devi KP, Sakthivel R, Nisha SA (2013) Eugenol alters the integrity of cell membrane and acts against the nosocomial pathogen proteus mirabilis. Arch Pharm Res 36:282–292

23. Arfa AB, Belloy LP, Chalier P et al (2007) Antimicrobial paper based on a soy protein isolate or modified starch coating including carvacrol and cinnamaldehyde. J Agric Food Chem 55:2155–2162
24. Nanasombat S, Wimuttigosol P (2011) Antimicrobial and antioxidant activity of spice essential oils. Food Sci Biotechnol 20:45–53
25. Becerril R, Gomez LR, Goni P et al (2007) Combination of analytical and microbiological techniques to study the antimicrobial activity of a new active food *E. coli* and *S. aureus*. Anal Bioanal Chem 388:1003–1011

Preparation of Dialdehyde Cellulose and Its Antibacterial Activity

Huanhuan Ge, Liming Zhang, Meng Xu, Jie Cao and Caicai Kang

1 Introduction

Microcrystalline cellulose (MCC) is a kind of purificatory partially depolymerized cellulose. MCCs are white, odorless, tasteless, porous particles composed of crystalline powder. They have different characteristics and application scope because of the different particle size. MCCs consists of aligned thread-like bundles of poly-β-(1, 4)-D-glucose molecules in extended-chain conformation. They have been widely used in pharmaceutical [1], cosmetic [2] and food industries [3].

Sodium periodate can selectively oxidize the hydroxyl groups in cellulose to 2, 3-dialdehyde functions. Many studies have been published regarding the periodate oxidation of cellulose [4]. Sodium periodate oxide of cellulose nanocrystal was used as a paper wet strength additive [5]. The periodate oxidation of cellulose, proceeds under mild aqueous conditions and is characterized by specific cleavage of the C_2-C_3 bond of the glucopyranoside ring that produces two aldehyde groups per unit [6]. Dialdehyde cellulose (DAC) is a cellulose derivate produced by a regioselective oxidation of cellulose using periodate as an oxidation agent. It is biodegradable and biocompatible and has a tremendous potential to be used in many applications [7, 8]. For example, the micron porous DAC membrane has been prepared by phase separation and was shown to be able to support cell adhesion and

H. Ge · L. Zhang (✉) · M. Xu · J. Cao · C. Kang
Key Laboratory of Industrial Fermentation Microbiology Ministry of Education, Tianjin University of Science and Technology, 300457 Tianjin, People's Republic of China
e-mail: zhlm@tust.edu.cn

H. Ge · L. Zhang · M. Xu · J. Cao · C. Kang
College of Bioengineering, Tianjin University of Science and Technology, 300457 Tianjin, People's Republic of China

H. Liu et al. (eds.), *Advances in Applied Biotechnology*, Lecture Notes in Electrical Engineering 444, https://doi.org/10.1007/978-981-10-4801-2_56

proliferation and hence exhibited its potential to be used as tissue engineering scaffold [9]. Recently, studies have found that direct immobilization of antibodies on dialdehyde cellulose film can be used as an electrochemical immunosensor [10].

DAC was a polymeric dialdehydye with the dialdehyde functionality that was similar to dialdehyde starch. At present, the antibacterial activity on dialdehyde starch has been studied [11], so we have examined the possibility of DAC with varied aldehyde contents becoming a polymeric biocide. In our study, DAC samples with varied aldehyde contents were prepared, and the physicochemical characterization of DAC and MCC were performed using Fourier transform infrared spectroscopy and thermogravimetric analysis.

2 Materials and Methods

2.1 Materials

Microcrystalline cellulose powder (MCC, particle size 90 μm, degree of polymerization: 500) was purchased from Ming Xin chemical products co., Ltd., Zhengzhou, Henan, China. Sodium periodate, absolute ethyl alcohol, peptone, agar powder and yeast extract powder were supplied by Tai airui Bio-Technology co., Ltd., Tianjin, China. Sulfuric acid, hydroxylamine hydrochloride, sodium chloride, sodium hydroxide, potassium bromide and ethylene glycol were supplied by Sinopharm Group Co., Ltd., Beijing, China. All aqueous solutions were prepared with doubly distilled water. The bacteria strains, *Staphylococcus aureus, Bacillus subtilis, Escherichia coli* and *Salmonella typhimurium* were maintained in the Key Laboratory of Industrial Fermentation Microbiology, Ministry of Education, Tianjin University of Science and Technology.

2.2 Sodium Periodate Oxidation of Microcrystalline Cellulose

The preparation of DAC can be found in our previous report [12]. An experimental procedure was performed as follows: 10.0 g of MCC was added to 500 mL of distilled water and the solution was adjusted to pH 2.0 by sulfuric acid (20%). Then, 13.2 g of sodium periodate was added. The mixture was stirred at 30 °C in the dark for 9 and 19 h. The remaining periodate was decomposed by adding excess ethylene glycol, the slurry was filtered and the product was washed with distilled water. Then anhydrous ethanol was used to remove water, and then the product was dried at 50 °C for 12 h and grinded into powder using mortar.

2.3 Determination of the Aldehyde Group Content

The determination of aldehyde contents of DAC was based on oxime reaction between aldehyde group and $NH_2OH{\cdot}HCl$ [13]. Briefly, 0.5 g DAC powder was dissolved in 25 mL deionized water in the flask. Hydroxylamine hydrochloride solution (20 mL, 50 g/L, adjusted to pH 5.0) was added to the DAC suspension solution. The mixture was stirred for 4 h in a water bath at 40 °C followed by titration of the released hydrochloric acid with 0.1 M NaOH. The endpoint was determined by the pH = 5 of the sample solution, and the consumption of sodium hydroxide solution was recorded as V_1 in mL. Moreover, the same weight of MCC was carried out a blank titration and its consumption of the alkali solution in mL was recorded as V_2. The percentage of the aldehyde content (AC %) was given by the following equation:

$$AC\% = \frac{M_{\mathrm{NaOH}}(V1-V2) \times 162}{m \times 1000} \times 100 \quad (1)$$

where M NaOH = 0.1 mol/L, m is the dry weight (g) of DAC, and 162 is the molecular weight of DAC. The experiments were performed in triplicate.

2.4 Fourier Transform Infrared (FTIR) Spectroscopy

The FT-IR analysis was performed using a Bruker Vector 22 FTIR spectrometer (Rheinstetten, Germany). The samples were prepared by blending about 1.0 mg MCC and DAC samples with 150 mg KBr and pressing the mixture into very thin tablets before measurement. All the spectra of measured samples were averaged from 32 scans ranging from 400 to 4000 cm^{-1} with a resolution of 4 cm^{-1}.

2.5 Thermogravimetric Analysis (TGA)

The thermal behavior of MCC and DAC were determined by TGA measurement on a TGA-Q50 thermal analyzer (Kyoto, Japan). Thermal analysis was performed between room temperature and 600 °C at a heating rate of 10 °C/min under nitrogen atmosphere. About 8–15 mg of sample was used in single runs.

2.6 Antibacterial Activity Test

The antibacterial activity of MCC and DAC was evaluated against *S. aureus*, *E. coli*, *B. subtilis* and *S. typhimurium* by the double dilution method (DDM), and the minimum inhibitory concentration (MIC) was determined as based on the National Committee for Clinical Laboratory Standards [14]. DDM and MIC provided qualitative and quantitative results, respectively. Before the start of any antibacterial activity, all the glassware and samples were autoclaved at 121 °C for 20 min.

2.6.1 Preparation of Bacteria Suspension

The Nutrient Agar plates were prepared for cell incubation at 37 °C for 24 h. The nutrient medium was inoculated with four types of bacteria (*S. aureus*, *E. coli*, *B. subtilis*, *S. typhimurium*) at 37 °C overnight with constant agitation under the aerobic condition. The four kinds of bacterial suspensions were diluted with a sterile 0.9% (w/v) saline solution. The turbidity of the bacterial suspension was adjusted to 1–2 × 10^8 CFU/mL based on the 0.5 McFarland standard. A total of 100 μL proofread cell culture was added to 2.9 mL lysogeny broth (LB) medium to prepare bacterial suspension spare and the concentration of the bacterial suspension was 5.0 × 10^6 CFU/mL.

2.6.2 Measurement of Minimal Inhibitory Concentration

A double dilution method was used to determine the MIC of the DAC aqueous suspension. Stock solutions of DAC aqueous suspension were prepared in an oil-bath at 100 °C for 1 h [15]. These suspensions were diluted to 12.00% and joined 5 mL LB culture medium into the first tube and blending. The concentration of DAC in the test tubes were, in order, 6.00, 3.00, 1.50, 0.75, 0.38 and 0.19% by using the method of DDM. At the same time, the tube did not add the DAC aqueous suspension was performed as the blank control. Finally, the bacterial suspension (0.1 mL, 5.0 × 10^6 CFU/mL) was added to each tube. All the test tubes were inoculated for 20 h with an agitation rate of 220 rpm at 37 °C. After 20 h, remove the tubes from the shaker and kept stationary for 5 min. Each of the tubes supernatant was measured absorbance at 600 nm using UV-2401 spectrophotometer (SHIMADZU, Japan). The MIC was defined as the lowest concentration of the DAC aqueous suspension at which the microorganism does not demonstrate visible growth [16]. Each assay was repeated three times.

3 Results and Discussion

3.1 DAC with Different Aldehyde Contents

According to the different stirring time, we have got two kinds of aldehyde content of DACs. The DAC with 60 and 93% aldehyde contents were obtained from 9 and

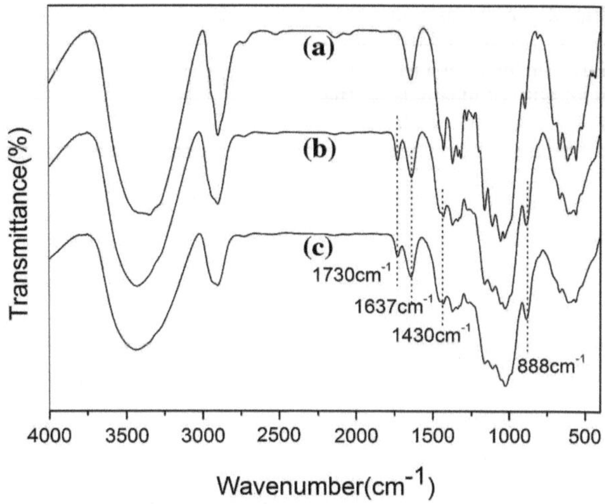

Fig. 1 Fourier transform infrared (FTIR) spectra of MCC and DAC. **a** MCC; **b** The DAC with 60% of aldehyde; **c** The DAC with 93% of aldehyde

19 h, respectively. The literature has reported that the oxidation degree 90% of DAS suspensions exhibited antibacterial activities [11]. In order to explore whether the low oxidation degree of DAC has the antibacterial activities, we chose the oxidation degree of 60% of the DAC for the experiment.

3.2 *FT-IR Analyses*

FTIR spectra analysis was performed to detect chemical changes of MCC and DAC (Fig. 1).

The band at 1637 cm^{-1} is the absorbed moisture of the sample, the characteristic absorption band of carbonyl groups is at 1730 cm^{-1} [17] for DAC (Fig. 1b and c). 1430 cm^{-1} is ascribed to the C–H stretching and bending modes of the methylene [18]. A slightly red shift of the hemiacetal vibration peak at 888 cm^{-1} implies the skeletal changed on the main chain. Furthermore, an increase in the intensity at 888 cm^{-1} (Fig. 1b and c) supported the conclusion that the aldehyde or hemiacetal groups were introduced into DAC by sodium periodate oxidation [5]. The resultant aldehyde or hemiacetal groups may be further modified to carboxylic acids or imines [19], thus serving as intermediates for cellulose-based functional products, such as drug carriers, adsorbents for heavy metals [20].

3.3 *Thermal Analyses*

The TGA curves (Fig. 2) show a characteristic change by MCC and oxidation of cellulose. The decomposition of MCC in nitrogen started at about 280 °C and rapid

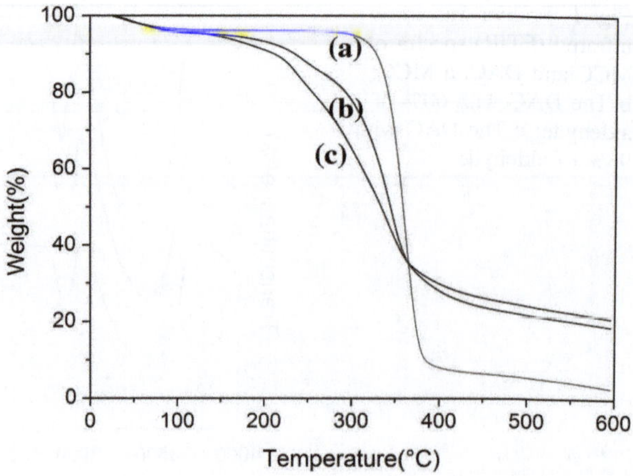

Fig. 2 Thermogravimetric *curves* (TGA) of MCC and DAC. **a** MCC; **b** The DAC with 60% of aldehyde; **c** The DAC with 93% of aldehyde

weight loss occurred at 300–370 °C. However, the decomposition temperature of DAC shifted to lower temperature side with increase in the degree of oxidation (Fig. 2b and c). We found that the chemical modification of cellulose is known to significantly change the thermal decomposition behaviour compared to unmodified cellulose. These changes are likely to be associated with facilitated thermal scission of covalent bonds leading to the generation of volatile substances [12].

3.4 Antibacterial Activity Analyses

The antibacterial activity of DAC with different contents and MCC against Gram-positive bacteria *S. aureus* and *B. subtilis* and Gram-negative bacteria *E. coli* and *S. typhimurium* were tested by the double dilution method. These bacterial species (*S. aureus*, *E. coli* and *S. typhimurium*) were selected because they are commonly known to cause human infections [21].

The MIC values of MCC were showed in Table 1 MCC did not show any antibacterial activity. The measured ultraviolet values have no obvious changes in different concentrations of DAC aqueous suspension. This proved that the raw material without aldehyde group showed no antibacterial effect. However, DAC aqueous suspension with 60 and 93% aldehyde contents showed clear antibacterial activity (Tables 2 and 3) and the ultraviolet values (OD_{600}) of these samples decreased coupled with the concentration of DAC aqueous suspension when compared with MCC.

Table 2 is the MIC results of DAC aqueous suspension with 60% aldehyde contents against four kinds of bacteria. The DAC aqueous suspension with 60% aldehyde contents exhibited a broad antibacterial spectrum against all of the test strains with MIC values between 1.50 and 3.00%. The MIC results of DAC aqueous suspension with 60% aldehyde contents against *S. aureus*, *E. coli*, *B. subtilis* and *S. typhimurium* were 1.50, 3.00, 3.00 and 3.00%, respectively. Table 3

Table 1 The MIC results of MCC against four kinds of bacteria

Sample concentration (%)	OD_{600}			
	S. aureus	*E. coli*	*B. subtilis*	*S. typhimurium*
0.00	1.769	1.848	1.816	1.839
0.19	1.740	1.992	1.739	1.805
0.38	1.692	1.945	1.802	1.787
0.75	1.737	1.862	1.785	1.679
1.50	1.764	1.805	1.752	1.638
3.00	1.683	1.803	1.717	1.746
6.00	1.721	1.831	1.613	1.712
MIC	–	–	–	–

– It indicates that the MCC did not have the effect of inhibiting bacterial growth. Data of OD_{600} were shown in average value (n = 3)

Table 2 The MIC results of DAC with 60% aldehyde contents against four kinds of bacteria

Sample concentration (%)	OD_{600}			
	S. aureus	*E. coli*	*B. subtilis*	*S. typhimurium*
0.00	1.924	1.868	1.941	1.975
0.19	1.514	1.639	1.772	1.780
0.38	0.812	1.175	1.208	1.637
0.75	0.794	1.038	1.062	1.528
1.50	0.045	0.577	0.709	0.869
3.00	0.042	0.038	0.039	0.045
6.00	0.040	0.038	0.037	0.043
MIC	1.50	3.00	3.00	3.00

Data of OD_{600} were shown in average value (n = 3)

Table 3 The MIC results of DAC with 93% aldehyde contents against four kinds of bacteria

Sample concentration (%)	OD_{600}			
	S. aureus	*E. coli*	*B. subtilis*	*S. typhimurium*
0.00	1.848	1.875	1.925	1.929
0.19	1.376	1.499	1.655	1.713
0.38	0.625	1.012	1.143	1.575
0.75	0.577	0.931	1.022	1.369
1.50	0.050	0.053	0.054	0.848
3.00	0.049	0.045	0.050	0.049
6.00	0.042	0.043	0.043	0.047
MIC	1.50	1.50	1.50	3.00

Data of OD_{600} were shown in average value (n = 3)

is the MIC results of DAC aqueous suspension with 93% aldehyde contents against four kinds of bacteria. The MIC results of DAC aqueous suspension with 93% aldehyde contents against *S. aureus*, *E. coli*, *B. subtilis* and *S. typhimurium* were 1.50, 1.50, 1.50 and 3.00%, respectively. The bacteriostatic effect of the DAC with 93% aldehyde contents aqueous suspension is superior to the DAC with 60% aldehyde contents aqueous suspension. It is seen that the bacteriostatic effect of *S. aureus* is the best one and the bacteriostatic effect of DAC aqueous suspension was increased with the increase of the degree of oxidation. These results indicate that DAC with 60 and 93% aldehyde contents showed antibacterial activity against both Gram-positive and Gram-negative bacteria because of the loading of the dialdehyde groups.

4 Conclusion

DAC with 60 and 93% aldehyde contents, prepared by the selective periodate oxidation of MCC with the introduction of the dialdehyde functions. The characterization of MCC and DAC by FTIR and TGA showed that the products were introduced two aldehyde groups and its thermal stability decreased. The results of this study indicate that the DAC shows high antibacterial properties. The higher DAC aldehyde content is, the better its bacteriostatic effect is. When DAC aldehyde content exceeded 60%, the MIC against *S. aureus*, *E. coli*, *B. subtilis* and *S. typhimurium* were 1.50, 3.00, 3.00 and 3.00%, respectively. When DAC aldehyde content exceeded 93%, the MIC of *S. aureus*, *E. coli*, *B. subtilis* and *S. typhimurium* reached 1.50, 1.50, 1.50 and 3.00%, respectively. Therefore, the findings in this study proved the potential applications of DAC as a new antibacterial agent.

References

1. Turbak AF, Snyder FW, Sandberg KR (1983) Microfibrillated cellulose, a new cellulose product: properties, uses and commercial potential. J Appl Polym Sci 37:815–827
2. Klemm D, Schumann D, Kramer F et al (2006) Nanocelluloses as innovative polymers in research and application. Adv Polym Sci 205:49–96
3. Dinand E, Chanzy H, Vignon MR (1999) Suspensions of cellulose microfibrils from sugar beet pulp. Food Hydrocolloids 13:275–283
4. Yang J, Han C, Duan J, Xu F, Sun R (2013) Mechanical and viscoelastic properties of cellulose nanocrystals reinforced poly (ethylene glycol) nanocomposite hydrogels. ACS Appl Mater Inter 5:3199–3207
5. Sun B, Hou QX, Liu ZH, Ni YH (2015) Sodium periodate oxidation of cellulose nanocrystal and its application as a paper wet strength additive. Cellulose 22:1135–1146
6. Browning BL (1976) Methods of wood chemistry, Vol. 2. Interscience, London, New York, pp 476, 477, 499

7. Kim UJ, Kuga S (2000) Reactive interaction of aromatic amines with dialdehyde cellulose gel. Cellulose 287–297
8. Ramirez H, Cao R, Fragoso A, Torres-Labandeira J, Dominguez A, Schacht E et al (2006) Improved anti-inflammatory properties for naproxen with cyclodextringrafted polysaccharides. Macromol Biosci 555–561
9. RoyChowdhury PV, Kumar J (2006) Biomed. Mater Res A 76, 300
10. Zhang X, Shen G, Sun S, Shen Y, Zhang C (2014) Direct immobilization of antibodies on dialdehyde cellulose film for convenient construction of an electrochemical immunosensor. Sensor Actuat B-Chem 200:304–309
11. Song L, Sang YJ, Cai LM, Shi YC et al (2010) The effect of cooking on the antibacterial activity of the dialdehyde starch suspensions. Starch-Starke 62(9):458–466
12. Kim UJ, Kuga S (2001) Thermal decomposition of dialdehyde cellulose and its nitrogen-containing derivative. Thermochim Acta 369:79–85
13. Sirvio J, Hyvakko U, Liimatainen H, Niinimaki J, Hormi O (2011) Periodate oxidation of cellulose at elevated temperatures using metal salts as cellulose activators. Carbohyd Polym 83:1293–1297
14. Standards NCFCL (2006) Performance standards for antimicrobial disk susceptibility tests: approved standards, 11th ed. National Committee for Clinical Laboratory Standards, USA
15. Kim UJ, Wada M, Kuga S (2004) Solubilization of dialdehyde cellulose by hot water. Carbohyd Polym 56(1):7–10
16. Yang CH, Chang HW, Lin HY, Chuang LY (2013) Evaluation of antioxidant and antimicrobial activities from 28 Chinese herbal medicines. J Pharmacogn Phytochem 2(1): 2278–4136
17. Gong R, Zhang J, Zhu J, Wang J, Lai Q, Jiang B (2013) Loofah sponge activated by periodate oxidation as a carrier for covalent immobilization of lipase. Korean J Chem Eng 30(8):1620–1625
18. Sheng Y, Wang QH, Xu XC, Jiang WY, Gan SC, Zou HF (2011) Oxidation of cornstarch using oxygen as oxidant without catalyst. LWT—Food Sci Technol 44:139–144
19. Kim UJ, Kuga S, Wada M, Okano T, Kondo T (2000) Periodate oxidation of crystalline cellulose. Biomacromol 1(3):488–492
20. Kim UJ, Kuga S (2001) Ion-exchange chromatography by dicarboxyl cellulose gel. J Chromatogr 919:29–37
21. Hanim SAM, Malek NANN, Ibrahim Z (2016) Amine-functionalized, silver-exchanged zeolite NaY: preparation, characterization and antibacterial activity. Appl Surf Sci 360: 121–130

Optimization of Microwave-Assisted Extraction of Total Flavonoids from China-Hemp Leaves and Evaluation of Its Antioxidant Activities

Jie Cao, Limin Hao, Liming Zhang, Meng Xu, Huanhuan Ge, Caicai Kang, Jianyong Yu and Zongzhen Wang

1 Introduction

Hemp (Cannabis sativa L.) is an annual herbaceous plant belongs to Cannabaceae family and originated in Central Asia, grown in many countries such as China, France, Chile, Russia, Turkey, United States and Canada. The plant has been cultivated widely for the purposes of fiber, food and medicine [1]. The hemp leaves is a by-product of hemp fiber and seed production, yield is rich. It is a rich source of flavonoids [2].

Flavonoids, a non-cannabinoid phenols, have received more and more attention by the biochemical and nutritional researchers because of their biological activities and health function in health-care food or medicine, especially their antioxidant, anti-ultraviolet radiation, and antibacterial effects [3]. Recently, natural antioxidants are becoming an attractive option as they are abundant and high safety [4, 5]. Therefore, most of the researches are currently focused on the extraction and antioxidant activities of flavonoids [6, 7].

Microwave assisted extraction (MAE) was relatively new method by which microwave energy is used to heat polar solvent in contact with solid samples and to partition compounds of interest between the sample and the solvent, reducing both extraction time and solvent consumption [8]. In this paper, by MAE, the effect of

J. Cao · L. Zhang (✉) · M. Xu · H. Ge · C. Kang
Key Laboratory of Industrial Fermentation Microbiology, Ministry of Education Tianjin University of Science and Technology, Tianjin 300457, People's Republic of China
e-mail: zhanglmd@126.com

L. Hao (✉) · J. Yu
The Quartermaster Equipment Institute of General Logistics Department of People's Liberation Army, Beijing 100010, People's Republic of China
e-mail: hlm2005@163.com

Z. Wang
Ningbo Hemp Biotechnology Co. Ltd, Ningbo 315153, People's Republic of China

H. Liu et al. (eds.), *Advances in Applied Biotechnology*, Lecture Notes in Electrical Engineering 444, https://doi.org/10.1007/978-981-10-4801-2_57

ethanol concentration, solvent-to-solid ratio, extraction time and temperature on the yield of total flavonoids from hemp leaves were investigated by RSM (response surface methodology). RSM is an effective statistical technique, which is used to find optimum processing parameters [9].The antioxidant activities of hemp leaves extracts were evaluated, including 1,1-diphenyl-2-picryl-hydrazyl (DPPH) free radical, reducing power and 2,2′-azino-di (3-ethylbenzthiazoline-6-sulfonate) (ABTS) radical cation inhibition antioxidant test. We hope this study will be helpful to further exploit and utilize this resource.

2 Materials and Methods

2.1 *Materials and Chemical Reagents*

China-hemp of leaves was collected from Mengwang County of Jinghong city, Yunnan province, China, and was identified by Professor Wenbin Hou, Tianjin Institute of Pharmaceutical Research, China. Collected leaves were dried in a forced-air oven at 40 °C to constant weight, and then ground using an electric grinder. The ground powder was passed through a standard 0.25 mm sieve and was collected and stored at 40 °C in airtight bags for further use.

Rutin, ascorbic acid (Vc) and DPPH were purchased from Sigma-Aldrich (St. Louis, MO, USA). Butylated hydroxytoluene (BHT), ABTS were from Merck (Darmstadt, Germany). All other chemical reagents used in experiments were of analytical grade and Millipore quality water was used throughout the experiment.

2.2 *Extraction and Quantification Total Flavonoids Content*

2.2.1 Microwave-Assisted Extraction

Flavonoids from powders of hemp leaves were extracted using a domestic microwave oven system. The apparatus was equipped with a digital control system for irradiation time, microwave powder (the latter was linearly adjustable from 100 to 1000 W), magnetic stirrer speed and temperature.

Tow gram of hemp leaves powder was stirred into aqueous ethanol by stirring in preparation for extraction using the MAE system. Parameters of the MAE extraction were ethanol proportion (40–80%), liquid-to-solid ratio (20–40 mL/g), extraction time (5–40 min) and extraction temperature (40–80 °C). Table 1 provides the experimental conditions respectively, where the influence of each parameter was in single-factor experiments. Each trial was carried out in triplicate. After MAE treatment, the extraction solution was separated from insoluble residue

Table 1 Experimental design with the observed responses of total flavonoids (TF) yield from China-hemp leaves using MAE

Run	X_1 ethanolm concentration (% v/v)	X_2 liquid-to-solid ratio (mL/g)	X_3 extraction time (min)	X_4 extraction temperature (° C)	Y yield of TF (%)
1	60.00	35.00	10.00	70.00	2.66
2	60.00	35.00	20.00	60.00	2.77
3	60.00	30.00	20.00	50.00	2.43
4	60.00	35.00	20.00	60.00	2.81
5	80.00	35.00	10.00	60.00	2.33
6	60.00	40.00	20.00	70.00	2.85
7	80.00	35.00	30.00	60.00	2.95
8	80.00	35.00	20.00	50.00	2.28
9	60.00	35.00	20.00	60.00	2.64
10	40.00	35.00	30.00	60.00	1.88
11	60.00	35.00	30.00	70.00	2.88
12	80.00	40.00	20.00	60.00	2.50
13	60.00	30.00	10.00	60.00	2.42
14	40.00	40.00	20.00	60.00	2.10
15	60.00	30.00	20.00	70.00	2.88
16	40.00	35.00	20.00	70.00	2.52
17	60.00	35.00	10.00	50.00	2.11
18	60.00	40.00	10.00	60.00	2.33
19	60.00	35.00	20.00	60.00	2.76
20	40.00	35.00	20.00	50.00	1.63
21	40.00	30.00	20.00	60.00	2.33
22	60.00	30.00	30.00	60.00	2.68
23	60.00	40.00	30.00	60.00	2.56
24	40.00	35.00	10.00	60.00	2.05
25	80.00	35.00	20.00	70.00	3.01
26	80.00	30.00	20.00	60.00	2.65
27	60.00	40.00	20.00	50.00	2.22
28	60.00	35.00	20.00	60.00	2.66
29	60.00	35.00	30.00	50.00	2.34

by centrifugation (8000 rpm for 10 min), and then the supernatant was collected in a volumetric flask for determination of FT content.

2.2.2 Determination of Flavonoids Contents

The flavonoid content of the extracts was estimated by the $Al(NO_3)_3$ method [10]. Briefly, 1 mL of 10-fold diluted supernatant was mixed with 0.3 mL of 5% $NaNO_2$.

The solutions were mixed thoroughly and incubated at room temperature for 6 min. And then 0.3 mL of 10% $Al(NO_3)_3$ solution was added and mixed. 6 min later, 4 mL of 4% NaOH solution was added and used 60% ethanol diluted to 10 mL. With 15 min standing, the absorbance of the solution was measured at 510 nm with UV-2401 spectrophotometer against the same mixture, without the sample as a blank. (The calibration curve: $y = 9.4975\,x + 0.0366$, $R^2 = 0.9996$).

2.3 Experimental Design and Statistical Analysis

Influence of the process parameter was investigate using a single-factor-test to determine the preliminary range of the extraction variables including X_1 (ethanol concentration), X_2 (liquid-to-solid ratio), X_3 (extraction time), X_4 (extraction temperature). As shown in Table 2, the four factors chosen for this study were designated as X_1, X_2, X_3, X_4, and prescribed into three levels, coded +1, 0, −1 for high, intermediate and low value, respectively. The four variables were coded according to the following equation

$$X_i = \frac{x_i - x_0}{\Delta x}, i = 1 - 4 \tag{1}$$

where X_i was a coded value of the variable; x_i was the actual value of the variable; x_0 was the actual value of the independent variable at the center point; Δx and was the step change of the variable.

Using a Box-Behnken design (BBD) [11], response surface methodology was conducted to determine the MAE optimized extraction process variables for maximum recovery of TF yield. Tables 1 and 2 represent the non-coded values of the experimental variables and 29 experimental points. On the basis of the experimental date, a second-order polynominal model corresponding to the BBD was fitted to correlate the relationship between the independent variables and the response to predict the optimized condition. The nonlinear computer-generated quadratic model was given as:

$$Y = \beta_0 + \sum_{i=0}^{4} \beta_i X_i + \sum_{j=0}^{4} \beta_{ii} X_i^2 + \sum_{i=0}^{4}\sum_{j=0}^{4} \beta_{ij} X_i X_j \tag{2}$$

Table 2 Variables and experimental design levels for response surface

Independent	Coded symbols	Levels		
		−1	0	1
Ethanol concentration (%)	X_1	40	60	80
Liquid-to-solid ratio (mL/g)	X_2	30	35	40
Extraction time (min)	X_3	10	20	30
Extraction temperature (°C)	X_4	50	60	70

where Y was the response function; β_0 was a constant; β_i β_{ii} and β_{ij} were the linear, quadratic and interactive coefficients, respectively; X_i was the coded levels of independent variables. The terms X_iX_j and X_i^2 represented the interaction and quadratic terms, respectively.

2.4 *Antioxidant Activity Determinations*

2.4.1 DPPH Radical Scavenging Ability Assay

Radical scavenging activity of the China-hemp leaves extract was evaluated using DPPH radicals based on the method by Xu and Chang [12]. DPPH radical scavenging activity was calculated by the following equation: scavenging effect (%) = $[1 - (A_1 - A_2)/A_0] \times 100$, where A_0 is the absorbance of the control, A_1 is the absorbance of the sample, and A_2 is the background absorbance of the sample.

2.4.2 Reducing Power

The determination of reducing power was carried out according to the method of Yen and Chen [13]. The reducing power was calculated as following: Reducing power = $A_1 - A_2$, where A_1 is the absorbance of the sample, A_2 the absorbance of the reagent blank without potassium ferricyanide.

2.4.3 ABTS Radical Cation Inhibition Antioxidant Assay

Determination of ABTS radical cation inhibition activity of sample extract was performed according to the methods of Biglari et al. [14]. The percentage radical inhibition activity was calculated as following: Radical inhibition activity (%) = $(1 - A_1/A_2) \times 100$, where A_1 was the absorbance of the sample and A_2 was the absorbance of the solvent control.

3 Results and Discussion

3.1 *The Effect of Ethanol Concentration on the Total Flavonoids Yield*

In this study, the effect of ethanol concentration on the extraction yield of TF from hemp leaves was investigated and prepared different concentrations of ethanol (40, 50, 60, 70, 80%), when other experimental parameters were set as follows: the ratio

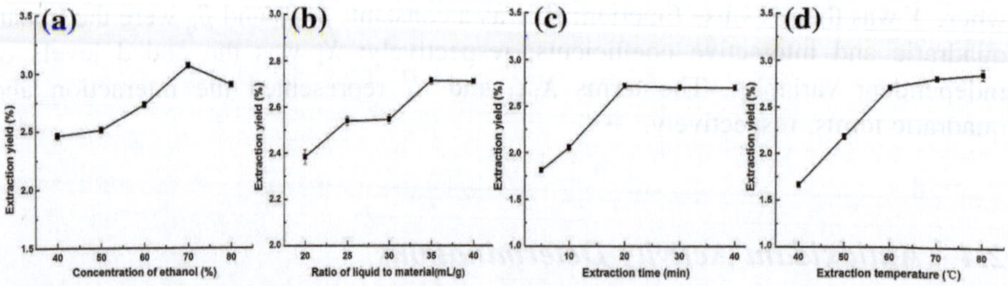

Fig. 1 Effect of different extraction parameters (**a** ethanol concentration, %; **b** liquid-to-solid ratio, mL/g; **c** extraction time, min; **d** extraction temperature, °C)

of liquid to material 30:1 (mL/g), extraction time 20 min, extraction temperature 60 °C and number of extraction 1. As shown in Fig. 1a, extraction yield of TF was highest when 70% ethanol and ethanol concentrations had important effects for it.

3.2 *The Effect of Extraction of Liquid-to-Solid Ratio on TF Yield*

The choice of liquid-to-solid ratio to raw material was another important step [15]. In this study, effect of different ratio of liquid to material (20:1, 25:1, 30:1, 35:1 and 40:1) on the extraction yield was investigated. The results were displayed in Fig. 1b. It was observed that the recovery of TF yield was maximized at a liquid-to-solid ratio of 35:1 (mL/g). A ratio of 30–40 (mL/g) was further used in the optimization of process parameters during MAE.

3.3 *The Effect of Extraction Time on the Total Flavonoids Yield*

Generally, by increasing the extraction time, the quantity of analytes is increased, although there is the risk of the degradation of extracted compounds. So the selection of extraction time (5, 10, 20, 30, 40 min) was tested for extraction yield of TF. The results were displayed in Fig. 1c. The yield of TF was increased markedly with the increase of extraction time from 5 to 20 min. Over 20 min, the yield decreased lightly. This might be due to the decomposition of active compounds during the prolonged extraction time [16].

3.4 *The Effect of Extraction Temperature on the Total Flavonoids Yield*

Extraction temperature was, respectively, set at 40, 50, 60, 70, 80 °C to examine the influence of different temperature on the yield of TF. Figure 1d indicated that the yield of TF rose gradually with the increase of temperature, reached the peak at 60 ° C, and finally dropped from 50 to 70 °C.

3.5 *Optimization of the Procedure*

3.5.1 Modeling and Fitting the Model Using Response Surface Methodology

The experiment design and corresponding response date for TF content from China-hemp leaves are presented in the Table 1.

All 29 of the designed experiments were conducted for optimizing the four individual parameters in the current BBD. By applying multiple regression analysis on the experimental date, the response variables and the best variables were related by the following second-order polynomial equation:

$$\begin{aligned} Y = {} & 2.73 + 0.27X_1 - 0.069X_2 + 0.12X_3 + 0.32X_4 + 0.020X_1X_2 \\ & + 0.20X_1X_3 - 0.040X_1X_4 - 7.500\text{E} - 033X_2X_3 + 0.045X_2X_4 \\ & - 2.500\text{E} - 033X_3X_4 - 0.28X_1^2 - 0.062X_2^2 - 0.16X_3^2 - 0.079X_4^2 \end{aligned}$$

where Y, X_1, X_2, X_3 and X_4 were the coded valued of yield of TF, ethanol concentration, liquid solvent to solid ratio, extraction time and extraction temperature, respectively.

The analysis of variance (ANOVE) for the experimental results of TF yield from China-hemp leaves were given in Table 3. It shows that the model is significant at F-value of 23.69. There is only a 0.01% chance that a "Model F-Value" this large could occur due to noises. The determination coefficient (R^2) was 0.9595, which implied that the sample variations of 95.95% for the MAE efficiency of TF were attributed to the independent variables, and only 4.05% of the total variations could not be explained by the model. However, a large value of R^2 does not always indicate that the regression model is a sound one. Here, the Adj.R^2 value was 0.919, which meant most variation (>91.9%) of TF yield could be predicted by the models, while only 8% variation could not be explained by the model.

Table 3 Estimated regression coefficients for the quadratic polynomial model and the analysis of variance (ANOVE) for the experimental results of total flavonoids yield from China-hemp leaves

Source	Sum of squares	Degree of freedom	Mean square	F-Value	*P*-Value (Prob>F)
Model	3.02	14	0.22	23.69	<0.0001
X_1	0.86	1	0.86	94.41	<0.0001
X_2	0.057	1	0.057	6.31	0.0249
X_3	0.16	1	0.16	17.7	0.0009
X_4	1.2	1	1.2	131.61	<0.0001
$X_1 X_2$	1.60E-03	1	1.60E-03	0.18	0.6813
$X_1 X_3$	0.16	1	0.16	17.16	0.001
$X_1 X_4$	6.40E-03	1	6.40E-03	0.7	0.4156
$X_2 X_3$	2.25E-04	1	2.25E-04	0.025	0.8773
$X_2 X_4$	8.10E-03	1	8.10E-03	0.89	0.3613
$X_3 X_4$	2.50E-05	1	2.50E-05	2.75E-03	0.9589
X_1^2	0.5	1	0.5	54.53	<0.0001
X_2^2	0.025	1	0.025	2.7	0.1228
X_3^2	0.16	1	0.16	17.47	0.0009
X_4^2	0.04	1	0.04	4.45	0.0534
Residual	0.13	14	9.10E-03		
Lack of fit	0.11	10	0.011	1.93	0.2757
Pure error	0.022	4	5.470E-003		
Cor total	3.14	28			
R^2	0.9595				
Adj.R^2	0.919				
Pred.R^2	0.7959				
Adequate precision	17.08				

3.5.2 Analysis of Response Surface

The 3D response surface and 2D contour profiles for flavonoids extraction from China-hemp leaves were presented in Figs. 2a–f and 3a–f. It has reported that the 3D response surface plots and 2D contour plots are able to reflect the effects of multiple independent variables and sensitiveness of response value toward the change of variables [17]. For example, if the response surface was more steeper, its influence on response value is extremely significant. The elliptical shape of contour plots was more significant than that of circular ones.

Figures 2a and 3a showed the combined effect of ethanol concentration (v/v) and the ratio of solvent volume to solid on the extraction yield of TF. The yield of TF increased with the increasing of ethanol concentration from 40 to 70%, then decrease when ethanol concentration continues to increase. The reason for this is that the appropriate concentration of ethanol could provide the most suitable

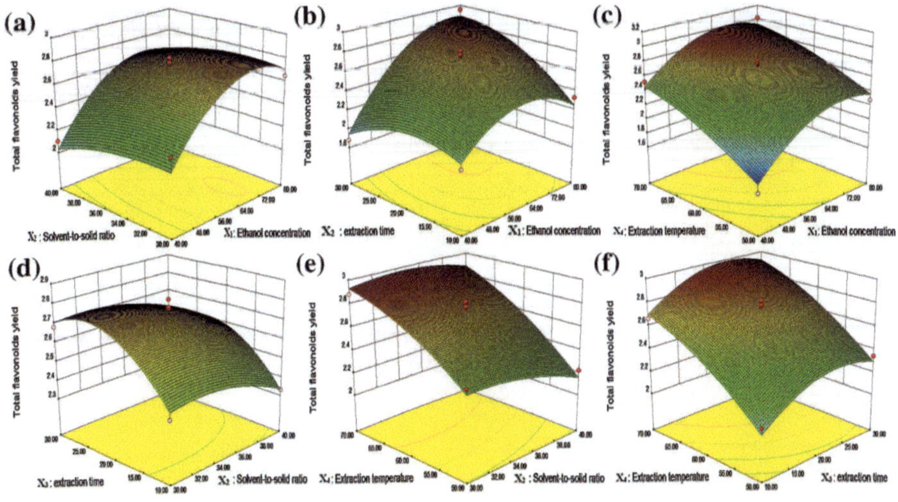

Fig. 2 Response surface plots showing the effect of X_1 (ethanol concentration, %), X_2 (solvent-to-solid ratio, mL/g), X_3 (extraction time, min), X_4 (extraction temperature, °C) to the total flavonoids on the response yield

polarity for flavonoid glucosides. The result was in consistent with the previous finding [18]. When the ratio of sovent to solid increased in the range from 34:1 to 40:1 mL/g, the yield of TF decreased. Therefore, the optimum ratio was 34:1 mL/L for the extraction of TF, this value was in consistance with the preliminary experimental result and could be determined for the accurate parameter.

Figures 2b and 3b showed the effects of ethanol concentration (v/v) and extraction time on the flavonoid extraction yield under the fixed other conditions. The extraction yield of TF increased with the increase of extraction time from 10 to 25 min, the yield was decreased when the extraction time was more than 25 min. The yield of TF was increased with the increase of ethanol concentration from 40 to 72%, further enhancing ethanol concentration led to the decrease of TF yield.

Figures 2c and 3c presented the effects of ethanol concentration and extraction temperature on the TF yield under the fixed other conditions. It could be seen that at lower ethanol concentration (below 50%), the TF yield changed little as extraction temperature improved. However, at higher ethanol concentration (50–60%), the TF yield increased significantly with the increase of extraction temperature.

Figures 2d and 3d depicted the effects of the ratio of solvent to solid and extraction time on the TF yield of TF under the fixed other conditions. It indicated that variation of extraction time has a marked effect on the TF yield. However, at low level of extraction time (from 10 to 15 min), the ratio of solvent to solid had little influence on the TF yield.

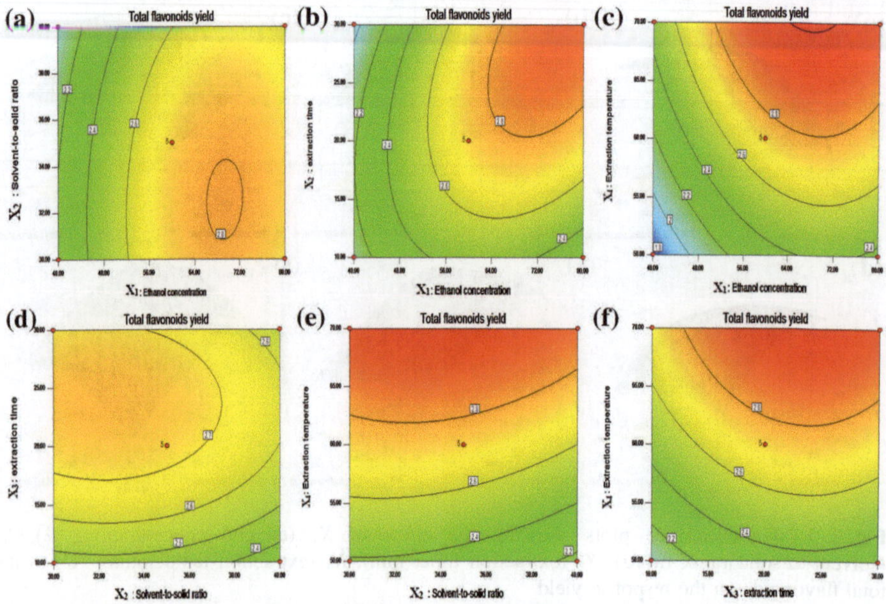

Fig. 3 Contour plots showing the effect of X_1 (ethanol concentration, %), X_2 (solvent-to-solid ratio, mL/g), X_3 (extraction time, min), X_4 (extraction temperature, °C) to the total flavonoids on the response yield

Figures 2e and 3e showed effect of the ratio of solvent to solid and extraction temperature on the yield of TF under the fixed other conditions. It revealed that the ratio of solvent to solid had little influence on the TF yield. However, the TF yield was linearly increased with the extraction temperature increment.

Figures 2f and 3f showed the effects of extraction temperature and extraction time on the yield of TF under the fixed other conditions. It can be seen that extraction time and extraction temperature exhibited a significant effect on TF yield. The maximum yield of TF was achieved when the process operated at extraction temperature of 70 °C for 20 min of extraction time.

3.5.3 Optimization of Extracting Parameters and Validation of the Model

In this study, the optimal microware extraction condition for obtain maximal yield of TF predicted by the quadratic model was as follows: ethanol concentration of 69.15%, solvent-to-solid ratio of 31.69 mL/g, extraction time of 25.14 min and extraction temperature of 69.96 °C. The predicted extraction yield of TF was 3.06%, which was consistent with the experimental yield of 3.04 ± 0.62%.

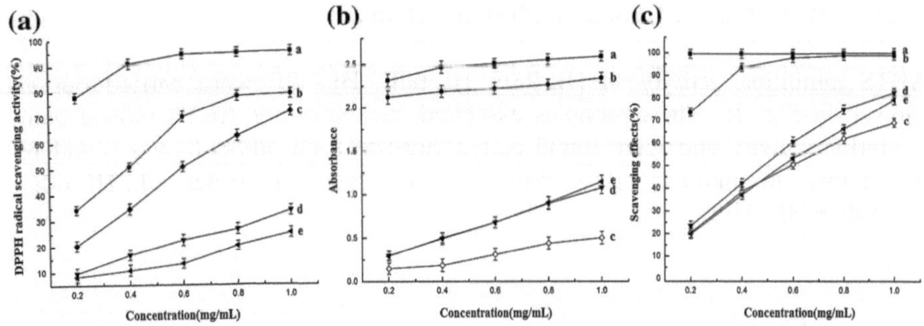

Fig. 4 Antioxidant activity assay: **a** scavenging effects on DPPH; **b** reducing power; **c** ABTS radical cation inhibition activity. **a** Vc; **b** BHT; **c** HL-95E; **d** HL-60E; **e** HL-30E. Each value is the mean ± SD of triplicate measurements

3.6 *Antioxidant Activity of Different Concentration of Ethanol Extract*

3.6.1 Scavenging Effects on DPPH Radicals

The hemp leaves powder (10 g) was respectively extracted by different concentration aqueous ethanol (95, 60, 30%) at 60 °C for 2 h. After filtration, the supernatants were concentrated and vacuum dried to obtain HL-95E (95% ethanol extraction), HL-60E (60% ethanol extraction), HL-30E (30% ethanol extraction), respectively.

The DPPH radical-scavenging activity of HL-95E, HL-60E, HL-30E were investigated at different concentrations (0.2–1.0 mg/mL) and the results were presented in Fig. 4a. In the results, Vc showed obvious scavenging activity and BHT, HL-95E, HL-60E, HL-30E showed scavenging activity on DPPH radical in a concentration-dependent manner. HL-95E, HL-60E and HL-30E showed a linear increase in scavenging activity with increase sample concentration, and exhibited under Vc and BHT. It was found that the ability to scavenge DPPH radical were in order of HL-95E > HL-60E > HL-30E.

3.6.2 Reducing Power

The reducing capacity of HL-95E, HL-60E and HL-30E always showed lower than Vc and BHT and indicated relatively low of its potential antioxidant activity. In Fig. 4b, the absorbance at 700 nm increased with the concentration of HL-95E, HL-60E and HL-30E. At 1.0 mg/mL, the reducing power of HL-95E, HL-60E and HL-30E was only 0.504, 1.079 and 1.146 respectively.

3.6.3 ABTS Radical Cation Inhibition Activity

ABTS inhibition activity of HL-95E, HL-60E, HL-30E were carried out and showed in Fig. 4c. These samples exhibited an interesting ABTS radical cation inhibition activity and were found active close to each other. It was found that the ability to inhibit ABTS radical cation were in order of HL-60E > HL-30E > HL-95E.

4 Conclusion

In this work, the microwave-assisted extraction (MAE) method was used to extract TF from China-hemp leaves. Based on single factor experiment and RSM analysis, the optimum technology parameters of MAE for extracting TF were obtained. The optimized results are as follows: ethanol concentration of 69.15%, solvent-to-solid ratio of 31.69 mL/g, extraction time of 25.14 min and extraction temperature of 69.96 °C. Under these conditions, a maximum TF yield was 3.04 ± 0.62%. The RSM could successfully model and optimize extraction process of TF from China-hemp leaves.

Additionally, the antioxidant activity of TF obtained from different concentration ethanol was evaluated in vitro by scavenging capacity of DPPH, ABTS, and reducing power. It is revealed that the ethanol concentration has significant effects on the antioxidant activity of TF. This result should be useful to further develop and apply China-hemp leaves resource.

Acknowledgements This research was financially supported by the Research Project of People's Liberation Army (No. AX110C002 and BX115C007).

References

1. Flores-Sanchez IJ, Verpoorte R (2008) Secondary metabolism in cannabis. Phytochem Rev 7:615–639
2. He J (2011) Advances in bioactive compounds in hemp leaves. J Chinese Inst Food Sci Tech 8(11):133–138
3. Wang WY, Zhou ZH, Huang NJ, Yun-lan Tang SY (2014) Study on the extraction of active ingredients in bamboo leaves by enzymatic coupling high pulse electric field. Food Ferment Tech 4(50):42–45
4. Bandyopadhyay M, Chakraborty R, Raychaudhuri U (2008) Antioxidant activity of natural plant sources in dairy dessert (Sandesh) under thermal treatment. LWT Food Sci Tech 41:816–825
5. Podsedek A (2007) Natural antioxidants and antioxidant capacity of Brassica vegetables: a review. LWT Food Sci Tech 40:1–11

6. Huang W, Xue A, Niu H, Jia Z, Wang JW (2009) Optimised ultrasonic-assisted extraction of flavonoids from Folium eucommiae and evaluation of antioxidant activity in multi-test systems in vitro. Food Chem 114:1147–1154
7. Meghashri S, Kumar HV, Gopal S (2010) Antioxidant properties of a novel flavonoid from leaves of Leucas aspera. Food Chem 122:105–110
8. Pérez-Serradilla JA, Castro LD (2011) Microwave-assisted extraction of phenolic compounds from wine lees and spray-drying of the extract. Food Chem 124:1652–1659
9. Liu J, Hu H, Xu J, Wen Y (2012) Optimizing enzymatic pretreatment of recycled fiber to improve its draining ability using response surface methodology. BioResources 7:2121–2140
10. It K, Vb V, Ml L (2011) Ultrasound-assisted extraction of total phenols and flavonoids from dry tobacco (Nicotiana tabacum) leaves. Nat Prod Commun 6:1855–1856
11. Ferreira SLC, Bruns RE, Ferreira HS, Matos GD, David JM, Brandão GC et al (2007) Box-Behnken design: an alternative for the optimization of analytical methods. Anal Chim Acta 597:179–186
12. Xu BJ, Chang SKC (2007) A comparative study on phenolic profiles and antioxidant activities of legumes as affected by extraction solvents. J Food Sci 72:S159–S166
13. Chen HY, Yen GC (2007) Antioxidant activity and free radical-scavenging capacity of extracts from guava (Psidium guajava, L.) leaves. Food Chem 101:686–694
14. Biglari F, AlKarkhi AFM, Easa AM (2008) Antioxidant activity and phenolic content of various date palm (Phoenix dactylifera) fruits for Iran. Food Chem 107:1636–1641
15. Luo C, Chen YS (2010) Optimization of extraction technology of Se-enriched Hericium erinaceum polysaccharides by Box-Behnken statistical design and its inhibition against metal elements loss in skull. Carbohyd Polym 82:854–860
16. Upadhyay R, Nachiappan G, Mishra HN (2015) Ultrasound-assisted extraction of flavonoids and phenolic compounds from Ocimum tenuiflorum leaves. Food Sci Biotechnol 24: 1951–1958
17. Yin G, Dang Y (2008) Optimization of extraction technology of the Lycium barbarum polysaccharides by Box-Behnken statistical design. Carbohyd Polym 74:603–610
18. Bimakr M, Rahman RA, Taip FS, Ganjloo A, Salleh LM, Selamat J et al (2011) Comparison of different extraction methods for the extraction of major bioactive flavonoid compounds from spearmint (Mentha spicata L.) leaves. Food Bioprod Process 89:67–72

Comparison of Different Protocols of Gradient Ammonium Sulfate Fractionation of Antibacterial Proteins/Peptides from *Clarias gariepinus* Wastes

Xiaomei Wang, Yan Wang, Yunxia Xu, Zhuanzhuan Li, Chengxun Chen, Jinwei Gao and Tao Li

1 Introduction

With the development of aquaculture, more and more intensive farming methods have instead of traditional fish farming mode. However, the high stocking density can result in deterioration of water quality and rapid growth of aquatic animal pathogenic bacteria, thereby causing disease outbreaks. To avoid economic loss caused by diseases, antibiotics are being used as prophylactic and therapeutic agent. The use of antibiotics can lead to the development of drug-resistant bacteria and residues of drugs in fish and in environment, which are potentially risky to human [1, 2]. Consequently, it is urgent to look for safer agent, instead of antibiotic, to fight the fish pathogens.

The antibacterial peptides (APMs) was first found in *Hyalophora cecropia*, and then named cecropin in 1981 by Steiner and Boman [3], thereafter, a variety of AMPs were reported in various organisms including bacteria, fungi, plants and animals and have been found to defend against invading bacteria, viruses and fungi [4]. AMPs, relatively low molecular mass, either inducible or constitutive [5], have also been found to display stable physical and chemical properties, broad-spectrum antimicrobial activity and do

X. Wang (✉) · Y. Wang · Z. Li · C. Chen · J. Gao · T. Li
Tianjin Key Laboratory of Aqua-Ecology and Aquaculture, College of Fisheries,
Tianjin Agricultural University, Tianjin 300384, People's Republic of China
e-mail: xiaomeiw@tjau.edu.cn

Y. Xu
Tianjin Diseases Prevention and Control Center of Aquatic Animals,
Tianjin 300221, People's Republic of China

H. Liu et al. (eds.), *Advances in Applied Biotechnology*, Lecture Notes
in Electrical Engineering 444, https://doi.org/10.1007/978-981-10-4801-2_58

not induce de novo bacterial resistance, which make AMPs a novel class of antimicrobials, particularly, an eco-friendly green antibacterial agents [6].

In teleost fish, innate immune system is very important and believed to be the first line of fish in opposing pathogenic organisms [7]. In innate immune system, endogenous AMPs are regarded as the earliest and fundamental immune molecules and exist widely in tissues and organs under physiologic and pathologic conditions [8]. African catfish, *Clarias gariepinus*, belonging to Siluriformes, Clariidae, *Clarias*, is rapid growth and high resistant to disease [9] and had become one of the most important fish in aquaculture in China since it was introduced to China in 1981 [10]. Owing to the fish muscle with high protein and low lipid content as well as absence of intramuscular bones [9], marketable size of the fish were supplied to processing industries. Fishery processing industries generate large amount of wastes such as viscera, skin, gill, blood and so on. To avoid wasting these fish wastes, various disposal methods have been applied in producing high added value products [11]. Interestingly, bio-active antibacterial compounds were obtained from *C. gariepinus* wastes with 70% saturation ammonium sulphate salting-out [12] and acidic extracts [13].

Ammonium sulphate [$(NH_4)_2SO_4$] fractionation, is generally utilized as the initial step in the isolation of proteins/peptides from various tissues, which provides a rapid and inexpensive method for concentrating large starting volumes. Stepwise precipitation with $(NH_4)_2SO_4$ is an effective way to fractionate proteins into similar molecular weight populations, and the higher the $(NH_4)_2SO_4$ saturation is, the lower molecular weight proteins/peptides will be obtained [14, 15].

Therefore, this study was carried out to investigate the effects of reducing steps of fractionated precipitation with $(NH_4)_2SO_4$ on the yields and antibacterial activities of proteins/peptides isolated from *C. gariepinus* wastes at 80 and 100% saturated ammonium sulphate solutions.

2 Materials and Methods

2.1 Tissues Collection

Healthy fish were obtained from Deren aquaculture center, Tianjin, China. The fish were reared at 180 kg/m^3 in static aerated concrete pond at temperature of 25 ± 1 °C. Fish body weight was 287.45 ± 69.46 g, body length was 32.06 ± 3.72 cm at the sampling time.

The *C. gariepinus* wastes, namely, gill, suprabranchial organ and viscera including hepatopancreas, spleen, kidney, head kidney and alimentary tract were collected and washed gently in cold sterilized 0.85% NaCl and intestinal contents were flushed out gently with same solution. After that, these tissues were pooled and stored at −20 °C for subsequent proteins/peptides extraction. The fish were anesthetized with MS-222 (tricaine methanesulphonate, sigma) before dissecting.

2.2 *Gradient Ammonium Sulphate Fractionation of Crude Antibacterial Proteins/Peptides*

A total of 100 g *C. gariepinus* wastes were homogenized in 500 ml PBS buffer (pH 6.0, 0.02 mol/L), heated to 80 °C and stirred for 30 min, then centrifuged at 8000 rpm for 20 min at 4 °C. The supernatant was dispensed in 3 equal volumes into 500 mL beakers, and termed as S1, S2 and S3. Solid $(NH_4)_2SO_4$ was slowly added to the S1, S2 and S3. The saturation level of $(NH_4)_2SO_4$ were adjusted stepwise to 20, 40, 60, 80 and 100% for S1; to 40, 60, 80 and 100% for S2 as well as to 60, 80 and 100% for S3, respectively. Each saturation solution was allowed to stand for 6 h at 4 °C except that 40% saturation of S2 was stood for 12 h and 60% saturation of S3 was stood for 18 h. The precipitated crude proteins/peptides were collected by centrifugation at 10,000 rpm for 15 min at 4 °C, corresponding precipitated fractions were named as S1P-20, S1P-40, S1P-60, S1P-80 and S1P-100 obtained from S1; S2P-40, S2P-60, S2P-80 and S2P-100 obtained from S2; S3P-60, S3P-80 and S3P-100 obtained from S3; respectively. The precipitates were separately dissolved in sterile deionized water and dialyzed in 1000 molecular weight cut-off dialysis bag against sterile deionized water for 24 h at 4 °C with changes of sterile deionized water per 2–3 h in order to remove $(NH_4)_2SO_4$. All of dialyzed crude proteins/peptides fractions were lyophilized, weighed and kept at −20 °C.

The data of the weight of dialyzed crude proteins/peptides fractions were expressed as means ± SD and subjected to one-way analysis of variance. After variance analysis, means were compared using Duncan test, with the differences considered to be statistically significant at $p < 0.05$. All statistical analyses were performed using the software SPSS version 11.5 for Windows.

2.3 *Antibacterial Activity Testing*

2.3.1 Preparation of Target Bacteria

Aeromonas hydrophila and *Escherichia coli* were stored in Tianjin Diseases Prevention and Control Center of Aquatic Animals (Tianjin, China), *Staphylococcus aureus* was purchased from Guangdong Microorganism Germplasm Resource Bank (Guangzhou, China). They were used as the test bacteria. Before the experiments, every bacterial strain was grown to logarithmic phase in LB media at an optimal temperature, harvested by centrifugation at 5000 rpm for 10 min at 4 °C, washed twice and resuspended with sterilized 0.85% NaCl, counted using a Neubauer hemocytometer, and then calculated its concentration.

2.3.2 Antibacterial Activity Assay

All of the lyophilized crude protein/peptide extracts were respectively redissolved in sterile water to a final concentration of 100 mg/mL. Antibacterial activities were examined using the agar diffusion method described by Wang et al. [12] with some modifications. Briefly, the sterile petri dish contained a layer of 20 mL 1.5% (w/v) agar LB medium, which was covered with a second layer of 5 mL 1.5% agar LB medium with a suspension of target bacterium (final concentration of approximately 10^6 cfu/mL) at 45 °C. After hardening, sterile Oxford cups (7.8 mm outer diameter, 6 mm inner diameter and 10 mm height) were placed on this bilayer medium. 100 μL of redissolved each sample was added to each Oxford cup, respectively, and then the petri dishes were placed in a 4 °C incubator for 12 h followed by incubation at an appropriate temperature (37 °C for *E. coli* and *S. aureus*, 30 °C for *A. hydrophila*) overnight for bacteria growth. The antibacterial activity was evaluated by measuring the diameter of the inhibition zone. 100 μL of sterile water and 100 μL of 25 μg/mL oxytetracycline hydrochloride were used as the negative and the positive control, respectively. All inhibition assays were carried out in duplicate.

2.4 SDS-PAGE Electrophoresis

All of the lyophilized crude proteins/peptides extracts were detected using SDS-polyacrylamide gel electrophoresis of Laemmli system according to the method of Xia et al. [16] with some modifications. 15% separating gel and 4% stacking gel were prepared, respectively, using acrylamide-bisacrylamide (30% T, 2.67% C) stock solution. 5 μL crude proteins/peptides extracts (10 μg/μL) and 5 μL protein loading buffer (4% SDS, 8% mercaptoethanol, 20% glycerol, 0.05% bromophenol blue, 100 mM Tris-HCl, pH 6.8) were mixed, boiled for 10 min, and then loaded the samples to each well of the gel. Start electrophoresis with an initial current of 11 mA and maintain at this current until the samples had completely entered the stacking gel. Next, a current of 24 mA could then be used until the bromophenol blue migrated to 1 cm from the bottom of the gel. After that, the samples were visualized directly in the gel with 0.05% Coomassie brilliant blue R-250 for 2 h and destained in 10% acetic acid until clear bands were observed. Transfer the gel to water for removing the acetic acid. The gel image was photographed with GeneSnap Version 6.0 software in the gel image system (GENEGENUS™, Syngene Company, UK).

3 Results

In total, 12 fractions of crude proteins/peptides were obtained in different saturation level of ammonium sulphatefrom from *C. gariepinus* wastes. The yields of the 12 samples were shown in Table 1.

In this paper, all of 100 μL of 12 samples at concentration of 100 mg/mL exhibited antibacterial activity against *A. hydrophila*, *E. coli* and *S. aureus*. However, all of the diameters of inhibition zone produced by the 12 samples were the same as outer diameter of Oxford cup and there were no differences in antibacterial activity between any two samples. The antibacterial activities of 12 samples against the 3 tested bacterial strains were shown in (Figs. 1, 2 and 3).

The profiles of 12 samples on SDS-PAGE gels showed a broad range of proteins/peptides. The electrophoretic patterns on the gel among the samples of S1P-80, S2P-80 and S3P-80 as well as among the samples of S1P-100, S2P-100 and S3P-100 had no obvious differences (Fig. 4).

4 Discussion

Fish possess both an innate and adaptive immune system, however, the innate immune system is an extremely important compared to adaptive immune system in fish [17]. Antimicrobial peptides, either inducible or constitutive [5], are key parameters and at the forefront of the innate immune system, which act against bacterial infection, killing bacteria by disrupting the cell membrane [18]. *C. gariepinus* possess high disease resistance at high stocking density [9], and in

Table 1 The yields of crude proteins/peptides fractions from *C. gariepinus* wastes

Saturation level of $(NH_4)_2SO_4$ (%)	Dialyzed lyophilized weight (g)		
	From supernatant S1	From supernatant S2	From supernatant S3
20	0.3953 ± 0.1283 (S1P-20)		
40	1.1750 ± 1.0711 (S1P-40)	1.5727 ± 0.7270 (S2P-40)	
60	0.2844 ± 0.1814 (S1P-60)	0.3739 ± 0.1081 (S2P-60)	1.8285 ± 0.0295 (S3P-60)
80	0.1128 ± 0.0574^{b} (S1P-80)	0.2814 ± 0.0916^{ab}(S2P-80)	0.3551 ± 0.1094^{a} (S3P-80)
100	0.1353 ± 0.0867 (S1P-100)	0.1867 ± 0.0703 (S2P-100)	0.1631 ± 0.0424 (S3P-100)

The contents in brackets are the names of samples. Values within the same rows sharing the same letters or no letters do not differ significantly ($P > 0.05$), whereas those with different letters differ significantly ($P < 0.05$)

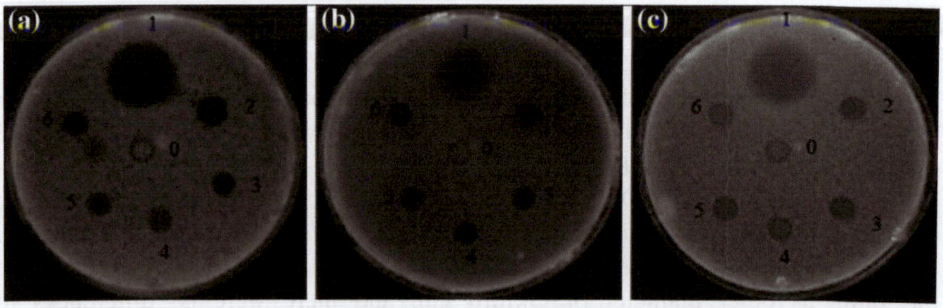

Fig. 1 Antimicrobial activity of crude proteins/peptides isolated from *C. gariepinus* wastes against *A. hydrophila* (**a**), *E. coli* (**b**) and *S. aureus* (**c**). Number *0* and *1*: negative (*0*) and positive control (*1*); Number *2–6*: samples of S1P-20, S1P-40, S1P-60, S1P-80 and S1P-100

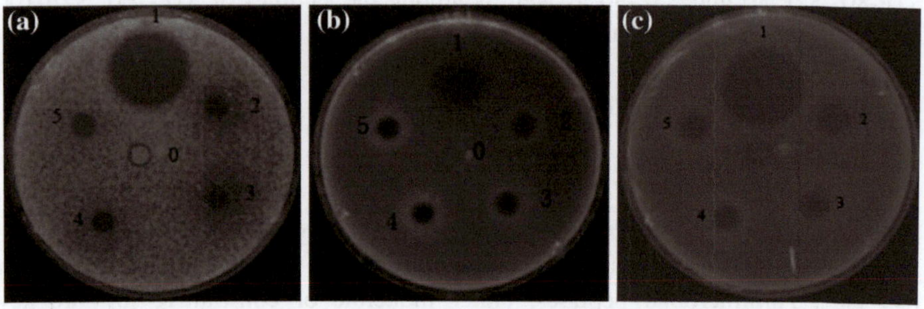

Fig. 2 Antimicrobial activity of crude proteins/peptides isolated from *C. gariepinus* wastes against *A. hydrophila* (**a**), *E. coli* (**b**) and *S. aureus* (**c**). Number *0* and *1*: negative (*0*) and positive control (*1*); Number *2–5*: samples of S2P-40, S2P-60, S2P-80 and S2P-100

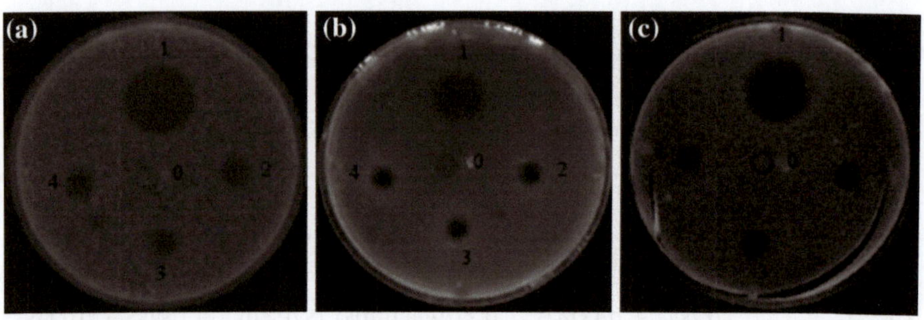

Fig. 3 Antimicrobial activity of crude proteins/peptides isolated from *C. gariepinus* wastes against *A. hydrophila* (**a**), *E. coli* (**b**) and *S. aureus* (**c**). Number *0* and *1*: negative (*0*) and positive control (*1*); Number *2–4*: samples of S3P-60, S3P-80 and S3P-100

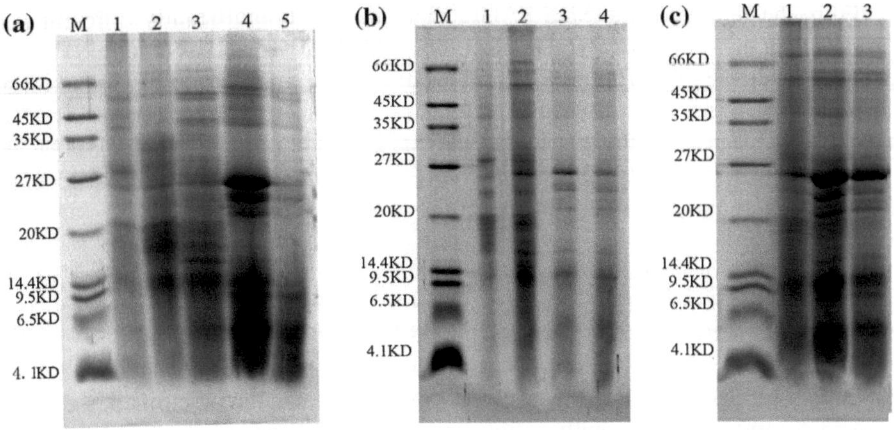

Fig. 4 Coomassie brilliant blue R-250 stained SDS-PAGE of crude proteins/peptides fractionated with $(NH_4)_2SO_4$ from *C. gariepinus* wastes. *Lane M* broad range protein markers. *Lane 1–5* in **a**: S1P-20–S1P-100; *Lane 1–4* in **b**: S2P-40–S1P-100; *Lane 1–3* in **c**: S3P-60–S3P-100. *Numbers* on the *left side* of the gels correspond to the molecular weight of the markers

healthy *C. gariepinus*, the bio-active antibacterial proteins/peptides were obtained from skin mucus, skin, gill, suprabranchial organ and intestine with ammonium sulphate of 70% saturation [12] and from the wastes with acidic extraction [13].

AMPs are normally considered small molecular mass proteins/peptides, on the other hand, relatively small molecular weight proteins/peptides could be better precipitated in higher saturation levels of $(NH_4)_2SO_4$ than in lower saturation of $(NH_4)_2SO_4$. Therefore, in present study, the yields of crude proteins/peptides obtained at 80% and 100% saturation of $(NH_4)_2SO_4$, which precipitated initially at 20%, 40% or 60% saturation respectively, were compared. As shown in Table 1, at 80% saturation level of $(NH_4)_2SO_4$, the highest yield of crude proteins/peptides precipitated initially at 60% saturation of $(NH_2)_4SO_4$ was found compared to the initial precipitation at 20% saturation of $(NH_2)_4SO_4$, however, there were no significantly differences between any two yields at 100% saturation with $(NH_4)_2SO_4$.

In this paper, all of 12 samples exhibited inhibitory activity against gram-negative bacteria (*A. hydrophila* and *E. coli*) and gram-positive bacterium (*S. aureus*), but the inhibitory activity among samples didn't show significant differences. In previous studies, the method of gradient ammonium sulphate fractionation for antibacterial peptides extraction was used, e.g. earthworms protein, extracted with $(NH_4)_2SO_4$ at 40, 50, 60 and 70% saturation, exhibited antibacterial activity against *E. coli* and *S. Aureus* [19]; antibacterial peptides were also obtained from cactus at 40%, 80% and 100% saturation of $(NH_4)_2SO_4$ [20] as well as from bamboo leaves at 30 and 60% saturation level of $(NH_4)_2SO_4$ [21]. These reports indicated that gadient ammonium sulphate salting-out technique was widely used and a valuable tool for obtaining antibacterial peptides from various tissues and organisms.

From the figures of the SDS-PAGE, there were not significantly different patterns among the samples of S1P-80, S2P-80 and S3P-80 as well as among S1P-100, S2P-100 and S3P-100. The protein profiles of these samples showed a broad range of proteins/peptides with molecular weights ranging from over 66 kDa to less than 6.5 kDa. It means that the molecular weights of crude proteins/peptides obtained with gradient $(NH_4)_2SO_4$ salting-out from *C. gariepinus* wastes were wide range. Abundant proteins/peptides with molecular weights of 26, 9.5 and 5.5 kD were observed in S3P-80 and S3P-100. The antibacterial proteins of 27 and 31 kD from the skin mucus of *Cyprinus carpio* [22], antibacterial proteins of 15.5 and 30 kD from the skin of *Ictalurus punctatus* [23], and various bio-active peptides with antimicrobial activities in fish were reported [24]. Therefore, the work on the 26, 9.5 and 5.5 kD proteins observed in this paper is being further studied in our laboratory for understanding their other characteristics. On the other hand, the crude proteins/peptides extracts with antibacterial activities from *C. gariepinus* wastes will be used as feed additive in order to test whether the extracts can enhance the fish resistance to pathogens infection.

In conclusion, in order to obtain relatively low-molecular weights antibacterial proteins/peptides from *C. gariepinus* wastes by using gradient ammonium sulphate precipitation technique, according to the results of yield, antimicrobial activity and protein profile on the SDS-PAGE, and simultaneously taking into account the convenience of operation, we propose that the fish wastes could be precipitated initially at 60% saturation. The data of the present and our previous studies [12, 13, 25] highlight the evidence that antimicrobial proteins/peptides in *C. gariepinus* consist of a range of molecular weight of proteins, and the wastes of the fish have a potential to be used as a valuable source for extracting antibacterial proteins/peptides.

Acknowledgements This work was supported by natural science foundation of Tianjin, China (No. 15JCZDJC33500), and partly supported by the innovative talent cultivation plan for young and middle-aged in Tianjin higher educational institutions (Fisheries resources utilization and ecological remediation).

References

1. Smith P, Hiney MP, Samuelesen OB (1994) Bacterial resistance to antimicrobial agent used in fish farming: a critical evaluation of method and meaning. Annu Rev Fish Dis 4:273–313
2. Alderman DJ, Hasting TS (1998) Antibiotic use in aquaculture: development of antibiotic resistance—potential for consumer health risks. Int J Food Sci Tech 33(2):139–155
3. Steiner H, Hultmark D, Engström Å et al (1981) Sequence and specificity of two antibacterial proteins involved in insect immunity. Nature 292(5820):246–248
4. Radek K, Gallo R (2007) Antimicrobial peptides: natural effectors of the innate immune system. Semin Immunopathol 29:27–43
5. Andreu D, Rivas L (1998) Animal antimicrobial peptides: an overview. Biopolymers 47 (6):415–433

6. Amy TYY, Shaan LG, Robert EWH (2011) Multifunctional cationic host defence peptides and their clinical applications. Cell Mol Life Sci 68:2161–2176
7. Whyte SK (2007) The innate immune response of finfish: a review of current knowledge. Fish Shellfish Immun 23:1127–1151
8. Lauth X, Shike H, Burns JC et al (2002) Discovery and characterization of two isoforms of moronecidin, a novel antimicrobial peptide from hybrid striped bass. J Biol Chem 277 (7):5030–5039
9. Prokešová M, Drozd B, Kouřil J (2015) Effect of water temperature on early life history of African sharp-tooth catfish, *Clarias gariepinus* (Burchell, 1822). J Appl Ichthyol 31(Suppl. 2):18–29
10. Li HY, Deng XH, Ye W et al (1984) Introduction of good freshwater fish—*Clarias gariepinus* (in Chinese). Freshwater Fisheries 1:7–12
11. Rebah FB, Miled N (2013) Fish processing wastes for microbial enzyme production: a review. Biotech 3:255–265
12. Wang X, Dai W, Xing K et al (2012) Antibacterial activities of antibacterial proteins/peptides isolated from organs and mucus of *Clarias gariepinus* reared at high stocking density. Adv Mater Res 455–456:455–460
13. Wang Y, Xu Y, Mei J et al (2015) Optimization of extraction conditions for crude antibacterial proteins/peptides from *Clarias gariepinus* by-products. LNEE 333:547–555
14. Barsukov AK, Barmin AV, Kuznetsov AI et al (2009) The isolation of immunoglobulin G and albumin protein standards and the study of their oligomerization and antigenity during storage in saturated ammonium sulfate solution. Appl Biochem Micro 45(3):343–348
15. Kent Ute M (1995) Purification of antibodies using ammonium sulfate fractionation or gel filtration. Methods Mol Biol 34:13–21
16. Xia Q, Zeng R (2004) Protein chemistry and proteomics (in Chinese). Science Press, Beijing
17. Magnadottir B (2010) Immunological control of fish diseases. Mar Biotechnol 12:361–379
18. Maier VH, Dorn KV, Gudmundsdottir BK et al (2008) Characterisation of cathelicidin gene family members in divergent fish species. Mol Immunol 45(14):3723–3730
19. Zheng J, Meng Q, Wang J (2008) Study on extraction of antimicrobial peptid and its antibacterial activity (in Chinese). Food Res Dev 29(10):51–53
20. Liu Y, Huang Q, Yang Z et al (2007) Screening of antibacterial peptides from cactuses and characterization of antimicrobial peptide MH-AMP-1 purified from *Mammillaria haw* (in Chinese). J Sichuan Agric Univ 25(9):271–276
21. Jiang H, Zhang G (2011) Purification and antimicrobial activity of a natural antimicrobial peptide from bamboo leaves (in Chinese). Food Ind 5:12–14
22. Lemaître C, Orange N, Saglio P et al (1996) Characterization and ion channel activities of novel antibacterial proteins from the skin mucosa of carp (*Cyprinus carpio*). Eur J Biochem 240(1):143–149
23. Robinette D, Wada S, Arroll T (1998) Antimicrobial activity in the skin of the channel catfish *Ictalurus punctatus*: characterization of broad-spectrum histone-like antimicrobial proteins. Cell Mol Life Sci 54:467–475
24. Altınelataman C, Torkova A, Tsentalovich M (2015) Fish derived bio-active peptides and their metabolic effects. Ege J Fish Aqua Sci 32(4):217–223
25. Wang X, Dai W, Chen Ch et al (2012) Analysis of antibacterial activities of antibacterial proteins/peptides isolated from serum of *clarias gariepinus* reared at high stocking density. LNEE 137:303–309

Screening of Microbial with the Ability of Epothilones Biotransformation

Meng Zhang, Lin Zhao, Xin Sun and Xin-li Liu

1 Introduction

Epothilones are antibiotics with 16-membered macrocyclic lactones. Its excellent anticancer activity have drawn more and more interests of researchers. Three main kinds of epothilones have been found up to now [1]. First, the natural products of epothilones which include epothilone A and B; second, the semi-synthetic derivatives of epothilones which include ixabepilone, 21-amino-epothilone, KOS-1584 and epothilone D; third, the total synthesis of derivatives of epothilones: sagopilone. Among them, the anticancer activity of ixabepilone and sagopilone ranked at the top [2]. However, the development of epothilones is limited by its high toxicity and super high price as chemical production [1]. What's more, ixabepilone can only be got through a rather complex chemical reaction with epothilone B as the raw material. And its productive rate is less than 40% [1], while sagopilone is less than 1% [3]. The further development of epothilones relies greatly on finding new and similar synthesized medicines with high productivity as well as low toxicity, which can be solved through microbial transformation.

Microbial transformation refers to a process in which certain microorganisms transform a substance into another one. In this process, the catalytic activity of certain microorganisms can be facilitated by its one or several special endoenzyme and extracellular enzyme [4]. *Rhizopusnigricans* was first used in 1950s to convert progesterone to 11α-hydroxy-progesterone, which help the scientists begin to notice the importance of structural modification of compounds with microorganisms [5]. From then on, microbial transformation has become an indispensable and important technology in transforming the structure of compounds in pharmaceutical industry.

M. Zhang · L. Zhao · X. Sun · X. Liu (✉)
Shandong Provincial Key Laboratory of Microbial Engineering
School of Bioengineering, QiLu University of Technology,
No. 3501, Daxue Road, Jinan 250353, People's Republic of China
e-mail: vip.lxl@163.com

H. Liu et al. (eds.), *Advances in Applied Biotechnology*, Lecture Notes
in Electrical Engineering 444, https://doi.org/10.1007/978-981-10-4801-2_59

In recent years, many researchers have devoted themselves to the research of microbial transformation of medicine, while rare attention was paid to the research of biotransformation of epothilones. Tang converted epothilone D to hydroxylated epothilones with equal activity by using *Amycolata autotrophica* [6]. Pandey RP and others successfully got the glycosylated product of epothilones by the method of glycosylation in vitro. The new product were better than epothilone B with lower toxicity and higher water-solubility [7, 8]. The biotransformation of epothilone compounds by using microorganisms or enzyme system may lead to some differences in the activity, polarity and toxicity of the derivatives [9]. By using the metabolism of microorganisms, this research aims to transform epothilone A and B to find better medicines with similar function as epothilones.

2 Materials and Methods

2.1 Microorganisms and Medium

Sorangium cellulosum is a microbial strain kept at our laboratory and is used for the fermentation and production of epothilone A and B. The wild type *Burkholderia* sp. and wild type *Bacillus megaterium* are isolated and purified in our laboratory. The PDA medium (200 g potatoes chopped into slices and boiled for 30 min, filtered, added with 20 g glucose and 15 g agar, diluted to a constant volume of 1 L using deionized water, mixed, adjusted pH to 7.2 with 1 mol/L NaOH, and sterilized at 115 °C for 30 min) and LB medium (10 g tryptone, 5 g yeast extract, 5 g NaCl and 15 g agar. Adjusted pH to 7.2 with 1 mol/L NaOH and diluted to a constant volume of 1 L with deionized water. Sterilized at 121 °C for 20 min) are chosen as the solid culture medium for microbial isolation and purification. The fermentation culture medium for *Sorangium cellulosum* is CNST culture medium (0.5 g KNO_3, 0.65 g Na_2HPO_4, 1.0 g $MgSO_4 \cdot 7H_2O$, 1 ml $FeCl_3$, 1 ml solution of trace elements, 15 g agar. Diluted to a constant volume of 1 L with deionized water. Adjusted pH to 7.2 with 1 mol/L NaOH. Sterilized at 121 °C for 20 min).

2.2 Method for Separating and Purifying Microorganisms in Soil

Adequate amounts of soil in different environments are obtained and suspended into sterile water. The supernatant is made into serial dilutions and is spread on PDA and LB culture plates. After incubating for 2 days at 33 °C, single colonies are picked and cultured. The purified microorganisms are preserved.

2.3 Preparation Method of Epothilones

The *Sorangium cellulosum* is inoculated on the solid CNST culture medium and incubated for 4 days. A layer of sterilized resin is laid on top and the incubation is continued for another 6 days. The resin is collected and dried at 30° C before desorbed by methanol to dissolve epothilone A and B. The methanol solution is subsequently dried with a rotary evaporator, and the obtained epothilone A and B are dissolved in anhydrous ethanol. The solution is stored at 4 °C for future use.

2.4 Screening and Identification of Microorganisms in Soil

Three experimental groups were setted for screening (Fig. 1). The purified microorganisms from 2.2 are inoculated in 25 mL liquid medium, while at the same time a 0.5 mL of ethanol solution containing epothilone is added into the medium (No. 3). Two experimental control groups are set up for comparison: one is 25 mL medium added with 0.5 mL ethanol solution of epothilones but with no microorganism inoculation (No. 1), and the other one is 25 mL medium added with 0.5 mL of anhydrous ethanol and inoculated with the purified microorganism. The three sets of groups are incubated simultaneously at 32 °C for 5 days on a 180 r/min shaker. The collected zymotic fluid is mixed with methanol by 1:1 and the supernatant is obtained by centrifugation in order remove the precipitating sugars and proteins in the culture medium. Subsequently, the supernatant is analyzed by HPLC. After comparing the three groups, the sample with new peak near the epothilone A and B in the No. 3 group was selected and the genome DNA of

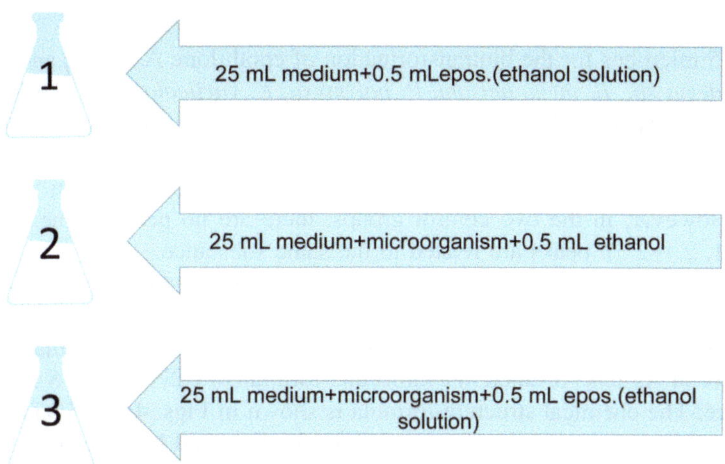

Fig. 1 The design of the screening of microorganisms in soil

Fig. 2 The chromatogram map of HPLC (1 medium + epothilones 3.1: midium + *Burkholderia* sp. + epothilones. 2.1: midium + *Burkholderia* sp. 3.2: midum + *B. megatherium* + epothilones. 2.2: midium + *B. megatherium.* 3.3: midium + *P. polymyxa* + epothilones. 2.3: midium + *P. polymyxa.* 3.4: midium + *A. minutus* + epothilones. 2.4: midium + *A. minutus.* 3.5: midium + *K. variicola* + epothilones. 2.5: midium + *K. variicola.*)

corresponding microorganismswas isolated. Using 27F and 1492R primers, the 16s rDNA fragment was amplified and compared with Blast in NCBI.

2.5 *Analytical Method for the Transformation Products*

The HPLC measurement was done with a Japanese SHIMADZU LC-20A HPLC machine. Column: Agilent column, RP-C18 filler, particle size of 5 μm, 4.1 × 250 mm; mobile phase: methanol: water = 6:4; flow rate: 1.0 mL/min; detection wavelength: 249 nm; column temperature: 35 °C; injection volume: 20 μL.

The LC-MS measurement was done using Agilent 6530 QTOF machine. LC Methods: Agilent column, ProeshellC18 filler, particle size of 2.7 μm, 3.0 × 100 mm; mobile phase: 0.1% formic acid solution in water/methanol gradient elution, flow rate: 0.3 mL/min; detection wavelength: 249 nm; column temperature: 35 °C, injection volume: 20 μL. MS conditions: Source: AJS ESI, Gas temp: 325 °C, Drying Gas: 6 L/min, Sheath gas: 350 °C, Sheath gas flow: 11 L/min, Nebulizer: 35 psig, Vcap: 3500 V, Nozzle voltage: 500 V.

3 Results and Discussion

We got 5 microbes for the biotransformation of epothilone A and B, and they are: *Burkholderia* sp., *B. megatherium*, *P. polymyxa*, *K. variicola*, and *A. minutus*.

As shown by the HPLC chromatogram map (Fig. 2), new peaks near epothilone A and B can be seen in all culture media for these five types of microbes (3.1–3.5) and the amount of epothilone A and B decreased significantly in experiment group. However, in the two control groups, there are no peaks at the same time points (1, 2.1–2.5), peaks are related to the same substance.

The new product and the epothilone B subsequently underwent MS/MS scanning, and the Mass chromatogram show that the structures of the two have a high degree of similarity (Fig. 3). The structure analysis by MSC (Molecular Structure Correlator of Agilent) resulted in a molecular formula of $C27H43NO7S$ for the new substance. The chemical structure formula is shown in Figs. 4 and 5 and it can be seen that the epoxy ring at positions C12–C13 opens and becomes two hydroxyl groups on the basis of epothilone B. It is suspected that the new substance is transformed from epothilone B, and its structural formula is speculated in Fig. 6.

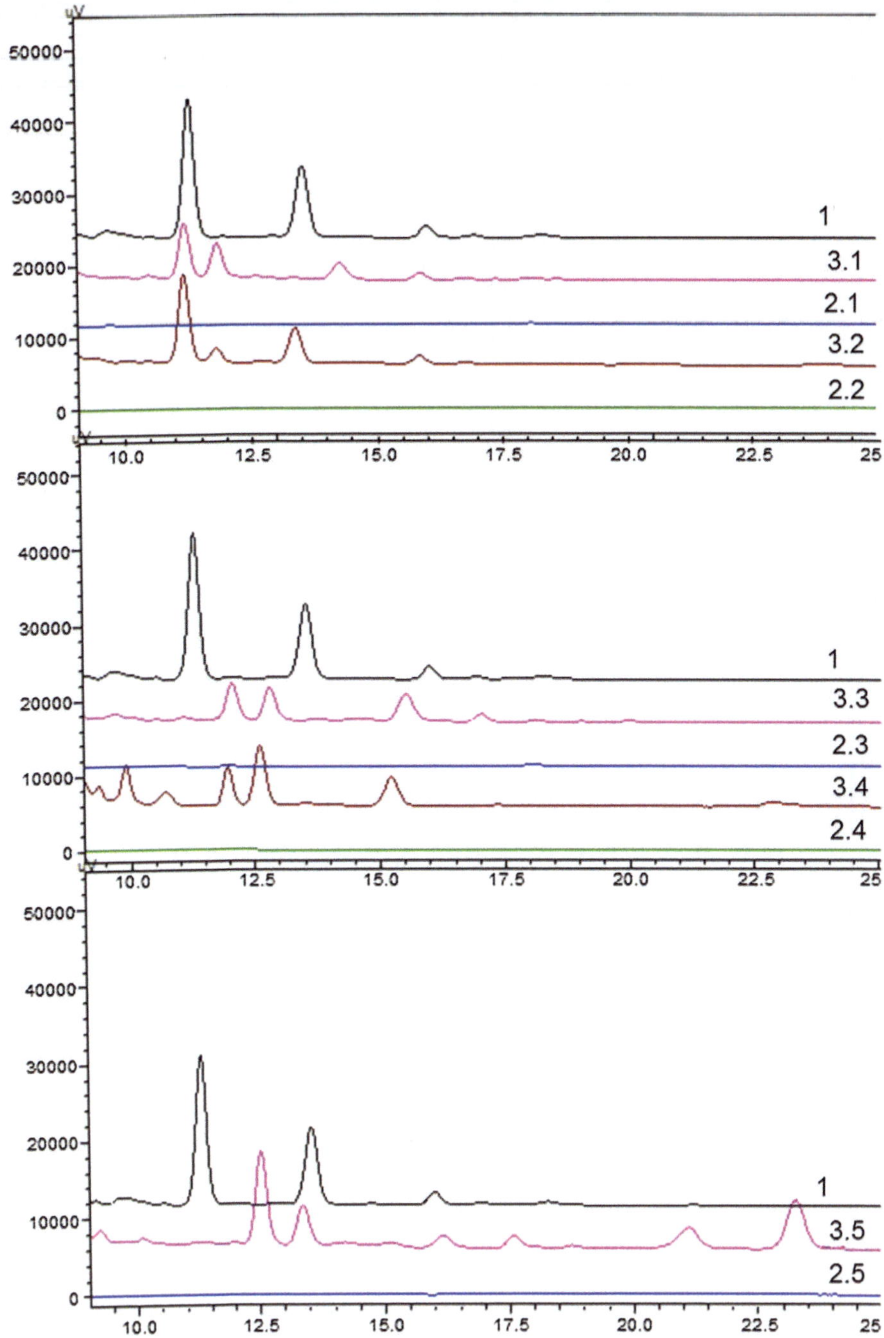
uV
50000
40000
30000
20000
10000
0
1
3.1
2.1
3.2
2.2
10.0
12.5
15.0
17.5
20.0
22.5
25
uV
50000
40000
30000
20000
10000
0
1
3.3
2.3
3.4
2.4
10.0
12.5
15.0
17.5
20.0
22.5
25
50000
40000
30000
20000
10000
0
1
3.5
2.5
10.0
12.5
15.0
17.5
20.0
22.5
25

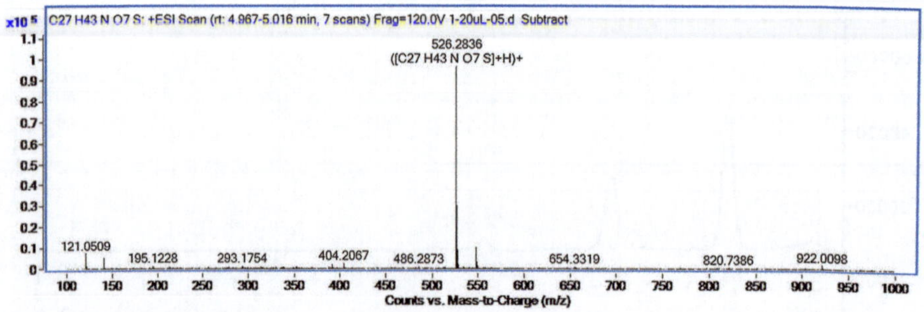

Fig. 3 Molecular weight of the new product measured by LC-MS

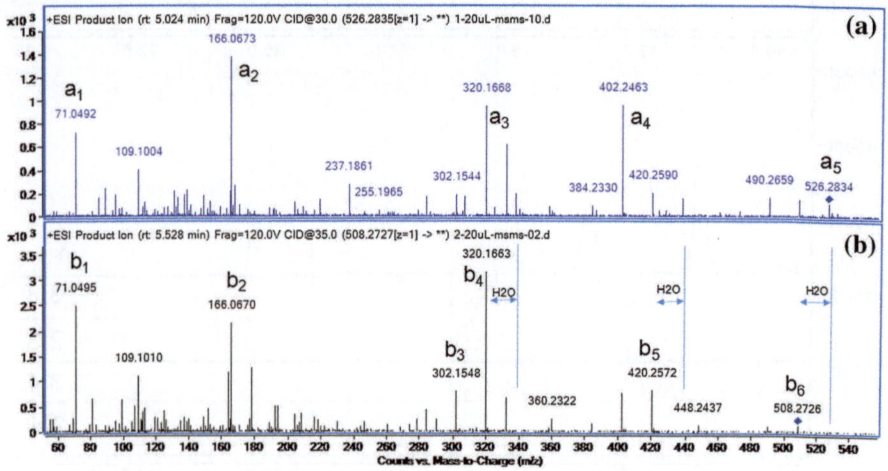

Fig. 4 Comparison of structure between new product and epothilone B, A is the MS/MS spectrometry of the new product. B is the MS/MS spectrometry of epothilone B

Sefkow et al. [10] has obtained a new analogue of epothilone by modifying the 12C–13C epoxy group of epothilone A with complex chemical methods. Karl-Heinz A et al. [11] have analyzed the anticancer activity of this new epothilone and the results showed that the activity is greatly reduced. But based on this structure, Altmann et al. [12] has synthesized an epothilone analogue with a better anticancer activity by chemical ketone reduction at the 12C–13C position. Altmann et al. [13] has pointed out that the opening of 12C–13C ring makes the conformation of epothilone analogues more flexible, thus making it possible to synthesize more epothilone analogs.

It is a new way to develop new medicines with similar functions as epothilones by transforming structure of epothilones with microorganisms. We will analyze the antitumor activity of the new medicines with similar functions as epothilones, and try to convert other drugs with the epoxy group *Burkholderia* sp. and B. megatherium so

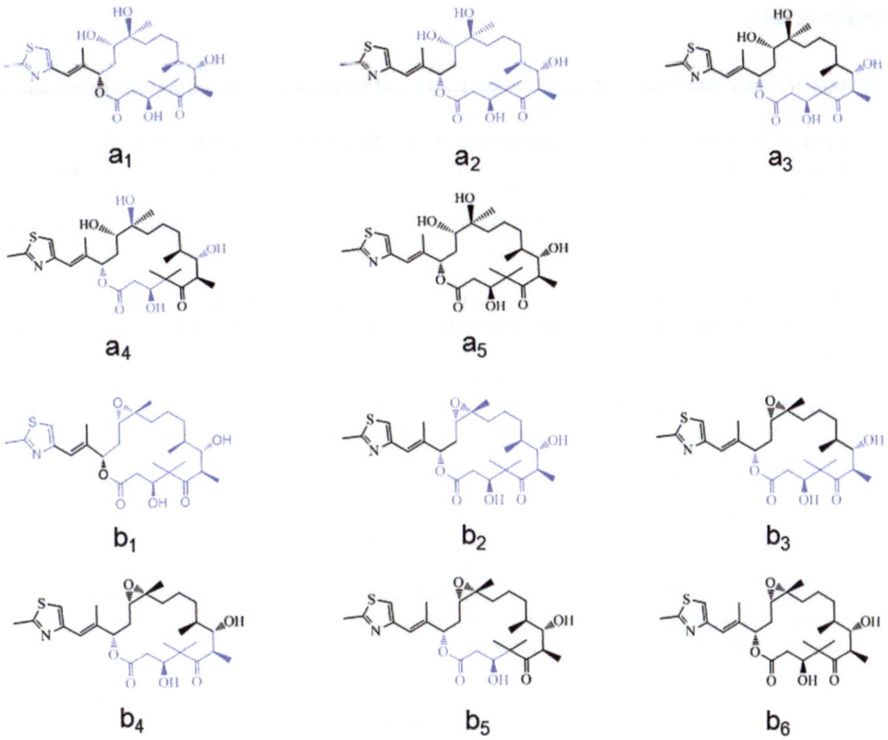

Fig. 5 The inferred structure of different molecular weight in Fig. 4

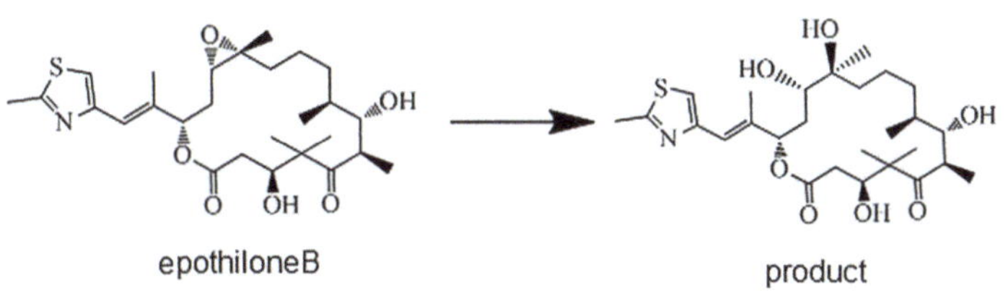

Fig. 6 The structure of the debris was inferred by MSC

as to get new more antibiotics. The transformation of natural compounds is not directional. However, compared with chemical synthesis, it can help us discover more derivatives and provide a guidance for the structural modification of the following functional molecular.

Acknowledgements The work was financially supported by Shandong Provincial Key Research and Development Program (No. 2015ZDZX05001, 2015ZDXX0502B01, 2015ZDXX0403B01, 2016GGB01409), National Natural Science Foundation (No. 31501396), Shandong Provincial Natural Science Foundation (No. ZR2012CQ027).

References

1. Yan JQ, Dai HM (2012) Review of the R&D and technology innovation of epothilones as new drugs. Chin J New Drugs 21:2241–2249
2. Kolman A (2005) Acitivity of epothilones. Curr Opin Ivestig Drugs 6:616–622
3. Klar U, Platzek J (2011) Asymmetric total synthesis of the epothilone sagopilone-from research to development. Acc 23:1291–1299
4. Hegazy MF, Mohamed TA, Eishamy AI et al (2015) Microbial biotransformation as a tool for drug development based on natural products from mevalonic acid pathway: a review. J A R 6:17–33
5. Holland HL (1998) Microbial transformations. Curr Opin Chem Biol 2:77–88
6. Tang L, Qiu RG, Katz L (2003) Generation of novel epothilone analogs with cytotoxic activity by biotransformation. J Antibot 56:16–23
7. Sohng JK, Kim BG, Ramesh PP et al (2016) Enzymatic synthesis of lactosylated and sialylated derivatives of epothilone A. Glycoconj J 33:137–146
8. Sohng JK, Parajuli P, Pandey RP (2014) Enzymatic synthesis of epothilone A glycosides. AMB Exp 4:31–44
9. Altmann KH, Pfeiffer B, Nicolaou KC (2007) The chemistry and biology of epothilones—the wheel keepsturning. Chem Med Chem 2:396–423
10. Sefkow M, Kiffe M, Hfle G (1998) Derivatization of the C12–C13 functional groups of epothilones A, B and C. Bioorg Med Chem Lett 8:3031–3036
11. Karl-Heinz A, Guido B, Giorgio C et al (2000) Epothilones and theiranalogs-potential new weapons in the fight against cancer. Chimia Int J Chem 54:612–621
12. Altmann KH (2009) Semisynthetic derivatives of epothilones. Fortschr Chem Org Naturst 90:135–156
13. Altmann KH, Gaugaz FZ, Schiess R et al (2011) Diversity through semisynthesis: the chemistry and biological activity of semisynthetic epothilone derivatives. Mol Divers 15: 383–399

Isolation, Screening and Evaluation of Potential Biocontrol Endophytes Against *Ralstonia solanacearum* on Ginger

Ning Zhou, Lin Zhao, Xin Sun and Xin-li Liu

1 Introduction

Ginger (*Zingiber officinale Rosc.*), a perennial herb, its extract has the activity of antibacterial, antioxidant, antiviral, anticancer, etc. Bacterial wilt caused by *R. solanacearum* is one of the major limiting factor in the cultivation of ginger. The peak of occurrence and prevalence about ginger wilt is from June to August, which has the characteristics of infecting easily and rapidly by *R. solanacearum* [1]. However, weight and bulk of the ginger tuber increase rapidly in the period, thus, the occurrence of diseases is easy to reduce the production of ginger. The common management strategies to control the ginger wilt include the adoption of resistant varieties, crop rotation and the use of synthetic chemical fungicides [2, 3]. However, considering the limitations of these methods, it seems appropriate to search for an alternative control strategy. Biocontrol approach using antagonistic microorganisms against the pathogen is now increasingly considered as an ideal treatment to manage plant diseases [4].

Plant endophyte can be defined as those microbe that colonize the internal tissue of the plant showing no external sign of infection or negative effect on their host [5, 6]. Plant endophyte are widely found in the root, stem, leaf, fruit and other organs, which have the characteristics of wide distribution, multiple species and abundant secondary metabolites. At present, scientists have proved that cotton, rice, potatoes, tomatoes, peppers and other plants [7] are host to one or more endophytes, and,

Ning Zhou and Lin Zhao have contribute equally to this paper.

N. Zhou · L. Zhao · X. Sun · X. Liu (✉)
Shandong Provincial Key Laboratory of Microbial Engineering,
School of Bioengineering, QiLu University of Technology,
No. 3501, Daxue Road, Jinan 250353, People's Republic of China
e-mail: vip.lxl@163.com

H. Liu et al. (eds.), *Advances in Applied Biotechnology*, Lecture Notes in Electrical Engineering 444, https://doi.org/10.1007/978-981-10-4801-2_60

some of these endophytes play a significant role to control plant diseases. A lot of studies have shown that niche exclusion, substrate competition, inducing plant resistance and production of secondary metabolites including antibiotics and siderophores make positive contribution in the control of pathogens by the antagonistic organism [2].

In this study, endophytes were isolated and screened from ginger (Shandong Laiwu, China). Our objectives were to assess the ability of different endophytes to control *R. solanacearum* in ginger, and provided a scientific basis for the prevention and treatment of ginger disease.

2 Materials and Methods

The materials and methods are listed as follows in this study.

2.1 Collection of the Ginger Material

The healthy tuber of ginger were collected from Ginger-growing Region in Laiwu, Shandong Province, China. The collected samples were placed in plastic bags and stored in a freezer at 4 °C.

2.2 Selected Pathogen

R. solanacearum was gifted from the Institute of Vegetables and flowers Chinese Academy of Agricultural Sciences.

2.3 Medium

① Bacteria isolation medium: beef extract peptone medium (NA), trypto-soya broth agar medium (TSA), King's B agar (KB), R_2A; ② Actinomycetes separation medium: gauserime synthetic agar medium; ③ Fungal selective culture medium: potato dextrose broth agar medium (PDA).

2.4 Determination of Surface Sterilized Condition

Freshly collected samples were thoroughly washed in washing liquid, followed by running tap water to rinse residue and two rinses in sterile double-distilled water. After thoroughly washing them, ginger tubers were surface sterilized with 75% ethanol for 1 min. Then, we explored the concentration of sodium hypochlorite and its disinfection time, these tubers sterilized respectively by 1, 3, 5, 10% sodium hypochlorite for 2, 5, 10 and 15 min. Then, these tubers were respectively washed in sterile double-distilled for five times to remove excess surface sterilants. To verify the effect of surface sterilization, two different methods were used. Firstly, the sterile double-distilled water of the final rinse were plated on NA and incubated in incubator, 28 °C, 2d. Secondly, the surface-sterilized samples were imprinted simultaneously for 5 min on NA and incubated under similar condition [8].

2.5 Isolation of Endophytes from the Tuber of Ginger

The surface-sterilized tuber was cut into a 3 cm × 3 cm square, placed in sterile mortar-pestle, crushed with the addition of sterile PBS buffer into a pulp, and stayed at room temperature for 30 min. After placement for 30 min, the tuber suspension was taken from the upper part and diluted to 10^{-1}–10^{-3} × in distilled PBS buffer. Serial dilutions from the suspensions obtained, were inoculated onto the plate of separation medium containing NA, TSA, KB, R_2A for bacteria, gauserime synthetic agar medium added with potassium dichromate (50 mg/L) for actinomycetes and PDA supplemented with streptomycin sulfate (100 mg/L) for fungal. All the petri dishes were incubated at 28 °C and were observed to know the growth condition of endophyte. These microbe were purified on the corresponding medium. To ensure the effect of surface sterilization, the sterile double-distilled water of the final rinse were plated on NA, and the surface-sterilized sample was imprinted simultaneously for 5 min on NA, then these plates were incubated under similar condition.

2.6 Screening for Antagonistic Strains in Vitro

The assessment of the inhibitory effects of strains on *R. solanacearum* was performed through the following procedure. Firstly, *R. solanacearum* was activated in NA Liquid medium at 28 °C for one day and tested strains were simultaneously recovered in NA liquid medium at 37 °C for one day. Secondly, the culture solution of *R. solanacearum* and melted NA medium were mixed and shook according to proportion of 1:100 to make the plates including pathogen. These plates were divided into 3 districts, and three holes with diameter of 6 mm were obtained in the center of each district. The culture solution of tested strains were added to the two hole, with the sterile water in another hole as the control. These plates were cultured in incubator at 28 °C for two days, and then, we observe and measure the size of inhibition zone.

2.7 Morphological, Biochemical Studies of Strain BJ-1, BJ-31

The morphological traits of strains BJ-1 and BJ-31 were observed under the microscope after dyeing. Twelve physiological and biochemical tests of strains BJ-1, BJ-31 for bacterial identification were implemented with the following experimental index: xylose, fructose, mannitol, glucose, V.P., citrate, gelatin hydrolysis, H_2S, nitric acid reduction, tyrosine hydrolysis, starch hydrolysis and catalase test [9].

2.8 16S RRNA Gene Amplification, Sequencing and Phylogenetic Analysis

The genomic DNA of endophytes: BJ-1 and BJ-31 were isolated using Ezup Column Bacteria Genomic DNA Purification Kit (Sangon Biotech, Shanghai, China) respectively. An approximately 1.5 kb sequence of the bacterial 16SrDNA gene was amplified using the primers 27f (5′-AGAGTTTGATCCTGGCTCAG-3′) and 1492r (5′-GGTTACCTTGTTACGACTT-3′). The PCR amplification was performed in a 50 μL reaction mixture containing 25 μL 2 × Easy pfu PCR SuperMix, 1 μL forward primer(10 μM), 1ul reverse primer(10 μM), 1 μL of template DNA, 22 μL of sterile double-distilled water. The PCR cycling protocol consisted of an initial denaturation at 94 °C for 4 min, followed by 30 cycles at 94 °C for 30 s, 55 °C for 30 s, 72 °C for 3.5 min, followed by a final step at 72 °C for 10 min. The PCR products were purified using SanPrep Column DNA Gel Extraction Kit (Sangon Biotech, Shanghai, China). These purified products were sequenced by Sangon Biotech (Shanghai, China). The 16S rDNA gene sequence of strains BJ-1 and BJ-31 were aligned with the reference sequences available in the Nucleotide Database of the National Center for Biotechnology Information (NCBI). Phylogenetic analysis was performed using the Neighbor-Joining (NJ) method with the MEGA software.

3 Results

The results were shown as follows:

3.1 *Surface Sterilization and Isolation of Endophytes from Ginger Tuber*

The effect of surface sterilization and the number of isolated endophytes are the basis of the separation experiment of plant endophytes. Results shown, when ginger tubers were disinfect with different concentration of sodium hypochlorite and different processing time, that 3% sodium hypochlorite for 2 min is the best disinfection condition for the isolation of ginger endophyte, and the number of isolated endophytes is maximum (Table 1). In addition, it also proved that the concentration of sodium hypochlorite and disinfection time have significant effect to the separation of plant endophyte.

In the study, a total of 39 endophytic bacterial isolates including ten strains of *Bacillus* (Table 2) and 2 endophytic actinomycetes were obtained, and fungal was not found, which is consistent with the result from Chu min [10]. In general, the method of tissue block is adopted for the isolation of endophytic fungi. Therefore, no fungi were found in this study, which may be associated with the separation method of endophytes.

Table 1 Surface sterilization rate of different sterilization (**a**) conditions and number of endophytes (**b**)

Concentration of sodium hypochlorite	Index	Processing time			
		2 min	5 min	10 min	15 min
1% NaClO	a	30.2%	75.6%	100%	100%
	b	0	0	0	20
3% NaClO	a	100%	100%	100%	100%
	b	32	28	20	15
5% NaClO	a	100%	100%	100%	100%
	b	23	18	11	7
10% NaClO	a	100%	100%	100%	100%
	b	8	5	0	0

Table 2 Morphological characteristics of 10 strains endophytic *Bacillus*

Number	Colonial morphology	Microbial morphology
BJ-1	Large, flat, moist and wrinkle-free	Rod, spore in the middle
BJ-2	Small, ridgy, dry and wrinkle	Rod, spore in the near middle
BJ-3	Small, flat, moist and wrinkle-free	Rod, spore in the middle
BJ-4	Small, ridgy, dry and wrinkle	Rod, terminal spore
BJ-24	Small, ridgy, moist and wrinkle	Rod, spore in the middle
BJ-26	Small, flat, moist and wrinkle-free	Rod, terminal spore
BJ-28	Large, flat, dry and wrinkle-free	Rod, spore in the near middle
BJ-29	Small, ridgy, dry and wrinkle	Rod, spore in the middle
BJ-30	Large, ridgy, dry and wrinkle	Rod, spore in the middle
BJ-31	Small, ridgy, dry and wrinkle	Rod, spore in the near middle

Table 3 Analysis of antagonistic activity of endophytes against *R. solanacearum*

Tested strains	Inhibition zone/cm	Tested strains	Inhibition zone/cm
BJ-1	3.54	BJ-26	0
BJ-2	2.35	BJ-28	3.00
BJ-3	2.95	BJ-29	3.10
BJ-4	1.95	BJ-30	3.23
BJ-24	2.17	BJ-31	3.54

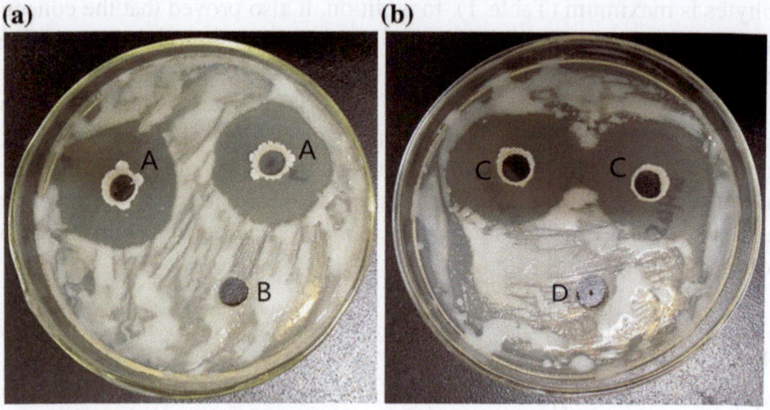

Fig. 1 Antagonistic effect of strains BJ-1, BJ-31 against *R. solanacearum.* **a** BJ-1, **b** Sterile water, **c** BJ-31, **d** Sterile water

3.2 Antagonistic Activity of the Isolates in Vitro

All the isolates were screened against *R. solanacearum* causing ginger wilt. We found that the strains BJ-1 and BJ-31 performed the highest and stable inhibition activity by analyzing the size of the diameter of the inhibition zone (Table 3). The antagonistic effect of BJ-1 and BJ-31 against *R. solanacearum* were shown in Fig. 1. From the Table 3, most of isolated endophytic *Bacillus* have obvious antagonistic activity against *R. solanacearum*, hence, the result revealed endophytic *Bacillus* used as potential biocontrol factor play an important role to control ginger wilt. We speculated that niche exclusion, substrate competition and the production of antibacterial substance may be positive factors in the inhibition to *R. solanacearum*.

3.3 Morphological, Biochemical Studies of Strain BJ-1, BJ-31

Strain BJ-1 has characteristics of larger, flat, moist, G +, short rod, spore-production on NA medium; strain BJ-31 is small, round, ridgy, dry, G+, short rod, spore-production on the NA medium (Fig. 2).

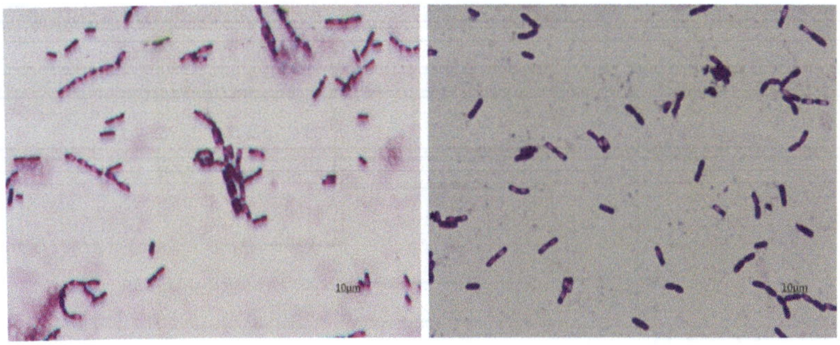

BJ-1 BJ-31

Fig. 2 Morphology of antagonistic Bacteria under microscope with 1000× condition

Table 4 Physiological and biochemical tests of strains BJ-1, BJ-31[Note 1]

Index	BJ-1	B-31
Glucose	+	+
Xylose	−	−
Fructose	+	−
H_2S	−	+
Nitric acid reduction	+	+
V.P.	+	+
Mannitol	−	+
Tyrosine hydrolysis	−	−
Citrate	−	+
Gelatin hydrolysis	+	+
Starch hydrolysis	+	+
Catalase test	ND	+

Note 1 "+" positive or can be used; "−" negative or cannot be used; "*ND*" not be tested

The results of physiological and biochemical tests (Table 4) revealed that strains BJ-1 and BJ-31 were similar with *Bacillus cereus* and *Bacillus subtilis* respectively.

3.4 Identification of Strains BJ-1, BJ-31 Based on Its 16S RDNA Gene Sequence

1500 bp PCR products from the 16S rDNA gene were amplified respectively from the genomic DNA of strains BJ-1 and BJ-31. Sequence analysis showed that strains BJ-1 and BJ-31 respectively shared 99% identity with *B. cereus* and *B. subtilis* in the NCBI. NJ trees using 16S rDNA sequences were constructed, and the results clearly showed that strains BJ-1, BJ-31 clustered with the genus *Bacillus* (Fig. 3).

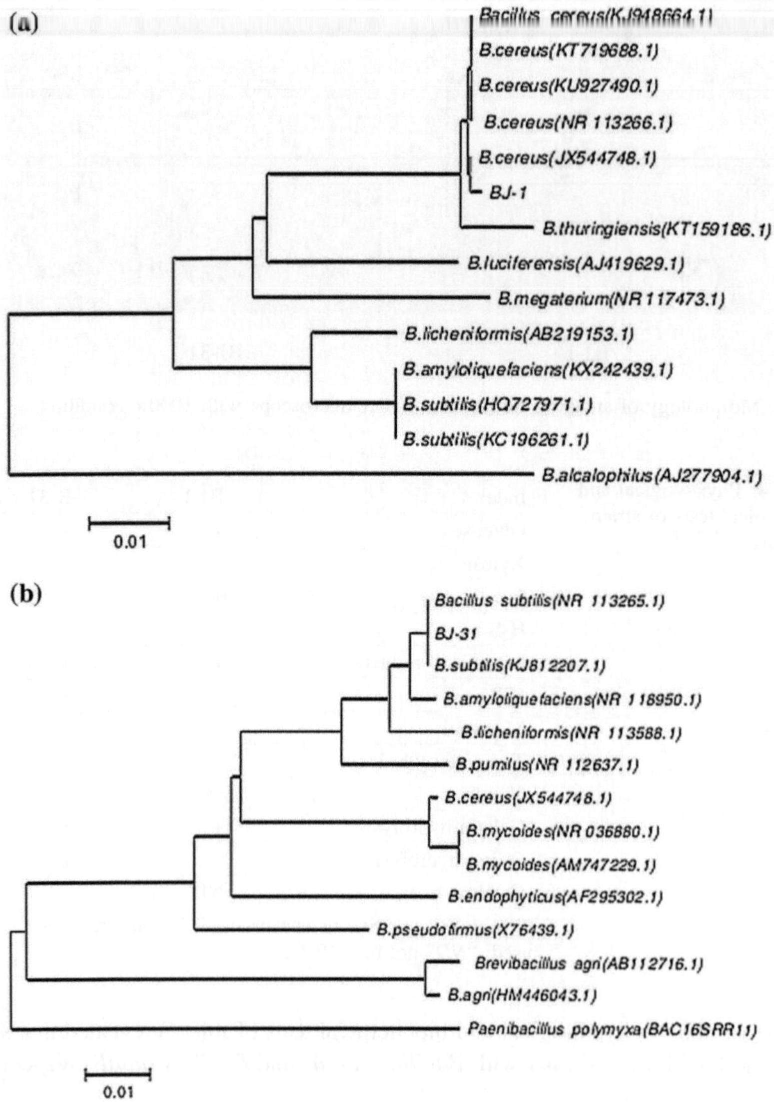

Fig. 3 Phylogenetic trees of strain BJ-1 (**a**) and BJ-31 (**b**) based on 16S rDNA gene sequences. The phylogenetic trees were constructed by the neighbor-joining (NJ) method using the MEGA 5.1 software

In a word, based on the consistency between the results of the morphological, physiological characterization and the 16S rDNA sequence analysis of strains BJ-1 and BJ-31 were respectively identified as *B. cereus* and *B. subtilis*.

4 Discussion

Plant endophytes considered as effective biological control agents (BCAs) play an enormously important role in plant disease control. Surface sterilization of sample is the key step in the separation of plant endophytes. However, we should consider the effect of disinfectant to the endophytes of explant on the basis of sterilization, otherwise, which will cause interference to isolate endophytes. The early experiments found that the disinfection of sodium hypochlorite is the key to the surface sterilization in this lab, therefore, this research adopts the different concentrations of sodium hypochlorite and processing time to treat ginger tuber. The separation of endophytes are related to many factors such as culture medium, culture conditions and so on. In this study, the method of separation of endophytes was relative, although endophytes were successful isolated from ginger tissue. In brief, various factors should be taken into account for the separation of endophytes.

In this study, 41 endophytic strains were obtained from the tuber of ginger, including 10 strains of *Bacillus*. Most of the isolated endophytic *Bacillus* showed antagonistic activity against *R. solanacearum*. The result is consistent with related literature [11, 12]. *Bacillus* agents are utilized for the biocontrol of diseases in various plants including rice, vegetables, and trees [11] and so on. Two strains named BJ-1 and BJ-31 performed the highest inhibition activity. BJ-1 and BJ-31 were respectively identified as *B. cereus* and *B. subtilis* from the 16 s rRNA gene analysis and morphological, physiological characterization.

In this experiment, the mechanism of antagonism about strains BJ-1 and BJ-31 against *R. solanacearum* may be the following two aspects [2]: ① nutritional competition. The rapid growth of endophytic bacteria occupy the nutrition space, so that the pathogen is inhibited because of the lack of nutrition; ② the production of active antimicrobial substances. Endophytes produce some active substances, such as antibiotics, antibacterial protein, degrading enzymes in cell wall.

In summary, our results demonstrated that strains BJ-1 and BJ-31 are effective for controlling *R. solanacearum* diseases on ginger, therefore, a further study would aim at the mechanism of biocontrol and greenhouse experiment to see how these organisms protect ginger.

Acknowledgements The work was financially supported by Shandong Provincial Key Research and Development Program (No. 2015ZDZX05001, 2015ZDXX0502B01, 2015ZDXX0403B01, 2016GGB01409), National Natural Science Foundation (No. 31501396), Shandong Provincial Natural Science Foundation (No. ZR2012CQ027).

References

1. Zhang CL, Zhao YQ, Xiao-Qing YU et al (2008) Screening and identification of antagonistic actinomyces strains against *Ralstonia solanacearum*. Acta Phytopathol Sin 38(4):414–419
2. Thomas P, Upreti R (2015) Testing of bacterial endophytes from non-host sources as potential antagonistic agents against tomato wilt pathogen *Ralstonia solanacearum*. Adv Microbiol 04 (10):656–666
3. Ohike T, Makuni K, Okanami M et al (2013) Screening of endophytic bacteria against fungal plant pathogens. J Environ Sci 25(25S1):S122–S126
4. Wang SL, Shih IL, Wang CH et al (2002) Production of antifungal compounds from chitin by bacillus subtilis. Enzym Microb Technol 31(3):321–328
5. Ryan RP, Germaine K, Franks A et al (2008) Bacterial endophytes: recent developments and applications. FEMS Microbiol Lett 278(1):1–9
6. Petrini O (1991) Fungal endophytes of tree leaves. Microbial ecology of leaves. Springer, New York
7. Zhu YJ, Chen L, Lan JL et al (2009) Isolation, identification and the biocontrol potential of endophyte in theasienensis (camellia sinensis). J Fujian Agric For Univ
8. Kusari P, Kusari S, Spiteller M et al (2014) Biocontrol potential of endophytes harbored in radula marginata (liverwort) from the new zealand ecosystem. Antonie Van Leeuwenhoek 106(4):771–788
9. Zhou D (1986) Microbiology experimental manual. Shanghai Science and Technology Press
10. Chu M, Zhang ZD, Wang W et al (2011) The diversity of endophytes in ginger and screening of the antagonism. Xinjiang Agric Sci
11. Hyakumachi M, Takahashi H, Matsubara Y et al (2014) Recent studies on biological control of plant diseases in japan. J Gen Plant Pathol 80(4):287–302
12. Wang Z, Wang Y, Zheng L et al (2014) Isolation and characterization of an antifungal protein from bacillus licheniformis, hs10. Biochem Biophys Res Commun 454(1):48–52

Isolation and Molecular Identification of *Lactobacillus brevis* from Spoilage Craft Beer in China

Zhu Liping

Craft beer is a consumed fermented alcoholic beverage produced with water, malted cereal grains (generally barley and wheat), hops, and yeast. Due to the inclusion of specialty malt and hop ingredients, a microbrew varies in aroma and flavor from commercially brewed beer. Craft beer is more fresh, nutritious and complex flavor than typical light lager beer because of the absence of sterilization and filtration of *Saccharomyces cerevisiae* after brewing. Therefore, craft beer is more favorable and popular than industry beer world widely in recent years. Thousands of craft beer firms have sprung up in the past several years in China and other countries. According to the data from Brewers Association, there are 3418 craft brewers in America in 2014, sales of craft beer was about 2,664,000 tons with up 19.6 billon dollars in yearly sales, accounting for 11% of total beer production.

Beer is considered an unfavorable substrate of growth for many microorganisms due to the presence of ethanol, hop, high content of carbon dioxide, low pH (3.8–4.7), extremely reduced content of oxygen and the presence of only traces of nutritive substances [1]. However, there are still a limited number of bacteria and wild yeasts which can grow in beer and result in the production of turbidity and off flavor. Beer spoilage bacteria reported include the genera *Lactobacillus*, *Pediococcus*, *Pectinatus* and *Megasphaera* [1, 2].

Moreover, craft beer seems to be easily spoiled because of without being pasteurized or sterile-filtered after fermentation in order to hold special flavor and nutrient. More important, a considerable amount of small-scale craft beer breweries are usually lack of good management and quality control system in China. Even if craft beer happens to spoil and result in turning a little acid, microbrewery perhaps cannot be found. We encountered several spoilage craft beer in microbrewery in

Z. Liping (✉)
School of Bioengineering, Qilu University of Technology, No. 3501, University Road, Changqing District, Jinan 250353, Shandong, People's Republic of China
e-mail: zhuliping2012@aliyun.com

H. Liu et al. (eds.), *Advances in Applied Biotechnology*, Lecture Notes in Electrical Engineering 444, https://doi.org/10.1007/978-981-10-4801-2_61

China in this summer. This study investigated on isolation and identification of *Lactobacillus brevis* from spoilage craft beer.

The extract of wheat malt is more than that of barley malt in quantity, and the starch in wheat is easier to be pasted. More important, wheat malt making beer tastes unique aroma different from barley beer, which brings about wheat beer is favored by consumers, therefore, wheat has been used as raw material for beer brewing production world wide for granted. However, due to carrying a large variety of microorganisms in wheat plica, more nutritious of the wheat malt extract, and more difficult on filtration, wheat craft beer is susceptible to spoilage or inferior quality of the finished beer product.

The manufacturing process of wheat craft beer is similar to that of conventional lager beer. The wheat craft beer in China usually brewed with 40% malted wheat and 60% malted barley instead of 100% barley. The addition of hop and *Saccharomyces cerevisiae* is respectively 0.05–0.07% and 0.8 L/L. The fermentation temperature is 12–18 °C, the riped craft beer don't need filter or sterilize after brewing. Qualified craft wheat beer is about 4% (v/v) alcohol, pH 4.4–4.5 and acidity of 1.6–1.8. However, when spoilage by some species of lactic acid bacteria, craft beer tastes slightly acid, its acidity changes from 1.6–1.8 to 2.1–3.0, and its pH value decreases from 4.5 to 4.0. Moreover, microscopic observation finds a lot of undesirable short-rod bacteria except yeast in beer.

1 Materials and Methods

1.1 Microbiological Analysis and pH Measurement of Spoilage Craft Beer

The spoilage beer sample was streaked or spread onto MRS agar plates (Difco), all the plates were incubated under anaerobic conditions (AnaeroPack Rectangular Jar, Mitsubishi Gas Chemicals, Japan) at 35 °C for 48–72 h. The inoculated MRS agar was examined every day for the presence of colony formation [3, 4]. Then picked a typical colony using sterilized toothpicks to inoculate MRS broth (pH 5.5) and cultured anaerobically at 35 °C for 24 h. Measured pH value of the bacterial media using a pH meter (SevenCompact™ S210, Mettler, Switzerland).

Applied a smear of bacteria onto a slide, aired dry, heat fixed by passing it through a flame a few times, stained the bacterial sample with Gram's Stain Kit, and then examined under microscope at both 400 × and 1000 × oil immersion.

1.2 Bacterial DNA Extraction

The isolate colony was inoculated in MRS broth media and grew at 35 °C for 24 h on a roller. Took 1 ml of pure bacterial culture into a 1.5 mL microfuge tube, centrifuged at 10,000×*g* for 2 min at 4 °C, resuspend the sediment in 1 ml of TE buffer by pipetting gently, pelleted the cells by centrifuging as before. Resuspend the pellet in 175 μL TE buffer (pH 8) and then mixed and centrifuged. The supernatant was removed, and then 175 μL of TE buffer and 10 μL of RNase A (Sigma-Aldrich) were added to resuspend the cell pellet, incubated at room temperature for 10 min. The cells were lysed by the addition of 20 μL of 10% sodium dodecyl sulphate and 5 μL of a proteinase K solution (20 mg/mL) (Sigma-Aldrich), followed by a 2 h incubation at 56 °C. Subsequently, 20 μL of 5 M NaCl and 500 μL of TE (pH 8) were added to the samples. The lysate was extracted once with 500 μL of equilibrated phenol, which was mixed for 10 min and then centrifuged. The supernatant was removed, and the aqueous phase was extracted once more with 500 μL of chloroform:isoamyl alcohol (24:1) and centrifuged. The supernatant was transferred to a sterile tube and 500 μL of isopropanol was added. The mixture was centrifuged and the pellet was washed with 10 μL of 70% cold ethanol and centrifuged as before. The pellet was air dried and resuspended in 100 μL of DNase-free water. Bacterial genomic DNA thus prepared was stored at −20 °C until used.

1.3 Amplification of Bacterial 16s rDNA Genes

Bacterial genomic DNA of the isolate was subjected to PCR analysis. The 16s rDNA polymerase chain reaction (PCR) amplifications of was performed with universal primers, 27F (5′-AGAGTTTGATCATGGCTCAG-3′) and 1492R (5′-TACGGTTACCTTGTTACGACTT-3′) [5, 6]. PCR was performed in a thermocycler (Mastercycler, Eppendorf, Germany).

The PCR amplifications were performed as follows: 2 μL of the PCR template was used in a 25 μL PCR mixture containing 1 × PCR buffer with 2 mM $MgCl_2$, 0.3 mM of dNTPs mixture, 0.2 μM of each primer, and 2.5 units of Taq DNA polymerase (Invitrogen). PCR amplification was conducted with the following temperature profile: an initial denaturation of 5 min at 94 °C, 31 cycles of 30 s at 94 °C, 60 s at 55 °C, and 90 s at 72 °C, followed by 5 min at 72 °C. PCR product was resolved by electrophoresis in a 1% (w/v) agarose gel and visualised by EB staining [7, 8].

1.4 DNA Sequencing and Heredity Evolution Analysis

PCR amplified product was purified using a PureLink Quick Gel Extraction Kit (Invitrogen) according to the manufacturer's instructions. The product was directly sequenced on an ABI-PRISM 310 Genetic Analyzer (Applied Biosystems, USA) according to the manufacturer's protocol. Ambiguous and incorrectly called bases were manually corrected using Chromas Lite software (version 2.01) and Seaview software (version 4.3.3).

The 16s rDNA gene sequence of the isolate was compared with the published sequences from GenBank DNA database (www.ncbi.nih.gov) for homology using BLAST algorithm. Sequences from the top BLAST hits were downloaded for further phylogenetic comparison. Multiple-aligned with 16s rDNA gene sequences of different strains for similarity using Clustal W 1.81 software program coupled with MEGA 5 software. A neighbor-joining method was employed to construct the phylogenic tree using MEGA 5. The stability or accuracy of the inferred topology was assessed via a bootstrap analysis of 1000 replicates. The identities of the sequences were determined based on the highest percentage (a minimum of 98%) of the total nucleotide match with sequences from GenBank.

2 Results and Discussion

2.1 Bacterial Isolation, Cultivation and Morphologic Identification

The isolated strain grew slowly on MRS agar plate, and it forms visible colonies two days later. Colonies are white, round, convex, regular and 0.5–1 mm in diameter after 2–3 days growing on MRS agar plate (Fig. 1).

A typical bacterial colony was inoculated in MRS broth media and grew at 35 °C for 24 h, white sediment was found at the bottom of bacterial media culture.

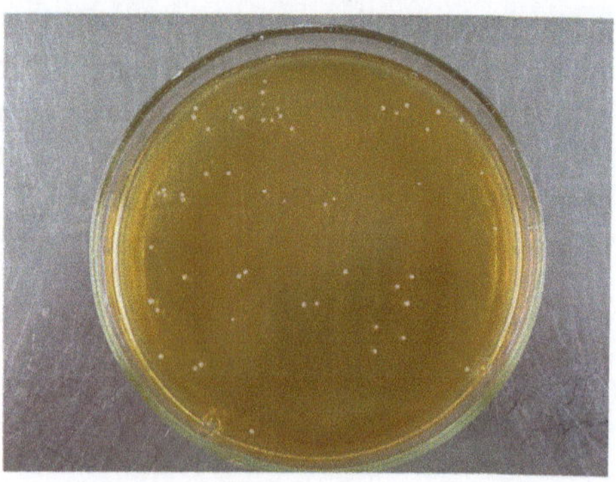

Fig. 1 Colony morphology of the isolate bacterium from spoilage craft beer

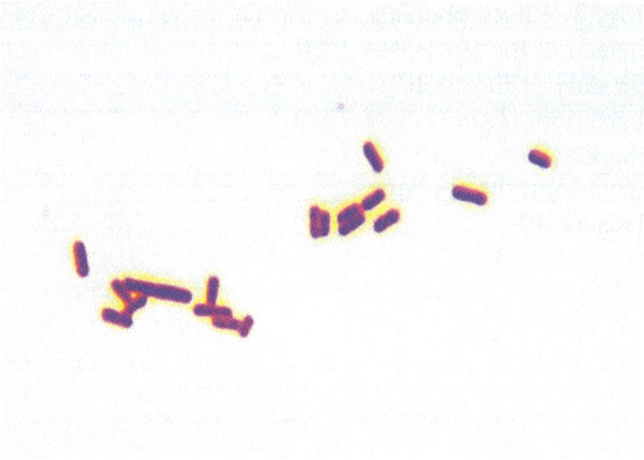

Fig. 2 Gram staining by microscopy of the spoilage bacterium

The isolate showed certain acidifying capacity. After growing 24 h, the pH value of the inoculated bacterial culture decreased from 6.4 to 3.8.

A loop of pure culture of the isolate was taken and plated on a slide, and then airdried, flame fixed, Gram stained, and microscopic observated. Figure 2 showed the isolate is Gram positive and short-rod shaped.

2.2 *Molecular Identification of the Beer Spoilage Bacterium Based on 16s rDNA Analysis*

Accurate identification of bacterial isolate depends on molecular characterizing of 16s rDNA sequencing, providing genus identification in >97% of cases and species in >99% [8, 9].

The full-length 16s rDNA of the isolate was PCR amplified using universal primers. Electrophoresis detection showed a about 1500-bp DNA fragment was obtained (Fig. 3), representing PCR product is consistent with the length of the objective gene fragment.

The purified 16s rDNA of the isolate was sequenced on an ABI-PRISM 310 Genetic Analyzer, the whole gene sequence was edited and validated, and then the full-length 16s rDNA sequence of the isolate was obtained. The sequence was submitted to the Genebank database on Sept 27th, 2016, and was deposited in GenBank under the accession number KX898576.1.

The 16s rDNA gene sequences of the isolate was 97%–98% similar to those of *Lactobacillus brevis* ATCC367 (Genbank accession number CP000416.1) and other 115 strains of *Lactobacillus brevis* available in the GenBank DNA database, suggesting the isolate belongs to the species *Lactobacillus brevis*.

Phylogenetic analysis was performed to confirm the identity of the 16s rDNA gene amplified from the isolate in this study. Sequences from the top BLAST hits were further phylogeneticly compared using the program Clustal W and a

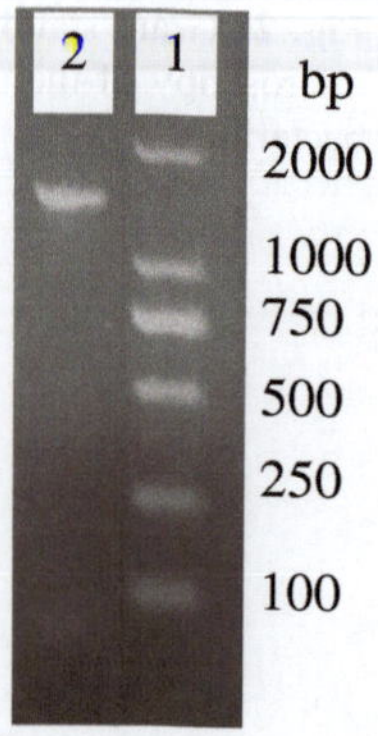

Fig. 3 Electrophoresis detection for 16s rDNA PCR product of the spoilage bacterium (*Lane 1*, molecular marker; *Lane 2*, 16s rDNA gene from the spoilage bacterium)

phylogenetic tree was constructed using Mega 5. A phylogenetic tree showing the relationship of the isolate with other *Lactobacillus brevis* species was shown in Fig. 4. Heredity evolution analysis showed the isolate belongs to different branch of *Lactobacillus brevis*, indicating potentially distinctive characteristics and function of the *Lactobacillus brevis* isolate to be further investigated in the future.

Craft beer contaminated by lactobacillus usually happened in microbrewing plants and was difficult to be realized by some microbreweries. We isolated, cultivated, Gram-staining microscopic examined an isolate bacterium from the spoilage craft beer, determined the pH value change of the isolate grown in MRS broth media, and further molecularly identified the isolate through 16s rDNA sequencing. Through the above research, the craft beer spoilage bacterium was isolated and identified as *Lactobacillus brevis*.

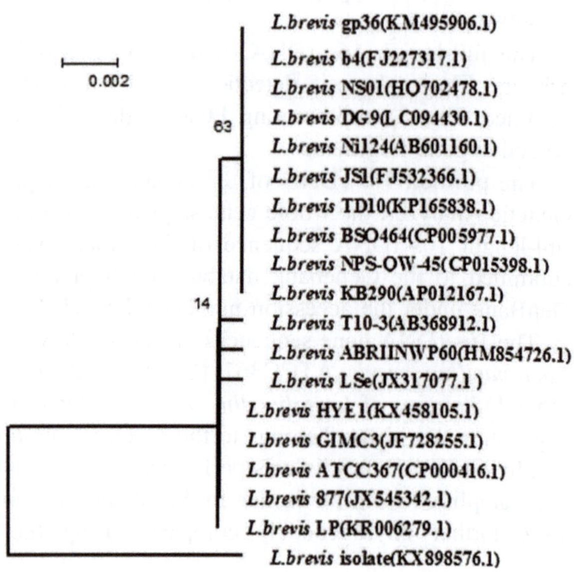

Fig. 4 Phylogenetic tree of 16s rDNA sequences (1.5 kb) of the isolated strain and other *L. brevis* strains derived from GenBank database, constructed using the neighbour-joining method

Judging the contaminated bacterium origined from the beer-making process or polluted *Saccharomyces cerevisiae*, we instructed microbreweries to take measures to strengthen sterilization in brewing process and stop utilizing passage *Saccharomyces cerevisiae* to successfully resolve the problem of craft beer spoilage.

References

1. Sakamoto K, Konings WN (2003) Beer spoilage bacteria and hop resistance. Int J Food Microbiol 89(2–3):105–124
2. Haakensen M, Ziola B (2008) Identification of novel hora-harbouring bacteria capable of spoiling beer. Can J Microbiol 54(4):321–325
3. Suzuki K, Asano S, Iijima K, Kuriyama H, Kitagawa Y (2008) Development of detection medium for hard-to-culture beer-spoilage lactic acid bacteria. J Appl Microbiol 104(5): 1458–1470
4. Bartowsky EJ, Xia D, Gibson RL, Fleet GH, Henschke PA (2003) Spoilage of bottled red wine by acetic acid bacteria. Lett Appl Microbiol 36(5):307–314
5. Wilson KH, Blitchington RB, Greene RC (1990) Amplification of bacterial 16s ribosomal dna with polymerase chain reaction. J Clin Microbiol 28(9):1942–1946
6. Frank JA, Reich CI, Sharma S, Weisbaum JS, Wilson BA, Olsen GJ (2008) Critical evaluation of two primers commonly used for amplification of bacterial 16s rRNA genes. Appl Environ Microbiol 74(8):2461–2470
7. Farimani RH, Najafi MBH, Bazzaz BSF, Edalatian MR, Bahrami AR, Flórez AB et al (2015) Identification, typing and functional characterization of dominant lactic acid bacteria strains from iranian traditional yoghurt. Eur Food Res Technol 242(4):517–526
8. Drancourt M, Bollet C, Carlioz A, Martelin R, Gayral JP, Raoult D (2000) 16S ribosomal DNA sequence analysis of a large collection of environmental and clinical unidentifiable bacterial isolates. J Clin Microbiol 38(10):3623–3630
9. Kim YW, Jeong YJ, Kim AY, Son HH, Lee JA et al (2014) *Lactobacillus brevis* strains from fermented Aloe vera survive gastroduodenal environment and suppress common food borne enteropathogens. PLoS ONE 9(3):e90866

Production of 5′-Inosinic Acid by Whole-Cell Biocatalyst Expressing a Mutated Acid Phosphatase/Phosphotransferase

Hui Yuan, Zi-fan Jia, Ju-hua He, Xiao-guang Fan and Ning Chen

1 Introduction

5′-Inosinic acid (Inosine 5′-monophosphate, 5′-IMP) is a dietary nucleotide that has been widely used in food industry, pharmaceuticals and agriculture [1]. Because of its characteristic umami taste and synergistic effect with monosodium glutamate, 5′-IMP has become a primary flavor potentiator in various food additives [2]. As inosine can be efficiently produced by fermentation, the progress of large-scale 5′-IMP production mainly depends on the improvement of phosphorylation process. The chemical phosphorylation process, which use phosphoryl chloride ($POCl_3$) or phosphorus pentoxide (P_2O_5) as phosphorylating reagent, requires complicated conditions and expensive raw material [3]. By contrast, enzymatic phosphorylation process using inosine kinase or acid phosphatase/phosphotransferase (AP/PTase) as biocatalyst has more advantage with respect to substrate specificity, product purity and reaction efficiency [4, 5].

AP/PTase is a novel selective nucleoside phosphorylating enzyme that consists of six subunits and possesses C-5′-position selective pyrophosphate-nucleoside phosphotransferase activity. AP/PTase has been purified and characterized from several microorganisms including *Morganella morganii* (*M. morganii*) and Enteric bacteria [6, 7]. Not like the inosine kinase, AP/PTase is an ATP-independent enzyme responsible for the biosynthesis of 5′-IMP by using inosine and tetrasodium pyrophosphate (TSPP) as the reactants [8]. However, the wild type AP/PTase exhibits not only phosphotransferase activity but also phosphatase activity.

H. Yuan · Z. Jia · J. He · X. Fan (✉) · N. Chen (✉)
Key Laboratory of Industrial Fermentation Microbiology, Ministry of Education,
Tianjin Key Lab of Industrial Microbiology, Tianjin University of Science
and Technology, Tianjin 300457, People's Republic of China
e-mail: xiaoguangfan@tust.edu.cn

N. Chen
e-mail: ningch@tust.edu.cn

H. Liu et al. (eds.), *Advances in Applied Biotechnology*, Lecture Notes
in Electrical Engineering 444, https://doi.org/10.1007/978-981-10-4801-2_62

The amounts of 5′-IMP produced were closely related to the Km value of AP/PTase for inosine [9, 10]. To enhance the molar yield of inosine to 5′-IMP, error-prone polymerase and site-directed mutation have been used for the improvement of phosphotransferase activity of AP/PTase [11–13]. With mutated AP/PTase, an improved phosphotrasferase reaction of inosine to 5′-IMP was obtained.

Although mutated AP/PTase has the potential for enzymatic synthesis of 5′-IMP, the high cost of enzyme still limit its large-scale industrial application. Recently, whole-cell biocatalyst has been applied into the production of many compounds in food and beverage industry [14, 15]. Compared with the purified enzymes, whole cell biocatalyst could simplify the process and be more stable to the external environment. Therefore, in the present study, the catalytic properties of whole-cell biocatalyst with mutated AP/PTase was studied and optimized.

2 Materials and Methods

2.1 Bacterial Strain, Plasmids and Cultural Conditions

A 690-bp nucleotide sequence of AP/PTase from *M. morganii* AP/PTase sequence (NCBI 10,466) with mutations of G92D and I171T was synthesized after optimization of the codons by gene designer software against *E. coli XL1-Blue* as host. The mutated AP/PTase gene defined as *phoC*^{M} was inserted into expression vector pQE30 (provided by Qiagen) between *BamH* I and *Pst* I restriction endonuclease sites, respectively. The recombinant plasmid pQE30-*phoC*^{M} was transformed into *E. coli XL1-Blue* by the classical heat shock procedure [16]. Transformation selection, colony PCR, plasmid isolation and agarose gel electrophoresis were operated according to standard protocol [17].

The seed culture medium (pH7.0) contained: 1.0% (w/v) tryptone, 1.0% (w/v) NaCl, 0.5% (w/v) yeast extract.

The composition of fermentation medium utilized for AP/PTase expression was as follows: 0.5% (w/v) typtone, 0.3% (w/v) yeast extract, 0.1% (w/v) KH_2PO_4, 0.25% (w/v) NaCl, 0.25% (w/v) $MgSO_4$, 2.0% (v/v) corn steep liquor and 0.3% (w/v) glucose. The initial pH of media was adjusted to 7.0 (Table 1).

Table 1 Bacterial strains and plasmids used in this study

Strains and plasmids	Charaters	Resources
E. coli XL1-Blue	lac I q, CmR	In this laboratory
E. coli XL1-Blue/pQE30	lac Iq, CmR, AmpR	In this paper
E. coli XL1-Blue/pQE30-phoCYM	lac Iq, CmR, AmpR, phoCYM	In this paper
pQE30	AmpR	In this laboratory
pQE30-phoCYM	AmpR, phoCYM	In this paper

2.2 Investigation of Mutated AP/PTase Expression Process

Recombinant *E. coli* was grown in seed culture medium for 9 h at 37 °C on a rotary shaker (200 rpm). The seed culture (5% (v/v)) was then inoculated into 5 L fermenter (Biotech-2002 Bioprocess controller, Baoxing, Shanghai, China) with a working volume of 3 L. The agitation speed, aeration rate, pH and initial temperature were maintained at 300 rpm, 1.0 vvm, 7.0 and 36 °C. IPTG was designed to be added with a final concentration of 0.1 mM at four different phases according to the growth curve of recombinant *E. coli*. The induction temperature was shifted to the set point ranged from 28 to 36 °C when IPTG was added to the cultures. Samples were taken hourly for enzyme activity assay after induction.

2.3 5′-Inosinic Acid Production by Whole-Cell Biocatalyst

The intact cells were harvested by centrifugation at 7000 × *g* for 10 min and washed twice with 50 mM phosphate buffer solution (pH = 7.0). The cell pellet was then resuspended in the same buffer solution containing with 20% glycerol, frozen in liquid nitrogen and stored at −80 °C.

The reaction was conducted in 250 mL Erlenmeyer flask containing 5 g/L (dry cell weight) whole-cell biocatalyst and 100 mM sodium acetate buffer solution (pH 5.2) and carried out in a shaking water bath (200 rpm, 35 °C) for 7 h. Different concentrations of inosine and disodium pyrophosphate were used as the substrates with a total reaction system volume of 100 mL.

2.4 Analysis Methods

AP/PTase activity was assayed in a standard reaction mixture comprising 0.2 mmol inosine, 0.2 mmol disodium pyrophosphate, 0.1 mmol acetate buffer (pH 5.0) and 10 mg (dry cell weight/DCW) whole-cell biocatalyst in a total volume of 2 mL. The reaction mixture was incubated at 35 °C and stopped after 20 min by adding 200 μl HCl (2 M). One unit of AP/PTase activity was defined as the amount of enzyme that generated 1 μmol of 5′-IMP per minute under the assay condition. Specific activity was defined as units per mg of protein. Protein concentration was determined using the method of Bradford with crystalline bovine serum albumin as a standard [18].

Quantitative determination of inosine and 5′-IMP was carried out by HPLC (Series 1200, Agilent Technologies, Santa Clara, CA, USA) with a diode array detection wavelength of 245 nm. The temperature of C18 column (4.6 × 150 mM, 3.5 μm; Agilent, USA) was maintained at 30 °C. The mobile phase was 10 mM potassium phosphate buffer (pH 2.8) and methanol (9:1, v/v), and the flow rate was 1.0 mL/min. Synthesized 5′-IMP was calculated as IMP · 2Na · $7.5H_2O$ (molecular weight, 527).

All experiments were carried out in triplicate and results represent mean standard ± error.

3 Results and Discussion

3.1 *Construction of Expression Plasmid for Mutated AP/PTase Gene*

Based on the reported nucleotide sequence, the AP/PTase gene with mutated sites from *M. morganii* was synthesized after optimization of codons. The mutated AP/PTase gene, named as *phoC*^{M}, was inserted into an expression vector pQE30 to construct the recombinant plasmid pQE30-*phoC*^{M}. The recombinant plasmid pQE30-*phoC*^{M} was transformed into the competent cells of *E. coli XL1-Blue*. The positive transformant was selected by colony PCR and double enzymatic digestion. The agarose gel electrophoresis analysis was shown in Fig. 1.

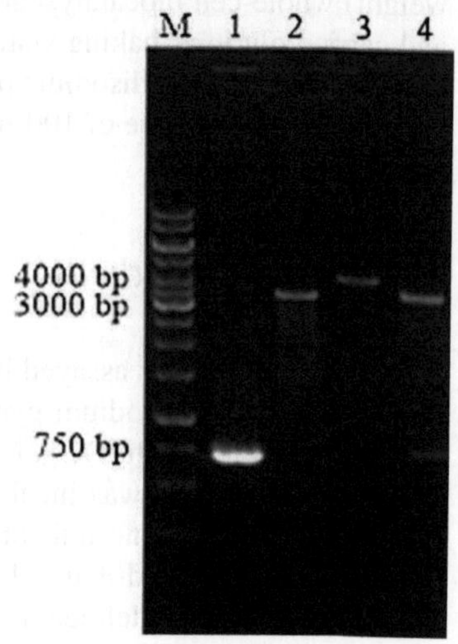

Fig. 1 The identification of recombinant plasmids pQE30-*phoCY*^{M} by digestion M: 1 kb DNA marker; *1* *phoCY*^{M} fragment; *2* pQE30 plasmid double enzyme digestion (*BamH* I/*Pst* I); *3* pQE30-*phoCY*^{M} plasmid single enzyme digestion (*BamH* I); *4* pQE30-*phoCY*^{M} plasmid double enzyme digestion (*BamH* I/Pst I)

3.2 Effect of Induction Initiation Time on Mutated AP/PTase Expression

The growth of *E. coli XL1-Blue* with pQE30-*phoC*^{M} was measured in 15 h cultivation without induction. According to the growth curve (data not shown), the microbe growth was divided into four phase: lag phase (dry cell weight, DCW between 0.17 and 0.73 g/L), exponential phase (DCW between 0.74 and 6.47 g/L), stationary phase (DCW between 6.48 and 7.33 g/L) and deceleration phase (DCW over 7.33 g/L). IPTG was added when the culture reached the DCW of 0.52, 1.03, 3.67 g/L at 36 C. The effect of induction initiation time on mutated AP/PTase expression was shown in Fig. 2. AP/PTase started to be expressed along with the addition of IPTG. When IPTG was added at the early stage of exponential phase (DCW = 1.03 g/L), the highest specific activity of mutated AP/PTase (5.59 U/mg) and total activity of mutated AP/PTase (25.27 U/mL) were obtained, but the biomass of strains was only 4.46 g/L. When IPTG was added at the middle stage of exponential phase (DCW = 3.67 g/L) and stationary phase (DCW = 6.47 g/L), the biomass was higher than 5.5 g/L, but the specific activity was lower than 3.3 U/mg. High cell density was a strategy for high production of recombinant protein, at the same time high cell density suffered from several drawbacks including limited availability of dissolved oxygen, decrease growth rates and stimulate acetate formation [19, 20]. Since the production of target protein was correlated with specific growth rate, the induction initiation time at different growth phase may greatly affect the activity of recombinant protein [21]. In addition, cells grew slowly when IPTG was added at the lag phase (DCW = 0.52 g/L), which resulted in low total enzyme activity during 12 h cultivation. Hossain found that overexpression of heterologous protein could perturb cell growth [22]. Therefore, the early stage of exponential phase was the best induction initiation time for mutated AP/PTase expression.

3.3 Effect of Induction Initiation Time on Mutated AP/PTase Expression

Furthermore, the effect of induction temperature on mutated AP/PTase expression was investigated. Temperature was adjusted to 28, 30, 32, 34 and 36 °C when IPTG was added at the early stage of exponential phase (DCW = 1.03 g/L). As shown in Fig. 3, the highest specific activity of mutated AP/PTase was 6.21 U/mg at 32 °C, which were 1.33 and 1.48 folds higher than that at 36 and 28 °C. The biomass of strains was reduced with the decrease of temperature, but the highest total activity of mutated AP/PTase (32.07 U/mL) was obtained at the temperature of 32 °C. It was found that inclusion body and incorrectly folding of synthesized peptides appeared when the temperature of expression was not preferable [23]. Lower temperature could increase the stability of the recombinant plasmid and prevent the

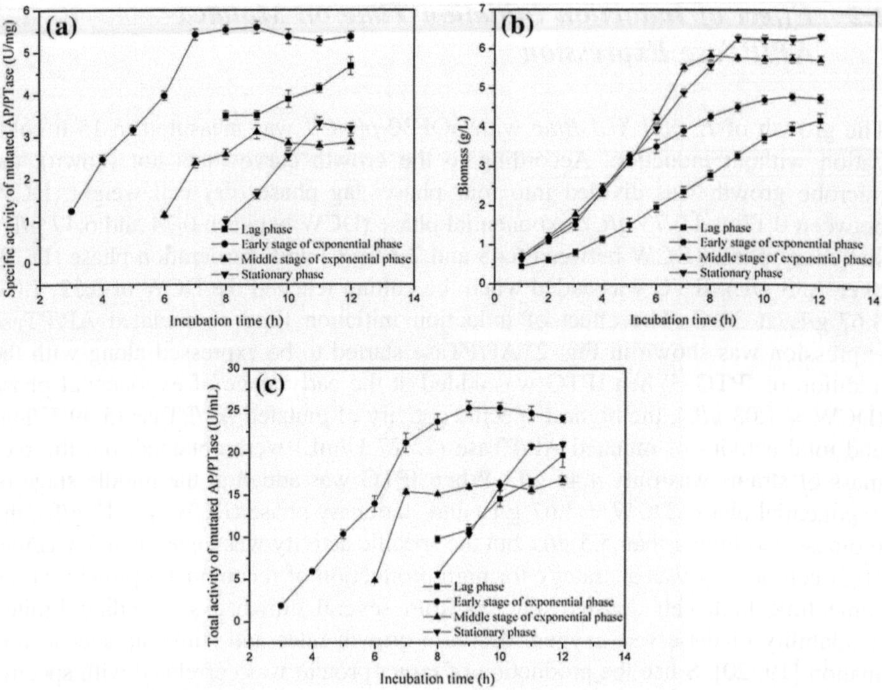

Fig. 2 The effect of induction initiation time on mutated AP/PTase expression. **a** The time course curve of unit enzyme activity with different induction initiation time; **b** The time course curve of biomass with different induction initiation time; **c** The time course curve of total enzyme activity with different induction initiation time

protein degradation and cell lysis [24]. However, when the mutated AP/PTase was incubated at a temperature of 28 °C, the total activity only remained 65%, indicating that the intracellular metabolism and enzyme secretion were weak at lower temperature [25].

3.4 Effect of Induction Duration on Mutated AP/PTase Expression

According to the above results, induction duration was evaluated at 32 °C when IPTG was added at the early stage of exponential phase (DCW = 1.03 g/L). As shown in Fig. 4, the specific activity of mutated AP/PTase approached to the same level of 6.1 U/mg after induction for 7 h. However, the total activity and biomass increased rapidly with the increase of induction duration and reached to the maximum of 32.08 U/mL at 9 h. In the late stage of induction phase (induction duration over 9 h), stresses could occur when the cell metabolism was subdued to high cell

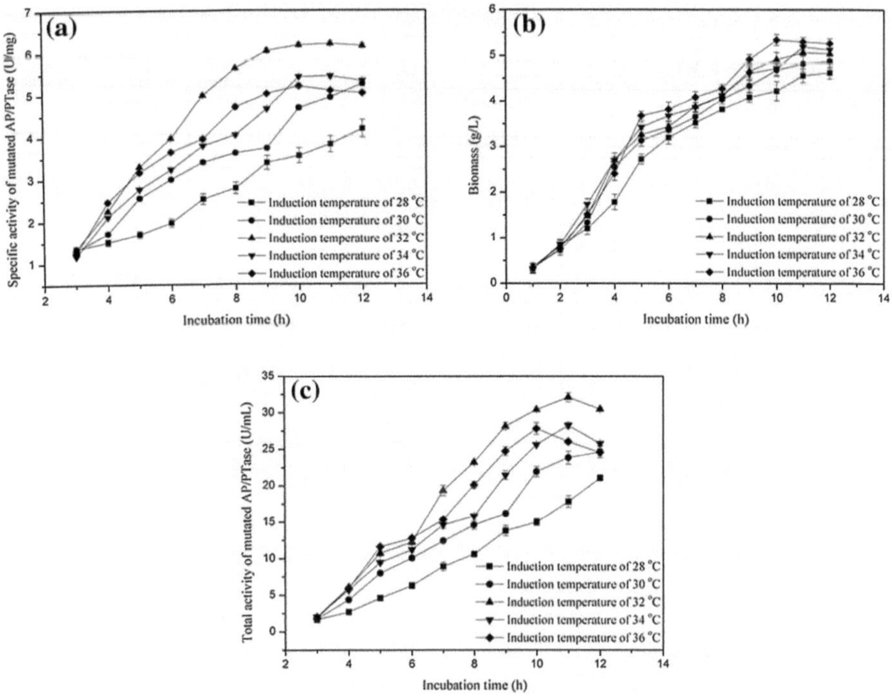

Fig. 3 The effect of induction temperature on mutated AP/PTase expression. **a** The time course curve of unit enzyme activity with different induction temperature; **b** The time course curve of biomass with different induction temperature; **c** The time course curve of total enzyme activity with different induction temperature

densities and undesirable culture conditions, which might cause the proteolysis by degradation of the cell membrane [26]. In the process of heterologous protein overexpression, total activity was the key factor for enzymatic reaction. Higher total activity resulted in lower protein dosage and more production formation. From an industrial point of view, 9 h was the best induction duration for mutated AP/PTase expression.

3.5 *Production of 5′-IMP by Whole-Cell Biocatalyst*

Using the whole-cell biocatalyst, the production of 5′-IMP from different concentrations of inosine and disodium pyrophosphate was examined. The reaction temperature and pH of this enzyme were set as 30 °C and 5.2 according to the previous study [7]. As shown in Fig. 5, 5′-IMP production decreased at lower substrate concentration because the superfluous whole-cell biocatalyst limited the mixing efficiency and oxygen transfer rate. Meanwhile, when the amount of substrate was

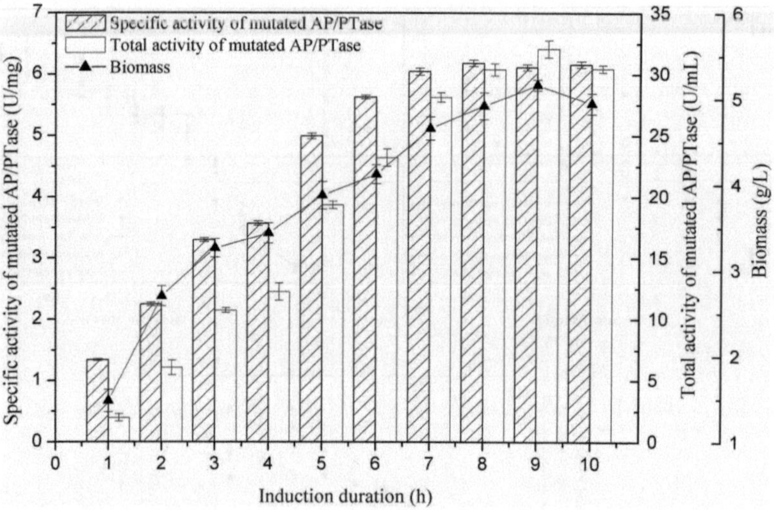

Fig. 4 The effect of induction duration on mutated AP/PTase expression

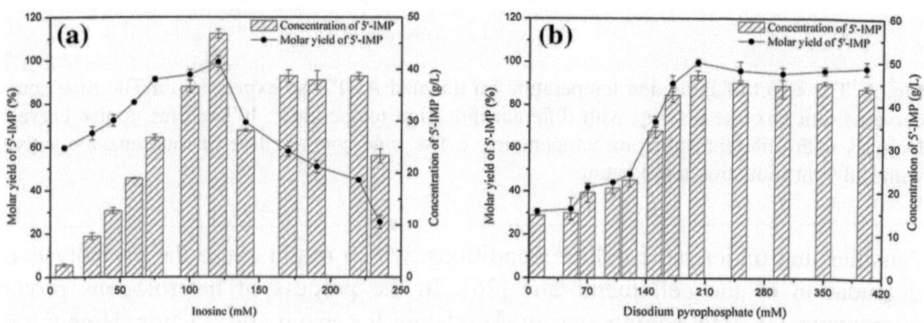

Fig. 5 Production of 5′-IMP by whole-cell biocatalyst under different concentrations of inosine (**a**) and disodium pyrophosphate (**b**)

greater than the required limit, the activities of the enzymes were inhibited, which in turn decreased the molar yield of 5′-IMP. The maximal molar yield and concentration of 5′-IMP were 99.65% and 46.89 g/L with 120 mM inosine and 200 mM disodium pyrophosphate. Yasuhiro Mihara et al. have developed several processes to produce 5′-IMP. Using mutated AP/PTase from *M. morganii*, 101 g/L of 5′-IMP production from inosine was achieved with a molar yield of 88% [13]. While using mutated AP/PTase from *E. blattae*, 156 g/L of 5′-IMP production from inosine was achieved with a molar yield of 79% [12]. The molar yield obtained in the present study was close to theoretical value, which was benefit for the separation and purification of the products. These results suggested that the whole-cell biocatalyst process described here might be an alternative choice for simple and efficient 5′-IMP production.

4 Conclusions

This paper presented a study of optimizing the induction conditions associated with whole-cell biocatalysis for 5′-IMP production. A successful extracellular expression of AP/PTase from *M. morganii* with mutations of G92D and I171T was reported. Results of single factor analysis indicated that the influence of induction initiation time and temperature on mutated AP/PTase expression was significant. 120 mM inosine and 200 mM disodium pyrophosphate could be converted to 5′-IMP during 7 h whole-cell biocatalysis with a molar yield of 99.65%. The approximate theoretical molar yield obtained in this study confirmed that the carbon efficiency value could be enhanced based on the optimization of mutated AP/PTase expression. Further study will be focused on the combination of inosine fermentation and immobilized whole-cell biocatalysis for 5′-IMP production.

Acknowledgements The authors are very grateful for financial support from Foundation for Key Laboratory of Industrial Fermentation Microbiology of Ministry of Education and Tianjin Key Lab of Industrial Microbiology (Grant No. 2016IM101).

References

1. Carver JD, Walker WA (1995) The role of nucleotides in human nutrition. J Nutr Biochem 6 (2):58–72
2. Ghirri A, Bignetti E (2012) Occurrence and role of umami molecules in foods. Int J Food Sci Nutr 63(7):871–881
3. Asano Y (2002) Overview of screening for new microbial catalysts and their uses in organic synthesis—selection and optimization of biocatalysts. J Biotechnol 94(1):65–72
4. Asano Y, Mihara Y, Yamada H (1999) A novel selective nucleoside phosphorylating enzyme from *Morganella morganii*. J Biosci Bioeng 87(6):732–738
5. Mori H, Iida A, Fujio T et al (1997) A novel process of inosine 5′-monophosphate production using overexpressed guanosine/inosine kinase. Appl Microbiol Biot 48(6):693–698
6. Asano Y, Mihara Y, Yamada H (1999) A new enzymatic method of selective phosphorylation of nucleosides[1]. J Mol Cata B: Enzy 6(6):271–277(7)
7. Mihara Y, Utagawa T, Yamada H et al (2001) Acid phosphatase/phosphotransferases from enteric bacteria. J Biosci Bioeng 92(1):50–54
8. Usuda Y, Shimaoka M, Kawasaki H et al (2002) Expression of the guanosine-inosine kinase gene from *Exiguobacterium*. World J Microb Biot 27(3):709–712
9. Ishikawa K, Mihara Y, Shimba N et al (2002) Enhancement of nucleoside phosphorylation activity in an acid phosphatase. Protein Eng Des Sel 15(7):539–543
10. Pugmire MJ, Ealick SE (2002) Structural analyses reveal two distinct families of nucleoside phosphorylases. Biochem J 361(1):1–25

11. Liu ZQ, Zhang L, Sun LH et al (2012) Enzymatic production of 5′-inosinic acid by a newly synthesised acid phosphatase/phosphotransferase. Food Chem 134(2):948–956
12. Mihara Y, Ishikawa K, Suzuki E et al (2004) Improving the pyrophosphate-inosine phosphotransferase activity of *Escherichia blattae* acid phosphatase by sequential site-directed mutagenesis. Biosci Biotech Bioch 68(5):1046–1050
13. Mihara Y, Utagawa T, Yamada H et al (2000) Phosphorylation of nucleosides by the mutated acid phosphatase from *Morganella morganii*. Appl Environ Microb 66(7):2811–2816
14. Carvalho CCCRD (2011) Enzymatic and whole cell catalysis: finding new strategies for old processes. Biotechnol Adv 29(1):75–83
15. Ishige T, Honda K, Shimizu S (2005) Whole organism biocatalysis. Curr Opin Chem Biol 9 (2):174–180
16. Seidman CE, Kevin S, Jen S et al (2001) Introduction of plasmid DNA into cells. Current protocols in molecular biology Chapter 1:A.4D.1-A.4D.2
17. Green MR, Green MR (2012) Molecular cloning: a laboratory manual, 4th edn. Three-Volume Set, Cold Spring Harbor Laboratory Press
18. Bradford MM (1976) Anal Biochem 72:248–254
19. Hannig G, Makrides SC (1998) Strategies for optimizing heterologous protein expression in *Escherichia coli*. Trends Biotechnol 16(2):54–60
20. Sivashanmugam A, Murray V, Cui C et al (2009) Practical protocols for production of very high yields of recombinant proteins using *Escherichia coli*. Protein Sci 18(5):936–948
21. Shokri A, Sandén A, Larsson G (2002) Growth rate-dependent changes in *Escherichia coli*, membrane structure and protein leakage. Appl Microb Biotechnol 58(3):386–392
22. Hossain GS, Li J, Shin HD et al (2014) Bioconversion of l-glutamic acid to α-ketoglutaric acid by an immobilized whole-cell biocatalyst expressing l-amino acid deaminase from *Proteus mirabilis*. J Biotechnol 169(Complete):112–120
23. Fu Z, Ng K, Lam TW (2005) Cell death caused by hyper-expression of a secretory exoglucanase in *Escherichia coli*. Protein Expres Purif 42(1):67–77
24. Baneyx F (1999) Recombinant protein expression in *Escherichia coli*. Curr Opin Biotechnol 10(5):411–421
25. Wang Y, Wang Z, Xu Q et al (2009) Lowering induction temperature for enhanced production of polygalacturonate lyase in recombinant *Pichia pastoris*. Process Biochem 44 (9):949–954
26. Mattanovich D, Gasser B, Hohenblum H, Sauer M (2004) Stress in recombinant protein yeasts. J Biotechnol 113(1–3):121–135

Bio-production of L-rhamnonate by *Pseudomonas taetrolens*

Shuhong Mao, Jianlin Wu, Lixia Zhang, Shuqi Gui and Fuping Lu

1 Introduction

Carbohydrate is one kind of important source of raw materials [1]. More and more attention was paid to the catalysis of the carbohydrate such as monosaccharides, disaccharides to their corresponding aldonic acids. These products have wide applications in the industry of food, cosmetics and medicine. For instance, gluconic acid is an oxidation product of glucose, in addition to its use as a food additive to regulate the acidity, its capability for chelating metals make them useful as the mineral supplement [2]. Calcium gluconate is on the World Health Organization's List of Essential Medicines. Lactobionic acid, which can be obtained by lactose oxidation, is a new important humectant and antioxidant due to its polyhydroxy acid structure [3]. Lactobionic acid can bind with water up to 14% to form a transparent matrix, and such a gel matrix is very useful for the skin as well as protection in wound healing [4].

The production of aldonic acid by microbial fermentation has become a reliable, cost-competitive, feasible way to overcome certain drawbacks associated with chemical synthesis or enzymatic catalysis [5]. In fact, microbial production of aldonic acid has been explored since 1950s. *Pseudomonas taetrolens* species such as *P. mucidolens*, *P. myxogenes* or *P. fluorescens* were found to be the main efficient aldonic acid producer [6, 7].

L-rhamnose is widely spread in plant biomass, and several microorganisms such as *Cry. ptococcus laurentii*, *Cry. neoformans* and *Pichiascolyti* are able to utilize

S. Mao · J. Wu · L. Zhang · S. Gui · F. Lu (✉)
Key Laboratory of Industrial Microbiology, Ministry of Education,
College of Biotechnology, Tianjin University of Science and Technology,
Tianjin 300457, People's Republic of China
e-mail: lfp@tust.edu.cn

S. Mao
e-mail: shuhongmao@tust.edu.cn

H. Liu et al. (eds.), *Advances in Applied Biotechnology*, Lecture Notes
in Electrical Engineering 444, https://doi.org/10.1007/978-981-10-4801-2_63

L-rhamnose as the carbon source to produce L-rhamnose acid after a long time fermentation with low yield but the by-product 1,2-propylene glycol was also obtained in this process [8], but little has been reported with *P. taetrolens* species.

This study investigated the pattern of L-rhamnose dissimilation by *P. taetrolens*, the isolation and identification of the product L-rhamnose acid as well as the effect of the temperature, cell dentistry and dissolved oxygen on the L-rhamnose acid production in a resting cell system was also studied.

2 Materials and Methods

2.1 *Microorganism and Culture Conditions*

P. taetrolens (TCCC 11892) was maintained at 28 °C on YPD medium containing 5 g/L yeast extract, 10 g/L peptone, 1 g/L glucose and 20 g/L agar. A loop of strain from the plate was inoculated in 50 mL YPD medium in a 250 mL flask. After grown for about 10 h on an orbital shaker at 220 rpm at 30 °C. The seed culture was inoculated into 50 mL YPD medium in 250 mL flask. The fermentation medium was performed for about 10 h at the same condition. The cells were harvested by centrifugation at 8000 rpm for 10 min and washed three times with stroke-physiological saline solution (pH 7). The washed cells were resuspended in 30 mL of the same solution with the final OD_{600} at about 10–50 [9–11].

2.2 *Biotransformation by Resting Cells*

Biotransformation of L-rhamnose using resting cells was carried out in 250 mL flasks with shaking at 100–220 rpm and 28–40 °C. Ca_2CO_3 (1 g) was used to adjust the acidity of the medium. The concentrations of the substrate and product were determined by HPLC.

2.2.1 Effect of Temperature on L-rhamnonic Acid Production

The transformation of L-rhamnose (1 g/L) by resting cells with the final cell concentration OD_{600} at about 20 was carried out under the shaker condition (180 rpm) at different temperature (28, 30, 32, 34, 37 and 40 °C).

2.2.2 Effect of Cell Concentration on L-rhamnonic Acid Production

The effect of different cell concentration at OD_{600} (10, 20, 30, 40 and 50) on L-rhamnonic acid was studied at 180 rpm and 37 °C.

2.2.3 Effect of Dissolved Oxygen Concentration on L-rhamnonic Acid Production

In order to investigated the effect of the dissolved oxygen concentration on L-rhamnonic acid production, the biotransformation of L-rhamnose was carried out using resting cells (OD_{600} = 40) at 37 °C under the different dissolved oxygen level controlled by orbital shaker (100, 140, 160, 180 and 220 rpm).

2.2.4 Time Course Profiles of L-rhamnonic Acid Production

The biotransformation of L-rhamnose was carried out using resting cells (OD_{600} = 40) at 37 °C, 180 rpm for about 48 h, and the samples was withdrawn at each 12 h and analyzed by HPLC.

2.3 *Separation and Identification of L-rhamnonic Acid*

After biotransformation, the medium was centrifugated at 8000 rpm, the supernatant was obtained and evaporated at 100 °C, the concentrated L-rhamnonate was then precipitated with equal volume of 95% ethanol, the precipitate was filtered and dried at 60° C for about 24 h, and the crude L-rhamnonate was then washed twice by water and filtered and dried at the same condition. The final product was dissolved in D_2O and analyzed by ^1H NMR.

2.4 *Analytical Methods*

HPLC was performed in an TSK-GEL NH2-100 column at 25 °C with acetonitrile: H_2O (60:40, v/v, Containing 0.1% of phosphoric acid) as the eluent, and the flow rate was 1 mL/min. The HPLC spectrum of L-rhamnonic acid produced by *P. taetrolens* was shown in Fig. 1.

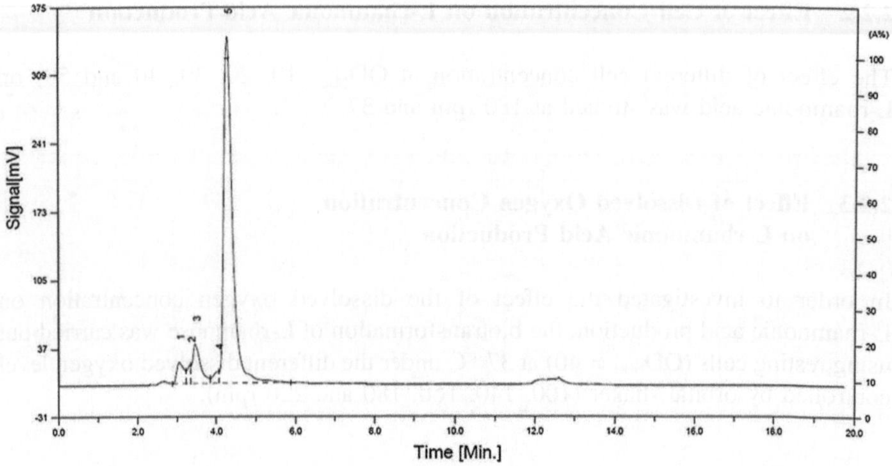

Fig.1 The HPLC spectrum of L-rhamnonic acid produced by *P. taetrolens*. The peak 4 is corresponded to L-rhamnonic acid

3 Results and Discussion

3.1 *Identification of L-rhamnonic Acid Obtained from Biotransformation*

The product obtain from the biotransformation by *P. taetrolens* should be L-rhamnonic acid because they had the same ^{1}H NMR spectrum as the standard sodium L-rhamnonate as shown in Figs. 2 and 3. *P. taetrolens* had shown for their ability of oxidizing disaccharides such as lactose, maltose and cellobiose to the corresponding aldobionic acids. Furthermore, *Pseudomonas* sp as an efficient microbial platform has gained much attention in the last decades because of the reliable and cost effective fermentation process. In this study, *P. taetrolens* was also found to be able to transform the L-rhamnose into L-rhamnonic acid, which has only been observed in several yeasts and the ability of *P. taetrolens* to transform monosaccharides to the corresponding aldobionic acids was reported for the first time (Fig. 4).

3.2 *Bio-production of L-rhamnonic Acid by* Pseudomonas taetrolens

Temperature has an important influence on the biotransformation efficiency. As shown in Fig. 5a, the production of L-rhamnonic acid increased with the temperature increased from 28 to 37 °C, when the temperature was more than 37 °C, the

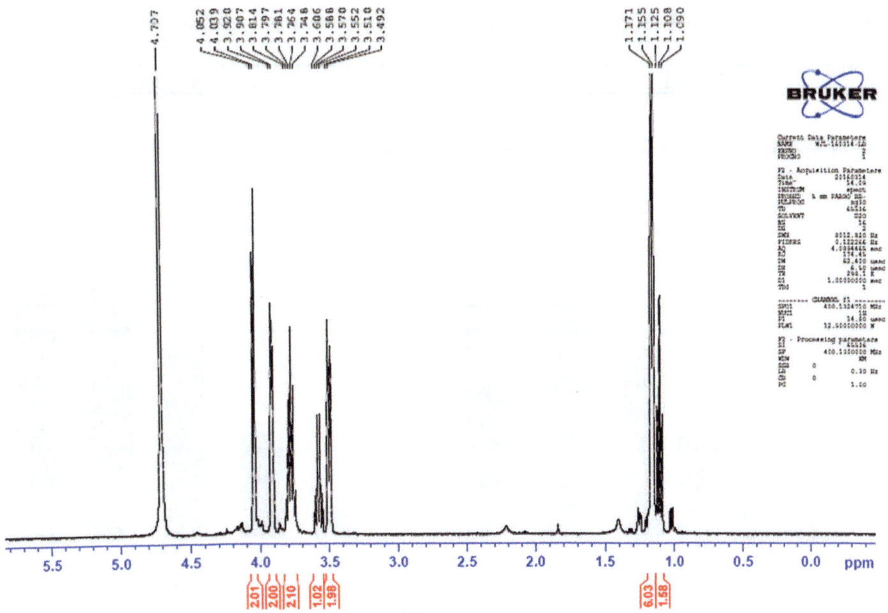

Fig. 2 The ^{1}H NMR of sodium L-rhamnonate produced by *P. taetrolens*

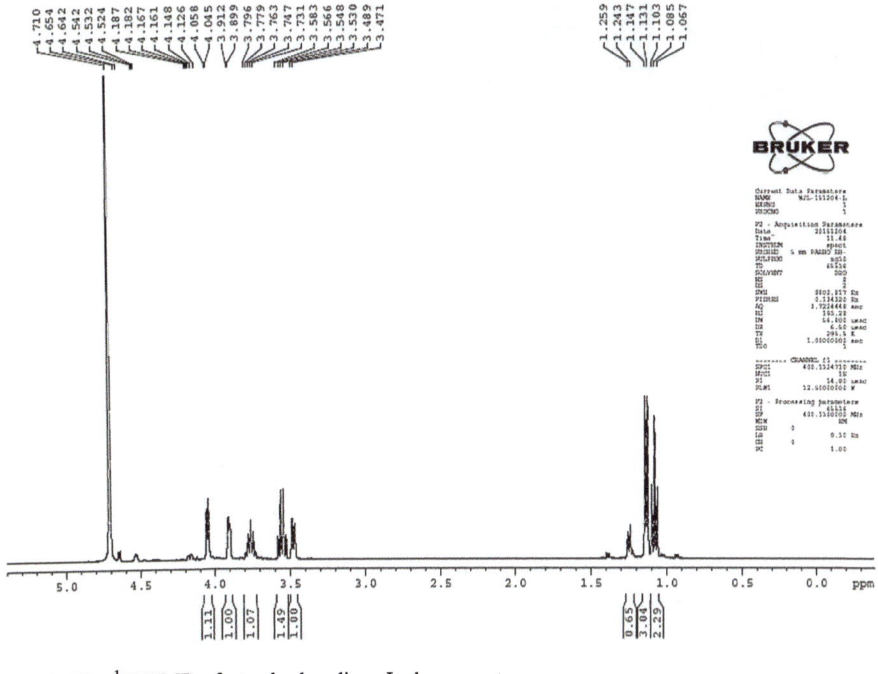

Fig. 3 The ^{1}H NMR of standard sodium L-rhamnonate

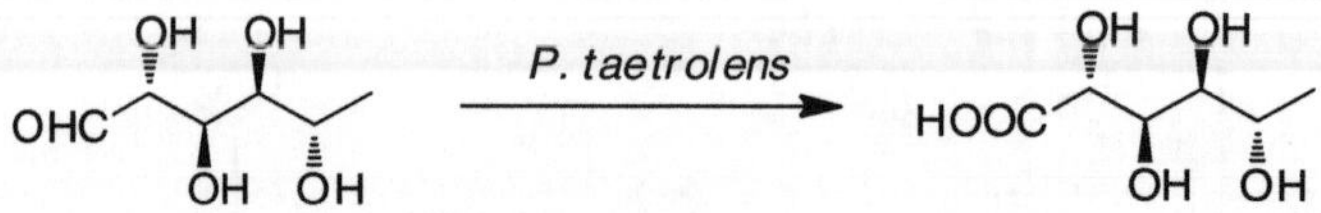

Fig. 4 Scheme of the L-rhamnonic acid produced by *P. taetrolens*

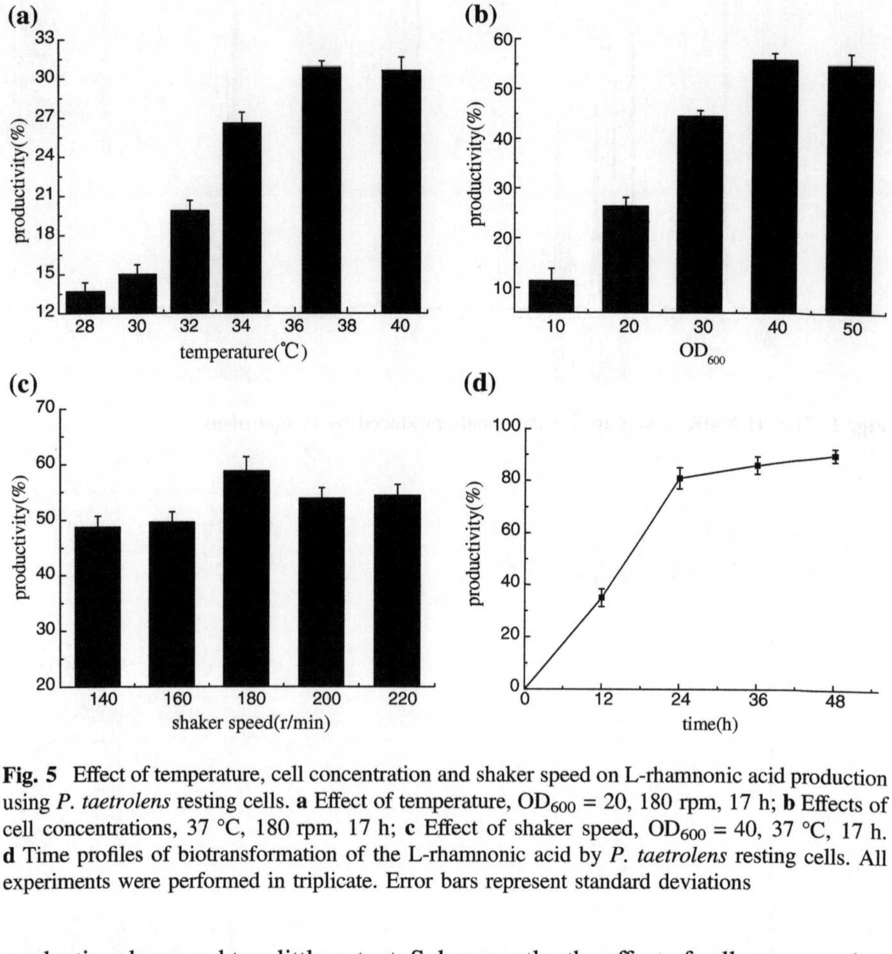

Fig. 5 Effect of temperature, cell concentration and shaker speed on L-rhamnonic acid production using *P. taetrolens* resting cells. **a** Effect of temperature, OD_{600} = 20, 180 rpm, 17 h; **b** Effects of cell concentrations, 37 °C, 180 rpm, 17 h; **c** Effect of shaker speed, OD_{600} = 40, 37 °C, 17 h. **d** Time profiles of biotransformation of the L-rhamnonic acid by *P. taetrolens* resting cells. All experiments were performed in triplicate. Error bars represent standard deviations

production decreased to a little extent. Subsequently, the effect of cell concentration on the production was also investigated, as can be seen in Fig. 5b, a gradual improvement of production was observed by increasing the cell concentration from OD_{600} 10 to OD_{600} 40, more addition of the cells didn't result in the higher production perhaps owing to the low dissolved oxygen in the system. Finally, the effect of the dissolved oxygen on the production was shown in Fig. 5c, and there is a slight increase in the production with increasing shake speed to 180 rpm, no more improvement in production was obtained although the shaker speed exceed 180 rpm, this is perhaps due to the maximum dissolved oxygen in the water.

4 Conclusions

The biotransformation of L-rhamnose to produce L-rhamnonic acid by *P. taetrolens* was first reported. The efficient L-rhamnonic acid bio-production process using resting cells was developed.

References

1. Robyt JF (1998) Essentials of carbohydrate chemistry. Springer, New York, p 399
2. Merfort M, Herrmann Ute, Stephanie BM, Hermann S (2006) High-yield 5-keto-d-gluconic acid formation is mediated by soluble and membrane-bound gluconate-5-dehydrogenases of *Gluconobacte*r oxydans. Appl Microbiol Biotechnol 73:443–451
3. Gutierrez LF, Hamoudi S, Belkacemi K (2012) Lactobionic acid: a high value-added lactose derivative for food and pharmaceutical applications. Int Dairy J 26:103–111
4. Gutierreza LF, Hamoudi S, Belkacemi K (2011) Selective production of lactobionic acid by aerobic oxidation of lactose over gold crystallites supported on mesoporous silica. Appl Catal A 402:94–103
5. Sauer M, Porro D, Mattanovich D, Paola B (2008) Microbial production of organic acids: expanding the markets. Trends Biotechnol 26:100–108
6. Tanimura R, Hamada A, Ikehara K, Iwamoto R (2003) Enzymatic synthesis of 2-keto-d-gluconate and 2-keto-d-galactonate from d-glucose and d-galactose with cell culture of *Pseudomonas fluorescens* and 2-keto-galactonate from d-galactono 1,4-lactone with partially purified 2-ketogalactonate reductase. J Mol Catal B Enzym 23:291–298
7. Vanschie BJ, Vandijken JP, Kuenen JG (1984) Non-coordinated synthesis of glucose dehydrogenase and its prosthetic group PQQ in and species. Sci Direct 24(1):133–138
8. Suzuki T, Onishi H (1968) Aerobic dissimilation of L-rhamnose and the production of L-rhamnonic acid and 1,2-propanediol by yeasts. Biol Chem 32(7):888–893
9. Grishko VV, Tarasova EV, Ivshina IB (2013) Biotransformation of betulin to betulone by growing and resting cells of the actinobacterium *Rhodococcus rhodochrous* IEGM 66. Process Biochem 48:1640–1644
10. Wang Z, Zhao F, Chen D, Li D (2006) Biotransformation of phytosterol to produceandrosta-diene-dione by resting cells of *Mycobacterium* in cloud point system. Process Biochem 41(3):557–561
11. Lang YJ, Bai L, Ren YN, Zhang LH, Nagata S (2011) Production of ectoine through a combined process that uses both growing and resting cells of *Halomonas salina* DSM 5928 (T). Extremophiles 15(2):303–310

Preparation and Stability Evaluation of Direct Vat Set *Lactobacillus helveticus* AAF1-5

Chuanxue Fu, Shuai Yang, Jun Mou, Jia Song, Min Wang and Yu Zheng

1 Introduction

Lactic acid bacteria (LAB) are well recognized worldwide, which are applied to food, health care, feed, agriculture etc. [1, 2]. Many strains of LAB are isolated and applied in production of fermented foods, such as cheese [3], sausage [4] and pickle [5]. Specially, LAB are important bacteria for Chinese traditional vinegar (CTV). LAB, including *Lactobacillus helveticus*, *Lactobacillus acetotolerans*, *Lactobacillus fructivorans* and *Lactobacillus pontis*, are dominant bacteria during fermentation process of Shanxi aged vinegar, a famous CTV, and produce abundant organic acids which give the vinegar unique taste and flavor [6]. To improve the property of CTV, fermentation reinforced with LAB that are isolated from itself is considered a potential strategy [7].

To make the use of LAB conveniently, the direct vat set (DVS) of LAB has been developed in spray-dried, or freeze-dried forms [8]. Within these forms, freeze-dried or lyophilized bacteria has been proven to be the most convenient and successful method for preserving bacteria, and have been used for several decades [9–11]. However, the freeze-drying process may be lethal to a large fraction of microorganisms, and the cytoplasmic membrane is considered the main site of dehydration damages [12]. Many studies focused on approaches to minimize the damage by adding protective agents. Different substances, including sugars (sucrose, lactose maltose trehalose), sugar alcohols (inositol, sorbitol), amino acids (sodium glutamate), have been tested for their protective effect during drying and storage [13, 14]. Sugars are reported to exhibit enhanced desiccation tolerance by forming hydrogen bonds to proteins during drying, which help to maintain the

C. Fu · S. Yang · J. Mou · J. Song · M. Wang · Y. Zheng (✉)
Key Laboratory of Industrial Fermentation Microbiology, Ministry of Education, College of Biotechnology, Tianjin University of Science and Technology, Tianjin 300457, P.R. China
e-mail: yuzheng@tust.edu.cn

H. Liu et al. (eds.), *Advances in Applied Biotechnology*, Lecture Notes in Electrical Engineering 444, https://doi.org/10.1007/978-981-10-4801-2_64

tertiary protein structure in the absence of water [15]. On the other hand, designing an appropriate fermentation medium is of crucial importance because medium composition can significantly affect the product yield of cells. MRS culture medium is commonly used for LAB culture, some other nutrient or enhancement factor such as tomato juice, $CaCO_3$ are added to improve the yield of cells [16–18].

In our previous work, *L. helveticus* AAF1-5 that appeared a potential application in reinforcing vinegar fermentation was isolated from *pei* of Shanxi aged vinegar. Thus, the objective of this work was to optimize the preparation of the DVS *L. helveticus* AAF1-5 and to evaluate the stability of this DVS.

2 Materials and Methods

2.1 *Microorganisms*

The strain *L. helveticus* AAF1-5 is conserved in Key Laboratory of Industrial Fermentation Microbiology, Ministry of Education. The strain is grown in the minimal medium for 12 h at 37 °C before experiments. The minimal medium is comprised (w/v) of glucose 2.0%, peptone 1.0%, beef extract 1.0%, yeast extract 0.5%, NaAc 0.5%, K_2HPO_4 0.2%, $MgSO_4$ 0.058%.

2.2 *Optimization of the Growth Factor*

Malt wort, tomato juice and corn steep liquor were chosen as growth factor. OD_{600} and the number of viable cells were chosen as the index to find out the suitable growth factor, and then the concentration (5, 10, 15, 20, 25, 30%) of malt wort was further optimized.

2.3 *Optimization of Preparing DVS* L. helveticus *AAF1-5*

The fermentation was ended after 24 h cultivation. The fermentation broth was treated by centrifugation at 5000×*g* for 5 min and the supernatant was discarded to harvest *L. helveticus* AAF1-5 cells. To protect the cell from low temperature and desiccation, lactose, sucrose, glucose, fructose and skim milk were chosen as protective agents. Each protective agent was added to the living cells (Bacterial mud: Protective agents (w/v), 1/17), and then cells were pre-frozen at 4 °C for 12 h and −20 °C for 24 h. Lyophilized was performed at −80 °C, 0.37 Pa for 24 h using a vacuum freeze dryer. DVS of *L. helveticus* AAF1-5 was spread on solid MRS media. The number of viable cells was calculated to select an appropriate protective agent. And then the concentration (5, 8, 11, 14%) was further optimized.

Prepared DVS of *L. helveticus* AAF1-5 was stored at 25 and 4 °C, respectively, to determine the survival rates of the strain.

2.4 Analytical Methods

Cell growth was monitored based on the OD value using a spectrophotometer (UVmini-1240, Shimadzu, Kyoto, Japan) at 600 nm. For assaying the concentration of available cells, 100 μL appropriate dilution of DVS was spread on solid MRS media. Plates were incubated at 37 °C for 48 h, and then the colony forming units (CFU) were counted.

Three replications were carried out for each experiment. The significance of each variable was determined by one-way variance analysis (ANOVA). Statistical significance was defined as $P < 0.05$.

3 Results and Discussion

3.1 Optimization of the Growth Factor

Growth factor, which is an important substance in the medium, is necessary for the growth of LAB. The growth factors, malt wort, tomato juice and corn steep liquor, were selected (Fig. 1a). Malt wort improved the growth of *L. helveticus* AAF1-5 ($P < 0.05$). Therefore, malt wort was selected to optimize the medium.

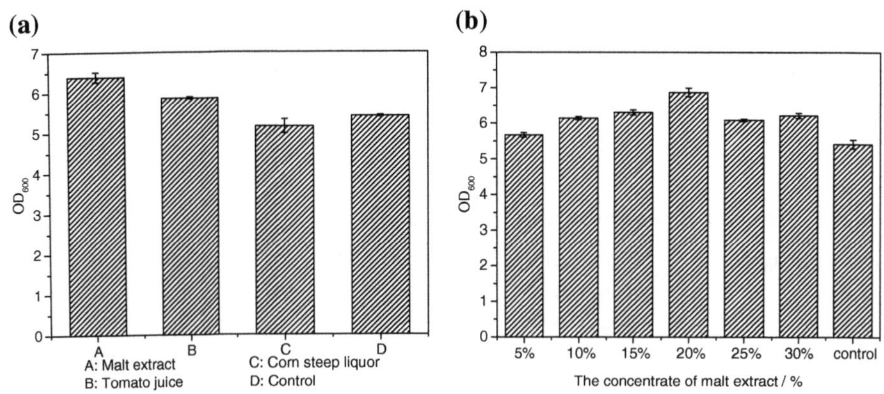

Fig. 1 The effect on the growth of *L. helveticus* AAF 1-5. Data are means ± SD (n = 3) ($P < 0.05$)

As show in Fig. 1b, 20% malt wort was the optimized concentration for *L. helveticus* AAF1-5 cultural. ($P < 0.05$), and the number of viable cells was $(2.35 \pm 6.00) \times 10^9$ CFU/mL.

3.2 *Optimization of Preparing DVS* L. helveticus *AAF1-5*

The kind of protective agent, which is a key factor that can significantly affect bacteria survival, serve as an anchor which play the role of protecting the structure of cells. Other researchers reported that the cell survival rate is approximate to 99.8% when adding the protective agent [19]. Thus, adding the protective agent is critical to the preparation of DVS *L. helveticus* AAF1-5. It was reported that *Micrococcus sp* played a reinforcement role in cheese production and could shorten the maturity of cheese [20].

Similarly, LAB appeared a potential application in reinforcing vinegar fermentation. To facilitate prepare and usage the DVS of *L. helveticus* AAF1-5 was prepared. The effeteness of skim milk and 4 sugars (lactose, sucrose, glucose and fructose) as protective agent was compared. As listed in Table 1, the highest viable cell was obtained when lactose was used as protective agent, and 8% was the optimal concentration for DVS preparation (Fig. 2). Lactose is commonly used in DVS preparation. Also, it is beneficial to the recovery of DVS when it was used for vinegar fermentation.

3.3 *Stability Evaluation of DVS L. Helveticus AAF1-5*

Temperature plays an important role during bacteria powder storage. It was reports that the cell membrane of freeze-dried powder of *Lactobacillus bulgaricus* often happen to oxidation during storage. To avoid this oxidation, freeze-dried powder of *L. bulgaricus* should be stored in low temperature [21]. Studies have also shown that, when it was preserved at 4 °C, the preservation period of freeze-dried product is much longer compared with room temperature preservation [22].

Table 1 The effect of protective agents on the property of DVS *L. helveticus* AAF1-5

Protective agent	Viable cells of DVS ($\times 10^{10}$ CFU/g)	Yield ratio (%)
Lactose	5.66 ± 0.07	87.5 ± 0.42
Sucrose	4.24 ± 0.02	26.5 ± 0.25
Glucose	4.97 ± 0.03	39.8 ± 1.47
Fructose	2.24 ± 0.01	13.4 ± 0.38
Skim milk	5.22 ± 0.05	83.7 ± 0.12

Data are means ± SD (n = 3) ($P < 0.05$)

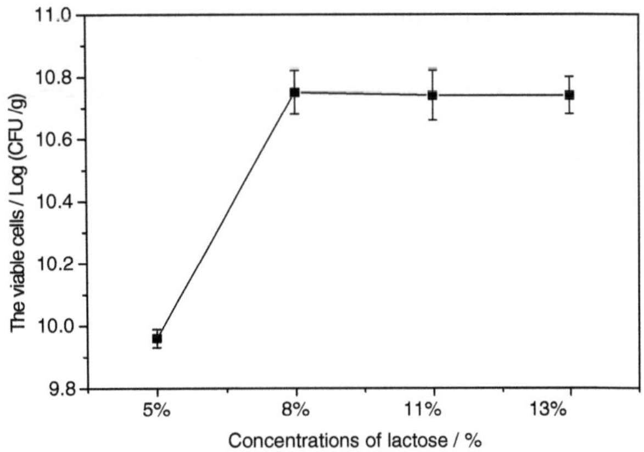

Fig. 2 The effect of lactose on the viable cells for DVS preparation. Data are means ± SD (n = 3)

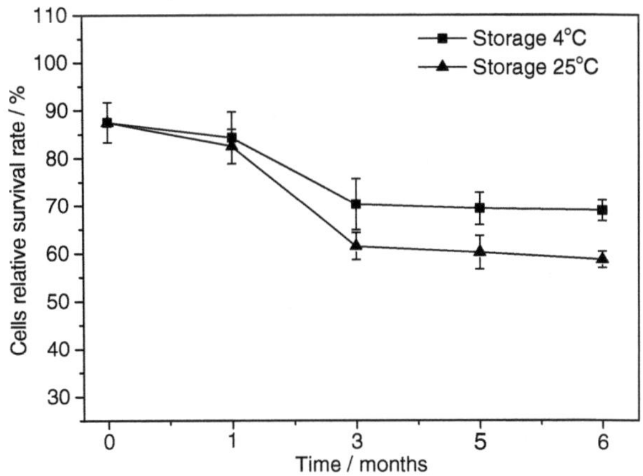

Fig. 3 Stability evaluation of DVS *L. helveticus* AAF1-5. Data are means ± SD (n = 3)

It must be recognized that the efficacy of DVS depends largely on the count of viable bacteria [23]. The allowable level of loss of viable bacteria in DVS should be generally less than 10%. As shown in Fig. 3, the viable cells was 84.3% and 82.5% of initial when it was stored at 4 and 25 °C, respectively, for 1 month. Then, the inactivation of *L. helveticus* AAF1-5 increased. After 6 months store at 4 and 25 °C, respectively, the viable cells was 69.0 and 58.7% of initial. Thus, the DVS should be used prior to 1 month.

4 Conclusions

The optimum medium for *L. helveticus* AAF1-5 cultural is consisted of glucose 2.0%, peptone 1.0%, beef extract 1.0%, yeast extract 0.5%, NaAc 0.5%, K_2HPO_4 0.2%, $MgSO_4$ 0.058%, malt wort 20%, the cell number is (2.35 ± 6.00) × 10^9 CFU/mL after 24 h cultivation at 37 °C. The DVS was prepared with the method of freeze-drying with 8% lactose as protective agent. Ant the DVS containing *L. helveticus* AAF1-5 of (5.66 ± 0.07) × 10^{10} CFU/g was obtained. The viable cells were 84.3 and 82.5% of initial after 1 month at 4 and 25 °C, which is satisfied for its usage.

Acknowledgements This work was supported by the Ministry of Science and Technology of the P.R. China (2016YFD0400505), Tianjin Municipal Science and Technology Commission (16YFZCNC00650).

References

1. Jiang Y, Guo J, Li Y et al (2010) Optimisation of lactic acid fermentation for improved vinegar flavour during rosy vinegar brewing. J Sci Food Agric 90(8):1334–1339
2. Yu MH, Im HG, Im NK et al (2009) Anti-hypertensive activities of lactobacillus isolated from kimchi. Korean J Food Sci Technol 41(4):428–434
3. Smit G, Smit BA, Engels WJM (2005) Flavour formation by lactic acid bacteria and biochemical flavour profiling of cheese products. FEMS Microbiol Rev 29(3):591–610
4. Junli Miao, Yourong Chen, Fenglan Qi et al (2005) The application of lactic acid bacteria in meat products, brewing industry etc. China Acad J Electron Publ House 111(2):49–53
5. Wenting LI, Zhenming CHE, Ji LEI et al (2011) Study on the quality and safety of pickle fermented by lactobacillus preparation. J Xihua Univ Nat Sci 30(3):97–112
6. Nie Z, Zheng Y, Du H et al (2014) Dynamics and diversity of microbial community succession in traditional fermentation of Shanxi aged vinegar. Food Microbiol 47:62–68
7. Yu YJ (2014) Organic acids analysis and enhanced lactic acid synthesis of Zhenjiang aromatic vinegar. Jiangnan University, Wuxi
8. Xiao-man HUA, Dong-mei JIA, Yu-shan WANG et al (2016) Current production and research development of direct vat set (DVS) for fermented dairy products. China Acad J Electron Publ House 44(7):23–27
9. Morgan C, Vesey G (2009) Encyclopaedia of microbiology (3rd ed). Academic Press, pp, 162–173
10. Berny JF, Hennebert GL (1992) Viability and stability of yeast cells and filamentous fungus spores during freeze-drying: effects of protectants and cooling rates. Mycological 83(6): 805–815
11. Garcia-Amezquita LE, Welti-Chanes J, Vergara-Balderas FT et al (2016) Freeze-drying: the basic process. Encyclopedia of food and health, pp 104–109
12. Lou CL, Verbreek MAM, Noomen A et al (1994) Mechanism of dehydration inactivation of *Lactobacillus plantarum*. Appl Microbiol Biotechnol 41(1):90–94
13. De Valdéz GF, De Giori GS, Aa DRH et al (1983) Protective effect of adonitol on lactic acid bacteria subjected to freeze-drying. Appl Environ Microbiol 45(1):302–304
14. Abadias M, Benabarre A, Teixidó N et al (2001) Effect of freeze drying and protectants on viability of the biocontrol yeast Candida sake. Int J Food Microbiol 65(65):173–182
15. Leslie SB (1995) Trehalose and sucrose protect both membranes and proteins in intact bacteria during drying. Appl Environ Microbiol 61(10):3592–3597
16. LIU Yun he (2003) The preparation of lactic acid bacteria powder starter. J Huaihai Inst Technol 12(4):60–62
17. Zhang L, Zhiqiang E, Wan H et al (2000) Selecting enhancement culture and dried protectors on lactic acid bacteria. China Dairy Ind 28(2):8–18

18. Gao H, Liu M, Liu J et al (2009) Medium optimization for the production of avermectin B1a by *Streptomyces avermitilis*, 14-12A using response surface methodology. Biores Technol 100(17):4012–4016
19. Li Hua, Luo Yaner (2002) Research progress of vacuum freeze-drying microorganisms. Microbiology 29(3):78–82
20. Morales P, Gaya P, Nuñez M (2005) Effect of the addition of Micrococcus sp. INIA 528 milk cultures and curds on cheese proteolysis and lipolysis. Milchwissenschaft-milk Sci Int 60 (4):410–414
21. Castro HP, Teixeira PM, Kirby R (1996) Changes in the cell membrane of *lactobacillus bulgaricus* during storage following freeze-drying. Biotechnol Lett 18(1):99–104
22. Matejtschuk P (2004) Freeze-drying of biological standards, freeze-drying of pharmaceutical and biological products. Marcel Dekker Inc, New York, pp 215–224
23. Champagne CP, Mondou F, Raymond Y et al (1996) Effect of polymers and storage temperature on the stability of freeze-dried lactic acid bacteria. Food Res Int 29(29):555–562

Overexpression of the Endo-inulinase Gene from Two Different Sources and Characteristics Analysis

Li-Kun Wei, Qing-Long Xin, Zhi-Mei Feng, Xue-Yan Xing, Wei Feng and Yu Li

1 Introduction

Endo-inulinases hydrolyze the internal β-2,1 fructofuranosidic linkages to produce oligofructose with the degree of polymerization (DP) of 2–7 using inulin as the substrate [1]. These oligofructoses are among the most popular functional oligoses because of low-calorie, bifidogenic, health-promoting and so on [2, 3]. They can also inhibit the bacterial growth, and have many applications in food, pharmaceutical industries and biotechnology [4]. Inulin, as a carbohydrate, was present in the roots and tubers of plants such as Jerusalem artichoke, yacon, chicory, and dahlia [5]. Recent years, it gets much more attention for inexpensive, renewable, and abundant sources as an excellent natural raw material.

Some strains were found to produce endo-inulinase, such as *Penicillium* sp. [6], *Aspergillus ficuum* [7], *Aspergillus niger* [8], *Arthrobacter* sp. [9]. There are numerous researches on high-level expression of exo-inulinase. However, the reports on high-level expression of endo-inulinase are quite fewer [10–13]. The current expression levels of endo-inulinases are still low and unsuitable for the large-scale production of oligofructose. Although there have been many reports on high-level conversion of inulin to oligofructose, these processes need long reaction time, large quantity of enzyme consumption. These disadvantages make product less profitable [14]. Therefore, gene engineering technology can provide an effective method to work with these problems.

L.-K. Wei · Q.-L. Xin · Z.-M. Feng · X.-Y. Xing · W. Feng · Y. Li (✉)
Key Laboratory of Industrial Microbiology, Ministry of Education,
College of Biotechnology, Tianjin University of Science and Technology,
Tianjin 300457, China
e-mail: liyu@tust.edu.cn

L.-K. Wei
e-mail: 1048158327@qq.com

H. Liu et al. (eds.), *Advances in Applied Biotechnology*, Lecture Notes in Electrical Engineering 444, https://doi.org/10.1007/978-981-10-4801-2_65

In this study, we expressed endo-inulinase *A. niger* TCCC41064 and *Penicillium* sp. TN-88 in *P. pastoris*. The characterizations of the recombinant endo-inulinases were investigated. The conditions of the biotransformation of inulin to oligose were optimized, including substrate concentration, pH, and temperature, aimed at establishing a conversion process for the industrial production of oligofructose.

2 Materials and Methods

2.1 *Strains, Vectors, Reagents, and Cultures*

The strain of *A. niger* TCCC41064, *Escherichia coli* JM109, *Pichia* GS115, the expression vector pPIC9K was preserved in the laboratory. The EZ-10 Spin Column Plasmid Mini-Prep kit, agarose gel DNA purification kit, restriction enzymes, PCR enzyme "LA Taq", DNA marker, Solution I DNA ligase, and cloning vector pMD19-T simple vector were purchased from TakaRa (Otsu, Japan). Synthesis of DNA primers and DNA sequencing were performed by AuGCT (Beijing, China). LB, YPD, MD, G418, BMGY, and BMMY cultures were prepared according to the Ref. [15]. Unless stated otherwise, all the chemicals were analytical grade and purchased from Sangon Biotech (Shanghai) Co. Ltd (Shanghai, China).

2.2 *Experimental Method*

2.2.1 Extraction of Targetting Gene DNA from Organisms

Genomic DNA of *A. niger* TCCC41064 and *Penicillium* sp. TN-88 were extracted as described in Ref. [16]. The gene encoding endo-inulinases were inuA, GenBank Accession No. CAA07345, and inuC, GenBank Accession No. AB041337.2. The DNA sequence of inuA was analyzed and designed using the analyzer Primer 5, including the addition of two restriction sites *EcoR I* and *Not I*, to the forward and reverse end, respectively. The PCR primers of inuA and inuC gene were synthesized by AuGCT (Beijing, China) (Table 1).

Table 1 The primers used in this study

Primer	Gene	Sequence (5′-3′)	Restriction	Size (bp)
P1	inuA-F	CGGAATTCATGTTGAATCCGAAGGTTG	*EcoR I*	27
P2	inuA-R	GCGGCCGCTCATTCAAGTGAAACACTC	*Not I*	27
P3	inuC-F	CGGAATTCATGATCTCCCAAGGACTTAC	*EcoR I*	28
P4	inuC-R	GCGGCCGCTCAAAGTGAGACTGAGGAC	*Not I*	27

PCR amplification conditions: 95 °C 5 min, 94 °C 45 s, 58 °C 40 s, 72 °C 1.5 min, 30 cycles; 72 °C extend 10 min.

2.2.2 Construction of Expression Plasmid

The full-length genes were amplified using the primers inuA-F, inuA-R, inuC-F, inuC-R and LA Taq DNA polymerase. The recombinant pMD19-T simple vector/inuA was obtained and transformed into *E. Coli* JM109 by the heat-shock method [17]. The recombinant pMD19-T simple vector/inuA and pPIC9 K were digested with *EcoR I* and *Not I*, gel-purified, and ligated into pPIC9K. Next, the recombination vector pPIC9K/inuA was extracted from *E. Coli* JM109 and then transformed into *P. pastoris* G418 using the electrical pulse method. The recombinant plasmid of gene inuC was performed as described above.

2.2.3 Transformation of *P. pastoris*, High Copy Recombinant Strains and Fermentation

The method of electroporation in *P. pastoris*, high copy recombinant strains and fermentation were described as Ref. [18].

2.2.4 Determination of Endo-inulinase Activity

200.0 μL of the crude endo-inulinase was mixed with 800 μL of 4.0% (w/v) inulin in 0.05 mol/L HAC-NaAC buffer (pH 4.6). The mixture was incubated at 50 °C for 15 min. After that, the mixture was immediately inactivated in the boiling water for 10 min. The reducing sugar released from inulin was measured using DNS method [19]. The same mixture with 200 μL of the fermentation supernatant heated at 100 °C for 10 min was used as the negative control. One unit of endo-inulinase activity was defined as the amount of enzyme causing release of reducing sugars equivalent to 1.0 μmol from inulin in 1 min under the assay conditions.

2.2.5 Effects of pH and Temperature on the Recombinant

The Optimum Temperature and Temperature Stability

Take appropriately diluted enzyme 200 μL, was added 800 μL 4% inulin (with 0.05 mol/L of HAC-NaAC buffer preparation), measure inulinase vitality at different temperatures (40, 45, 50, 55, 60, 65, 70, 75 °C) conditions. The enzyme was saved at 50, 55, 60, 65 °C at per hour and then rapidly cooled, measuring residual enzyme activity.

The Optimum Value of pH and pH Value Stability

The optimum pH was measured by 0.05 mol/L of HAC-NaAC buffer (pH 3.5–11) compounded in 4% of inulin, inulin determination of endo-enzyme activity under different pH conditions.

The enzyme was placed in buffers with different pH, at room temperature for 60 min, after acetate buffer diluted 0.1 mol/L pH 4.6, the remaining enzyme activity was measured at 50 °C

2.2.6 Hydrolytic Analysis by HPLC

The products of hydrolyzed inulin were analyzed by HPLC (Agilent, America) using a TSK-GEL Amide-80 Series column (4.6 × 100 mm, 5 μm) and an evaporative light scattering detector. The eluent was a mixture of 60% acetonitrile and 40% water, and the flow rate was 1 mL/min.

Using the oligosaccharide standard sample curves, we took the group of sugar peak area into the curve, to determine the content of the degree of polymerization at the oligofructose of DP 3–5.

$$\begin{aligned} \text{DP}\,(\text{x}) &= \text{concentration} * \text{volume} * \text{ dilution factor}\,(\text{x} = 3, 4, 5) \\ \text{FOSyield}\,(\%) &= \text{DP3}{-}\text{5quality}/\text{initial inulin quality} * 100\% \end{aligned} \tag{1}$$

2.2.7 Hydrolytic Analysis by TLC

2 μL sample of the reaction product was carried out on the TLC plate for qualitative analysis, The developing solvents were n-butanol–isopropanol–water–acetic acid (volume ratio of 7:5:4:2). After each developing layer was dried with a hair dryer, exhibited the floor for twice. And next, added liquid on the TLC plate(water, 235 mL, ammonium molybdate 12 g, cerium ammonium molybdate 0.5 g, concentrated sulfuric acid 15 mL), dried at 95 °C for 10 min, observation result lastly.

3 Results and Discussion

3.1 Identification of the Expression Vector pPIC9K-inuA and pPIC9K-inuC

A. niger TCCC 41064 strain DNA was used as a template for gene amplification, PCR amplification by using primers P1, P2 according to the Method Sect. 2.2.2. A single band DNA of the target gene fragment at the same size in the vicinity of 1.5 kb was detected on 8% agarose gel electrophoresis. The PCR product T-inuA and T-inuC were sequenced by Beijing BGI Ltd. The PCR product sequence similarity was 99.67% compared with those published on NCBI.

The amplified endo-inulinase gene fragment was inserted into the expression vector pPIC9K, constitute recombinant plasmid pPIC9K-inuA, using *EcoR I* and *Not I* for double enzyme digestion. Small gene fragments about 1500 bp is inuA, a large fragment about 9000 bp is the pPIC9K carrier. It is indicated that the gene has been inserted into pPIC9K carrier. Similarly as above, gene inuC double digestion with *EcoR I* and *Not I*, was inserted into the expression vector PMD19T-simple and pPIC9K successively.

3.2 Fermentation of Recombinant Pichia

After the transformants of electroporation was certified, and then detected the activity of the recombinants fermentation supernatants. The activity of culture supernatant of inuA and inuC were 13.5 U/mL and 69.3 U/mL.

3.3 SDS-PAGE Analysis and Protein Purification

SDS-PAGE showed that the supernatant from the transformant carrying the inuA and inuC gene exhibited one band compared with the control group with a molecular mass of about 66.2 kDa. This is caused by glycosylation process, in which the folded protein was secreted into the extracellular. The recombinant enzyme was purified as described in Ref. [20]. After purification, this unique band was the His-tagged protein of the recombinant endo-inulinase (Fig. 1a, b). InuC specific activity reached 792.09 U/mg after purified, which was higher than inuA (350.00 U/mg).

3.4 The Optimum Temperature and Temperature Stability

The temperature data showed that under assay conditions used, the temperature for optimal activity of the inuA and inuC endo-inulinase gene were determined at 60, 50 °C (Fig. 2a, b), which were in accordance with the optimum inulinases temperatures of 45–60 °C from other molds [21]. In addition, the enzyme of gene inuA retained 83.2% of its activity after 9 h at 55 °C, and the enzyme of gene inuC retained 75.4% activity after 9 h at 40 °C (Fig. 2c, d), which displayed good thermostability among the inulinase family [22, 23]. Since temperatures of inuA over 60 °C, temperatures of inuC temperatures over 50 °C ensure proper solubility

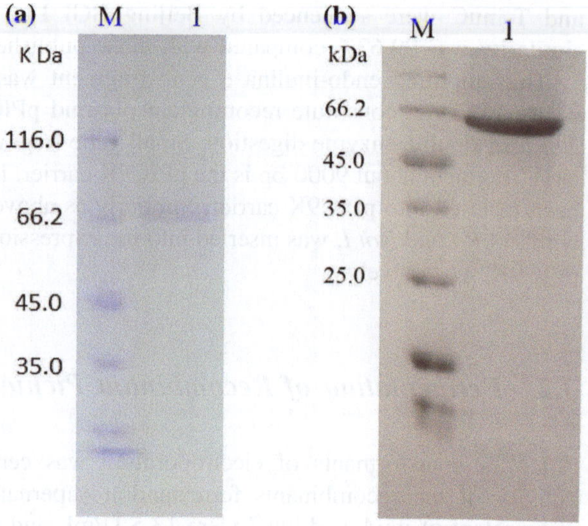

Fig. 1 InuA and inuC gene SDS-PAGE analysis of the purified endo-inulinase. Note: *M* standard protein; **a** purified endo-inulinase of inuA; **b** purified endo-inulinase of inuC

of inulin and prevent microbial contamination [24], thermostability of the endoinulinase at 55, 40 °C is definitely beneficial for industrial applications. Compared with the two genes, the inuC gene optimum temperature is lower than inuA.

3.5 The Optimum pH and pH Stability

As shown in Fig. 3, the effects of pH on the activity and stability of the recombinant endo-inulinase were also determined. The enzyme inuA and inuC exhibited the highest activity at pH 5.0, 4.0. The enzyme inuA was stable over a pH range of 4.5–5.5, retaining >80% of its initial activity. Meanwhile, the enzyme inuC was stable over a pH range of 3.0–6.0, retaining >85% of its initial activity (Fig. 3a, b). These results showed that the recombinant endo-inulinase inuC was potentially to be effectively in the preparation of product. The pH stability of recombinant enzymes were similar, the two enzyme were stable over a pH range of 3.0–6.0, retaining >85% of its initial activity (Fig. 3c, d).

3.6 Hydrolytic Analysis by HPLC

The combination and mode of action for inulin and endo-inulinase were different, the distribution of polymerization degree of the products were different, the

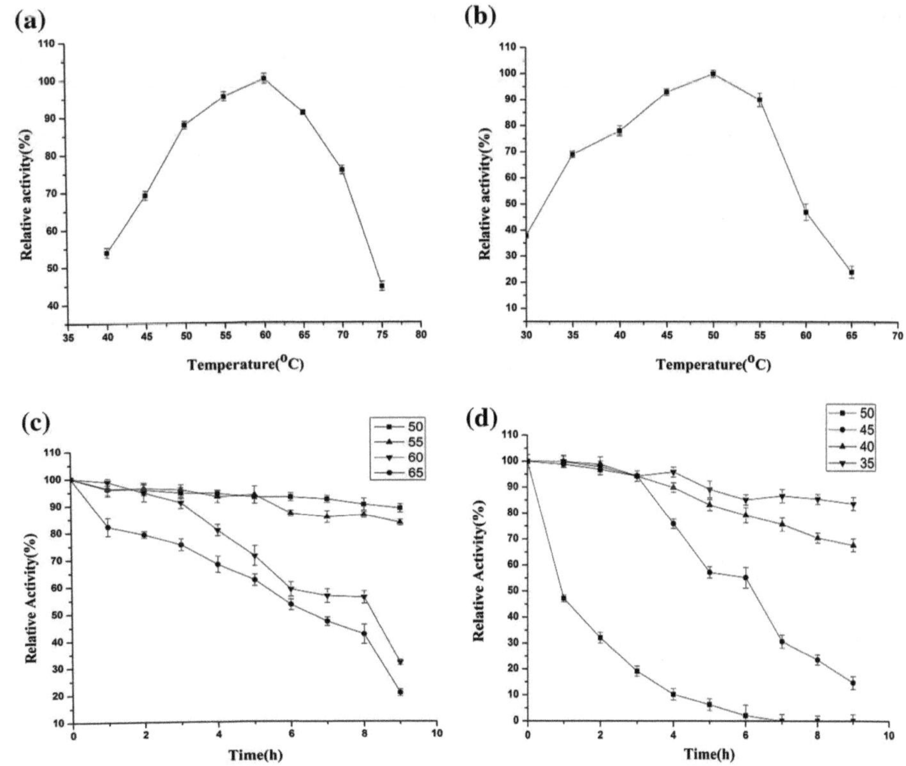

Fig. 2 The optimum temperature and thermostability of the recombinant enzymes **a** the optimum temperature of inuA **b** the optimum temperature of inuC **c** the thermostability of inuA **d** the thermostability of inuC

distribution of the degree of polymerization of the resulting product, and the products were oligofructose with DP 3–5, which are far more functional than the counterparts of other DP [22].

During the first 120 min, endo-inulinase inuA hydrolyze inulin to oligosaccharide as shown in Fig. 4. The increase rate of trisaccharide is much higher than several others. Inulin ultimately was hydrolyzed to the oligosaccharides with DP 3–5, with a small amount of disaccharide.

In contrast, retention time trisaccharide hydrolyzed by inuC was 10 min, and major products were trisaccharides, tetrasaccharides. The product of trisaccharides accounted for 74.8% (1.1), which is much higher than inuA. Therefore, inuC is more conducive for the production (Fig. 5).

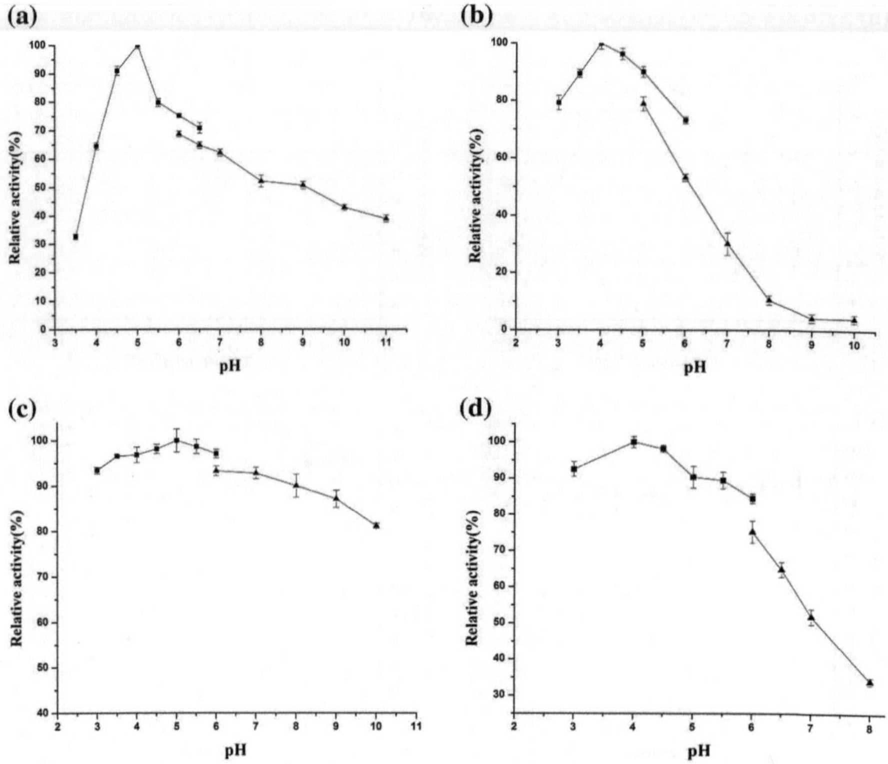

Fig. 3 The optimum pH and pH stability of recombinant enzymes **a** the optimum pH of inuA **b** the optimum pH of inuC **c** the pH stability of inuA **d** the pH stability of inuC

3.7 Hydrolytic Analysis by TLC

The TLC analysis result of inuA was shown in Fig. 6. The long-chain inulin was gradually hydrolyzed to short-chain. After 8 h, inulin was completely hydrolyzed to short-chain oligosaccharides.

Hydrolytic product by inuC was initial detected, shows in Fig. 7. Long-chain inulin was gradually hydrolyzed to low molecular weight saccharide [25]. After continuous reaction of 20 h, inulin is mostly converted to trisaccharide as the main ingredient and other oligosaccharide mixtures. The long-chain inulin random was hydrolyzed to relatively short molecules (lanes 4, 5) (DP > 5), and then turned them to the lower degree of polymerization oligosaccharides. It also shows that endo-inulinase cleavage the internal glycosidic bond was random.

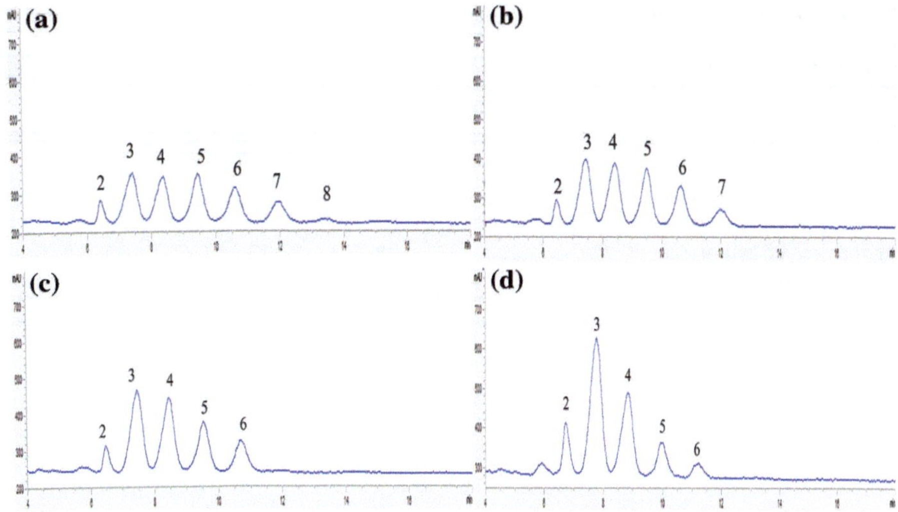

Fig. 4 The hydrolyzd products in different periods using inulin as the substrate by HPLC analyze **a** hydrolysis 30 min; **b** hydrolysis 120 min; **c** hydrolysis 240 min; **d** Hydrolysis 480 min *peak 2–8*: F2 (fructose), F3, F4, F5, F6, F7, F8

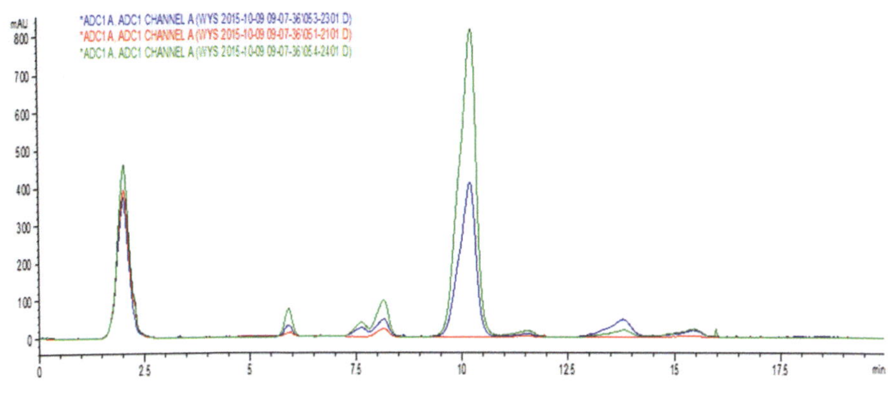

Fig. 5 The hydrolyzd products of *inuC* in different periods with HPLC analyze *red line* original inulin substrate, *green line* hydrolysis 480 min; *blue line* hydrolysis 240 min

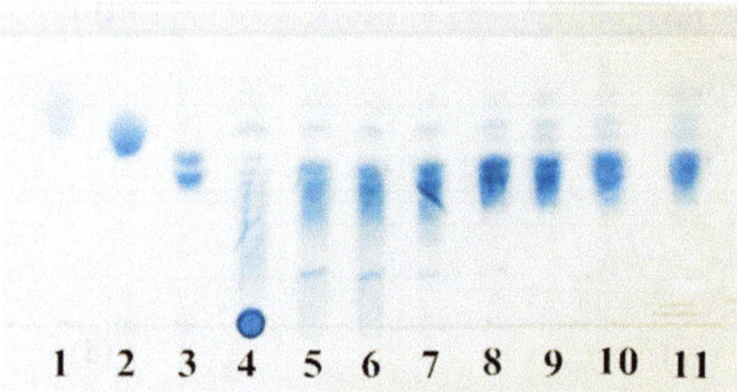

Fig. 6 TLC analysis of the hydrolysate of inulin by inuA *Note 1* glucose; *2* sucrose (GF); *3* from top to bottom are glucose, sucrose (GF), kestose (GF2), fungitetraose (GF3); *4* inulin substrate + inactivated fermentation supernatant; *5–11* the time of hydrolysate of inulin 10, 30, 120, 240, 480, 720, 1200 min

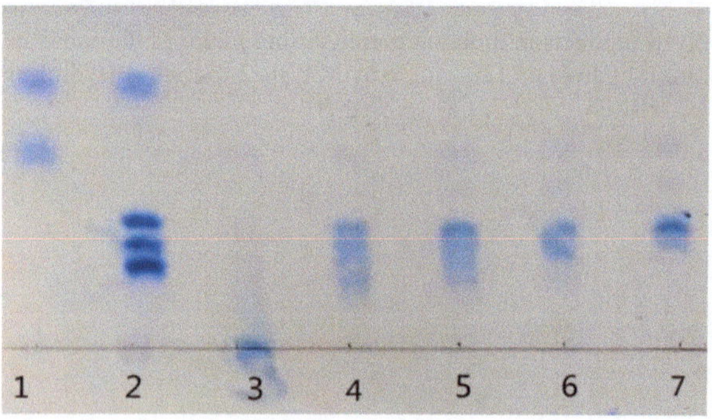

Fig. 7 TLC analysis of the hydrolysate of inulin by inuC *Lane 1* from top to bottom are glucose, sucrose; *Lane 2* from top to bottom are glucose, kestose (GF2), fungitetraose (GF3), F-fructofuranosylnystose (GF4); *Lane 3*: inulin substrate + inactivated fermentation supernatant; *Lane 4–7* the time of hydrolysate of inulin 240, 480, 720 min

4 Conclusions

In this study, we cloned two endo-inulinase genes (inuA, inuC) from *A. niger* TCCC41064 and *Penicillium* TN-88, and over-expressed in *P. pastoris* GS115. The molecular weight of the purified inuA and inuC were both about 66.2 kDa. The optimal pH and temperature of the purified inuA were 5 and 60 °C, respectively. The enzyme of inuA retained 83.2% of its activity after 9 h at 55 °C, and exhibited higher thermal stability. After 8 h, inulin substantially could be hydrolyzed to the short-chain oligosaccharides, (inulin-disaccharide 32.5%, inulin-trisaccharide 36.7%, and inulin-tetrasaccharide 30.7%). After 11 h, FOS conversion yield was 63.5%. For the gene inuC, the optimal pH and temperature of the purified were

4 and 50°C. The enzyme of inuC retained 75.4% of its activity after 9 h at 40 °C. Major products of inulin hydrolyzed by inuC were trisaccharide (74.8%), inulin tetrasaccharide (21.6%), but also produce a small quantity of inulin-disaccharides and monosaccharides.

Comparing with these two genes, inuC showed the higher enzyme activity. Secondly, the temperature of hydrolysis at 40 °C is lower, which can save more energy by compared with inuA. The hydrolyze inulin by inuC can generate more fructo-oligosaccharides, which are better for health, more easily absorbed and can be used by probiotics [20]. The endo-inulinase inuC derived from *Penicillium* has more advantages, providing important guidance for the use of inulin-oligofructose industrial production. Further research will focus on the optimum process conditions of inuC for preparing fructo-oligosaccharides.

References

1. Respondek F, Swanson KS, Belsiro KR et al (2008) Short-chain fructooligosaccharides influence insulin sensitivity and gene expression of fat tissue in obese dogs. J Nulrilion 138 (9):1712–1718
2. Morita T, Kasaoka S, Ilase K et al (1991) Oligo-L-methionine and resistant protein promote cecal butyrate production in rats fed resistant starch and fructooligosaccharide. J Nulrilion 129 (7):1333–1339
3. Li D, Kim JM, Jin Z et al (2008) Prebiotic effectiveness of inulin extracted from edible burdock. Anaerobe 14(1):29–34
4. Respondek F, Goachet AG, Julliand V (2008) Effects of dietary short-chain fructooligosaccharides on the intestinal microflora of horses subjected to a sudden change in diet. J Anim Sci 86(2):316–323
5. Li Y, Liu G-L et al (2012) Overexpression of the endo-inulinase gene from Arthrobacter sp. S37 in Yarrowia lipolytica and characterization of the recombinant endo-inulinase. J Mole Catal B Enzym 74:109–115
6. Nakamura T, Shitara A, Matsuda S et al (1997) Production, purification and properties of an endoinulinase of Penicillium sp.TN-88 that liberates inulotriose. J Ferment Bioeng 84(4): 313–318
7. Uhm TB, Chae KS, Lee DW et al (1998) Cloning and nucleotide sequence of the endoinulinase-encoding gene, inu2, from *Aspergillus ficuum*. Biotech Lett 20(8):809–812
8. Ohta K, Akimoto H, Matsuda S et al (1998) Molecular cloning and sequence analysis of two endoinulinase genes from *Aspergillus ficuum*. Biosci Biotechnol Biochem 62(9):1731–1738
9. Kang SII, Chang YJ, Oh SJ et al (1998) Purification and properties of an endo-inulinase from an Arthrobacter sp. Biotech Lett 20(10):983–986
10. Kumar GP, Kunamneni A, Prabhakar T, Ellaiah P (2005) Optimization of process parameters for the production of inulinase from a newly isolated *Aspergillus niger* AUP19. World J Microb Biotechnol 21:1359–1361
11. Li Y, Liu GL, Wang K, Chi ZM, Madzak C (2011) Overexpression of the endo-inulinase gene from Arthrobacter sp. S37 in Yarrowia lipolytica and characterization of the recombinant endo-inulinase. J Mol Catal B-Enzym 74:109–115
12. Nakamura T, Shitara A, Matsuda S, Matsuo T, Suiko M, Ohta K (1997) Production, purification and properties of an endoinulinase of Penicillium sp. TN-88 that liberates inulotriose. J Ferment Bioeng 84:313–318

13. Wang JH, Teng D, Yao Y, Yang YL, Zhang F (2004) Expression of *Aspergillus niger* 9891 endoinulinase in *Pichia pastoris*. High Tech Lett 10:52–56
14. He M, Dan W, Jing W, Chen Jian (2014) Enhanced expression of endoinulinase from *Aspergillus niger* by codon optimization in *Pichia pastoris* and its application in inulooligosaccharide production. J Ind Microbiol Biotechnol 41:105–114
15. Hu S, Luo S, Zhang J, Zhangsun D (2007) *Pichia pastoris* expression system and strategies for high-level expression. Biotechnology 17(6):79–83
16. Cho Y, Yun JW (2002) Purification and characterization of an endoinulinase from Xanthomonas oryzae No. 5. Process Biochem 37(11):1325–1331
17. Shi XL, Feng MQ, Shi J (2007) High-level expression and purification of recombinant human catalase in *Pichia pastoris*. Protein Expr Purif 54(2):193–203
18. Kango N, Jain SC (2011) Production and properties of microbial inulinases: recent advances. Food Biotechnol 25:165–212
19. Chen HQ, Chen XM, Li Y, Wang J, Jin ZY, Xu XM, Zhao JW, Chen TX, Xie ZJ (2009) Purification and characterisation of exo-and endo-inulinase from *Aspergillus ficuum* JNSP5-06. Food Chem 115:1206–1212
20. Chi Z, Zhang T, Liu G, Yue L (2009) Inulinase-expressing microorganisms and applications of inulinases. Appl Microbiol Biotechnol 82:211–220
21. Nguyen QD, Rezessy-Szabó JM, Czukor B, Hoschke Á (2011) Continuous production of oligofructose syrup from Jerusalem artichoke juice by immobilized endo-inulinase. Process Biochem 46:298–303
22. Cho YJ, Yun JW (2002) Purification and characterization of an endoinulinase from Xanthomonas oryzae No. 5. Process Biochem 37:1325–1331
23. Zhang T, Gong F, Chi Z et al (2009) Cloning and characterization of the inulinase gene from yeast *Pichia guillermondii* and its expression in *Pichia pastoris*. Antione van Leeuwenhoek 95(1):13–22
24. Ohta K, Akimoto H, Moriyama S (2004) Fungal inulinase: enzymology, molecular biology and biotechno logy. J Appl Glycosci 51:247–254
25. Kolida S, Tuohy K, Gibson GR (2002) Prebiotic effects of inulin and oligofructose. Br J Nutr 87(S2):S193–S197

Production of Ethyl Acetate Catalyzed by Activated Carbon-Based Solid Acid Catalyst

Jia Li, Yan Li and Hua Zhao

1 Introduction

Aroma components in Chinese baijiu are nearly 1000. But the main aroma components are esters, including ethyl acetate, ethyl lactate, ethyl hexanoate and ethyl butyrate esters, which cover more than 90% of the total ester content. Esters are produced by dehydration of acid and alcohol, and the esterification reaction is a typical acid catalytic reaction. Concentrated sulfuric acid or other liquid acid was used as catalyst during esterification reaction in traditional industrial production. However, there are many shortcomings, such as a lot of by-products and acid waste water, difficulties in the separation of reaction products, serious environment pollution and the equipment corroded [1]. Compared with the traditional liquid acid, the solid acid catalyst has the advantages of high activity, easy separation, no corrosion to the equipment, less pollution to the environment and easy recovery and reuse. Therefore, solid acid catalyst has been widely used as a substitute for liquid acid.

In recent years, carbon-based solid acid catalysts with high activity and good stability, which are produced by carbohydrate compound or woody biomass containing carbon, have attracted wide attention from domestic and overseas scholars [2–4]. Carbon-based solid acid is a new type of solid acid material. The solid acid catalyst is prepared by using the biomass carbon as the carrier and loading acid substance. Wood materials with high cellulose content are abundant renewable resource. The advantage of low cost and environmental protection makes them

J. Li
Tianjin University of Science and Technology, Tianjin, China

Y. Li · H. Zhao (✉)
Key Laboratory of Industrial Fermentation Microbiology,
Ministry of Education, Tianjin Key Laboratory of Industrial Microbiology,
College of Biotechnology, Tianjin University of Science and Technology,
Tianjin 300457, China
e-mail: zhaohua@tust.edu.cn

H. Liu et al. (eds.), *Advances in Applied Biotechnology*, Lecture Notes
in Electrical Engineering 444, https://doi.org/10.1007/978-981-10-4801-2_66

great importance to the development and utilization in the current age of resource scarcity and pollution [5]. The carbon-based solid acid catalyst prepared by high carbon content of cellulose can be widely used in the process of esterification, hydrolysis and alkylation [3, 6–8].

Ren [9] used expanded starch as the carrier, and combined it with methyl benzene sulfonic acid to produce a new carbon-based solid acid catalyst, which was applied in the esterification of oleic acid and ethanol. The results showed that the catalytic activity of the new type of carbon based solid acid catalyst was very strong, and the esterification rate of oleic acid ethyl ester was 83.78%. The high stability of the solid acid catalyst could be reused more than 6 times. Kastner [10] used peanut shell, sawdust and concentrated sulfuric acid to prepare a carbon-based solid acid catalyst. It showed high activity in the esterification of both palm acid and stearic acid, and the conversion of palm acid could be close to 100%. Wu [11], through the process of carbonization and sulfonation, used bamboo charcoal as raw materials to prepare solid acid catalyst and catalytic esterification reaction of spend jatropha seed oil with methanol. The results showed that the esterification efficiency was improved significantly, and the esterification rate of the seed oil could reach 90.02%.

In this study,activated carbon-based solid acid catalyst was prepared from baijiu vinasse. The effects of solid acid catalyst amount, acid alcohol ratio, esterification time and esterification temperature on the esterification reaction were investigated. The optimum conditions of the reaction were obtained, and the stability of the solid acid catalyst was also investigated to further evaluate the catalyst.

2 Experimental

2.1 *Preparation of Activated Carbon-Based Solid Acid Catalyst*

One biomass carbon-based solid acid catalysts was prepared by a carbonization–sulfonation method from baijiu vinasse precursor. The prepared activated carbon was crushed to 120 mesh for use. A total of 3 g active carbon and 15 mL concentrated H_2SO_4 (98%) were put into a magnetic stirrer controlled at 80 °C and 700 rpm for 24 h. When the reactor was cooled to room temperature, the mixture was filtered, washed with deionized water to remove the H_2SO_4 solution until the pH of the washing solution reached neutral. Finally, the prepared catalyst was dried at 121 °C for 12 h.

2.2 Catalyst Characterization

Structural information was gained for carbon materials by Fourier Transform Infrared Spectroscopy (FT-IR). FT-IR spectra were carried out on a NICOLET-6700 Fourier Transform Infrared spectrophotometer and anhydrous KBr was used to prepare the sample [12].

2.3 Solid Acid Catalyze Esterification

The activity of catalysis for the prepared solid acid was conducted using the esterification rate of ethyl acetate, which was produced through the reaction of ethanoic acid and ethanol. Ethanoic acid and ethanol with certain mass ratios were first mixed in a 100 mL flask with a reflux condenser. And the catalyst with different dosages was added into the reaction system. Then, the mixture was heated in a water bath on certain temperature for different time. The esterification rate was measured after the reaction.

The esterification rate can be calculated by the formula:

$$\text{Esterification rate}\,(\%) = \left(1 - \frac{\text{amount of substance of acid}}{\text{amount of substance of initial ethanoic acid}}\right) * 100\%$$

3 Results and Discussion

3.1 Characterization

It can be seen from the infrared spectrum that the 3300–3600 cm^{-1} has a wide spectrum band, and the low frequency absorption peak is due to the stretching vibration of O–H. The absorption peak near 1592 cm^{-1} is caused by the C=C stretching vibration. The absorption peak near 2850 cm^{-1} is caused by C–H stretching vibration. Characteristic peak of carbon based solid acid catalyst existed in 500–1500 cm^{-1}; C–S (573 cm^{-1}), S–O (752 cm^{-1}), $-SO_3H$ (1031 cm^{-1}), O=S=O (1185 cm^{-1}), the S functional groups in these stretching vibrational peaks show that the solid acid catalyst is prepared by using activated carbon as the carrier and sulfonate with the concentrated sulfuric acid. At the same time, the catalytic performance of the activated carbon-based solid acid is also showed (Fig. 1).

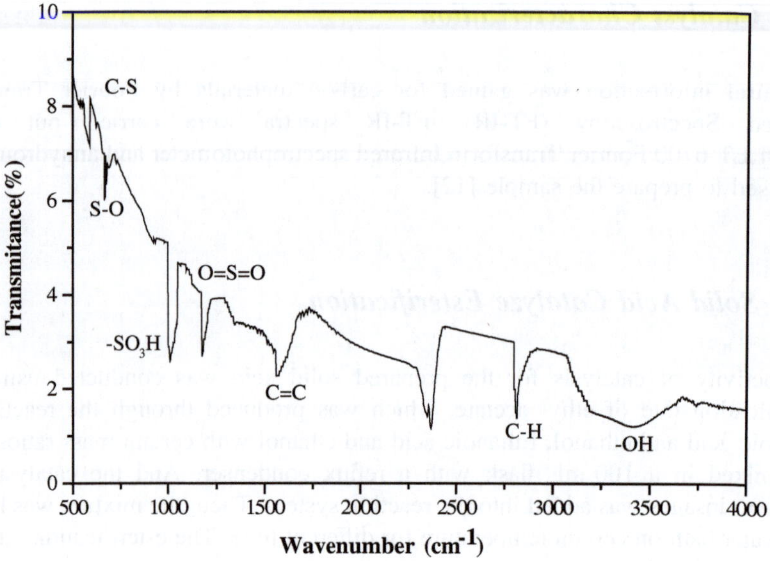

Fig. 1 FT-IR spectra of solid acid

3.2 *Effect of Activated Carbon Based Solid Acid Catalyst on Esterification Reaction*

Four factors, that were catalyst dosage, ethanoic acid/ethanol mass ratio, reaction time and reaction temperature, were chosen in order to evaluate catalytic activity of activated carbon-based solid acid catalyst prepared at optimum conditions.

3.2.1 Effect of Catalyst Dosage on Esterification Rate

The effect of the catalyst dosage on esterification rate was observed in our work. The results are showed in Fig. 2.

It can be seen that esterification rate was obviously influenced before the dosage of catalyst was 3%. When the addition amount of solid acid catalyst was 3%, the esterification rate reached the maximum value of 85.4%. After that, it was not improved significantly. Since in a certain range of catalyst concentration, the amount of catalyst was increased, the more active groups were provided, the catalytic performance was enhanced, and the esterification rate was increased. However, when the added amount of the catalyst reaches a certain value, the amount of the active group can no longer change the thermodynamic effect in esterification reaction. Therefore, the catalyst dosage of 3% was used in the following experiments.

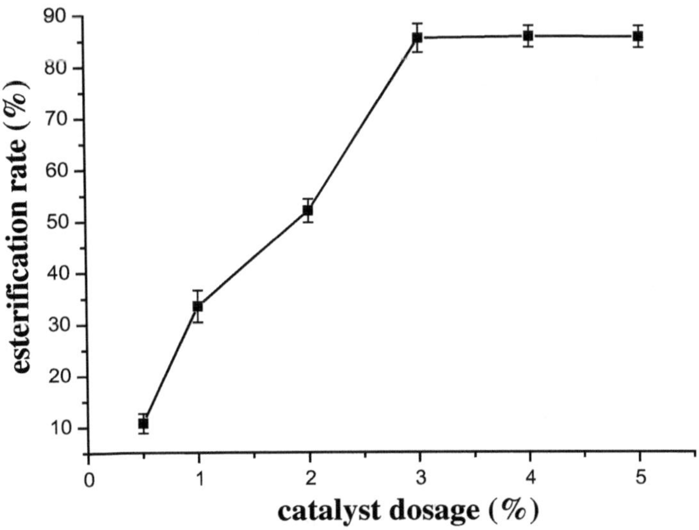

Fig. 2 Effect of catalyst dosage on esterification rate at the conditions of ethanoic acid/ethanol mass ratio of 1:4, reaction time of 4 h and reaction temperature of 80 °C

3.2.2 Effect of Acid Alcohol Molar Ratio on Esterification Rate

The effect of the molar ratio of ethanoic acid to ethanol on esterification rate was observed in our work. The results are showed in Fig. 3.

As shown in Fig. 3, with the increase of ethanol content, the conversion of acetic acid increased, and the esterification rate increased. In the esterification process, theoretical molar ratio of acetic acid and ethanol amount is 1:1. But the esterification reaction is reversible. Increasing the concentration of a reactant or separating the birth of a product can promote the reaction to the positive direction and increase the efficiency of the reaction. Ethanol volatilizes easily, increase its content can improve the rate of esterification. When the ratio of acid to alcohol was 1:5, the esterification rate reached 85.9%. However, it cannot increase owing to the limitation of equilibrium condition of reaction if molar ratio of ethanoic acid to ethanol was further increased. In the following experiments, the mass ratio of 3:1 for ethanol to ethanoic acid was used.

3.2.3 Effect of Esterification Time on Esterification Rate

It can be seen from Fig. 4 that the esterification rate is low when the esterification time short, because the esterification reaction is not complete. With the extending of reaction time, the esterification rate increases. When the esterification time was 4 h, the esterification rate was 86%. The esterification reaction is reversible, the reaction system can reach the dynamic equilibrium when the time exceeds 4 h, and the esterification rate is no longer increased. Therefore, the esterification time of 4 h was chosen as the best reaction time.

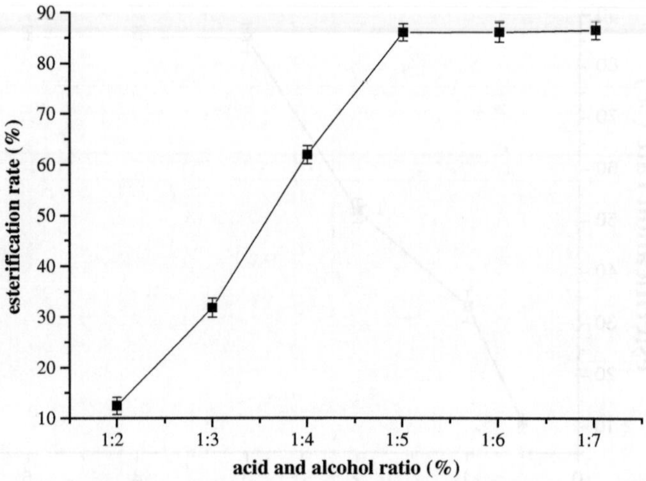

Fig. 3 The effects of acid and alcohol ratio on the esterification rate of ethyl acetate at the conditions of catalyst dosage was 3%, the esterification time was 4 h, the esterification temperature was 80 °C

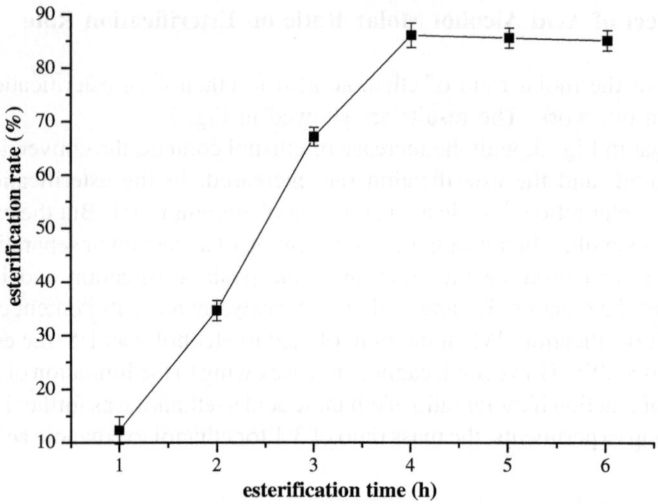

Fig. 4 The effects of esterification time on the esterification rate of ethyl acetate at the conditions of catalyst dosage of 3%, ethanoic acid/ethanol molar ratio of 1:5 and esterification temperature of 80 °C

3.2.4 Effect of Esterification Temperature on Esterification Rate

From Fig. 5, we can see with the increase of the reaction temperature, the esterification rate rises rapidly. And the esterification temperature reaches 80 °C, the esterification rate reaches a maximum value of 86.1%. The reaction of ethyl acetate

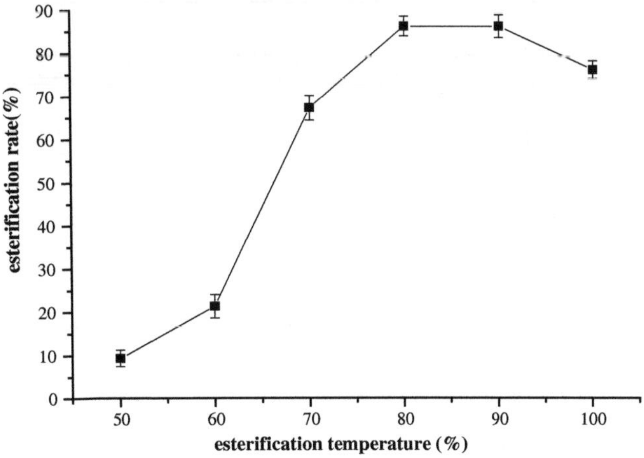

Fig. 5 The effects of esterification temperature on the esterification rate of ethyl acetate at the conditions of catalyst dosage of 3%, ethanoic acid/ethanol molar ratio of 1:5 and esterification time of 4 h

needs a certain temperature, with the increase of esterification temperature, the esterification rate increases. But the esterification reaction is reversible reaction. When the esterification rate reaches the maximum value, system will produce a reversible decomposition reaction. The reverse reaction is endothermic reaction, with the increase of temperature, the reverse reaction is strengthened, and the esterification rate is reduced. The optimum esterification temperature of ethyl acetate was selected at 80 °C.

3.3 Characterization of Infrared Spectra of Solid Acids for Recycling

The stability of solid acid depends on the acidic groups inside the carbon. Fourier transform infrared spectroscopy was used to characterize the internal structure of the carbon based solid acid recovered after each reaction. The results were showed in Fig. 6.

Figure 6 shows that with the increase of the number of applications, the peak of the intrinsic characteristics group of carbon-based solid acid gradually weakened, the most significant for $-SO_3H$ (1000–1180 cm^{-1}), O=S=O (1185–1380 cm^{-1}). When use times reaches 5 times, C–S, $-SO_3H$ peak basically disappeared, when it reaches sixth times, the O=S=O peak basically disappeared. C–S, $-SO_3H$, O=S=O groups are provided by sulfuric acid, but also are the characteristic peaks of carbon-based solid acid with catalytic activity. With the extension of the use of carbon-based solid acid, the active carbon groups attached to solid acids with

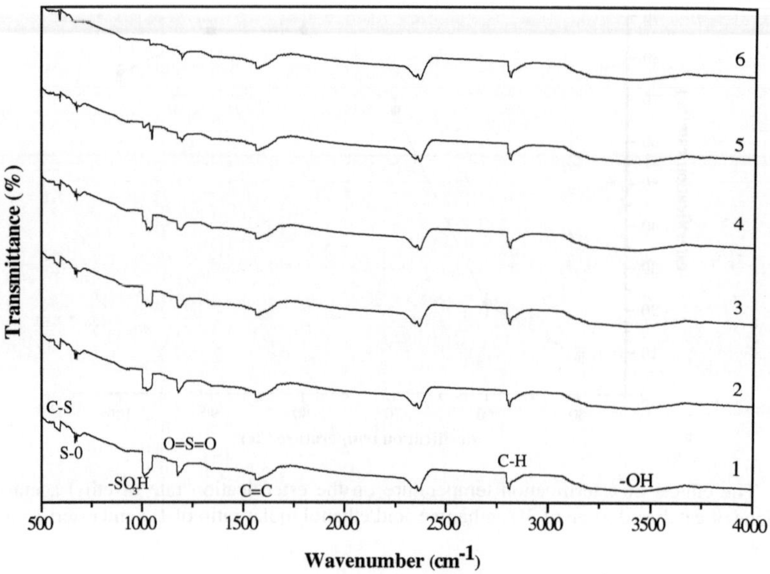

Fig. 6 The IR spectrogram of the repeated use of solid acid

catalytic properties gradually shed, the catalytic activity gradually decreased. Therefore, the use of carbon-based solid acid catalyst has a certain life. However, the active carbon carrier can be recycled and can be used again after sulfonated.

4 Conclusions

The esterification reaction of ethyl acetate catalyzed by solid acid catalyst suggested that the esterification rate of ethyl acetate with 85.4% was obtained under the best conditions of the single factor tests, which were catalyst dosage of 3%, ethanoic acid/ethanol molar ratio of 1:5, reaction time of 4 h, reaction temperature of 80 °C. At the same time, the stability of the activated carbon solid acid catalyst was analyzed by the esterification rate and the infrared spectrum. Finally, it was concluded that the activated carbon-based solid acid catalyst prepared by this experiment can be reused for 6 times.

References

1. Danlin Z, Min S, Yali M (2012) Recent developments in biomass carbon-based solid acid catalyst. Pet Proc Petrochem 43(11):63–68
2. Issariyakul T, Dalai AK (2010) Biodiesel production from green seed canola oil. Energy Fuels 24(9):4652–4658

3. Takagaki A, Toda M, Okamura M, Hara M (2006) Esterification of higher fatty acids by a novel strong solid acid. Catal Today 116(2):157–161
4. Geng L, Wang Y, Yu G, Zhu Y (2011) Efficient carbon-based solid acid catalysts for the esterification of oleic acid. Catal Commun 13(1):26–30
5. Roberto R, Regina P, Ferdi S (2008) Depolymerization of cellulose using solid catalysts in ionic liquids. Cell Convers 47:8047–8050
6. Hara M, Yoshida T, Takagaki A (2004) A carbon material as a strong protonic acid. Angew Chem Int Ed 43(22):2955–2958
7. Nakajima K, Tomita I, Hara M et al (2005) A stable and highly active hybrid mesoporous solid acid catalyst. Adv Mater 17(15):1839–1842
8. Nakajima K, Tomita I, Hara M et al (2006) Development of highly active SO_3H-modified hybrid mesoporous catalyst. Catal Today 116(2):151–156
9. Ren LJ, Yu JW, Zhang XY et al (2009) Study on catalytic performance of new carbon based solid acid catalyst in esterification reaction. Pet Proc Petrochem 40(5):38–42
10. Lou WY, Guo Q, Chen WJ et al (2012) A highly active bagasse-derived solid acid catalyst with properties suitable for production of biodiesel. Chemsuschem 5(8):1533–1541
11. Wu XH, Zhang F, Fang Z et al (2015) Study on the esterification reaction of the seed oil by the carbon based solid acid catalyst. J Yunnan Univ (Nat Sci Edn) 37(1):111–116
12. Chin LH, Abdullah AZ, Hameed BH (2012) Sugar cane bagasse as solid catalyst for synthesis of methyl esters from palm fatty acid distillate. Chem Eng J 183:104–107

The Preparation of Chitosan from Corncob Hydrolyzate by *Actinomucor elegans*

Yan Li, Jia Li and Hua Zhao

1 Introduction

Lignocellulose, an important component of raw material, the most abundant on Earth, which is the cheapest renewable resources [1]. It consists of three forms polymer, cellulose, hemicellulose and lignin [2, 3]. Wherein hemicellulose easily be hydrolyzed into xylose relatively [4]. Many natural microorganisms have the capable of utilizing xylose, including bacteria, fungi and yeasts. Most of them are able to convert xylose into other products [5–7]. Sulfuric acid hydrolysis pretreatment is common and effective treatment of lignocellulose. Unfortunately, during the hydrolysis process, various sugars and microorganisms inhibitors produced, these components include xylose, glucose, mannose, acetic acid, furfural [8]. It is of practical significance to study the utilization of lignocellulose by hemicellulose hydrolysis and detoxification of hydrolyzate.

Chitosan, a linear hydrophilic polysaccharides linked by β-1,4-glucosidic linkages [9]. Which has practical applications in agriculture, cosmetics, food and pharmaceutical industries [10]. The traditional methods of producing chitosan are extracted from shrimp shells [11]. Due to seasonal reasons and the use of a large number of strong alkali at high temperatures result in the nature of chitosan varied, quality decreased and costs increased and environmental pollution, for industrial production is a potential limitation [12, 13]. Chitosan is also an important component of fungal cell wall [14], extracted from the cell walls of fungi can avoid the

Y. Li
Tianjin University of Science and Technology, Tianjin, China

J. Li · H. Zhao (✉)
Key Laboratory of Industrial Fermentation Microbiology, Ministry of Education,
Tianjin Key Laboratory of Industrial Microbiology, College of Biotechnology,
Tianjin University of Science and Technology, Tianjin 300457, China
e-mail: zhaohua@tust.edu.cn

H. Liu et al. (eds.), *Advances in Applied Biotechnology*, Lecture Notes in Electrical Engineering 444, https://doi.org/10.1007/978-981-10-4801-2_67

use of acid and alkali, reduction of industrial pollution on the environment, Which will become a common alternative in the future.

In this paper, the hydrolytic conditions of corncob and hydrolyzate detoxification were studied, as well as the components of medium to obtain chitosan. The aim of this study was to investigate the potential of mycelia to produce chitosan from hemicellulose.

2 Materials and Methods

2.1 *Preparation of Corncob Hydrolysate*

After the corncob (hemicellulose 32%) was crushed, respectively using sulfuric acid with the concentration of 0.5, 0.8, 1.0, 1.5 and 2.0% as the solid-liquid ratio of 1:10, hydrolyzing at 100 and 120 °C for 2 h.

2.2 *Hydrolysate of Detoxification*

2.2.1 Initial Detoxification

After addition of $Ca(OH)_2$ powder to the corncob hydrolyzate, adjusted the pH to 10 and centrifuged, The hydrolyzate was obtained by regulating the supernatant pH to 7 with 10% H_2SO_4.

2.2.2 Secondary Detoxification

$Ca(OH)_2$ Detoxification: The procedure is the same as the initial detoxification.

Active Carbon Detoxification: Thel supernatant was added to activated carbon at a solid-liquid ratio of 1:20, placed in a shaker at 120 r/min for 1 h at 50 °C, and centrifuged to obtain hydrolyzate

Macroporous resin Detoxification: Thel supernatant was added to NKA-2 macroporous at a ratio of solid-liquid 1:10, placed in a shaker at 120r/min for 1 h at 30 °C, and then centrifuged.

2.3 *Mycelium Cultivation*

The spores were obtained by using PDA medium, then inoculated into the seed medium, thereafter cultured in fermentation medium at 6% inoculum, placed in a shaker at 200 r/min for 48 h at 30 °C.

2.4 *Cell Biomass and Chitosan Yield*

2.4.1 Cell Biomass Determination

The mycelium was collected by suction using a Buchner funnel and washed with distilled water until clarified, then dried in an oven at 50 °C.

2.4.2 Determination of Chitosan Yield

Weigh the sample 1 g, with concentrated hydrochloric acid hydrolysis at 100 °C for 6 h, after adjusting pH to neutral with sodium hydroxide, dilute the measured absorbance, glucosamine hydrochloride from the standard curve of the sample in the quality of glucosamine hydrochloride. According to the formula obtained chitosan content

$$\text{Chitosan yield} = \frac{m_1 \times 10^{-3}}{m_2 \times 10^{-2}} \times 0.8309 \times 100\% \quad (1)$$

Note: m_1: The quality of glucosamine hydrochloride was determined from the working curve (mg); m: The mass of the sample (g); 0.8309: Glucosamine hydrochloride is converted to glucosamine by a factor.

3 Results and Discussion

3.1 *Effect of Hydrolysis Conditions on Xylose Yield*

The corncob was hydrolyzed by solid-liquid ratio 1:10 under different conditions for 2 h, and the xylose content was determined. The hydrolysis condition of each sample is shown in Table 1. The total yield of xylose is shown in Table 2.

Table 2 shows that the yield of xylose increases obviously with the increase of acid concentration and temperature. When the concentration of sulfuric acid was 2.0% at 120 °C, the yield of xylose was the highest, reaching 87.79%.

Table 1 Hydrolysis conditions for each sample

T (°C)/sulfuric acid concentration (%)	0.50	0.80	1.00	1.50	2.00
100	1	2	3	4	5
120	1°	2°	3°	4°	5°

Table 2 The effect of hydrolysis conditions on xylose yield

Hydrolysis	1	2	3	4	5	1°	2°	3°	4°	5°
Yield of xylose (%)	49.33	68.04	72.83	75.12	77.88	82.26	84.62	85.44	86.91	87.79

Table 3 The effects of hydrolysis conditions on mycelium growth and chitosan yield after initial detoxification

Hydrolysis	Mycelium concentration (g/L)	Chitosan yield (g/L)
1	11.6	1.5
2	17.42	2.26
3	17.27	2.24
4	17,24	2.24
5	17.12	2.22
1°	18.79	2.44
2°	17.55	2.28
3°	16.27	2.16
4°	No growth	–
5°	No growth	–

3.2 *Effects of Hydrolysis Conditions of Corncob on Mycelium Concentration and Chitosan Yield*

3.2.1 Initial Detoxification

The corncob was hydrolyzed as shown in Table 1. The hydrolyzate obtained by detoxification of $Ca(OH)_2$ was used as carbon source, addition to 7.0% corn steep liquor, 0.2% $MgSO_4 \cdot 7H_2O$ and 0.2% KH_2PO_4, initial pH 7, the inoculation amount of 6%, and the temperature of 30 °C for 60 h. The mycelium concentration and chitosan yield were measured, and results are shown in Table 3.

As shown in Table 3, when the hydrolysis condition is 0.5% sulfuric acid concentration, the mycelium concentration is the highest, and the corresponding chitosan yield is the highest, reaching 2.44 g/L. This may be due to the relatively high concentration of sugar in the hydrolyzate and the relatively low concentration of the inhibitor in this hydrolytic condition, the mycelium can resist the external environment and grow well, and further study on the deep detoxification can be carried out. With 1.0% sulfuric acid, the mycelia could resist the inhibitor and grow well, the concentration reached 16.27 g/L and the yield of chitosan was 2.16 g/L. While the hydrolysis condition with 2.0% sulfuric acid, the concentration of the inhibitor in the hydrolyzate was too high to make mycelium to grow.

3.2.2 Effect of Secondary Detoxification on Mycelium Growth and Chitosan Yield

The hydrolyzate with 0.5 and 1.0% sulfuric acid undergo deep detoxification experiments. After be detoxified by $Ca(OH)_2$, then $Ca(OH)_2$, activated carbon and NKA-2 macroporous adsorption resin used for further detoxification. The mycelia concentration and the chitosan yield were measured with the hydrolyzate as carbon source. The results are shown in Table 4.

From Table 4, it was found that $Ca(OH)_2$ had no significant effect on the mycelial growth after two treatments, whereas activated carbon and NKA-2 macroporous resin detoxification hydrolyzate is more conducive to the growth of mycelium. It may be that the latter causes the concentration of sugar in the hydrolyzate to rise and the amount of the inhibitor to be reduced. The amount of chitosan produced by activated carbon detoxification solution and NKA-2 macroporous resin was 2.74 and 2.78 g/L, respectively. As the cost of activated carbon is low and easy to recover, therefore, the ideal detoxification method for is $Ca(OH)_2$ detoxification with the deep detoxification by activated carbon.

Table 4 The effects of secondary detoxification on mycelium growth and chitosan yield

Sulfuric acid concentration (%)	Detoxification	Mycelium concentration (g/L)	Chitosan yield (g/L)
0.50	$Ca(OH)_2$	18.8	2.46
	Activated carbon	19.21	2.55
	NKA-2	19.29	2.56
1.00	$Ca(OH)_2$	16.3	2.17
	Activated carbon	20.6	2.74
	NKA-2	22.1	2.78

3.3 *Effects of Medium Components on Biomass and Chitosan Yield*

3.3.1 Effect of Nitrogen Source

Nitrogen sources were mainly composed of cell material and nitrogen metabolites. With 2% peptone, beef extract, yeast extract, corn steep liquor, 0.58% sodium nitrate and 0.73% ammonium nitrate as nitrogen source in the basal medium, respectively (Inorganic nitrogen source was converted to the same nitrogen content as peptone). The effects of different nitrogen sources on mycelial growth and chitosan yield were investigated using 3% xylose as carbon source.

As shown in Table 5, the yield of chitosan was 0.67 g/L with corn steep liquor as nitrogen source, which probably attribute to the rich soluble protein, inorganic salt, biotin, and necessary nutrients in corn steep liquor [15], effectively promote cell biomass synthesis. The yield of chitosan by yeast powder was 0.78 g/L, but it's price was expensive, therefore, using corn steep liquor as the medium nitrogen is better.

3.3.2 Effects of Corn Syrup Addition

After the detoxification of the hydrolyzate, 0.2% $MgSO_4{\cdot}7H_2O$ and 0.2% KH_2PO_4 were added. The results are shown in Fig. 1.

As shown in Fig. 1, the mycelial concentration increased with increasing corn syrup, and the yield of chitosan increased from 1.58 to 2.74 g/L. However, when the addition of corn steep liquor was over 7%, the yield of chitosan decreased with the decrease of mycelium concentration due to the inhibition of substrate. Therefore, the addition of corn steep liquor in the medium was determined to be 7%.

Table 5 The effect of nitrogen sources on chitosan production

Nitrogen	Peptone	Beef extract	Yeast	Corn syrup	NH_4NO_3	$NaNO_3$
Mycelium concentration (g/L)	10.81	10.8	11.91	10.35	3.26	1.86
Chitosan yield (g/L)	0.71	0.71	0.78	0.67	0.21	0.12

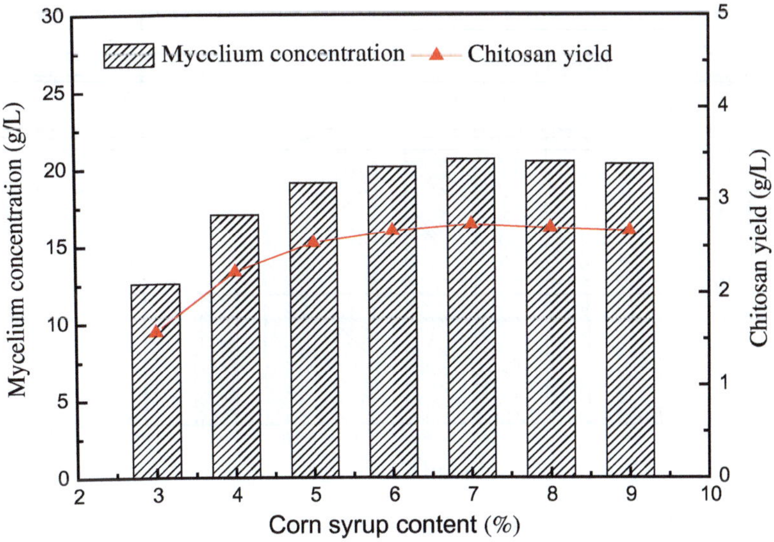

Fig. 1 The effects of corn syrup contents on mycelium concentration and chitosan production

3.3.3 Effect of Initial pH on Mycelium Concentration and Chitosan Yield

The initial pH of the fermentation medium was adjusted to different levels. The fermentation temperature was controlled at 30 °C, shaking speed was 200 r/min, inoculation amount was 6%. After 60 h, the mycelial concentration, chitosan yield and degree of deacetylation were measured (see Fig. 2).

As shown in Fig. 2, when the medium pH was controlled at 4–5, the mycelium concentration and the yield of chitosan were the highest. When the initial pH of the culture medium was controlled from 6 to 7, the mycelium concentration and chitosan production decreased, but the difference was not significant, which indicated that the optimum pH of the *Actinomucor elegans* was 4–5. The degree of deacetylation of chitosan increased with the initial pH of the culture medium. When the pH was 7, the degree of deacetylation was the highest, and the degree of deacetylation began to decrease after the pH was over 7, which mostly because the deacetylase activity is highest under neutral conditions. Taking into account, the optimal pH of the culture medium was 7.

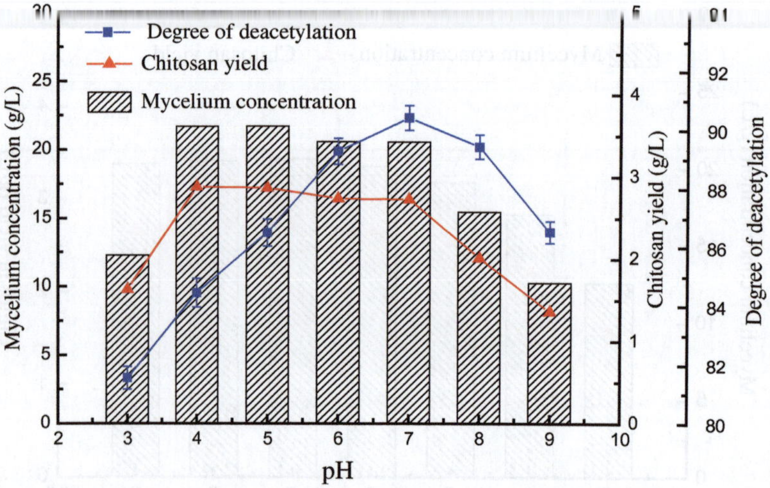

Fig. 2 The effects of initial pH on mycelium growth, chitosan production and deacetyiation

3.4 Dynamic Fermentation of Elegant Mucor

The fermentation temperature was controlled at 30 °C, pH 7, control air flow rate 1.667 vvm, stirring speed 100 r/min. The pH, xylose concentration, glucose concentration, cell biomass and chitosan content in the fermentation broth were measured every 4 h.

As shown in Fig. 3, the pH did not change much throughout the fermentation. The xylose and glucose in the fermentation broth were first utilized by the mycelium, and the glucose was almost completely utilized at 20 h. When the fermentation started, the mycelium grew slowly, entered the logarithmic phase at 24 h, stabilized at 52 h, reached a maximum of 2.87 g/L, and the time of culture was 6 h earlier than in the shaker, mainly depend on ventilation and stirring improve the dissolved oxygen in the fermentation, cell growth faster, so that the fermentation cycle shortened. Prolonged fermentation time, chitosan production began to decline, which may depend on increased cell harmful metabolites and dissolved oxygen, mycelium cell wall autolysis.

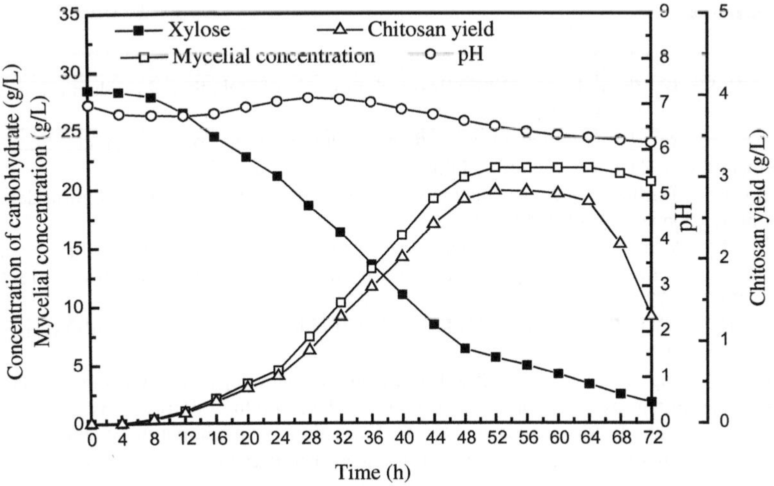

Fig. 3 The fermentation course of cell culture

4 Conclusion

In this paper, the hydrolysis conditions of corncob and the optimization of the process for the production of chitosan by fermentation of *Actinomucor elegans* were studied. The results are as follows:

(1) The best hydrolysis condition of corncob was hydrolyzed by 0.5% H_2SO_4 at the ratio of 1:10 and 120 °C for 2 h, the best detoxification method was $Ca(OH)_2$ and activated carbon.

(2) The optimal medium components of *Actinomucor elegans* were determined as following: the detoxification corncob hydrolysate, 7% corn syrup, 2% $MgSO_4 \cdot 7H_2O$ and 2% KH_2PO_4, initial pH 6.5, inoculation amount 6% under 60 h, the yield of chitosan increased to 2.87 g/L.

(3) The feasibility of fermentative preparation of Chitosan by corncob hydrolysate was proved by fermentation experiment. However, the toxic components in the hydrolysate have a great impact on the growth of M. elegans. The detoxification process needs a high cost, and the domestication bacteria is needed to enhance the tolerance of toxic components from the hydrolyzate.

References

1. Du FG, Shi JP, Zhang L et al (2007) Research progress of fuel ethanol production in cellulose production. Chin J Hemorheol 29(1):72–85
2. Dong XM, Revol J, Gray DG (1998) Effect of microcrystallite preparation conditions on the formation of colloid crystals of cellulose. Cellulose 5(1):19–32
3. Slamenova Darina (2002) Reduction of carcinogenesis by bio-based lignin derivatives. Biomass Bioenergy 23(8):153–159
4. Lloyd TA, Wyman CE (2005) Combined sugar yields for dilute sulfuric acid pretreatment of corn stover followed by enzymatic hydrolysis of the remaining solids. Biores Technol 96 (18):1967–1977
5. Fan YC, Xu QQ, Zhuang YJ (2004) Extraction of chitin and ChITOSAN FROM *Aspergillus niger*. Biotechnology 10(2):2–3
6. Liu HW, Wang AM, Feng GN (2009) Study on the technology of producing saccharomyces cerevisiae protein by corn corn hydrolyzate. Food Sci Technol 34(6):48–51
7. Zamani A, Edebo L, Sjostrom B et al (2007) Extraction and precipitation of chitosan from cell wall of zygomycetes fungi by dilute sulfuric acid. Biomacromol 8(12):3786–3790
8. Herrera A, Luis SJ, Cabriales JJ et al (2004) Effect of the hydrochloric acid concentrationon the hydrolysis of sorghum straw at atmospheric pressure. J Food Eng 63(1):103–109
9. Jiang TD (2009) Chitosan (2nd edn). Chemical Industry Press; Beijing, p 4
10. Aranaz I, Mengibar M, Harris R, et al (2009) Functional characterization of chitin and chitosan. Curr Chem Biol 3(2):203–230
11. Chatterjee S, Adhya M, Guha AK et al (2005) Chitosan from Mucor rouxii: production and physico-chemical characterization. Process Biochem 40(1):395–400
12. Dhillon GS, Brar SK, Kaur S et al (2013) Green synthesis approach: extraction of chitosan from fungus mycelia. Crit Rev Biotechnol 33(4):379–403
13. Cheung RC, Ng TB, Wong JH et al (2014) Chitosan: an update on potential biomedical and pharmaceutical applications. Mar Drugs 13(8):5156–5186
14. Chen X, Lai XH, Yuan YY et al (2000) Preparation of chitosan from filamentous. Fine Chem 17(3):132–134
15. Li WY, Zhao XM (2006) Study on L-lactic acid fermentation of corn steep liquor as organic nitrogen source. Chem Ind Times 20(9):61–63

High-Efficiency Separation and Purification of Taq DNA Polymerase

Hao Zhou, Yujie Zhang, Zhiyin Hu, Ai Mu and Xiangchao Gu

1 Introduction

The thermostable Taq DNA polymerase was firstly isolated by Ms. Qian Jiayun from the YT-1 strain of the aquatic thermophilic bacteria (Thermus aquatics) [1–3]. Its high specificity, yield and sensitivity make it widely used in PCR [4, 5] and other related techniques for many years. Taq polymerase has a half-life of 9 min at 97.5 °C and its optimum temperature is 75–80 °C. It can replicate a DNA fragment of 1000 bp within 10 s at 72 °C. Its higher enzyme activity is temperature dependent. At low temperature, the enzyme activity of replicating DNA was significantly lower comparing to that at higher temperature. However, its activity on DNA synthesis is also quickly reduced when temperature is higher than 90 °C.

At present, the main task in optimizing Taq DNA polymerase is to improve its yield by genetic engineering as well as simplifying the purification and shortening the production process [6, 7]. Large-scale production of efficient and cost-effective Taq DNA polymerase relys on new engineering method, and therefore the PCR technology still has much room to improve [8].

Engelke et al. constructed recombinant plasmid pTTQ18 and expressed the recombinant Taq DNA polymerase in Escherichia coli. It was purified by denaturing heat- hybrid protein and PEI. Then used BioRex 70 Ion Exchange Chromatography to obtain recombinant Taq DNA polymerase. However, the yield, purity and enzyme activity of this method are not very high. Qinchuan et al. used ammonium sulfate precipitation and freezing and thawing to extract Taq DNA polymerase. It showed that ammonium sulfate precipitation is relatively simple and low in cost. The activity of

H. Zhou (✉) · Y. Zhang · Z. Hu · A. Mu · X. Gu
Key Laboratory of Industrial Microbiology, Ministry of Education and Tianjin City, College of Biotechnology, Tianjin University of Science and Technology, Tianjin 300457, People's Republic of China
e-mail: zhouhao@tust.edu.cn

H. Liu et al. (eds.), *Advances in Applied Biotechnology*, Lecture Notes in Electrical Engineering 444, https://doi.org/10.1007/978-981-10-4801-2_68

prepared Tap DNA polymerase can be effectively amplified DNA fragments. Freeze-thawing method did not sufficiently purify protein. The activity of Taq DNA polymeraseis low. PCR amplification effect of this method is not ideal.

In this study, we established a highly efficient and rapid method for the expression and purification of Taq DNA polymerase in an engineering strain of Taq DNA polymerase. This method optimized various conditions to elevate the production and purity of the enzyme, providing high quality Taq DNA polymerase for routine molecular biology experiments. His6 refers to a fusion tag consisting of six histidine residues that can be inserted at the C-terminus or N-terminus of the protein. It can constitute an epitope that facilitates purification and detection, it also can format the unique structural features (binding ligands) to facilitate purification. So the constructed plasmid of Recombinant Escherichia coli strain uesd His-tag. The Taq DNA polymerase Recombinant Escherichia coli strain has the advantage of producing large quantities of this enzyme in E. Coli using IPTG as an inducer. The technique does not require tedious work, and purified enzyme is effective and can be used with relatively low amount. Therefore it reduces the cost of each experiment to a certain extent [9]. Taq DNA polymerase is an important biotechnology tool, and is widely used in diagnosis and treatment of infectious diseases, drug mechanism and other medical fields. Therefore, this study not only has clear potential in applications, but also has some theoretical significance for the research of polymerase expression, separation and purification.

2 Materials and Methods

2.1 *Materials*

Taq DNA polymerase Recombinant Escherichia coli strain preserved in our laboratory. Chemicals and culture medium: Luria-Bertani (LB, 1% Tryptone, 0.5% Yeast extract, 1% NaCl), 50× TAE Buffer (Tris, acetate, EDTA, pH 8.5), DNA extracts (100 mmol/L Tris-Cl pH 8.0, 50 mmol/L EDTA pH 8.0, 500 mmol/L NaCl, 1% SDS, 10 mmol/L βHere) [10]. Buffer A (500 ml): 0.6057 g Tris (pH 7.9), 0.2033 g $MgCl_2$, 0.1982 g $(NH_4)\ _2SO_4$, 0.99 g Glucose, 1.86 g KCl.

2.2 *Definition of Taq DNA Polymerase*

The enzyme activity of one unit of Taq polymerase was defined by the amount of enzyme required to incorporate 10 nmol deoxynucleotides into acid-insoluble matter at 74 °C for 30 min using activated salmon sperm DNA as template/primer [11].

2.3 Selection and Activation of Culture

The *E. coli* strain of this experiment contains an exogenously expressed Taq DNA polymerase gene. To purity Taq DNA polymerase, the strains were streaked on LB plates (Amp 100 μg/mL), cultured for 16–20 h at 37 °C. Single colony is picked and inoculated into 6 ml LB medium containing ampicillin (100 μg/mL), and cultured overnight at 37 °C.

When the bacteria reached their semi-growth phase(OD660 = 0.4), induction of Taq DNA polymerase was performed using different concentrations of IPTG and samples were shook for 12 h at 37 °C.

2.4 Separation, Purification and Concentration of Taq DNA Polymerase

Using lysozyme, freezing-thawing and heparin-Sepharose affinity chromatograph to separate and purify Taq DNA polymerase. Lysozyme, also known as the cell wall enzyme or N-acetyl cell wall polysaccharides hydrolase, is an alkaline enzyme that can effectively hydrolyze the peptidoglycan of bacterial cell wall [12]. Freezing and thawing method cooled brakes cells to −15 to −80 °C, and then melt quickly at 30–40 °C, so repeated freezing and thawing many times, the formation of ice particles in the cell to make the remaining fluid solution of salt Concentration increased caused by cell swelling and fragmentation. Repeated freezing and thawing method is more gentle, avoiding the destruction of high temperature on the active substance.

The procedures of lysing bacteria and denaturing by high temperature to remove some miscellaneous proteins are (1) Adding 5 μL Buffer A to the centrifuge tube. After the cells were completely dissolved in buffer A, added 200 μL lysozyme, placed at 70 °C for 1 h. (2) Lysed cells were placed at −80 °C for 10 min, remove the cells and place in water at 80 °C for 5 min. After repeated freezing and thawing for three times, transfer the bacteria into water with a temperature of about 75–80 °C for 30 min, mix upsiding down the bacteria culture every 5 min. (3) Centrifuging at 8000 rpm for 30 min at 4 °C, transferring the supernatant to a clean tube, and placed on ice. (4) Filtering with 0.45 nm filter, taked 1 mL as a crude enzyme.

In this study, the heparin affinity chromatography was used to further purify. Finally the high concentration of Taq DNA polymerase was obtained. Heparin affinity chromatography medium (Heparin QZT 6FF) was prepared by conjugating heparin to an agarose gel using a self-made cross-linked agarose as the matrix. In this paper, Taq DNA polymerase was concentrated by dialysis.

2.5 Detection of Taq DNA Polymerase

Taq DNA polymerase were separated by SDS-PAGE [13–15]. A 12% polyacrylamide gel was used for electrophoresis and Coomassie blue R-250 (0.1%) was used for gel staining. The Odyssey Infrared Imaging System, and the corresponding application software version 3.0 from Li-Cor Biosciences [16].

2.6 DNA Extraction and PCR Analysis

Total DNA was extracted by using the Kit (Solarbio) according to the manufacturer's instructions. Using extracted cell genome as PCR template through culturing cell. Three kinds of primers were designed and verified by PCR [17]. The length of the target fragment was 2200, 750 and 600 bp respectively. It can verify the amplify efficiency of PCR.

3 Results

3.1 Optimal Screening of Taq DNA Polymerase Strain

Drawing three section line [18] with a small amount of Taq DNA polymerase solution. The treated plate was then incubated overnight at 37 °C and the growing rounded single colony was picked as the candidate Taq DNA polymerase strain.

3.2 Induction of Taq DNA Polymerase

Recombinant bacteria were inoculated into LB medium. After incubating at 37 °C for 12 h, 5 mL bacteria solution was inoculated into 100 mL culture medium (Containing 100 μg/mL ampicillin). Adding IPTG in the latter part of the growth curve [19]. The final concentration of IPTG is 0.5 mmol/L. After inducting for 12 h, the bacteria were collected by centrifugating at 5000 rpm for 10 min. Using SDS polyacrylamide gel electrophoresis to examine the induced effect of protein of the crude Taq DNA polymerase.

It can be seen from Fig. 1 that the level of protein expression of Taq DNA polymerase increased significantly after IPTG induction for 12 h compared to the strain that was not induced by IPTG. The production of Taq DNA polymerase increased 50% by the gray level analysis through Quantity One.

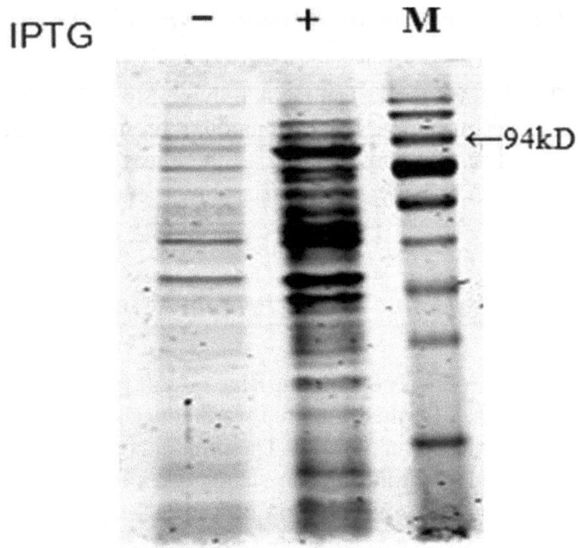

Fig. 1 Induction of Taq DNA polymerase

3.3 Purification of Taq DNA Polymerase

The crude Taq DNA polymerase was obtained by repeated freezing and thawing cycle as previously described [20] with constantly switching high and low temperature. The protein was purified by affinity chromatography. The purified Taq DNA polymerase that was obtained after the elution and detected the purification effect of protein by SDS polyacrylamide gel electrophoresis [21].

As shown in Fig. 2, compared with the purchased Taq DNA polymerase (control), the crude enzyme solution contains a small amount of hybrid protein, but the purity of Taq DNA polymerase has been greatly improved after affinity chromatography. The purity of Taq DNA polymerase is up to about 90% by the gray level analysis through Quantity One.

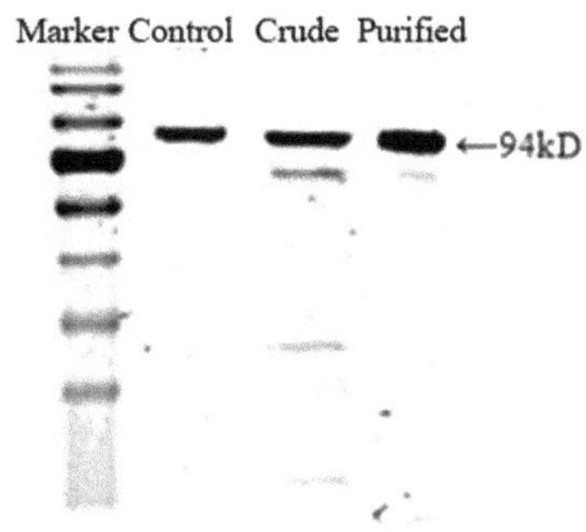

Fig. 2 Affinity purification of Taq DNA polymerase

3.4 Detection of Residual Nucleic Acids

The obtained crude Taq DNA polymerase and purified enzyme were checked to see whether there are residual amount of nucleic acid by using agarose gel electrophoresis. As shown in Fig. 3, there was no ribonucleic acid bands in the obtained Taq DNA polymerase extracts, thus it could be used for PCR experiment.

3.5 Detection of the Protein Concentration

200 μL Coomassie Brilliant Blue G-250 Formulation Solution and 5 μL Taq DNA Polymerase Purified Enzyme were added to each well of a cleaning 96-well plate, reading the OD at 595 nm. The protein concentration of Taq DNA polymerase was calculated according to the standard curve [22]. As shown in Fig. 4, after purification, protein concentration of Taq DNA polymerase had increased.

3.6 Detection Heat Tolerance

Toleration of the enzyme to high temperature is critical for PCR, therefore we determined the thermal stability of the enzyme. As shown in Fig. 5, the Taq DNA polymerase was heated at 95 °C for 5, 10, 15 and 20 min, respectively. PCR and agarose gel electrophoresis were used to detect the heat tolerance of Taq DNA polymerase.

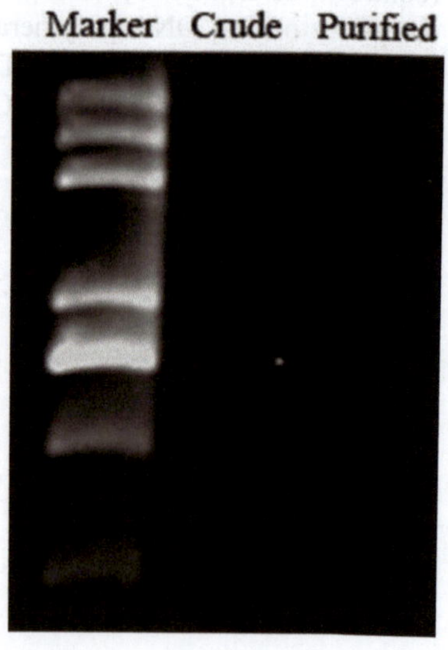

Fig. 3 Detection of residual nucleic acid

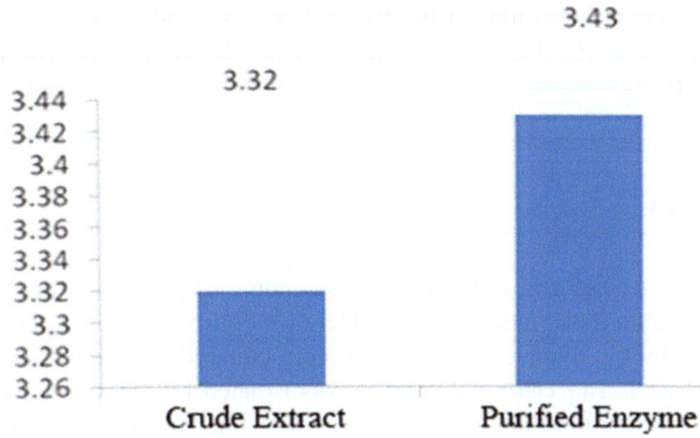

Fig. 4 Protein concentration of Taq DNA polymerase (μg/mL)

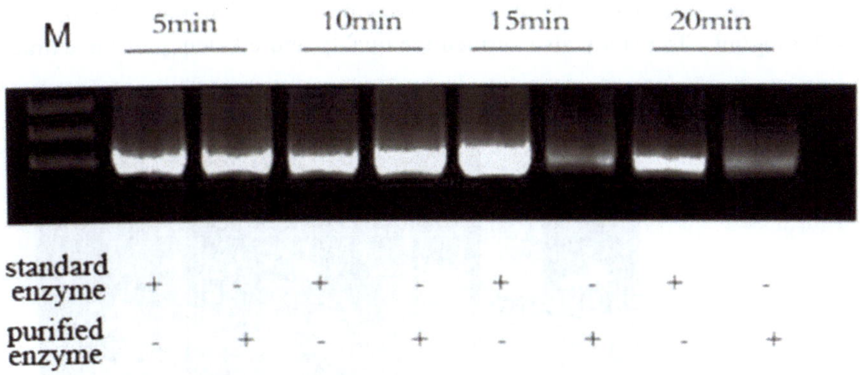

Fig. 5 Detection of Taq DNA polymerase heat resistance

It can be found that the purified enzyme of Taq DNA polymerase was heated to 15 min. PCR target band did not differ from the standard enzyme, which can meet the high temperature conditions during PCR. Thus it could prove that the extracted Taq DNA polymerase has good heat resistance.

3.7 The Amplification Efficiency of Taq DNA Polymerase

Using PCR to analyze the efficiency of the obtained Taq DNA polymerase. The length of the target fragment was 2200, 750 and 600 bp respectively. The results indicated that our purified Taq DNA polymerase not only showed the high efficiency on PCR as that of the standard enzyme, but even more powerful in amplifying short length DNA template as it revealed higher specificity. Our purified

Taq DNA polymerase was not do the transformed, so fidelity will not have any increase or decrease. Fidelity was not significantly different from the purchased standard (Fig. 6).

4 Discussion

This study characterized and optimized the procedure of produce Taq DNA polymerase using a strain developed in our lab. Here we optimized the induction time of IPTG for Taq DNA polymerase, and the protein content after induction was significantly increased. Taq DNA polymerase was isolated by the method of preliminary protein extraction combined with freeze-thawing and thermal denaturation, followed by further purified using heparin affinity chromatography. The production of Taq DNA polymerase was in high yield that reached ~3.43 mg/mL. Taq DNA polymerase was not do the transformed, so enzymatic activity will not have any increase or decrease. The yield of purified enzyme was 5.15 mg and the final yield was 2.6 mg/mL. The study also verified the quality and efficiency of the extracted

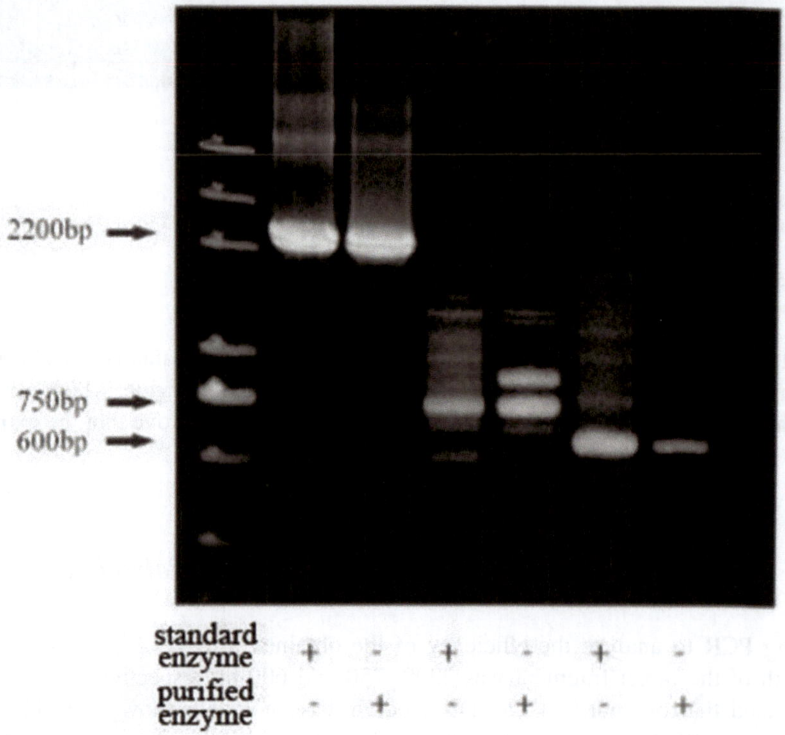

Fig. 6 Effect of Taq DNA polymerase amplification

Taq DNA polymerase, and our results showed that Taq DNA polymerase could amplify the DNA fragment of different length with even higher specificity than that of the standard commercial enzyme.

Acknowledgements This work was financially supported by the Laboratory Open Foundation of Tianjin University of Science and Technology (1304A303).

References

1. Zhan Q, Zhan W, Liu Z, Zhu K (2013) Preparation and purification of Taq DNA polymerase. Hunan Agric Sci 13(10–11):15
2. Ding Y, Liu S, Qi Q (2011) Agricultural biotechnology. Agric Sci Technol 12(3):375–378
3. Yu Y, Li C, Chen S (2012) Rapid purification of Taq DNA polymerase by anion exchange column. Fujian Agric J 27(7):734–738
4. Ding Y, Liu S, Li Q (2011) Preparation of Taq DNA polymerase by thermal purification. Anhui Agric Sci 39(17):10153–10155
5. Ishmael Faoud T, Stellato C (2008) Principles and applications of polymerase chain reaction. Basic Sci Pract Phys 101(4):437–443
6. Wang JC, LI RG (2004) Expression and purification of recombinant Taq DNA polymerase in *Escherichia coli*. J QiangDao Univ (E&T)
7. Arezi B, Xing W, Sorge JA, Hogrefe HH (2003) Amplification efficiency of thermostable DNA polymerases. Anal Biochem 321(2):226–35
8. Moazen F, Rastegari A, Hoseini SM, Panjehpour M, Miroliaei M, Sadeghi HM (2012) Optimization of Taq DNA polymerase enzyme expression in *Escherichia coli*. Adv Biomed Res, vol 82
9. Sadeghi HM, Rabbani M, Moazen F (2006) Amplification and cloning of Taq DNA polymerase gene from *Thermus aquaticus* strain YT-1. Res Pharm Sci 1:49–52
10. Bu Z, Biehl R, Monkenbusch M (2005) Coupled protein domainmotionin Taq polymerase revealed by neutron spin-echo spectroscopy. Natl Acad Sci USA 102(490):17646–17651
11. Lawyer FC1, Stoffel S, Saiki RK, Myambo K, Drummond R, Gelfand DH (1989) Isolation, characterization, and expression in Escherichia coli of the DNA polymerase gene from Thermus aquaticus.The Journal of Biological Chmistry 264(11):6427–37
12. Arabski M, Konieczna I, Wąsik S, Relich I, Zając K (2015) The use of lysozyme modified with fluorescein for the detection of Gram-positive bacteria. Microbiol Res 170:242–247
13. Urh M, Simpson D, Zhao K (1990) Affinity chromatography: general methods. Methods Enzymol 463:417–438
14. Zhang C, Long AM, Swalm B, Charest K, Wang Y, Hu J, Schulz C, Goetzinger W, Hall BE (2016) Development of an automated mid-scale parallel protein purification system for antibody purification and affinity chromatography. Protein Expr Purif 128:29–35
15. Krieg RC, Dong Y, Schwamborn K, Knuechel R (2015) Protein quantification and its tolerance for different interfering reagents using the BCA-method with regard to 2D SDS PAGE. J Biochem Biophys Method 65(1):13–9
16. Wiechelman K, Braun RD, Fitzpatrick JD (1988) Investigation of the bicinchoninic acid protein assay: identification of the groups responsible for color formation. Anal Biochem, vol 175(1) (McCarthy)
17. McCarthy MW, Walsh TJ (2016) PCR methodology and applications for the detection of human fungal pathogens. Expert Rev Mole Diagn 16(9):1025–36
18. Si Z, Zhu J, Wang W, Huang L, Wei P, Cai J, Xu Z (2016) Novel and efficient screening of PQQ high-yielding strains and subsequent cultivation optimization. Appl Microbiol Biotechnol

19. Marbach A, Bettenbrock K (2012) Lac operon induction in *Escherichia coli*: Systematic comparison of IPTG and TMG induction and influence of the transacetylase LacA. J Biotechnol 157(1):82–88
20. Pluthero FG (1995) Rapid purification of high- activity Taq DNA polymerase Nucl Acids Res 21:4850–4851
21. Ying Y, Zhao L, Kong L, Kong X, Hua Y, Chen Y (2015) Solubilization of proteins in extracted oil bodies by SDS: a simple and efficient protein sample preparation method for Tricine-SDS-PAGE. Food Chem 181:179–185
22. Brunelle JL, Green R (2014) Coomassie blue staining. Methods Enzymol 541:161–167

Determination of Activity and Extraction of Thrombin from the Porcine Blood

Tianjun Li, Heng Li, Hong Pan, Tao Li and Jun Shi

1 Introduction

Thrombin (Thrombin, EC 3.4.21.5), a kind of serine proteolytic enzymes, is important in the system of blood coagulation and is first known as one of coagulation factors [1, 2]. In the late 19th century, thrombin was discovered and nominated by physiologist in Scotland. The research and acquaintance of thrombin are deeper and deeper after being purified and sequenced in 1951. In the middle 20th century, the application was more and more widely due to its prominent cure effect after the thrombin was applied to clinic. It is a kind of quick and local hemostatics in recent years in China [3].

Thrombin is apt to stop local bleeding for the all kinds of blood capillary and a few of visceral organs. It has been applied widely to trauma, surgery, oral cavity, ear-nose-throat, urinary and alimentary canal. Thrombin is safe and convenient, and has no obvious adverse effect for long time and high dose. Its styptic effect is better than classic hemostatic such as the 4-amino benzoic acid and tranexamic acid. In clinical practice, thrombin is usually applied to the wound and operation in the form of the power or solution, which can be used to control the exudation of capillary blood, and sometimes, it can be used as oral medicine in the stomach and duodenum. Not only directly as hemostatic drug, but also thrombin can be combined with other drugs to obtain a new dosage form, can also be used as an important raw material for a variety of topical hemostatic agents.

Application of animal thrombin in the human body does not appear the antigenicity of thrombin, oral and topical did not produce the antibodies of allergic reaction and other adverse reactions, which is currently used by thrombin preparations derived mostly from animal plasma. Although a lot of researches are concentrated in the preparation of thrombin in the domestic and foreign, few researches

T. Li (✉) · H. Li · H. Pan · T. Li · J. Shi
Department of Basic Science, Tianjin Agriculture University, Tianjin 300384, China
e-mail: litianjun65@163.com

H. Liu et al. (eds.), *Advances in Applied Biotechnology*, Lecture Notes
in Electrical Engineering 444, https://doi.org/10.1007/978-981-10-4801-2_69

reported on porcine thrombin. Porcine production in China ranked first in the world, a total of about 1 billion kilograms of porcine was got every year. At present, most of the blood in the form of sewage is discharged out, not only a waste of valuable biological resources and cause serious environmental pollution. Therefore, the extraction of thrombin from porcine blood can make full use of blood resources, has great market value.

Thrombin is extracted by the methods of isoelectric precipitation and ammonium sulfate fractionation precipitation from fresh porcine blood. The solution of crude thrombin which was activated by the 0.05 mol/L Ca^{2+} solution were obtained and purified through chromatographic column of cellulose DEAE-52. The measurement of activity of the obtained thrombin was carried.

2 Materials and Methods

2.1 *Materials*

The fresh porcine blood is obtained from Yanchen porcine farm in Wuqing distinct Tianjin. Before sampling, the 3.8% sodium citrate solution as anticoagulation was shook and centrifuged under 4000 r/min for 20 min. The adtevak (the supernatant) is packed in small tubes and saved under −20 °C.

Fibrinogen standard products (1000 μ, dissolved in 1 mL NS, and then packed into 100 tubes, each tube is 10 μ) were supplied by the microbiology laboratory of Tianjin University.

3.8% sodium citrate solution, 2% acetate acid, 30% ammonium sulfate, Tris-HCl buffer solution(0.05 mol/L, pH = 7.2), 0.05 mol/L $CaCl_2$ solution, NS, 0.1 mol/L NaCl solution

2.2 *Experimental Methods*

2.2.1 Extraction of Thrombin

Isoelectric precipitation Two tubes of adtevak were taken out and unfrozen under −4 °C, and then centrifuged under 4000 r/min for 10 min. The lower precipitation, fibrinogen and other miscellaneous protein, was removed, and the supernatant was collected. The supernatant was diluted 10 times with distilled water. The pH was adjusted to 5.0 with 2% acetate acid. Then the solution was placed overnight under −4 °C. The supernatant was removed by siphon method. The precipitation was centrifuged under 4000 r/min for 10 min.

Ammonium sulfate fractionation precipitation The saturated solution 30% ammonium sulfate was added to the precipitation. The precipitation was dissolved completely. This solution was placed overnight. The above-mentioned solution was centrifuged under 4000 r/min for 10 min in the next day. Ammonium sulfate was added to the supernatant until the solution reach to its degree of saturation of 65%. This solution was centrifuged under 4000 r/min for 20 min after being placed for 10 h. The supernatant was removed and the precipitation (prothrombin) was collected.

Dialysis The above-mentioned precipitation was dissolved with 0.05 mol/L Tris-HCl solution with pH = 7.2. The dialysis bag was boiled for 10 min and checked its leaks. The solution was transferred into dialysis bag. After this dialysis bag was full of 60% and the air was removed, the solution was dialyzed for 4 h under −4 °C. During the process, the solvents was changed once time. Then the solution was dialyzed against buffer solution overnight. This solution was centrifuged under 3000 r/min for 20 min. The supernatant was the prothrombin of porcine blood.

2.2.2 Activation of Prothrombin

0.4 mL 0.25 mol/L $CaCl_2$ was added to the 19.6 mL solution of prothrombin. The final concentration of Ca^{2+} was 0.05 mol/L. The prothrombin was activated during 1.5 h. The activated prothrombin was the crude thrombin.

2.2.3 Purification of Prothrombin

The thrombin was purified by ion-exchange chromatography [4]. First, the suspension including 30 g cellulose DE-52 was boiled for 10 min to swell it. This suspension was added into column and balanced with the 0.05 mol/L, pH = 7.2 Tris-HCl solution. The sample of thrombin was eluded by 0.1 mol/L, 0.1 mol/L NaCl, respectively after the sample was dissolved with 0.05 mol/L, pH = 7.2 Tris-HCl solution. The constant pump flow rate is set as 20 m L/h, 4 mL/tube. The collected solution was the thrombin solution which was dialyzed against water to remove the ion of Na^+ and Cl^- and frozen drying. The product was thrombin.

2.2.4 Determination of Activity of Thrombin

The activity for thrombin was determined according to the method provided by the Pharmacopoeia of the People's Republic of China [5]. The standard curve was drawn with different concentrations of thrombin standard vs the reaction time of different fibrinogen standard solution. The activity of purified thrombin could be obtained by the standard curve after determining the reaction time between thrombin and fibrinogen.

Determination of protein content was done using UV absorption method, with bovine serum albumin as the standard with 752 spectrophotometer.

3 Results and Discussions

3.1 Purification of Thrombin

After the prothrombin extracted from blood was activated, the thrombin was purified with ion-exchange chromatography. The column was balanced with the 0.05 mol/L, pH = 7.2 Tris-HCl solution overnight before the 2 mL crude thrombin solution was pour into the column. The sample was eluted with 0.05 mol/L, pH = 7.2 Tris-HCl solution and 0.1 mol/L NaCl, respectively. The flow rate was 20 mL/h. The sample was collected onto automatic collector with 4 mL/tube. The number of the tubes was 1–10. The elution solution was changed into 0.2 mol/L NaCl after the sample collection was completed. The flow rate and the number of the tubes were unchanged. The number of these tubes was 11–20. The adsorption of these samples was measured with UV–Vis spectrophotometer at 280 nm. The reference solutions were 0.1 mol/L NaCl for the 1–10 tube and 0.2 mol/L NaCl for 11–20 tube, respectively. The protein content solution of each sample was shown in Fig. 1. At the same time, 50 μL solution, added the same volume solution of fibrinogen (0.25 g/mL), was taken out from each tube in a small centrifuge tube. There was precipitation in tube to account for the existence of thrombin activity. So the thrombin activity was measured qualitatively.

As shown in Fig. 1, a small amount protein was eluted by 0.1 mol/L NaCl while a large amount protein was eluted by 0.2 mol/L NaCl. The results showed that there was thrombin activity in the 12–17 tube, not in other tubes. So the protein eluted by

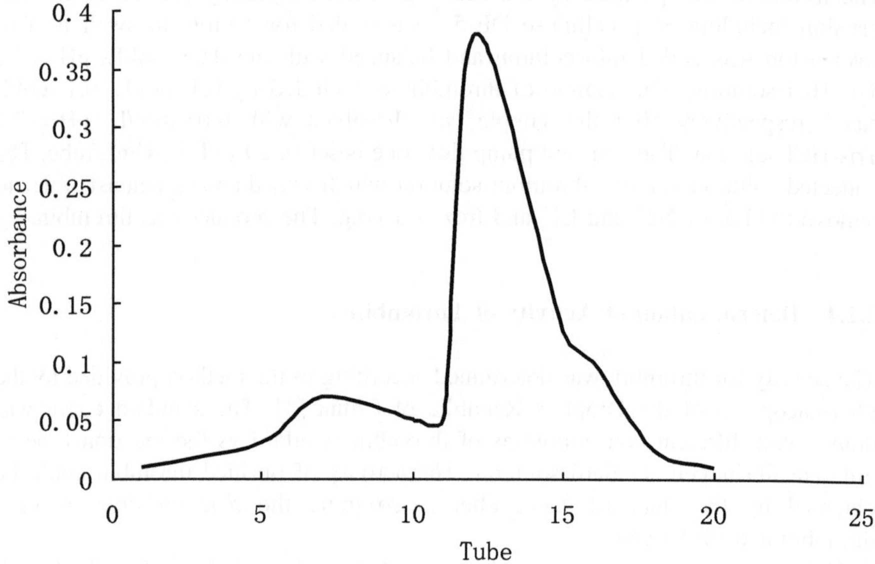

Fig. 1 Elution curve of the thrombin from porcine blood

0.2 mol/L NaCl was the thrombin protein. The solution in 12–17 tubes was merged to obtain 24 mL solution. The 4 mL solution was taken out for subsequent measurement and the surplus 20 mL solution was dialyzed against water to remove the Na^+ and Cl^- overnight. The dialysis solution was frozen-drying overnight.

There were different methods in every step during the separation and purification of the thrombin. So the purity and yield of thrombin obtained by different combination methods were different. Thrombin extraction is mainly divided into four steps: obtained plasma, extraction of prothrombin, prothrombin activation and thrombin purification. The different dilution ration of the blood plasma had a certain effect on the extraction of thrombin when the method of the isoelectric point precipitation was used. The obtained blood plasma was frozen for 12 h under −20 °C to remove the fibrinogen and other proteins of the blood plasma. The prothrombin, which was extracted with the combined method of the isoelectric point precipitation and ammonium sulfate precipitation, was diluted 10 times, activated by 0.05 mol/L Ca^{2+} and purified by DEAE-52 cellulose ion exchange chromatography. After the combination of the above methods, the extracted thrombin showed that the protein content was similar to that of the thrombin standard, and the product was fully purified. 5 mg thrombin was extracted from the 100 mL blood plasma. The activity of thrombin was low. So the different method combination is used to achieve the best results in the future experiments.

3.2 *Measurement of Protein Content for Thrombin*

The purified water was added to the 25 mg bovine serum albumin to 25 mL. The concentration of the bovine serum albumin solution was 1 mg/L. A variety of reagents was added and mixed into the test tubes numbered 1–8 according to the Table 1. The value of adsorption (OD280) was measured at 280 nm with the solution in 1 tube as reference.

Figure 2 show that the standard curve of adsorption with protein concentration was drawn. As shown in Fig. 2, the equation of protein concentration was y = 0.1805x. The adsorption of thrombin measured with 752 UV–Vis spectrometer was 0.139. The calculated protein concentration was 0.77 mg/mL.

Table 1 Standard series of protein with UV–Vis spectrophotometer

Tube number	Standard solution of protein (mL)	Concentration of protein (mg/mL^{-1})	Purified water	OD_{280}
1	0.0	0.000	4.0	0
2	0.5	0.125	3.5	0.086
3	1.0	0.250	3.0	0.177
4	1.5	0.375	2.5	0.268
5	2.0	0.500	2.0	0.364
6	2.5	0.625	1.5	0.454
7	3.0	0.750	1.0	0.548
8	4.0	1.000	0.0	0.716

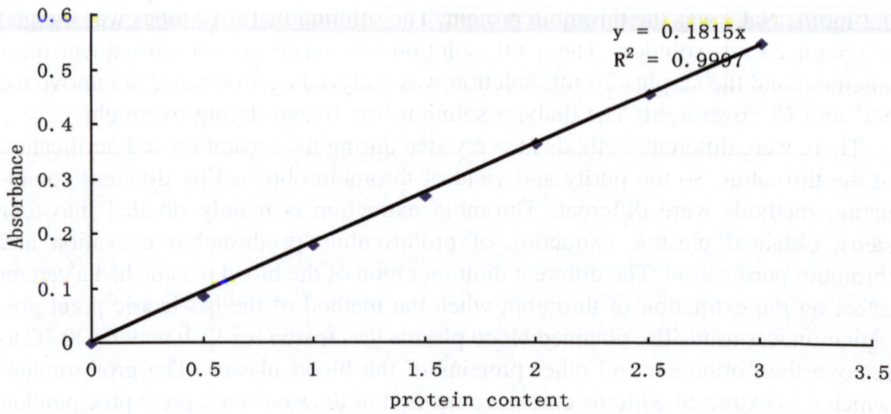

Fig. 2 Standard curve of protein content with UV–Vis spectrometer

3.3 *Measurement of Activity of Thrombin*

5.0 U thrombin standard solutions (5 μL) was added 0.9% NaCl to the final volume of 1 mL, which was the thrombin standard solution. 0.1 mL thrombin standard solution was added 0.9 mL fibrinogen standard solution with different concentrations which heated at 37 °C for 5 min. The reaction time was controlled about 1 min by adjusting the concentration of fibrinogen standard solution. The reaction time was about 80 s when the fibrinogen standard solution was 0.25 mg/mL. So 0.25 mg/mL of fibrinogen standard solution was chosen during the process of drawing the standard curve.

The thrombin standard solution was dissolved with 0.9% NaCl to obtain a serial of solution with 5.0, 6.4, 8.0, 10.0 and 12.0 U. 0.1 mL above-mentioned solutions and 0.9 mL fibrinogen standard solution with concentration of 0.25 mg/mL was quickly mixed in five tubes respectively at 37 °C for 5 min. At the same time, the time started. These tubes was shaken well and placed in a water bath at 37 °C. The initial setting time of fibrinogen was recorded. The measurements were repeated three times to obtain the average value.

The reaction time of the thrombin standard solution reacted with fibrinogen standard was shown in Table 2.

In the double logarithmic, changes in the setting times with the average value of actual activity units was shown in Fig. 3.

As shown in Fig. 3, the equation of activity of thrombin was $y = -0.115x + 14.406$. The initial setting time of thrombin was measured with the same method. The times were 75, 73, 76 s respectively. The average value was 74.7 s. The calculated activity of thrombin was 5.8 U/mL and the specific activity was 29.9 U/mg.

According to the literatures [4], the activity of thrombin was affected by the temperature. The lower the temperature was, the higher the activity of thrombin

Table 2 The time of the thrombin standard samples

Times (s)	Activity unit (U)				
	5.0	6.4	8.0	10.0	12.0
1	81	67	53	36	22
2	87	73	51	39	25
3	79	76	50	37	24
Average value	82.3	72	51.3	37.3	23.7

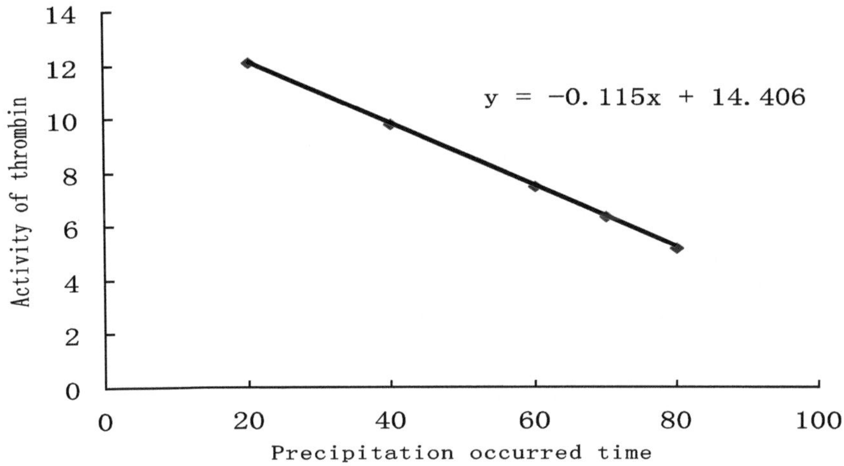

Fig. 3 The standard curve of activity for thrombin

was. The activity of thrombin decreased remarkably when the temperature was higher than 25 °C. The activity of the thrombin decreased to 15% with the increase of the temperature. So, the extraction should be carried out at low temperature in the process of extracting thrombin. The experiment was carried out at room temperature, which may cause effects on the activity of thrombin with extent.

The concentration of $CaCl_2$ and the activated time have effect on the activity of thrombin. The higher the concentration of $CaCl_2$ was, the higher the activity of thrombin was. The activity of the thrombin kept a constant when the concentration of $CaCl_2$ was increased to a certain value. So the effect of the different temperature and concentration on the activity of thrombin should be considered in the future research.

4 Conclusions

There is rich porcine blood in China. The extraction of thrombin from porcine blood may make fully use of the porcine blood resource. The thrombin has great application value in clinical due to its good hemostatic effect. In our experiment, the porcine blood was used as raw material to extract the thrombin. Thrombin is extracted by the methods of isoelectric precipitation and ammonium sulfate fractionation precipitation from fresh porcine blood. The solution of crude thrombin which was activated by the 0.05 mol/L Ca^{2+} solution were obtained and purified through chromatographic column of cellulose DEAE-52. The active component, as pure solution of thrombin, was frozen-drying to obtain the white and pure thrombin. The specific activity was 29.9 U/mg by measuring the activity and purity of the extracted thrombin.

References

1. Wenjing Zhou, Jiali Lv (2005) The improvement of method for the extraction and purification of thrombin from the porcine blood. Res Dev Food 26(2):93–95
2. Changfa Xu, Yulan Wang, Defu Liu et al (1994) Derternination, separation and purification of thrombin. J Beijing Union Univ 8(1):45–48
3. Yi Lai, Yang Liu, Fangzhao Lin (2009) Overview of thrombin research. Thrombosis Hemostasis 15(3):142–144
4. Wanqiong Ren (2008) The effect of the conditions for the separation and extraction of thrombin. J Med Sci 29(23):53–54
5. Chinese Pharmacopoeia Commission (2005) The third Pharmacopoeia of the People's Republic of China. Chemcal Industry Press, Beijing, pp 193–194

An Efficient Method for Isolation and Separation of Pigments from *Streptomyces alboflavus* TD-1

Xiaoyue Gu, Yali Zhang, Laifeng Lu, Zhenjing Li and Changlu Wang

1 Introduction

Colorants derived from natural sources, such as plant and microorganisms, are believed to be a good alternative to artificial pigments because they are non-toxic, non-carcinogenic and biodegradable [1, 14]. The isolation and application of microbial pigments have been investigated by various researchers [13]. Urgent problems that need to be solved are improving extraction and separation process of pigment produced by microorganisms expanding the scope of natural pigment application and reducing the production cost of natural pigment [7, 15]. For microbial pigments, their solubility, coloring, and stability are important indices of their applicability in industry [18]. Microbial pigments are also attractive for its wide range of biological activities, including activities as antimalarial, antifungal, immunosuppressant, and antibiotic agents, which make them an excellent target for dual or multifunctional application [8].

Despite a long history of studies on *Streptomyces* species, they still have many biochemical mysteries to be elucidated [5]. Our previous studies indicated that *Streptomyces alboflacus* TD-1 could significantly inhibit various fungi, such as *Fusarium moniliforme*, *Fusarium oxysporum*, *Fusarium solani*, *Aspergillus flavus*, *Aspergillus niger*, *Aspergillus oryzae*, *Aspergillus ochraceus*, *Aspergillus nidulans*, *Penicillium verrucosum* and *Penicillum citrinum* [17, 19]. Interestingly, *S. alboflacus* TD-1 was found to produce amounts of pigments which are supposed to have anti-bacterial property. Necessities both in extraction and isolation from microbial pigments have increased as new research areas of their characteristic and antibacterial activities have been developed [16]. However, there still be many problems

X. Gu · Y. Zhang · L. Lu · Z. Li · C. Wang (✉)
Key Laboratory of Food Nutrition and Safety, Ministry of Education,
College of Food Engineering and Biotechnology, Tianjin University
of Science and Technology, Tianjin 300457, People's Republic of China
e-mail: clw123@tust.edu.cn

H. Liu et al. (eds.), *Advances in Applied Biotechnology*, Lecture Notes
in Electrical Engineering 444, https://doi.org/10.1007/978-981-10-4801-2_70

limiting the large-scale preparation of natural pigments in the process of separation and purification processes [10].

The conventional methods for the separation and identification of pigments are silica gel chromatography and high-performance liquid chromatography (HPLC) [9]. Nowadays, high-speed countercurrent chromatography (HSCCC) is widely used to extract and separate natural products. But, to our best knowledge, there are no reports of extracting and using HSCCC to isolate pigments from *S. alboflacus* TD-1. Therefore, the objective of this study was to find an optimum extraction and isolation method for pigments from *S. alboflacus* TD-1, and isolation the pigments by HSCCC.

2 Materials and Methods

2.1 Bacterial Strains and Culture Medium

The strain *S. alboflavus* TD-1 was isolated from the soil around the grain storage silo of the Feed Mill of Baodi District in Tianjin City and preserved in the China General Microbiological Culture Collection Center (CGMCC No. 4666). The spore suspension was inoculated in Gause's synthetic medium on a gyratory shaker at 30 rpm s^{-1} for 7 days at 30 °C. The Gause's synthetic medium consisted of 20.0 g of soluble starch, 1.0 g of KNO_3, 0.5 g of K_2HPO_4, 0.5 g of NaCl, 0.5 g of $MgSO_4 \cdot 7H_2O$, and 0.01 g of $FeSO_4 \cdot 7H_2O$ in 1 L of tap water (pH 7.2). The fermentation broth was filtered, and the mycelium of *S. alboflavus* TD-1 was used for pigment extraction.

2.2 Extraction of Microbial Pigments

Dissolve exactly 0.200 g of grinding bacteria powder in 10 mL of ethanol, methanol or hexane in a centrifuge tube. Extraction of pigment goes three times with assisting ultrasonic unit for 30 min was each. After ultrasonic extraction, 58.3 rpm s^{-1} centrifugation for 20 min is performed [4]. The supernatants extracted three times are collected into the rotary steam bottle. A total of 30 mL of pigment solution is extracted to be concentrated with the vacuum rotary steam equipment at 40 °C and 1.13 rpm s^{-1}. The 10 mL of concentrated liquid is preserved under the condition of a low temperature and away from sunlight for next experiment.

2.3 *Pigments Separation*

2.3.1 Measurement of Partition Coefficient (K)

In the present study, two-phase solvent systems with four different solvents were selected for the pigment separation. Each solvent system was full equilibrium at room temperature after thorough vigorous shaking. Approximately 0.1 mg of *S. alboflavus* TD-1 pigment powder was added in 4 mL upper phase of the two-phase solvent system. Taken 1 mL solution as the value of K_1, the 3 mL remaining upper phases and lower phase were mixed, then taken 1 mL lower phase of mixture solution as the value of K_2 [2]. Equal volumes of the upper and lower phases were evaporated to dryness, respectively. The sample was dissolved with 100% methanol to equal volumes and analysed by HPLC to obtain the partition coefficient (K) of *S. alboflavus* TD-1 pigment [12].

2.3.2 High-Speed Countercurrent Chromatography Separation of Pigments

Intensively mix four organic reagents of a specific volume, then stand for 30 min, to reach stratification between the liquid phase and the stationary phase. Select upper phase as the stationary phase and the lower one as the liquid phase, perform ultrasonic degaussing for 30 min, to avoiding air bubbles, respectively [3]. Turn down the pump at 10 mL min^{-1} in the stationary phase. Turn off the pump when it flows out of the column outlet [20]. Change the pump head and put it into the liquid phase. Then, adjust the speed of countercurrent chromatography of the host, select mode of reverse connection and forward mode, and slowly modulates the revolving speed to 14.1 rpm s^{-1}. After the host speed is stable, pump it into the lower phase at a speed of 2 mL min^{-1}. Make sure that the stationary phase and the liquid phase are balanced, and then recording the volume of the upper and lower phase, and continue to pump into the lower phase [11]. Injecting the pigments into the sample with a syringe, closing the injection program, switching the button to load mode. Finally, collect the outflow of liquid and stop the host's rotation. Adjust the rotational speed slowly to 15 rpm s^{-1} under forward positive connection mode, collect the outflow of liquid, and record the time.

3 Results

3.1 *Optimization of Extraction Conditions of Pigments*

3.1.1 Fully Wavelength Scanning

Spectrums of *S. alboflavus* TD-1 extracts in different solvents between 290 and 1100 nm are shown in Fig. 1. The absorbance of *S. alboflavus* TD-1 extracts in ethanol and methanol show the similar pattern between 290 and 1100 nm, a maximum absorption appears at 360 and 530 nm, whereas the hexane extract only has a maximum absorption at 530 nm.

3.2 *Extraction Method*

3.2.1 Effects of Material-to-Liquid Ratio on Pigments

The *S. alboflavus* TD-1 mycelium sample was mixed with different solvents at a ratio of 1:10 (v:v), 1:50 (v:v), 1:100 (v:v), respectively. Its absorption value was then measured under absorbance of 530 and 360 mm.

As it is shown in Fig. 2a, absorbency of *S. alboflavus* TD-1 methanol extracts at 360 nm is higher than ethanol and hexane extracts. The methanol fully extract the secondary metabolites at the solid-liquid ratio of 1:10 (v:v); ethanol extracts the secondary metabolites fullest at the solid-liquid ratio of 1:50 (v:v); but the

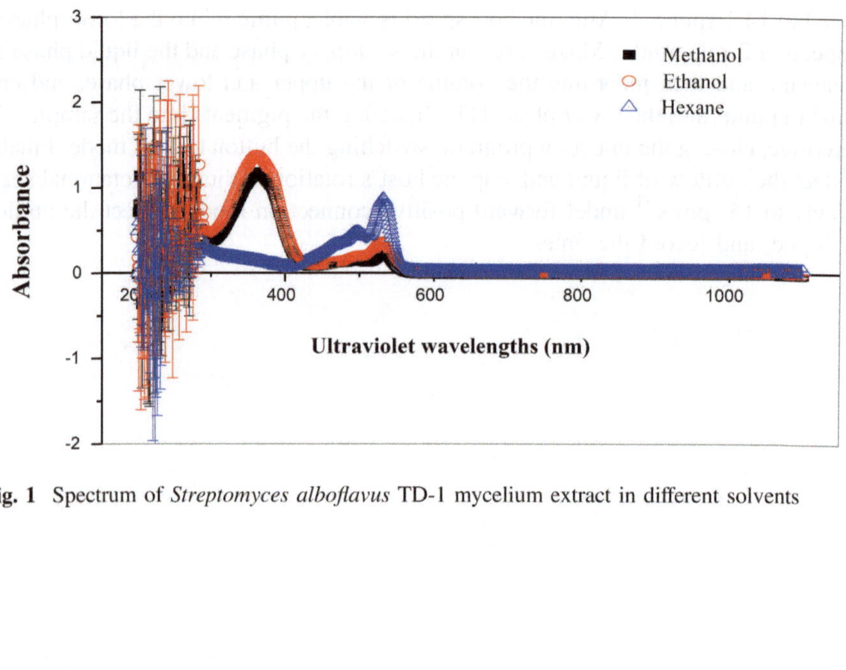

Fig. 1 Spectrum of *Streptomyces alboflavus* TD-1 mycelium extract in different solvents

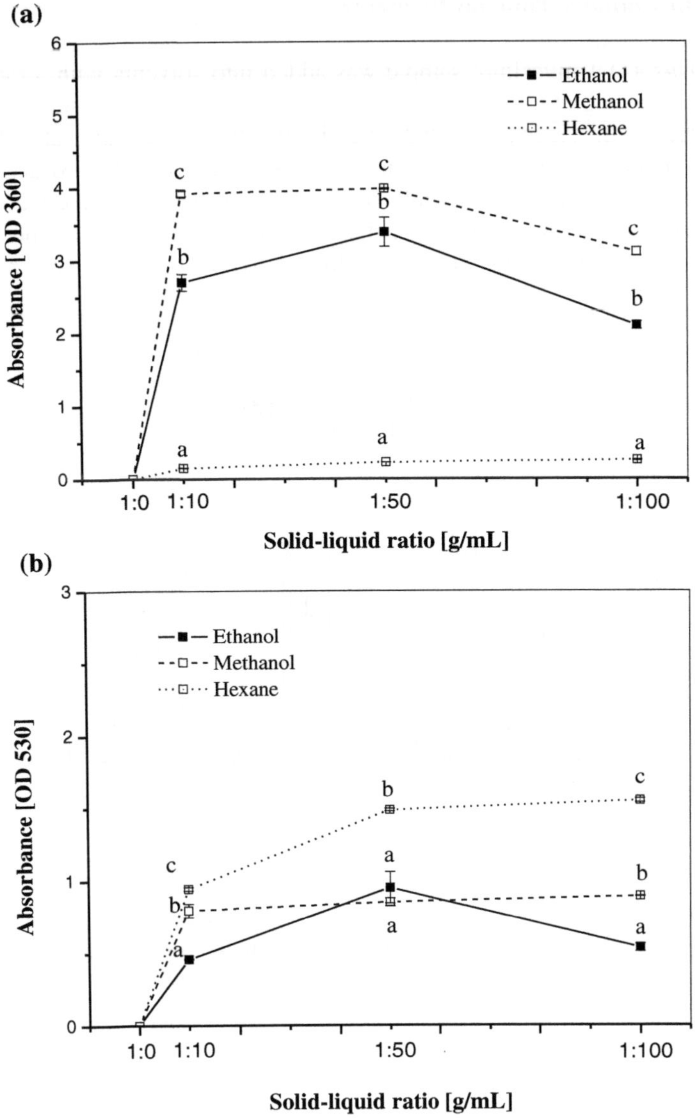

Fig. 2 Effects of material-to-liquid ratio on pigment extraction

absorbance of secondary metabolites in hexane is generally low. Figure 2b presented the absorbance of secondary metabolites in hexane is higher than in ethanol and methanol; hexane extracts the secondary metabolites fullest at the solid-liquid ratio of 1:50 (v:v).

3.2.2 Ultrasound's Time on Pigments

S. alboflavus TD-1 mycelium sample was added into solvents with a ratio of 1:50 (v:v), then assisting of ultrasound for 10, 30, and 60 min. As it is shown in Fig. 3a, the absorbency of secondary metabolites is highest in methanol at 360 nm, followed by ethanol and lower in hexane. It takes 10 min to fully extract secondary metabolites by methanol and 30 min by ethanol and hexane at the wavelength of 360 nm. As it is shown in Fig. 3b, the absorbency of secondary metabolites is highest when extracting by hexane, followed by methanol, lowest by ethanol;

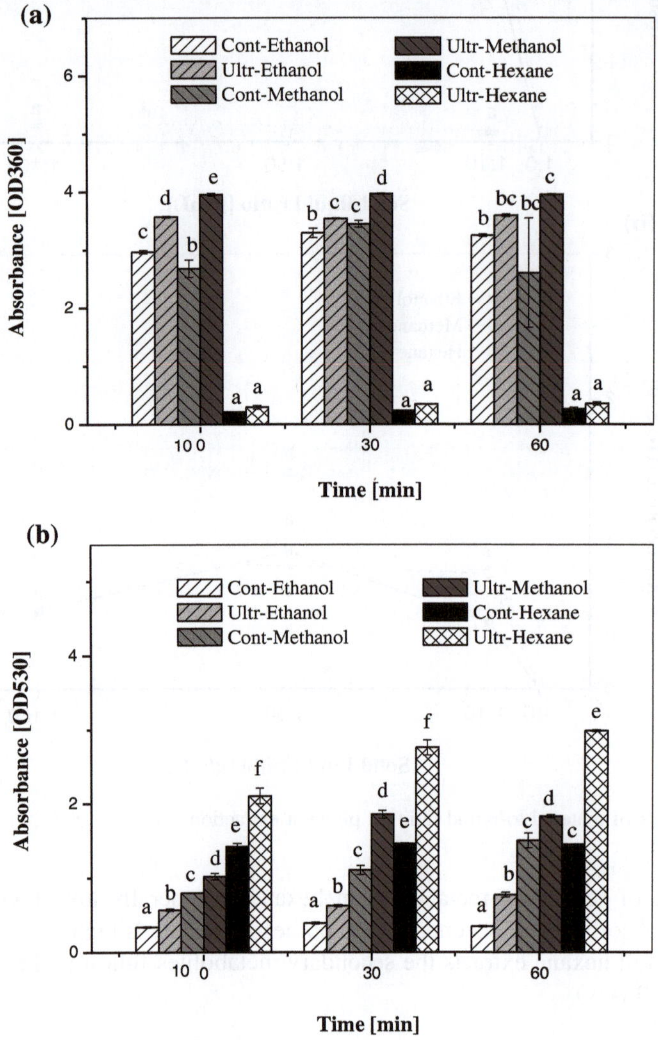

Fig. 3 Effect of ultrasound time on pigment extraction

hexane fully extract the secondary metabolites for 30 min and the absorbency of secondary metabolites are lower without ultrasonic than under the condition of ultrasonic.

3.2.3 Times of Extraction on Pigments

In Fig. 4, the absorbance rate of 30-fold diluted extraction was measured four times after each time of extraction, respectively.

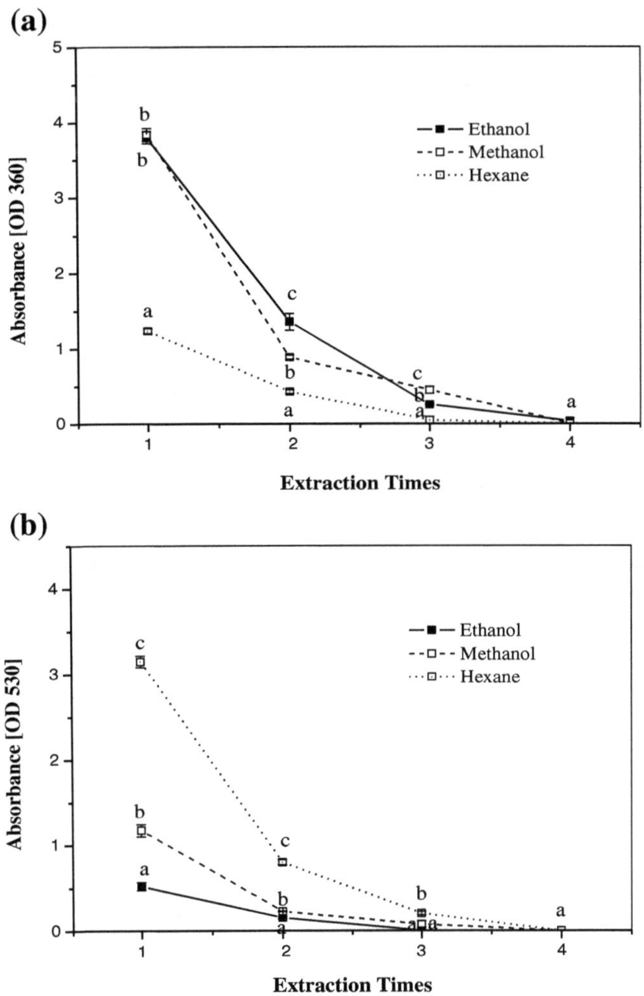

Fig. 4 Effect of extraction times' on pigment extraction

The absorbance of extract in methanol and ethanol is higher than hexane and decrease significantly during the fourth extraction with the wavelength of 360 nm. The absorption value of hexane at 360 nm was low. The high absorption value was at 530 nm, and the secondary metabolites were fully extracted the third time, and the secondary metabolites of methanol and hexane in the third extraction were completely adequate.

3.3 The Separation of Microbial Pigments with High-Speed Countercurrent

3.3.1 Determination of Solvent System of High-Speed Countercurrent

A suitable two-phase solvent system that contains n-hexane:ethyl acetate:methanol: water (4:6:5:5; 5:5:5:5; 6:4:7:3; 7:3:7:3) was chosen for separating pigments in extracts of *S. alboflavus* TD-1 mycelium. The result shows that since the solubility of orange partion in n-hexane is excellent, the proportion of n-hexane in a solvent system is an important indicator when one chooses a high-speed countercurrent solvent system for the separation of four kinds of pigments. The greater the proportion of n-hexane intakes, the easier to separate jacinth pigments from rose pigments (Table 1).

3.3.2 Separation of Four Pigments of *S. alboflavus*

N-hexane:ethyl acetate:methanol:water (6:4:7:3) were chosen as the solvent system for conducting the high-speed countercurrent separation. A total of 200 mg of pigment mixture was dissolved in 15 mL of the upper phase. In the forward-rotating and reverse connection mode, the lower phase was used as the stationary phase and the upper phase was used as the liquid phase with the flow rate of 2 mL min^{-1}. Then, the mode was switched to the forward-rotating and forward connection mode to separate three pigments. Peaks appeared in 12–24 min, 25–54 min, and 59–62 min. Peaks also appeared in 30–42 min in the forward-rotating and forward connection mode. A high-performance liquid was used to detect the four kinds of separated pigments. The results showed the peak area of the compositions starting from the first one was 95.6, 99.8, 94.1, and 96.7%, respectively.

Table 1 Partition coefficients from four pigments in four solvent systems

Hexane:ethyl acetate:methanol:water	4:6:5:5	5:5:5:5	6:4:7:3	7:3:7:3
Orange yellow	0.73	0.61	0.52	–
Orange red	0.4	0.52	0.71	0.97
Rose Bengal	4.2	29	5.6	0.75
Yellow	0.4	0.1	0.02	–

4 Discussion

Microorganism pigments showed potential in various applications, but the pigment production is limited due to high cost. This study has mainly focused on the high pigments yield strain *S. alboflavus* TD-1 with Gause 1 culture medium, and it has optimized the best condition for extracting pigments by n-hexane, ethyl acetate and methanol ensured that their uses. Hexane extracts are in rose red and the maximum absorption wavelength is at 530 nm. To determine the ultrasound assisted extraction, the ratio of the solution and material being 1:50 (v:v), extracting 3 times and 30 min extraction time to extracting rose pigment. The extraction liquid from methanol and ethanol is the orange one. There are absorption peaks at 360, 530 nm. To determine the ultrasound assisted extraction, the ratio of the solution and material being 1:10 (v:v), extracting 3 times and 10 min extraction time to extracting orange pigment.

In the present study, we find used different solvent in different polar can extract different substances from *S. alboflavus* TD-1 mycelium. The extracts in methanol and ethanol solvent has shown antimicrobial activity against *botrytis cinerea* (the inhibition rate was 87.17%), while the hexane extract is red pigment which was supposed to be used in dyeing [6]. The results form present study provided the optimum conditions for the economical large-scale extraction of *S. Alboflavus* TD-1 pigment, to obtain substantial amounts of pigments for application in the antibacterial and dyeing area.

In partition coefficient (K) system, the proportions of n-hexane and ethyl acetate are important indicators for separating orange pigments, and the proportion of n-hexane is too small in the system, which is mainly the most difficult to separate saffron yellow pigments and yellow pigments. In the lower phase, with different proportions of methanol, the polarities of the upper and lower phase systems also changed. In order to obtain the best proportion for the separation of orange, nacarat, and rose red pigments, the allocation proportion of the hexane-ethyl acetate-methanol-water system was designated after defining the scope of the

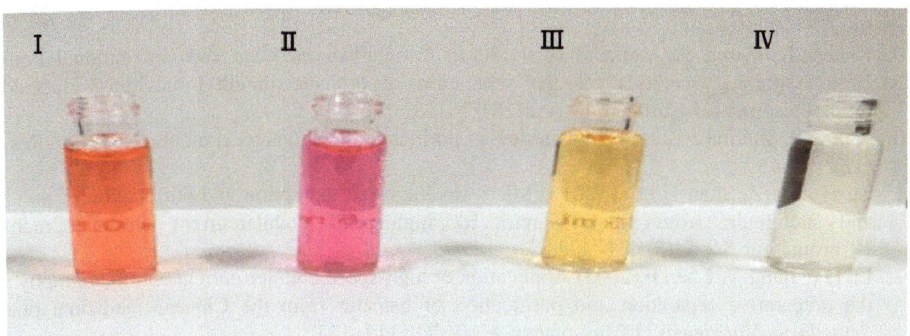

Fig. 5 High-speed countercurrent result using n-hexane:ethyl acetate:methyl alcohol:water (6:4:7:3) solvent system

balance value K in high-speed reflux. The liquid phase, with the Solvent system ratio is 6:4:7:3, can separate the yellow pigments (Fig. 5I) by using the corotation and positive connection mode and orange red, rose red, and orange pigments (Fig. 5II–IV) by using the forward-rotating and reverse connection mode.

In conclusion, four kinds of pigments were separated from the *S. alboflavus* TD-1 extracts by HSCCC, the four kinds of pigment with moderate polarity could be more suitable used in dyeing industry. What's more, *S. alboflavus* TD-1 pigment showing excellent antibacterial properties which give it higher values in colorants industrial. However, the metabolic processes of pigment production in *S. alboflavus* TD-1 still needs further research.

References

1. Ahmad WA, Venil CK, Aruldass CA (2015) Production of violacein by chromobacterium violaceum grown in liquid pineapple waste: current scenario. In: Beneficial microorganisms in agriculture, aquaculture and other areas, vol 29, pp 45–58
2. Chandrasekhar J, Sonika G, Madhusudhan M et al (2015) Differential partitioning of betacyanins and betaxanthins employing aqueous two phase extraction. J Food Eng 144:156–163
3. Claessens H, Van Straten M (2004) Review on the chemical and thermal stability of stationary phases for reversed-phase liquid chromatography. J Chromatogr A 1060(1):23–41
4. Gao J, Huang C, Lin Y et al (2016) In situ solvothermal synthesis of metal–organic framework coated fiber for highly sensitive solid-phase microextraction of polycyclic aromatic hydrocarbons. J Chromatogr A 1436:1–8
5. Hwang KS, Kim HU, Charusanti P et al (2014) Systems biology and biotechnology of Streptomyces species for the production of secondary metabolites. Biotechnol Adv 32 (2):255–268
6. Keck Wilhelm A, Kratz E, Mildau G et al (2015) Chemical analysis and risk assessment of prohibited colouring agents in face paint with special regard to CI 15585 (D&C Red No. 9, Pigment Red 53: 1). J Cosmet Sci 37(2):187–195
7. Korumilli T (2014) Studies on pigment production by microorganisms using raw materials of agro-industrial origin. National Institute of Technology Rourkela
8. Kramar A, Ilic-Tomic T, Petkovic M et al (2014) Crude bacterial extracts of two new streptomyces sp. isolates as bio-colorants for textile dyeing. World J Microbiol Biotechnol 30 (8):2231–2240
9. Kučera L, Fanali S, Aturki Z et al (2016) Comparison of nano and conventional liquid chromatographic methods for the separation of (+)-catechin-ethyl-malvidin-3-glucoside diastereoisomers. J Chromatogr A 1428:126–133
10. Kumar P, Sharma S (2015) An overview of purification methods for proteins. Int J Appl Res 1 (12):450–459
11. Lü H, Sun Z, Shan H et al (2016) Microwave-assisted extraction and purification of arctiin and arctigenin from fructus arctii by high-speed countercurrent chromatography. J Chromatogr Sci 54(3):472–478
12. Lu HT, Jiang Y, Chen F (2003) Application of high-speed counter-current chromatography to the preparative separation and purification of baicalin from the Chinese medicinal plant *Scutellaria baicalensis*. J Chromatogr A 1017(1):117–123
13. Pearce CI, Lloyd JR, Guthrie JT (2003) The removal of colour from textile wastewater using whole bacterial cells: a review. Dyes Pigm 58(3):179–196

14. Shirsath SR, Sonawane SH, Gogate PR (2012) Intensification of extraction of natural products using ultrasonic irradiations—a review of current status. Chem Eng Process 53:10–23
15. Venil CK, Zakaria ZA, Ahmad WA (2013) Bacterial pigments and their applications. Process Biochem 48(7):1065–1079
16. Verma B, Kumar P, Dhanasekaran D et al (2015) Gas chromatography-mass spectrometry analysis and antibacterial activity of bluish-green pigment from pseudomonas sp. JJTBVK (KF836502). Braz Arch Biol Tech 458(4):628–635
17. Wang C, Wang Z, Qiao X et al (2013) Antifungal activity of volatile organic compounds from *Streptomyces alboflavus* TD-1. FEMS Microbiol Lett 341(1):45–51
18. Wang Y, Lu Z, Lv F et al (2009) Study on microencapsulation of curcumin pigments by spray drying. Eur Food Res Technal 229(3):391–396
19. Wang Z, Wang C, Li F et al (2013) Fumigant activity of volatiles from *Streptomyces alboflavus* TD-1 against fusarium moniliforme sheldon. J Microbio 51(4):477–483
20. Yang Y, Aisa HA, Ito Y (2009) Mathematical model of computer-programmed intermittent dual countercurrent chromatography applied to hydrostatic and hydrodynamic equilibrium systems. J Chromatogr A 1216(35):6310–6318

Part IV
Progress of Biotechnology

Establishment of the Method for Screening Small Molecule Inhibitors Blocking the Interaction Between PD-1 and Its Ligand PD-L1

Lei Jing, Fushan Yan, Yingchun Wang, Bo Jiang, Li Chang, Cheng Cheng, Yuyin Li and Aipo Diao

1 Introduction

In the last decades, cancer biology studies have attempted to stimulate the antitumor immune response to fight cancer. However, those attempts were not very successful because the cancer cells could evade the immune system by modulating the "immune checkpoints". Until recent years, the antibodies targeted the "immune checkpoints" were developed [1], the immunotherapies had a fast development in clinical treatment for some types of solid tumors [2], especially in melanoma [1, 3–5].

Programmed cell death-1 (PD-1) protein is a type I transmembrane receptor that was originally isolated from T-cell hybridoma [6]. It belongs to Ig super-family that contains a Ig-like-V domain in its extracellular region and has been found to be expressed on macrophages, mature B cells and T cells following activation [7–9]. PD-1 has been shown to negatively regulate T-cell receptor signaling, which lead to reduced T-cell cytolysis, proliferation and cytokine production [10, 11]. Its ligand PD-L1 also known as B7-H1 is highly over-expressed in many solid tumors. In addition, PD-L1 was also found to be expressed on various human tumor cell lines containing colon, breast and lung cell lines [12]. IFN-γ up-regulates the expression of PD-L1 on human tumor cells [13]. These observations have led to the founding that tumors escape immune system via PD-1/PD-L1 interaction by negative the host immune responses.

The PD-1/PD-L1 immune checkpoint is a key target in the immune treatment of cancer, and mostly inhibitors on PD-1/PD-L1 checkpoint are monoclonal antibodies targeting PD-1 or PD-L1. These antibody-based drugs have inherent drawbacks, such as poor tumor penetrance due to their large size and activated cytotoxic immune responses through NK cells and macrophages [14, 15], which limit their

L. Jing · F. Yan · Y. Wang · B. Jiang · L. Chang · C. Cheng · Y. Li · A. Diao (✉)
School of Biotechnology, Key Lab of Industrial Fermentation Microbiology of the Ministry of Education, Tianjin University of Science and Technology, Tianjin 300457, China
e-mail: diaoaipo@tust.educn

H. Liu et al. (eds.), *Advances in Applied Biotechnology*, Lecture Notes in Electrical Engineering 444, https://doi.org/10.1007/978-981-10-4801-2_71

effectiveness. In this study, we established a method to measure the interaction of PD-1/PD-L1 using competitive ELISA assay for screening of small molecule inhibitors blocking the interaction between PD-1 and its ligand PD-L1.

2 Materials and Methods

2.1 *Construction of Plasmids*

The DNA fragments encoding the extracellular domain of human PD-1 and PD-L1 protein were amplified by PCR using the primers listed in Table 1, which cloned from the cDNA plasmids pMD-PD-1 and pMD-PD-L1 (Sino Biological). The PCR product of amplified *PD-1* gene was digested by using *Bam*HI and *Eco*RI and inserted into the pET32a (GE Healthcare) vector. The PCR product of amplified *PD-L1* gene was purified and digested with *Eco*RI and *Xho*I and inserted into the pGEX-4T-2 (GE Healthcare) vector.

2.2 *Protein Expression and Purification*

Plasmid encoding GST-PD-L1 (AA26-236) fusion protein was transformed into *E. coli* BL21-CodonPlus (DE3)-RIL (Stratagene). Cells were induced with 0.2 mM IPTG overnight at 16 °C, and lysed in lysis buffer (20 mM Tris-Cl pH 7.4, 150 mM NaCl, 2 mM DTT) containing protease inhibitors. The recombinant protein was purified using Glutathione-Sepharose beads (GE Healthcare). After sonication, the cell extracts were clarified by centrifugation. The recovered supernatants were allowed to bind to Glutathione Sepharose beads for 3 h at 4 °C. After washed three times with lysis buffer, the beads was eluted with elution buffer (100 mM Tris-Cl pH 7.4, 150 mM NaCl, 20 mM reduced glutathione). His-Trx-PD-1(AA21-163) was expressed in *E. coli* BL21-CodonPlus (DE3)-RIL (Stratagene) cells that transformed with pET32a-His-Trx-PD-1 plasmid, and induced with 0.2 mM IPTG

Table 1 Primers used for PCR experiments

Target gene	Primer sequences
PD-1	
Forward: PD-1-*Bam*HI-21	5′-CGGGATCCATGCCAGGATGGTTCTTAGACTCC-3′
Reverse: PD-1-*Eco*RI-163	5′-GGAATTCTCAGGCTGGCCTGGGTGAGGGGCTG-3′
PD-L1	
Forward: PD-L1-*Eco*RI-26	5′-GGAATTCCCGACCTATATGTGGTAGAGTATGG-3′
Reverse: PD-L1-*Xho*I-236	5′-CCGCTCGAGTCAATTTGGAGGATGTGCCAGAGGTAG-3′

overnight at 16 °C. The His-tagged protein was purified using Ni-Sepharose beads after lysis of the bacteria in Na-P lysis buffer (50 mM Phosphate Buffer pH 8.0, 150 mM NaCl, 5 mM β-mercaptoethanol) plus protease inhibitors. After sonication, the extracts were clarified by centrifugation. The recovered supernatants were allowed to bind to Ni-Sepharose beads (GE Healthcare) for 3 h at 4 °C. After washed three times with Na-P lysis buffer containing 50 mM imidazole, the beads was eluted with elution buffer (50 mM Phosphate Buffer pH 8.0, 150 mM NaCl, 5 mM β-mercaptoethanol and 250 mM imidazole).

2.3 *GST Pull-Down Assays and Western Blot*

For the GST pull-down assays, 5 μg GST-fusion proteins were incubated with the Glutathione-Sepharose beads for 2 h in rotation at 4 °C, then washed three times with cold GST pull-down buffer (20 mM Tris-Cl pH 7.4, 150 mM NaCl, 1% Triton-100, 2 mM DTT). 0.5 μg protein of His-Trx-PD-1 was resuspended in the GST pull-down buffer and clarified by centrifugation at 14,000 rpm for 10 min, the supernatant was added to the GST-beads and incubated for 2 h in rotation at 4 °C, then washed three times with cold GST pull-down buffer and washed once with 1 × PBS. Finally, the beads were resuspended and boiled in 40 μl 2 × SDS loading buffer for 5 min, the supernatants were collected and separated by 10% SDS-PAGE and analyzed by Western blot using anti-His-tag antibody.

2.4 *Competitive ELISA Assay and Small Molecule Inhibitors Screening*

100 μl of recombinant human His-Trx-PD-1(1 μg/ml) in carbonate-bicarbonate buffer (pH 9.6) was coated onto a 96-well plate overnight at 4 °C. The unbound sites were blocked with 5% non-fat milk in PBS. Each well was washed with PBST followed by incubating with the recombinant human GST-PD-L1 (2 μg/ml) and small molecule chemicals (5 μM) for 2 h at 37 °C. The unbound GST-PD-L1 and small molecule chemicals were removed by three times washing with PBST. 100 μl of a 1:10,000 dilution of anti-GST antibody (Rabbit, sigma) was added and incubated for 2 h at 37 °C, then washed three times. 100 μl of a 1:20,000 dilution of secondary antibody conjugated-HRP (thermo scientific) was added and incubated for 45 min at 37 °C. The plates were washed four times with PBST. The color reaction was developed with TMB as substrate, then the reaction was stopped with 2 M H_2SO_4. The inhibiting of small molecules blocking PD-1/PD-L1 interaction was quantified by determining the absorbance at 450 nm using the Infinite 200Pro microplate reader (Tecan). The L1300-Selleck-FDA-Approved-Drug-Library (Selleck) was used to screen the inhibitors for blocking the PD-1/PD-L1 interaction.

3 Result

3.1 *Expression and Purification of the His-Trx-PD-1 and GST-PD-L1 Fusion Proteins*

PD-1 protein contains 288 amino acids, including an extracellular Ig-like-V domain, a transmembrane region and an intracellular tail in C-terminal (Fig. 1a). Its ligand PD-L1 is a type I transmembrane protein containing 290 amino acids, which includes extracellular a Ig-like-V domain and a Ig-like-C2 domain followed by a transmembrane region and an intracellular tail in C-terminal (Fig. 1b).

The pET32a-His-Trx-PD-1 and pGEX-4T-2-PD-L1 plasmids were transformed into *E. coli* BL21-CodonPlus (DE3)-RIL. The recombinant His-Trx-PD-1 (AA21-163) and GST-PD-L1 (26-236) proteins were expressed under IPTG induction overnight at 16 °C. SDS-PAGE and coomassie blue staining analysis indicated that expressed His-Trx-PD-1 and GST-PD-L1 proteins were partially soluble in the cell lysate and largely in the pellets (Fig. 2a, b). The molecular weight of two fusion proteins was about 50 and 32 kDa respectively. The His-Trx-PD-1 (AA21-163) and GST-PD-L1 (26-236) fusion proteins were purified and detected by SDS-PAGE and coomassie blue staining (Fig. 2c). The results indicated that the highly pure recombinant proteins His-Trx-PD-1 (AA21-163) and GST-PD-L1 (26-236) were obtained successfully.

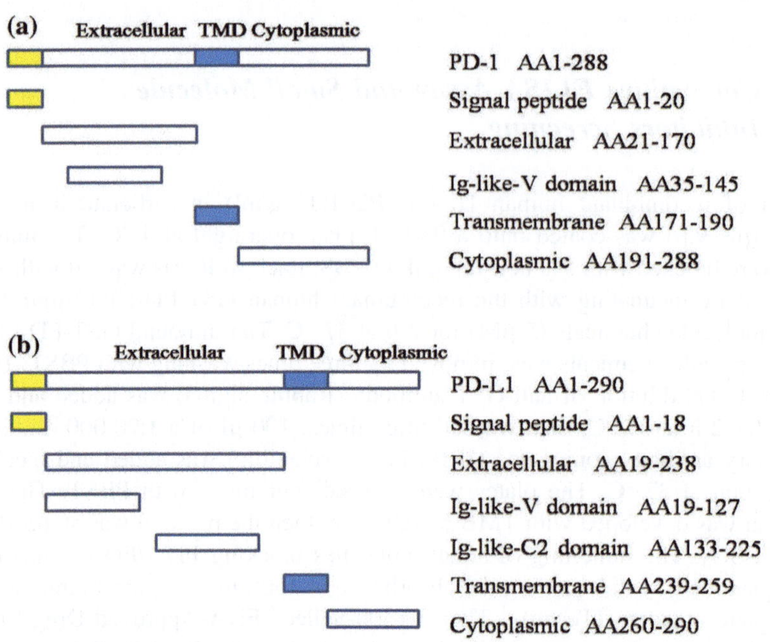

Fig. 1 The schematic structure of human PD-1 (**a**) and its ligand PD-L1 (**b**). Amino acid positions of individual domains are denoted with corresponding numbers. *TMD* transmembrane domain

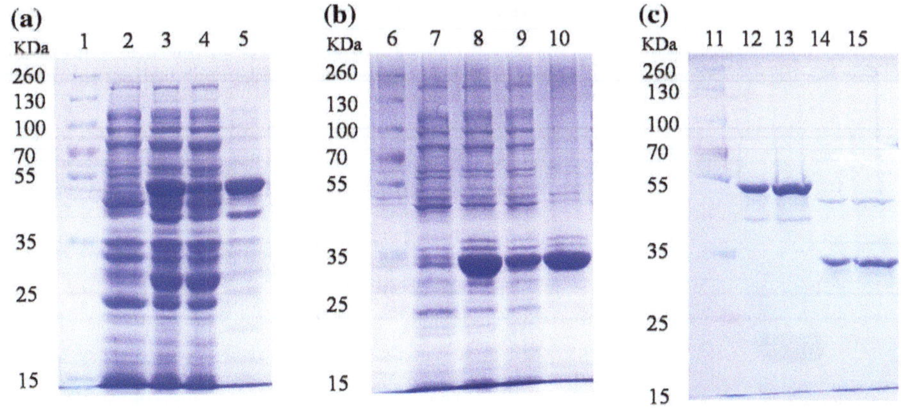

Fig. 2 Expression and purification of the GST-PD-L1 (**a**, **c**) and His-Trx-PD-1 (**b**, **c**) fusion protein induced by 0.2 mM IPTG. *1*, *6*, *11* protein makers; *2*, *7* culture before IPTG induction; *3*, *8* the total cell lysate; *4*, *9* the soluble cell lysate after centrifugation; *5*, *10* the cell pellet after centrifugation; *12*, *13* the purified GST-PD-L1 protein; *14*, *15* the purified His-Trx-PD-1 protein

3.2 *Measuring the PD-1/PD-L1 Interaction Using Competitive ELISA Assay*

GST pull-down experiment was performed to verify whether PD-1 interacts with PD-L1. As shown in Fig. 3a, GST-PD-L1, but not GST, pulled down purified His-Trx-PD-1, which indicates that GST-PD-L1 interacts with His-Trx-PD-1 directly. To largely screen the small molecule inhibitors blocking PD-1/PD-L1 interaction, the competitive ELISA assay was used to measure the PD-1/PD-L1 interaction. As shown in Fig. 3b, the interaction of GST-PD-L1 with His-Trx-PD-1 had a high absorbance value than the GST control, the ratio of the binding of GST-PD-L1 with His-Trx-PD-1 to the GST control was optimized with the GST-PD-L1, His-Trx-PD-1 and GST proteins were at 200 ng/well separately (Fig. 3c). These results showed that competitive ELISA assay could be used measure PD-1/PD-L1 interaction for inhibitor screening.

3.3 *Screening of the Small Molecule Inhibitors Blocking the PD-1/PD-L1 Interaction*

The L1300-Selleck-FDA-Approved-Drug-Library was used to screen the small molecule chemicals inhibiting the PD-1/PD-L1 interaction, about 320 small molecules were screened using the competitive ELISA assay, and several chemicals

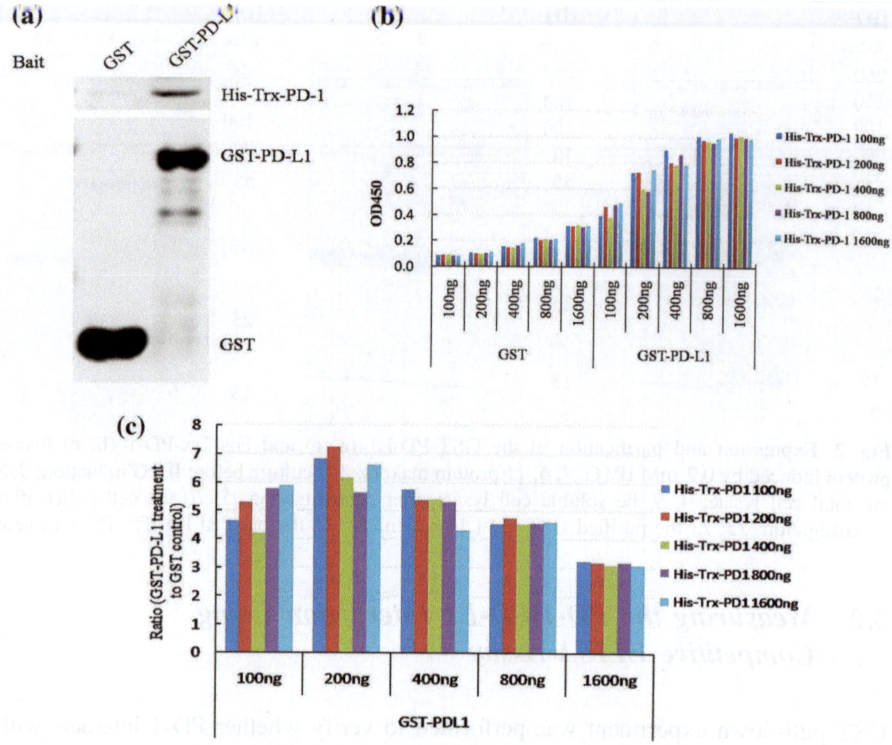

Fig. 3 Measuring the PD-1/PD-L1 interaction. **a** Direct interaction between PD-1 and PD-L1 determined by the GST pull-down assay; **b** measure the interaction between PD-1 and PD-L1 determined using the ELISA assay, GST protein was used as a negative control. **c** The optimized combination of different concentration of adding GST-PD-L1, His-Trx-PD-1 or GST proteins. Y-axis: the ration of absorbance value for GST-PD-L1 with His-Trx-PD-1 to GST control

had an obviously inhibitory effect, the chemicals S2468 (Fenbendazole) and S2084 (Duloxetine HCl) had the most inhibitory effect with the inhibition rate 0.49 and 0.52 respectively (Fig. 4a). The chemical structure of Fenbendazole and Duloxetine HCl were shown in Fig. 5. During the screening, the chemicals S2485 (Mitoxantrone Hydrochloride) and S1957 (Sulfamethizole) had a positive effect with the inhibition rate 1.8 and 1.6 respectively (Fig. 4a). The chemical structure of Mitoxantrone hydrochloride and Sulfamethizole were shown in Fig. 5.

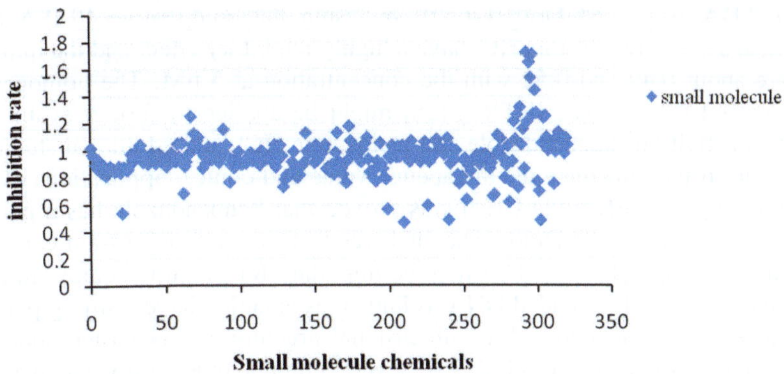

Fig. 4 Screening of the small molecule chemicals inhibiting the PD-1/PD-L1 interaction using competitive ELISA assay. Inhibition rate, the ratio between the treatments added small molecule and the treatment without. Value < 1, inhibitory effect. Value > 1, positive effect

Fig. 5 The chemicals structure of small molecules

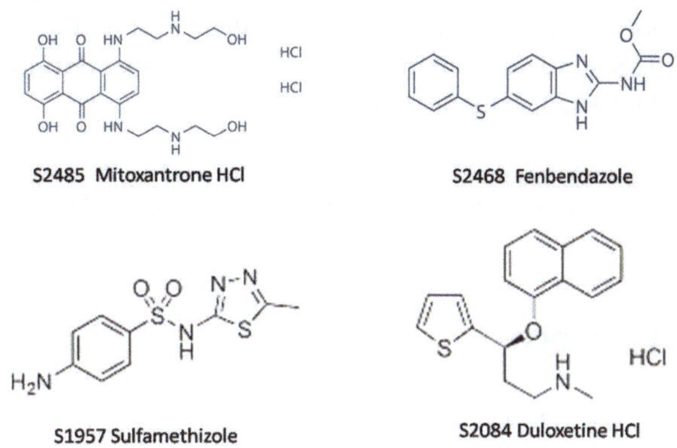

4 Discussion

It has been reported that blocking immune checkpoint has a positive role on the anti-tumor treatment. Most monoclonal antibodies, targeting immune checkpoint are widely used on clinical trials now, such as pembrolizumab [16], nivolumab [4], MDX-1105 [17], MDX-1106 [2]. Although antibody-based inhibitors for blocking immune checkpoints, such as PD-1, PD-L1, CTLA-4, have been significantly developed recent years, almost all antibodies have immunogenicity and have different therapeutic effect to different patients. These deficiency limit their application on clinical trial.

Blocking the interaction of PD-1 protein with its ligand PD-L1 was proved to be an effective way for suppressing the development of tumors. Thus, screening the small molecule inhibitors blocking the interaction of PD-1 with PD-L1 became an efficient strategy. Here, we screened 320 small molecule chemicals from L1300

Selleck FDA Approved Drug Library by using the competitive ELISA assay. Fenbendazole and Duloxetine HCl have a highly inhibitory effect and the inhibition rates are about 0.49 and 0.52 with the concentration at 5 μM. The homogeneous time-resolved fluorescence (HTRF) [18] might be applied to further evaluate the inhibitive activity of Fenbendazole and Duloxetine HCl, and additional studies are required to further elucidate the therapeutic value and clinical applications of these inhibitors targeting PD-1/PD-L1. It was proved that Fenbendazole has a potential anti-cancer activity while Duloxetine HCl was just as a anti-depressant combining with chemotherapy [19–21]. This may be important for immunotherapy. In recent years, the short single stranded DNA as known as aptamer, have a strong plasticity that can be formed the unique three dimensional structures and be easier modified to change their pharmacokinetics [22, 23]. Several immune modulatory aptamers have been developed to block immune checkpoint such as the DNA aptamer blocking the murine PD-1/PD-L1 interaction [24, 25]. Based our established screening model for PD-1/PD-L1 interaction inhibitors, this method also can be used to screen more efficient DNA aptamers inhibiting the PD-1/PD-L1 interaction.

In conclusion, we established a method for screening small molecule inhibitors blocking the interaction between PD-1 and its ligand PD-L1 using the competitive ELISA assay, which can be used for the development of inhibitors blocking PD-1/PD-L1 interaction.

Acknowledgements This research is supported by the program for Tianjin University of Science and Technology student laboratory Innovation Fund (1504A302X).

References

1. Hodi FS, O'Day SJ, Mcdermott DF et al (2010) Improved survival with ipilimumab in patients with metastatic melanoma. N Engl J Med 363(8):711–723
2. Brahmer JR, Drake CG, Wollner I et al (2010) Phase I study of single-agent anti-programmed death-1 (MDX-1106) in refractory solid tumors: safety, clinical activity, pharmacodynamics, and immunologic correlates. J Clin Oncol 28(19):3167–3175
3. Hamid O, Robert C, Daud A et al (2013) Safety and tumor responses with lambrolizumab (anti-PD-1) in melanoma. N Engl J Med 369(2):134–144
4. Tsai KK, Daud AI (2014) Nivolumab plus ipilimumab in the treatment of advanced melanoma. J Hematol Oncol 8(1):1–4
5. Wolchok JD, Kluger H, Callahan MK et al (2013) Nivolumab plus ipilimumab in advanced melanoma. N Engl J Med 369(2):122–133
6. Ishida Y, Agata Y, Shibahara K et al (1992) Induced expression of PD-1, a novel member of the immunoglobulin gene superfamily, upon programmed cell death. EMBO J 11(11): 3887–3895
7. Nishimura H, Honjo T (2001) PD-1: an inhibitory immunoreceptor involved in peripheral tolerance. Trends Immunol 22(5):265–268
8. Carreno BM, Collins M (2002) The B7 family of ligands and its receptors: new pathways for costimulation and inhibition of immune responses. Annu Rev Immunol 20(20):29–53
9. Agata Y, Kawasaki A, Nishimura H et al (1996) Expression of the PD-1 antigen on the surface of stimulated mouse T and B lymphocytes. Int Immunol 8(5):765–772

10. Rodig N, Ryan T, Pang H et al (2003) Endothelial expression of PD-L1 and PD-L2 down-regulates CD8+ T cell activation and cytolysis. Eur J Immunol 33(11):3117–3126
11. Freeman GJ, Long AJ, lwai Y et al (2000) Engagement of the PD-1 immunoinhibitory receptor by a novel B7 family member leads to negative regulation of lymphocyte activation. J Exp Med 192(7):1027–1034
12. Dong H, Strome SE, Salomao DR et al (2002) Tumor-associated B7-H1 promotes T-cell apoptosis: a potential mechanism of immune evasion. Nat Med 8(8):793–800
13. Wintterle S, Schreiner B, Mitsdoerffer M et al (2003) Expression of the B7-related molecule B7-H1 by glioma cells: a potential mechanism of immune paralysis. Can Res 63(21): 7462–7467
14. Lee CM, Tannock IF (2010) The distribution of the therapeutic monoclonal antibodies cetuximab and trastuzumab within solid tumors. BioMed Central Cancer 10(1):1–11
15. Scott AM, Wolchok JD, Old LJ (2012) Antibody therapy of cancer. Nat Rev Cancer 12(4):278–287
16. Khoja L, Butler MO, Kang SP et al (2015) Pembrolizumab. J Immunother Cancer 3:36
17. Taneja SS (2012) Re: safety and activity of anti-PD-L1 antibody in patients with advanced cancer. J Urol 188(6):2148–2149
18. Liu A, Dong L, Wei XL et al (2016) Development of amino and dimethylcarbamate substituted resorcinol as programmed cell death-1 (PD-1) inhibitor. Eur J Pharm Sci Official J Eur Fed Pharm Sci 88:50–58
19. Duan Q, Liu Y, Rockwell S (2013) Fenbendazole as a potential anticancer drug. Anticancer Res 33(2):355–362
20. Aycock-Williams AN, Pham LK, Liang M et al (2011) Effects of fenbendazole and vitamin E succinate on the growth and survival of prostate cancer cells. J Cancer Res Exp Oncol 3(9):115–121
21. Hershman DL, Lacchetti C, Dworkin RH et al (2014) Prevention and management of chemotherapy-induced peripheral neuropathy in survivors of adult cancers: American Society of Clinical Oncology clinical practice guideline. J Oncol Pract 10(6):421–424
22. Ku TH, Zhang T, Luo H et al (2015) Nucleic acid aptamers: an emerging tool for biotechnology and biomedical sensing. Sensors 15(7):16281–16313
23. Sun H, Zhu X, Rosato RR et al (2014) Oligonucleotide aptamers: new tools for targeted cancer therapy. Mol Therpy Nucleic Acids 3(8):e182–e182
24. Prodeus A, Abdulwahid A, Fischer NW et al (2015) Targeting the PD-1/PD-L1 immune evasion axis with DNA aptamers as a novel therapeutic strategy for the treatment of disseminated cancers. Mol Therapy Nucleic Acids 4(4):1–10
25. Hervasstubbs S, Soldevilla MM, Villanueva H et al (2015) Identification of TIM3 2′-fluoro oligonucleotide aptamer by HT-SELEX for cancer immunotherapy. Oncotarget 7(4): 4522–4530

Construction of the *PD-L1* Promoter-Luciferase Reporter Expressing Vector for Small Molecule Inhibitors Screening

Bo Jiang, Zhichen Shi, Ali Wang, Yuyin Li, Qiurong Zhang, Lei Jing and Aipo Diao

1 Introduction

T cell mediated immune response plays an important role in modulating the growth, proliferation and recurrence of tumor cells. Activation of T cell requires binding of the T cell receptor (TCR) with a cognate peptide presented on the MHC of an APC and a co-stimulatory signal, mainly generated between members of the B7 ligand family on the APC and the CD28 receptor family on the T cell. On the contrast, the inhibitory signaling between these two families acts to down-grade T cell activation, resulting in T cell exhaustion, deletion or tolerance [1]. Antigen-specific T cell responses are regulated by the balance between co-stimulatory and co-inhibitory signals [2, 3], and modulation of such co-signal pathways is beneficial when immune intervention is necessary [4].

PD-1 (programmed cell death-1) and its ligand PD-L1 (programmed cell death-ligand 1) belonged to the immunoglobulin super-family of CD28/B7, are co-stimulatory molecules which have a negative regulation effect. The hypothesis that PD-1/PDL-1 interaction mediated immune evasion from tumor specific T cells was founding due to the widely expression of PD-L1 in human various cancer tissues or cell lines [5]. Thus, blocking of such negative immune regulatory signals on tumor cells has given rise to hope that their manipulation may lead to enhanced tumor-specific CD8+ T-cell immunity in vivo [6]. To extend survival further, novel type's targets for PD-L1 have been developed and achieved good clinical efficacy [7]. In recent years, immunotherapy involving drugs such as humanized monoclonal antibodies against PD-1 and its ligand PD-L1, have been applied to the

B. Jiang · Z. Shi · A. Wang · Y. Li · Q. Zhang · L. Jing (✉) · A. Diao (✉)
School of Biotechnology, Key Lab of Industrial Fermentation Microbiology of the Ministry of Education, Tianjin University of Science and Technology, Tianjin 300457, China
e-mail: shudaozhinan@sina.cn

A. Diao
e-mail: diaoaipo@tust.edu.cn

H. Liu et al. (eds.), *Advances in Applied Biotechnology*, Lecture Notes in Electrical Engineering 444, https://doi.org/10.1007/978-981-10-4801-2_72

clinical treatment of cancers, with improved treatment effects observed for melanoma, lung cancer, and other cancers. Pembrolizumab and nivolumab are the first of this anti-PD-1 pathway family of checkpoint inhibitors to gain accelerated approval from the US Food and Drug Administration (FDA) for the treatment of ipilimumab-refractory melanoma [8].

Previous studies have demonstrated that PD-L1 was expressed in various human tumor cell lines [9, 10]. Therefore, discovery of safe and effective anti-PD-L1 drugs was required. In this study, we established a method to measure the transcriptional activity of *PD-L1* promoter using luciferase reporter gene assay system to screen small molecule inhibitors down-regulating the expression of PD-L1.

2 Materials and Methods

2.1 Plasmid Construction

Human genomic DNA was isolated from MCF-7 cells. The 5′-untranslated region (from nt −818 bp to +134 bp) of *PD-L1* was amplified by PCR using the primers (forward primer: 5′-GGGGTACCTAGAAGTTCAGCGCGGGATAATAC-3′ and reverse primer: 5′-CCGCTCGAGCTGCAGGCGGACAGAAGCGCGGCTG-3′). The PCR product of amplified *PD-L1* promoter region was digested by *Kpn*I and *Xho*I and inserted into pGL4 vector.

2.2 Cell Culture and Reagents

HeLa cells were cultured in Dulbecco's modified Eagle's medium (DMEM), containing 10% fetal bovine serum (FBS), 2 mM glutamine, 100 U/ml penicillin and 100 mg/ml streptomycin (GIBCO-BRL, Gaithersburg, MD) at 37 °C in a humidified incubator with 5% CO_2. The small molecule chemicals were used from L1300-Selleck-FDA-Approved-Drug-Library (Selleck).

2.3 Transiently Transfection and Luciferase Reporter Assay

All transfections were performed using TurboFect Transfection Reagent (thermo scientific) according to the manufacturer's protocol. HeLa cells transfected the pGL4-PD-L1 and pCMV-β-galactosidase plasmids for 12 h were treated with different small molecule chemicals (1 μM) or DMSO (0.1% v/v, the negative control) in DMEM containing 3% FBS (v/v) for 24 h. Luciferase assays were performed using the luciferase reporter assay system (Promega) according to the manufacturer's protocol [11]. All firefly luciferase values were normalized to β-galactosidase in order to minimize the influence of variation in transfection efficiency [12]. All luciferase assays were repeated in three independent experiments.

2.4 RT-PCR

Total RNA was extracted from MGC-803 cells treated with drugs for 48 h by using Trizol Reagent (thermo scientific) according to the manufacturer's protocol. One microgram of total RNA was reverse transcribed by oligo(dT)$_{18}$ primers using the Reverse Transcription System (thermo scientific). The single-stranded cDNA was amplified by PCR using PD-L1 specific primers (forward primer: 5′-GGAATTCATGGACCTATATGTGGTAGAGTATGG-3′, reverse primer: 5′-CCGCTCGAGTCAATTTGGAGGATGTGCCAGAGGTA-3′) and β-actin primers (forward primer: 5′-TGACGAGGCCCAGAGCAAGA-3′, reverse primer: 5′-ATGGGCACAGTGTGGGTGAC-3′). The PCR products were analyzed by electrophoresis on a 0.8% agarose gel.

2.5 Western Blot Assay

MGC-803 cells treated with drugs were washed twice with PBS before lysed with RIPA lysis buffer. Equal amounts of protein were loaded into each well and separated by 10% SDS-PAGE gel, followed by transferring onto PVDF membranes. The membranes were blocked in PBST buffer containing 5% nonfat dry milk for 1 h at room temperature. The blots were then incubated overnight at 4 °C with primary antibodies. Secondary antibodies were incubated for 1 h at room temperature. Finally, the Odyssey infrared laser imaging system (LI-COR Biosciences) was used to image the experimental results.

2.6 Statistical Analysis

All the presented data and results were confirmed in at least three independent experiments. Data were expressed as mean ± SD.

3 Result

3.1 Construction of the PD-L1 Promoter-Luciferase Reporter Expressing Vector

The PCR product of the 5′-flanking non-coding region of *PD-L1* promoter was amplified and separated by agarose gel electrophoresis, the result showed that the size of the amplified DNA fragment is consistent with the expectation which was 952 bp (Fig. 1a). The recombinant plasmid was digested with *Kpn*I and *Xho*I and the products were detected using agarose gel electrophoresis, the recombinant plasmid was digested into two fragments of 952 and 5599 bp which represented the inserted DNA fragment and pGL4 vector respectively (Fig. 1b). The recombinant plasmid was further verified by DNA sequencing. These data showed that *PD-L1* promoter luciferase reporter expressing vector pGL4-PD-L1 was constructed successfully.

3.2 Measuring the Activity of PD-L1 Promoter Using the Luciferase Reporter Assay System

The pGL4-PD-L1 and pCMV-β-galactosidase plasmids were transiently transfected in HeLa cells for 24 h and the activity of *PD-L1* promoter was then detected using luciferase reporter assay system, the pGL4 plasmid was used as a negative control. The results showed that the relative luciferase activity of the pGL4-PD-L1 was significantly higher than that of the control (Fig. 2). Therefore, the constructed *PD-L1* promoter-luciferase reporter expressing vector can be used to screen the inhibitors deregulating the promoter activity of human *PD-L1* gene.

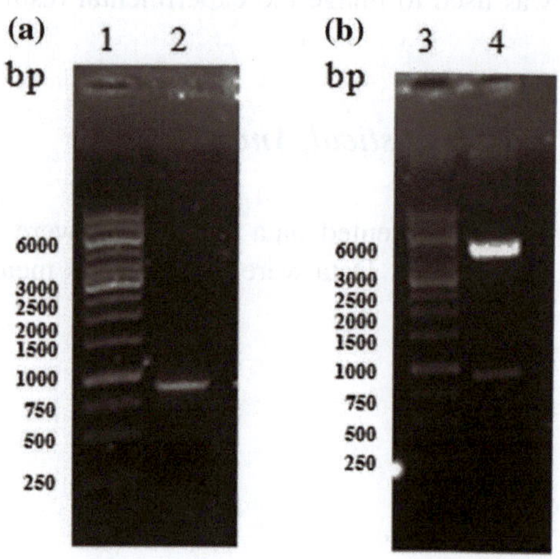

Fig. 1 Construction of the *PD-L1* promoter luciferase reporter expressing vector. **a** 5′-flanking non-coding region of *PD-L1* promoter was amplified by PCR. **b** Detection of the recombinant plasmid by double digestion with *Kpn*I and *Xho*I. *1*, *3* DNA Marker; *2* PCR product; *4* Double digested recombinant plasmid pGL4-PD-L1

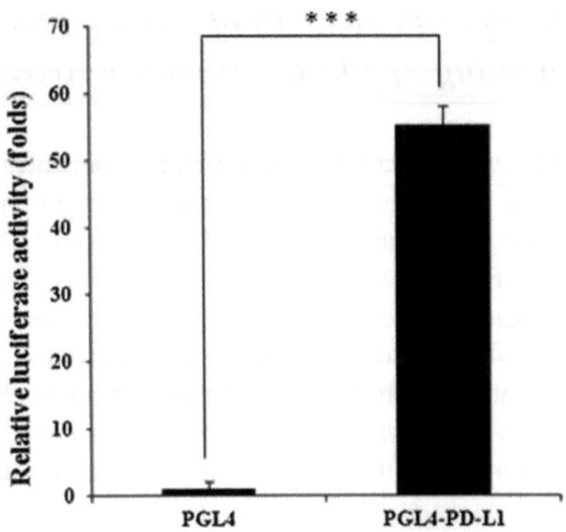

Fig. 2 Detecting the activity of *PD-L1* promoter in HeLa cells. pGL4-PD-L1 plasmid was transfected into HeLa cells and cultured for 24 h, the empty pGL4 was as a negative control. Relative luciferase activity was measured (N = 3, ***$P < 0.001$)

3.3 Screening the Small Molecule Inhibitors Down-Regulating PD-L1 Promoter Activity

To obtain potential small molecule inhibitors for suppressing the translational activity of *PD-L1* promoter, over 200 compounds were screened at the concentration of 1 μM by using luciferase reporter assay. The results showed that Fludarabine Phosphate had a largest inhibition and the inhibitory rate was 0.6 (Fig. 3a). The chemical structure of Fludarabine Phosphate was shown in Fig. 3b.

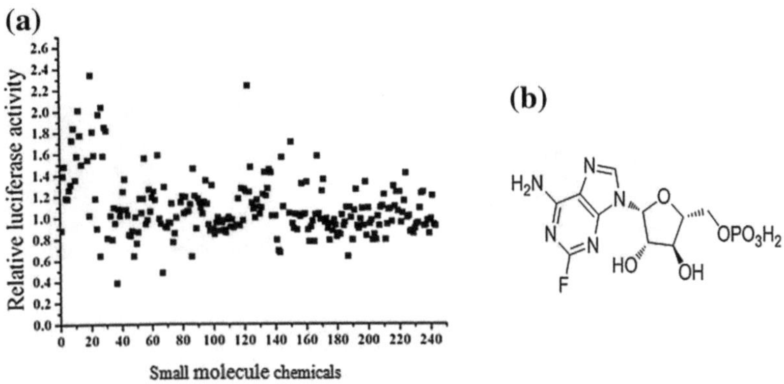

Fig. 3 Screen the small molecule inhibitors down-regulating activity of *PD-L1* promoter. **a** Over 200 compounds were used to screen the inhibitors of *PD-L1* promoter. **b** The chemical structure of Fludarabine Phosphate

3.4 Detecting the Specificity of Fludarabine Phosphate Induced Inhibiting of PD-L1 Promoter Activity

Transfected HeLa cells were treated with different concentration of Fludarabine Phosphate for 24 h and then measured by luciferase reporter assay. The results showed that Fludarabine Phosphate inhibited the activity of *PD-L1* promoter in a dose dependent manner (Fig. 4a). To detect the specificity of Fludarabine Phosphate on *PD-L1* promoter activity, the pGL4-SM22 and pGL4-Beclin1 were also detected by Fludarabine Phosphate at the concentration of 1 μM. The data indicated that Fludarabine Phosphate had a strong inhibitory effect for the *PD-L1* promoter activity, but not for *Beclin1* or *SM22* promoters (Fig. 4b), which indicates that Fludarabine Phosphate inhibits *PD-L1* promoter activity specifically.

3.5 Fludarabine Phosphate Inhibited the Expression of PD-L1

Since Fludarabine Phosphate inhibited *PD-L1* promoter activity, we further investigated whether it down-regulated PD-L1 expression in MGC-803 cells expressing high levels of endogenous PD-L1. The expression levels of PD-L1 were analyzed by RT-PCR and western blot analysis and showed that Fludarabine Phosphate down-regulated PD-L1 expression at both mRNA (Fig. 5a) and protein (Fig. 5b) levels in MGC-803 cells.

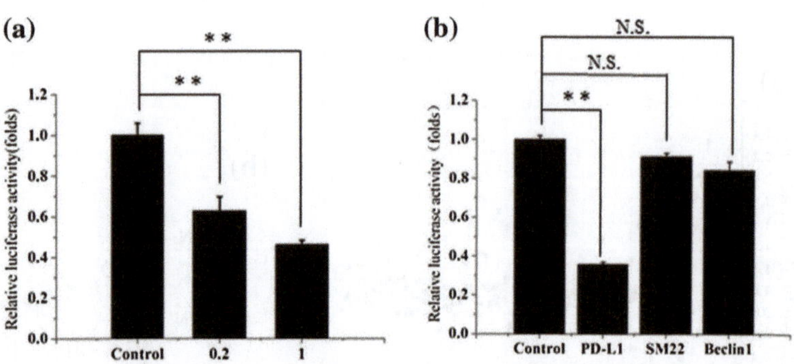

Fig. 4 Detecting the specificity of Fludarabine Phosphate induced inhibiting of *PD-L1* promoter activity. **a** HeLa cells transfected with pGL4-PD-L1and pCMV-β-galactosidase plasmids were treated for 24 h with different concentrations of Fludarabine Phosphate (0.2, 1 μM), 0.1% (v/v) DMSO was used as a negative control. **b** pGL4-PD-L1, pGL4-SM22 or pGL4-Beclin1 plasmids were co-transfected with pCMV-β-galactosidase plasmid and then1 μM Fludarabine Phosphate were treated for 24 h. Relative luciferase activity was measured (N = 3, $**p < 0.01$, $***P < 0.001$, *N.S.* indicates not significant)

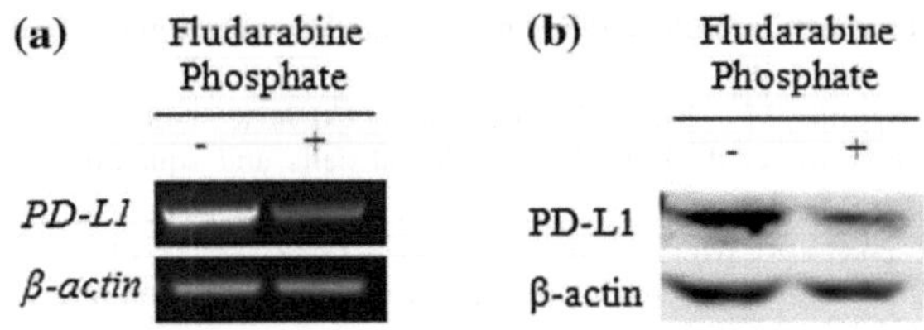

Fig. 5 Fludarabine Phosphate down-regulated PD-L1 expression in MGC-803 cells. **a** MGC-803 cells were treated with Fludarabine Phosphate (1 μM) for 48 h, and 0.1% (v/v) DMSO was used as a negative control. *PD-L1* mRNA levels were analyzed by RT-PCR. *β-actin* was used as an internal control. **b** MGC-803 cells were treated with Fludarabine Phosphate (1 μM) for 48 h, and 0.1% (v/v) DMSO was used as a negative control. Cells were then lysed and PD-L1 expression was analyzed by Western blot

4 Discussion

Immune inhibitory checkpoints, such as *PD-1* and *PD-L1*, allow the tumor to evade immune monitoring, which has become a new hallmark of cancer cells. Activation of *PD-1* signaling pathway by *PD-L1* has been shown to inhibit T cells activity. Balancing with co-stimulatory signals is crucial to maintain peripheral tolerance and prevent excessive damage to tissues during clearing infection. Two monoclonal antibodies against PD-1 have been approved for the treatment of unresectable and metastatic melanoma (nivolumab and pembrolizumab). As for the anti-PD-L1 mAbs, they have already shown some promising results on advanced incurable cancer treatment [13], which suggests that PD-L1 plays a crucial role in tumorigenesis and a novel drug screening target for cancer therapies.

PD-1/PD-L1 monoclonal antibodies have a significant efficacy in clinical trials. However, it also appears to the adverse reactions associated with T cell excessive activation and clinically observable autoimmune damage. In this study, we successfully constructed *PD-L1* promoter luciferase reporter expressing plasmids and found that Fludarabine Phosphate down-regulated the transcriptional activity of *PD-L1* promoter. We have verified that Fludarabine Phosphate inhibited the expression of *PD-L1* in MGC-803 cells, which suggests Fludarabine Phosphate is a potential drug to activate the anti-tumor immune response of T cells.

Acknowledgements This research is supported by the program for College students' innovative entrepreneurial training plan (201510057058).

References

1. Barach YS, Lee JS, Zang X (2011) T cell coinhibition in prostate cancer: new immune evasion pathways and emerging therapeutics. Trends Mol Med 17:47–55
2. Greenwald RJ, Freeman GJ, Sharpe AH (2005) The B7 family revisited. Annu Rev Immunol 23:515–548

3. Zou W, Chen L (2008) Inhibitory B7-family molecules in the tumour microenvironment. Nat Rev Immunol 8:467–477
4. Ritprajak P, Azuma M (2015) Intrinsic and extrinsic control of expression of the immuno-regulatory molecule PD-L1 in epithelial cells and squamous cell carcinoma. Oral Oncol 51:221–228
5. Yamamoto R, Nishikori M, Tashima M et al (2009) B7-H1 expression is regulated by MEK/ERK signaling pathway in anaplastic large cell lymphoma and Hodgkin lymphoma. Cancer Sci 100:2093–2100
6. Blank C, Gajewski TF, Mackensen A (2005) Interaction of PD-L1 on tumor cells with PD-1 on tumor-specific T cells as a mechanism of immune evasion: implications for tumor immunotherapy. Cancer Immunol Immunother 54:307–314
7. Shien K, Papadimitrakopoulou VA, II Wistuba (2016) Predictive biomarkers of response to PD-1/PD-L1 immune checkpoint inhibitors in non-small cell lung cancer. Lung Cancer 99:79–87
8. Mahoney KM, Freeman GJ, McDermott DF (2015) The next immune-checkpoint inhibitors: PD-1/PD-L1 blockade in melanoma. Clin Ther 37:764–782
9. Padda SK, Riess JW, Schwartz EJ et al (2015) Diffuse high intensity PD-L1 staining in thymic epithelial tumors. J Thorac Oncol 10(3):500–508
10. Katsuya Y, Fujita Y, Horinouchi H et al (2015) Immunohistochemical status of PD-L1 in thymoma and thymic carcinoma. Lung Cancer 88(2):154–159
11. Holets LM, Carletti MZ, Kshirsagar SK et al (2009) Differentiation-induced post-transcriptional control of B7-H1 in human trophoblast cells. Placenta 30:48–55
12. Qin K, Rosenfield RL (2005) Characterization of the basal promoter element of the human type 5 17beta-hydroxysteroid dehydrogenase gene. Biochim Biophys Acta 1728:115–125
13. Silva R, Gullo I, Carneiro F (2016) The PD-1: PD-L1 immune inhibitory checkpoint in *Helicobacter pylori* infection and gastric cancer: a comprehensive review and future perspectives. Porto Biomed J 1:4–11

Construction and Functional Analysis of BNP Promoter Luciferase Reporter Plasmid

Jian Zhang, Nan Wang, Yanzhong Liu, Man Li, Huiqin Gong, Hongpeng He and Tongcun Zhang

1 Introduction

Cardiac hypertrophy is an adaptive response of the heart to mechanical, hemodynamic, hormonal and pathologic stimuli [1]. Long-term hypertrophy can lead to reduced coronary flow reserve, increased incidence of cardiac ischemia and arrhythmias, which may eventually lead to heart failure and sudden death [2]. Myocardin is a cardiac and smooth muscle-specific co-activator of SRF that potently binds to CArG box which are within the promoters of cardiac and smooth muscle target genes, such as brain natriuretic peptide (BNP), one of the most sensitive markers of hypertrophic signaling, and activates its transcription. Previous study showed that overexpression of myocardin can induce the cardiomyocyte hypertrophy and fetal cardiac gene expression [3].

Neuron-restrictive silencer element (NRSE), also known as repressor element 1 (RE-1), was originally identified as a repressive DNA regulatory element in gene. A transcriptional repressor, neuron-restrictive silencer factor (NRSF), also known as RE-1 silencing transcription factor (REST) [4, 5] binds to NRSE to inhibit expression of neuronal genes in non-neuronal cell types and in undifferentiated neuronal cells [6, 7]. NRSF is widely expressed in non-neuronal tissues, including the adult heart, and acts as the transcriptional repressor of BNP gene expression [8, 9]. BNP promoter contains a NRSE-like sequence by Bioinformatics analysis. Thus, the purpose of this paper is to investigate the effect of NRSF and myocardin in regulating the transcription of BNP.

J. Zhang · N. Wang · Y. Liu · M. Li · H. Gong · H. He · T. Zhang (✉)
A Key Laboratory of Industrial Microbiology, Ministry of Education and Tianjin City, College of Biotechnology, Tianjin University of Science and Technology, Tianjin 300457, People's Republic of China
e-mail: tony@tust.edu.cn

H. Liu et al. (eds.), *Advances in Applied Biotechnology*, Lecture Notes in Electrical Engineering 444, https://doi.org/10.1007/978-981-10-4801-2_73

2 Materials and Methods

2.1 Plasmid Construction

The BNP promoter containing CArG boxes and NRSE was amplified from genomic DNA of MCF-7 cells by PCR and cloned into a pGL3 luciferase reporter vector. The primers used in PCR reactions are as followed: forward primer-ACTA CGCGTGAGACGGGGTTTCACCGT, reverse primer-TTATCTCGAGTGAACC GGGGCTGCCCAG. Amplification conditions for PCR are as follows: pre-degeneration for 5 min at 95 °C, denaturation for 45 s at 95 °C, annealing for 45 s at 55 °C and extension for 1 min 30 s at 72 °C. PCR reaction was carried out for 35 cycles.

The PCR products and pGL3-Basic vector were digested with restriction enzyme *Hind*III and *Xho*I at 37 °C for 1 h and then the gene fragment was linked to pGL3-Basic vector using T4 DNA ligase at 16 °C for 12 h. The recombinant plasmids were transformed into *E. coli DH5α*. The correct construction of BNP promoter reporter plasmids was confirmed by restriction enzyme digestion and DNA sequencing.

2.2 Cell Culture and Transfection

COS-7 cells were derived from the American Type Culture Collection (ATCC). COS-7 cells were cultured in Dulbecco's Modified Eagle Medium Nutrient Mixture F-12(Ham)(1:1)(DMEM/F-12)(GIBCO) supplemented with 10% fetal bovine serum (FBS) at 37 °C in a 5% CO_2 incubator.

For transfection experiments, COS-7 cells were plated in DMEM/F-12 medium without antibiotics. Transfection was carried out using transfection reagent Turbofect following the manual of the manufacturer (Thermo), when cells were cultured to 70–80% confluence. The myocardin expression construct contained a cDNA of mouse myocardin encoding amino acids 129–935. The NRSF ORF was amplified from pHR′-NRSF-CITE-GFP, which was a gift from Jay Nadeau (Addgene plasmid # 21310) and then cloned to pcDNA3.1 vector.

2.3 Luciferase Reporter Assays

Luciferase activity assay was performed using the Luciferase Assay System (Promega) according to the instructions. Briefly, after transfection 24 h, the transfected cells were lysed in Cell Culture Lysis Reagent. 20 μL of cell lysate and 100 μL of luciferase assay reagent were added into a 96-well plate and then luciferase activity was measured by on a SynergyTM 4 (Bioteck). All experiments were performed at least three times with different preparations of plasmids and primary cells, producing qualitatively similar results.

2.4 Statistical Analyses

All experiments were performed at least three times. Data are analyzed with unpaired *t*-test. Values of $P < 0.01$ was represented by **. Data are presented as means ± SD.

3 Results

3.1 Construction of Human BNP Luciferase Reporter Plasmid

The schematic structure of human BNP promoter containing CArG boxes and NRSE was shown in Fig. 1. The BNP promoter sequence from −1046 to +189 was cloned from the genomic DNA of the MCF-7 cell line.

To estimate the PCR amplification of BNP promoter, agarose gel electrophoresis was performed. As shown in Fig. 2a, one band emerged at the site of 1235 bp, which represent PCR product of BNP promoter.

As shown in Fig. 2b, the PCR products and pGL3-basic vector were digested with restriction enzyme *Hind*III and *Xho*I. Purified vector fragments and PCR fragments were used in a DNA ligation reaction to generate recombinant plasmids.

Next, the recombinant plasmid was analyzed with 1% agarose gel electrophoresis. The size of recombinant plasmid (lane 2) was larger than pGL3-basic vector (lane 1) (Fig. 3). Next, the recombinant plasmid was digested with *Hind*III and *Xho*I at 37 °C for 1 h and 1% agarose gel electrophoretic was used to analyze the DNA products. As shown in Fig. 3, analysis of these digests revealed two major bands: a 1235 bp band could represent BNP gene promoter; a 4818 bp band could represent a pGL3-Basic vector. Finally, BNP promoter was successfully constructed, confirmed by plasmid sequencing.

Fig. 1 A schematic structure of BNP promoter luciferase reporter plasmid. "+1" represents the transcription start site

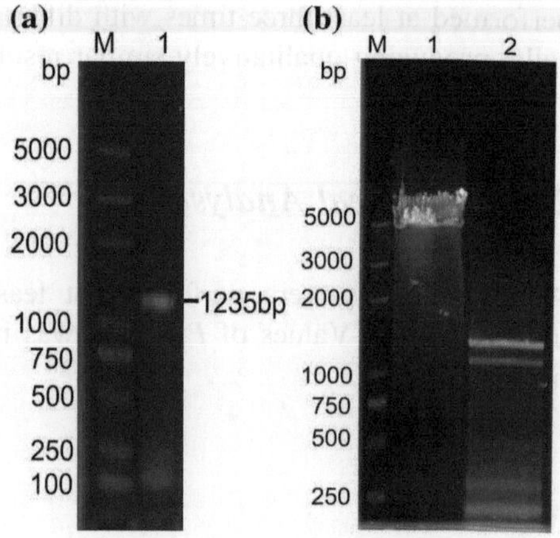

Fig. 2 Agarose gel electrophoretic analysis of PCR products and restriction enzyme digestion analysis of PCR products. **a** PCR product of BNP promoter (*M* DL2000 Plus Marker; *1* PCR products). **b** Restriction enzyme digestion of pGL3-basic vector and PCR product (*M* DL2000 Plus Marker; *1* Double digestion of pGL3-basic vector; *2* Double digestion of BNP promoter)

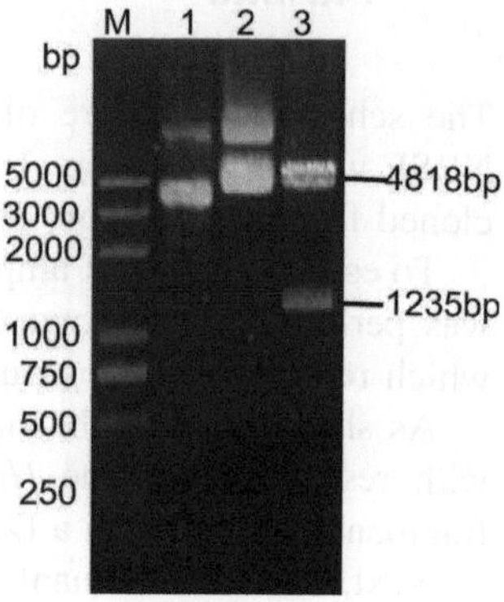

Fig. 3 Agarose gel electrophoretic analysis of recombinant plasmids. *M* DL2000 Plus Marker; *1* pGL3-basic vector; *2* BNP recombinant plasmid; *3* double digestion of BNP recombinant plasmid

3.2 Luciferase Activity Assay

To analyze whether the activity of BNP promoter is affected by myocardin or NRSF, the BNP promoter plasmids were transfected into COS7 cells together with pcDNA3.1-Myocardin or pcDNA3.1-NRSF or pcDNA3.1-Myocardin and pcDNA3.1-NRSF using Turbofect for 24 h. As shown in Fig. 4, the activity of BNP promoter was inhibited significantly by over-expression of NRSF.

It has been reported that myocardin, the member of SRF coactivators, can activate the transcription of BNP and promote its expression. In our results, the activity of BNP promoter was increased 18-fold in myocardin-transfected cells, compared with the control group, whereas in NRSF and myocardin co-transfected group, NRSF inhibited significantly the myocardin-induced activation of BNP promoter.

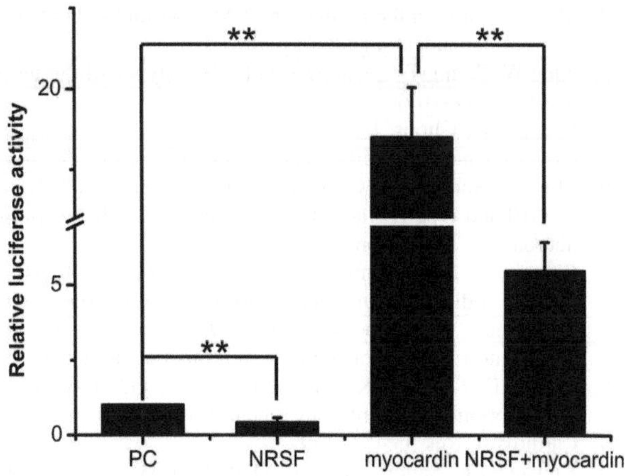

Fig. 4 NRSF inhibited the luciferase activity of BNP promoter activated by myocardin (**, $P < 0.01$, n = 4; Data are presented as means ± SD)

4 Discussion

Heart failure is a common heart disease and one of the most common causes leading to death. Some transcription factors contribute to the development and progression of heart failure by promoting cardiac hypertrophy. It has been reported that myocardin induces cardiac hypertrophy by activating the transcription of hypertrophic marker genes including ANF and BNP, etc. Generally, the re-expression of a fetal cardiac gene program, including BNP, leads to the development of cardiac hypertrophy. In this study, we found that myocardin activated BNP promoter activity and BNP was the target gene of myocardin.

NRSF inhibits the expression of a variety of genes as a transcriptional activity inhibitor. Previous studies showed the role of NRSF/REST in the regulation of cardiac gene expression and function [10]. But about transcription factors which can target myocardin and regulate its transcription and expression are poorly understood. Whether NRSF can inhibit myocardin-stimulated hypertrophic marker gene express is not clarity. Our results preliminarily demonstrated that NRSF inhibited the transcription activity of BNP activated by myocardin. However, further study is needed to illuminate the role of NRSF in cardiac hypertrophy.

Acknowledgements This work was financially supported by the National Natural Science Foundation of China (31171303, 31301073 and 31470816).

References

1. Pipes GCT, Creemers EE, Olson EN (2006) The myocardin family of transcriptional coactivators: versatile regulators of cell growth, migration, and myogenesis. Genes Dev 20:1545–1556

2. Man L, Study on the regulation of Myocardin modulated cardiomyocyte Hypertrophy via the calcium-dependent signaling pathway
3. Xing W, Zhang T-C, Cao D et al (2006) Myocardin induces cardiomyocyte hypertrophy. Circ Res 98:1089–1097
4. Kraner SD, Chong JA, Tsay HJ et al (1992) Silencing the type II sodium channel gene: a model for neural-specific gene regulation. Neuron 9:37–44
5. Mori N, Schoenherr C, Vandenbergh DJ et al (1992) A common silencer element in the SCG10 and type II Na+ channel genes binds a factor present in nonneuronal cells but not in neuronal cells. Neuron 9:45–54
6. Chong JA, Tapia-Ramirez J, Kim S et al (1995) REST: a mammalian silencer protein that restricts sodium channel gene expression to neurons. Cell 80:949–957
7. Schoenherr CJ, Anderson DJ (1995) The neuron-restrictive silencer factor (NRSF): a coordinate repressor of multiple neuron-specific genes. Science 267:1360–1363
8. Ogawa E, Saito Y, Kuwahara K et al (2002) Fibronectin signaling stimulates BNP gene transcription by inhibiting neuron-restrictive silencer element-dependent repression. Cardiovasc Res 53:451–459
9. Kuwahara K, Saito Y, Takano M et al (2003) NRSF regulates the fetal cardiac gene program and maintains normal cardiac structure and function. EMBO J 22:6310–6321
10. Kuwahara K (2013) Role of NRSF/REST in the regulation of cardiac gene expression and function. Circ J 77:2682–2686

HPV18 E6 and E7 Influence the Expression of Cancer Related LncRNAs in HeLa Cells

Xiang Liu, Yongwei Lai, Hailin Yao, Mengmeng Zhang, Hao Zhou, Tongcun Zhang and Hongpeng He

1 Introduction

Cervical cancer which is identified as the second most common type of cancer in women worldwide, is responsible for more than 275,100 mortalities each year and associated with high-risk human papillomavirus, including HPV18 [1]. HPV18 DNA contains two vital oncogenes, E6 and E7, which have crucial roles in malignant transformation in cervical cancer. The E6 promotes p53 degradation through E6AP interaction [2], and interfere with other pro-apoptotic proteins [3], E7 interacts with the PDZ domain of cellular proteins, such as retinoblastoma protein (pRb) to cause cellular transformation, leading to neoplastic progression [4].

Previous studies have revealed that LncRNAs (long non-coding RNAs), which are non-coding transcripts that are more than 200nt in length, are important regulators in various biological processes such as proliferation, differentiation and metastasis, and took vitally an function on the oncogenesis and progression of cervical cancer [5]. It was revealed that there was a close correlation between HPV16 E7 and LncRNA HOTAIR expression in cervical cancer, whereas the detailed mechanism was not identified [6]. So far, the mechanism by which LncRNAs are regulated in cervical cancer is mainly obscure. As an important causative factor of cervical cancer [7], demonstrating how high-risk HPVs regulated the expression of cancer-related LncRNAs would help to understand the mechanism of cervical cancer development.

X. Liu · Y. Lai · H. Yao · M. Zhang · H. Zhou · T. Zhang · H. He (✉)
Key Laboratory of Industrial Microbiology, Ministry of Education and Tianjin City, College of Biotechnology, Tianjin University of Science and Technology, Tianjin 300457, People's Republic of China
e-mail: hehongpeng@tust.edu.cn

H. Liu et al. (eds.), *Advances in Applied Biotechnology*, Lecture Notes in Electrical Engineering 444, https://doi.org/10.1007/978-981-10-4801-2_74

Through a review of literatures regarding LncRNAs associated with various malignant tumors [8], particularly cervical cancer [9], we selected 11 LncRNAs, including 9 previously reported ones MALAT1 [10], TMPOP2 [11], TUSC8 [12], H19 [13], UCA1 [14], CRNDE [15], GAS5 [16], NPTN-IT1 and MEG3 [17], and two function-unknown ones Lnc-01468 and Lnc-00657, for investigation in the present study. MALAT1, TMPOP2, UCA1, H19 and CRNDE which were up-regulated in cervical cancer compared with adjacent tissues were proposed to be correlated with the size, FIGO stage, vessel invasion, lymphatic diffusion, and other features of cervical cancer. TUSC8, GAS5, NPTN.IT1 and MEG3 were reported to be downregulated in cervical cancer, and may play a role in tumor suppression. Lnc-01468 and Lnc-00657 are newly identified LncRNAs with unknown function in cervical cancer. The expression of these 11 LncRNAs in HeLa, a cervical cancer cell line carrying HPV18 DNA in cellular genome, was measured with RT-qPCR method and a comparison was made between HeLa and HaCaT, a nonmalignant epithelial cell line. Additionally, the expression of these LncRNAs was also examined in other cancer or noncancer cell lines. Further, ectopic expression of HPV18 E6 and E7 was introduced into HeLa and HaCaT cells to investigate the effects of HPV18 E6 and E7 on the regulation of lncRNA expression.

2 Materials and Methods

2.1 *Cell Culture*

Human cervical cancer cell line HeLa (adenocarcinoma, containing an integrated HPV-18 E6-E7 coding sequence) was grown in DMEM (low glucose, Gibco) supplemented with 10% fetal bovine serum (FBS, PAA), penicillin (100 U/mL) and streptomycin (100 U/mL). Human epidermal cell line HaCaT was grown in MEM supplemented with 15% FBS. Breast cancer cell line T47D was cultured in RPMI 1640 medium supplemented with 10% FBS. HEK293T (human embryonic kidney cell), HUVEC (Human umbilical vein endothelial cells) and MCF7 (Breast cancer cell line) were cultured in DMEM-F12 supplemented with 10% FBS. All cells were grown at 37 °C in a humidified incubator with 5% CO_2 atmosphere.

2.2 *Cell Transfection*

HeLa or HaCaT cells cultured in 6-well plates at about 80% confluency were transfected with 2.5 μg of plasmids, namely PCMV-tag2B, PCMV-tag2B-E6,

PCMV-HA-E7 and PCMV-tag2B-E6-E7 for 48 h, using Lipofectamine 2000 as instructed by the manufacture (Invitrogen).

2.3 Extraction of Total RNA and RT-PCR

Total RNA was extracted with Trizol reagent (Invitrogen). Reverse transcription was performed with M-MLV reverse transcriptase (Promega). Complementary DNA was quantified by Semi-quantitative PCR using EasyTaq DNA Polymerase kit (TransGen Biotech). PCR primers were designed with NCBI online software Primer-BLAST and synthesized by Invitrogen. Sequences of the primers were: 18s rRNA-forward: 5′-CACGGGAAACCTCACCCGGC-3′; 18s rRNA-reverse: 5′-CGGGTGGCTGAACGCCACTT-3′; HPV18 E6-forward: 5′-CCAGAAACCG TTGAATCCAG-3′; HPV18 E6-reverse: 5′-G TTGGAGTCGTTCCTGTCGT-3′; HPV18 E7-forward: 5′-TGAAATTCCGGTTGACCTTC-3′; HPV 18E7-reverse: 5′-TCGGGCTGGTAAATGTTGAT-3′; TMPOP2-forward: 5′-CTGGAATAACTG GGA CTGA-3′; TMPOP2-reverse: 5′-CTCGCTGAAATTACTGCTC-3′; GAS5-forward: 5′-TAATGACCAC AACAAGCAA-3′; GAS5-reverse: 5′-CCATCAGGCAGTCTACAA-3′; MEG3-forward: 5′-CTGCCCA TCTACAC CTCACG-3′; MEG3-reverse: 5′-AGCTGGCTGGTCAGTTCC-3′; NPTN.IT1-forward: 5′-C CTGTGGTGCTGTATCCT-3′; NPTN.IT1-reverse: 5′-CACCAGC CACTCACTCTAC-3′; TUSC8-Forward: 5′-GGAGGTGAACTGCTTCTTAT-3′; TUSC8-reverse: 5′-CAGGTCCCTTTCTTTGAGAA-3′; U-CA1-forward: 5′-TAGTGGCTGAAGACTGATGC-3′; UCA1-reverse: 5′-GAGGCTGTAGAGT TTGA-3′; CRNDE-forward: 5′-GCTGAAATTCATCCCAAGGC-3′; CRNDE-reverse: 5′-CCTCCTTCCAAT AGCCAGTA-3′; H19-forward: 5′-TCGGTGC CTCAGCGTTCG-3′; H19-reverse: 5′-CTGGCGTCTT GGCCTTCG-3′; MALAT1-forward: 5′-GTGATGCGAGTTGTTCTCCG-3′; MALAT1-reverse: 5′-CTG GCTGCCTCAATGCCTAC-3′; Lnc-01468-forward: 5′-TGAGTCTCAA CTCTGCTGCTTT-3′; Lnc-014-68-reverse: 5′-CACTGCGATTAGATCTTC ACGA-3′. cDNA was quantified by real-time qPCR, with BiosyStems StepOne™ Real-Time PCR and Fast SYBR Green MasterMix (Applied Biosystems).

2.4 Statistical Analysis

The data are representative of three independent experiments, and the GraphPad Prism 6 was used for analysis. The data are presented as the mean ± standard deviation and were compared using Student's t-test or one-way ANOVA test, as appropriate. Statistical significance was defined as $P < 0.05$.

3 Results and Discussion

3.1 *The Different Expression of Selective LncRNAs in HeLa and HaCaT Cell Lines*

To figure out potential LncRNAs correlated with cervical cancer development, the difference in LncRNA expression between HeLa cervical cancer cell line and HaCaT nonmalignant epithelial cell line was investigated. Total RNA was extracted from freshly cultured cells and the expression levels of LncRNAs were determined with RT-qPCR. As shown in Fig. 1a, LncRNAs MALAT1, TMPOP2 and GAS5 were more highly expressed in HeLa cells than that in HaCaT cells. Additionally, the expression of LncRNA TUSC8 was only detected in HeLa cells but not in HaCaT cells (data not shown). The above results suggested that expression of these four LncRNAs could be positively correlated with cervical cancer development. By contrast, the expression of Lnc-01468, H19, UCA1 and Lnc-00657 in HeLa were lower than that in HaCaT (Fig. 1b), indicating that these four LncRNAs might be repressers of carcinogenesis. The expression of MEG3, NPTN.IT1 and CRNDE were undetectable in HeLa (data not shown).

3.2 *The Expression Profiles of the 11 LncRNAs in Different Cell Lines*

To narrow down the list of interesting LncRNAs for further investigation, the expression of above mentioned LncRNAs were investigated in human embryonic epithelial cell line HEK293T, in human breast cancer cell lines MCF7 and T47D, and in human blood vessel endothelial cell line HUVEC. The results of RT-qPCR showed that, comparing with the expression in HaCaT cell line, the RNA level of TMPOP2 was nearly 90-fold higher in HEK293T cells, about 25- to 30-fold higher

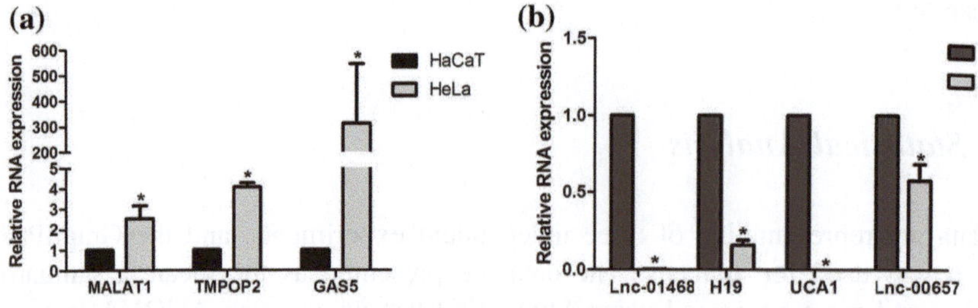

Fig. 1 The expression levels of LncRNAs in HeLa and HaCaT cell lines. **a** The expression of MALAT1, TMPOP2 and GAS5 in HeLa were higher than that in HaCaT. **b** The expression of Lnc-01468, H19, UCA1 and Lnc-00657 in HeLa were lower than that in HaCaT

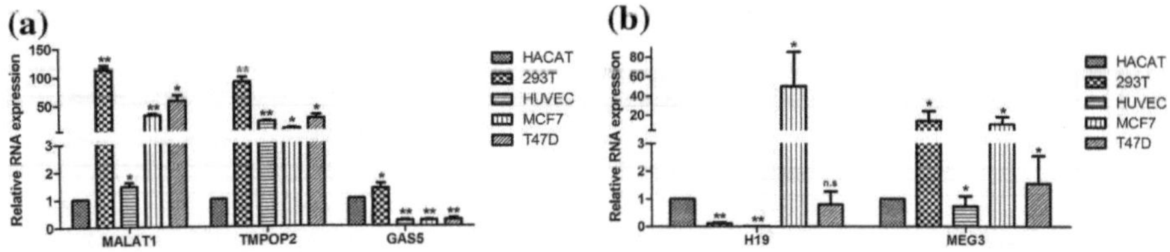

Fig. 2 Expression of lncRNAs varied from cell line to cell line. **a** The expression of TMPOP2, MALAT1 and GAS5 in 293T, MCF7, HUVEC and T47D were basically higher than that in HaCaT. **b** The expression of, H19 and MEG3 in 293T, MCF7, HUVEC and T47D had different varieties comparing with that in HaCaT

in HUVEC and T47D cells, and about 5-fold higher in MCF7 cells, suggesting that TMPOP2 expression was up-regulated in embryonic and cancer epithelial cell lines (Fig. 2a). Also, in non-malignant blood vessel endothelial cell HUVEC, TMPOP2 was moderately up-regulated (Fig. 2a). Unlike the relative high expression in HeLa cells (Fig. 1a), GAS5 expression was slightly higher in HEK293T cells but was lower in MCF7, HUVEC and T47D cells (Fig. 2a), suggesting that the high expression of GAS5 in HeLa cells might be cervical cancer specific.

In parallel, the lncRNAs that were relatively low expressed in HeLa were also determined. And as shown in Fig. 2b, the expression of H19 was obviously lower in 293T and HUVEC cells, on the contrary, it was surprisingly high in MCF7 cells. LncRNA MEG3 was highly expressed in 293T and MCF7 cells but not much varied in HUVEC and T47D cells. The rest of the LncRNAs examined in this study were expressed at too low level to be detected herein.

Taken together, these results suggested that the expression of LlncRNAs was different from cell line to cell line and, among the 11 LncRNAs tested in this study, TMPOP2, MALAT1 and GAS5 were most possible to be correlated with cervical cancer due to their highly expression in HeLa cells.

3.3 *The Effects of HPV18 E6, E7 Over-Expression on LncRNA Expression in HeLa Cervical Cancer Cells*

As shown in Figs. 1 and 2, lncRNA expression profiles were various in different cells. In HeLa, a cervical cancer cell line, MALAT1, TMPOP2 and GAS5 were highly expressed. Especially GAS5 was highly expressed only in HeLa cells but not in other cell lines used in this study. In HeLa cells, HPV18 E6 and E7 genes were constitutionally expressed, altering the expression pattern of protein-coding genes directly or indirectly. To test the effects of HPV18 E6 and E7 on expression of cancer-corrected LncRNAs, plasmids encoding either HPV18 E6, HPV18 E7 or HPV18 E6-E7 were transfected into HeLa cells. Forty-eight hours after

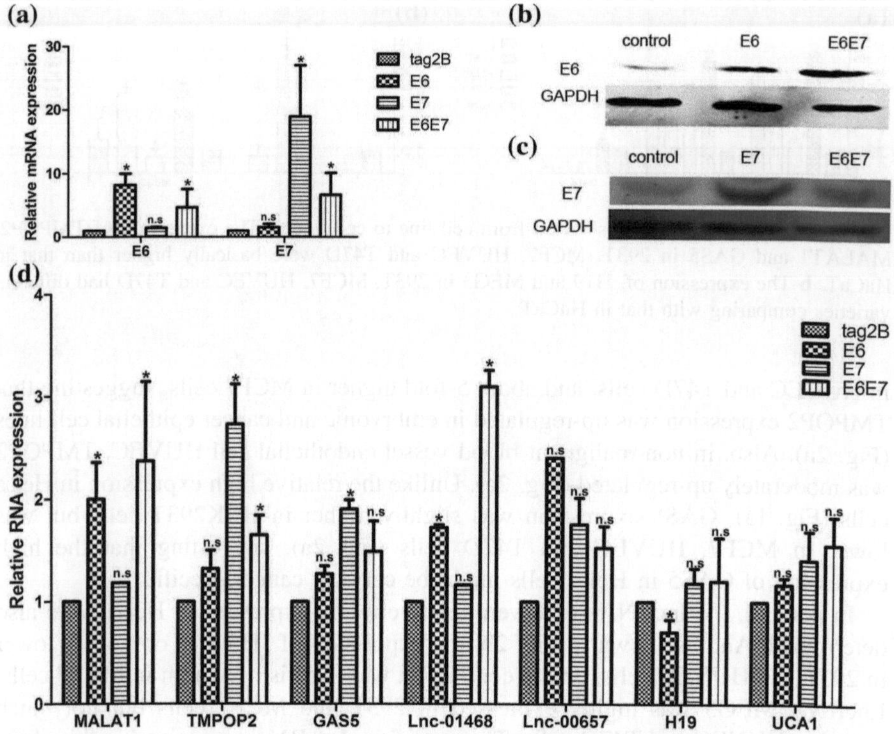

Fig. 3 Over-expression of HPV18 E6 E7 altered the expression levels of lncRNAs in HeLa cells. **a** The mRNA levels of E6 and E7 in HeLa transfected with PCMV-tag2B, PCMV-tag2B-E6, PCMV-HA-E7 or PCMV-tag2B-E6-E7 plasmids were tested with RT-qPCR. **b, c** Western blot assay detected E6 expression in HeLa treated with PCMV-tag2B, PCMV-tag2B-E6 or PCMV-tag2B-E6-E7 plasmids, and E7 expression in HeLa cells treated with PCMV-tag2B, PCMV-HA-E7 or PCMV-tag2B-E6-E7 plasmids. **d** The RNA levels of H19, TMPOP2, MALAT1, Lnc-01468, UCA1, GAS5, Lnc-00657 and MEG3 in HeLa were determined 48-h post transfection with PCMV-tag2B, PCMV-tag2B-E6, PCMV-HA-E7 or PCMV-tag2B-E6-E7 plasmids

transfection, total RNA was extracted and analyzed with RT-qPCR. Protein expression was analyzed with Western-blot. Figure 3a shows the overexpression of E6 and E7 at mRNA level. Figure 3b, c show the overexpression of E6 and E7 at protein level. As shown in Fig. 3d, TMPOP2 and GAS5 RNA levels were increased in E7- and E6-E7-overexpressed cells, indicating that E7 enhanced the expression of TMPOP2 and GAS5. MALAT1 and lnc-01468 RNA levels were boosted by E6 overexpression or E6-E7 co-overexpression, suggesting a positive role of E6 in the regulation of MALAT1 and Lnc-01468 expression (Fig. 3d). Expression of LncRNA MEG3 was repressed by E6 and E7 over-expression, indicating that HPV18 E6 and E7 played a negative role in MEG3 expression (Fig. 3d). RNA levels of lncRNA H19 and UCA1 were not significantly affected by E6 or E7 overexpression (Fig. 3d).

3.4 *The Effects of HPV18 E6, E7 Overexpression on LncRNA Expression in Nonmalignant HaCaT Cells*

Due to the basal expression of HPV18 E6 and E7 in HeLa cells, the relation between HPV18 E6 E7 oncogenes and LncRNAs needed further confirmation. We transfected PCMV-tag2B, PCMV-tag2B-E6, PCMV-HA-E7 and PCMV-tag2B-E6-E7 plasmids into the HPV-negative nonmalignant HaCaT cells, and then measured the expression of the LncRNAs. The successful expression of HPV18 E6 and E7 in HaCaT cells was demonstrated with RT-qPCR as shown in Fig. 4a. The results for TMPOP2, MALAT1, UCA1, Lnc-01468, GAS5 and MEG3 in HaCaT were basically consistent with that in HeLa (Fig. 4b), while the RNA level of the H19 in HaCaT was significantly negative correlated with HPV18 E6, and the RNA level of the Lnc-00657 in HaCaT had significantly positive correlation with HPV18 E7 and E6-E7 (Fig. 4b).

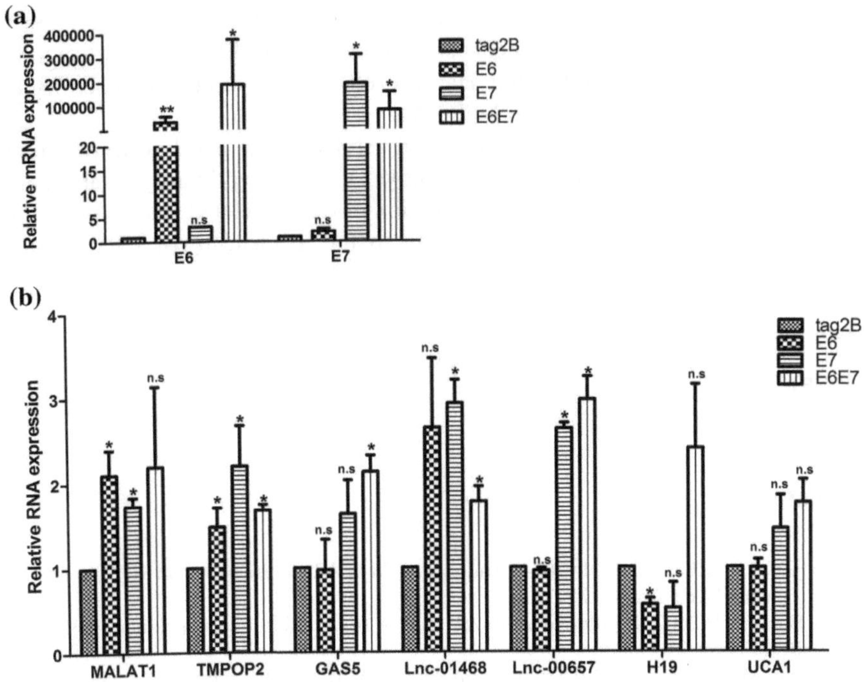

Fig. 4 Expression of HPV18 E6 E7 in HaCaT cells affected the expression of LncRNAs. **a** The mRNA levels of E6 and E7 in HaCaT transfected with PCMV-tag2B, PCMV-tag2B-E6, PCMV-HA-E7 or PCMV-tag2B-E6-E7 plasmids were tested with RT-qPCR. **b** The RNA levels of H19, TMPOP2, MALAT1, Lnc-01468, UCA1, GAS5, Lnc-00657 and MEG3 in HaCaT were determined 48-h post transfection with PCMV-tag2B, PCMV-tag2B-E6, PCMV-HA-E7 or PCMV-tag2B-E6-E7 plasmids

4 Discussion

To investigate the relation between HPV18 E6/E7 and the LncRNAs in cervical cancer, the RNA levels of 11 LncRNAs in HeLa, HaCaT, HEK293T, MCF7, HUVEC and T47D were tested by RT-qPCR with 18S rRNA served as endogenous control. We found that the lncRNA expression profile was cell type dependent. Basically, TMPOP2 and MALAT1 were highly expressed in embryonic or cancer cell lines comparing with nonmalignant HaCaT cells. Intriguingly, GAS5 was highly expressed in HeLa but not in other cancer cell lines tested herein, suggesting that GAS5 might be specifically upregulated in HeLa cells. Introduction of ectopically expressed HPV18 into HeLa cells further boosts the expression of TMPOP2, MALAT1 and GAS5. Similar effects were observed in HaCaT cells. Taken together, our results suggested that HPV18 E6 E7 oncoproteins stimulated the expression of cervical cancer-related LncRNA TMPOP2, MALAT1 and GAS5. The mechanism by which HPV18 E6 and E7 regulate the expression of these lncRNA was worth of further study.

The expression of lncRNA H19, UCA1, Lnc-01468 and Lnc-00657 were relatively lower in HeLa cells than that in HaCaT cells, indicating that they might be negatively correlated with cervical cancer, whereas HPV18 E6 E7 did not suppress their expression in both HeLa and HaCaT cells. There might be other factors involved in the regulation of these lncRNAs.

HPV infection was highly correlated with cervical caner and LncRNAs irregulation was a novel "hot spot" of cancer study. Our results primarily established a link between HPV18 E6 E7 and the expression of cancer related LncRNAs, which would contribute to a better understanding to the mechanism of HPV induced cervical cancer development.

Acknowledgements This work was supported by National Natural Science Foundation of China (31301073) and Applied Basic Science and Frontier Technology Program of Tianjin (13JCYBJC38000).

References

1. Jiang Y, Li Y, Fang S et al (2014) The role of MALAT1 correlates with HPV in cervical cancer. Oncol Lett 7:2135–2141
2. Zur Hausen H (2002) Papillomaviruses and cancer: from basic studies to clinical application. Nat Rev Cancer 2:342–350
3. Thomas M, Banks L (1999) Human papillomavirus (HPV) E6 interactions with Bak are conserved amongst E6 proteins from high and low risk HPV types. J Gen Virol 80(Pt 6):1513–1517
4. Ramakrishnan S, Partricia S, Mathan G (2015) Overview of high-risk HPV's 16 and 18 infected cervical cancer: pathogenesis to prevention. Biomed Pharmacother 70:103–110
5. He JH, Han ZP, Li YG (2014) Association between long non-coding RNA and human rare diseases (Review). Biomed Rep 2:19–23
6. Sharma S, Mandal P, Sadhukhan T et al (2015) Bridging links between long noncoding RNA HOTAIR and HPV oncoprotein E7 in cervical cancer pathogenesis. Sci Rep 5:11724

7. Nohata N, Abba MC, Gutkind JS (2016) Unraveling the oral cancer lncRNAome: identification of novel lncRNAs associated with malignant progression and HPV infection. Oral Oncol 59:58–66
8. Yan X, Hu Z, Feng Y et al (2015) Comprehensive genomic characterization of long non-coding RNAs across human cancers. Cancer Cell 28:529–540
9. Peng L, Yuan X, Jiang B et al (2015) LncRNAs: key players and novel insights into cervical cancer. Tumour Biol
10. Sun R, Qin C, Jiang B et al (2016) Down-regulation of MALAT1 inhibits cervical cancer cell invasion and metastasis by inhibition of epithelial-mesenchymal transition. Mol BioSyst 12:952–962
11. Sun NX, Ye C, Zhao Q et al (2014) Long noncoding RNA-EBIC promotes tumor cell invasion by binding to EZH2 and repressing E-cadherin in cervical cancer. PLoS ONE 9: e100340
12. Liao LM, Sun XY, Liu AW et al (2014) Low expression of long noncoding XLOC_010588 indicates a poor prognosis and promotes proliferation through upregulation of c-Myc in cervical cancer. Gynecol Oncol 133:616–623
13. Feigenberg T, Gofrit ON, Pizov G et al (2013) Expression of the h19 oncofetal gene in premalignant lesions of cervical cancer: a potential targeting approach for development of nonsurgical treatment of high-risk lesions. ISRN Obstet Gynecol 2013:137509
14. Hong HH, Hou LK, Pan X et al (2016) Long non-coding RNA UCA1 is a predictive biomarker of cancer. Oncotarget
15. Ellis BC, Molloy PL, Graham LD (2012) CRNDE: a long non-coding RNA involved in CanceR, Neurobiology, and DEvelopment. Front Genet 3:270
16. Qian X, Xu C, Zhao P et al (2016) Long non-coding RNA GAS5 inhibited hepatitis C virus replication by binding viral NS3 protein. Virology 492:155–165
17. Zhang J, Yao T, Wang Y et al (2016) Long noncoding RNA MEG3 is downregulated in cervical cancer and affects cell proliferation and apoptosis by regulating miR-21. Cancer Biol Ther 17:104–113

Construction of Tip60-Encoding Plasmid and the Effect of Tip60 on the Expression of HPV18 Genes in HeLa Cells

Yongwei Lai, Xuena Liu, Yijie Wang, Yunpeng Yue, Xiang Liu, Hao Zhou, Nan Wang, Xue-Gang Luo, Wenjian Ma, Tong-Cun Zhang and Hongpeng He

1 Introduction

Tip60 is a cellular protein belonging to the MYST family of histone acetyl transferases (HATs) and it is universally expressed in various types of cells. Tip60 was originally isolated as an HIV Tat-interactive protein with molecular weight of 60 kDa thus was named Tip60 [1]. Tip60 functions as a transcriptional co-factor in the regulation of HIV gene expression. Later on, the **h**istone **a**cetyl **t**ransferase (HAT) activity of Tip60 was identified [2]. In addition to histone, Tip60 was also found to be a lysine acetyl transferase therefore it is also known as KAT5. ER, a well-known transcription activator, is a typical target of Tip60 for acetylation [3].

Previous studies suggest that Tip60 plays important roles in many biological processes, such as cellular signalling, DNA damage repair, cell cycle regulation and apoptosis [4]. In response to DNA double strand breaks, Tip60 is recruited to DNA lesions where it participates both in the initial as well as the final stages of DNA repair [5]. Tip60 is part of a multiprotein complex, NuA4, which is recruited by many transcription factors to their target promoters, where it is thought to participate in histone acetylation and transcriptional regulation [6]. The role of Tip60 in target gene transcription could be active or repressive depending on the context of chromatin [7, 8].

Human papillomavirus (HPV) is a causative factor of cervical cancer which threatens the health of females worldwide. In cervical cancer cells, HPV DNA was broken into DNA fragments and integrated into cellular genome thereby being regulated by cellular transcriptional factors. It was reported that Tip60 is involved

Y. Lai · X. Liu · Y. Wang · Y. Yue · X. Liu · H. Zhou · N. Wang · X.-G. Luo · W. Ma · T.-C. Zhang · H. He (✉)
Key Laboratory of Industrial Microbiology, Ministry of Education and Tianjin City, College of Biotechnology, Tianjin University of Science and Technology, Tianjin 300457, People's Republic of China
e-mail: hehongpeng@tust.edu.cn

H. Liu et al. (eds.), *Advances in Applied Biotechnology*, Lecture Notes in Electrical Engineering 444, https://doi.org/10.1007/978-981-10-4801-2_75

in E2-mediated HPV16 E6-E7 transcriptional repression [9]. Additionally, EP400 was figured out to repress HPV16 E6-E7 in an E2-dependent manner [10]. In cervical cancer cells, HPV E2 gene is usually lost during HPV DNA integration, which raises the question about the roles of Tip60 and EP400 in HPV E6-E7 regulation in the E2-deficient cervical cancer cells, such as HeLa. In this study, we constructed a Tip60-expressing plasmid, furthermore, we overexpressed Tip60 and EP400 in HeLa cells to explore the effects of Tip60 and EP400 in HPV18 E6-E7 gene transcription.

2 Materials and Methods

2.1 *Plasmid Construction*

The encoding sequence of Tip60 was amplified with RT-PCR method using total RNA extracted from HeLa cells as template. Pfu DNA polymerase (Thermo Scientific, Waltham, USA) was used in PCR reaction. Restriction enzyme sites *EcoR*I and *Xho*I were introduced into forward or reverse PCR primers, respectively. Primer sequences for Tip60 cloning were: Tip60-Fd (*EcoR*I): AGAT**GAATTC** GCGGAGGTGGTGAGTCCGGTG, Tip60-Rv (*Xho*I): GGTCT**CTCGAG**TC ACCACTTCCCCCTCTTGCTC. pCMV-Tag2B was used as expressing vector and Tip60-encoding sequence was inserted into the vector with the help of T4-DNA ligase (Promega, Madison, USA). The correct Tip60 constructs were primarily identified by double RE-digestion and finally confirmed by plasmid sequencing.

2.2 *Cell Culture and Transfection*

HeLa cells were cultured in DMEM (Gibico, Waltham, USA) supplemented with 10% FBS (Kangyuan, Tianjin, China) at 37 °C in a humidified incubator with 5% CO_2. Tip60-expressing plasmid was constructed in this study and EP400-expressing plasmid is a gift from Prof. Robert Roeder (USA) [8]. Empty vector pCMV-Tag2B served as negative control. Plasmids were transfected into HeLa cells with TurboFect (Thermo Scientific, Waltham, USA) following the manufacturer's instruction.

2.3 *Reverse Transcription-PCR and QPCR*

Total RNA was extracted from Hela cells with Trizol (Invitrogen, Waltham, USA). cDNA was synthesized by reverse transcription with M-MLV reverse transcriptase (Promega, Madison, USA). cDNA was quantified by either regular PCR using

EasyTaq DNA Polymerase kit (TransGen Biotech, Beijing, China) or realtime qPCR using Biosystems StepOne™ Real-Time PCR system and SYBR Green Master Mix (ABI, Waltham, USA). PCR products were visualized by electrophoresis using agarose gels.

2.4 Western-Blotting

HeLa cell lysates were separated with SDS-PAGE using a 15% gel and then transferred to nitrocellulose membrane (Millipore, Darmstadt, Germany). Primary antibodies included HPV18 E6, HPV18 E7, β-actin and GAPDH (Santa Cruz, Santa Cruz, USA). Secondary antibodies were IRDye-conjugated donkey anti-mouse or anti-rabbit IgG LI-COR Biosciences, Lincoln USA) and membranes were visualized with Odyssey Infrared Imaging System (LI-COR Biosciences, Lincoln USA).

3 Results and Discussion

3.1 Construction of Tip60-Expressing Plasmid

Tip60 is universally expressed in various types of human cells including HeLa cervical cancer cells. Total RNA was isolated from HeLa cells. The encoding sequence of Tip60 was amplified with RT-PCR. As shown in Fig. 1a, the PCR fragment is about 1600 bp in length which is consistent with the length of Tip60 ORF.

Tip60 PCR fragment was cloned into pCMV-Tag2B expressing vector utilizing multiple cloning sites *Eco*RI and *Xho*I. The insertion of Tip60 was identified with double restrictive enzyme digestion (Fig. 1b). At last, the correct construction of Tip60 expressing plasmid was confirmed with sequencing.

3.2 Over-Expression of Tip60 in HeLa Cells

To test whether the recombinant Tip60 plasmid would be expressed in human cells, pCMV-Tag2b-Tip60 was transfected into HeLa cells. Fourty-eight hours post transfection, cells were harvested for mRNA and proteins analysis. mRNA level was determined with RT-realtime qPCR and the results showed that Tip60 mRNA level was dramatically boosted with Tip60-plasmid transfection (Fig. 2a). Results of Western-blotting showed that the protein level of Tip60 increased in pCMV-Tag2b-Tip60 transfected cells, demonstrating the successful overexpression of Tip60 in HeLa cells (Fig. 2b).

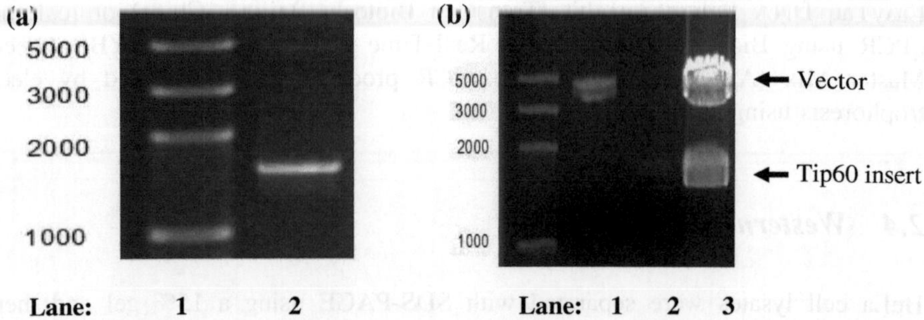

Fig. 1 Construction of Tip60-expressing plasmid (**a**). RT-PCR amplification of Tip60-encoding sequence. *Lane 1* DNA marker. *Lane 2* Tip60 PCR fragment. **b** *EcoR*1 and *Xho*1 digestion of candidate plasmids. *Lane 1* DNA marker. *Lane 2* digested pCMV-Tag2b vector. *Lane 3* digested pCMV-Tag2b-Tip60

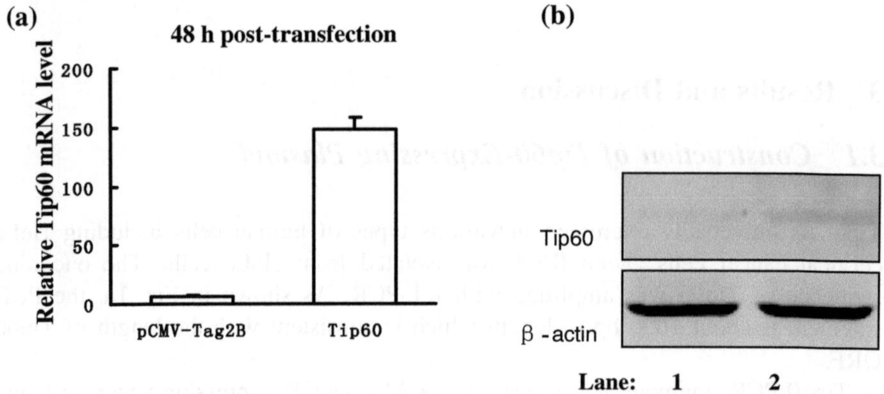

Fig. 2 Over-expression of Tip60 in HeLa cells. **a** The mRNA level of Tip60 was increased by about 150 folds in pCMV-Tag2b-Tip60 transfected cells. The mRNA levels were measured with realtime qPCR. 18S rRNA was used as endogenous control. **b** The protein level of Tip60 was increased in Tip60 overexpressing cells. *Lane 1* HeLa cells transfected with empty vector. *Lane 2* HeLa cells transfected with pCMV-Tag2b-Tip60 plasmids. β-actin served as loading control

3.3 The Repressive Effect of Tip60 on HPV18 E6 and E7 Genes

Tip60 was previously found to repress HPV E6 and E7 gene transcription in a HPV E2-dependent manner [9]. To examine the effect of Tip60 over-expression on HPV18 E6 and E7 gene regulation in the E2-defective HeLa cells, mRNA levels of HPV18 E6 and E7 were measured with RT-realtime qPCR. The results showed that both E6 and E7 mRNA levels were reduced by roughly 60% (Fig. 3), indicating

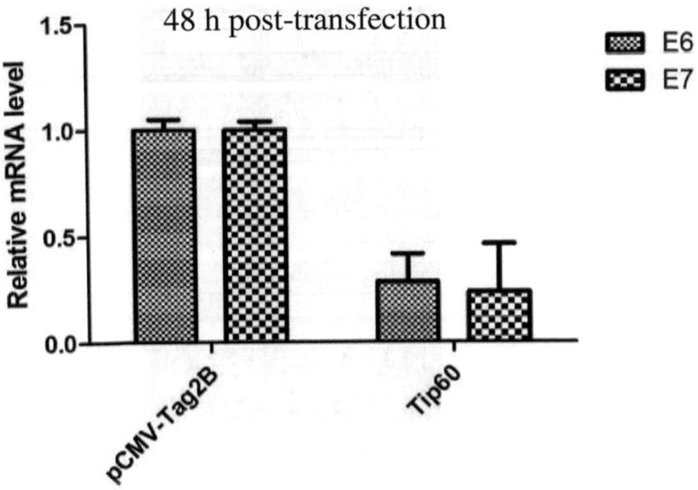

Fig. 3 Tip60 repressed HPV18 E6 and E7 gene in E2-defective HeLa cells

that Tip60 is able to repress HPV18 E6 and E7 gene transcription without the mediation of HPV E2.

3.4 *Tip60 and EP400 Worked Synergistically in the E2-Independent Repression of HPV18 E6 and E7 Genes*

EP400 is a chromatin remodelling factor with ATPase activity. It was identified as an E2-dependent HPV E6-E7 repressor from a genome-wide siRNA screening [10]. To explore the relationship of Tip60 and EP400 in the regulation of HPV18 E6 and E7 genes in HeLa cells, Tip60 and EP400 plasmids were co-transfected into HeLa cells. The results of RT-PCR showed the successful overexpression of EP400 and Tip60. Meanwhile, the mRNA levels of E6 and E7 were further decreased upon EP400 and Tip60 co-overexpression (Fig. 4, compare lane 4 with lanes 1–3)

The results of Western-blotting, as shown in Fig. 5, is in agreement with RT-PCR, showing that the protein levels of HPV18 E6 and E7 decreased in EP400 or Tip60 overexpressing HeLa cells and this inhibitory effect was more dramatic when EP400 and Tip60 were co-overexpressed. Taken together, these results suggest that Tip60 and EP400 work synergistically in the regulation of HPV18 E6 E7 genes which is HPV E2-independent in HeLa cells.

To conclude, in the present study, a Tip60-expressing plasmid was constructed and the correct expression was demonstrated. In addition, the HPV E2-independent repressive effect of Tip60 on HPV18 E6 and E7 genes was observed and its synergistic effect with EP400 in the regulation of HPV18 E6 and E7 was identified

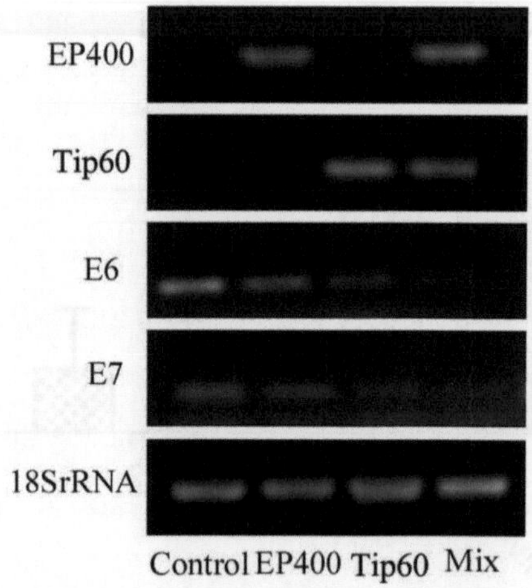

Fig. 4 EP400 and Tip60 co-overexpression repressed the transcription HPV18 E6 and E7 more significantly than single-overexpression of either one

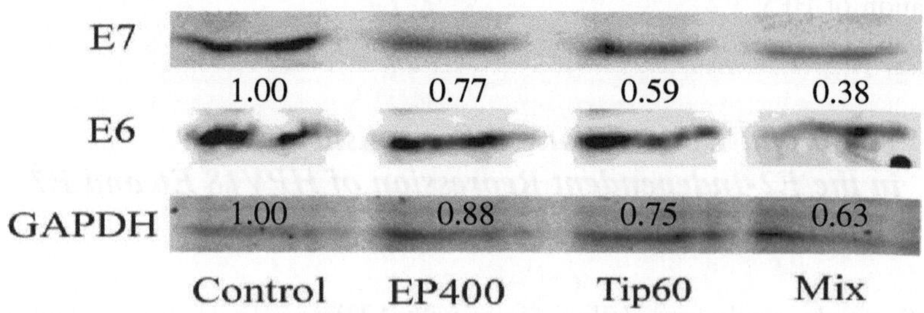

Fig. 5 EP400 and Tip60 worked synergistically in the repression of HPV18 E6 and E7 in HeLa cells. Density of bands was analyzed with Image J software. GAPDH served as a loading control. The density of E6 or E7 bands was normalized with that of GAPDH bands and then calibrated by control. Protein expression levels in control were set as 1 and the *numbers* under the bands indicate the expression levels of various proteins

for the first time. EP400 is a chromatin-remodeling factor with ATPase activity which could facilitate a more condensed chromatin structure to reduce the accessibility of HPV18 promoter to the transcription machinery. Tip60 and EP400 were previously demonstrated to form a protein complex [8], which might contribute to their synergistic effect in HPV18 gene repression. The detailed mechanism will be further studied in the future.

References

1. Kamine J, Elangovan B, Subramanian T et al (1996) Identification of a cellular protein that specifically interacts with the essential cysteine region of the HIV-1 Tat transactivator. Virology 216:357–366
2. Kimura A, Horikoshi M (1998) Tip60 acetylates six lysines of a specific class in core histones in vitro. Genes Cells 3:789–800
3. Gaughan L, Logan IR, Cook S et al (2002) Tip60 and histone deacetylase 1 regulate androgen receptor activity through changes to the acetylation status of the receptor. J Biol Chem 277:25904–25913
4. Sapountzi V, Logan IR, Robson CN (2006) Cellular functions of TIP60. Int J Biochem Cell Biol 38:1496–1509
5. Ikura M, Furuya K, Fukuto A et al (2016) Coordinated regulation of TIP60 and Poly (ADP-Ribose) Polymerase 1 in damaged-chromatin dynamics. Mol Cell Biol 36:1595–1607
6. Squatrito M, Gorrini C, Amati B (2006) Tip60 in DNA damage response and growth control: many tricks in one HAT. Trends Cell Biol 16:433–442
7. Ravens S, Yu C, Ye T et al (2015) Tip60 complex binds to active Pol II promoters and a subset of enhancers and co-regulates the c-Myc network in mouse embryonic stem cells. Epigenetics Chromatin 8:45
8. Park JH, Sun XJ, Roeder RG (2010) The SANT domain of p400 ATPase represses acetyltransferase activity and coactivator function of TIP60 in basal p21 gene expression. Mol Cell Biol 30:2750–2761
9. Smith JA, Haberstroh FS, White EA et al (2014) SMCX and components of the TIP60 complex contribute to E2 regulation of the HPV E6/E7 promoter. Virology 468–470:311–321
10. Smith JA, White EA, Sowa ME et al (2010) Genome-wide siRNA screen identifies SMCX, EP400, and Brd4 as E2-dependent regulators of human papillomavirus oncogene expression. Proc Natl Acad Sci U S A 107:3752–3757

Investigation of Aquatic Pathogens and Diversity Analysis of *Aeromonas* Isolates

Zhaoyuan Jing, Yang He, Qian Li, Bo Zhang and Hongjiang Yang

1 Introduction

China is a country in the production and consumption of a large amount of aquatic products, accounting for about two-thirds of the total amount in the world's total. In recent years, aquaculture constantly tended to operate in the ways with high density, industrialization and intensification, for increasing the output of aquatic products. However, the water rich in nutrients led to pathogenic bacteria blooming in aquaculture. Annual losses from aquaculture diseases accounts for about 30% of production, while pathogenic bacterial infection diseases accounts for about half.

The major aquatic bacterial pathogens mainly comprise multiple genera of both Gram-negative and Gram-positive bacteria, such as *Vibrio*, *Aeromonas*, *Escherichia coli*, *Streptococcus iniae*, *Salmonella*, *Edwardsiella tarda*, *Pleisionomas*, *Flavobacterium*, *Acinetobacter*, *Pasteurella*, *Pseudomonas*, *Staphylococcus aureus*, *Mycobacterium*, *Clostridium* and others [1]. *Aeromonas*, including 25 different species, is the important pathogen of fish, shrimp and other animals [2, 3], causing serious economic losses [4].

In this work, fish of Tianjin and Xiamen, Fujian, China, *Penaeus vannamei* shrimp seeds of Hangu, Tianjin, China, and seawater samples of Jiyun River and Bohai Bay were collected in Tianjin. The epidemiology of aquatic pathogens was isolated with the selective media. The isolates were identified with the 16S rRNA

Z. Jing · Y. He · Q. Li · H. Yang (✉)
Key Laboratory of Industrial Fermentation Microbiology,
Ministry of Education, Tianjin Key Laboratory of Industrial Microbiology,
College of Biotechnology, Tianjin University of Science & Technology,
Tianjin 300457, China
e-mail: hongjiangyang@tust.edu.cn

B. Zhang
College of Marine & Environmental Science, Tianjin University
of Science & Technology, Tianjin 300457, China

H. Liu et al. (eds.), *Advances in Applied Biotechnology*, Lecture Notes
in Electrical Engineering 444, https://doi.org/10.1007/978-981-10-4801-2_76

sequencing analysis [5]. Since 16S rRNA gene sequence is highly homologous among most *Aeromonas* species, the *Aeromonas* isolates in this study were also analyzed with the multilocus sequence typing (MLST) method for further discrimination of the *Aeromonas* isolates [6].

2 Materials and Methods

2.1 Samples, Media and Bacteria Isolation

Freshwater fishes were obtained from five local markets and two fish farms in Tianjin, China from March through May 2012. The fishes belonged to 11 different species, including *Hypophthalmichthys molitrix* (4), *Ctenopharyngodon idella* (4), *Carassius auratus* (4), *Hypophthalmichthys nobilis* (4), *Megalobrama amblycephala* (2) and *Ephippus orbis* (2). *Penaeus vannamei* shrimp seeds (2) and Saltwater fish *Epinephelus awoara* (12) were respectively obtained from Hangu, Tianjin, and Xiamen, Fujian, China, during March 2012. Seawater samples (4) were obtained from in Tianjin, China from March through May 2012.

The gastrointestinal tract was removed from fish and the content was collected and stored at 4 °C. The content sample was serially diluted in phosphate buffered saline and spread on three types of agar plates, including SS (*Salmonella Shigella* agar), TCBS (thiosulfate citrate bile salts sucrose agar) [7, 8], MAC (MacConkey agar) [9], CIN-1 (Cepulodin Irgasan Novobiocin Agar). L-agar plates containing ampicillin (20 μg/ml), erythromycin (20 μg/ml), and kanamycin (20 μg/ml) were also used for screening the possible bacterial pathogens in the gut samples.

2.2 The Identification of Isolated Strains

The genomic DNA was extracted from the isolated strains. The DNA was used as template for PCR amplification of the 16S rRNA gene with primers described previously [10]. The PCR product was subjected to sequencing directly. The obtained sequences were analyzed by the BLAST program on the NCBI website. MEGA 5 was used to construct the phylogenetic trees [11].

2.3 MLST Analysis

Multilocus sequence typing (MLST) was proposed as a universal method to characterize bacteria based on sequence polymorphisms within internal fragments of housekeeping genes. The primers for amplifying a set of six housekeeping genes, namely, *gyrB*, *groL*, *gltA*, *metG*, *ppsA*, and *recA*, was synthesized for MLST of the *Aeromonas* isolates [12]. The sequences of distinct alleles were deposited in the *Aeromonas* MLST database (http://pubmlst.org/aeromonas). Allele number and sequence type (ST) number were determined by described previously [12].

3 Results

3.1 Strain Separation

120 bacteria strains were isolated from 37 different samples with various media. Morphology of bacterial colonies grown on the TCBS plate were mostly yellow and green, smooth, and moist; on the SS plate were mostly yellow, pink, colorless with black center; on the MAC plate mostly were pink and red, flat; on the CIN-1 plate mostly were pink, big, and smooth.

3.2 Identification of Isolated Strains

120 isolated strains were identified with the comparative sequence analysis of the 16S rRNA gene. The 16S rRNA sequences were deposited in GenBank with accession no KC210757-KC210826, KC210827- KC210872 and KC252599-KC252602. All the isolates were tentatively classified into 15 genera of seven families, including *Aeromonas* spp. (53), *Vibrio* spp. (19), *Proteus* spp. (5), *Citrobacter* spp. (9), *Hafnia* spp. (8), *Providencia* spp. (6), *Pseudomonas* spp. (3), *Kluyvera* spp. (2), *Enterobacter* spp. (2), *Bacillus* spp. (2), *Pantoea* sp. (1), *Leclercia* sp. (1), and *Acinetobacter* sp. (1). *Aeromonas* strains were not isolated from all the samples except from *Epinephelus awoara* from samples of Xiamen, Fujian, China. The *Vibrio* isolates include *V. parahaemolyticus* (2), *V. alginolyticus* (5), *V. azureus* (2), *V. anguillarum* (9), and *V. fluvialis* (1), found in the samples from Tianjin and Xiamen, Fujian, China (Table 1).

3.3 Phylogenetic Analysis

Based on the 16S rRNA gene sequences, the phylogenetic tree of all isolated strains was constructed. As shown in Fig. 1, *Aeromonas* isolates, *Enterobacteriaceae* isolates, *Vibrio* isolates, and *Shewanella* isolates were in grouped in one cluster, respectively; 9 *C. freundii* isolates were also grouped into two clades, one comprising of 4 members and the other comprising of 5 members. The phylogenetic

Table 1 Comparative analysis of 16S rRNA gene sequences of the isolated

Genus	Number	Host[a]	Medium[b]
Vibrio	19	Ea, Pv, Eo	MAC, TCBS, CIN-1
Proteus	5	Ea, Hn	TCBS, Tc^{20}
Citrobacter	9	Ea, Hm, Ma, Hn, Cau	SS, MAC, TCBS, CIN-1, Km^{20}, Em^{20}
Hafnia	8	Ea, Cc	SS, MAC, TCBS, Em^{20}
Providencia	6	Ea, Ci	CIN-1, TCBS
Kluyvera	2	Ea	CIN-1
Morganella	2	Cau, Ci	TCBS
Pantoea	1	Ci	SS
Enterobacter	2	Ea, Hn	TCBS, Km^{20}
Leclercia	1	Ea	Em^{20}
Shewanella	6	Hn, Eo, Hm, Pv	CIN-1, TCBS
Pseudomonas	3	seawater, Pv	SS
Acinetobacter	1	Pv	SS
Bacillus	2	Ea	SS, TCBS
Aeromonas	53	seawater, Hn, Eo, Cau, Hm, Pv, Cc, Ci, Ma	TCBS, MAC, SS, CIN-1

[a]Hm *Hypophthalmichthys molitrix*; Ci *Ctenopharyngodon idella*; Cau *Carassius auratus*; Hn *Hypophthalmichthys nobilis*; Ma *Megalobrama amblycephala*; Eo *Ephippus orbis*; Pv *Penaeus vannamei*; Ea *Epinephelus awoara*; Cc *Cyprinus carpio*
[b]Em^{20}, LB medium contain 20 μg/ml erythromycin; Tc^{20}, LB medium contain 20 μg/ml tetracycline; Kn^{20}, LB medium contain 20 μg/ml kanamycin

analysis result indicated that the isolated strains within the same genus or species might belong to different lineages (Fig. 1).

Since 16S rRNA gene sequence is highly homologous among most *Aeromonas* species, the *Aeromonas* isolates in this study were identified only at genus level [6].

3.4 MLST Analysis

The fragments of the six housekeeping genes were successfully amplified and sequenced in all 53 *Aeromonas* strains. The sequence alignment results showed only 21 out of the 318 alleles was known alleles in the database, including *gyrB*22 (LSB2), *gyrB*22 (MY-1), *gyrB*97 (MJ-4), *gyrB*22 (MY), *groL*42 (MJ-1), *groL*96 (CH13), *groL*96 (SH11), *gltA*56 (MY-1), *gltA*56 (MJ-4), *gltA*51 (ML-1), *gltA*56 (MY), *gltA*56 (SJ1), *gltA*51 (SL-1), *gltA*99 (TB2), *metG*64 (TY-4), *ppsA*67 (LSJ1), *ppsA*40 (MJ-1), *ppsA*40 (SL1), *ppsA*40 (TL1), *ppsA*40 (TY-2) and *recA*42 (MJ-1). The remaining 297 sequences represented the novel alleles. Correspondingly, only 8 STs were known sequence types and 45 STs were novel. The results suggested the highly genetic diversity of the *Aeromonas* isolates in our work.

Fig. 1 Phylogenetic tree based on 16S rRNA gene sequences of the isolates

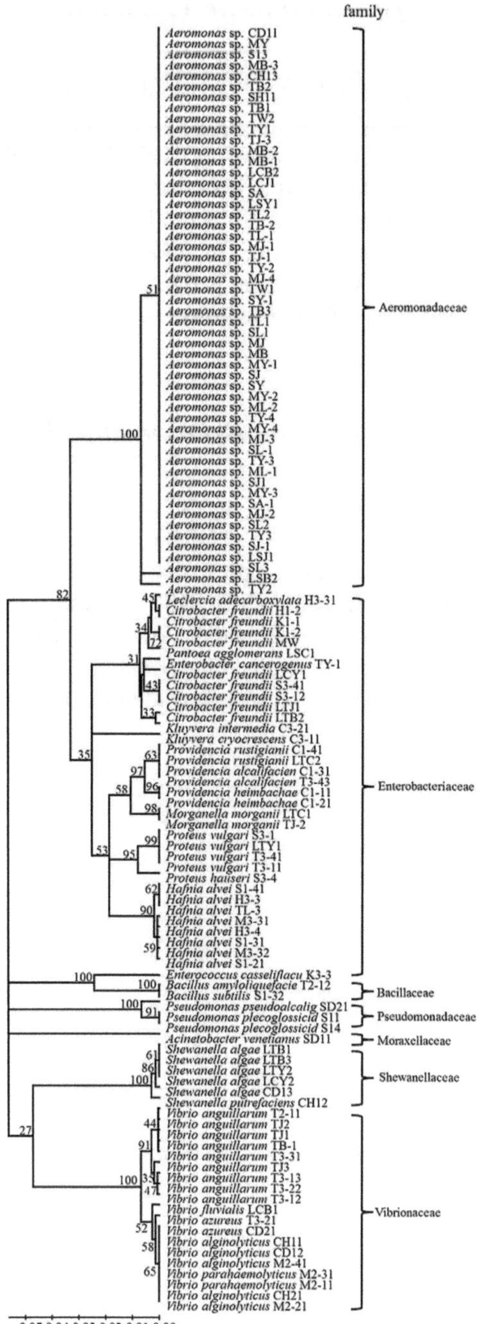

3.5 Distribution of Aeromonas in Worldwide

By analyzing the database of all 398 *Aeromonas* STs, *Aeromonas* strains geographic isolation was found. As shown in Fig. 2a, the Asia, Europe, North America and South America were respectively discovered 195, 181, 8 and 1 unique STs, only one ST is the North American and Asian shared, the ST is famous ST 251 [13].

As shown in Fig. 2b, a total of 171 unique ST were found in the Chinese region, Beijing, Tianjin, Jilin, Guangdong and Zhejiang as the main source of the data, and was found to also have geographic isolation; no duplicate ST occurs among strains isolated in different regions.

3.6 Aeromonas Relationship Between the STs

Relationships among the STs deposited in the *Aeromonas* MLST database were analyzed by using the eBURST algorithm [9]. The eBURST analysis revealed that 42 irrelevant clonal complexes (CCs) and 81 singletons were identified. Figure 3 is a schematic diagram of eBURST analysis, the identified 45 STs distributed in the 8 of 42 clonal complexes, and the low correlation between strains, indicating a high degree of biological diversity of *Aeromonas* strains. Additionally, in this study only one allele number difference was identified between ST127 (MY-1) and ST 140 (MY).

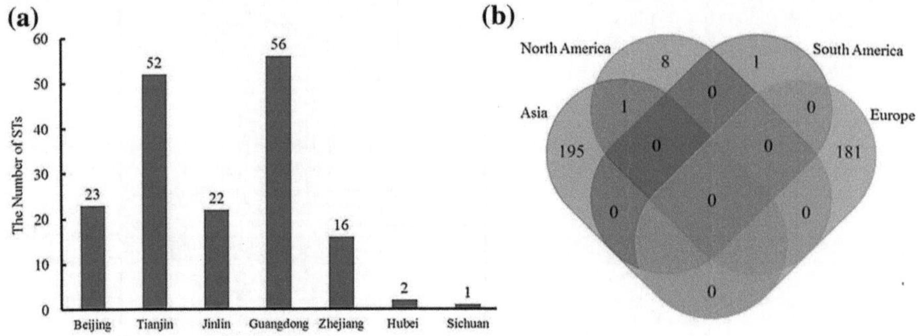

Fig. 2 Distributions of *Aermonas* STs across China (**a**) and the world (**b**)

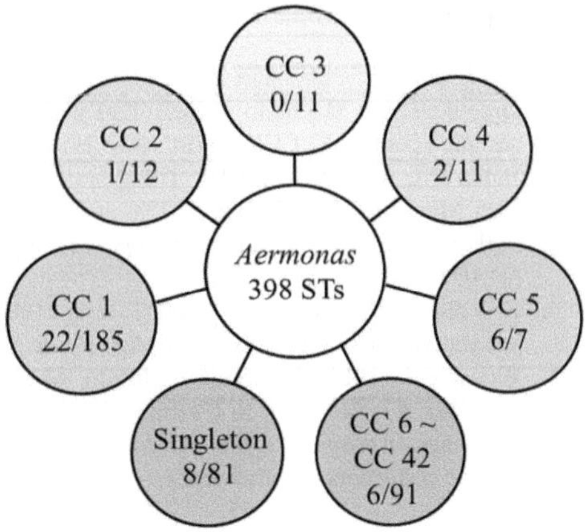

Fig. 3 eBURST algorithm analysis of 398 *Aermonas* STs. *CC* clonal complexes; Transparent circle is 398 *Aermonas* STs, *grey circle* is grouped by eBURST algorithm analysis 398 *Aermonas* STs; Fraction = (Number of *Aermonas* CCs in this study and CCs)/(Number of *Aermonas* STs in this CCs)

4 Discussion

Aquatic pathogenic bacteria can cause fish diseases, leading to the death of aquatic animals, bringing economic losses to farmers. Deng et al. [14] collected samples from a variety of aquatic animals from 64 aquaculture farms in Guangdong, China, and isolated 112 *Aeromonas* strains with multidrug-resistant phenotype and Class I integrons. Adesoji et al. [15] isolated 108 tetracycline resistant strains from water distribution systems in southwestern Nigeria, including *Aermonas* spp., *Alcaligenes* spp., *Bacillus* spp., *Klebsiella* spp., *Leucobacter* spp., *Morganella* spp., and *Proteus* spp..

MLST was performed to further discriminate the 53 *Aeromonas* isolates. Totally 297 new alleles and 45 novel STs were identified in our study. The data indicated the strains of the *Aeromonas* genus from different sources were highly genetically diversified [16, 17]. By eBURST algorithm analysis 398 *Aeromonas* STs were found belonging to 42 CCs, further suggesting the diversity of all *Aeromonas* isolates across the world. That may be due to the small number of *Aeromonas* isolates in the MLST database, and more deposited strains will better unveil the distribution patterns of *Aeromonas* strains.

This study has identified a number of novel alleles and STs of the *Aeromonas* spp. The relevant information has been deposited in the *Aermonas* MLST database. The data obtained in our study will provide more support for the study of the *Aeromonas* epidemiology.

Acknowledgements This work was partly supported by The National Natural Science Foundation of China (Grant 31370205 and 30970114).

References

1. Greenlees KJ, Machado J, Bell T et al (1998) Food borne microbial pathogens of cultured aquatic species. Vet Clin North Am Food Anim Pract 14(1):101–120
2. Janda JM, Abbott SL (2010) The genus *Aeromonas*: taxonomy, pathogenicity, and infection. Clin Microbiol Rev 23(1):35–73
3. Parker JL, Shaw JG (2011) *Aeromonas* spp. clinical microbiology and disease. J Infect 62 (2):109–118
4. Greenlees KJ, Machado J, Bell T et al (1998) Food borne microbial pathogens of cultured aquatic species. Vet Clin North Am Food Anim Pract 14(1):101–112
5. Woese CR (1987) Bacterial evolution. Microbiol Res 51(2):221–271
6. Martinez AJ, Benlloch S, Collins MD (1992) Phylogenetic interrelationships of members of the genera *Aeromonas* and *Plesiomonas* as determined by 16S ribosomal DNA sequencing: lack of congruence with results of DNA-DNA hybridizations. Int J Syst Bacteriol 42(3):412–421
7. Uchiyama H (2000) Distribution of vibrio species isolated from aquatic environments with TCBS agar. Environ Health Prev Med 4(4):199–204
8. Feil EJ, Li BC, Aanensen DM et al (2004) eBURST: inferring patterns of evolutionary descent among clusters of related bacterial genotypes from multilocus sequence typing data. J Bacteriol 186:1518–1530
9. Mishra S, Nair GB, Bhadra RK et al (1987) Comparison of selective media for primary isolation of *Aeromonas* species from human and animal feces. J Clin Microbiol 25(11):2040–2043
10. Weisburg W, Barns S, Pelletier D et al (1991) 16S ribosomal DNA amplification for phylogenetic study. J Bacteriol 173(2):697–703
11. Tamura K, Peterson D, Peterson N et al (2011) MEGA5: molecular evolutionary genetics analysis using maximum likelihood, evolutionary distance, and maximum parsimony methods. Mol Biol Evol 28(10):2731–2739
12. Elena MM, Luca F, Filomena M (2011) Determination of microbial diversity of *Aeromonas* strains on the basis of multilocus sequence typing, phenotype, and presence of putative virulence genes. App Environ Microbiol 77(14):4986–5000
13. Pang M, Jiang J, Xie X et al (2015) Novel insights into the pathogenicity of epidemic *Aeromonas hydrophila* ST251 clones from comparative genomics. Sci Rep 5:9833
14. Deng Y, Wu Y, Jiang L et al (2016) Multi-Drug resistance mediated by class 1 integronsin *Aeromonas* isolated from farmed freshwater animals. Front Microbiol 7:935
15. Adesoji AT, Ogunjobi AA, Olatoye IO et al (2015) Prevalence of tetracycline resistance genes among multi-drug resistant bacteria from selected water distribution systems in southwestern Nigeria. Ann Clin Microbiol Antimicrob 14:35
16. Neo E, La T, Phillips ND et al (2013) The pathogenic intestinal spirochaete Brachyspira pilosicoli forms a diverse recombinant species demonstrating some local clustering of related strains and potential for zoonotic spread. Gut Pathog 5(1):24
17. Cui SY, Sun QH, Li JX et al (1997) Study on the phenotypic characters and pathogenic factors of 260 *Aeromonas* spp. strain. Microbiol China 24(4):227–230

Expression of Transcription Factor EB (TFEB) Promotes Cancer Cell Proliferation, Migration and Invasion

Wei Li, Yang Liu, Min Hao, Meng Yang, Shuang Zhao, Zhenxing Liu and Aipo Diao

1 Introduction

Transcription factor EB (TFEB), a member of the bHLH leucine-zipper family of transcription factors, drives the expression of autophagic and lysosomal genes [1, 2]. TFEB is localized to the cytoplasm in normal conditions and translocated to nuclear under specific conditions, such as nutrient starvation [3]. TFEB positively regulates the expression of lysosomal genes, controls lysosome population and promotes cellular degradation of lysosomal substrates [4–6]. Lysosome plays a key role in autophagic process in which autophagic substrates are degraded [7]. TFEB may link autophagy with lysosomal biogenesis via activating multiple genes encoding lysosomal proteins [8]. Thus, identification of TFEB as a target regulator may provide a novel therapeutic strategy for modulating lysosomal function in human disease [9].

Because the significant effects of TFEB in autophagy and lysosomal biogenesis, the physiological roles of TFEB have been studied in recent years. The effects of high-fat diet enhanced by lipid degradation was prevented by TFEB overexpression [10]. Cells overexpressing TFEB have a larger number of lysosomes and an enhanced degradative capability against lysosomal and autophagy substrates [4]. In HeLa cells, the expression of several genes involved in lysosomal biogenesis were

W. Li · Y. Liu · M. Hao · M. Yang · S. Zhao · Z. Liu (✉) · A. Diao (✉)
Key Lab of Industrial Fermentation Microbiology of Ministry
of Education, College of Bioengineering, Tianjin University
of Science and Technology, Tianjin 300457, China
e-mail: liuzx@tust.edu.cn

A. Diao
e-mail: diaoaipo@tust.edu.cn

W. Li
College of Basic Medical, Inner Mongolia Medical University,
Hohhot 010110, Inner Mongolia, China

H. Liu et al. (eds.), *Advances in Applied Biotechnology*, Lecture Notes
in Electrical Engineering 444, https://doi.org/10.1007/978-981-10-4801-2_77

increased and HIF-1 reporter activity was inhibited due to the overexpression of TFEB [11]. In addition, increasing TFEB expression in RAW 264.7 cells resulted in the generation of osteoclast-like cells that were more efficient in resorbing a mineralized ECM [12].

It has been shown that overexpression of TFEB leads to the generation of new lysosomes and increased the number of autophagosomes in variety of cell types [5, 7, 13]. Stable and transient overexpression of TFEB in HeLa cells significantly increased the number of autophagosomes [7]. However, whether the expression of TFEB inflect the proliferation, migration and invasion of cancer cells hasn't been investigated. In this study, we investigated whether expression of TFEB was involved in tumor cells proliferation, migration and invasion, in an attempt to gain further insights on the role of TFEB in tumorigenesis.

2 Materials and Methods

2.1 Reagents

Fetal bovine serum (FBS) was purchased from Tianhang Biological Technology (Zhejiang, China). Rabbit anti-TFEB antibody was obtained from Santa Cruz Biotech (USA). Mouse anti-β-actin and anti-Flag antibodies, goat anti-mouse 680 and anti-rabbit 680 secondary antibodies were purchased from Sungene Biotech (Tianjin, China). MTT (3-[4,5-dimethylthiazol-2-yl]-2,5-diphenyl tetrazolium bromide) was purchased from Solarbio Biotech (Beijing, China). All other biological or chemical reagents were from Thermo Scientific or Sigma Aldrich unless otherwise stated.

2.2 Cell Culture and Transfection

The HeLa cells were cultured in Dulbecco's modified Eagel's medium (DMEM) supplemented with 10% FBS, 100 unit/mL penicillin and 100 μg/mL streptomycin in humidified incubator at 37 °C and 5% CO_2.

The *TFEB* gene was amplified by PCR and the product was purified and digested with *Eco*RI and *Bam*HI restriction enzyme, and then inserted into pcDNA3.1$^+$ vector with flag tag on its upstream and HA tag on its downstream. The recombinant plasmid and empty vector (EV) were transiently transfected into HeLa cells. In brief, transient transfection were performed when cells grown to about 70–80% confluence using 2 μg plasmid DNA, 4 μL TurboFect transfection reagent (Thermo, USA) and 200 μL Opti-MEM per 35 mm dish according to the manufacturer's instructions.

2.3 RNA Interference

The pLKO.1-TRC2 cloning vectors expressing TFEB shRNA (TRCN0000013108: 5′-CCGGCCCAC TTTGGTGCTAATAGCTCTCGAGAGCTATTAGCACCAAA GTGGGTTTTTG-3′) and non-targeting shRNA control vector (SHC002) were purchased from Sigma and the knockdown level was tested by Western blot. Lentiviruses were produced using 293T cells according to the manufacturers' manual. Cells were infected with lentiviruses for 3 days prior to the start of the experiments.

2.4 Western Blot

Transfected/infected and control HeLa cells were incubated in the condition described in Sect. 2.2 and the protein level of TFEB was detected by Western blot. The cells were collected and proteins were extracted using RIPA buffer on ice for 30 min. The lysates were centrifugated and mixed with 5× SDS supplemented with DTT (SERVA, Germany). After denatured, 8% SDS-PAGE was used to separate the proteins with 30 μg protein in each lane and subsequently transferred onto the PVDF membrane using transmembrane instrument (DYCP-40E, Liuyi instrument company, Beijing, China). After blocking for 1 h with 5% nonfat milk at room temperature, the membrane was incubated with primary antibodies against either TFEB (1:200), flag or β-actin overnight at 4 °C, and then with goat anti-mouse or anti-rabbit secondary antibody at a dilution of 1:5000 for 2 h at room temperature. Proteins were visualized on ODYSSEY infrared imaging (Li-cor Biosciences, USA) western blotting detection.

2.5 Cell Proliferation Assay

The methyl thiazolyl tetrazolium (MTT) assay was used to assess cell proliferation [14]. HeLa cells were seeded in 96-well plates at a density of 8000 cells per well, then incubated at 37 °C for 24, 48, and 72 (h), respectively. After incubation, 20 μL of MTT (5 mg/mL) was added and the cells were cultured for additional 4 h to format the formazan salt crystals. The reaction was terminated by removal of the supernatant and addition of 200 μL of DMSO. The 96-well plate was shaken for 10 min at room temperature gently. A Microplate Reader (Thermo, USA) was used to measure the optical density (OD) of each well at 490 nm. The results were plotted as means ± SD of three independent experiments.

2.6 *Wound Healing Assay*

The wound healing assay was performed to evaluated the migration ability of tumor cells [15]. HeLa cells were cultured in 35 mm dishes and grown to approximate 100% confluence. A clear area was scraped in the monolayer with a 200 μL pipette tip. After washing three times with DPBS, cells were incubated at 37 °C in DMEM medium containing 3% FBS. The wounded gap sizes were evaluated at 0, 24, 48 and 72 (h) with an inverted microscope and photographed. The healing rate was quantified with the gap sizes compared to the control (0 h). Three different areas in each assay were chosen to measure.

2.7 *Ttranswell Invasion Assay*

For the invasion assay [16], matrigel solution (BD Biosciences) was prepared in serum-free cell culture medium at a dilution of 1:8, and used to coat the 24-well transwell chambers overnight at 37 °C before cell seeding. The cells were cultured in the chambers with serum-free media at 1×10^5 cells per well. After 48 h, the insert were removed by wiping with a cotton swab. Cells that migrated to the bottom surface of the insert were fixed with 4% paraformaldehyde, stained with 0.1% crystal violet, and counted in five random fields.

2.8 *Statistical Analysis*

Experiments were independently repeated three times. Student's t-test and One-way analysis of variance (ANOVA) were used to investigate the difference between the transfected/infected cells and control cells. SPSS18.0 statistical software package was used for all statistical analysis. $P < 0.05$ was considered to indicate statistical significance.

3 Results

3.1 *Expression of TFEB in HeLa Cells*

The *TFEB* gene with flag- and HA-tag was inserted into a pcDNA3.1$^+$ plasmid. The empty vector (EV) and *TFEB* gene were transiently transfected into HeLa cells, named as pcDNA3.1$^+$-EV and pcDNA3.1$^+$-flag-TFEB-HA, respectively. TFEB expression in HeLa cells was analyzed by Western blot. As shown in Fig. 1, Western blot analysis confirmed that only endogenous TFEB protein was detected in control

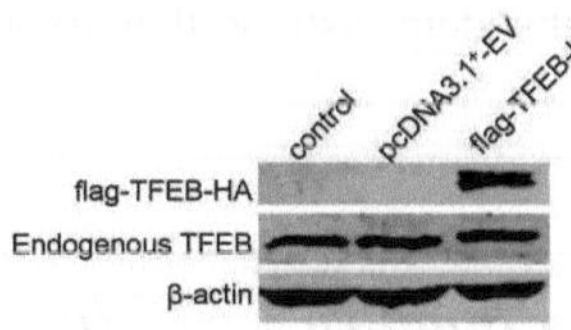

Fig. 1 Expression of TFEB in the transiently transfected HeLa cells. The protein level of TFEB in control or transfected HeLa cells (pcDNA3.1$^+$-EV and pcDNA3.1$^+$-flag-TFEB-HA) was detected by Western blot. The untransfected HeLa cells were used as control. β-actin was used as the loading control

cells and cells transfected with pcDNA3.1$^+$-EV plasmid, while TFEB fusion protein was detected in the cells transfected with pcDNA3.1$^+$-flag-TFEB-HA, which indicated that exogenous TFEB protein was expressed in HeLa cells.

3.2 TFEB Expression Promoted Cell Growth

To assess whether TFEB promotes or inhibits the growth of HeLa cells, the MTT assay was performed. The empty plasmid was transient transfected serving as control, and pcDNA3.1$^+$-flag-TFEB-HA was transfected into HeLa cells under the same experimental circumstances. As shown in Fig. 2a, cell growth rate between pcDNA3.1$^+$-flag-TFEB-HA and pcDNA3.1$^+$-EV cells had significant difference ($P < 0.05$) at 72 h post-transfection. Moreover, TFEB was depleted in HeLa cells infected lentiviruses expressing TFEB-shRNA (named shTFEB), HeLa cells expressing TFEB-shRNA showed efficient silencing of TFEB expression, as determined by Western blot (Fig. 2b). MTT assay showed that knockdown of TFEB expression significantly decreased HeLa cell viability compared to the

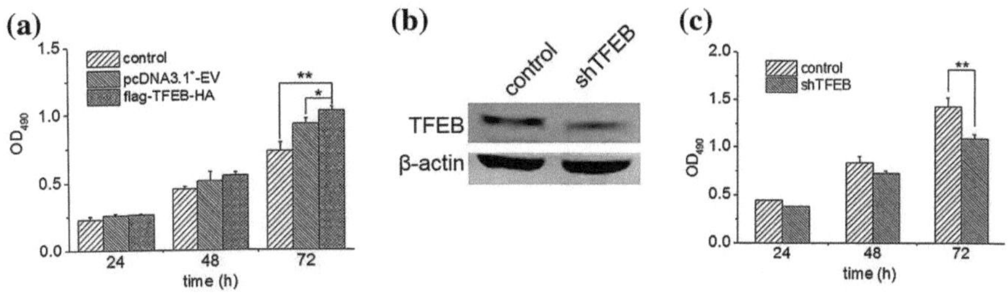

Fig. 2 The effect of TFEB expression on the proliferation of HeLa cells. **a** MTT assay were performed and the growth rates in transiently transfected HeLa cells expressing TFEB were analyzed compared to control cells (*$P < 0.05$, **$P < 0.01$). The untransfected HeLa cells were used as control. **b** TFEB expression was inhibited in HeLa cells infected lentiviruses expressing TFEB-shRNA. **c** The proliferation of HeLa cells depleted TFEB was assayed by MTT compared to control cells (**$P < 0.01$)

control (Fig. 2c). These results demonstrate that expression of TFEB promotes HeLa cells proliferation.

3.3 TFEB Expression Promoted Cell Migration

To investigate the effect of TFEB on cell migration, wound healing assay was performed in our study. The transfected HeLa cells were incubated for 24 h and then scraped with a 200 μL pipette tip. The cells were incubated continuously in DMEM medium with 3% FBS under the same condition and the wound size were measured at 0, 24, 48 and 72 (h). The results showed that the gaps were gradually reduced from 24 to 72 h. The pcDNA3.1^{+}-flag-TFEB-HA transfected HeLa cells had a much faster wound-healing rate compared with the control pcDNA3.1^{+}-EV transfected Hela cells ($P < 0.05$), which suggests that TFEB expression increases the migration ability of HeLa cells (Fig. 3a, b).

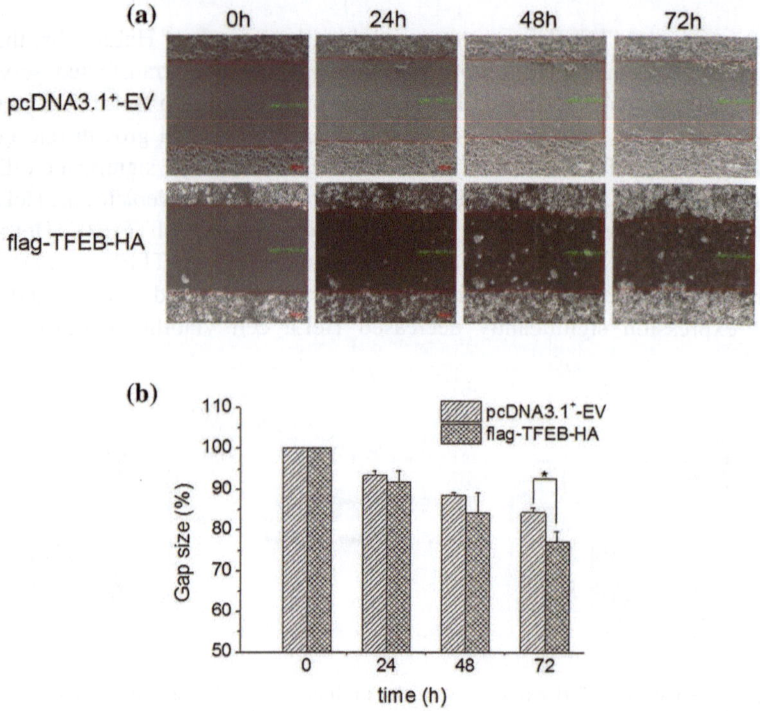

Fig. 3 The effect of TFEB expression on HeLa cell migration was determined using wound healing assay. **a** Observations were carried out at 0, 24, 48 and 72 (h) after the start of culture. *Scale bar* 100 μm. **b** The gap size was measured and plotted as the percentage of the original time point (0 h), (*$P < 0.05$)

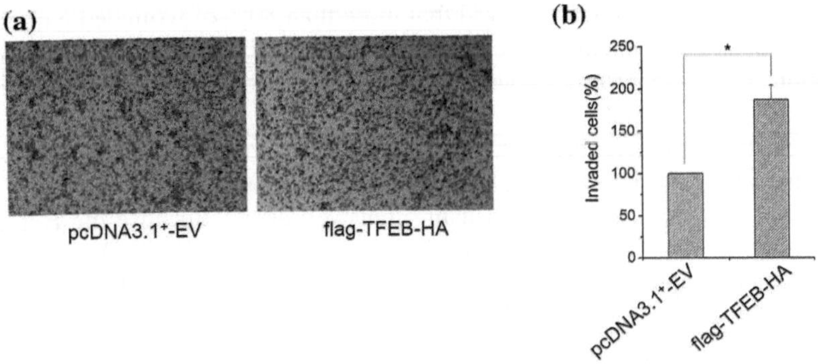

Fig. 4 Expression of TFEB promoted MGC-803 cells invasion. **a** MGC-803 cells overexpressing TFEB were used to determine the effect of TFEB on cell invasion by a transwell invasion assay. **b** The invasive cells of MGC-803 cells overexpressing TFEB were calculated and then normalized with that of pc-DNA3.1^{+}-EV transfected cells (*$P < 0.05$)

3.4 TFEB Expression Promoted Cell Invasion

To further assess the role of TFEB on the invasion of cancer cells, transwell invasion assay was performed. The pcDNA3.1^{+}-flag-TFEB-HA plasmid was transiently transfected into MGC-803 cells, the empty vector were used as control. The results indicated that TFEB expression promoted MGC-803 cells invasion through matrigel by about 1.8 times compared to the control (Fig. 4a, b).

4 Discussion

TFEB is considered as a master control gene of autophagy and exocytosis. Stable or transient overexpression of TFEB in HeLa cells increases the number of autophagosomes [7, 17]. Moreover, TFEB overexpression increases the degradation of protein that causes Huntington's disease [5]. But at the present, whether expression of TFEB inflects the cancer cell proliferation, migration and invasion remains to be elucidated. In our study, *TFEB* gene was successfully transfected into HeLa cells or MGC-803 cells. Then we measured the cell proliferation, migration and invasion ability via MTT, wound healing and transwell assay respectively.

TFEB overexpression increased proliferation of NSC-34 cells [18], and knockdown of TFEB reduced the expansion rates of MEFs and HeLa cells [19]. But down-regulation of TFEB expression in A549 cells has no particular effect on cell proliferation [20]. Our results suggested that overexpression of TFEB promoted the proliferation of HeLa cells (Fig. 2a) and knockdown of TFEB significantly decreased HeLa cell viability (Fig. 2c). The results indicated that TFEB is associated with cell proliferation of HeLa cell lines.

Giatromanolaki et al. [20] showed that more than 80% of wounded gap was not refilled after TFEB silencing, which suggested that TFEB had influences on the cells migration. The present study suggested the overexpression of TFEB could accelerate the healing rate of HeLa cells, which indicated that TFEB might increase the migration ability of HeLa cells. Our results further indicate that TFEB expression is connected with cancer cells migration. In addition, we detected the effects of TFEB expression on cancer cell invasion using transwell assay. Because the HeLa cells have a weak ability of cell invasion, we overexpressed *TFEB* gene in MGC-803 cells and investigated the effect of TFEB on the invasion of cancer cells. Our findings indicated that expression of TFEB promoted the invasion of MGC-803 cells.

Collectively, this study provides a novel evidence that TFEB expression promotes the cancer cell proliferation, migration and invasion. Therefore, our findings implicate TFEB is a potential therapeutic target for cancer treatment.

References

1. Napolitano G, Ballabio A (2016) TFEB at a glance. J Cell Sci 129(13):2475–2481
2. Martina JA, Diab HI, Brady OA, Puertollano R (2016) TFEB and TFE3 are novel components of the integrated stress response. EMBO J 35(5):479–495
3. Roczniak-Ferguson A, Petit CS, Froehlich F, Qian S, Ky J, Angarola B et al (2012) The transcription factor TFEB links mTORC1 signaling to transcriptional control of lysosome homeostasis. Sci Signal 5(228):ra42
4. Palmieri M, Impey S, Kang H, di Ronza A, Pelz C, Sardiello M et al (2011) Characterization of the CLEAR network reveals an integrated control of cellular clearance pathways. Hum Mol Genet 20(19):3852–3866
5. Sardiello M, Palmieri M, di Ronza A, Medina DL, Valenza M, Gennarino VA et al (2009) A gene network regulating lysosomal biogenesis and function. Science 325(5939):473–477
6. Settembre C, Medina DL (2015) TFEB and the CLEAR network. Methods Cell Biol 126:45–62
7. Settembre C, Di Malta C, Polito VA, Garcia-Arencibia M, Vetrini F, Erdin S et al (2011) TFEB links autophagy to lysosomal biogenesis. Science 332(6036):1429–1433
8. Medina DL, Fraldi A, Bouche V, Annunziata F, Mansueto G, Spampanato C et al (2011) Transcriptional activation of lysosomal exocytosis promotes cellular clearance. Dev Cell 21 (3):421–430
9. Settembre C, Fraldi A, Medina DL, Ballabio A (2013) Signals from the lysosome: a control centre for cellular clearance and energy metabolism. Nat Rev Mol Cell Biol 14(5):283–296
10. Settembre C, De Cegli R, Mansueto G, Saha PK, Vetrini F, Visvikis O et al (2013) TFEB controls cellular lipid metabolism through a starvation-induced autoregulatory loop. Nat Cell Biol 15(6):647–658
11. Hubbi ME, Hu H, Kshitiz, Ahmed I, Levchenko A, Semenza GL (2013) Chaperone-mediated autophagy targets hypoxia-inducible factor-1α (HIF-1α) for lysosomal degradation. J Biol Chem 288(15):10703–10714
12. Ferron M, Settembre C, Shimazu J, Lacombe J, Kato S, Rawlings DJ et al (2013) A RANKL–PKCβ–TFEB signaling cascade is necessary for lysosomal biogenesis in osteoclasts. Genes Dev 27(8):955–969
13. Settembre C, Zoncu R, Medina DL, Vetrini F, Erdin S, Erdin S et al (2012) A lysosome-to-nucleus signalling mechanism senses and regulates the lysosome via mTOR and TFEB. EMBO J 31(5):1095–1108

14. Zhang T, Chen X, Qu L, Wu J, Cui R, Zhao Y (2005) Chrysin and its phosphate ester inhibit cell proliferation and induce apoptosis in Hela cells. Bioorg Med Chem 12(23):6097–6105
15. Zhang H-Y, Sun H (2010) Up-regulation of Foxp3 inhibits cell proliferation, migration and invasion in epithelial ovarian cancer. Cancer Lett 287(1):91–97
16. Lv T, Yuan D, Miao X, Lv Y, Zhan P, Shen X et al (2012) Over-expression of LSD1 promotes proliferation, migration and invasion in non-small cell lung cancer. PLoS ONE 7(4): e35065
17. Mizushima N, Komatsu M (2011) Autophagy: renovation of cells and tissues. Cell 147(4):728–741
18. Chen Y, Liu H, Guan Y, Wang Q, Zhou F, Jie L et al (2015) The altered autophagy mediated by TFEB in animal and cell models of amyotrophic lateral sclerosis. Am J Transl Res 7(9):1574–1587
19. Pena-Llopis S, Vega-Rubin-de-Celis S, Schwartz JC, Wolff NC, Tran TAT, Zou L et al (2011) Regulation of TFEB and V-ATPases by mTORC1. EMBO J 30(16):3242–3258
20. Giatromanolaki A, Kalamida D, Sivridis E, Karagounis IV, Gatter KC, Harris AL et al (2015) Increased expression of transcription factor EB (TFEB) is associated with autophagy, migratory phenotype and poor prognosis in non-small cell lung cancer. Lung Cancer 90(1):98–105 (Amsterdam, Netherlands)

Research Progress of Squalene Synthase on Function and Application

Dengyue Sun, Qianqian Guo, Zhangliang Zhu, Songtao Li, Jian-Wen Wang, Yu-Fu Zhang, Lijun Guan, Hui-Min Qin and Fuping Lu

1 Introduction

Triterpene and sterols play an important role in organisms such as the fungi, animals and plants, and they have potential application value in pharmaceuticals industry [1], therefore synthesis of triterpene and sterols draw more and more researchers' attention [2]. Squalene (SQ) is a direct descendent of SQS, and SQS has a wide range of application value in our daily life [3, 4].

SQ is a key intermediate product in the synthesis of cholesterol, which is the composition of various organs and tissues, including cell membranes and liver fat [5–7]. In addition, SQ could be used as oil phase raw materials for cosmetics [8], which has function of potent free radical scavenger and antioxidant capability [9, 10]. SQ is effective in cancer treatment since it has the strong ability supplying

D. Sun · Q. Guo · Z. Zhu · S. Li · J.-W. Wang · Y.-F. Zhang · H.-M. Qin (✉) · F. Lu
College of Biotechnology, Tianjin University of Science and Technology,
Tianjin, China
e-mail: huiminqin@tust.edu.cn

H.-M. Qin · F. Lu (✉)
Key Laboratory of Industrial Fermentation Microbiology, Ministry of Education,
Tianjin, China
e-mail: lfp@tust.edu.cn

H.-M. Qin · F. Lu
Tianjin Key Laboratory of Industrial Microbiology, Tianjin, China

H.-M. Qin · F. Lu
National Engineering Laboratory for Industrial Enzymes, Tianjin 300457,
People's Republic of China

L. Guan
Food Processing Institute, Heilongjiang Academy of Agricultural Science,
368# Xuefu Road, Nangang District, Harbin 150086, China

H. Liu et al. (eds.), *Advances in Applied Biotechnology*, Lecture Notes
in Electrical Engineering 444, https://doi.org/10.1007/978-981-10-4801-2_78

oxygen in organisms [6, 11]. So, SQ has curative effect to many kinds of cancers, especially for the ovarian cancers and the lung cancers [12, 13]; One hand, SQ could promote the blood circulation and prevent high or low blood pressure caused by poor circulation, on the other hand, SQ lowers the risk of myocardial infarction and coronary heart disease [14]; Besides, SQ was also used for anti-infection cure as it possesses stronger antibacterial action on most parts of the bacterium [15].

2 The Structure and Function of SQS

2.1 The SQS Structure

The SQS has a transmembrane domain at its C-terminal with a quadruple spiral to anchor at cell membrane [16]. And it has a large N-terminal catalytic domain facing the cytosol. This orientation would enable SQS to accept water soluble farnesyl pyrophosphate (FPP) and NADPH from the cytosol, and release lipophilic squalene to the membrane of the endoplasmic reticulum, thus reflecting the enzyme's unique position in the cholesterol biosynthetic pathway [17]. Mammalian SQS include 416 amino acids with a molecular mass of 47 kDa. The crystal structure of the human SQS has been identified that the overall structure has a single domain, which is composed of an α-helical bundle fold. In hSQS, one side of the central cavity is buttressed by two mobile segments, the AB flap and the JK loop and helix αK, which are assumed to regulate the binding of substrates (prenyl donor and acceptor) and NADPH (Fig. 1) [18]. hSQS catalyzes the first step of cholesterol biosynthesis, in which two farnesyl pyrophosphate (FPP) molecules are condensed in a 'head-to-head' or '1′–1' coupling to form SQ. The catalytic process involves two half-reactions: firstly, two FPP molecules are condensed to form presqualene diphosphate (PSPP), an intermediate with a cyclopropane C1′–C2–C3 ring structure; secondly, PSPP undergoes a NADPH-dependent rearrangement and reduction to generate the end product SQ. The active site is lined to both sides by conserved aspartate and arginine residues. Those two residues are important for the two half-reactions, which are testified by mutagenesis experiments [19].

2.2 Biological Function of SQS

SQS catalyzes the first reaction of the branch of the isoprenoid metabolic pathway committed specifically to sterol biosynthesis [17]. And sterols serve as hormone precursors and secondary messengers play crucial role in developmental signalling [20]. SQS also would compete with several other enzymes to produce precursor of various terpenoid using FPP. The FPP conversion to the cholesterol will be inhibited when the SQS activity reduced, and that leads to increasing the generation

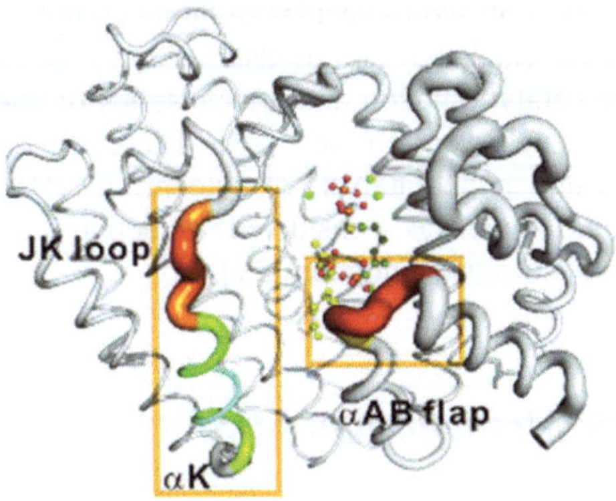

Fig. 1 The tertiary structure of the human squalene synthase, and overall structure of the hSQS–FSPP–Mg^{2+} complex [18]. The two molecules of FsSPP, S1 FSPP (*green*) and S2 FSPP (*yellow*), bound to the active site are shown as ball-and-stick models. The *two golden boxes* highlight the flexible regions in ligand binding. The *coloured ribbon* model is based on B factors, with *red* representing a high B factor and *deep blue* representing a lower B factor (the B factor stands for conformational states of protein crystals)

of non-steroidal including ubiquinone, dolichol, heme [21]. However, studies of the SQS gene in knock-out mice have shown that SQS-deficient was fatal influence on growth of central nervous system [22].

3 Synthesis and Catalytic Mechanism of SQS

3.1 The Metabolic Pathways Involved in Squalene Synthesis

SQS is a key enzyme in isoprenoid synthesis and it catalyzes two molecules of FPP into 30-carbon linear SQ [23]. The isoprenoid metabolic pathway is essential in the biosynthesis of plant sterols and triterpenoid saponins. A large amount of triterpenoid saponin and phytosterol were produced when SQS was overexpressed in *Panax ginseng*, and it indicated that SQS played an important role in the synthesis of phytosterol [24, 25]. In addition, the overexpression of SQS in the bupleurum (*Bupleurum falcatum*) would cause the increase of phytosterol and saikosaponin, meanwhile the squalene epoxidase and cycloartenol synthase mRNA expression also increased [26].

The synthetic pathway of steroidal and saponin [27] is as follow: (1) the mevalonic acid is converted into isoprene pyrophosphate (IPP) with the participation of dimethylallyl diphosphate (DMAPP). IPP is normally catalyzed by long geranyl pyrophosphate synthase to generate geranyl diphosphate (GPP); (2) GPP is catalyzed by famesyl diphosphate synthetase (FPPS), which react with IPP to produce famesyl diphosphate (FPP); (3) SQS is a bifunctional enzyme that catalyzes the condensation of two molecules of farnesyl diphosphate (FPP) to form the SQ, meanwhile SQ is catalyzed into oxidation SQ by the SQS, in the end, with oxidation SQ enzyme catalytic action and after a series of chemical reactions, the phytosterols and triterpenes were formed [28]. These triterpenes produce many types of ginsenosides under the catalysis of cytochrome P450 and glycosyltransferases catalytic action, which is considered as active ingredients of ginseng showing distinctive physiological function [29].

3.2 *The Catalytic Mechanism of SQS*

SQS catalyzes two molecules of FPP to form pre-squalene diphosphate, and then it is converted to SQ in a two-stage reaction [30, 31]. Initially, two molecules of FPP are condensed to form presqualene diphosphate (PSPP), Secondly; PSPP is rearranged and reduced by NADPH to form SQ. Research of yeast SQS enzyme activity in vitro showed PSPP would be transformed to SQ directly with binding to the active site when NADPH is present in the reaction buffer [23].

Farnesyl diphosphate (FPP) is formed in the process of generating SQ which is C15 allylic compound [22]. Its reaction is shown in Fig. 2 [22]. It is two FP that are connected together, forming a carbon bond in the position of the C-terminal leading to the reaction of biosynthetic terpenoid. The divalent cations are required to take part in reaction regularly so as to facilitate the FPP phosphate group [32].

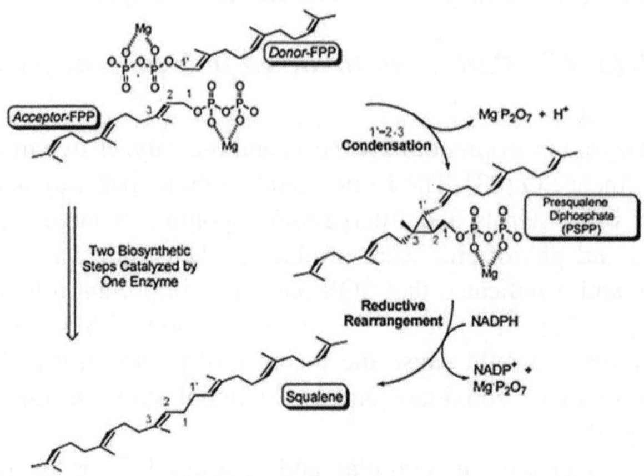

Fig. 2 The proposed reaction mechanisms for the two half-reactions catalyzed by SQS. The two molecules of FPP appear to bind sequentially with different affinities to two distinct regions of the enzyme, prior to the binding of NADPH, and prior to dissociation of the first diphosphate leaving group. The binding site for the donor FPP is more selective for FPP than is the acceptor FPP binding site. The PSPP binding site either overlaps with, or is the same as, the FPP binding sites [22]

In the first half reaction, two FPP molecules are combined to SQS by the CIS acting. One FPP molecule is got by pyrophosphate group cleavage when FPP molecules bind to different regions of the enzyme with different binding affinity [33]. By the catalytic reaction of SQS, a head-to-head condensation reaction (10-2, 3-linked) between two molecules of FPP forms a cyclopropylcarbinyl intermediate, presqualene diphosphate (PSPP) [34].

In the second step of SQS-catalyzed reaction, PSPP moves to the second reaction sites in SQS, which is located in the central channel, and would be exactly avoid water when reaction happen [35]. Then the cyclopropyl is opened, PSPP is rearranged, and the NADPH is reduced to a linear end product SQ, while the pyrophosphate, H^+, and $NADP^+$ is generated, then SQ is released into the endoplasmic reticulumomentum [23].

3.3 *The Sequence Analysis of Genes Encoding SQS*

Genes encoding SQS have been identified from a variety of eukaryotic sources, the fungi, animals and plants included. We conducted a database search for proteins from fungi, animals, and plants, to biochemically characterized SQS and found hits in two DXXED motifs as the substrate-binding sites and residues involved in NADPH recognition [36].

Multiple sequence alignment of SQS proteins indicated (Fig. 3, I–IV) that there existed four highly conserved regions among these organisms (animals, plants and fungi). Regions I, II, and III are related to the first half-reaction, the condensation of two molecules of FPP to produce PSPP. The binding of the diphosphate units in FPP via bridging Mg^{2+} is likely involved in regions I and III [35, 37]. While the region IV is considered to be required for the rearrangement of PSPP to SQ. The amino acid sequence of animals and fungi shares highest similarity in regions I and II and drops considerably in regions III and IV, which is a little high compared with that from plants [38].

Meanwhile the AB-flap (SRSF) was also found, which were thought to regulate the binding of NADPH, respectively. The distinctive domains distributed between the conserved domains were also shown by the amino acid sequence alignment [36].

4 The Research Progress of SQS

4.1 *Applications of SQS/SQ in Biosynthetic Pathways*

The first committed steps of steroid/hopanoid pathways involve SQS that are constitute two of the major sub-groups of isoprenoids, and it is the biosynthesis of squalene from two farnesyl diphosphates (FPPs), thus, some synthetic biology

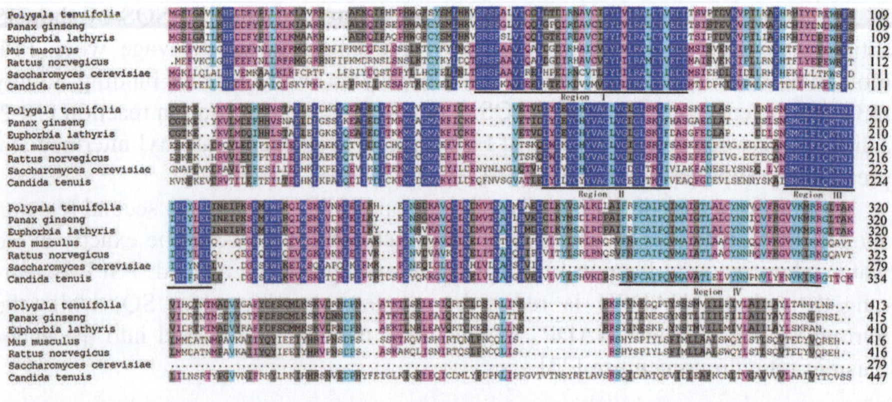

Fig. 3 Amino acid sequence alignments for representative SQS proteins from *Polygala tenuifolia*, *Panax ginseng*, *Euphorbia lathyris*, *Mus musculus*, *Rattus norvegicus*, *Saccharomyces cerevisiae*, *Candida tenuis* [36]. The ClustalW algorithm was used to generate the alignment (www.ebi.ac.uk/Tools/clustalw/index.html). *Dashes* indicate gaps that were introduced to maximize the alignment. The four conserved regions (I–IV) are indicated below the sequences

products can use this pathway to realize overexpression. Since these reactions in pathway are all placed at the branching position, the precursors enter pathway are determined by its enzyme activities. Once some special pathway formed, SQ could be exclusively used.

Here, some experiments showed that SQS could synthesize detectable amounts of dehydrosqualene in the absence of NADPH [23, 31]. Therefore, SQS could generate the carotenoid dehydrosqualene, and this could be a theoretically possible route for carotenoid biosynthesis [39]. It is showed the carotenoid pigments could be synthesized mainly via the desaturation of SQ rather than the direct synthesis of dehydrosqualene which had a low production. This also revealed the possible existence of a "SQ route" for C30 carotenoids, using SQS in carotenoid biosynthetic pathways [39].

In addition, SQS could control sterol biosynthesis that has been well established using SQS mutants in animals [40] and yeast [41]. The upstream part of the withanolide biosynthesis pathway is about isoprenoid biosynthesis and that diverges into sterol and triterpene biosynthesis. In consequence, to enhance withanolide production which is held in high repute in traditional Indian medicine can make overexpression of SQ in *Withania somnifera* [42].

Gene overexpression strategies are commonly employed for the upregulation of metabolic pathways in engineered hosts [43]. It suggests that the enhanced coproduction of SQ, ergosterol, oleic acid and biomass in yeast due to the overexpression of genes in the squalene biosynthetic (SB) pathway by newly characterized constitutive promoter. Hence, the regulation of complementary SB pathways to achieve a large-scale SQ production can be a useful strategy for engineering of the terpenoid pathway in *Saccharomyces cerevisiae co-overproduces* SQ [44]. According to overexpression genes of the SB pathway, production of SQ could increase 10-fold in the *M1EG* strain of *Saccharomyces cerevisiae* compared to the wild-type strain [45].

The research of SQS/SQ in biosynthetic pathways is essential and also helpful to realize large-scale production.

4.2 Potential Applications of SQS/SQ in Human Health

SQ was extensively studied in medical and pharmaceutical research for its chemical effectiveness to inhibit chemically induced colon, lung, and skin cancers; possess bactericidal and antifungal properties; significantly increases cellular immune and reduce total cholesterol of plasma levels [46].

SQS has been investigated as a key enzyme, because it plays an important role in the biosynthesis of SQ, which could be as an intermediate for the production of sterols, hopanoids or other triterpenoids. SQS catalyzes the first committed step of the sterol branch of the cholesterol biosynthesis pathway, which could regulate target of cholesterol level. Studies have shown that the increase of SQS expression constitutes an important approach to improve the level of cholesterol in mice [22]. Therefore, researchers are interest in SQS function of inhibiting high cholesterol and coronary heart disease (CHD) [32]. But others believe that the variation of the enzyme may be associated with hypercholesterolemia genetic disorders [47].

The hybrid human-yeast SQS was expressed by genetic engineering, which can be as a tool for anti-cholesterol drug assessment [48]. HMG-CoA reductase inhibitors are safe and effective in reducing cardiovascular disease risk, and also can be tolerated well for the majority of patients [49], but some of SQS failed to pass clinical phase I/II trials because of their hepatoxicity [50]. However, SQS is still regarded as a potential for therapeutic molecules that could decrease cholesterol level without affecting other isoprenoids [48]. In summary, SQS inhibitors have shown great promise and potential as cholesterol-lowering agents in human health, and both as alternatives to statins.

4.3 Research Achievements of SQS

The fungal SQS was the first active enzyme to be purified from yeast [16], and then Rat liver active SQS is also purified from mammals, which has a transmembrane structure binding in the endoplasmic reticulum [51, 52]. And using similar research method of mouse liver, human liver SQS gene was cloned [53]. And then SQS genes were researched widely. SQS genes have been cloned from a variety of organisms including animals (such as, human, mouse, *Trypomosoma cruzi*, etc.) [54, 55], plants (*Euphorbia tirucalli*, *Taxus chinensis*, *Arabidopsis thaliana*, etc.) [30, 56, 57], fungi (*Ganoderma lucidum*, yeast, etc.) [58, 59] and *Thermosynechococcus elongates* [60].

Through the study of SQS in these organisms, the results revealed that all the SQS had a transmembrane region in the enzyme C terminal anchored the cell membrane by a short helix, and its N terminal is located in the cytoplasm. This structure of SQS prevents the researchers from purifying the soluble wild SQS. Compared to the eukaryotic SQS, investigations of prokaryotic SQS are very limited and only one example from was reported. Previous studies have shown that the full length of SQS gene was cloned and expressed in *Thermosynechococcus elongatus*. The bacterial SQS protein structure analysis showed that was not similar to the eukaryotic SQS protein for it lacked the C-terminal transmembrane heli [60].

The studies found that the SQS catalytic activity was not affected in vitro when the amino acids of the C-terminal transmembr were knocked out. Some scholars confirmed that the purification of the *Trypanosoma cruzi* SQS still had catalytic activity when the N-terminal and C-terminal amino acids were knockout [61, 62]. Human squalene synthetase crystal has been crystallized, composed mostly of spiral Q. Later, on the basis of the human SQS structure, the new identification of SQS was performed by researchers [57, 62].

5 Conclusions and Perspectives

Due to SQS crucial functions in vivo and significant pharmaceutical application, SQS studies have attracted increasing interest in recent years. The understanding of the reaction mechanism, metabolic pathways, the functional domains of the enzyme, and transcriptional regulation of the SQS gene is being actively explored and referenced. In future, SQS research is bound to get more exploration and will drive the successful development of a series of industry, and it will also contribute to the human health.

References

1. Silva AAS, Morais SM, Falcão MJC et al (2014) Activity of cycloartane-type triterpenes and sterols isolated from *Musa paradisiaca* fruit peel against *Leishmania infantum chagasi*. Phytomedicine 21(11):1419–1423
2. Lukić M, Lukić I, Krapac M et al (2013) Sterols and triterpene diols in olive oil as indicators of variety and degree of ripening. Food Chem 136(1):251–258
3. Batista I, Nunes ML (1992) Characterisation of shark liver oils. Fish Res 14(4):329–334
4. Hye JC, Taylor W, Timothy P et al (2013) Vibrational spectra and DFT calculations of squalene. J Mol Struct 1032:203–206
5. Harivardhan LR, Squalene CP (2009) A natural triterpene for use in disease management and therapy. J Adv Drug Delivery Rev 61(15):1412–1426
6. Kin S, Karadeniz F (2012) Biological importance and application of squalene and squalane. Adv Food Nutr Res 65:223–233
7. Pragst F, Auwärter V, Kiessling B et al (2004) Wipe-test and patch-test for alcohol misuse based on the concentration ratio of fatty acid ethyl esters and squalene C FAEE/C SQ in skin surface lipids. Forensic Sci Int 143(2):77–86

8. Rissmann R, Oudshoorn MH, Kocks E et al (2008) Lanolin-derived lipid mixtures mimic closely the lipid composition and organization of vernix caseosa lipids. Biochim Biophys Acta (BBA)-Biomembr 1778(10):2350–2360
9. Preez HD (2008) Squalene-antioxidant of the future. Nat Med 33:106–112
10. Kelly GS (1999) Squalene and its potential clinical uses. Altern Med Rev 4(1):30–36
11. Newmark HL (1997) Squalene, olive oil and cancer risk review and hypothesis. Cancer Epidemol 889(1):193–203
12. Chinthalapally VR, Newmark HL, Bandaru SR (1998) Chemopreventive effect of squalene on colon cancer. Carcinogenesis 19(2):287–290
13. Desai KN, Wei H, Lamartiniere CA (1996) The preventive and therapeutic of the squalene-containing compound, Roidex on tumor promotion and regression. Cancer Lett 101(1):93–96
14. Nowicki R, Baraska-Rybak W (2007) Shark liver oil as a supporting therapy in atopic dermatitis. Polski Merkuriusz Lekarski 22(130):312–313
15. Lewkowicz N, Lewkowicz P, Kurnatowska A et al (2006) Biological action and clinical application of shark liver oil. Polski merkuriusz lekarski: organ Polskiego Towarzystwa Lekarskiego 20(119):598–601
16. Jennings SM, Tsay YH, Fisch TM et al (1991) Molecular cloning and characterization of the yeast gene for squalene synthetase. Proc Natl Acad Sci U S A 88(16):6038–6042
17. Tansey TR, Shechter I (2000) Structure and regulation of mammalian squalene synthase. Biochim Biophys Acta 1529(23):49–62
18. Liu CI, Jeng WY, Chang WJ et al (2014) Structural insights into the catalytic mechanism of human squalene synthase. Acta Crystallogr A 70(2):231–241
19. Gu P, Ishii Y, Spencer TA et al (1998) Function-structure studies and identification of three enzyme domains involved in the catalytic activity in rat hepatic squalene synthase. J Biol Chem 273(16):12515–12525
20. Kourounakis AP, Katselou MG, Matralis AN et al (2011) Squalene synthase inhibitors: an update on the search for new antihyperlipidemic and antiatherosclerotic agents. Curt Med Chem 18(2):4418–4439
21. Paradise EM, Kirby J, Chan R et al (2008) Redirection of flux through the FPP branch-point in *Saccharomyces cerevisiae* by down-regulating squalene synthase. Biotechnol Bioeng 100 (5):371–378
22. Okazaki H, Tazoe F, Okazaki S et al (2006) Increased cholesterol biosynthesis and hypercholesterolemia in mice overexpressing squalene synthase in the liver. J Lipid Res 47 (3):1950–1958
23. Jarstfer MB, Zhang DL, Poulter CD (2002) Recombinant squalene synthase. Synthesis of non-head-to-tail isoprenoids in the absence of NADPH. J Am Chem Soc 124(11):8834–8845
24. Lee MH, Jeong JH, Seo JW et al (2004) Enhanced triterpene and phytosterol biosynthesis in *Panax ginseng* overexpressing squalene synthase gene. Plant Cell Physiol 45(3):976–984
25. Seo JW, Jeong JH, Shin CG et al (2005) Overexpression of squalene synthase in *Eleutherococcus senticosus* increases phytosterol and triterpene accumulation. Phytochemistry 66(5):869–877
26. Kim YS, Cho JH, Park S et al (2011) Gene regulation patterns in tfiterpene biosynthetic pathway driven by overexpression of squalene synthase and methyl jasmonate elicitation in *Bupleurum falcatum*. Planta 233(13):343–355
27. Yang Y, Laval S, Yu B (2014) Chemical synthesis of saponins, Chap 2. Adv Carbohydrate Chem Biochem 71:137–226
28. Zhao R-Y, Xiao W, Cheng H-L et al (2010) Cloning and characterization of squalene synthase gene from *Fusarium fujikuroi* (Saw.) Wr. J Ind Microbiol Biotechnol 37(2):171–1182
29. Kim YJ, Zhang D, Yang DC (2015) Biosynthesis and biotechnological production of ginsenosides. Biotechnol Adv 33(6):717–735
30. Nakashima T, Inoue T, Oka A et al (1995) Cloning, expression, and characterization of cDNAs encoding Arabidopsis thaliana squalene synthase. Proc Natl Acad Sci U S A 92(7): 2328–2332

31. Blagg BS, Jarstfer MB, Rogers DH, Poulter CD (2002) Recombinant squalene synthase. A mechanism for the rearrangement of presqualene diphosphate to squalene. J Am Chem Soc 124(30):8846–8853
32. Davidson MH (2007) Squalene synthase inhibition: a novel target for the management of dyslipidemia. Curt Atheroscler Rep 9(2):78–80
33. Mookhtiar KA, Kalinowski SS, Zhang D et al (1994) Yeast squalene synthase. A mechanism for addition of substrates and activation by NADPH. J Biol Chem 269(21):1201–1207
34. Lin FY, Liu CL, Liu YL et al (2010) Mechanism of action and inhibition of dehydrosqualene synthase. Proc Natl Acad Sci U S A 107(11):21337–21342
35. Pandit J, Danley DE, Schulte GK et al (2000) Crystal structure of human squalene synthase. A key enzyme in cholesterol biosynthesis. J Biol Chem 275(8):30610–30617
36. Huang D, Yao Y, Zhang H et al (2015) Directed optimization of a newly identified squalene synthase from *Mortierella alpine* based on sequence truncation and site directed mutagenesis. J Ind Microbiol Biotechnol 42:1341–1352
37. Pandit J, Danley DE, Schulte GK et al (2000) Crystal structure of human squalene synthase. A key enzyme in cholesterol biosynthesis. J Biol Chem 275:30610–30617
38. Lee Sungwon, Dale Poulter C (2008) Cloning, solubilization, and characterization of squalene synthase from *Thermosynechococcus elongatus* BP-1. J Bacteriol 190(11):3808–3816
39. Furubayashi M, Li L, Katabami A et al (2014) Construction of carotenoid biosynthetic pathways using squalene synthase. FEBS Lett 588:436–442
40. Tozawa R, Ishibashi S, Osuga J et al (1999) Embryonic lethality and defective neural tube closure in mice lacking squalene synthase. J Biol Chem 274:30843–30848
41. Karst F, Lacroute F (1977) Ergosterol biosynthesis in *Saccharomyces cerevisiae* mutants deficient in early steps of pathway. Mol Gen Genet 154:269–277
42. Grover A, Samuel G, Bisaria VS et al (2013) Enhanced withanolide production by overexpression of squalene synthase in *Withania somnifera*. J Biosci Bioeng 115(6):680–685
43. Paddon CJ, Westfall PJ, Pitera DJ, Benjamin K et al (2013) High-level semi-synthetic production of the potent antimalarial artemisinin. Nature 496:528–532
44. Rasool A, Zhang G, Li Z et al (2016) Engineering of the terpenoid pathway in *Saccharomyces cerevisiae* co-overproduces squalene and the non-terpenoid compound oleic acid. Chem Eng Sci 152:457–456
45. Rasool A, Ahmed MS, Li C (2016) Overproduction of squalene synergistically downregulates ethanol production in *Saccharomyces cerevisiae*. Chem Eng Sci 152:370–380
46. Kim SK, Karadeniz F (2012) Biological importance and applications of squalene and squalane. Adv Food Nutr 65:223–233
47. Do R, Kiss RS, Gaudet D et al (2009) Squalene synthase: a critical enzyme in the cholesterol biosynthesis pathway. Clin Genet 75(5):19–29
48. Goral WM, Wysocka-Kapcinska M et al (2016) Genetic engineering and molecular characterization of yeast strain expressing hybrid human-yeast squalene synthase as a tool for anti-cholesterol drug assessment. J Appl Microbiol 120:877–888
49. Armitage J (2007) The safety of statins in clinical practice. Lancet 370:1781–1790
50. Liao JK (2011) Squalene synthase inhibitor lapaquistat acetate: could anything be better than statins. Circulation 123:1925–1928
51. Shechter I, Klinger E, Rucker ML et al (1992) Solubilization, purification, and characterization of a truncated form of rat hepatic squalene synthetase. J Biol Chem 267(21): 8628–8635
52. Stamellos KD, Shackelford JE, Shechter I et al (1993) Subcellular localization of squalene synthase in rat hepatic cells. Biochemical and immunochemical evidence. J Biol Chem 268 (11):12825–12836
53. Jiang G, McKenzie TL, Conrad DG et al (1993) Transcriptional regulation by lovastatin and 25-hydroxycholesterol in HepG2 cells and molecular cloning and expression of the cDNA for the human hepatic squalene synthase. J Biol Chem 268(13):12818–12824

54. Summers C, Kant F, Charles AD (1993) Cloning, expression and characterisation of the cDNA encoding human hepatic squalene synthase, and its relationship to phytoene synthase. Gene 136(9):185–192
55. Soltis DA, McMahon G, Caplan SL et al (1995) Expression, purification, and characterization of the human squalene synthase: use of yeast and baculoviral systems. Arch Biochem Biophys 316(26):713–723
56. Uchida H, Yamashita H, Kajikawa M et al (2009) Cloning and characterization of a squalene synthase gene from a petroleum plant. *Euphorbia tirucalli* L. Planta 229(19):1243–1252
57. Huang Z, Jiang K, Pi Y et al (2009) Molecular cloning and characterization of the yew gene encoding squalene synthase from Taxus cuspidate. J Biochem Mol Biol 40(2): 625–635
58. Zhao MW, Liang WQ, Zhang DB et al (2007) Cloning and characterization of squalene synthase (SQS) gene from *Ganoderma lucidum*. J Microbiol Biotechnol 17(2):1106–1112
59. Lograsso PV, Soltis DA, Boettcher BR (1993) Overexpression, purification, and kinetic characterization of a carboxyl-terminal-truncated yeast squalene synthetase. Arch Biochem Biophys 307(12):193–199
60. Lee S, Pouiter CD (2008) Cloning, solubilization, and characterization of squalene synthase from *Thermosynechococcus elongatus* BP-1. J Bacteriol 190(15):3808–3816
61. Sealey-Cardona M, Cammerer S, Jones S et al (2007) Kinetic characterization of squalene synthase from *Trypanosoma cruzi*: selective Inhibition by quinuclidine derivatives. Antimicrob Agents Chemother 51(3):2123–2129
62. Shang N, Li Q, Ko T-P et al (2014) Squalene synthase as a target for Chagas disease therapeutics. PLoS Pathogen 10(5):1178–1191

Review in Metabolic Modulation of Higher Alcohols in Top-Fermenting Yeast

Zhongguan Sun and Dongguang Xiao

1 Introduction

Wheat beer is a Bavarian and Austrian specialty beer, which is produced by wheat malt, non-germinated malt occasionally, as the major raw material, usually requires the amount of wheat malt accounts for at least 50% of the total raw material internationally [1]. All the traditional wheat beer that is produced by top fermentation has pure white exquisite froth, mellow and smooth taste, slight bitterness, high calorie and nutritious [2–4].

In 80 s last century, China's Beer Enterprises has imported wheat beer, but the awareness of this beer in the consumers is not high for many years. With the current consumer demand becoming increasingly diversified, differentiation, personalized, and in recent years, the shortage of global barley resources leading to rise in price, all beer enterprises actively adjust the product structure, by the way of developing high-quality, diversified and personalized products. Wheat malt is favored by the brewer, with its excellent brewing characteristics [5, 6]. Among them, Weiss and Bitter are the two most famous top-fermented beer styles. The Weiss style is characterized by a very fruity aroma (usually banana) and a strong phenolic (usually clove) content; on the other hand, the Bitter beer has a malt aroma and low-moderate fruity and hop aroma [3].

One of the main characteristic of fermentation and beer flavor maturation is the adjustment of the concentration of all volatile compounds. Among them, fusel or higher alcohols are the major abundant organoleptic compounds in wheat beer, contributing along with the related acetate esters to the overall flavor of beer.

Z. Sun · D. Xiao (✉)
Key Laboratory of Industrial Fermentation Microbiology,
Ministry of Education, Tianjin Industrial Microbiology
Key Laboratory, College of Biotechnology, Tianjin University
of Science and Technology, 300457 Tianjin, People's Republic of China
e-mail: xiao99@tust.edu.en

H. Liu et al. (eds.), *Advances in Applied Biotechnology*, Lecture Notes
in Electrical Engineering 444, https://doi.org/10.1007/978-981-10-4801-2_79

However, due to higher concentration of higher alcohols, drinking wheat beer will cause thirst, headache and other symptoms, and is harmful to the health of the drinkers, which seriously affects the development of wheat beer.

2 Effect of Higher Alcohols on Beer Flavor

Through the whole brewing process, fermenting cultures of the yeast produce many low-molecular-weight flavor compounds, along with ethanol and carbon dioxide. These higher alcohols, esters, aldehydes, phenols, organic acids, organic sulfides and other carbonyl substances have a strong impact on product quality [7, 8]. Higher alcohols have the function of giving the beer taste harmonious and full-bodied, but if the concentration of higher alcohols over the taste threshold, not only affect the quality of beer flavor, but also bring the body of the drinkers obvious side effects [9]. The optimum concentration of higher alcohols in 12.0°P beer, brewed by bottom fermentation, is 70–100 mg/L, as for beer with lighter taste and lower wort concentration, the concentration of higher alcohols should be even lower. Top fermented beer with wheat wort as the main raw material, because of its plumper aroma, is allowed a little higher concentration of higher alcohols [10]. However, the concentration of higher alcohols in wheat beer is as high as 300 mg/L or even higher, while generally below 100 mg/L in barley beer and about 50 mg/L in high-quality beer. Too much higher alcohols in wheat beer will cause obvious side effects after drinking, and it is one of the primary reasons for inhibiting the development of wheat beer. The major higher alcohols are listed in Table 1 with their characteristic flavor, taste threshold and fluctuation range of normal content in lager.

In the higher alcohol groups, it is reported that amyl alcohol is the most quantitatively significant aromatic compound. Active amyl and isoamyl alcohols are sometimes considered as one and represented simply as amyl alcohol. Active amyl

Table 1 List of beer flavors associated with the major higher alcohols [2, 7]

Higher alcohols	Flavor in beer	Organoleptic threshold (mg/L)	Range of normal content (mg/L)
n-propanol	Bitterness	25	5–15
Isobutyl alcohol	Alcohol	75	5–15
Isoamyl alcohol	Alcohol, banana, sweetish, aromatic	75	30–35
Active amyl alcohol	Alcohol, banana, edicinal, solvent	50	15–50
β-Phenyl ethanol	Roses, sweetish, perfumed	75	15–50
Total	–	–	70–100

alcohol is generally one-fifth to one-quarter of the total amyl alcohol, but it affects drinkability because beer flavor is considered heavier as amyl alcohol concentration increases. Isobutyl alcohol has an undesirable effect on beer quality when its concentration exceeds 20% of the total concentration of three alcohols: n-propanol, isobutyl, and amyl [7].

3 Effects of Fermentation Conditions on Higher Alcohols and the Expression of Related Genes

The higher alcohol in beer is produced by during brewing, including *n*-propanol, isobutyl alcohol, isoamyl alcohol, active amyl alcohol, β-phenyl ethanol and et al. In *Saccharomyces cerevisiae*, two metabolic pathways could be used to product higher alcohols. One is Harris pathway which is from glycolysis and another one is from amino acids called Ehrlich pathway [7, 8]. During the brewing process, the expression of yeast genes involved in higher alcohols metabolism are easily influenced by the fermentation conditions and the nature of raw materials, especially the brewing water, the types and addition rate of hops and the blend and types of malt [11–17], resulting in concentration and composition differences in higher alcohols and esters important for favor and taste.

The amino acid composition of wort affects the synthesis of higher alcohols in fermented beverages. Procopio et al. [18] using partial least square regression analyzed the impact of significant amino acids on higher alcohols during yeast fermentation and reached the conclusion that the most important explanatory variables affecting the synthesis of higher alcohols of *S. cerevisiae* were leucine, isoleucine, valine, histidine, glutamine and surprisingly proline. Lei et al. [14] found that wort gravity and nitrogen level had a significant effect on the accumulation of higher alcohols in the brewing process. High gravity resulted in lower expression levels of *ATF1*, *BAP2*, *BAT1*, *HSP12*, and *TDH*, whereas the higher nitrogen level caused higher expression levels for these genes. Furthermore, the lower nitrogen level resulted in increases in the levels of higher alcohols at high wort gravity.

Not only the blend and types of malt but also the fermentation conditions affect the accumulation of higher alcohols in brewing process. These parameters also affect many aspects of beer brewing, including yeast growth, carbon dioxide production, oxidation-reduction reaction, production kinetics, and other aspects. Marcia et al. [19] observed that a decrease in the concentration of higher alcohols occurred at the higher temperature during aging. Verbelen et al. [20] discovered that when the oxygen concentration reached to 8 mg/L, the expression level of *BAP2* gene and the accumulation of isobutyl alcohol and isoamyl alcohol all increased. Saerens et al. [21] studied the influence of wort gravity and fermentation temperature on the accumulation of the higher alcohols, the expression level of *BAT1* gene and the accumulation of higher alcohol all enhanced with the increasing of wort gravity and fermentation temperature.

Thus, the study on the effect of fermentation conditions on the expression of genes (enzyme) related to higher alcohols metabolism in top-fermenting yeast, and the ascertain of biological basis for the regulation of the key enzymes in the metabolism of higher alcohols, will be of great significance to reduce the accumulation of higher alcohols, and to improve the quality of wheat beer.

4 Directional Modification of Higher Alcohols Metabolic Pathway

The importance of aromatic yeast metabolites in the beer brewing has been a driving force for the screening of mutant strains in order to find those capable of producing compounds with desirable aromas. Higher alcohols are major determinants of the flavor of yeast fermented alcohols.

In recent years, there are many research reports about the modification of higher alcohols metabolism gene in *S. cerevisiae* [22–26]. Branched chain amino acids (BCAAs) are regarded as the key substrates in the formation of higher alcohols. The first step in the catabolic BCAA degradation is a transaminase step, catalyzed by a branched-chain amino acid transaminase (BCAAT). *Saccharomyces cerevisiae* possesses a mitochondrial and a cytosolic BCAAT, *Bat1*p and *Bat2*p, respectively [24]. Eden et al. [27] found out that the deletion of the *BAT2* gene encoding amino transferase had a great influence on the production of isobutyl alcohol and isoamyl alcohol.

It has been studied how yeast can convert branched chain amino acids to higher alcohols. Carbon skeletons are derived from the catabolism of BCAAs through the Ehrlich pathway as branched-chain oxoacids. And also, in *S. cerevisiae*, production of 2-ketoisovalerate (2-KIV), one of the branched chain oxoacids, from pyruvate occurs in the mitochondrial matrix by the consecutive actions of *ILV2* (ALS), *ILV5* (KARI), and *ILV3* (DHAD). 2-KIV is further degraded to isobutanol via several 2-keto acid decarboxylases and alcohol dehydrogenases in yeast. Conversion of 2-KIV to 2-ketoisocaproate (2-KIC) is catalyzed by three enzymatic steps involving Leu4p (2-isopropylmalate synthase), Leu1p (isopropylmalate isomerase), and Leu2p (3-isopropylmalate dehydrogenase), and 2-KIC can be either converted to leucine by *Bat2* or degraded to 3-methyl-1-butanol via Ehrlich pathway [28]. Chen et al. [29] came to the conclusion that overexpression of genes *ILV2*, *ILV3*, *ILV5*, and *BAT2* in valine metabolism led to an increase in isobutanol production in *S. cerevisiae* while additional overexpression of *ILV6* in the *ILV2*, *ILV3*, *ILV5* overexpression strain had a negative effect. Park et al. [28] obtained a mutant strain overexpressing *LEU2* and $LEU4^{D578Y}$ showed a 34-fold increase in 3-methyl-1-butanol synthesis compared with the original strain.

In our laboratory, Zhang et al. [30] constructed two engineered strains *S*5-2 and *S*5-4, which feature partial *BAT2* allelic genes replaced by the constructed *ATF1* overexpression cassette. The experiment results showed that the engineered strains *S*5-2 and *S*5-4, respectively, produced about 65 and 51% of higher alcohols produced by the parental strain. Wang Dan studied the effect of *BAT2* gene deletion on the production of higher alcohols in the top-fermenting yeast *S*-17, the research results indicated that the higher alcohols content of the *BAT2* gene deletion mutant was reduced by 34.08%, compared with the starting strain, the largest decline in higher alcohols was isobutanol, which decreased by 48.26%, and more importantly, the fermentation performance of the mutant was basically unchanged.

5 Outlook

Higher alcohols represent the most important aroma-active substances in wheat beer produced by top-fermenting yeast due to their essential aroma impressions. Great advance has been made in elucidating their biochemical pathways and their regulatory mechanisms. However, there are still many problems about the genetic modification of top-fermenting yeast strains. First of all, in the process of brewing wheat beer by using top-fermenting yeast, the effect of different fermentation factors on the accumulation of the higher alcohols is not clear enough. Secondly, the metabolic regulation mechanism of the higher alcohols in top-fermenting yeast is not yet unequivocal [31], which will cause us unable to achieve the desired results when we carry out genetic modification of industrial yeast strains in according with the standard brewing yeast gene map. In our laboratory, Hao Xin et al. [32] proposed to reduce the production of higher alcohols in industrial brewing yeast by knocking out a pyruvate decarboxylase-like gene *YDL080C*, however, the experimental results showed that there is no significant change in the production of higher alcohols. Thirdly, how to achieve the genetic regulation of higher alcohols content without affecting the production of other flavor compounds, especially their harmony, is still uncertain. Finally, the vast majority of the mutant strains are obtained by introducing exogenous genes, especially resistance genes, which will bring security risks to the mutants [33]. Because of these reasons, at present most of the research on the metabolism of *Saccharomyces cerevisiae* flavor substances remains in the experimental stage, and there are rare commercial production reports.

Acknowledgements This work was supported by the National Natural Science Foundation of China (31471724).

References

1. Langos D, Granvogl M, Schieberle P (2013) Characterization of the key aroma compounds in two bavarian wheat beers by means of the sensomics approach. J Agric Food Chem 61:11303–11311
2. Yu AH, Geng JW, Mi XT et al (2011) Research progress on higher alcohols control in top-fermentation wheat beer production. China Brewing 1:21–25

3. Liguori L, De Francesco G, Russo P et al (2015) Quality attributes of low-alcohol top-fermented beers produced by membrane contactor. Food Bioprocess Technol 9:191–200
4. Cui YQ, Wang AP, Zhang Z et al (2015) Enhancing the levels of 4-vinylguaiacol and 4-vinylphenol in pilot-scale top-fermented wheat beers by response surface methodology. J Inst Brew 121:129–136
5. Filip D, Wouter G, Johan M et al (2001) Contribution of wheat and wheat protein fractions to the colloidal haze of wheat beers. J Am Soc Brew Chem 59:135–140
6. Phiarais BPN, Mauch A, Schehl BD et al (2010) Processing of a top fermented beer brewed from 100% buckwheat malt with sensory and analytical characterisation. J Inst Brew 116:265–274
7. Kobayashi M, Shimizu H, Shioya S (2008) Beer volatile compounds and their application to low-malt beer fermentation. J Biosci Bioeng 106:317–323
8. Hazelwood LA, Daran JM, Van Maris AJ et al (2008) The Ehrlich pathway for fusel alcohol production: a century of research on *Saccharomyces cerevisiae* metabolism. Appl Environ Microbiol 74:2259–2266
9. Kobayashi M, Nagahisa K, Shimizu H et al (2006) Simultaneous control of apparent extract and volatile compounds concentrations in low-malt beer fermentation. Appl Microbiol Biotechnol 73:549–558
10. Vanderhaegen B, Neven H, Coghe S et al (2003) Evolution of chemical and sensory properties during aging of top-fermented beer. J Agric Food Chem 51:6782–6790
11. Landaud S, Latrille E, Corrieu G (2001) Top pressure and temperature control the fusel alcohol/ester ratio through yeast growth in beer fermentation. J Inst Brew 107:107–117
12. Molina AM, Swiegers JH, Varela C et al (2007) Influence of wine fermentation temperature on the synthesis of yeast-derived volatile aroma compounds. Appl Microbiol Biotechnol 77:675–687
13. Sigler K, Matoulkova D, Dienstbier M et al (2009) Net effect of wort osmotic pressure on fermentation course, yeast vitality, beer flavor, and haze. Appl Microbiol Biotechnol 82:1027–1035
14. Lei HJ, Zhao HF, Yu ZM et al (2012) Effects of wort gravity and nitrogen level on fermentation performance of brewer's yeast and the formation of flavor volatiles. Appl Biochem Biotechnol 166:1562–1574
15. Chen D, Vong WC, Liu SQ (2015) Effects of branched-chain amino acid addition on chemical constituents in lychee wine fermented with *Saccharomyces cerevisiae*. Int J Food Sci Technol 50:2519–2528
16. Rollero S, Bloem A, Camarasa C et al (2015) Combined effects of nutrients and temperature on the production of fermentative aromas by *Saccharomyces cerevisiae* during wine fermentation. Appl Microbiol Biotechnol 99:2291–2304
17. Procopio S, Sprung P, Becker T (2015) Effect of amino acid supply on the transcription of flavour-related genes and aroma compound production during lager yeast fermentation. LWT—Food Sci Technol 63:289–297
18. Procopio S, Krause D, Hofmann T et al (2013) Significant amino acids in aroma compound profiling during yeast fermentation analyzed by PLS regression. LWT—Food Sci Technol 51:423–432
19. Mascia I, Fadda C, Karabín M et al (2016) Aging of craft durum wheat beer fermented with sourdough yeasts. LWT—Food Sci Technol 65:487–494
20. Verbelen PJ, Saerens SM, Van Mulders SE et al (2009) The role of oxygen in yeast metabolism during high cell density brewery fermentations. Appl Microbiol Biotechnol 82:1143–1156
21. Saerens SM, Verbelen PJ, Vanbeneden N et al (2008) Monitoring the influence of high-gravity brewing and fermentation temperature on flavour formation by analysis of gene expression levels in brewing yeast. Appl Microbiol Biotechnol 80:1039–1051
22. Friden P, Schimmel P (1987) *LEU3* of *Saccharomyces cerevisiae* encodes a factor for control of RNA levels of a group of leucine-specific genes. Mol Cell Biol 7:2708–2717

23. Friden P, Schimmel P (1998) *LEU3* of *Saccharomyces cerevisiae* activates multiple genes for branched-chain amino acid biosynthesis by binding to a common decanucleotide core sequence. Mol Cell Biol 8:2690–2697
24. Schoondermark-Stolk SA, Tabernero M, Chapman J et al (2005) Bat2p is essential in Saccharomyces cerevisiae for fusel alcohol production on the non-fermentable carbon source ethanol. FEMS Yeast Res 5:757–766
25. Abe F, Horikoshi K (2005) Enhanced production of isoamyl alcohol and isoamyl acetate by ubiquitination-deficient *Saccharomyces cerevisiae* mutants. Cell Mol Biol Lett 10:383–388
26. Lilly M, Bauer FF, Styger G et al (2006) The effect of increased branched-chain amino acid transaminase activity in yeast on the production of higher alcohols and on the flavour profiles of wine and distillates. FEMS Yeast Res 6:726–743
27. Eden A, Van Nedervelde L, Drukker M et al (2001) Involvement of branched-chain amino acid aminotransferases in the production of fusel alcohols during fermentation in yeast. Appl Microbiol Biotechnol 55:296–300
28. Park SH, Kim S, Hahn JS (2014) Metabolic engineering of *Saccharomyces cerevisiae* for the production of isobutanol and 3-methyl-1-butanol. Appl Microbiol Biotechnol 98:9139–9147
29. Chen X, Nielsen KF, Borodina I et al (2011) Increased isobutanol production in *Saccharomyces cerevisiae* by overexpression of genes in valine metabolism. Biotechnol Biofuels 4:21
30. Zhang CY, Liu YL, Qi YN et al (2013) Increased esters and decreased higher alcohols production by engineered brewer's yeast strains. Eur Food Res Technol 236:1009–1014
31. Procopio S, Qian F, Becker T (2011) Function and regulation of yeast genes involved in higher alcohol and ester metabolism during beverage fermentation. Eur Food Res Technol 233:721–729
32. Hao X, Xiao DG, Zhang CY (2010) Effect of *YDL080C* gene deletion on higher alcohols production in *Saccharomyces cerevisiae* haploids. Acta Microbiologica Sinica 50:1030–1035
33. Zhang YY, Xiao DG, Zhang CY et al (2012) Construction and crossing trial of *BAT2* deletion mutants lacking antibiotic resistant gene in *Saccharomyces cerevisiae*. Food Ferment Ind 38:36–40

Research Progress of Aldehyde Ketone Reductase for Asymmetric Catalysis of Chiral Compounds

Songtao Li, Zhangliang Zhu, Jian-Wen Wang, Qianqian Guo, Panpan Xu, Dengyue Sun, Hui-Min Qin and Fuping Lu

1 Introduction

AKRs exist in nearly all phyla (Table 1), they are mainly monomeric soluble proteins (34–37 kDa) [1]. These enzymes reduce carbonyl substrates such as sugar aldehydes, keto-steroids, keto-prostaglandins, retinals, quinones, and lipid peroxidation by-products. AKR family members owns the form of an $(\alpha/\beta)_8$ TIM barrel fold consisting of a cylindrical core of eight parallel β-strands, forming the β-barrel, surrounded by eight α-helices running antiparallel to the strands [2] (Fig. 1). The carboxy ends of the β-strands are connected to amino ends of the α-helices by loops of varying lengths forming the active site. These loops are most flexible and variable in length, so that they enable the enzymes to accommodate substrates of varying shapes and sizes, and to control many molecular events such as catalytic action [3]. Sequence alignment in AKR family enzymes showed a conserved quadruplet, which is the key characteristic to the enzymes responsible for the

S. Li · Z. Zhu · J.-W. Wang · Q. Guo · P. Xu · D. Sun · H.-M. Qin (✉) · F. Lu (✉)
College of Biotechnology, Tianjin University of Science and Technology,
Tianjin, China
e-mail: huiminqin@tust.edu.cn

F. Lu
e-mail: lfp@tust.edu.cn

H.-M. Qin · F. Lu
Key Laboratory of Industrial Fermentation Microbiology, Ministry of Education,
Tianjin, China

H.-M. Qin · F. Lu
Tianjin Key Laboratory of Industrial Microbiology, Tianjin, China

H.-M. Qin · F. Lu
National Engineering Laboratory for Industrial Enzymes, Tianjin 300457,
People's Republic of China

H. Liu et al. (eds.), *Advances in Applied Biotechnology*, Lecture Notes
in Electrical Engineering 444, https://doi.org/10.1007/978-981-10-4801-2_80

Table 1 Members of the aldo-keto reductase superfamily

AKR		Enzyme	Species
AKR1	AKR1A1-AKR1A4	Aldehyde reductases	Homo sapiens, Sus scrofa et al.
	AKR1B1-AKR1B19	Aldose reductases	Homo sapiens, Rattus norvegicus et al.
	AKR1C1-AKR1C35	Hydroxysteroid dehydrogenases	Homo sapiens, Rattus norvegicus et al.
	AKR1D1-AKR1D3	Steroid 5b-reductases	Homo sapiens, Oryctolagus cuniculus et al.
AKR2	AKR2A1-AKR2A2	Manose	Apium graveolens et al.
	AKR2B1-AKR2B9	Xylose reductases	*Pichia stipitis*, Candida tenuis et al.
AKR3	AKR3A1-AKR3A3	Yeast AKRs	*Saccharomyces cerevisiae*, Escherichia coli et al.
AKR4	AKR4A1-AKR4A4	Chalcone	Medicago sativa et al.
	AKR4B1-AKR4B10	Codienone reductases	*Sesbania rostrata*, Oryza sativa et al.
AKR5	AKR5A1-AKR5A2	Gluconic acid reductases	Leishmania major et al.
AKR6	AKR6A1-AKR6A14	b-subunits of the potassium gated voltage channels	Bos taurus, Rattus norvegicus et al.
AKR7	AKR7A1-AKR7A5	Aflatoxin dialdehyde	Homo sapiens et al.
	AKR7B1	Succinic semialdehyde reductases	Homo sapiens
AKR8	AKR8A1-AKR8A2	Pyridoxal reductases	*Schizosaccharomyces pombe*
AKR9	AKR9A1-AKR9A3	Aryl alcohol dehydrogenases	*Aspergillus nidulans*, Haloferax volcanii et al.
AKR10	AKR10A1	Streptomyces AKRs	*Streptomyces glaucescens*, Streptomyces bluensis
AKR11	AKR11A1	Bacillus AKRs	*Bacillus subtilis*, E. coli et al.
AKR12	AKR12A1	Streptomyces sugar aldehyde reductases	*Streptomyces fradiae*, Streptomyces avermitilis et al.
AKR13	AKR13A1	Hyperthermophilic bacteria reductases	*Xylella fastidiosa*, Xylella fastidiosa et al.
AKR14	AKR14A1	*E. coli* reductases	*Escherichia coli*
AKR15	AKR15A1	Mycobacterium reductases	Microbacteriun luteolum
AKR16	AKR16A1	*V. cholera* reductases	*vibrio cholerae*

substrate catalytic amino acids. The quadruplet is composed of aspartic acid, tyrosine, lysine and histidine. AKRs catalyze an ordered bi-bi kinetic mechanism, in which NAD(P)H cofactor binds first and leaves last. Following cofactor binding, the steroid is bound and transformed through a "push–pull" mechanism. Therefore, AKRs catalytic enantioselective reduction reaction to show high levels of the chemical selectivity, antipodal selectivity and regioselectivity.

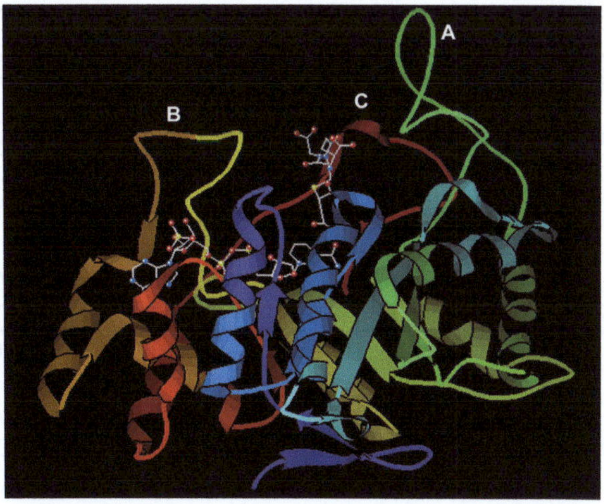

Fig. 1 Crytal structure of aldose reductase. The ribbon drawing is a side-view representation of the protein with NAD(P)H bound to the active site. The active-site loops are marked *A*, *B* and *C*. The ball-and-stick at loop B is the NAD(P)H molecule and the structure at loop C is the glutathione analog 1,2-dicarboxyethyl, glutathione, which binds to the substrate binding site of the enzyme

Now AKRs from a variety of microorganisms have been used in the synthesis of chiral compounds. Here, we reviewed the research progress about improving AKRs as industrial catalysts.

2 Recent Progress About AKRs for Designing Industrial Catalysts

2.1 Modified Aldehyde Ketone Reductase Protein

Usually, the wild-type AKR has poor catalytic efficiency. Therefore, it needs to be further modified to improve the catalytic properties of these enzymes. Studies have shown that AKRs have some important amino acids directly or indirectly affecting the process of enzyme catalysis [4]. Therefore, it is necessary to solve the crystal structure, and elucidate the activity center of the enzymes and the catalytic mechanism of AKR. It is also important to find the key amino acids related to the enzyme catalytic reaction to direct transformation enzyme protein or to create a new enzyme protein.

Widboom et al. [5] used the method of homologous modeling reported the first structure of an oxygenase class in complex with a bound substrate mimic. The results resolve the unique and complex chemistry of DpgC, a key enzyme in biosynthetic pathway of antibiotics. Furthermore, mechanistic parallels exist between DpgC and cofactor-dependent flavoenzymes, providing information regarding the general mechanism of enzymatic oxygen activation. Now, it has become the research focus. To simulate the structure of the enzyme targeting for the retional design, Gregory Zeikus et al. [6] developed a catalyst, which was able to produce 1-phenyl-2-propanol from phenylacetone using TeSADH as target enzyme.

They designed a catalytic site by point mutation W110A, which makes TeSADH active on phenylacetone, producing 1-phenyl-2-propanol. They also showed that W110A in TeSADH is active on benzylacetone, and it is specific for (S)-4-phenyl-2-butanol and (S)-1-phenyl-2-propanol. Stephanie Majkowicz et al. [7] changed the enzyme stereoselectivity and significantly improved the optical purity of the product through enzymes from yeast by site-directed mutation. Recently, some researches on aldehyde ketone reductases from the genus *Candida* by site-directed mutation have made a lot of progress. The wild type NADPH-dependent carbony1 reductase from *Candida magnoliae* could not utilize NADH as a coenzyme. Surprisingly, Souichi et al. [8] designed mutants, which are activities toward NADH-dependency and disability to utilize NADPH as a coenzyme. A short-chain carbonyl reductase (SCR) from *Candida parapsilosis* catalyzes an anti-prelog reduction of 2-hydroxyacetophenone to (S)-1-phenyl-1,2-ethanediol (PED), and exhibits coenzyme specificity for NADPH over NADH. Rongzhen Zhang et al. [9] designed the mutants with different combinations of Ser67Asp, His68Asp, and Pro69Asp substitutions inside or adjacent to the coenzyme binding pocket by site-directed mutagenesis. The S67D/H68D mutant produced (R)-PED with high optical purity and yield in the NADH-linked reaction. Huimin Qin et al. [10] found the enzyme, Thr25, Lys26 and Ser260 play a key role in combination with coenzyme through analyzing the crystal structure of the CPR-C1 derived from *Candida*, which providing the key information for further reforming the enzyme.

2.2 *Using Genome Database Mining for Development of New Aldehyde Ketone Reductase*

At present, the resources and application information about aldehyde ketone reductase are still limited. Genome database mining plays an important role in exploring and developing new stereoselectivity aldehyde ketone reductase, enriching the aldehyde ketone reductase resources. Furthermore, it has gradually developed into effective means as the rapid development of genomics, proteomics and bioinformatics, using existing data resources to exploit potential biological catalyst [11]. A variety of microorganisms have been developed as a catalytic asymmetric reduction enzyme donor in the field of medicine and industry, such as *C. parapsilosis*, *Mucor ambiguus*, and *Saccharomyces cererisiae*. *C. parapsilosis* shows excellent catalytic performance, in which the enzymes of conjugated polyketone reductases (CPR), CPR-C1 and CPR-C2 have highly stereoselectivity catalytic performance [12]. Therefore, the whole genome of *C. parapsilosis* can help to developing the new reductase.

As early as 1999, eighteen key reductases from baker's yeast have been overproduced in *Escherichia coli* and evaluated. It demonstrates that individual yeast reductases can be used to supply important chiral building blocks [13]. Guo et al.

[14] identified eight new type of aldehyde ketone reductase using genome information from *C. parapsilosis* to catalytize N,N-dimethyl-3-ketone-3-(2-thiophene)-1-propanamine (DKTP) for generation of (S)-N,N-dimethyl-3-hydroxy-3-(2-thiophene)-1-propanamine (DHTP). Genome mining can effectively develop new biocatalysts for pharmaceutical and fine chemical industry.

2.3 Research of Co-enzyme Regeneration

Aldehyde ketone reductases are redox enzyme that dependent on cofactors. It needs a certain amount of coenzyme NADPH as electron transfer for catalytic asymmetric reduction reaction process [15]. Therefore, it is necessary to add extra coenzyme in the catalytic reaction to maintain the reaction. However, the coenzyme NADPH is cost, and not suitable for the industrial utilization of chiral compounds. Therefore, it is important to build an efficient and economic system of co-enzyme regeneration cycle. In recent years, a series of methods, such as electrochemical, chemical, photochemical methods and enzymatic regeneration system, and enzymatic regeneration, have been developed in order to solve the problem of co-enzyme regeneration. Enzymatic regeneration has two methods: double enzymes of aldehyde ketone reductase and regeneration enzyme catalyzedcoupling reaction to build aldehyde ketone reductase and regeneration enzyme into the same expression vector (Fig. 2). Kataoka et al. [16] have applied ethanol, in place of glucose for coenzyme factor, to the catalytic reduction reaction of asymmetric carbonyl compounds, but the utilization rate of ethanol is extremely low. Furthermore, high concentration of ethanol is toxic to which cells. The researchers found that excessive glucose as the electron donor resulted in a high rate of alcoholic fermentation, oxygen consumption and biomass formation, and therefore causing low efficiency of glucose utilization. Controlling the supply of the electron donor of glucose at the highest rates applied prevented alcoholic fermentation but still resulted in biomass formation and a high oxygen requirement. Low supply rates of

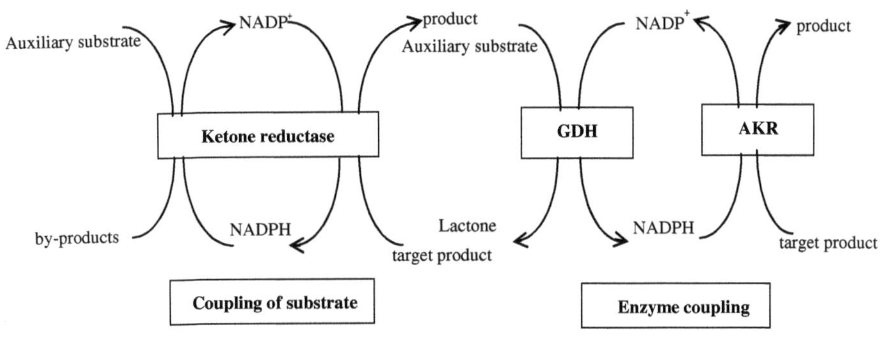

Fig. 2 Coenzyme regeneration process

ethanol resulted in biomass decrease while low supply rates of glucose provided the most efficient strategy for electron donor provision and yielded a high products. Therefore low supply rates of ethanol prevented by-product formation and biomass increase, and resulted in a low oxygen requirement [17]. In 1989, Makino et al. [18] used glucose dehydrogenase as coenzyme factor regeneration enzyme, because both NADH and NADPH can be generated by glucose oxidation of glucose dehydrogenase. All kinds of aldehyde ketone reductase can show the independent cofactor specificity. The sources of glucose dehydrogenase are extensive and the product of lactone can be quickly converted to other acids, which can be jointly with the aldehyde ketone reductase coenzyme regeneration cycle system for large-scale chiral synthesis. Yasohara et al. [19] constructed a recombinant bacterium with two genes of aldehyde ketone reductase and glucose dehydrogenase. These two enzymes coordinated expression of chiral product with 85% of production rate. Ni Yan et al. coexpressed the aldehyde ketone reductase gene from *Bacillus subtilis* and glucose dehydrogenase gene into the same plasmid for preparation of (R)-CHBE. The final yield was 91.7% [20].

3 Conclusion

Compared with chemical synthesis, biosynthesis of chiral compounds has the advantages of mild reaction condition, high selectivity and environmental protection. It has a broad development prospects. As the development of biotechnology, gene engineering and enzyme engineering, the aldehyde ketone reductase catalytic activity, stereoselectivity and tolerance will be improved significantly. At the same time, the molecular mechanics model and the computer simulation technology will become more matured. It will have a more profound understanding of enzyme reaction mechanisms guiding the rational design for asymmetric synthesis.

References

1. Penning TM (2015) The aldo-keto reductases (AKRs): overview. Chem Biol Interact 234:236–246
2. Ren D, Villeneuve NF, Jiang T et al (2011) Brusatol enhances the efficacy of chemotherapy by inhibiting the Nrf2-mediated defense mechanism. Proc Natl Acad Sci U S A 108(4): 1433–1438
3. Couture JF, Legrand P, Cantin L et al (2004) Loop relaxation, a mechanism that explains the reduced specificity of rabbit 20α-hydroxysteroid dehydrogenase, a member of the aldo-keto reductase superfamily. J Mol Biol 339:89–102
4. Sanli G, Blaber M (2001) Structural assembly of the active site in an aldo-keto reductase by NADPH cofactor. J Mol Biol 309(5):1209–1218
5. Widboom PF, Fielding EN, Liu Y et al (2007) Structural basis for cofactor-independent dioxygenation in vancomycin biosynthesis. Nature 447(7142):342–345

6. Ziegelmann-Fjeld KI, Musa MM, Phillips RS et al (2007) A *Thermoanaerobacter ethanolicus* secondary alcohol dehydrogenase mutant derivative highly active and stereoselective on phenylacetone and benzylacetone. Protein Eng Des Sel 20(2):47–55
7. Zhu D, Yang Y, Majkowicz S et al (2008) Inverting the enantioselectivity of a carbonyl reductase via substrate—enzyme docking-guided point mutation. Org Lett 10(4):525–528
8. Morikawa S, Nakai T, Yasohara Y et al (2005) Highly active mutants of carbonyl reductase s1 with inverted coenzyme specificity and production of optically active alcohols. Biosci Biotechnol Biochem 69(3):544–552
9. Zhang R, Xu Y, Sun Y et al (2009) Ser67Asp and His68Asp substitutions in *Candida parapsilosis* carbonyl reductase alter the coenzyme specificity and enantioselectivity of ketone reduction. Appl Environ Microbiol 75(7):2176–2183
10. Qin HM, Yamamura A, Miyakawa T et al (2013) Crystal structure of conjugated polyketone reductase (CPR-C1) from *Candida parapsilosis* IFO 0708 complexed with NADPH. Proteins 81:2059–2063
11. Fewer M, Martínez-Abarca F, Golyshin PN (2005) Mining genomes and metagenomes for novel catalysts. Curr Opin Biotechnol 16(6):588–593
12. Hidalgo A, Akond MA, Kita K et al (2001) Isolation and primary structural analysis of two conjugated polyketone reductases from *Candida parapsilosis*. Biosci Biotechnol Biochem 65(12):2785–2788
13. Kaluzna IA, Matsuda T, Sewell AK et al (2004) Systematic investigation of *Saccharomyces cerevisiae* enzymes catalyzing carbonyl reductions. J Am Chem Soc 126(40):12827–12832
14. Guo RY, Nie Y, Mu XQ et al (2014) Preparation of asymmetric transformation of (S)-N, N-dimethyl-3-hydroxyl-3-(2-thienyl)-1-propanamine(DHTP) by the novel AKR. Cheml Ind Eng Progr 33(9):2344–2349
15. Zhao H, Donk WA (2003) Regeneration of cofactors for use in biocatalysis. Curr Opin Biotechnol 4(6):583–589
16. Kataoka K, Kita M, Wada Y, Yasohara J et al (2003) Novel bioreduction system for the production of chiral alcohols. Appl Microbiol Biotechnol 62:437–445
17. Chin-Joe I, Straathof AJ, Pronk JT et al (2001) Influence of the ethanol and glucose supply rate on the rate and enantioselectivity of 3-oxo ester reduction by baker's yeast. Biotechnol Bioeng 75:29–38
18. Makino Y, Ding JY, Negoro S et al (1989) Purification and characterization of a new glucose dehydrogenase from vegetative cells of *Bacillus megaterimn*. J Ferment Bioeng 67:374–379
19. Kizaki N, Yasohara Y, Hasegawa J et al (2001) Synthesis of optically pure ethyl(S)-4-chloro-3-hydroxybutanoate by *Escherichia coli* transformant cells coexpressing the carbonyl reductase and glucose dehydrogenase genes. Microbiol Biotechnol 55(5):590–595
20. Ni Y, Li CX, Wang LJ et al (2011) Highly stereoselective reduction of prochiral ketones by a bacterial reductase coupled with cofactor regeneration. Org Biomol Chem 9(15):5463–5468

Population Structure and Genetic Diversity Analysis of Peanut (*Arachis hypogaea* L.) Using Molecular Markers

Xiu Rong Zhang, Feng Zhen Liu, Kun Zhang and Yong Shan Wan

1 Introduction

Peanut, one of the most important oilseed crops, is grown in more than 100 countries in the world [1, 2]. Cultivated peanut (*Arachis hypogaea* L.) is a highly selfing allotetraploid species (2n = 4x = 40, AABB), covering subspecies *fastigiata* and *hypogaea*. The ssp. *hypogaea* can be divided into two botanical types var. *hypogaea* and var. *hirsuta*, while ssp. *fastigiata* including four botanical types var. *fastigiata*, var. *vulgaris*, var. *peruviana* and var. *aequatoriana* [1, 3]. Furthermore, many breeding programs in china have developed a lot of varieties or lines with high yield and quality and strong resistance to environmental stress through cross breeding between subspecies or different types, known as the "irregular type" in the world [4]. Compared with other countries, peanut germplasm resources belonging to irregular type are more abundant in China [4].

Improvement of yield, quality and resistance in peanut is mostly rely on germplasm resources with distinct characters by breeding ways. However, varieties or lines, developed from breeding parents with narrow genetic basis, may be remarkably similar to each other in many aspects [5]. Thus, it is very difficult to make breakthrough in peanut breeding due to their closely related genetic relationship. Morphological characters were always used in artificial selection for millennia, as well as in selection of breeding parents. Most characters of economic importance, such as yield and quality, are quantitative traits, influenced by numerous genes that often have individually small effects throughout the genome [6]. These traits are also susceptible to environment, making breeding and selection

X.R. Zhang · F.Z. Liu (✉) · K. Zhang · Y.S. Wan (✉)
State Key Laboratory of Crop Biology, Shandong Key Laboratory of Crop Biology, Shandong Agricultural University, Tai'an 271018, China
e-mail: liufz@sdau.edu.cn

Y.S. Wan
e-mail: yswan@sdau.edu.cn

H. Liu et al. (eds.), *Advances in Applied Biotechnology*, Lecture Notes in Electrical Engineering 444, https://doi.org/10.1007/978-981-10-4801-2_81

more difficult. Molecular markers evaluate genetic variation on DNA level, providing reliable evidence for germplasm enhancement [1, 3, 7]. A peanut integrated consensus map was constructed by Shirasawa [8] in 2013, covering 2651 cM with 3693 marker loci which was anchored to 20 linkage groups corresponding to peanut genomes. The integrated map supplies valuable data resources to increase the genetic and genomic understanding and facilitate molecular breeding in peanut [8].

There are two most commonly used tools for dissecting complex quantitative traits in crops, linkage analysis and association mapping [9]. Linkage analysis has been always performed with experimental populations which are derived from a bi-parental cross, and individuals shared inheritance of polymorphisms within families of known ancestry [10]. Association mapping has been typically conducted using a collection of individuals often with unobserved ancestry, which is called association mapping population or natural population [9, 10]. When compared with linkage analysis, association mapping has many advantages including higher mapping resolutions, shorter research time and greater number of alleles investigated at the same time [10].

Abundant phenotypic variation and genetic diversity among individuals are necessary for association mapping. However, the confounding effects of population structure can lead to spurious results if not controlled during association analysis. We collected hundreds of peanut germplasm resources including preserved germplasm by our lab and cultivars (lines) supported by other breeding program in China. Most of the collections were from different areas in China, and the rest were originally from other countries. A number of phenotypic traits related to agronomy and quality of all varieties had been identified for many years. Based on phenotypic data, a collection of 367 peanut germplasm was developed, which represented the phenotypic variation of all materials. In the present study, genetic variation of the population was evaluated using a total of 101 SSR and transposable element (TE) markers from peanut integrated map [8]. The results from genetic diversity and population structure analysis will provide important information to understand the peanut genetic variation worldwide and broaden the genetic base of peanut varieties. In addition, it could be used for constructing association mapping population and doing further marker-trait association analysis.

2 Materials and Methods

2.1 Plant Materials

A total of 367 peanut germplasm resources were used as materials, and the number of resources for botanical type irregular type, var. *hypogaea*, var. *hirsuta*, var. *vulgaris,* var. *fastigiata* and var. *aequatoriana* is 109, 89, 34, 106, 26 and 3, respectively (Supplemental file, Table S1).

2.2 DNA Extraction and Genotyping with Markers

Young leaves of all peanut lines were sampled and put in liquid nitrogen immediately for DNA extraction. Total genomic DNA was extracted using the hexadecyl trimentyl ammonium bromide (CTAB) method as described in previous study [2]. The DNA quality and concentration of each samples were respectively estimated by 1% agarose gel and NanoDrop2000 spectrophotometer (Thermo, USA). Finally, the DNA concentration was normalized to about 50 ng μl^{-1} for polymerase chain reaction (PCR).

A set of 101 molecular markers from peanut integrated map [8] were selected in our research, distributed on to the 20 linkage groups (Supplemental file, Table S2). PCR amplification was performed in a total volume of 20 μl comprising of 2 μl genomic DNA as template, 10 × *Taq* PCR buffer, 0.25 μM of each forward and reverse primer, 0.2 mM dNTPs and 1.0 U *EasyTaq* polymerase. Touchdown PCR was carried out in ABI 2720 Cycler (ABI, USA) using the following protocol: pre-denaturation at 94 °C for 5 min, followed by denaturation at 94 °C for 45 s, annealing (from 60 to 50 °C, −0.5 °C for each cycle) for 45 s, and elongation at 72 °C for 45 s, and then 35 cycles of 94 °C for 45 s, 55 °C for 45 s, 72 °C for 45 s and post-elongation at 72 °C for 7 min.

The PCR products were size separated in non-denaturing polyacryl-amide gel electrophoresis (PAGE) using 8% (w/v) polyacryl-amide gel at 150 V for 2.5 h in 1 × TBE buffer using DYCZ-24B gel rigs (Beijing, China). Based on the expected product size, the size of the most intensely amplified bands around the expected product size for each marker was determined using identified concentrations of standard molecular markers after the bands were visualized by silver staining.

2.3 Data Analysis

Based on whether the unique polymorphic band is present or absent in each germplasm, allele score was denoted presence as 1 and absence as 0, respectively. A 1/0 matrix was constructed for 101 markers to be analyzed for genetic diversity and population structure. According to different requirements of analytic software, the genotype data was changed to different formats to perform analysis.

PowerMarker v3.25 [11] was used to calculate major allele frequency (MAF), gene diversity (GD), number of alleles per locus and polymorphism information content (PIC). A dendrogram was constructed by using unweighted pair group method with arithmetic mean (UPGMA) with 1000 replications of bootstrapping based on *Nei*'1983 genetic distance matrix. Population structure analysis was performed using the Bayesian model-based clustering method in program STRUCTURE v2.3.4 [12] using the admixture model, correlated allele frequency option. The program was run 5 times for each K (the number of subpopulations) value, ranging from 1 to 20, using a burn-in period of 10,000 and 100,000 Markov

Chain Monte Carlo (MCMC) replications during analysis. The program Structure Harvester [13] was used to determine the optimal K. Out of the 5 runs for K, the run with the highest log likelihood value, known as Ln P(D), was selected to assign the membership coefficients (Q) to each sample. When Q is 0.50 or higher, samples were clustered to one subpopulation, otherwise they were clustered to mixed group. Program NTSYSpc v2.20 [14] was used to perform principal component analysis (PCA) to identify amounts of variance, eigenvectors, and cumulatively explained variances per component using the *Dice* coefficient of similarity.

3 Results

3.1 Molecular Marker Polymorphism

In total, the 101 markers produced 503 polymorphic bands. The number of alleles per locus, ranging from 2 to 15 with an average of 4.98, varied widely among the markers (Supplemental file, Table S2). The major allele frequency (MAF) varied from 0.213 to 0.975 with an average of 0.618. The mean value of gene diversity (GD) was 0.495, ranging from 0.048 to 0.857. The polymorphism information content (PIC) for all the markers ranged from 0.047 to 0.845 with an average of 0.432 (Supplemental file, Table S2).

3.2 Genetic Diversity Analysis

Among all the peanut lines, the genetic distance ranged from 0.033 to 0.846 with an average of 0.470. The largest genetic distance 0.846 was between Qinghua505 from China and Ganyinxuan1003 from India. Shanhua15 and 98H101 showed the least genetic distance, as they were both selected from one breeding institute in Shandong, China. Overall 42% peanut lines had a genetic distance greater than 0.5. These results indicated that peanut germplasm resources were abundant in genetic variation.

The UPGMA (unweighted pair group method with arithmetic average) dendrogram was constructed from DNA marker data based on *Nei*'1983 genetic distance. Clustering analysis showed 367 peanut lines were obviously grouped into 2 branches (Fig. 1a), denoted as G1 and G2. G1 contained 119 peanut lines and grouped into G1a, G1b, G1c and G1d, while G2 contained 248 peanut lines and grouped into G2a, G2b, G2c and G2d (Fig. 1a). Interestingly, G1b contained 24 peanut lines, of which 3 were var. *aequatoriana* and 19 were var. *fastigiata*. G1c and G1d mainly contained var. *vulgaris*, and represented peanut varieties from northern and southern China, respectively. G2a contained 24 var. *hirsuta* and 47 var. *hypogaea*. G2b mainly contained var. *hypogaea*. G2d, the largest group,

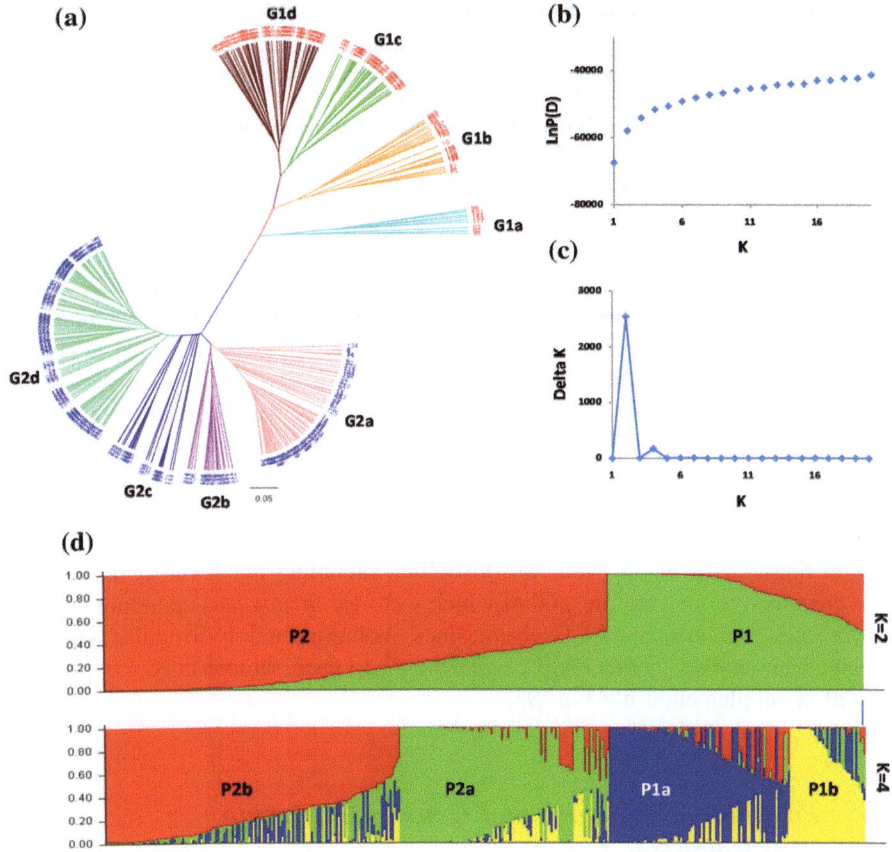

Fig. 1 The UPGMA clustering and population structure. **a** UPGMA dendrogram based on *Nei*'1983 genetic distance, **b** the changes of likelihood value Ln P(D) against K value, **c** the changes of Delta K against K value and **d** the distribution of sub-populations from population structure analysis at K = 2 and K = 4

contained 119 peanut lines, 83 of which were irregular type. Therefore, G1 mainly contained peanut lines from var. *fastigiata*, var. *vulgaris* and var. *aequatoriana*, while G2 mainly contained peanut lines from var. *hypogaea*, var. *hirsuta* and irregular type. G1 and G2 was in accordance with ssp. *fastigiata* and ssp. *hypogaea*, respectively. The detailed information of each peanut line, classified into each group (G), was presented in supplemental file Fig. S1.

3.3 Population Structure Analysis

When K was set from 1 to 20, the log likelihood score did not plateau at a single K value, instead, it continued to increase at relatively constant increments (Fig. 1b). In this case, Delta K score based on calculation from Structure Harvester [13] was used to find the optimal K. Figure 1c showed that a sharp peak of Delta K appeared at K = 2, followed by that at K = 4.

Population structure analysis showed 367 peanut lines were divided into 2 subpopulations at K = 2, designated as P1 and P2 respectively (Fig. 1d). Among the two groups, P1 contained 124 peanut lines, and P2 contained 243 peanut lines (the membership coefficients Q to each sample at K = 2 was presented in supplemental file Fig. S2). There are 77 var. *vulgaris*, 27 var. *fastigiata* and 3 var. *aequatoriana* in P1 group, while 100 irregular type, 84 var. *hypogaea* and 29 var. *hirsuta* in P2 group. P1 and P2 was in accordance with ssp. *fastigiata* and ssp. *hypogaea* as the same as G1 and G2. Furthermore, at K = 4, these peanut lines were classified into 4 groups, named as P1a, P1b, P2a and P2b. P1a contained 83 peanut lines, 69 of which were var. *vulgaris* mainly from China. P1b contained 34 peanut lines, 27 of which were var. *fastigiata*, including the 3 var. *aequatoriana*. P2a contained 95 peanut lines, 60 of which were var. *hypogaea*, including other 25 var. *hirsuta*. P2b contained 131 peanut lines, including mainly irregular type from China. The detailed membership coefficients Q to each sample at K = 4 was presented in supplemental file Fig. S3.

3.4 Population Structure and Genetic Clustering with Botanical Type

Among all the materials, there were six botanical varieties that were classified based on morphological characters. The distribution frequency of each botanical type within each group and the frequency of each group within each botanical type were showed in Fig. 2a, b, respectively. From Fig. 2a, ssp. *fastigiata*, including var. *fastigiata*, var. *vulgaris* and var. *aequatoriana*, accounted for 86.44% of G1 group, while 86.46% as to P1 group. The irregular type accounted for 41.22 and 37.36% of G2 and P2 group, and ssp. *hypogaea* (var. *hypogaea* and var. *hirsuta*) accounted for 45.71 and 53.85%, respectively. The distribution frequency of each group within each botanical type showed that 91.30% of the var. *hypogaea* and 82.35% of the var. *hirsuta* were grouped into G2, while 81.43% of the var. *hypogaea* and 75.59% of the var. *hirsuta* were grouped into P2. Most of the var. *vulgaris* and var. *fastigiata* were grouped into G1 (72.82, 82.76%) and P1 (56.33, 79.31%). The 3 var. *aequatoriana* were grouped into G1 and P1. In spite of discrepancies, the population structure and genetic clustering is obviously associated with the botanical types.

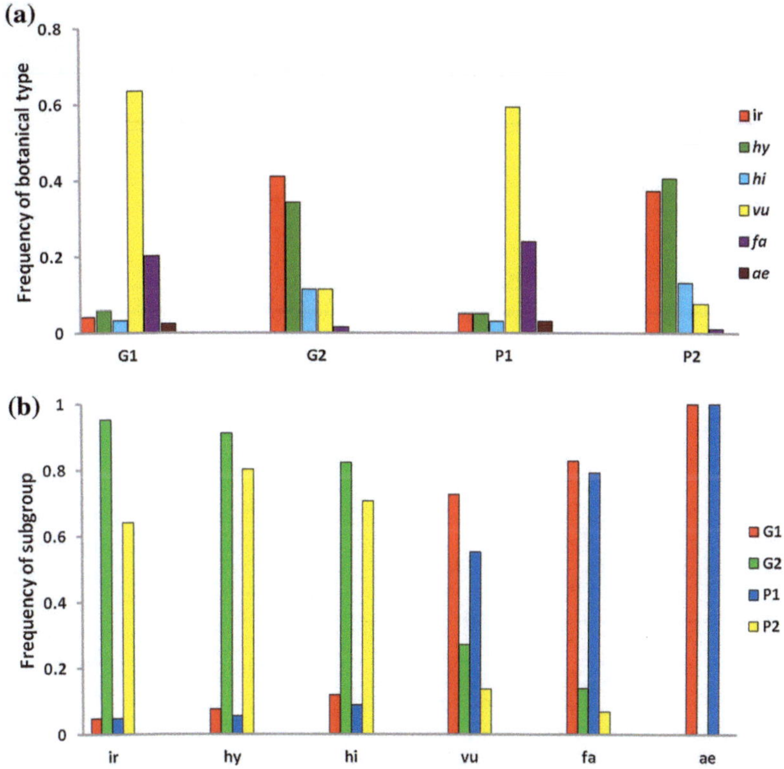

Fig. 2 The distribution frequency of each botanical type within each group (**a**) and the frequency of each group within each botanical type (**b**). *Note* the botanical types were abbreviated as *ir* for irregular type, *hy* for var. *hypogaea*, *hi* for var. *hirsuta*, *fa* for var. *fastigiata*, *vu* for var. *vulgaris* and *ae* for var. *aequatoriana*

3.5 *Principal Component Analysis*

Principal component analysis (PCA) was performed using *Dice*'s coefficient of similarity based on marker-genotype data. The eigenvectors of the first three components were 32.00, 14.06 and 4.91, accounting for 19.10, 8.39 and 2.93% of the total molecular variance, respectively. The first three components explained 30.43% of the total variation cumulatively. The two-dimensional scatter plots were plotted using the first two components to perform factorial analysis (Fig. 3).

Results from factorial analysis confirmed that the clustering was closely related to botanical type. The var. *fastigiata* and var. *aequatoriana* were mainly clustered on quadrant "I", in contrast most of the var. *vulgaris* were clustered on quadrant "IV", and the var. *hypogaea* and var. *hirsuta* were mainly clustered on quadrant "II" (Fig. 3a). Besides, almost all the irregular type were clustered on quadrant "III" (Fig. 3a). The 367 peanut lines were collected from different countries

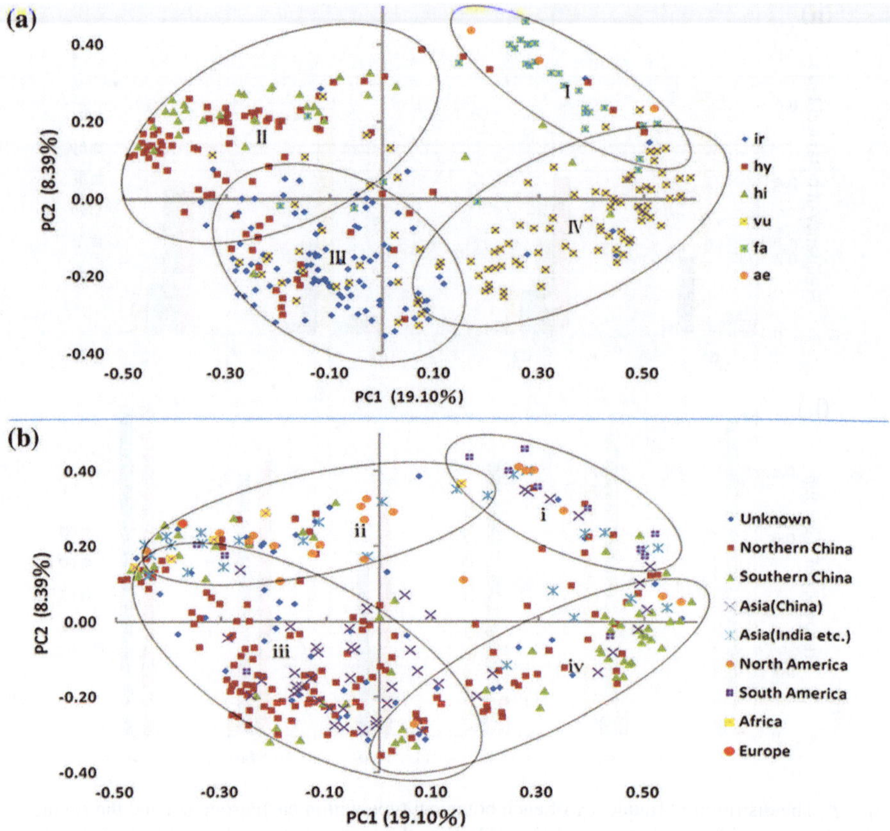

Fig. 3 Factorial analysis of the *peanut lines* based on markers. **a** Clustering related to botanical type, **b** clustering related to geographical origin. *Note* the botanical types were abbreviated as *ir* for irregular type, *hy* for var. *hypogaea*, *hi* for var. *hirsuta*, *fa* for var. *fastigiata*, *vu* for var. *vulgaris* and *ae* for var. *aequatoriana*. The botanical type and origin of each peanut materials were listed in Table S1

(Supplemental file, Table S1). Factorial analysis showed that the majority of peanut lines from South America were clustered on the quadrant "i", and peanut lines from North America, Africa and Europe were mostly clustered on the quadrant "ii" (Fig. 3b). The peanut lines from Asian countries such as China and India were widely distributed. Took Chinese peanut lines as an example, these varieties from northern China including Shandong, Henan and Hebei were mainly distributed on quadrant "iii", while these varieties from southern China including Guangdong, Guangxi and Fujian were mainly clustered on quadrant "iv" (Fig. 3b). Therefore, these results showed that the clustering was closely related to geographical origin.

4 Discussion and Conclusions

The cultivated peanut is divided into two subspecies, ssp. *fastigiata* and ssp. *hypogaea*, including different botanical types. There were six botanical types using in the present study. For one thing, the ssp. *fastigiata* (var. *fastigiata*, var. *vulgaris* and var. *aequatoriana*) accounted for more than 80% cumulatively of both G1 and P1 group, and factorial analysis confirmed the results as most of the ssp. f*astigiata* were clustered on quadrant "I" and "IV". Besides, over 80% of the var. *hypogaea* and var. *hirsuta* were grouped into G2 and P2 group, which was in accordance with factorial analysis that those peanut lines were clustered on quadrant "II". Those results were consistent with previous studies [1, 15].

There are obvious phenotypic and genetic differences among peanut lines from different areas due to geographical location, climate, natural and artificial selection [6]. Those germplasm from other countries were mostly clustered on quadrant "i" and "ii", while germplasm from China were mainly clustered on quadrant "iii" and "iv". Those results were also consistent with previous studies [1]. Among the tested materials, more than 70% of peanut lines were from China, including local varieties, market types and other breeding lines (Supplemental file, Table S1). Those varieties from northern and southern China were clustered into different groups based on UPGMA and factorial analysis, indicating that peanut varieties from different regions showed obvious ecological adaptability on genetic level [16, 17].

Compared with other crops, molecular marker techniques in peanut is developing slowly. There is a long way to go on molecular breeding, while conventional cross-breeding is still the main method at present. Breeding parents with obvious genetic background difference within a certain range are more likely to generate an elite variety in the cross-breeding program. Therefore, understanding of genetic relationship is critically important for crop improvement [18]. In the present study, analysis showed that 42% of peanut lines had a genetic distance greater than 0.5, indicating that variation were abundant on genetic level. Genetic distance between Chinese and some foreign peanut lines or between subspecies were larger than others. Thus, it is essential to choose breeding parents with larger genetic distance from different countries, subspecies or types to conduct cross-breeding programs in peanut.

The power of association mapping relies on a large population of individuals with abundant phenotypic and genetic variation. From our study, the 101 molecular markers produced 503 alleles, and the mean values of GD and PIC were 0.495 and 0.432, which were much higher than a previous study by Ren et al. [16] but similar to a recent study by Wang [1]. The results suggested that the 367 peanut lines were rich in genetic variation with a higher level of allelic diversity. However, there were still some varieties found to be extremely similar in both phenotypic and genetic characters. In order to perform marker-trait association analysis, the natural population maybe need to be further optimized. Furthermore, clustering results from UPGMA, population structure and PCA showed the peanut lines grouped into subpopulations based on molecular data, and suggested that population structure must be taken into consideration when conducting association analysis.

To summarize, genetic variation of a collection comprising 367 peanut lines were assessed by 101 molecular markers, indicating that abundant genetic diversity existed in the population. Results from population structure and genetic clustering analysis were closely related to botanical types, consistent with results from factorial analysis. The results will help us to understand genetic variation of peanut germplasm resources and can be used by breeding programs to broaden the genetic base of peanut varieties. In addition, it provided important information for constructing association mapping population and doing further marker-trait association analysis in peanut.

Acknowledgements We would like to thank Oil Crops Research Institute of the Chinese Academy of Agricultural Sciences, Hebei Agricultural University, Hebei Academy of Agriculture and Forestry Sciences, Henan Academy of Agricultural Sciences and Shandong Peanut Research Institute for providing valuable peanut germplasm resources. This research was supported by the earmarked fund for China Agriculture Research System (CARS-14), the Peanut Seed Industry Project in Shandong province of China, and the earmarked fund for Agriculture Research System in Shandong province of China (SDAIT-04-03).

References

1. Hui W, Khera P, Huang B, Mei Y, Katam R, Zhuang W, Harris-Shultz K, Moore KM, Culbreath AK, Zhang X (2016) Analysis of genetic diversity and population structure of peanut cultivars and breeding lines from China, India and the US using simple sequence repeat markers. J Integr Plant Biol 25(5):S149
2. Zhang XR, Wan Q, Liu FZ, Zhang K, Sun A, Luo B, Sun L, Wan YS (2015) Molecular analysis of the chloroplast Cu/Zn-SOD gene (*AhCSD2*) in peanut. Crop J 43(3):246–257
3. Huang L, He H, Chen W, Ren X, Chen Y, Zhou X, Xia Y, Wang X, Jiang X, Liao B, Jiang H (2015) Quantitative trait locus analysis of agronomic and quality-related traits in cultivated peanut (*Arachis hypogaea* L.). TAG Theor Appl Genet (Theoretische und angewandte Genetik) 128(6):1103–1115
4. Jiang H-F (2008) Peanut core collection established in China and compared with ICRISAT mini core collection. Acta Agron Sin 34(1)
5. Kochert G, Halward T, Branch WD, Simpson CE (1991) RFLP variability in peanut (*Arachis hypogaea* L.) cultivars and wild species. Theor Appl Genet 81(81):565–570
6. Lande R, Thompson R (1990) Efficiency of marker-assisted selection in the improvement of quantitative traits. Genetics 124(3):743–756
7. Moretzsohn MC, Gouvea EG, Inglis PW, Leal-Bertioli SC, Valls JF, Bertioli DJ (2013) A study of the relationships of cultivated peanut (*Arachis hypogaea*) and its most closely related wild species using intron sequences and microsatellite markers. Ann Bot 111(1):113–126
8. Shirasawa K, Bertioli DJ, Varshney RK, Moretzsohn MC, Leal-Bertioli SC, Thudi M, Pandey MK, Rami JF, Fonceka D, Gowda MV, Qin H, Guo B, Hong Y, Liang X, Hirakawa H, Tabata S, Isobe S (2013) Integrated consensus map of cultivated peanut and wild relatives reveals structures of the A and B genomes of Arachis and divergence of the legume genomes. DNA Res: Int J Rapid publ Rep Genes Genomes 20(2):173–184
9. Yu J, Buckler ES (2006) Genetic association mapping and genome organization of maize. Curr Opin Biotechnol 17(2):155–160

10. Yang X, Yan J, Shah T, Warburton ML, Li Q, Li L, Gao Y, Chai Y, Fu Z, Zhou Y, Xu S, Bai G, Meng Y, Zheng Y, Li J (2010) Genetic analysis and characterization of a new maize association mapping panel for quantitative trait loci dissection. TAG Theor Appl Genet (Theoretische und angewandte Genetik) 121(3):417–431
11. Liu K, Muse SV (2005) POWERMARKER: integrated analysis environment for genetic marker data. Bioinformatics 21(9):2128–2129
12. Hubisz MJ, Falush D, Stephens M, Pritchard JK (2009) Inferring weak population structure with the assistance of sample group information. Mol Ecol Resour 9(5):1322–1332
13. Earl DA, Vonholdt BM (2012) STRUCTURE HARVESTER: a website and program for visualizing STRUCTURE output and implementing the Evanno method. Conserv Genet Resour 4(2):1–3
14. Rohlf FJ (2008) NTSYSpc: numerical taxonomy system, Version 2.20. Exeter Publishing, Ltd, Seatauker, NY
15. Wang ML, Sukumaran S, Barkley NA, Chen Z, Chen CY, Guo B, Pittman RN, Stalker HT, Holbrook CC, Pederson GA, Yu J (2011) Population structure and marker-trait association analysis of the US peanut (*Arachis hypogaea* L.) mini-core collection. TAG Theor Appl Genet (Theoretische und angewandte Genetik) 123(8):1307–1317
16. Ren X, Jiang H, Yan Z, Chen Y, Zhou X, Huang L, Lei Y, Huang J, Yan L, Qi Y (2014) Genetic diversity and population structure of the major peanut (*Arachis hypogaea* L.) cultivars grown in China by SSR markers. PLoS One 9(2):e88091
17. Yan M, Zhang X, Han S, Huang B, Dong W, Liu H, Sun Z, Zhang Z, Tang F (2015) Genome-wide association study of agronomic and yield traits in a worldwide collection of peanut (*Arachis hypogaea*) germplasm. Chin Bull Bot
18. Xiao Y, Cai D, Yang W, Ye W, Younas M, Wu J, Liu K (2012) Genetic structure and linkage disequilibrium pattern of a rapeseed (*Brassica napus* L.) association mapping panel revealed by microsatellites. TAG Theor Appl Genet (Theoretische und angewandte Genetik) 125 (3):437–447

Research on γ-Polyglutamic Acid Fermentation with Extract from Waste Beer Yeast

Jun Yu, Lin Zhao, Song Li, Xin Sun and Xin-li Liu

1 Introduction

γ-polyglutamic acid (γ-PGA) is a kind of polypeptide composed of glutamic acid monomer [1]. It has been widely used in cosmetics, food, environmental protection and other industries by its good biocompatibility, such as environmental friendliness, strong water absorption and moisture retention performance and so on [2]. However, the price of γ-PGA remains at high level (the price of cosmetic-grade PGA is over 400 yuan/kg in the Chinese market in 2016) due to the high costs in fermentation and extraction which seriously restricted its application in certain extents. Several papers reported that yeast extract played a critical role in the fermentation of γ-PGA [3], and it had a large proportion in the cost composition.

Beer yeast is nutritionally valuable, containing a great amount of protein, ribonucleic acid, glutathione, coenzyme A and B-vitamins and other nutrients [4]. Waste beer yeast is always supplied at lower price as by-product of brewery [5]. Therefore, raw material costs of γ-PGA can save almost half by utilizing yeast extract from waste beer yeast. Development and utilization of waste beer yeast could create huge benefits not only in economic, but also in the environment.

In this paper, the feasibility of γ-PGA fermentation with extract of waste beer yeast was investigated. Composite method for preparation of yeast extract from waste beer yeast was constructed, and the yeast extract was employed into γ-PGA fermentation [3].

Jun Yu, Lin Zhao and Song Li are contributed equally to this paper.

J. Yu · L. Zhao · S. Li · X. Sun · X. Liu (✉)
Shandong Provincial Key Laboratory of Microbial Engineering,
School of Bioengineering, QiLu University of Technology,
No. 3501 Daxue Road, Jinan 250353, People's Republic of China
e-mail: vip.lxl@163.com

H. Liu et al. (eds.), *Advances in Applied Biotechnology*, Lecture Notes
in Electrical Engineering 444, https://doi.org/10.1007/978-981-10-4801-2_82

2 Materials and Methods

2.1 Microorganisms and Medium

Bacillus lincheniformis was kept at our laboratory and used for the fermentation of γ-PGA. Fresh waste beer yeast (moisture content 50%) was provided by China-Germany Brewing Technical Center of Qilu University of Technology. The fermentation medium for *B. lincheniformis* is PP medium (10.0 g glucose, 10.0 g yeast powder, 5.0 g peptone, 10.0 g sodium glutamate, 1.0 g K_2HPO_4, 0.5 g $MgSO_4 \cdot 7H_2O$, diluted to a constant volume of 1 L with deionized water, adjusted pH to 7.2 with 1 mol/L NaOH, sterilized at 115 °C for 20 min).

2.2 Fermentation

Batch fermentation was carried out in a 50 L fermenter (Shanghai Baoxing Bio-Engineering Equipment Co., Ltd.) with a working volume of 35.0 L. The agitation speed was set at 300 rpm, and the aeration rate was controlled at 1.5 m^3/h. The temperature was kept at 37.0 ± 0.2 °C. The pH was measured by the pH electrode and adjusted at 6.50 ± 0.30 with $NH_3 \cdot H_2O$ and phosphoric acid. Samples were taken at the indicated time points and analyzed for γ-PGA production.

2.3 Reagent and Enzyme Preparation

Related enzymes, dextranase (10,000 μ/g), cellulose (40,000 μ/g) and complex protease were bought from Ji'nan De-Ke Bio-Tech Co., Ltd., alkaline protease (400,000 μ/g) and papain (800,000 μ/g) were bought from Nan'ning Doing-Higher Bio-Tech Co., Ltd. Other reagents were analytical pure. Raw materials used in the fermentator were industrial grade or above.

2.4 Preparation Process of Waste Yeast Extract

The waste beer yeast was sieved and washed with water to remove the impurities and residual alcohol, and then the yeast paste was washed with 0.5% sodium hydrogen carbonate solution for 30 min to remove the bitterness. The treated yeast paste was prepared to yeast cream at a concentration of 10%, and a pH value of 7.0.

First, the accelerant dosage experiment was performed. Yeast autolysis is an enzymatic reaction process, and a plasmolysis process. It is reported that sodium chloride, ethanol, ethyl acetate, toluene and other substances can be useful in promoting the autolysis of yeast, among them, sodium chloride is the most safe and economic promoter [6]. Sodium chloride dosage was optimized to determine the optimum dosage. The autolysis conditions were listed as follows: 55 °C, 24 h.

Then, the external enzyme dosage experiment was performed. By external enzyme addition, it was available to shorten the time of yeast autolysis and improve the rate of yeast autolysis. We studied independent or compound application of dextranase, cellulose, papain, alkaline protease and complex protease, as well as the effect of such application approaches on the yield of yeast extract solids content. The dosage was 2‰, the enzymolysis conditions were listed as follows: 55 °C, 24 h.

After the optimal enzyme preparation was determined, the enzymolysis temperature and dosage of enzyme preparation were optimized. The experiment was conducted at 45, 50, 55, 60, and 65 °C respectively. The experimental dosage was set to 1, 2, 3, 4 and 5‰.

2.5 *Experimental Process on Determining the Impact of Extract Dosage on γ-PGA Production*

The impact of yeast extract dosage on γ-PGA production was studied with the extract prepared in Sect. 2.4. The experiment contained five groups at a dosage of 0, 2.5, 5, 7.5 and 10% respectively.

2.6 *Analytic Methods*

Experiments above were repeated three times. After autolysis and enzymatic hydrolysis, the hydrolysate was centrifuged at 10,000 rpm for 10 min. Then the supernatant was removed, the solution was heated and dried at 110 °C until constant weight [7]. After fermentation, 100 mL of fermentation broth was taken, three times the volume of fermentation broth (300 mL) of absolute ethyl alcohol was added, and stirred evenly. The liquid was centrifuged at 3000 rpm for 5 min. After the supernatant was removed, the white precipitate was PGA crude extract [8]. The precipitate was dried at 65 °C until constant weight, and the yield of PGA was calculated finally.

3 Results and Discussion

3.1 *The Impact of Sodium Chloride on the Yield of Yeast Autolysis*

Sodium chloride could affect the autolysis. The yields of soluble solids increased by 18.49, 20.64, 28.79, 29.21 and 13.36% with different sodium chloride addition (as shown in Fig. 1). Meanwhile, the soluble solids yield of 4% and 5% conditions

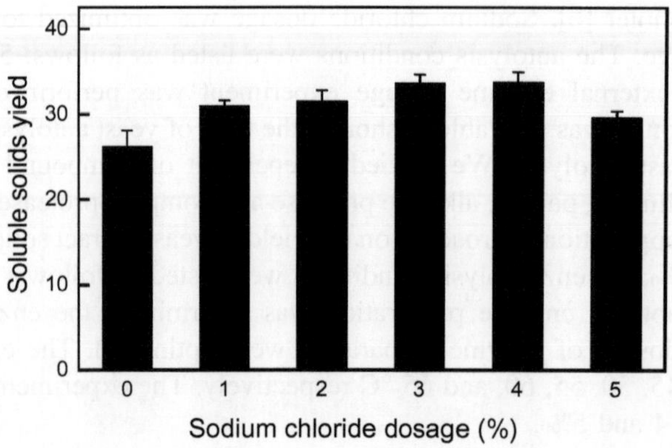

Fig. 1 Impact of sodium chloride dosage on the yield of the soluble solids of yeast extract

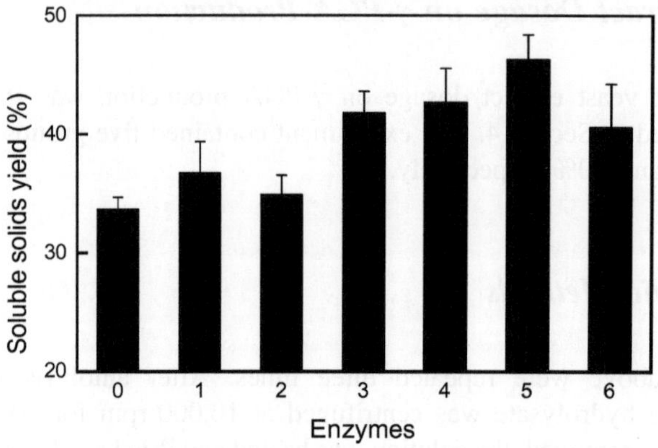

Fig. 2 Enzymes addition effect on yeast autolysis. *0* 3.0% NaCl, *1* 3.0% NaCl + dextranase, *2* 3.0% NaCl + cellulose, *3* 3.0% NaCl + dextranase +papain, *4* 3.0% NaCl + dextranase + alkaline protease, *5* 3.0% NaCl + dextranase + complex protease and *6* dextranase + papain + alkaline protease + complex protease

were lower than 3% condition, which meant that the high concentration could inhibit the activity of enzymes in vivo [9]. Therefore, 3% sodium chloride was chosen for autolysis in follow-up experiments (Fig. 2).

3.2 The Impact of Enzyme Preparation on the Yield of Yeast Autolysis

Yeast cell wall was mainly composed of dextran and cellulose [10]. When the dextranase and cellulase were added, the soluble solid yields increased by 9.22 and

3.63% than control. Furthermore, the complex protease showed significant effect, but the compound application of three proteases did not significantly improve the yield of solids, which may be due to the interaction of enzymatic reactions. According to the above results, the yield of soluble solid of yeast extract could be effectively improved by adding dextranase and complex protease.

3.3 Optimization on Addition Conditions of External Enzyme

The optimal temperature of complex protease was 55 °C according to Fig. 3, which was also consistent with the appropriate temperature of endogenous enzyme of yeast in vivo [11]. In this paper, 55 °C was chosen for the temperature of enzymatic hydrolysis in follow-up experiments.

It can be seen from Fig. 4 that the solid content increased with the increase of compound protease dosage, and the best yield of 48.64% was obtained at the

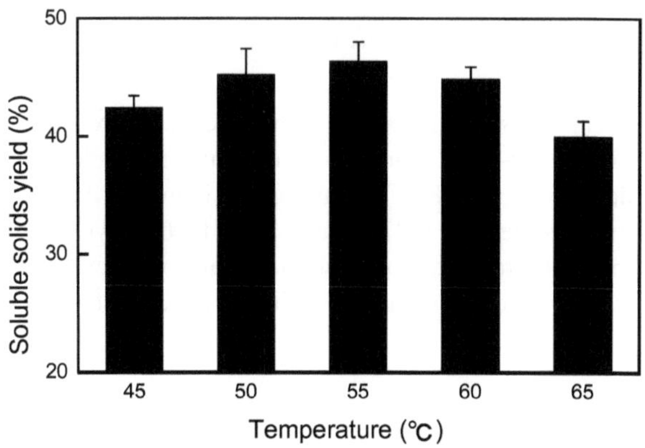

Fig. 3 Temperature optimization of complex protease

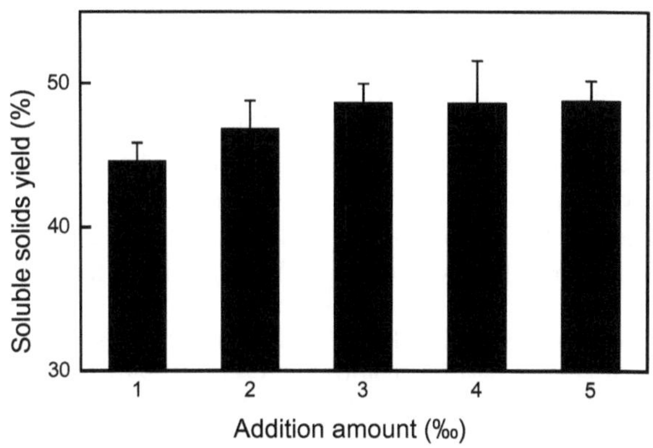

Fig. 4 Dosage optimization of the complex protease

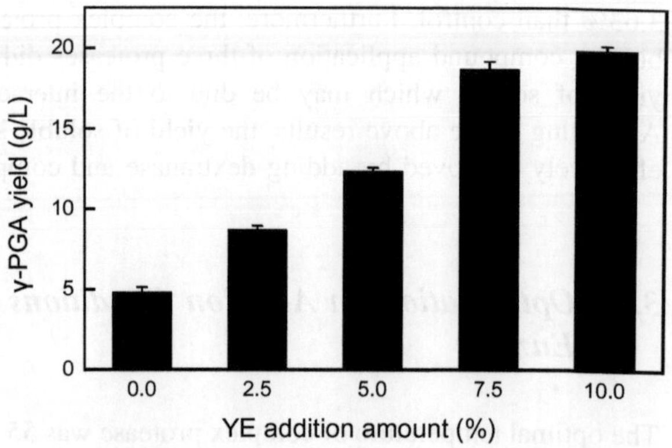

Fig. 5 Impact of addition of yeast extract on γ-PGA yield

dosage of 3‰. Continuing to increase protease dosage contributed little in improving solid content. Therefore, in this research, the follow-up experiment was performed at a dosage of 3‰.

3.4 Impact of Addition of Yeast Extract on γ-PGA Yield

It can be seen from Fig. 5 that the yeast extract could significantly improve the yield of γ-PGA, and the second highest yield of 18.69 g/L was obtained at the dosage of 7.5%. Continuing to increase extract dosage contributed a minor improvement of 6.31%, indicating that the yeast extract had become almost saturated due to culture conditions, strain productivity and other factors. In this paper, 7.5% was chosen for the optimum dosage of yeast extract for cost considerations.

4 Conclusions

In this paper, the feasibility of γ-PGA fermentation with yeast extract from waste beer yeast was investigated and composite method for preparation of yeast extract from waste beer yeast was constructed. The optimized condition was: 24-hours autolysis at 55 °C with 2% NaCl, 2‰ dextranase and 3‰ complex protease addition, the highest extracting yield reached 48.64% (w/w). Then the yeast extract was employed into γ-PGA fermentation process, and the γ-PGA yield increased greatly by 290% from 4.8 to 18.69 g/L, with adding 7.5% waste beer yeast extract. These results showed a significant valuable for industrial applications of γ-PGA.

Acknowledgements The work was financially supported by Shandong Provincial Key Research and Development Program (No. 2015ZDZX05001, 2015ZDXX0502B01, 2015ZDXX0403B01, 2016GGX107003), National Natural Science Foundation (No. 31501396), Shandong Provincial Natural Science Foundation (No. ZR2012CQ027).

References

1. Sugimoto H, Takeuchi H, Yokotsuka T (1976) Process for autolysis of yeast: US, US 3961080
2. Yamashita S, Kado H, Masuda T (2007) Beer yeast cell bodies containing high ribonucleic acid content and their production method: US, US 20070031968 A1
3. Jazwinski SM (1990) Preparation of extracts from yeast. Methods Enzymol 182(6):154–174
4. Shih IL, Van YT (2001) The production of poly-(gamma-glutamic acid) from microorganisms and its various applications. Bioresour Technol, 79(3):207–25
5. Ding MS, Ji LI, Xiao DG (2001) Research on reclaim and utilization in beer yeast mud. Liq-Mak Sci Technol 2001(4):73–75 (in Chinese)
6. Li X, Peng L, Wang Y et al (2001) The study of yeast autolysis. China Brewing 2001(5):17–19 (in Chinese)
7. Chae HJ, Joo H, In MJ (2001) Utilization of brewer's yeast cells for the production of food-grade yeast extract. Part 1: effects of different enzymatic treatments on solid and protein recovery and flavor characteristics. Bioresour Technol 76(3):253–258
8. Sugimoto H (1974) Synergistic effect of ethanol and sodium chloride on autolysis of baker's yeast for preparing food-grade yeast extracts. J Food Sci 39(5):939–942
9. Obst M, Steinbüchel A (2004) Microbial degradation of poly(amino acid)s. [J]. Biomacromol 5(4):1166–1176
10. Toshio T, Osamu H, Takafumi F et al (2014) Purification and characterization of poly (γ-glutamic acid) hydrolase from a filamentous fungus, myrothecium sp. TM-4222. Biosci Biotech Biochem 20(3):536–561
11. Knorr D, Shetty KJ, Hood LF et al (2006) An enzymatic method for yeast autolysis. J Food Sci 44(5):1362–1365

Isolation and 16SrDNA Identification of Bacteria from Traditional Kazak Dairy Products

Wenyuan Sun, Yanli Fan, Jing Li, Gaoshaer Kayierhali, Xuejiao Liu, Yun Hao, Yirong Hou, Yajian Song and Tongcun Zhang

1 Introduction

Lactic acid bacteria (LAB), which are a group of bacteria found as gastrointestinal symbionts in humans and animals, are recognized as probiotics. These bacteria are also found in fermented foods as the prime fermenting microorganism [1]. The concept of the LAB as a group of organisms developed at the beginning of the 1900s, preceded by pioneering scientific and technical developments during the latter part of the 19th century [2]. LAB are commonly thought to be having health-promoting effects on human [3]. Currently, this group comprises the following genera: *Carnobacterium*, *Enterococcus*, *Lactobacillus*, *Lactococcus*, *Leuconostoc*, *Oenococcus*, *Pediococcus*, *Streptococcus*, *Tetragenococcus*, *Vagococcus* and *Weissella* [4]. They can be found in different habitats such as soil, water, animal and human gastrointestinal tract, as well as in food and fermented products [5].

Probiotics are live microorganisms which confer a health benefit on the host when administered in adequate amounts. Most probiotics are from LAB and other genera such as *Bacillus* [6]. The health benefits they displaying includes: maintain the microbial balance of intestinal microflora, improve product flavor, promote the maturity of fermented products, improve nutrient utilization, promote nutrient absorption, control of endotoxin, lower cholesterol and et al. [7]. Probiotics can be consumed directly as part of a dietary supplement or as a component of a health-promoting food [8]. They are usually available as culture concentrates in dried or deep-freeze form to be added to a food for industrial or home uses [9].

W. Sun · Y. Fan · J. Li · G. Kayierhali · X. Liu · Y. Hao · Y. Hou · Y. Song (✉) · T. Zhang
Key Laboratory of Industrial Fermentation Microbiology,
Ministry of Education and Tianjin City, College of Biotechnology,
Tianjin University of Science and Technology, Tianjin 300457,
People's Republic of China
e-mail: songyajian@tust.edu.cn

H. Liu et al. (eds.), *Advances in Applied Biotechnology*, Lecture Notes in Electrical Engineering 444, https://doi.org/10.1007/978-981-10-4801-2_83

In recent years, the health effects of probiotics attracted more and more attentions of domestic and foreign scholars. Numerous researches about LAB diversity of dairy products in Inner Mongolia Autonomous Region, Xinjiang Uygur Autonomous Region and other places have been investigated. Xinyuan County in Yili Kazakh Autonomous Prefecture, Xinjiang Uygur Autonomous Region, China is one of the main regions inhabited by Kazakh people. So far, LAB in the traditional dairy products of this region has not been studied. In this study, we collected dairy samples from this region and studied its biodiversity by isolating and identifying the bacteria in the samples.

2 Materials and Methods

2.1 Sample Collection and Preservation

All samples used in this study were collected from Kazak herdsman families of Xinyuan County, Yili Kazakh Autonomous Prefecture, Xinjiang Uygur Autonomous Region, China. These samples are all dairy products including ghee, dry yogurt, butter, yogurt milk cheese and red knots made by traditional methods. All the samples were immediately transported to the laboratory by vacuum packaging after collection and preserved at 4 degree to ensure the freshness of the samples.

2.2 Isolation, Purification and Preservation of the Strains

0.1 g sample was taken and diluted gradually by normal saline. The dilutions were plated onto Man Rogosa and Sharp (MRS) agar (Beijing AoBoXing Bio-Tech Co. Ltd, China). Isolates were picked based on colony morphology and purified by successive streak plate method. The strains were cultivated in MRS liquid medium and collected in mid-logarithmic phase. 20% glycerol was used to preserve the culture and the mixture was stored at −80 °C for further analysis.

2.3 16SrDNA Sequencing

The genomic DNA of the LAB was extracted using DNA Extraction Kit (Axygen, Hangzhou, China) following the manufacture's protocol. The quality of the product was determined by electrophoresis in 1.5% agarose gel with 1 × TAE (40 mM Tris-acetate, 1 mM EDTA, pH 8.0) at 130 V. The 16SrDNA gene was amplified by the universal prime pairs 27F and 1492R and using genome as template [10]. The quality of PCR product was determined by gel electrophoresis.

2.4 *Sequence Analysis*

The PCR product was sequenced by GENEWIZ company (GENEWIZ, Hangzhou, China), and assembled by DNAMAN software (Lynnon Biosoft, USA) [11]. The obtained sequences were analyzed using the BLAST algorithm at the NCBI server (http://blast.ncbi.nlm.nih.gov/Blast.cgi). Phylogenetic relationships of the 22 strains were analyzed by Neighbor-joining method using MEGA software.

2.5 *Bile Salt and Acid Tolerance of the Selected Strains*

The strains were cultivated to mid-logarithmic phase and collected by centrifugation at 5000*g*. The sediment was diluted to 10^8 CFU by MRS medium with 0.3% bile salt and MRS medium adjusted to pH 3.0 respectively. The dilutions were incubated at 37 °C for 4 h, and their viability was determined by MTT assay [12].

3 Results and Discussion

3.1 *Sample Collection and Strain Isolation*

We collected 6 samples from Kazak herdsman families. Ghee from two different families was labeled as A1 and A2 respectively. Dry yogurt, butter, yogurt milk cheese and red knots were labeled as B1, C1, D1, E1 respectively. After separation and purification, 22 strains were isolated according to the colonial morphology.

3.2 *Identification of the Strains by Sequence Alignment of 16SrDNA*

16SrDNA genes of all the strains isolated were successfully amplified and sequenced. The alignment of the sequences with Genbank database was conducted using the BLAST algorithm. 22 strains were identified and belong to 5 genera including *Lactobacillus*, *Enterococcus*, *Leuconostoc*, *Acetobacter*, and *Bacillus*. Strains in *Lactobacillus* are clarified into 2 species, *Lactobacillus casei and Lactobacillus plantarum.* Strains in *Enterococcus* also belong to 2 species,

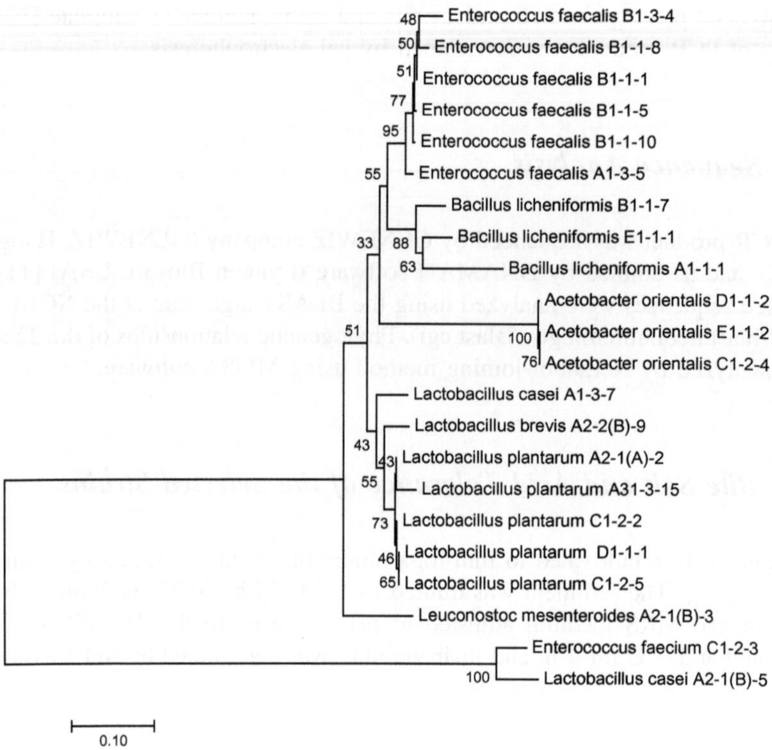

Fig. 1 Phylogenetic relationships of the 22 strains by neighbor-joining analysis. *Numbers* on branches represent the bootstrap support for each node in 1000 replicates

Enterococcus faecalis and Enterococcus faecium. Each other genera contains only one species, and they were Acetobacter orientalis, Bacillus licheniformis, and Leuconostoc mesenteroides. Phylogenetic relationship of the 22 strains was constructed by Neighbor-joining analysis as shown in Fig. 1. All the strains are LAB except *Bacillus licheniformis* and *Acetobacter orientalis.* No pathogenic bacterium was detected in the samples.

3.3 The Distribution of Different Species in Dairy Samples

All the samples in this study were found to contain two or more strains (Table 1). The isolation strains from two ghee sample A1 and A2 showed dramatic difference. It indicated that the environment of different family influence the microorganism composition even though the process method was identical. There are 6 different strains found in dry yogurt sample identified as *E. faecalis*, as is likely to be related to the special process method and flavor of dry yogurt. *L. plantarum and*

Table 1 Distribution of the species in different samples

Sample	Species	Number
A1 sample (ghee)	*Bacillus licheniformis*	A1-1-1
	Enterococcus faecalis	A1-3-5
	Lactobacillus casei	A1-3-7
	Lactobacillus plantarum	A1-3-15
A2 sample (ghee)	*Lactobacillus plantarum*	A2-1(A)-2
	Leuconostoc mesenteroides	A2-1(B)-3
	Lactobacillus casei	A2-1(B)-5
	Lactobacillus brevis	A2-2(B)-9
B1 sample (dry yogurt)	*Enterococcus faecalis*	B1-1-1, B1-1-5, B1-1-8, B1-1-10, B1-3-4, B1-4-10
	Bacillus licheniformis	B1-1-7
C1 sample (butter)	*Lactobacillus plantarum*	C1-2-2, C1-2-5
	Acetobacter orientalis	C1-2-4
	Enterococcus faecium	C1-3-2
D1 sample (yogurt milk cheese)	*Lactobacillus plantarum*	D1-1-1
	Acetobacter orientalis	D1-1-2
E1 sample (red knot)	*Bacillus simplex*	E1-1-1
	Acetobacter orientalis	E1-1-2

A. orientalis are the most widespread strains, which could be found in three different kinds of dairy products. *L. plantarum* was isolated in ghee, butter, and yogurt milk cheese, while *A. orientalis* was isolated from butter, yogurt milk cheese and red knots. The results above demonstrated that the traditional dairy products exhibited good biodiversity and the microorganism composition was dramatically influence by environment and process method.

3.4 Bile Salt and Acid Tolerance of LAB

6 LAB were selected to test the bile salt and acid tolerance, including *L. plantarum* A1-3-15 and C1-2-5, *E. faecalis* B1-1-1, and B1-1-5, and *B. licheniformis* A1-1-1 and B1-1-7. Cell viability of the strains was determined by MTT method after treating by 0.3% bile salt as well as under pH 3.0. *L. plantarum* A1-3-15 and C1-2-5 were found to have better bile salt tolerance than other strains (Fig. 2a). *L. plantarum* A1-3-15 also showed good acid tolerance (Fig. 2b). It was suggested that this strain is likely to have good viability in human digestive tract, and is a candidate probiotic to be further studied.

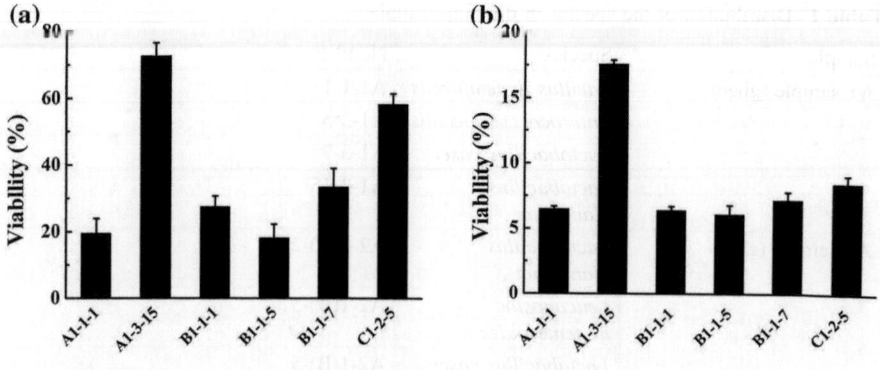

Fig. 2 Viability of 6 LAB after treated by 0.3% bile salt (**a**) and pH 3.0 (**b**)

4 Conclusion

In this study, we collected 6 traditional dairy samples from Kazak herdsman families in Xinyuan County. Total 22 strains were isolated and identified. The strains belong to 5 genera, including *Lactobacillus*, *Enterococcus*, *Leuconostoc*, *Acetobacter, and Bacillus*. Among them, *L. plantarum* A1-3-15 exhibits good bile salt and acid tolerance. This study indicated that the traditional Kazak dairy product exhibits good biodiversity and is a good resource of probiotics. The health function of the strains isolated is worth to be further studied.

References

1. Damodharan K, Palaniyandi SA (2015) In vitro probiotic characterization of Lactobacillus strains from fermented radish and their anti-adherence activity against enteric pathogens. J Can J Microbiol 61:837–850
2. Stiles ME, Holzapfel WH (1997) Lactic acid bacteria of foods and their current taxonomy. J Int J Food Microbiol 36:1–29
3. Zhang L, Wang L et al (2016) A novel antimicrobial substance produced by *Lactobacillus rhamnous* LS8. J Food Control. doi:10.1016/j.foodcont.2016.09.028
4. García-Hernández Y, Pérez-Sánchez T et al (2016) Isolation, characterization and evaluation of probiotic lactic acid bacteria for potential use in animal production. J Res Vet Sci 108:125–132
5. Holzapfel WH, Haberer P et al (2001) Taxonomy and important features of probiotic microorganisms in food and nutrition. J Am J Clin Nutr 73:365S–373S
6. Sánchez B, Ruiz L et al (2013) Omics for the study of probiotic microorganisms. J Food Res Int 54:1061–1071
7. Olnood CG, Beski SSM et al (2015) Delivery routes for probiotics: Effects on broiler performance, intestinal morphology and gut microflora. J Animal Nutr 1:192–202
8. Garrigues C, Johansen E et al (2013) Pangenomics—an avenue to improved industrial starter cultures and probiotics. J Curr Opin Biotechnol 24:187–191

Research on Extracting Technology of Chlorogenic Acid from *Honeysuckle*

Yang Sun, Ye-Min Yu, Hong-Bo Suo, Ying-Lan Zhu, He Huang and Yi Hu

Honeysuckle, the flower of *Lonicera japonica* Thunb with sweet flavor and cold nature, is commonly used as traditional Chinese medicine and tea beverage [1]. It has effects of antifebrile and detoxifcation, anti-bacteria and dephlogisticate, detumescence, liver protection and cholagogue and so on. *Honeysuckle* is widely distributed in China and as a kind of both food and medicinal plants, it has a long medicinal history and it is one of the 60 kinds of Chinese herbal medicines and the first batch issued by national ministry of health [2]. Its chemical composition contains chlorogenic acid, isochlorogenic acid, flavonoid compounds, linalool and so on.

Chlorogetic acid is recognized as the "gold plant" [3, 4] because of its main active ingredient in *honeysuckle*. Chlorogetic acid is styrene acrylic class of secondary metabolites produced in shikimic acid pathway of aerobic respiration process. Its molecular structure contains three instable parts, including ester bond, unsaturated double bond and polyphenol. The pharmacological functions of chlorogetic acid have been studied extensively. Potentially beneficial properties to human such as antimicrobial, anti-inflammatory, antioxidant and anticancer have been received more and more attention [5]. Li et al. [6] reported that chlorogetic acid could reduce blood pressure acutely, which would benefit cardiovascular health. Because of these functions, chlorogetic acid has been recorded officially in the National Pharmacopoeia of China, and which is widely applied to food, medicine, cosmetics and other industries [3, 7].

A series of methods have been reported for the extraction bioactive chlorogetic acid from plant materials, such as water extraction [8], ethanol refluxing method [9]. As two traditional methods for the extraction of chlorogenic acid, they take long time and the extraction rate of chlorogenic acid is lower. According to the

Y. Sun · Y.-M. Yu · H.-B. Suo · Y.-L. Zhu · H. Huang · Y. Hu (✉)
State Key Laboratory of Material-Oriented Chemical Engineering,
College of Pharmacy Nanjin Tech University, Nanjin 211816, China
e-mail: huyi@njtech.edu.cn

H. Liu et al. (eds.), *Advances in Applied Biotechnology*, Lecture Notes
in Electrical Engineering 444, https://doi.org/10.1007/978-981-10-4801-2_84

demand, the emergence of new technologies in industrial production were needed. In recent years, PEM and UEM have been widely used in the extraction of natural products [10–12]. Cao et al. [13] optimized the extraction process of chlorogenic acid in *honeysuckle* and found the extraction yield of chlorogenic acid was 5.62%, while the key factors on ultrasonic power and the solid-liquid ratio were not investigated. Fu et al. [14] optimized the extraction of chlorogenic acid in *honeysuckle* by cellulase and found extraction yield of chlorogenic acid was only 2.39%.

In this paper, PEM and UEM method were developed to extract chlorogetic acid respectively and orthogonal experiment was designed to select the optimum parameters for providing a reference on the extraction of chlorogenic acid from *honeysuckle*.

1 Materials and Methods

Honeysuckle was purchased from Xinyi *honeysuckle* Agricultural Development co. (Xin yi, China). The standards of chlorogenic acid was purchased purchased Aladdin (Shanghai, China) and their their purities was $\geq 98\%$. Ethanol of analytical grade for extraction was purchased from Tianjin. KQ-KDB 400 type ultrasonic cleaners was purchased from Kunshan ultrasonic instrument co. (Kunshan, China). N_4 ultraviolet-visible spectrophotometer was purchased from Shanghai instrument analysis instrument co. (Shanghai, China). BSA224S type electronic balance was purchased from Sartorius Scientific Instrument Co. (Beijing, China). RCT basic type of magnetic stirring heater were purchased from IKA. (Shanghai, China). Pectinase was purchased from Jiangsu Rui Yang biotechnology co. (Jiangsu China).

2 Methods and Results

2.1 Standard Curve Drawing of Chlorogenic Acid

The 7.5 mg Chlorogenic acid standard was dissolved in the capacity of 25 mL and then 1, 1.5, 2, 2.5, 3 mL were accurately quantified to placed in capacity of 10 mL. Chlorogenic acid standard was determinated by using UV-visible spectrophotometer at the wavelength of 328 nm. Standard curve of chlorogenic acid was drew using absorbance (Y) as the vertical coordinate and the mass concentration (X, μg/mL) as the horizontal coordinate, the curve equation was $Y = 00651X + 0.0045$, $R^2 = 0.9998$.

9. Tripathi MK, Giri SK (2014) Probiotic functional foods: Survival of probiotics during processing and storage. J Funct Foods 9:225–241
10. Tongjie L, Yun L et al (2016) Prevalence and diversity of lactic acid bacteria in Chinese traditional sourdough revealed by culture dependent and pyrosequencing approaches. J LWT-Food Sci Technol 68:91–97
11. Katrolia P, Jia H et al (2012) Characterization of a protease-resistant α-galactosidase from the thermophilic fungus Rhizomucor miehei and its application in removal of raffinose family oligosaccharides. J Bioresour Technol 110:578–586
12. Muñoz-Atienzaa E, Araújo C et al (2015) Different impact of heat-inactivated and viable lactic acid bacteria of aquatic origin on turbot (*Scophthalmus maximus* L.) head-kidney leucocytes. J Fish Shellfish Immun 44:214–223

2.2 Determination of Chlorogenic Acid Content

The 15 mL samples obtained by different extraction methods were diluted 625 times, Using 100% methanol solution as the blank control, samples were measured at the absorbance wavelength of 328 nm. According to the linear regression equation, the mass concentration was calculated, then the extraction rate of chlorogenic acid was calculated according to formula (1).

$$\text{Rate of chlorogenic acid}\,(\%) = (C \times n \times V)/(M \times 10^6) \tag{1}$$

In this formula: C—the mass concentration calculated by the standard curve/(μg/mL); V—total volume of the extraction/mL; n—dilution ratio; M—quality of raw materials/g.

2.3 Study on PEM

2.3.1 Single Factor Investigation

(1) Study on the dosage of enzyme *Honeysuckle* was 2 g, solid-liquid ratio was 1:50 (g/mL), the dosage of enzyme were 0, 0.5, 1, 1.5, 2% respectively (proportion of raw materials), Enzymatic hydrolysis temperature was 45 °C, Enzymatic hydrolysis time was 90 min, Enzymatic hydrolysis pH was 4, the extraction rates were 2.01, 3.59, 3.51, 3.43, 3.4% respectively. It showed that with the increase of the dosage of enzyme, the extraction rate of the chlorogenic acid increased gradually in Fig. 1, when the dosage of enzyme increased to 0.5%, the extraction yield of the chlorogenic acid decreased slightly.

Fig. 1 Effect of enzyme dosage on the yield of chlorogenic acid

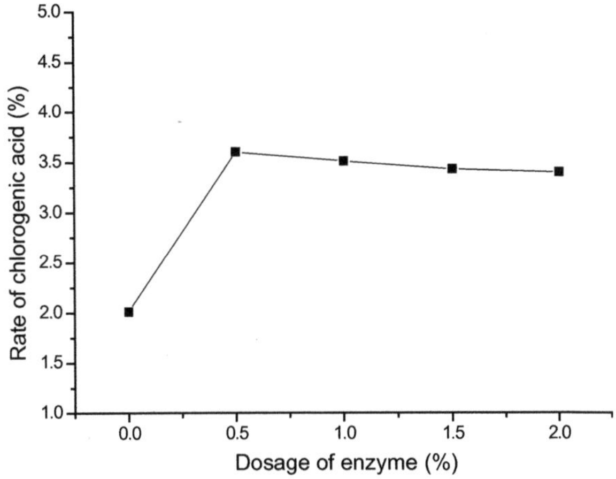

(2) Study on enzyme temperature *Honeysuckle* was 2 g, solid-liquid ratio was 1:50 (g/mL), Enzymatic hydrolysis temperature were 35, 40, 45, 50, 55 °C respectively, the dosage of enzyme were 0.5% (proportion of raw materials), enzymatic hydrolysis time was 90 min, enzymatic hydrolysis pH was 4, the extraction rates were 3.41, 3.43, 3.59, 3.52, 3.11% respectively. Figure 2 showed that with the increase of the enzyme temperature, the extraction rate of the chlorogenic acid increased gradually, when enzyme temperature increased to 45 °C, the extraction rate of the chlorogenic acid significantly decreased. The possible main reason is that the optimum temperature of enzymatic reaction is favorable to the diffusion of the extraction medium. High temperature not only would make the enzyme protein denaturation, it would also lead different levels of chlorogenic acid oxydic.

(3) Study on enzymatic hydrolysis time *Honeysuckle* was 2 g, solid-liquid ratio was 1:50 (g/mL), Enzymatic hydrolysis time were 60, 75, 90, 105, 120 min respectively, the dosage of enzyme were 0.5% (proportion of raw materials), enzymatic hydrolysis temperature were 45 °C, Enzymatic hydrolysis pH was 4, the extraction rates were 3.41, 3.53, 3.59, 3.54, 3.51% respectively. Figure 3 showed that with the increase of the enzymatic hydrolysis time, the extraction rate of the chlorogenic acid increased gradually, when the enzymatic hydrolysis time was 90 min, the extraction rate of chlorogenic acid reached the maximum, and the extraction rate of chlorogenic acid decreased gradually with the time.

(4) Study on enzymatic hydrolysis pH *Honeysuckle* was 2 g, solid-liquid ratio was 1:50 (g/mL), Enzymatic hydrolysis pH were 2, 3, 4, 5, 6 respectively, the dosage of enzyme were 0.5% (proportion of raw materials), Enzymatic hydrolysis temperature were 45 °C, Enzymatic hydrolysis time was 90 min, the extraction rates were 3.32, 3.48, 3.59, 3.53, 3.50% respectively. It showed that proper pH can maintain the activity of enzyme and promote the combination between enzyme and substrate in Fig. 4. Different pH could make

Fig. 2 Effect of enzymatic hydrolysis temperature on the yield of chlorogenic acid

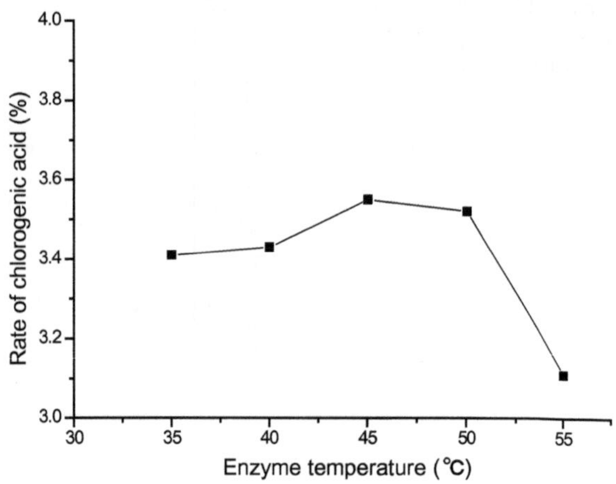

Fig. 3 Effect of enzymatic hydrolysis time on the yield of chlorogenic acid

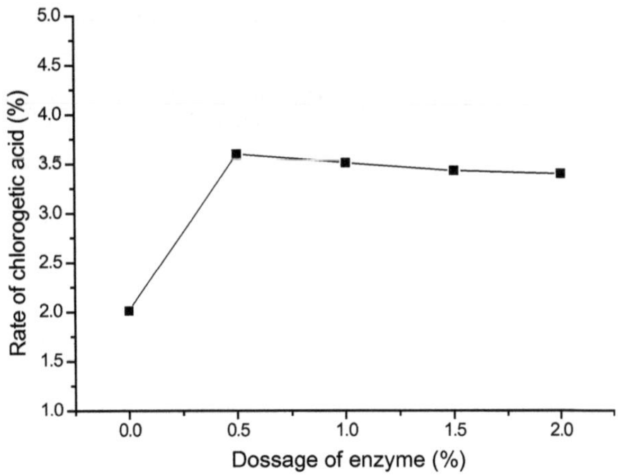

Fig. 4 Effect of enzymatic hydrolysis pH on the yield of chlorogenic acid

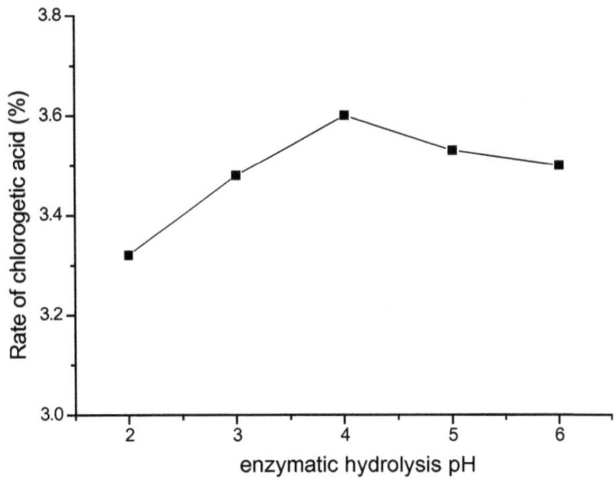

different affinity between enzyme and substrate, which lead to different catalytic rate. With the increasing of pH, the extraction rate of chlorogenic acid was increasing, and when it was 4, the extraction yield of chlorogenic acid reached the maximum and the extraction rate was decreased.

2.3.2 Orthogonal Experiment Results and Analysis

The round-bottom flask containing 2 g dried material and certain solvent were placed in oil bath, and the dispersion tool was put in the solvent at the same time. When the extraction was ended, the extraction solvent was filtered through a filtration net into the collection pot and concentrated liquid was added to 15 mL,

Table 1 Orthogonal experiment level table of the chlorogenic acid

Level	A	B	C	D
	Enzyme dosage (%)	Enzymatic hydrolysis temperature (°C)	Enzymatic hydrolysis time (min)	Enzymatic hydrolysis pH
1	0.5	40	60	3
2	1	45	90	4
3	1.5	50	120	5

and then according to the method of 2.2, the extraction rate of chlorogenic acid was calculated.

On the basis of single factor investigation, Enzyme dosage (A), enzymatic hydrolysis temperature (B), enzymatic hydrolysis time (C), and pH (D) were used as the factors of investigation, the extraction rate of chlorogenic acid was used as an index of investigation and orthogonal experiment table ($L_9(3^4)$) was designed to carry out the experiment. The factor levels were shown in Table 1, orthogonal experimental results were shown in Table 2, the results of analysis of variance were shown in Table 3.

Significant test: variance analysis showed that factor B was highly significant, factors A and D were significant, factors C was not significant, influence order of each factor on the experimental indexes was BDAC. The Optimal conditions were $A_1B_3C_3D_3$.

Verification tests: according to the orthogonal experiment, the optimal process conditions were repeated for 3 times. When the solid-liquid ratio was 1:50 (g/mL), the enzyme hydrolysis temperature was 50 °C, the enzymatic hydrolysis pH was 5, the dosage of enzyme was 0.5%, the time of enzymatic hydrolysis was 120 min, the extraction rate of chlorogenic acid reached 4.75%.

2.4 *Study on UEM*

2.4.1 Single Factor Investigation

(1) Study on ultrasonic time *Honeysuckle* was 2 g, ultrasonic temperature was 45 °C, ultrasonic time were 40, 60, 80, 100, 120 min respectively, ethanol concentration was 70%, solid-liquid ratio was 1:30 (g/mL), the extraction rates were 5.62, 5.92, 6.16, 6.02, 5.96% respectively. It showed that with the increase of the ultrasonic time, the extraction rate of the chlorogenic acid increased gradually in Fig. 5, when ultrasonic time was 80 min, the extraction rate of chlorogenic acid reached the maximum and the extraction rate was decreased after that.

Table 2 Results and analysis of the orthogonal test

Runs	A	B	C	D	Rate (%)
1	1	1	1	1	3.54
2	1	2	2	2	3.37
3	1	3	3	3	4.63
4	2	1	2	3	3.39
5	2	2	3	1	3.55
6	2	3	1	2	3.50
7	3	1	3	2	2.88
8	3	2	1	3	3.79
9	3	3	2	1	4.17
1	3.85	3.27	3.61	3.75	
2	3.48	3.57	3.64	3.25	
3	3.61	4.1	3.69	3.94	
R	0.37	0.83	0.08	0.69	

Table 3 Results of analysis of variance

Factors	SS	df	s^2	F	F_a	Significance
A	0.207	2	0.1035	41.4	$F_{0.05}$ (2, 4) = 6.94	**
B	1.060	2	0.53	212	$F_{0.01}$ (2, 4) = 18.0	**
C	0.009	2	0.0045	0.9		
D	0.758	2	0.379	75.8		**
Error	0.01	4	0.0025			

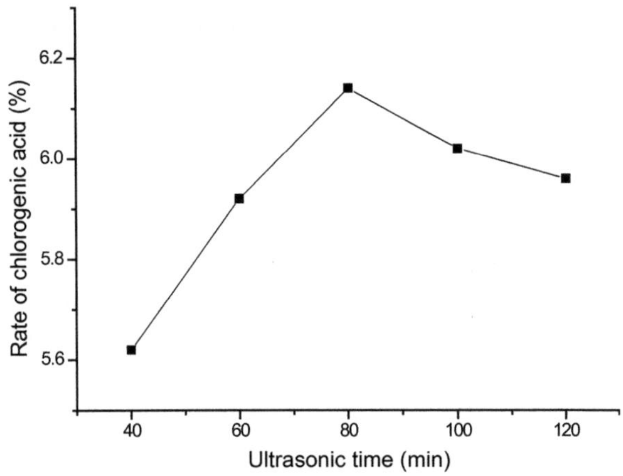

Fig. 5 Effect of ultrasonic extraction time on the yield of chlorogenic acid

Fig. 6 Effect of ultrasonic power on the yield of chlorogenic acid

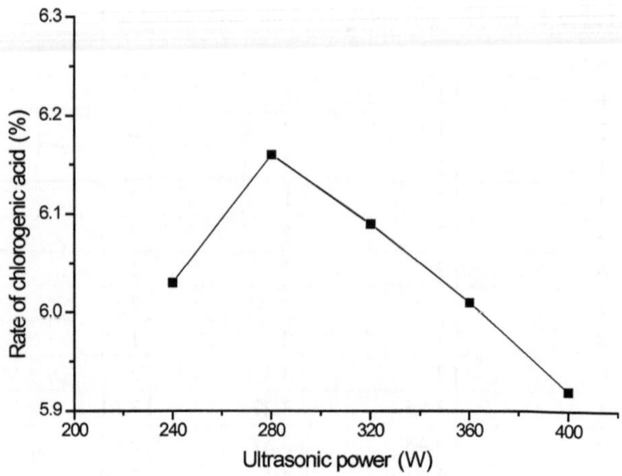

(2) Study on ultrasonic power *Honeysuckle* was 2 g, ultrasonic temperature was 45 °C, ultrasonic power were 240, 280, 320, 360, 400 W respectively, ultrasonic time was 80 min, ethanol concentration was 70%, solid-liquid ratio 1:30 (g/mL), the extraction rates were 6.03, 6.16, 6.09, 6.01, 5.92% respectively. Figure 6 showed that with the increase of the ultrasonic power, the extraction rate of the chlorogenic acid increased gradually, the result was better when the ultrasonic power was 280 W, it was because the ultrasonic power was so high that it destroyed the structure of *honeysuckle*. When the ultrasonic power exceeded 280 W, the tendency of extraction rate decreased instead, as may be attributed to the possibility that higher ultrasonic power would lead to destroy the structure of phenolic hydroxyl.

(3) Study on ethanol concentration *Honeysuckle* was 2 g, ultrasonic temperature was 45 °C, ethanol concentration was 50, 60, 70, 80, 90% respectively, ultrasonic time was 80 min, ultrasonic power was 280 W, solid-liquid ratio was 1:30 (g/mL), the extraction rates were 5.69, 6.00, 6.16, 5.90, 5.68% respectively. It showed that ethanol concentration had a great influence on the extraction rate of chlorogenic acid from Fig. 7. With the increase of ethanol concentration, the extraction rate of the chlorogenic acid increased gradually, when ethanol concentration was 70%, the extraction rate of chlorogenic acid reached the maximum and the extraction rate was decreased after that.

(4) Study on solid-liquid ratio *Honeysuckle* was 2 g, ultrasonic temperature was 45 °C, ultrasonic solid-liquid ratio were 1:20, 1:30, 1:40, 1:50, 1:60 (g/mL), respectively, ultrasonic time was 80 min, ultrasonic power was 280 W, ethanol concentration was 70%, and extraction rate were 6.05, 6.16, 6.05, 5.98, 5.92% respectively. it showed that when the solid-liquid ratio was 1:30, the extraction rate of chlorogenic acid reached the maximum in Fig. 8. Taking the effective dissolution of chlorogenic acid into account and not a waste of solvent, it was appropriate for us to select solid-liquid 1:30 (g/mL).

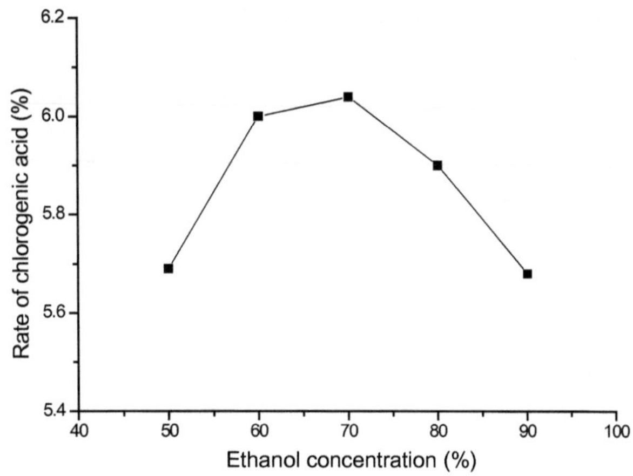

Fig. 7 Effect of ethanol concentration on the yield of chlorogenic acid

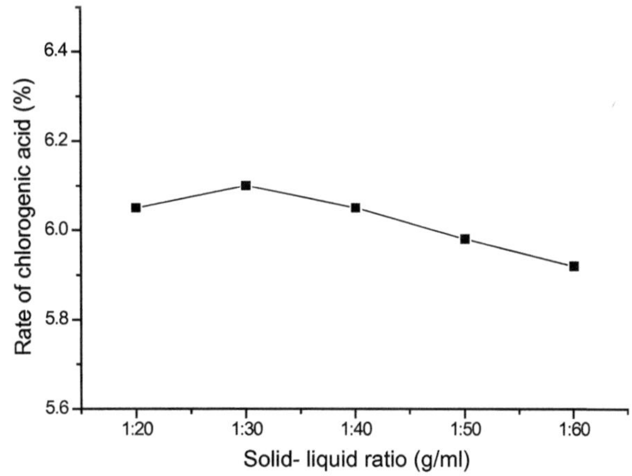

Fig. 8 Effect of solid-liquid ratio on the yield of chlorogenic acid

2.4.2 Orthogonal Experiment Results and Analysis

The round-bottom flask containing 2 g dried material and certain solvent were placed in oil bath, and the dispersion tool was put in the solvent at the same time. When extraction process ended, the extraction solvent was filtered through a filtration net into the collection pot and concentrated liquid was added to 15 mL, and then according to the method of 2.2, the extraction rate of chlorogenic acid was calculated.

On the basis of single factor investigation, ultrasonic time (A), ultrasonic power (B), ethanol concentration (C), and solid-liquid ratio (D) were used as the factors of investigation, extraction yield of chlorogenic acid was used as an index of investigation and orthogonal experiment table ($L_9(3^4)$) was designed to carry out the experiment. The factor levels were shown in Table 4, orthogonal experimental

Table 4 Orthogonal experiment level table of the chlorogenic acid

Level	A	B	C	D
	Ultrasonic time (min)	Ultrasonic power (W)	Ethanol concentration (%)	Solid-liquid ratio (g/mL)
1	40	240	50	1:30
2	60	280	60	1:40
3	80	320	70	1:50

Table 5 Results and analysis of the orthogonal test

Runs	A	B	C	D	Rare (%)
1	1	1	1	1	5.54
2	1	2	2	2	6.02
3	1	3	3	3	5.43
4	2	1	2	3	5.47
5	2	2	3	1	6.23
6	2	3	1	2	5.44
7	3	1	3	2	6.04
8	3	2	1	3	6.19
9	3	3	2	1	5.81
1	5.663	5.683	5.723	5.903	
2	5.757	6.190	5.767	5.833	
3	6.013	5.560	5.943	5.697	
R	0.350	0.630	0.220	0.206	

Table 6 Results of analysis of variance

Factors	SS	df	s^2	F	F_a	Significance
A	0.197	2	0.0985	5.6285	$F_{0.05}$ (2, 4) = 6.94	
B	0.669	2	0.3345	19.1143	$F_{0.01}$ (2, 4) = 18.0	**
C	0.081	2	0.0405	2.3142		
D	0.066	2	0.033			
Error	0.07	4	0.0175			

results were shown in Table 5, the results of analysis of variance were shown in Table 6.

Significant test: variance analysis showed that factor B was highly significant, factors A, C, D were not significant, the order of influence of each factor on the experimental indexes was BACD. The Optimal conditions are $A_3B_2C_3D_1$.

Verification tests: according to the orthogonal experiment, the optimal process conditions were repeated for 3 times. When ultrasonic temperature was 40 °C, ultrasonic power was 280 W, ultrasonic time was 80 min, ethanol concentration was 70%, ultrasonic solid- liquid ratio was 1:30 (g/mL), and under this condition, the extraction rate of chlorogenic acid reached 6.26%.

Table 7 Comparison of different extraction methods of chlorogenic acid

Extraction methods	Rate (%)
Water extraction[a]	3.15
Ethanol refluxing method[b]	3.34
PEM	4.75
UEM	6.26

[a]Through the single factor experiments on time, temperature, pH and solid-liquid ratio, orthogonal experiment was designed to optimize conditions: time 60 min, temperature 70 °C, pH = 3, solid-liquid ratio 1:40 (g/mL), the extraction yield of the chlorogenic acid reached 3.15%

[b]Through the single factor experiments on solid-liquid ratio, temperature, ethanol concentration and time, orthogonal experiment was designed to optimize conditions: solid-liquid ratio 1:30 (g/mL), temperature 70 °C, ethanol concentration 70%, time 80 min, the extraction yield of the chlorogenic acid reached 3.34%

Compared with the traditional methods such as water extraction, ethanol refluxing method, pectinase method and ultrasonic extraction method had a relatively efficient extraction effect. Moreover, compared with previous reports, the extraction yield had been greatly improved [13, 14]. Comparison results were shown in Table 7.

3 Conclusions

Through above experimental research, in pectinase method, with the change of enzymolysis temperature, pH, dosage of enzyme and enzymatic hydrolysis time, the extraction yield of chlorogenic acid in *honeysuckle* was changing. The optimal condition was that solid-liquid ratio was 1:50 (g/mL), enzyme hydrolysis temperature was 50 °C, enzymatic hydrolysis pH was 5, dosage of enzyme was 0.5%, enzymatic hydrolysis time was 120 min, and the extraction yield of the chlorogenic acid was 4.75%. In the ultrasonic extraction method, the optimal condition was ultrasonic temperature was 40 °C, ultrasonic power was 280 W, ultrasonic time was 80 min, ethanol concentration was 70%, ultrasonic solid-liquid ratio was 1:30 (g/mL), and the extraction yield of chlorogenic acid reached 6.26%.

Compared with the extraction yield of chlorogenic acid reported in related literatures, thus optimized pectinase method and ultrasonic extraction method had improved significantly. It's a new direction to match pectinase method with ultrasonic extraction method, and to study the optimal extraction conditions of pectinase-ultrasonic extraction methods.

Acknowledgments This research was supported by National Natural Science Foundation of China (21676143), Qing Lan Project of JiangSu Province and Program for Innovative Research Team in University of Jiangsu Province.

References

1. Xiang Z, Ning Z (2008) Scavenging and antioxidant properties of compound derived from chlorogenic acid in South-China honeysuckle. LWT-Food Sci Technol 41(7):1189–1203
2. Li M, Wang YX, Meng J (2014) Determination of eight components in Lonicerae Japonicae Flos by HPLC. China Tradit Herb Drugs 45(7):1006–1010
3. Marques V, Farah A (2009) Chlorogenic acids and related compounds in medicinal plants and infusions. Food Chem 113(4):1370–1376
4. Du YB, Qian AY (2006) Bioactivity, resources, extraction and purification of chlorogenic acid. Mod Food Sci Technol 22(2):250–252
5. Seo ON, Kim GS, Park S et al (2012) Determination of polyphenol components of *Lonicera japonica* Thunb. using liquid chromatography–tandem mass spectrometry: contribution to the overall antioxidant activity. Food Chem 134(1):572–577
6. Li J, Jin S, Zu YG et al (2014) Rapid preparative extraction and determination of major organic acids in honeysuckle (*Lonicera japonica* Thunb.) tea. J Food Compos Anal 33 (2):139–145
7. Saleh IA, Vinatoru M, Mason TJ et al (2016) A possible general mechanism for ultrasound-assisted extraction (UAE) suggested from the results of UAE of chlorogenic acid from *Cynara scolymus* L. (artichoke) leaves. Ultrason Sonochem 31:330–336
8. Zheng XQ, Jiang JF, Liu X et al (2006) Extraction of chlorogenic acid from sunflower seeds by water solution and alcohol sedimentation. Food Sci 27(1):159–161
9. Wu L, Zhang ZS (2005) Extraction and examination of chlorogenic acid from Flos Lonicerae. Food Sci 26(6):130–134
10. Xu WJ, Zhai JW, Cui Q et al (2016) Ultra-turrax based ultrasound-assisted extraction of five organic acids from honeysuckle (*Lonicera japonica* Thunb.) and optimization of extraction process. Sep Purif Technol 166:73–82
11. Li AM, Yao YZ, Guo Y et al (2013) Optimization of ultrasonic-assisted extraction process of chlorogenic acid from Sambucus chinensis Lindl. Leaves. Sci Technol Food Ind 31(6):15–19
12. Chen YS, Li ZG, Dong JG (2008) Study on extraction process of sunflower meal by enzyme treatment. Food Sci 29(2):202–204
13. Cao Y, Li CJ, Xia ZN (2008) Optimization of ultrasound extraction technology of chlorogenic acid in Flos Lonicerae. Lishizhen Med Mat Med Res 19(12):2857–2858
14. Fu SL, Zhong JP, Li HX et al (2007) Study on extracting process of chlorogenic acid from Flos Lonicerae. J Guangdong Ocean Univ 27(4):70–73

Construction and Functional Analysis of Luciferase Reporter Plasmid Containing Vimentin Gene Promoter

Cheng-Xi Yu, Yuan Xiang, Xing-Hua Liao, Xiao-Yu Zhang, Hui Li, Jia-Peng Li and Tong-Cun Zhang

1 Introduction

Breast cancer is the most common malignancy of women [1]. The major cause of death from breast cancer is due to metastases that are resistant to conventional therapies [2]. Tumor metastasis is a multi-step, multi-stage, multi-channel complex process involving multiple gene variations. The Epithelial to mesenchymal transition (EMT) process, in which epithelial cells are converted into mesenchymal cell, is frequently activated during cell invasion and migration, and facilitates metastasis in multiple carcinoma types [3]. Vimentin is an important marker gene for mesenchymal cell. It was discovered that abnormal expression or abnormal activation of certain related proteins during the occurrence and metastasis of cancer [4].

It is reported that the persistent activation of signal transducer and activator of transcription 3 (STAT3) in prostate cancer [5]. And STAT3-expressing cells had decreased E-Cardherin levels, and enhanced migratory capacities compared to control-expressing cells [6]. STAT3 convey signal from numerous cytokines and growth factors to the nucleus. The expression and activity have been shown to be perturbed in a variety of malignancies including breast cancer [7].

The myocardin protein family, myocardin-related transcription factor MRTF-A is co-activators of serum response factor (SRF). MRTF-A bind to SRF and strongly activate transcription from promoter by binding to serum response element (SRES), also known as CArG box, in their promoter [8]. It is reported that the Rho-A pathway appears to activate MRTF-A by altering MRTF-A binding to actin and causing MRTF-A translocation from cytoplasm to the nucleus [9]. Previously

C.-X. Yu · Y. Xiang · X.-H. Liao (✉) · X.-Y. Zhang · H. Li · J.-P. Li · T.-C. Zhang (✉)
Institute of Biology and Medicine, Wuhan University of Science and Technology, Wuhan 430081, China
e-mail: xinghualiao@hotmail.com

T.-C. Zhang
e-mail: zhangtongcun@wust.edu.cn

H. Liu et al. (eds.), *Advances in Applied Biotechnology*, Lecture Notes in Electrical Engineering 444, https://doi.org/10.1007/978-981-10-4801-2_85

reported MRTF-A play dual roles in EMT, direct regulation of slug transcription and reorganization of actin cytoskeleton [10].

In this study, it is our objective whether MRTF-A and STAT3 have interaction in the regulation of EMT during breast cancer [11]. The Vimentin promoter was amplificated from human genome by PCR and inserted into pGL-3 basic vector (luciferase vector). Furthermore, luciferase assays were performed in MDA-MB-231 cells.

2 Materials and Methods

2.1 Cell Culture

Human breast cancer cell line MDA-MB-231 was obtained from Boster Immunoleader. MDA-MB-231 cells were cultured in Dulbecco's modified Eagle's medium (GIEMO) containing 10% fetal bovine serum (FBS) and incubated in a 5% CO_2 incubator at 37 °C.

2.2 Plasmid Construction

The human genome was extracted from MDA-MB-231 cells. The Vimentin promoter fragment (−326 to +83) was PCR amplified by the following primer:

F: 5′-CGGGGTACCCATTTGTGTTACATAATTG-3′
R: 5′-GGCGAGCTCTTTTAATAACTCGCTAAAGC-3′

PCR conditions are as follows: 95 °C pre-denaturation 5 min, 95 °C denaturation 30 s, annealing at 55 °C for 30 s, extension at 72 °C for 2 min. The PCR reaction was carried out for 30 cycles, and the amplification product was subjected to 1% agarose gel stained with ethidium bromide under UV lamp. PCR products and pGL3-Basic vector were digested with restriction endonucleases *KpnI* and *XhoI* at 37 °C for 1 h. The PCR product was mixed with the pGL3-Basic vector plasmid with 2 μl of T4 ligase buffer and 1 μl of T4 DNA ligase, and water was added to 20 μl. Incubated at 16 °C for 2 h and then transformed into *E. coli* cells. Monoclonal colonies were isolated and cultured in 3 ml of LB medium containing ampicillin and incubated overnight at 37 °C. Plasmids were extracted using the plasmid kit (Axygene) according to the manufacturer's instructions and sequenced.

2.3 Transfection

MDA-MB-231 cells were seeded in 24-well plates at a density of 1.5×10^4 cells/cm^2. Cells were transfected with 0.25 μg of pcdna3.1-mrtf-a plasmid and the pcDNA3.1-STAT3 plasmid through the TurboFect reagent (Fermentas). pcDNA3.1-MRTF-A and pcDNA3.1-STAT3 plasmid was transfected into MDA-MB-231 cells meanwhile. And Transfection with pcDNA3.1 plasmid to be used as a negative control. 1 μg of plasmid was incubated with 2 μg of the transfection reagent for 20 min and then added to serum-free medium. After transfection for 6 h, serum-free medium was replaced with medium containing 10% FBS. The Vimentin promoter luciferase reporter gene plasmid was respectively transfected into MDA-MB-231 cells.

2.4 *Luciferase Reporter Assay*

After 48 h of transfection, 800 μl of protein (200 μl/well) was extracted for detection of luciferase activity, which were measured by a luciferase reporter assay system on SpectraMax i3.

3 Results

Construction of Vimentin Promoter Luciferase Reporter Plasmid. The PCR amplification of the Vimentin gene promoter was estimated by agarose gel electrophoresis. We know that the fragment size is 427 bp. As shown in Fig. 1, we can observe a single fragment of about 420 bp, which represents the PCR product of the Vimentin promoter. The PCR product was then cloned into pGL3-Basic vector by double digestion.

Recombinant plasmids were purified and analyzed by agarose gel electrophoresis. Figure 2 shows the size of the purified recombinant plasmid. Lane 2 and 3 represents a recombinant plasmid, and Lane 4 represents the vector plasmid. To confirm whether the recombinant plasmid was correct, we digested with double digestion and then electrophoresed by agarose gel.

As shown in Fig. 3, a band at about 420 bp represents the vimentin promoter, and the other band is about 4800 bp, representing the pGL3-Basic vector. Vimentin promoter luciferase reporter gene sequencing results were further confirmed. DNA sequence alignment results showed that the vimentin promoter luciferase reporter gene plasmid was successfully constructed.

Luciferase Assay. Luciferase Reporter Assays were performed after MDA-MB-231 cells were transfected with MRTF-A and STAT3 and Vimentin promoter. As shown in Fig. 4, Contrasting with control group transfected with pcDNA3.1, MRTF-A showed a significant effect to enhance the transcription activity of Vimentin promoter. STAT3 upregulation of transcriptional activity was not obviously. MRTF-A and STAT3 co-transfection also can significantly enhance the transcriptional activity of the vimentin promoter.

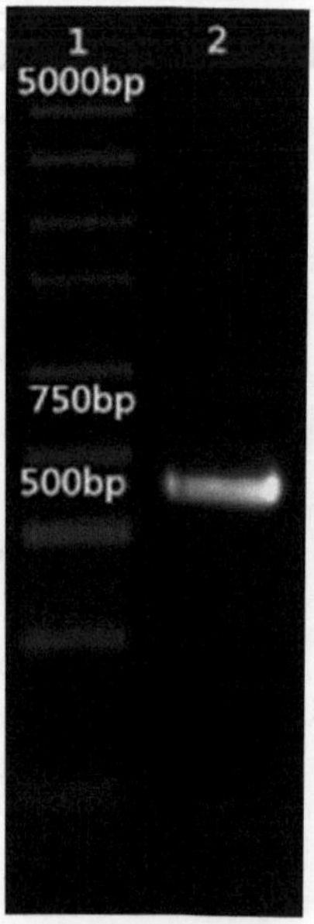

Fig. 1 Agarose gel electrophoretic analysis of PCR product; *1* 500 bp DNA marker; *2* Vimentin gene promoter

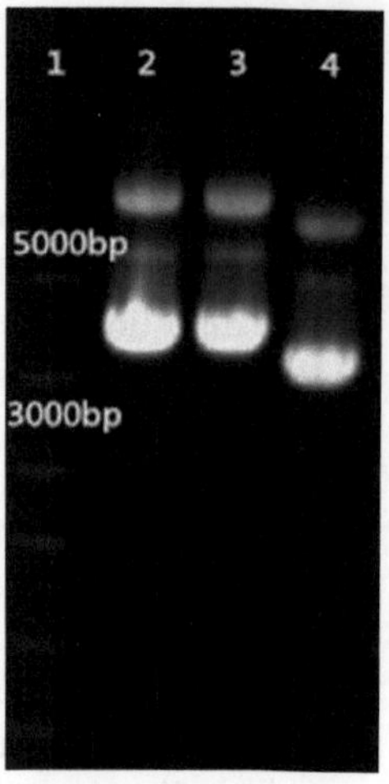

Fig. 2 Agarose gel electrophoretic analysis of recombinant plasmids; *1* 500 bp DNA marker; *2–3* recombinant plasmid; *4* pGL3-basic vehicle plasmid

4 Discussion

Previous studies found that Vimentin was closely related to breast tumor migration [12]. Vimentin is known as an important marker gene for epithelial-mesenchymal transition (EMT). EMT is a key process that occurs during embryonic development and fibrosis and tumor development [13]. Myocardin related transcription factor A (MRTF-A), belongs to the MRTF family member, also known as MKL1, has been

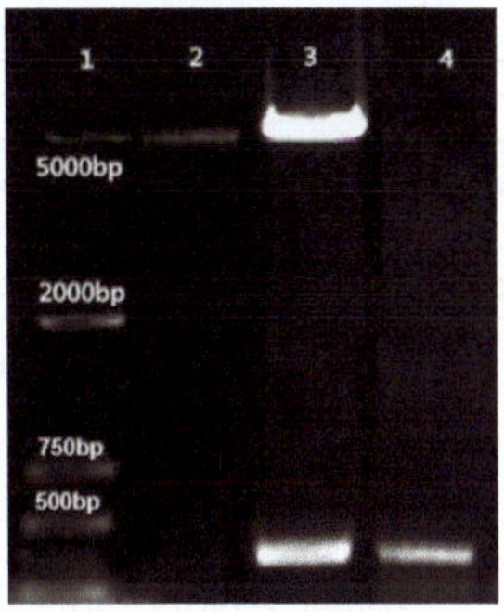

Fig. 3 Agarose gel electrophoretic analysis of digested plasmids. *1* 500 bp DNA marker; *2* pGL3-basic vehicle plasmid; *3* digested recombinant plasmid; *4* digested vimentin gene promoter

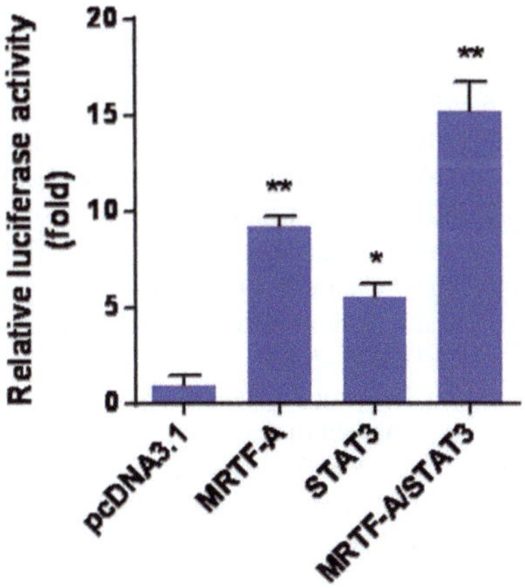

Fig. 4 The relative luciferase activity of vimentin by transfected with MRTF-A and STAT3

reported as a transcriptional co-activator of the serum response factor (SRF) [14]. It has been found that STAT3 and MRTF-A are involved in the process of epithelial mesenchymal transition [15]. Therefore, the construction of Vimentin luciferase reporter plasmid is necessary, which will testify us whether STAT3 and MRTF-A related to Vimentin. Our studies found that MRTF-A showed a significant effect to enhance the transcription activity of Vimentin promoter, STAT3 has a weak ability to enhance the activity of the Vimentin promoter. Based on our results, we speculate that MRTF-A have directly related to Vimentin promoter, and STAT3 and MRTF-A may exist protein-protein interactions. The next step, we will continue to study the mechanism of interaction between MRTF-A and STAT3, including determining their interaction binding sites. And we will continue to study which the signal pathway to achieved transcriptional regulation.

References

1. Colditz GA, Bohlke K (2014) Priorities for the primary prevention of breast cancer. CA Cancer J Clin 64:186–194
2. Zhang XH-F, Giuliano M, Trivedi MV, Schiff R, Osborne K (2013) Metastasis dormancy in estrogen receptor-positive breast cancer. Clin Cancer Res: Off J Am Assoc Cancer Res 19 (23). doi:10.1158/1078-0432
3. Chaffer CL, Brennan JP, Slavin JL, Blick T, Thompson EW, Williams ED (2006) Mesenchymal-to-epithelial transition facilitates bladder cancer metastasis: role of fibroblast growth factor receptor-2. Cancer Res 66(23):11271–11278
4. Vuoriluoto K, Haugen H, Kiviluoto S, Mpindi JP, Nevo J, Gjerdrum C, Tiron C, Lorens JB, Ivaska J (2011) Vimentin regulates EMT induction by Slug and oncogenic H-Ras and migration by governing Axl expression in breast cancer. Oncogene 30(12):1436–1448
5. Lee HT, Xue J, Chou PC, Zhou A, Yang P, Conrad CA, Huang S (2015) Stat3 orchestrates interaction between endothelial and tumor cells and inhibition of Stat3 suppresses brain metastasis of breast cancer cells. Oncotarget 6(12):10016–10029
6. Ko HS, Choi SK, Kang HK, Kim HS, Jeon JH, Park IY, Shin JC (2013) Oncostatin M stimulates cell migration and proliferation by down-regulating E-cadherin in HTR8/SV neo cell line through STAT3 activation. Reprod Biol Endocrinol RB&E, 11, 93
7. Klampfer L (2008) The role of signal transducers and activators of transcription in colon cancer. Front Biosci 1(13):2888–2899
8. Li S, Chang S, Qi X, Richardson JA, Olson EN (2006) Requirement of a myocardin-related transcription factor for development of mammary myoepithelial cells. Mol Cell Biol 26 (15):5797–5808
9. Ni J, Dong Z, Han W, Kondrikov D, Su Y (2013) The Role of RhoA and cytoskeleton in myofibroblast transformation in hyperoxic lung fibrosis. Free Radical Biol Med 0:26–39
10. O'Connor JW, Gomez EW (2013) Cell adhesion and shape regulate TGF-Beta1-induced epithelial-myofibroblast transition via MRTF-A signaling. PLoS One 8(12):e83188
11. Liao XH, Wang N, Liu LY, Zheng L, Xing WJ, Zhao DW, Sun XG, Hu P, Dong J, Zhang TC (2014) MRTF-A and STAT3 synergistically promote breast cancer cell migration. Cell Signal 26(11):2370–2380
12. Lehtinen L, Ketola K, Mäkelä R, Mpindi J-P, Viitala M, Kallioniemi O, Iljin K (2013) High-throughput RNAi screening for novel modulators of vimentin expression identifies MTHFD2 as a regulator of breast cancer cell migration and invasion. Oncotarget 4(1):48–63
13. Wang Y, Zhou BP (2011) Epithelial-mesenchymal transition in breast cancer progression and metastasis. Chin J Cancer 30(9):603–611
14. Morita T, Mayanagi T, Sobue K (2007) Dual roles of myocardin-related transcription factors in epithelial–mesenchymal transition via slug induction and actin remodeling. J Cell Biol 179 (5):1027–1042
15. Lamouille S, Xu J, Derynck R (2014) Molecular mechanisms of epithelial–mesenchymal transition. Nat Rev Mol Cell Biol 15(3):178–196

STAT5A and MKL-1 Activate the Activity of Luciferase Reporter Plasmid Containing FOXP3 Gene Promoter

Jia-Peng Li, Hui Li, Xiao-Yu Zhang, Cheng-Xi Yu, Yuan Xiang, Ze Yin, Xing-Hua Liao and Tong-Cun Zhang

1 Introduction

Adult idiopathic thrombocytopenic purpura (ITP) is a organ-specific autoimmune disease [1]. The thrombocytopenia of ITP is mainly ascribe to the early destruction of platelets by the reticuloendothelial system, following their sensitization by ant platelet autoantibody [2]. Other mechanisms also contribute, such as complement mediated lysis [3], ineffective thrombopoiesis [2], or direct T-cell cytotoxicity [4]. The disordered cellular immunity is also crucial in the pathophysiology of ITP [5–7]. Decreased number and function of Treg cells might play a role on the immune regulation dysfunction in ITP [8].

T regulatory (Treg) cells comprise a population of cells enriched in $CD4^+CD25^+T$ cells that attenuates immune responses against self-antigens [9–11]. Several studies have been demonstrated the role of Treg cells in the pathogenesis of human autoimmune disease [12, 13]. More recently, several groups discovered that a subset of $CD4^+CD25^+T$ cells expresses the transcription factor *FOXP3*, which is necessary and sufficient for Treg cell development and function [14–17]. *FOXP3* is highly conserved in mice and humans, and can be used as a reliable marker for the Treg lineage.

Stat5a and Stat5b are two closely related proteins that have overlapping functions with respect to lymphoid development and differentiation [18, 19]. Gene

J.-P. Li (✉) · H. Li · X.-Y. Zhang · C.-X. Yu · Y. Xiang · Z. Yin · X.-H. Liao (✉) · T.-C. Zhang (✉)
Institute of Biology and Medicine, Wuhan University of Science and Technology, No. 2 Huangjiahu, Wuhan 430065, China
e-mail: 13277973853@163.com

X.-H. Liao
e-mail: xinghualiao@hotmail.com

T.-C. Zhang
e-mail: zhangtongcun@wust.edu.cn

H. Liu et al. (eds.), *Advances in Applied Biotechnology*, Lecture Notes in Electrical Engineering 444, https://doi.org/10.1007/978-981-10-4801-2_86

targeting of Stat5a and Stat5b (collectively referred to as Stat5), results in impairment in the development of T, B, and natural killer (NK) cells [20]. The essential role of Stat5a/b in regulating Treg cells has been proved. Stat5a/b was also found bound directly to the *FOXP3* gene [21].

Myocardin and myocardin-related transcription factors (MRTFs) act as powerful transcriptional coactivators of SRF in mammalian cells [22]. It has been proved there may be a direct protein interaction between MKL-1 and STAT5A, and the *FOXP3* promoter sequence containing a specific bonding site of MKL-1, which prompt that MKL-1 may also bound directly to the *FOXP3* gene.

In this study, luciferase reporter with *FOXP3* promoter was successfully constructed. Then the activation of the *FOXP3* promoter was detected in 293T cells by luciferase reporter assay after transfected one of the expression vector of human MKL-1 and human STAT5A or both. The result showed that transfected each of the expression vectors can enhance the transcriptional activity of *FOXP3*, and this activity effect can be enhancing when the two vectors were cotransfected. Our research will further reveal the proliferation and differentiation of Treg cells and may provide theoretical foundations for methods to cure ITP.

2 Materials and Methods

2.1 Cell Culture

Human renal epithelial cell line 293T was obtained from American Type Culture Collection. 293T cells were cultured in Dulbecco's modified Eagle's medium (DMEM) (GIBCO) supplemented with 10% fetal bovine serum (FBS) at 37 °C in an atmosphere of 5% CO_2 incubator.

2.2 Plasmid Construction

The human genome was extracted from 293T cells. The cDNA was obtained by reverse transcription used RNA extracted from Jurkat cell line. FOXP3 promoter fragment from −1847 to +10 bp is amplified by PCR using the primers (F: 5′-CCTAGCTAGCCACACCCAAGCCATTTTGG-3′; R: 5′-GCATCTCGAGCTC GAGTAGTCCAGCAGCTGATAA-3′), PCR condition is as follows: pre-degeneration for 5 min at 95 °C, denaturation for 1 min at 95 °C, annealing for 30 s at 55 °C, and extension for 2 min at 72 °C. PCR reaction was carried out for 32 cycles and PCR product was visualized in 1% agarose gels stained with ethidium bromide under UV transillumination.

The PCR product of *FOXP3*-pGL3-Promoter vehicle plasmid were digested with restriction enzyme NheI and XhoI at 37 °C for 2 h. These fragments of *FOXP3* PCR product and pGL3-Promoter vehicle plasmid were mixed with 2 μL T4 ligase buffer and 1 μL T4 DNA ligase, incubated at 16 °C for 24 h, and then transformed into competent *E. coli*. A single colony was picked and cultured in LB which contains ampicillin. This plasmid was extracted and sequenced.

2.3 Transfection

293T cells at logarithmic phase were seeded in 24-well plates at the density of 1×10^5 cells/well in an atmosphere of 5% CO_2 at 37 °C. The 60–70% confluent 293T cells were cotransfected with 1 μg of total plasmid containing 0.2 μg *foxp3* promoter luciferase reporter plasmid and MKL-1 or/and STAT5A expression plasmids or control plasmids (pcDNA3.1) using TurboFect reagent (Fermentas) according to the instructions of the manufacturer. Mainly, the overexpression vector plasmid pcDNA3.1 was added to make sure that every group had a same quality of total plasmid, 1 μg total plasmids were added into 2 μL Turbo reagents, and, and then added to the medium without serum. incubated for 20 min. After transfection for 6 h, the medium without serum were replaced with the medium containing 10% fetal calf serum. Transfected with comparable quality of pcDNA3.1 vector plasmids can be used as a negative control. And a green fluorescent protein (GFP) plasmid was used to test the efficiency of transfection.

2.4 Luciferase Reporter Assays

Firefly luciferase activity was determined using the Luciferase Assay system (Promega) after cells were lysed with passive lysis buffer according to the instructions of the manufacturer. The content of total protein was measured with a BCA protein assay kit (Beyotime).The firefly luciferase activities were measured by Microplate Reader (Molecular Devices). Relative luciferase activity was calculated by normalizing the firefly luciferase activity (*FOXP3*-pGL3-Promoter) to the internal control content of total protein. All experiments were performed at least three times with different preparations of plasmids and primary cells, producing qualitatively similar results. Columns represent the means of three independent experiments expressed relative to the luciferase activities obtained for untreated cells. The error bars represent the standard errors of the mean. The data was detected using t-test.

3 Result

3.1 The FOXP3 Promoter Contained Key Sites

Based on the results of fragment competition test and homology, the FOXP3 promoter contained some key sites which located at base pairs 1827–1837 (5′-CCATTTTTGG-3′), 1476–1482 (5′-TCTTTC-3′) (Fig. 1)

3.2 Construction of FOXP3 Promoter Luciferase Reporter Plasmids

Agarose gel electrophoresis was performed to estimate the PCR amplification of FOXP3 promoter. As shown in Fig. 2a, a specific band emerged at the site of 1877 bp, which represented nucleotide fragment of FOXP3 promoter (between upstream 2000 and downstream 1500 bp).

Then, the PCR nucleotide fragment was digested by 2 restriction enzyme *Nhe*I and *Xho*I and cloned to the pGL3-Promoter vector. Extracted and purified the recombinant plasmids and agarose gel electrophoretic was performed. Figure 2b represented agarose gel electrophoretic analysis of recombinant plasmids.

We digest the recombinant plasmid with the 2 respective cloning restriction enzymes *Nhe*I and *Xho*I and then electrophoresed through agarose gel to confirm these recombinant plasmids. As shown in Fig. 2c, the recombinant plasmid was cut into two bands. The first lane contained two bands, about 4999 and 1877 bp, which respectively represented pGL3-Promoter vehicle plasmid and *FOXP3* promoter nucleotide fragment. This luciferase reporter plasmid of *FOXP3* gene promoter was finally confirmed by sequencing. DNA sequence alignments showed that we successfully constructed the luciferase reporter plasmid containing *FOXP3* promoter nucleotide fragment.

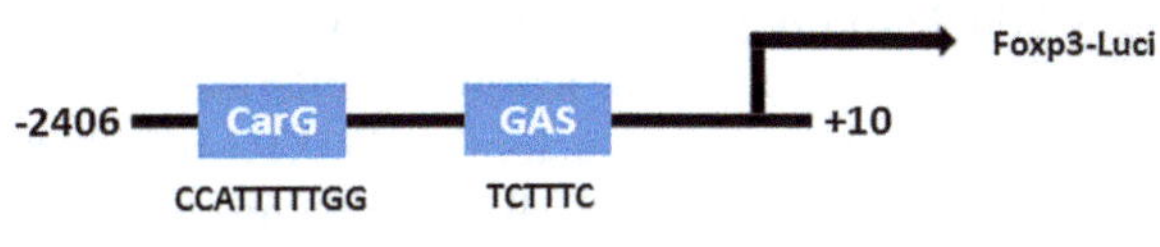

Fig. 1 The *FOXP3* promoter contained key sites. The *foxp3* promoter contained a MKL-1 and a STAT5A bInding sequence

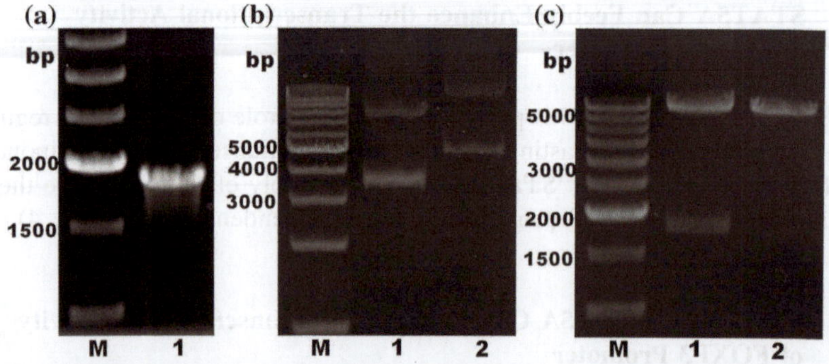

Fig. 2 Agarose gel electrophoretic analysis of FOXP3 promoter luciferase reporter plasmids. **a** Agarose gel electrophoretic analysis of PCR product. *M* 500 bp DNA marker; *1* Foxp3 gene promoter. **b** Agarose gel electrophoretic analysis of recombinant plasmids. *M* 1 kb DNA marker; *1* pGL3-Promoter vehicle plasmid; *2* Foxp3 recombinant plasmid. **c** Agarose gel electrophoretic analysis of recombinant plasmids by digestion. *M* 500 bp DNA marker; *1* foxp3 recombinant plasmid; *2* pGL3-Promoter vehicle plasmid

3.3 Luciferase Assay

3.3.1 The Efficiency of Transfection

The transfection efficiency was demonstrated using an EGFP (enhanced green fluorescent protein) expression plasmid. The transfection efficiency was approximately 70%, demonstrating that DNA was transfected efficiently into 293T cells.

3.3.2 MKL-1 Can Obviously Enhance the Transcriptional Activity of FOXP3 Promoter

Luciferase Reporter Assay was performed to test the role of ML-1 in regulating FOXP3 transcription. Contrasting with control group transected with equiponderant pcDNA3.1 vehicle plasmid, MKL-1 showed a significant effect to enhance the transcription activity of FOXP3 promoter in a dose-dependent manner (Fig. 3).

Fig. 3 The transfection efficiency of 293T. Fluorescence microscope excitation glowing green fluorescent protein

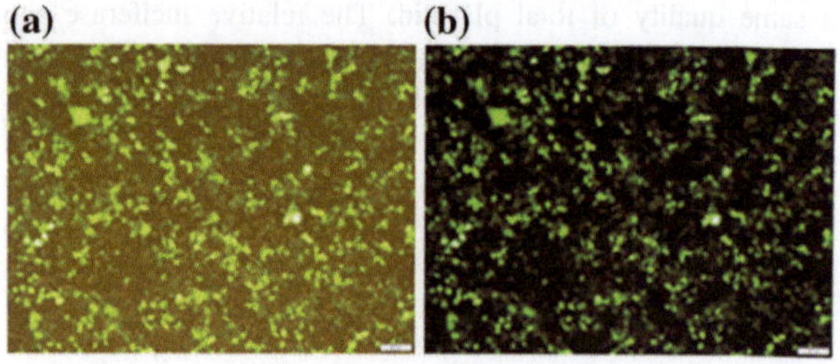

3.3.3 STAT5A Can Feebly Enhance the Transcriptional Activity of FOXP3 Promoter

Luciferase Reporter Assay was performed to test the role of STAT5A in regulating FOXP3 transcription. Contrasting with control group transected with equiponderant pcDNA3.1 vehicle plasmid. STAT5A showed a feebly effect to enhance the transcription activity of FOXP3 promoter in a dose-dependent manner (Fig. 4).

3.3.4 MKL-1 and STAT5A Can Enhance the Transcriptional Activity of FOXP3 Promoter

Luciferase Reporter Assay was performed to test the role of MKL-1 and STAT5A in regulating FOXP3 transcription. Contrasting with control group transected with equiponderant pcDNA3.1 vehicle plasmid, MKL-1 showed a significant effect to enhance the transcription activity of FOXP3 promoter STAT5A showed a feebly effect to enhance the transcription activity of FOXP3 promoter, and the transcription activity of FOXP3 promoter can be enhanced when MKL-1 and STAT5A were cotransfected (Figs. 5 and 6).

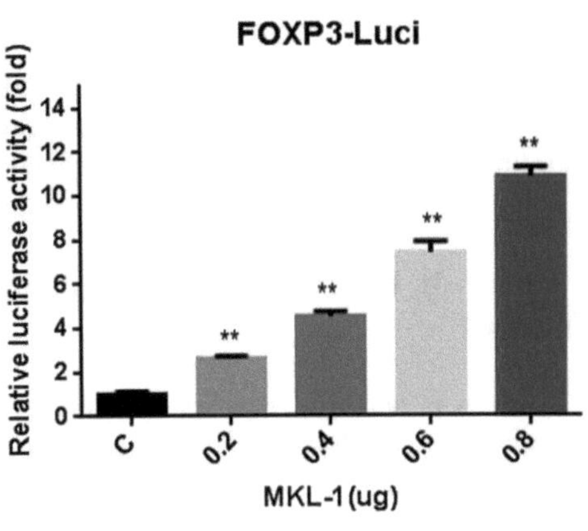

Fig. 4 MKL-1 enhanced the transcriptional activity of FOXP3 promoter. A gradient concentration quality of MKL-1 overexpression plasmid were cotransfected with *foxp3* promoter luciferase plasmid, and the overexpression vector plasmid pcDNA3.1 was added to make sure that every had a same quality of total plasmid. The relative luciferase activity increased with the increase of the amount of MKL-1 overexpression plasmid. All groups have significant difference when compared with control ($p < 0.001$)

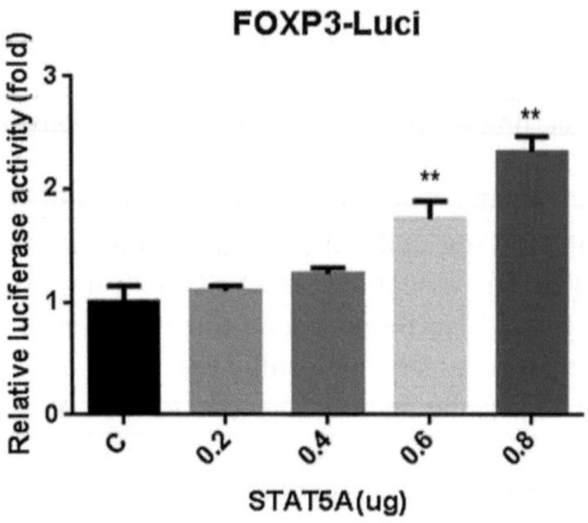

Fig. 5 STAT5A enhanced the transcriptional activity of FOXP3 promoter. A gradient concentration quality of STAT5A overexpression plasmid were cotransfected with *foxp3* promoter luciferase plasmid, and the overexpression vector plasmid pcDNA3.1 was added to make sure that every had a same quality of total plasmid. The relative luciferase activity increased with the increase of the amount of STAT5A overexpression plasmid. The relative luciferase activity of transfected 0.6 and 0.8 μg STAT5A overexpression plasmid has a significant difference when compared with control ($p < 0.001$)

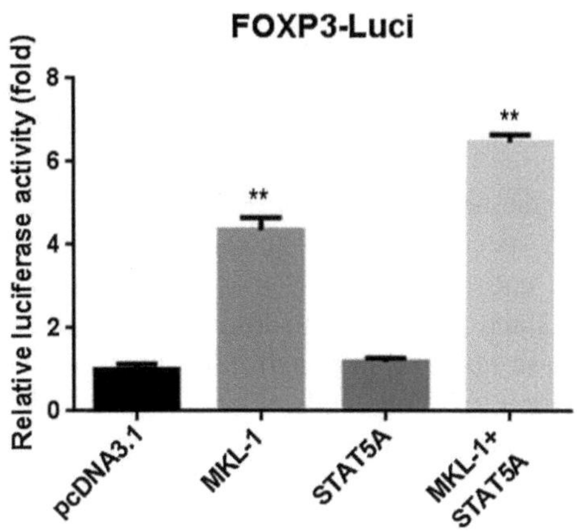

Fig. 6 MKL-1 and STAT5A enhanced the transcriptional activity of *foxp3* promoter. A same quality of 0.4 μg MKL-1 or STAT5A overexpression plasmid were cotransfected with foxp3 promoter luciferase plasmid, the relative luciferase activity increased when transfected MKL-1 overexpression plasmid ($p < 0.001$) or cotransfected MKL-1 and STAT5A overexpression plasmids ($p < 0.001$). And the fold of cotransfected MKL-1 and STAT5A overexpression plasmids is higher than transfected MKL-1 overexpression plasmid alone

4 Discussion

The abnormal cellular immunity is crucial in the pathophysiology of ITP [5–7]. T regulatory (Treg) cells comprise a population of cells enriched in $CD4^{+}CD25^{+}$ T cells that attenuates immune responses against self-antigens [9–11]. Decreased number and function of Treg cells might play a role on the immune regulation dysfunction in ITP [8]. The essential role of Stat5a/b in regulating Treg cells has been proved. Stat5a/b was also found bound directly to the FOXP3 gene [21], the expression of which is necessary and sufficient for Treg cell development and function [14–17]. However, the definite transcription mechanism of FOXP3 gene is unclear.

It has been proved there may be a direct protein interaction between MKL-1 and STAT5A, and there is a specific bonding site of MKL-1 in *FOXP3* promoter sequence, which prompt that MKL-1 may also bound directly to the *FOXP3* gene.

In this study, luciferase reporter with *FOXP3* promoter was successfully constructed. Then the activation of the *FOXP3* promoter was detected in 293T cells by luciferase reporter assay after transfected one of the expression vectors of human MKL-1 and human STAT5A or both, the transfection efficiency was demonstrated using an EGFP (enhanced green fluorescent protein) expression plasmid. The result showed that MKL-1 could enhanced the transcriptional activity of *FOXP3* promoter obviously, STAT5A could feebly enhanced the transcriptional activity of *FOXP3* promoter. There might have a weekly binding between STAT5A and the non-classical binding sequence in *FOXP3* promoter. And the activation effect could be enhanced when the expression vectors of MKL-1 and STAT5A vectors were cotransfected.

In the next phase, the activation of the *FOXP3* promoter with the mutation of MKL-1 and STAT5A specific binding site can be detected in 293T cells by luciferase reporter assay, and a CHIP assay may also be used to prove the definite binding site of MKL-1 and STAT5A. In addition, developments of new reagents to further understand these remarkable transfactors' function and reveal the proliferation and differentiation of Treg cells. Our research may provide theoretical foundations for methods to cure ITP.

References

1. Cines DB, Blanchette VS (2002) Immune thrombocytopenic purpura. N Engl J Med 346:995–1008
2. McMillan R (2000) The pathogenesis of chronic immune (idiopathic) thrombocytopenic purpura. Semin Hematol 37:5–9
3. Hed J (1998) Role of complement in immune or idiopathic thrombocytopenic purpura. Acta Paediatr Suppl 424:37–40
4. Olsson B, Andersson PO, Jernås M et al (2003) T-cell-mediated cytotoxicity toward platelets in chronic idiopathic thrombocytopenic purpura. Nat Med 9:1123–1124

5. Zhou B, Zhao H, Yang RC et al (2005) Multi-dysfunctional pathophysiology in ITP. Crit Rev Oncol Hematol 54:107–116
6. Semple JW, Freedman J (1991) Increased antiplatelet T helper lymphocyte reactivity in patients with autoimmune thrombocytopenia. Blood 78:2619–2625
7. Ware RE, Howard TA (1993) Phenotypic and clonal analysis of T lymphocytes in childhood immune thrombocytopenic purpura. Blood 82:2137–2142
8. Liu B, Zhao H, Poon MC et al (2007) Abnormality of $CD4^{+}CD25^{+}$ regulatory T cells in idiopathic thrombocytopenic purpura. Eur Haematol 78(2):139–143
9. Thornton AM, Shevach EM (1998) $CD4^{+}CD25^{+}$ immunoregulatory T cells suppress polyclonal T cell activation in vitro by inhibiting interleukin 2 production. J Exp Med 188:287–296
10. Shevach EM (2000) Regulatory T cells in autoimmmunity. Annu Rev Immunol 18:423–449
11. Sakaguchi S (2004) Naturally arising $CD4^{+}$ regulatory t cells for immunologic self-tolerance and negative control of immune responses. Annu Rev Immunol 22:531–562
12. van Amelsfort JM, Jacobs KM, Bijlsma JW et al (2004) $CD4^{+}CD25^{+}$ regulatory T cells in rheumatoid arthritis: differences in the presence, phenotype, and function between peripheral blood and synovial fluid. Arthritis Rheum 50:2775–2785
13. Balandina A, Lecart S, Dartevelle P et al (2005) Functional defect of regulatory $CD4^{+}CD25^{+}$ T cells in the thymus of patients with autoimmune myasthenia gravis. Blood 105:735–741
14. Fontenot JD, Gavin MA, Rudensky AY (2003) FOXP3 programs the development and function of $CD4^{+}CD25^{+}$ regulatory T cells. Nat Immunol 4:330–336
15. Fontenot JD, Rasmussen JP, Williams LM et al (2005) Regulatory T cell lineage specification by the forkhead transcription factor FOXP3. Immunity 22:329–341
16. Hori S, Nomura T, Sakaguchi S (2003) Control of regulatory T cell development by the transcription factor FOXP3. Science 299:1057–1061
17. Khattri R, Cox T, Yasayko SA et al (2003) An essential role for Scurfin in $CD4^{+}CD25^{+}$ T regulatory cells. Nat Immunol 4:337–342
18. Feldman GM, Rosenthal LA, Liu X et al (1997) STAT5A-deficient mice demonstrate a defect in granulocyte-macrophage colony-stimulating factor-induced proliferation and gene expression. Blood 90:1768–1776
19. Imada K, Bloom ET, Nakajima H et al (1998) Stat5b is essential for natural killer cell-mediated proliferation and cytolytic activity. J Exp Med 188:2067–2074
20. Yao Z, Cui Y, Watford WT et al (2006) Stat5a/b are essential for normal lymphoid development and differentiation. Proc Natl Acad Sci U S A 103:1000–1005
21. Yao Z, Kanno Y, Kerenyi M et al (2007) Nonredundant roles for stat5a/b in directly regulating FOXP3. Blood 109(10):4368–4375
22. Han Z, Li X, Wu J et al (2004) A myocardin-related transcription factor regulates activity of serum response factor in drosophila. Proc Natl Acad Sci U S A 101(34):12567–12572

Study the Role of E-selectin and Its Ligand sLeX in the Adhesion Between THP-1 Cells and HUVEC Cells

Qian Zhang, Huan Liu, Chaoran Yao, Tingshen Li, Xuehui Li, Li Zhang, Zhen Liu, Peng Yu and Yuou Teng

1 Introduction

It is all known that inflammation plays an important role in occurrence and development of tumor. In the process of the formation of tumors, inflammatory factors provide a suitable micro-environment that promotes the growth and metastasis of tumor. At the end of tumor formation, tumor cells are facilitated by the destruction of the immune function which may be the metastasis mechanism of tumor cells in inflammation. Meanwhile, the injured endothelial cells not only make inflammation aggravated but also can induce malignant tumors to diffuse. E-selectin is not constitutively expressed by endothelial cells [1–3] but its expression is stimulated by inflammatory molecules such as TNF-α, IL-β and LPS. Therefore, E-selectin has been associated with tumor angiogenesis and metastasis in a variety of cancers [4, 5].

E-selectin (64 kDa), the member of selectin family, is represented by three receptors composed of calcium-dependent type I transmembrane glycoproteins [1, 6]. Relative molecular weight values of 100 and 115 kDa have been detected for different glycosylated forms [7]. It is constituted by a series of protein fragments including an amino terminal lectin-like domain, followed by an epidermal growth factor (EGF)-like domain and six repeated motifs (about 60 amino acids each)

Q. Zhang · H. Liu · C. Yao · T. Li · X. Li · L. Zhang · Z. Liu · P. Yu · Y. Teng (✉)
Key Laboratory of Industrial Fermentation Microbiology, Tianjin Key Laboratory of Industrial Microbiology, Sino-French Joint Lab of Food Nutrition/Safety and Medicinal Chemistry, College of Biotechnology, Tianjin University of Science and Technology, Tianjin 300457, People's Republic of China
e-mail: tyo201485@tust.edu.cn

H. Liu et al. (eds.), *Advances in Applied Biotechnology*, Lecture Notes in Electrical Engineering 444, https://doi.org/10.1007/978-981-10-4801-2_87

similar to those found in some complement-binding proteins [1]. The selectin has variety of ligand, such as E-selectin ligand-1 (ESL-1), p-selectin glycoprotein ligand-1 (PSGL-1), glycosylation-dependent cell adhesion molecule-1 (GlyCAM-1). However, the sialyl Lewis X (sLex) and sialyl Lewis A (sLeA) are the minimum unit of the ligand recognized by the selectin. The lectin with conservative sugar-binding domain, ligand-binding domain of E-selectin, can combine with sLex specificity [8]. It's worth noting that both of leukocytes and tumor cells can express sLex. While tumor cells themselves have ability to induce endothelial cells to express E-selectin [5].

As reported, modulation of Ca^{2+} by engagement of E-selectin receptor starts signal transduction pathways that affect cell spreading, tyrosine phosphorylation signaling, and cancer cell motility [9]. Therefore, we use Calcein-AM to stain THP-1 cells.

The model of THP-I adhesion to HUVEC we constructed is indispensable ingredient in building up a system that consents us to evaluate the ability of targeted compound to inhibit the progress.

2 Experimental

2.1 *Material*

Calcein-AM was purchased from Molecular probes. Flow cytometry was purchased from the company of BD, USA. Microplate reader was purchased from SYNERGY, BioTek, USA

2.2 *Method*

2.2.1 Cell Culture

THP-1 cells (purchased from the Shanghai Institutes of Biological Sciences, China) were grown in suspension culture in RPMI 1640 medium supplemented with 10% fetal calf serum, 100 U/mL glutamine and 100 mg/mL penicillin (all from Biological Industries, Kibbutz Beit-Haemek, Israel) at 37 °C under an atmosphere consisting of 95% air and 5% CO_2. Human umbilical vein endothelial cells (HUVEC, purchased from the Shanghai Institutes of Biological Sciences, China) were grown in F-12 medium supplemented with 10% fetal calf serum, 100 U/mL glutamine and 100 mg/mL penicillin/streptomycin (all from Biological Industries, Kibbutz Beit-Haemek, Israel) at 37 °C under an atmosphere consisting of 95% air and 5% CO_2. The cells we used are in the logarithmic growth phase in the experiments.

2.2.2 The Optimization of Inflammatory Cytokines and Stimulation Time for the Expression of E-selectin

HUVEC cells (1 × 10^5 cells/mL) were seeded into 12-well plates for 24 h at 37 °C. The cells were then activated with 20 ng/mL TNF-α, 1 μg/mL LPS, 1 ng/mL IL-β for 0, 2, 4, 6, and 8 h, respectively. The cells were harvested by centrifugation at 1000 g for 5 min. 3% paraformaldehyde was used to sustain the cell at room temperature for 30 min, and then the cells were rinsed by cold PBS and fixed with 5% BSA for 30 min at room temperature. Then the cells were washed by cold PBS and fixed with anti-E-selectin (1:50) at 4 °C overnight. Control cells were washed only by PBS. HUVEC were rinsed by cold PBS and added goat-anti-E-selectin contained 5% BSA (1:50) at 4 °C for 30 min in the dark. The cells were washed by cold PBS again and analyzed by flow cytometry.

2.2.3 The Optimization of Adhesion Conditions Between THP-1 Cells and HUVEC

Poly-L-lysine (100 μg/mL) was coated onto 96-well plates (Costar 9600) at 12 h at 37 °C. HUVEC cells were seeded at a density of 1 × 10^5 cells/mL, 2 × 10^5 cells/mL, and 3 × 10^5 cells/mL, respectively at 36 h at 37 °C onto the coated 96-well plates. The cells were then activated with 20 ng/mL TNF-α, 1 μg/mL LPS, 1 ng/mL IL-β in medium for 0, 2, 4, 6, and 8 h, respectively. Control cells were left untreated. Meanwhile, THP-1 cells were collected by centrifugation at 1000 *g* for 5 min, washed with 5 mL of PBS, and stained with calcein-AM (10 μM) for 45 min at 37 °C. After removing the treatments, HUVECs were incubated with 1 × 10^6 cells/mL of stained THP-1 cells for 30 min at 37 °C. Then HUVECs were washed three times with PBS to remove unbound THP-1 cells. Then the plates were kept at −80 °C overnight, thawed at 37 °C, and the fluorescence of cell lysates was read with a microplate reader (ex, 485 nm; em, 535 nm) [10].

2.2.4 Statistical Analysis

All data were expressed as mean ± S.D. Results were analyzed by one-way analysis of variance (ANOVA), and significant differences were determined by post hoc Tukey test using SPSS 21.0 software.

3 Results and Discussion

3.1 *The Determination of Inflammatory Factor for E-selectin Expression*

E-selectin is a transmembrane protein with a molecular weight of 64 KDa. E-selectin, only expressed on activated endothelial cells, is now considered as an association with development of cancer. We intend to optimize the model of an adhesion between HUVEC and THP-1 through the stimulation of inflammatory factors. According to the result of flow cytometry, the concentration of E-selectin reached the maximum when TNF-α stimulated HUVEC for 6 h (Fig. 1). Meanwhile, when HUVEC was activated by different doses of TNF-α, the number of THP-1 cells adhesion to HUVEC and concentration of TNF-α was dose-dependent. As shown in Fig. 2, the concentration of E-selectin is higher with 20 ng/mL TNF-α than 10 ng/mL TNF-α did in 6 h. Unfortunately, the value of lower concentration of inflammatory factor has few effect to HUVEC compared with the negative group.

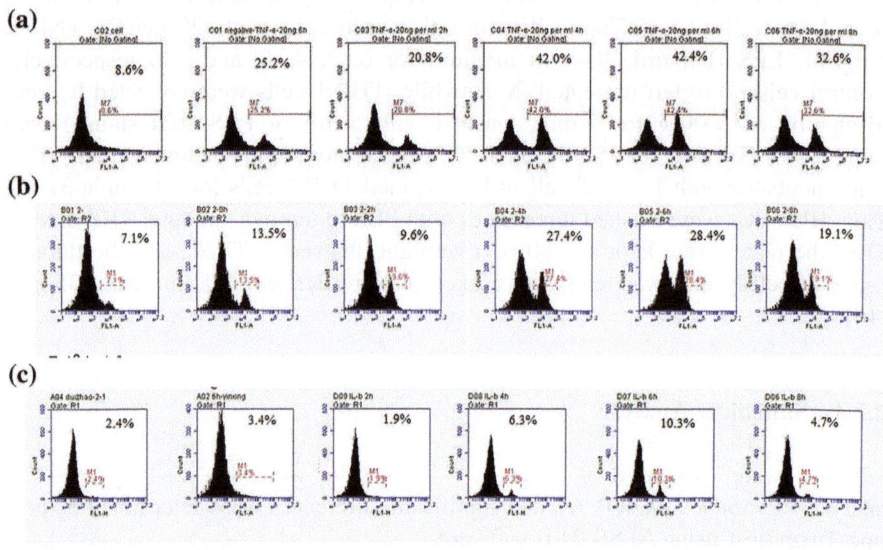

Fig. 1 Effects of three kinds of inflammatory factors on HUVEC at 2, 4, 6, 8 h respectively. At the time points indicated, cells labeled with anti-E-selectin was determined using FL1-A analysis. Data are presented as Mean ± S.D. of three separate experiments. HUVEC cells were treated with 20 ng/mL TNF-α (**a**), 1 μg/mL LPS (**b**) or 1 ng/mL IL-β (**c**) for 0, 2, 4, 6, 8 h, respectively

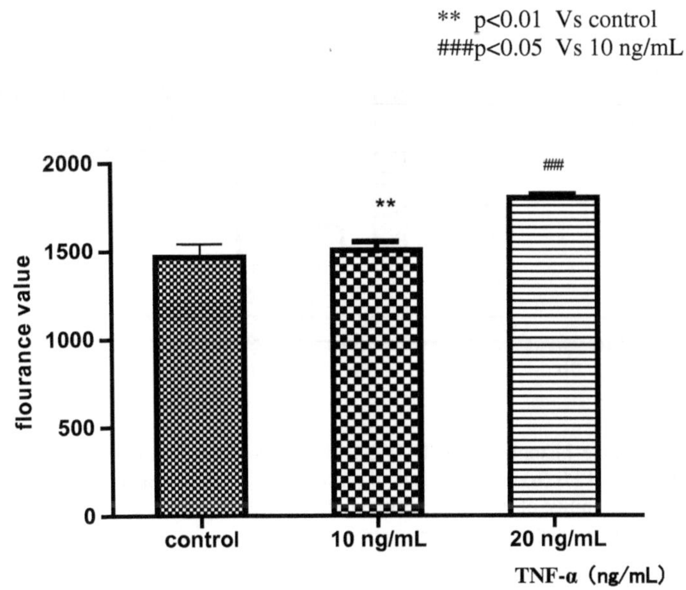

Fig. 2 Representative histograms of the expression of E-selectin induced by TNF-α (10 or 20 ng/mL) for 6 h. The fluorescence value presented the ability of adhesion between HUVEC and THP-1. The data was presented as mean ± S.D. of three independent experiments

3.2 The Determination of Incubation Time for E-selectin Expression

The fluorescence value of TNF-α, LPS, IL-β are 5213.39, 420.83, and 180.00 at 4 h. In addition, the data got to 9538.00, 2233.00, and 4444.00 at 6 h. Compared with three inflammatory factors, we observed that TNF-α with 20 ng/mL reached the highest fluorescence value when the HUVEC was incubated with it in 6 h. E-selectin peaked at 6 h when stimulated by three inflammatory cytokines (Fig. 3), which consistent with others study [11]. However, we perceive the fluorescence value of LPS keep on increasing while the value of other inflammatory factors is decreased, especially IL-β.

3.3 The Determination of Cell Density for E-selectin Expression

The density of HUVEC was also evaluated in the experiment. During the experiment, we designed three kind of density of HUVEC: 1×10^5, 2×10^5, 3×10^5 cells/mL. All of them were stimulated with TNF-α in 6 h. As shown in

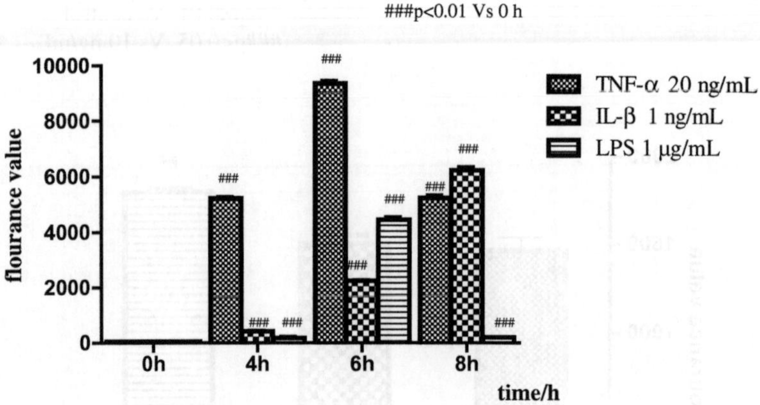

Fig. 3 Representative histograms for HUVEC binding to THP-1 at 0, 4, 6, 8 h by 20 ng/mL TNF-α (**a**), 1 μg/mL LPS (**b**) or 1 ng/mL IL-β (**c**) for 0, 2, 4, 6, 8 h, respectively. The fluorescence value presented the ability of adhesion between HUVEC and THP-1. The data was presented as mean ± S.D. of three independent experiments

Fig. 4, both of the groups of 20 ng/mL were higher than the 10 ng/mL ones. Thus, the value of 3×10^5 cells/mL group is lower than negative. It is probably related to the density of HUVEC, which is so high resulted the unenough THP-1 adhesion to it.

3.4 Discussion

Recent finding demonstrated that selectins mediate the adhesion by interacting with specific carbohydrate ligands on the cell surface [12]. Tumor cell adhesion to endothelial cells is an important part of tumor metastasis. Adhesion is not only the basis of directional metastasis of tumor cells, but also is the mainly method to help tumor immune evasion. E-selectin is not constitutively expressed by endothelial cells but expressed by inflammatory molecules stimulating such as tumor necrosis factor (TNF-α), interleukin-1 (IL-1) and bacterial lipopolysaccharide (LPS) [1–3, 6]. sLeX contained oligosaccharides, is a group of carbohydrate antigens with sugar esters and glycoprotein in the surface of tumor cells. sLex antigens, the ligand of E-selectin, is expressed on the surface of cells mostly [13, 14]. Therefore, it had a specificity of the interaction between leukocytes or tumor cells and human umbilical vein endothelial cells. This study found that the number of THP-1 cells adhesion to HUVEC significantly increased after HUVEC cells cultivated with TNF-α for 6 h. Some recent researches had shown that expression of E-selectin reached to the maximal after cytokine stimulation for 4–6 h and then decreases rapidly [11]. Meanwhile, the fluorescence value is the highest with stimulation of 20 ng/mL TNF-α. In the course of clinical treatment of cancer, the lack of tumor

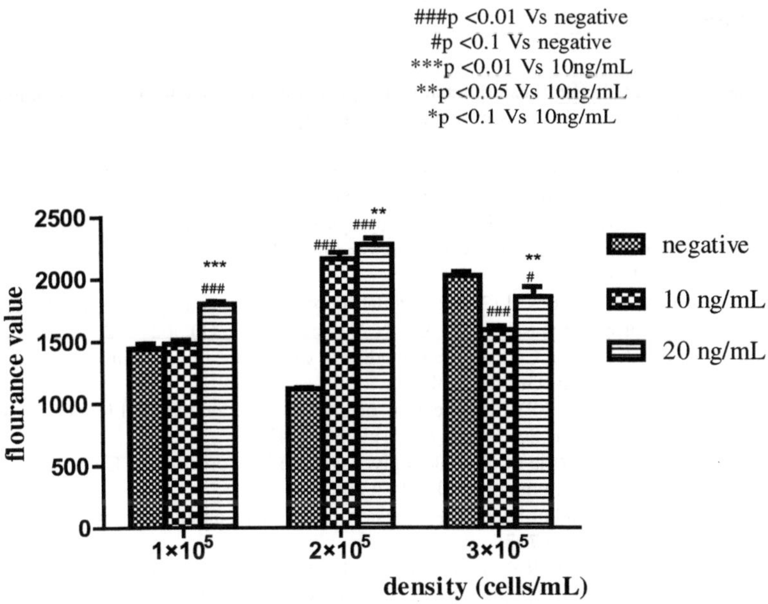

Fig. 4 The fluorescence value at different densities of HUVEC induced by TNF-α (20 ng/mL). The densities of HUVEC are 1×10^5, 2×10^5, 3×10^5 cells/mL respectively. All of them were stimulated with TNF-α for 6 h. The fluorescence value presented the ability of adhesion between HUVEC and THP-1. The data was presented as mean of three independent experiments

specificity displayed by low-molecular-weight drugs for cancer treatments often results in significant toxicity to non-cancerous tissues [15]. Therefore, E-selectin and its ligand sLex lead to the rolling of leukocytes on inflamed endothelial cells which is the first step of transmigration to the surrounding tissue [16].

4 Conclusion

In our study, we optimized the model of THP-1 adhesion to HUVEC including plating density and the time needed that E-selectin can express when inflammatory factors stimulate to endothelial cells. We demonstrated that the number of THP-1cells adhesion to HUVEC increased significantly after HUVEC cells cultivated with TNF-α for 6 h. Compared with three inflammatory molecules, the expression of E-selectin had a significant increase by TNF-α.

This study provides a foundation which can support targeted delivery of chemotherapeutics and imaging agents to tumor vasculature for therapeutic and diagnostic applications.

References

1. Bevilacqua MP et al (1989) Endothelial leukocyte adhesion molecule 1: an inducible receptor for neutrophils related to complement regulatory proteins and lectins. Science 243 (4895):1160–1165
2. Montgomery KF et al (1991) Activation of endothelial-leukocyte adhesion molecule 1 (ELAM-1) gene transcription. Proc Natl Acad Sci 88(15):6523–6527
3. Wong D, Dorovinizis K (1996) Regulation by cytokines and lipopolysaccharide of E-selectin expression by human brain microvessel endothelial cells in primary culture. J Neuropathol Exp Neurol 55(2):225–235
4. Julie L, François H, Jacques H (2002) Regulation of the metastatic process by E-selectin and stress-activated protein kinase-2/p38. Ann N Y Acad Sci 973(1):562–572
5. Kannagi R et al (2004) Carbohydrate-mediated cell adhesion in cancer metastasis and angiogenesis. Cancer Sci 95(5):377–384
6. Bevilacqua MP et al (1987) Identification of an inducible endothelial-leukocyte adhesion molecule. Proc Natl Acad Sci U S A 84(24):9238–9242
7. Påhlsson P et al (1995) Role of N - linked glycosylation in expression of E-selectin on human endothelial cells. Eur J Immunol 25(9):2452–2459
8. Bevilacqua MP Endothelial-leukocyte adhesion molecules—annual review of immunology. Annu Rev Immunol 11(1):767
9. D'Amato M et al (2003) Role of calcium in E-selectin induced phenotype of T84 colon carcinoma cells. Biochem Biophys Res Commun 301(4):907–914
10. Amoozgar Z et al (2013) Development of quinic acid-conjugated nanoparticles as a drug carrier to solid tumors. Biomacromol 14(7):2389–2395
11. Asmuth EJUV et al (1992) Evidence for endocytosis of E-selectin in human endothelial cells. Eur J Immunol 22(10):2519–2526
12. Kansas GS (1996) Selectins and their ligands: current concepts and controversies. Blood 88 (9):3259–3287
13. Walz G et al (1990) Recognition by ELAM-1 of the sialyl-Lex determinant on myeloid and tumor cells. Science 250(4984):1132–1135
14. Srinivas U, Påhlsson P, Lundblad A (1996) E-selectin: sialyl Lewis, a dependent adhesion of colon cancer cells, is inhibited differently by antibodies against E-selectin ligands. Scand J Immunol 44(3):197–203
15. Shamay Y et al (2009) E-selectin binding peptide–polymer–drug conjugates and their selective cytotoxicity against vascular endothelial cells. Biomaterials 30(32):6460–6468
16. Olofsson AM et al (1994) E-selectin mediates leukocyte rolling in interleukin-1-treated rabbit mesentery venules. Blood 84(8):2749–2758

The Application of Microbial Technology in Harbor Engineering: The Impact of Extracellular Polymeric Substances on the Sedimentation and Properties of Fluid Mud

Xiaohua Chen, Qixiu Pang, Mengnan Li, Baojiang Sun, Ruibo Zhang, Shiru Jia and Peipei Han

1 Introduction

Extracellular polymeric substances (EPS), a complex high-molecular-weight mixture of polymers, were secreted by a variety of microorganisms in particular environment, which was mainly composed of carbohydrate and protein, accounting for about 70–80% of the total EPS [1–3]. EPS has a significant influence on the physicochemical properties of microbial aggregates [4]. Because of diversity of property, EPS are widely used in various fields, such as the field of wastewater treatment, metallurgical industries, the field of food processing and the biomedical industry [5]. It was found that the bacteria and their EPS played an important role in the formation of stable fluid mud layers consisting mainly of silt and clay-sized sediments [6], so it is also applied in some silt harbors to delaying the sedimentation of fluid mud.

There are many large-medium silt harbors around the world, most of them have a problem that nautical thickness decreased due to the serious silting and sedimentation of fluid mud, bring difficult to voyage. Moreover, regular dredging brings great economic pressure to the Port enterprises [7]. With years of exploration, the

X. Chen · M. Li · B. Sun · S. Jia · P. Han (✉)
Key Laboratory of Industrial Fermentation Microbiology,
Ministry of Education, College of Biotechnology, Tianjin University
of Science & Technology, Tianjin 300457, China
e-mail: pphan@tust.edu.cn

Q. Pang · R. Zhang
Tianjin Research Institute of Water Transport Engineering, Key Laboratory
of Engineering Sediment of Ministry of Transport, Tianjin 300456,
People's Republic of China

H. Liu et al. (eds.), *Advances in Applied Biotechnology*, Lecture Notes
in Electrical Engineering 444, https://doi.org/10.1007/978-981-10-4801-2_88

Emden Port in Germany effectively decreased the settling velocity by microorganisms in fluid mud. The EPS secreted by these bacteria could be bond to the surface of sediment and decreased the settling velocity and therefore maintained the fluid mud a stable state for several months. However, the mechanism that the EPS secreted by microorganisms delayed the sedimentation of fluid mud still remains unclear.

Based on the previous findings, the aim of the study was to investigate the properties of EPS secreted by microorganisms, explore its influence on the sedimentation of fluid mud, and reveal the underlying mechanism for EPS delaying the sedimentation of fluid mud. The EPS produced by *Burkholderia* sp. TKS1 in the study was isolated and characterized. The influence of EPS on the sedimentation of fluid mud was studied and the underlying mechanism delaying the sedimentation of fluid mud was subsequently investigated.

2 Materials and Methods

2.1 *Strain and Culture Conditions*

The strain used to produce EPS was *Burkholderia* sp. TKS1 isolated from fluid mud from one muddy port in China. The culture conditions followed our early report [5], and medium composition was as follows: 21.0 g/L sucrose, 1.1 g/L NH_4Cl, 0.2 g/L $MgSO_4$, 1.0 g/L KH_2PO_4, $Na_2HPO_4 \cdot 12H_2O$ 2.0 g/L, $CaCO_3$ 1.5 g/L.

2.2 *Extraction and Analysis of Chemical Constituents of EPS*

After incubation for 3 days, the medium was centrifuged at 4000 r/min for 15 min after diluting with 4 volumes of distilled water. A total of 3 volumes of pre-chilled ethanol is added to the supernatant to collect the precipitate of crude EPS. The crude EPS was dissolved in distilled water, dialyzed by 2000 Da dialysis bag at 4 ° C for 48 h, and then preserved at 4 °C after being freeze-dried to obtain purified EPS, which was used in the subsequent experiment for testing the effects of EPS on sedimentation of fluid mud.

Carbohydrate content in the EPS was measured by phenol-sulfuric acid method using glucose as standard [8]. Protein content was measured according to the method of Coomassie brilliant blue [9]. Nucleic acid content was measured by spectrophotometer method (BioSpectrometer basic, Eppendorf, Germany). Lipid content was determined by vanillin assay and cholesterol was used as the standard [10].

Carbohydrate in the EPS was purified by ion exchange resin DEAE-650M [11]. The column was eluted with deionized water first, and then with a linear gradient of 0–0.5 mol/L NaCl solution at a flow rate of 1.0 mL/min. The fractions were collected and the carbohydrates were monitored via phenol-sulfuric acid method. Carbohydrate-positive fractions were merged and lyophilized.

The purity of carbohydrate was measured by the High Performance Liquid Chromatography (HPLC, Dionex, USA) equipped with RI-101 refractive index detector (RID) and TSK G4000PWXL sugar column (7.8 mm × 30.0 cm). The column oven was set at 30 °C and 20 μL of sample was injected. The mobile phase is 0.9% NaCl solution with a flow rate of 0.5 mL/min.

Monosaccharide composition was analyzed using 1-Phenyl-3-methyl-5-pyrazolone-High Performance Liquid Chromatography (PMP-HPLC, Dionex, USA). The separation was carried out using C18 column (250 mm × 4.6 mm i.d., 5 μm, ZORBAX SB-C18, Aglient, USA). The wavelength of 250 nm was selected for HPLC analysis and 10 μL of injection volume was injected. During measurements the column oven was set to 30 °C and the mobile phase consist of 17% acetonitrile and 0.05 mmol/L phosphate buffer solution (KH_2PO_4-NaOH, pH 6.9) with a flow rate of 1 mL/min.

2.3 *Measurement of Effects of EPS on the Properties of Fluid Mud*

The mud used in the study was collected from the fluid mud layer of one muddy port in China. The sedimentation of fluid mud after addition of EPS with different concentrations was evaluated by settled sediment volume, which was determined by measuring the volume of settled sediment in the 100 mL graduated cylinder. The initial mud sample was diluted to a density of 1,050 g/L, and adequate amount of EPS was added to make the final concentration at 0.1, 0.2, 0.3, 0.4, 0.5, 0.6, 0.7 and 0.8 g/L, respectively. After thoroughly mixed for 12 h, the mixture was transferred to the glass cylinder and then allowed to settle to measure the settled sediment volume. Rotational viscosimeter (NDJ-79, Changji, China) was used for measuring viscosity. The surface charge density of fluid mud was measured according to the method of colloid titration [12]. Mixed liquor suspended solids (MLSS) was determined in conformity with the standard methods [13]. The sediment particle size distribution of fluid mud was measured by Laser Scattering Particle Size Distribution Analysis (LSPSDA, LS13 320, Beckman, USA).

3 Results

3.1 EPS Composition

The chemical composition of EPS produced by *Burkholderia* sp. TKS1 was characterized and presented in Table 1, indicating that polysaccharides might be the key component influencing the characteristics of EPS.

3.2 Properties of Polysaccharides

The collected EPS was redissolved in deionized water, loaded on a DEAE-650 cellulose column and eluted with a linear gradient of 0–0.5 mol/L NaCl, in order to obtain the purified polysaccharides. The carbohydrate content of the eluent was determined by the phenol-sulfuric acid method during the period of collection and result was presented in Fig. 1. It was in a single and symmetrical peak, suggesting that homogeneous polysaccharide was obtained. The purity of polysaccharides was further determined by HPLC-RID. As shown in Fig. 2, the peak area of purified polysaccharides accounted for 92.9%, which met the required purity.

1-Phenyl-3-methyl-5-pyrazolone-High Performance Liquid Chromatography (PMP-HPLC) was applied to identify PMP-derivatized monosaccharides

Table 1 The chemical composition of EPS produced by *Burkholderia* sp. TKS1

Components	Polysaccharides	Proteins	Nucleic acids	Lipids
Content (%/EPS)	47.2	5.9	6.9	Not detected

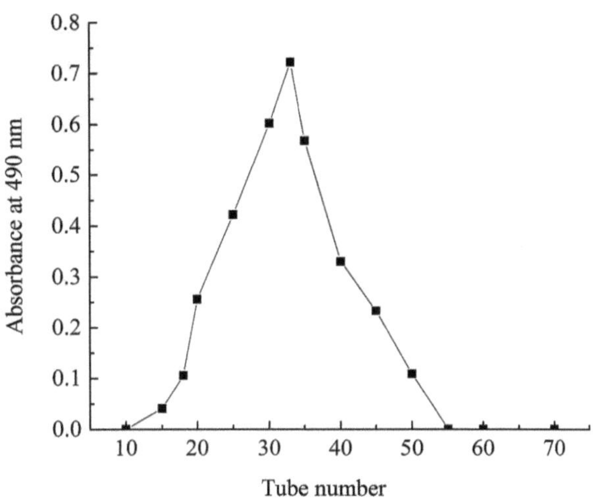

Fig. 1 Elution profile of purified polysaccharides on DEAE-650M chromatography column with gradient of NaCl solution (0 ~ 0.5 mol/L)

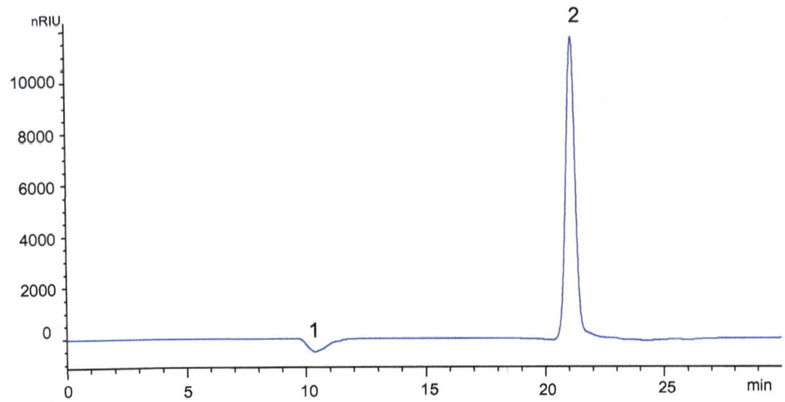

Fig. 2 Purified degree of polysaccharide by HPLC-RI

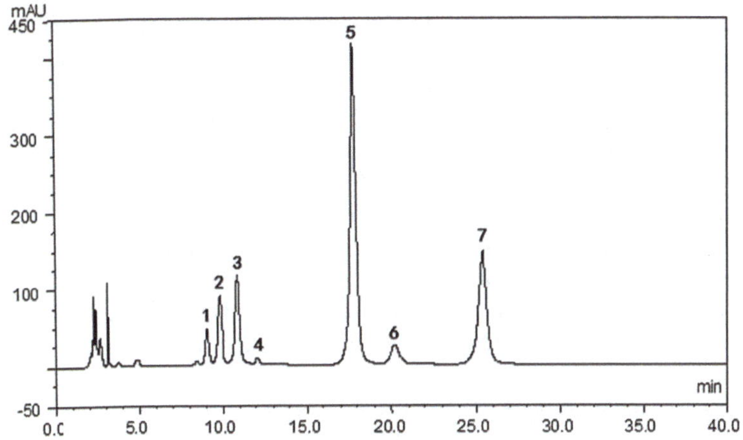

Fig. 3 High performance liquid chromatography (HPLC) separation of 1-phenyl-3-methyl-5-pyrazolon (PMP) derivatives of acid hydrolysates of purified polysaccharide. Peaks 1, 2, 3, 4, 5, 6, 7 represent mannose, reagent peak, glucuronic acid, galacturonic acid, glucose, galactose and fucose, respectively

unambiguously by comparing the retention time and mass data with those of standard monosaccharides and the result was presented in Fig. 3. The result showed that polysaccharide was heteropolysaccharide composed mainly of mannose, glucuronic acid, glucose, galactose and fucose at the molar ratio of 0.06:0.19:1.00:0.09:0.48, and also included a little galacturonic acid, the main component of which was glucose.

Intrinsic viscosity is a parameter reflecting the hydrodynamic volume occupied by polymers, which is defined as the reduced viscosity when concentration of polymer solution closes to zero [14]. It was determined in dilute solution and

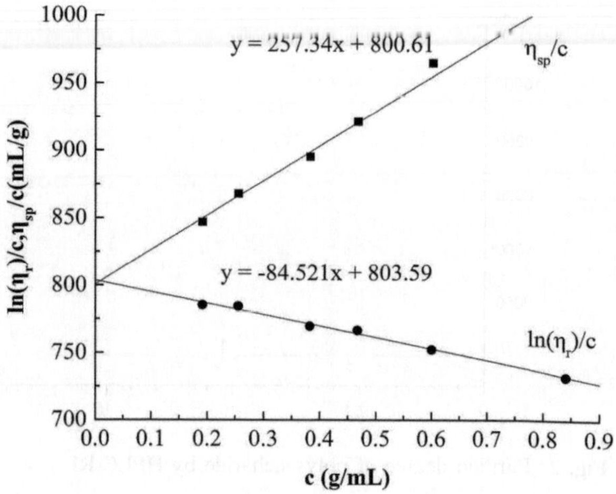

Fig. 4 Intrinsic viscosity of purified polysaccharide in 0.1 mmol/L NaCl solution at 25 °C by combined Huggins and Kraemer extrapolations

mainly depends on the molecular size and chain stiffness of the polymer, as well as solvent quality [15]. The intrinsic viscosity of polysaccharide was determined in 0.1 M NaCl using isoionic dilution. The value of intrinsic viscosity ([η]) was reported as the mean of both intercepts by plotting Huggins and Kraemer against concentration and extrapolating each linear trend line to zero concentration [16]. Sometimes, the intercept of the combined Huggins and Kraemer extrapolations might not meet at C = 0, which is the situation in this case, thus [η] was presented as the average of the Huggins and Kraemer intercepts. The intrinsic viscosity of polysaccharide of *Burkholderia* sp. TKS1 at 25 °C is 802.1 mL/g, as presented in Fig. 4. Xanthan gum is a polysaccharide secreted by the bacterium *Xanthomonas campestris*, used as a food additive, rheology modifier, and food thickening agent due to its high intrinsic viscosity [17] of 541 mL/g, which is significantly lower than the polysaccharide of *Burkholderia* sp. TKS1.

3.3 Effects of EPS on Fluid Mud Sedimentation

Settled sediment volume was used to characterize the sedimentation of different fluid mud samples. The larger value of settled sediment volume indicates slower sedimentation of the fluid mud, i.e. lower settling velocity of sediments. The impact of EPS with different concentrations on the settling velocity of sediments was presented in Fig. 5. The results showed that there were two stages during sediment settling, i.e. firstly fast and then slow sedimentation. The fast sedimentation stage prolonged significantly when the fluid mud was added with more EPS. The control group (0 g/L) entered the slow sedimentation stage after 2 h; the group with EPS concentration at 0.2 g/L entered the slow sedimentation stage after 5 h; while the group added with 0.4 g/L EPS entered the slow sedimentation stage after 10 h.

the study. The phenomenon that the amount and properties of EPS played a predominant role on settling properties was also found in activated sludge. The settling velocity of sludge would be deceased with more EPS content [19]. For flocs in the sludge, a high or excessive polysaccharide content was viewed as being detrimental to settling and dewatering properties owing to higher water content associated with polysaccharide rich EPS [20]. These studies indicated that polysaccharide in EPS might play a role in delaying the sedimentation and the composition of EPS had significant influences on the settling properties besides EPS contents. The results agreed well with previous reports, and the composition of EPS consisting of 47.2% polysaccharides and only 5.9% proteins revealed that polysaccharides were the key component delaying the sedimentation of fluid mud. It was found that the loosely bound EPS (LB-EPS) had a negative effect on bioflocculation and sludge-water separation. Although EPS was essential to sludge floc formation, excessive EPS in the form of LB-EPS could weaken cell attachment and the floc structure, resulting in poor bioflocculation, greater cell erosion and retarded sludge-water separation [21]. Additionally, it has been reported that the surface properties, hydrophobicity, and surface charge of EPS affected bioflocculation in the sludge and then influenced its settling properties [12]. Therefore, based on the results of the effects of EPS on fluid mud properties such as sediment particle density, viscosity and surface charge, the possible mechanism for EPS delaying the sedimentation of fluid mud was proposed as shown in Fig. 7. With the presence of EPS, the sediment density decreased due to the join of EPS, according to Stokes law, the decrease of sediment particle density caused the reduced particle settling velocity, the positively charged EPS could interact with sediments in fluid mud, which was negatively charged, to form a stable electric double layer structure, and steric repulsive energy was generated as a result of the interaction between the organic substances adsorbed onto the particle surfaces [22]. Due to the repulsion between particles, the flocculation between fine sediment particles by closing to each other and collision to form larger particles was inhibited, therefore dispersing the sediments in the mutually single-particle state. Moreover,

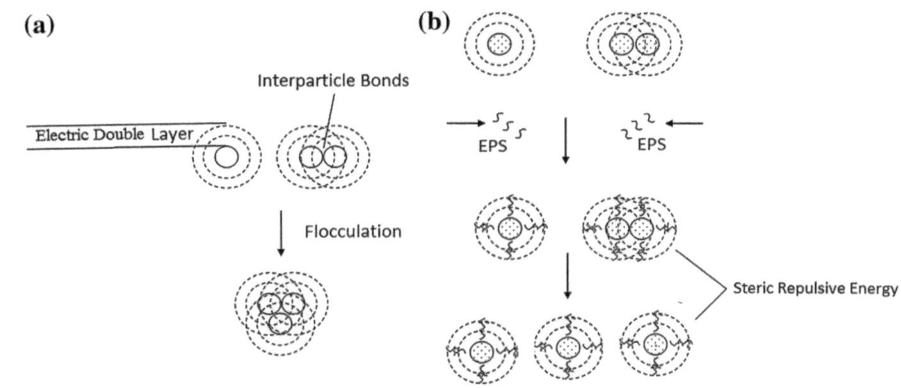

Fig. 7 The proposed mechanism for EPS delaying the sedimentation of fluid mud. **a** Represents under natural conditions (without EPS addition) and **b** represents under the presence of EPS

owing to the EPS with higher molecular weight and rich in polysaccharides (47%), in the presence of EPS, the water conditions of fluid mud system changed and the viscosity notably increased. According to Stokes law, the increase of viscosity caused the reduced particle settling velocity. Therefore, under the combined effects of viscosity and surface charges, the settling velocity of sediments was decreased and sedimentation of fluid mud was significantly delayed. The results showed that EPS could effectively delay the sedimentation of fluid mud. The application of EPS are successfully applied the biotechnology to the harbor engineering and save a lot of money for the port.

5 Conclusions

In this study, one EPS-producing strain *Burkholderia* sp. TKS1 isolated from fluid mud of a muddy port in China was utilized to investigate the impact of EPS on the sedimentation of fluid mud and explore the underlying mechanism. The analysis of chemical constituents of EPS showed the main component was a heteropolysaccharide and had high intrinsic viscosity. The effects of EPS with different concentrations on the sedimentation of fluid mud were investigated and the results showed that EPS could effectively delay the sedimentation of fluid mud. The addition of EPS significantly changed the viscosity, sediment particle size distribution and surface charge density of fluid mud. Based on the results, the possible mechanism for EPS delaying the sedimentation of fluid mud was finally proposed. The results would have important scientific significance to elucidate the mechanism of interaction between EPS and the fluid mud and to promote the use of the microbial technology in harbor engineering.

Acknowledgements The authors are very grateful for the financial support from Tianjin Research Institute for Water Transport Engineering M.O.T. Research Innovation Funds of China (TKS160201) and Changjiang Scholars and Innovative Research Team in University (IRT1166), Ministry of Education, China.

References

1. Liu H, Fang HH (2002) Extraction of extracellular polymeric substances (EPS) of sludges. J Biotechnol 95(3):249–256
2. Bo F, Palmgren R, Keiding K et al (1996) Extraction of extracellular polymers from activated sludge using a cation exchange resin. Water Res 30(8):1749–1758
3. Sponza DT (2002) Extracellular polymer substances and physicochemical properties of flocs in steady and unsteady-state activated sludge systems. Process Biochem 37(9):983–998
4. Sheng GP, Yu HQ, Li XY (2010) Extracellular polymeric substances (EPS) of microbial aggregates in biological wastewater treatment systems: a review. Biotechnol Adv 28(6):882–894

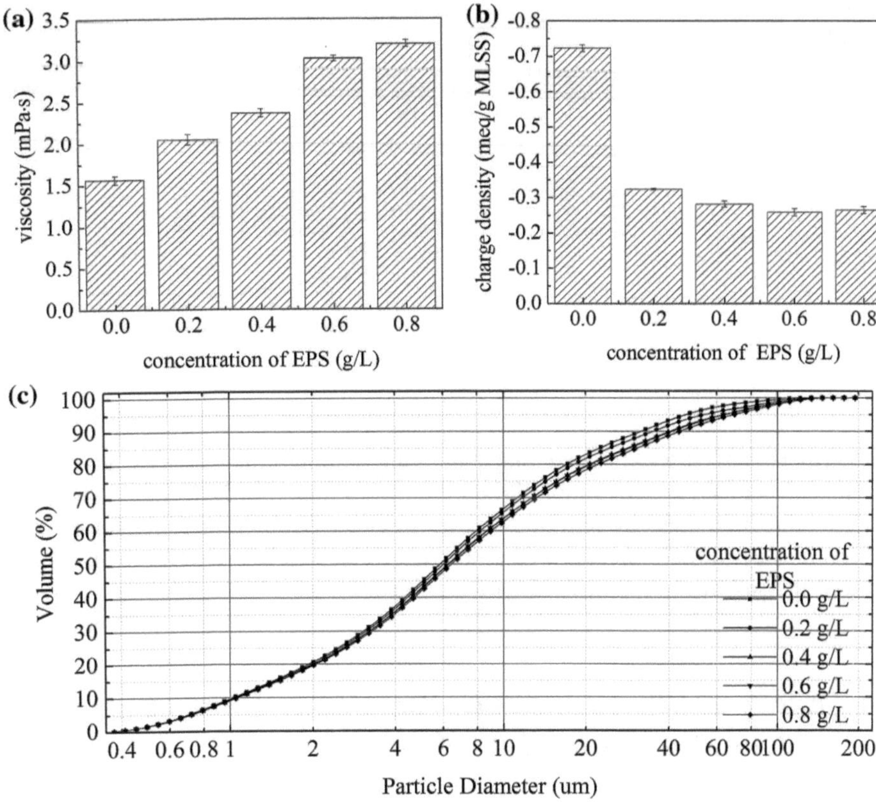

Fig. 6 The changes of viscosity (**a**), surface charge density (**b**) and sediment particle size (**c**) of fluid mud after adding EPS with different concentrations

was 6 μm, and after adding EPS the proportion of sediment particles with size larger than 6 μm significantly increased with the increase of EPS concentration, and it was speculated as the result of adsorption of EPS to the particle surface.

4 Discussion

Researches about the effects of EPS on sedimentation mainly focused on the process of wastewater treatment with activated sludge, while studies on its effects on the sedimentation process of fluid mud are very few at present. The sedimentation of fluid mud with EPS of different concentrations was measured and the results indicated that EPS could delay the sedimentation of fluid mud very effectively and the effect was dependent on the EPS concentration in the range of 0.1–0.8 g/L tested in

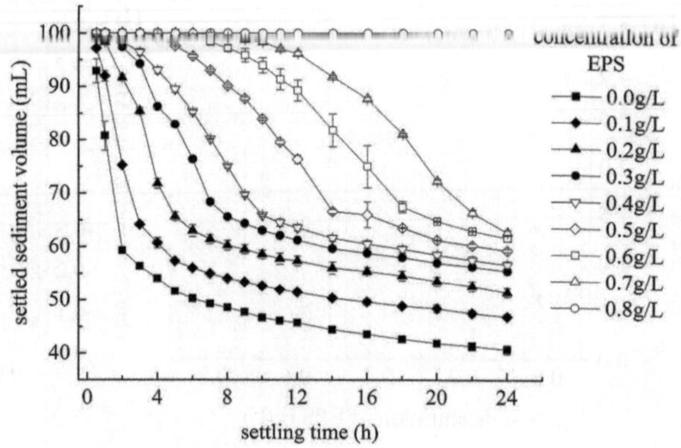

Fig. 5 The impact of EPS with different concentrations on the sedimentation of fluid mud

Furthermore, settled sediment volume at the same settling time increased along with the increase of EPS concentration. When the concentration of EPS was higher than 0.8 g/L, the fluid mud system showed no obvious sediment settling during the observation period. Compared to the control group (0 g/L), addition of EPS has delayed the sedimentation of fluid mud significantly.

3.4 *Effects of EPS on Viscosity, Particle Size Distribution and Surface Charge of Fluid Mud*

The main factors influencing the settling velocity of sediments include salinity, sediment concentration, organic matter etc., mainly by flocculation. The strong flocculation would make the particles condense into larger particles more easily, which was conducive to settlement. Therefore, studies on these factors affecting the flocculation could help us explore the mechanism of EPS delaying the settling velocity of sediments.

The changes of viscosity, particle size distribution and surface charge density of fluid mud samples with different EPS concentrations were measured. The results of viscosity and surface charge density were presented in Fig. 6a, b. The results showed that the viscosity increased from 1.5 to 3.2 mPa·S with the increase of EPS concentration. There were reports that the polysaccharide in EPS could form colloidal system to change the viscosity of the system [18]. The surface charge density of EPS was |0.38 meq/g EPS|, which could confirm that EPS was positively charged. The surface charge density of fluid mud without EPS was |−0.7 meq/g MLSS|, and the value decreased after adding EPS, and would remain stable until the surface charge density fell to |−0.3 meq/g MLSS|. It showed that the interaction between EPS molecules and sediment particles resulting in the absorption of EPS onto the particle surfaces therefore forming a stable electric double layer structure. As shown in Fig. 6c, the median sediment particle size of fluid mud without EPS

5. Sun BJ (2015) Preliminary study of microbial extracellular polymeric substances delaying the sedimentation of fluid mud and its mechanism. M.E. Dissertation, Tianjin University of Science & Technology, Tianjin 300457, China
6. Greiser N, Wurpts R (2008) Microbiological impact on formation and rheological properties of fluid mud. In: Zanke U (ed) Chinese-German joint symposium on hydraulic and ocean engineering. Eigen-Verlag, Darmstadt, pp 369–371
7. Kirby R (2011) Minimising harbour siltation—findings of PIANC working group 43. Ocean Dyn 61(2):233–244
8. Dubois M, Gilles KA, Hamilton JK et al (2002) Colorimetric method for determination of sugars and related substances. Anal Chem 28(3):350–356
9. Bradford MM (2015) A rapid and sensitive method for the quantitation of microgram quantities of protein utilizing the principle of protein-dye binding. Anal Biochem 72(s 1–2):248–254
10. Frings CS, Dunn RT (1970) A colorimetric method for determination of total serum lipids based on the sulfo-phospho-vanillin reaction. Am J Clin Pathol 53(1):89–91
11. Al-Sheraji SH, Ismail A, Manap MY et al (2012) Purification, characterization and antioxidant activity of polysaccharides extracted from the fibrous pulp of *Mangifera pajang* fruits. Lebensmittel-Wissenschaft und-Technologie 48(2):291–296
12. Liao BQ, Allen DG, Droppo IG et al (2001) Surface properties of sludge and their role in bioflocculation and settleability. Water Res 35(2):339–350
13. Eaton AD, Franson MA (2005) Standard methods for the examination of water and wastewater, 21th edn. American Public Health Association, American Water Works Association and Water Environment Federation, Washington, USA
14. Eich A, Wolf BA (2011) Intrinsic viscosities of polyelectrolytes: determination and modeling of the effects of extra salt. ChemPhysChem 12(15):2786–2790
15. Moresi M, Lo PS, Mancini M (2001) Rheology of scleroglucan dispersions. J Food Eng 50 (4):235–245
16. Qian KY, Cui SW, Wu Y et al (2012) Flaxseed gum from flaxseed hulls: extraction, fractionation, and characterization. Food Hydrocolloids 28(2):275–283
17. Nie LH, Zhou RJ, Ning ZX (2003) Focus on Xanthan Gum. China Food Addit 3:82–85
18. Yin JY, Nie SP, Li J et al (2012) Mechanism of interactions between calcium and viscous polysaccharide from the seeds of *Plantago asiatica* L. J Agric Food Chem 60(32):7981–7987
19. Wang LF, Wang LL, Li WW et al (2014) Surfactant-mediated settleability and dewaterability of activated sludge. Chem Eng Sci 116:228–234
20. Basuvaraj M, Fein J, Liss SN (2015) Protein and polysaccharide content of tightly and loosely bound extracellular polymeric substances and the development of a granular activated sludge floc. Water Res 82:104–117
21. Li XY, Yang SF (2007) Influence of loosely bound extracellular polymeric substances (EPS) on the flocculation, sedimentation and dewaterability of activated sludge. Water Res 41(5):1022–1030
22. Sato T, Ruch R (1980) Stabilization of colloidal dispersions by polymer adsorption. Surfactant Science Series Volume 9. Marcel Dekker Ltd, New York

Polyvalent Effect Enhances Anti-influenza Virus Activity

Haipeng Liu, Haojie He, Zhaoliang Yang, Peng Yu and Kui Lu

1 Introduction

Influenza accounts for approximately 20,800 hospitalizations and 1 death in November 2016 alone in China [1]. Zanamivir (**ZA**) and Oseletamivir (**OS**) are the two FDA approved antiviral drugs as the inhibitors of Neuraminidase (NA) [2] in the United States. They can competitively bind to NA and prevent the release of the newly assembled virus from the infected host cells [3]. However with the emerging of the drug resistance strain [4], it is still urgent to develop new kind of antiviral agents.

The efficient way to solve the drug-resistance is introducing chemical modifications to the existing **ZA** or **OS** [5]. Recently, a new kinds of 2, 3-fluorosialosyl fluorides (difluorosialic acids, **DFSA**s) that exhibit better inhibitory activity compared to **ZA** especially for drug-resistant strains have been prepared [6]. These leading compounds encouraged researchers to further enhance its potency for the antiviral application.

Structural studies have found that an estimated 50 copies of tetrameric NA are present on the surface of the influenza viral particle. This multiple presentation can significantly enhance the binding affinity between NA and its ligands which called "cluster effects" [7, 8]. Therefore, covalently attaching multicopies of NA inhibitor to a polymeric backbone which can interact with several NAs or NA's subunits is a promising method to get potent antiviral conjugates [9]. It has been reported that

H. Liu · H. He · Z. Yang · P. Yu · K. Lu (✉)
Key Laboratory of Industrial Fermentation Microbiology, Tianjin Key Laboratory of Industrial Microbiology, Sino-French Joint Lab of Food Nutrition/Safety and Medicinal Chemistry, College of Biotechnology, Tianjin University of Science and Technology, No. 29, 13th Street, Tianjin Economic-Technology Development Area, Tianjin 300457, People's Republic of China
e-mail: lukui@tust.edu.cn

H. Liu et al. (eds.), *Advances in Applied Biotechnology*, Lecture Notes in Electrical Engineering 444, https://doi.org/10.1007/978-981-10-4801-2_89

attaching **ZA** on the polymers can dramatically enhance antiviral activity not only against drug-sensitive but also drug-resistant strains [10, 11].

Our previous work [12] has shown that when attaching the **DFSA**s on octavalent scaffolds, the increased binding affinities of this conjugate to NA was achieved with IC_{50} values is maximal 145-fold to the monomer. With these promising results, polyvalent **DFSA**s were synthesized and their antiviral activity against H7N9 virus like particle (VLP) was evaluated in this paper. This report represents an initial attempt for the development of polyvalent **DFSA**s conjugates as strong anti-influenza inhibitors.

2 Experimental

2.1 Material

All solvents were dried prior to use. Thin-layer chromatography (TLC) was purchased from EMD Co. Ltd. (German). All compounds were stained with 5% H_2SO_4 in ethanol followed by heating. Detection with UV light was employed when possible. Flash column chromatography was performed on silica gel 200–300 mesh. NMR spectra were recorded on Bruker AVANCE III (400 MHz) instruments. Chemical shifts (d) were reported in parts per million downfield from TMS, the internal standard; J values were given in Hertz. Fluorescence intensity was measured using a Synergy™ H1/H1MF microplate reader (BioTek Instruments, Inc. USA) with excitation and emission wavelengths of 360 and 440 nm, respectively.

2.2 Method

2.2.1 Biological Assay

The neuraminidase inhibition assay was performed as previously reported with slight modification. The NA activity was measured using 4-methylumbelliferyl-α-D-N-acetylneuraminic acid sodium salt (4-MUNANA) in the enzyme buffer (100 mM sodium acetate pH 5.5 and 10 mM $CaCl_2$) as the substrate. All compounds were dissolved in buffer and diluted to the corresponding concentrations. The enzyme inhibitory assay was conducted in 384-well plates containing 15 μL diluted H7N9 VLP supernatant (containing active NA), 15 μL compounds in the buffer. The mixture was incubated for 30 min at 37 °C, and then added with 30 μL 4-MUNANA substrate per well in the buffer. The enzymatic reactions were carried out for 2 h at 37 °C. The fluorescent signal was monitored using the kinetics function of Synergy™ H1/H1MF microplate reader with excitation and emission wavelengths of 360 and 440 nm, respectively. The IC50 was calculated as the concentration of inhibitor resulting in a 50% reduction in fluorescent units (FU) compared to the control.

2.2.2 Chemical Synthesis

See Scheme 1.

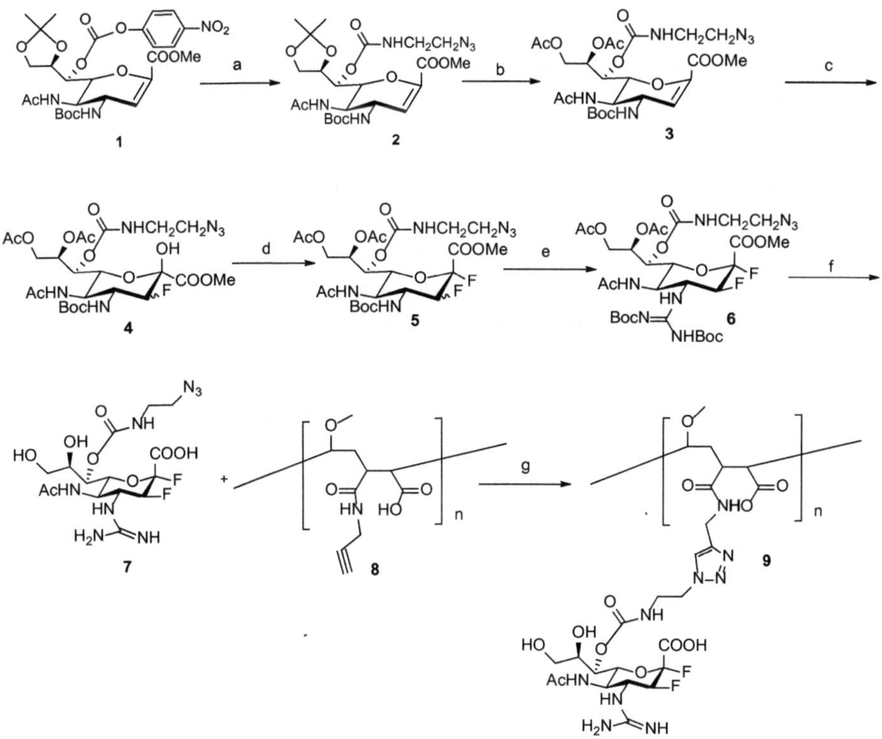

Scheme 1 Reagents and conditions: (*a*) 2-azidoethanamine, Py, DMAP, 90%; (*b*) (i) 80% HAc; (ii) Ac_2O/Py, 80% over two steps; (*c*) CH_3NO_2, H_2O, Selectfluor®, rt, 50%; (*d*) CH_2Cl_2, DAST, −40 °C, 79%; (*e*) (i) TFA/CH_2Cl_2 1:1; (ii) MeSC(=NBoc)NHBoc, $HgCl_2$, Et_3N, CH_2Cl_2, rt, 12 h, 83% over two steps, (*f*) (i) CH_3OH, CH_3ONa, rt, 2 h (ii) NaOH (aq) rt, 3 h, (iii) TFA/CH_2Cl_2 1:1, rt, 2 h, 45% over three steps; (*g*) $CuSO_4{\cdot}5H_2O$, VcNa, THF/H_2O, rt, 12 h, 76%

3 Results

3.1 *Synthesize and Characterize Difluorinated Zanamivir Analogs*

Methyl [5-acetamido-2,6-anhydro-7-*O*-(2-azidoethylcarbamoyloxy)-8,9-*O*-isopropylidene-4-*N*-tert-butyloxycarbonyl-3,4,5-trideoxy-D-glycero-D-galacto] non-2-enonate (2). To a solution of 2-azidoethanamine (932.17 mg, 10.83 mmol) in Py (10 mL), DMAP (660 mg, 5.41 mmol) was added. The solution was stirred at rt for 30 min then **1** [12] (3.30 g, 5.41 mmol) was added and continue stirring for 3 h. The progress of the reaction was monitored by TLC. Upon completion, the reaction mixture was washed by HCl (1 M, 25 mL), extracted by DCM (3 × 20 mL), the organic phases were combined and dried over Na_2SO_4. Solvent was removed under vacuum and the product was purified by column chromatography using PE:EtOAc (1:1) to give compound **2** (2.6 g, 86.3%).

^{1}H NMR (400 MHz, $CDCl_3$) δ 6.37 (d, J = 9.6 Hz, 1H, N*H*Ac), 5.91 (d, J = 2.1 Hz, 1H,), 5.43 (t, J = 5.6 Hz, 1H), 5.29 (d, J = 4.1 Hz, 1H), 5.10 (d, J = 9.0 Hz, 1H), 4.51 (t, J = 9.0 Hz, 1H), 4.34 (dt, J = 16.7, 8.3 Hz, 2H), 4.20–4.07 (m, 2H), 4.04–3.93 (m, 1H), 3.77 (s, 3H, CH_3O), 3.44 (dd, J = 12.4, 6.2 Hz, 2H), 3.36–3.25 (m, 2H), 1.93 (s, 3H), 1.40 (s, 9H), 1.37 (s, 3H), 1.35 (s, 3H).

Methyl [5-acetamido-8,9-di-*O*-acetyl-2,6-anhydro-7-*O*-(2-azidoethylcarbamoyloxy)-4-*N*-tert-butyloxycarbonyl-3,4,5-trideoxy-D-glycero-D-galacto] non-2-enonate (3). Compound **2** (2.6 g, 4.67 mmol) was dissolved in 20 mL 80% acetic acid, then stirred at rt for 4 days. Then the reaction mixture was washed by saturated $NaHCO_3$, extracted by DCM (3 × 30 mL), the organic phases were combined and dried over Na_2SO_4. Solvent was removed under vacuum and the residue was dissolved in 10 mL Py. Acetic anhydride (3 mL) was added and the reaction was stirred overnight. The reaction mixtrue was washed by HCl (1 M, 50 mL), extracted by DCM (3 × 20 mL), the organic phases were combined and dried over Na_2SO_4. DCM was removed *in vacuo* and the product was purified by column chromatography using PE:EtOAc (1:1) to give compound **3** (2.1 g, 86.29%).

^{1}H NMR (400 MHz, $CDCl_3$) δ 6.14 (d, J = 8.8 Hz, 1H, N*H*Ac), 5.91 (d, J = 1.9 Hz, 1H), 5.45–5.25 (m, 3H), 4.99 (d, J = 9.0 Hz, 1H), 4.65 (d, J = 11.7 Hz, 1H), 4.51 (t, J = 9.0 Hz, 1H), 4.28 (d, J = 10.2 Hz, 1H), 4.15 (ddd, J = 27.3, 14.9, 6.9 Hz, 2H), 3.78 (s, 3H), 3.57–3.18 (m, 4H), 2.06 (d, J = 7.6 Hz, 6H), 1.94 (s, 4H), 1.42 (s, 9H).

Methyl 5-acetamido-8,9-di-*O*-acetyl-7-*O*-(2-azidoethylcarbamoyloxy)-4-*N*-tert-butyloxycarbonyl-3,4,5-trideoxy-3-fluoro-D-*erythro*-β-L-gluco-non-2-ulopyranosonate (4) compound **3** (3 g, 5 mmol), nitromethane (20 mL), water (4 mL) and Selectfluor® (5.3 g, 15 mmol, 3 eq.) was stirred for 4 days at room temperature. The reaction was quenched with saturated $NaHCO_3$ (100 mL), extracted with EtOAc (4 × 200 mL). The organic phase was washed with saturated $NaHCO_3$ (300 mL) and brine (300 mL), dried over Na_2SO_4. After evaporation, the resulting residue was purified by flash column chromatography (Acetone/PE = 1/2.5) to give the desired compound **4** (1.5 g, 50%).

^{1}H NMR (400 MHz, CDCl3) δ 6.95 (d, J = 9.9 Hz, 1H), 6.57 (d, J = 9.0 Hz, 1H), 5.43 (dd, J = 21.0, 15.2 Hz, 3H), 5.29–5.02 (m, 6H), 4.84–4.45 (m, 3H), 4.33–3.90 (m, 9H), 3.79 (d, J = 22.7 Hz, 7H), 3.56–3.09 (m, 9H), 2.84 (s, 2H), 2.09–1.72 (m, 19H), 1.36 (d, J = 11.5 Hz, 19H).^{19}F NMR (376 MHz, $CDCl_3$) δ −196.90 (s), −203.47 (s).

Methyl 5-acetamido-8,9-di-*O*-acetyl-7-*O*-(2-azidoethylcarbamoyloxy)-4-*N*-tert-butyloxycarbonyl-2,3,4,5-tetradeoxy-2,3-difluoro-*D-erythro-α-L*-gluco-non-2-ulopyranosonate (5). To the compound of **4** (1.5 g, 9.4 mmol) dissolved in dry DCM (20 mL) DAST (1.36 mL, 13.9 mmol, 1.5 eq.) was added drop-wise at −40 °C. After addition, the reaction mixture was stirred for 0.5 h at −40 °C, and then warmed up to −10 °C. The reaction was quenched with saturated $NaHCO_3$, diluted with DCM (200 mL) and washed with brine (150 mL). The water phase was extracted again with EtOAc (3 × 200 mL) and washed with brine (300 mL). The combined organic phase was dried over $MgSO_4$. After evaporation, the resulting residue was purified by flash column chromatography (PE: Acetone = 3:1) to give product **5** as a white solid (1.2 g, 79.75%).

^{1}H NMR (400 MHz, CDCl3) δ 6.05 (d, J = 9.8 Hz, 1H), 5.65 (d, J = 8.4 Hz, 1H), 5.52–4.85 (m, 11H), 4.75–4.02 (m, 14H), 3.88 (d, J = 5.3 Hz, 7H), 3.64–3.13 (m, 10H), 2.27–1.86 (m, 26H), 1.44 (d, J = 9.0 Hz, 23H).^{19}F NMR (376 MHz, $CDCl_3$) δ −115.58 (d, J = 13.3 Hz), −123.08 (d, J = 10.6 Hz), −196.33 (d, J = 13.3 Hz), −214.67 (d, J = 10.5 Hz).

Methyl 5-acetamido-8,9-di-*O*-acetyl-7-*O*-(2-azidoethylcarbamoyloxy)-4-(bis-*N,N'*-tert-butyloxycarbonyl)-guanidino-2,3,4,5-tetradeoxy-2,3-difluoro-*D-erythro-α-L*-gluco-non-2-ulopyranosonate (6). To a solution of **5** (1.2 g, 1.88 mmol) in DCM (5 mL), TFA (5 mL) was added, the reaction was stirred at rt for 1 h. DCM was removed *in vacuo*, Et_3N (1 mL) was added. The solution was stirred at rt for 30 min. $HgCl_2$ (405.45 mg, 0.93 mmol) and 1, 3-Bis(tert-butoxycarbonyl)-2-methyl-2-thiopseudourea (647.14 mg, 2.23 mmol) was added. The reaction was stirred at rt for 12 h. The reaction mixture was washed with HCl (1 M, 25 mL), extracted with DCM (3 × 10 mL) and the product was purified by column chromatography using Acetone: PE (3.5:1) to give the product **6**. (500 mg, 39%).

^{1}H NMR (400 MHz, $CDCl_3$) δ 11.31 (s, 1H), 8.57 (d, J = 8.1 Hz, 1H), 6.41 (d, J = 9.3 Hz, 1H), 5.18 (dt, J = 23.9, 7.1 Hz, 3H), 4.99–4.50 (m, 2H), 4.47–3.94 (m, 4H), 3.82 (s, 3H), 3.52–3.13 (m, 4H), 2.00 (d, J = 19.1 Hz, 6H), 1.83 (s, 3H), 1.42 (s, 18H). ^{19}F NMR (376 MHz, $CDCl_3$) δ −115.18 (d, J = 12.8 Hz), −195.46 (d, J = 13.0 Hz).

5-Acetamido 7-*O*-(2-azidoethylcarbamoyloxy)-2,3,4,5-tetradeoxy-2,3-difluoro-4-guanidino-*D-erythro-α-L*-gluco-non-2-ulopyranosonic acid (7). To a solution of compound **6** (100 mg) dissolved in methanol (8 mL) $NaOCH_3$ (18 mg, 185 μmol) was added and stirred at rt for 2 h. The reaction mixture was neutralized by H^+ resin adjusts to pH 7.0, then the suspension was filtered. The liquid phase was collected and dried *in vacuo* to give colorless oil. Then a solution of the deacetylation compound in methanol (4 mL) and water (4 mL) was treated with

0.1 M sodium hydroxide solution (1 mL, 1 mmol). After 3 h, the reaction mixture was neutralized by H^+ resin adjust to pH 7.0, the solvent was evaporated and the residue was stirred in 50% trifluoroacetic acid (10 mL) solution in DCM at room temperature for 2 h and evaporated to dryness. The resulting syrup was dissolved in distilled water, after removing the copper complex by CupriSorb® if needed, and purified by Sephadex® LH-20 then lyophilisation to afford the compound 7 (33 mg, 45%).

^{1}H NMR (400 MHz, D_2O) δ 4.76 (d, J = 10.1 Hz, 2H), 4.63 (t, J = 11.8 Hz, 1H), 4.42–4.26 (m, 1H), 4.11 (dd, J = 22.3,11.6 Hz, 1H), 3.94–3.72 (m, 1H), 3.55 (d, J = 12.0 Hz, 1H), 3.44–3.14 (m, 5H), 1.89 (s, 3H).

Polymer 9. 7 (20 mg, 0.00028 mmol) was dissolved in the buffer (pH = 7.2), the polyvalent alkynyl backbone **8** (50 mg, 0.014 mmol), $CuSO_4{\cdot}5H_2O$ (2 mg, 1 eq.) and sodium ascorbate (4 mg, 1 eq.) were added. The mixture was stirred at room temperature overnight 16 h. The solvent was evaporated to dryness, the copper complex was absorbed by CupriSorb® if needed and the crude products were purified by Sephendex LH-20. Then lyophilisation to afford the compound 8 (40 mg, 76%).

^{1}H NMR (400 MHz, D_2O) δ 8.38, 7.93, (2 s, 2H), 4.55–2.26 (m, 12H), 2.14–0.68 (m, 5H). ^{19}F NMR (376 MHz, D_2O) δ −111.82 (s), −122.28 (s), −198.70 (s).

3.2 H7N9 Virus like Particle (VLP) Neuraminidase Inhibition Assay

See Table 1.

4 Conclusions

We have successfully attached the 2,3-difluorinated zanamivir analogs to the polymer via "Click Chemistry". The polyvalent conjugate showed improved anti-influenza virus activity with IC_{50} value about 1000 fold better than that of the corresponding monomer. This glycopolymer even showed better inhibitory activity compared with Zanamivir. A cell based cytopathic effect (CPE) reduction assay of the synthesized glycopolymer will be studied for the further inhibition evaluation and we expect this kind of polyvalent conjugates will be further used as potent anti-influenza agents.

Table 1 The inhibition of H7N9 VLP by the different valent DFSAs

Compound	Zanamivir	Monomer **7**	Polymer **9**
IC_{50} (nM)	1.92	506.67	<0.54

Acknowledgements This work was financially supported by the Natural Science Foundation of China (21402140, 21502139) and the International Science & Technology Cooperation Program of China (2013DFA31160). The authors are thankful to the Research Centre of Modern Analytical Technology, Tianjin University of Science and Technology for NMR measurements.

References

1. National Health and Family Planning Commission of the People's Republic of China, Reports of nationally notifiable infectious diseases. Available online: http://www.moh.gov.cn/jkj/s3578/201612/8a84c914186441d9a444e81751f58863.shtml
2. Itzstein MV et al (1993) Rational design of potent sialidase-based inhibitors of influenza virus replication. Nature 363(363):418–423
3. Wagner R, Matrosovich M, Klenk HD (2002) Functional balance between haemagglutinin and neuraminidase in influenza virus infections. Rev Med Virol 12(3):159–166
4. Hay AJ, Hayden FG (2013) Oseltamivir resistance during treatment of H7N9 infection. Lancet 381(9885):2230–2232
5. Kim JH et al (2013) Mechanism-based covalent neuraminidase inhibitors with broad-spectrum influenza antiviral activity. Science 340(6128):71–75
6. Das K et al (2010) Structures of influenza A proteins and insights into antiviral drug targets. Nat Struct Mol Biol 17(5):530–538
7. Lundquist JJ, Toone EJ (2002) The cluster glycoside effect. Chem Rev 102(2):555–578
8. Cecioni S, Imberty A, Vidal S (2015) Glycomimetics versus multivalent glycoconjugates for the design of high affinity lectin ligands. Chem Rev 115(1):525–561
9. Cheng CK et al (2014) From neuraminidase inhibitors to conjugates: a step towards better anti-influenza drugs? Future Med Chem 6(7):757–774
10. Weight AK et al (2011) Attaching zanamivir to a polymer markedly enhances its activity against drug-resistant strains of influenza a virus. J Pharm Sci 100(3):831–835
11. Haldar J et al (2010) Bifunctional polymeric inhibitors of human influenza A viruses. Pharm Res 27(2):259–263
12. Yang ZL, Zeng XF, Liu HP, Yu Q, Meng X, Yan ZL, Fan ZC, Xiao HX, Iyer SS, Yang Y, Yu P (2016) Tetrahedron Lett 57(24):2579–2582

Heidelberg, Germany

MIX
Papier aus verantwortungsvollen Quellen
Paper from responsible sources
FSC® C105338

If you have any concerns about our products,
you can contact us on
ProductSafety@springernature.com

In case Publisher is established outside the EU,
the EU authorized representative is:
Springer Nature Customer Service Center GmbH
Europaplatz 3, 69115 Heidelberg, Germany

Printed by Libri Plureos GmbH
in Hamburg, Germany